DATA ANALYSIS METHODS IN PHYSICAL OCEANOGRAPHY

Second and Revised Edition

Some other Related Elsevier Titles of Interest

P. MALANOTTE-RIZZOLI
Modern Approaches to Data Assimilation in Ocean Modelling
Elsevier Oceanography Series, 61

J.H. STEL
Operational Oceanography
Elsevier Oceanography Series, 62

D. HALPERN
Satellites, Oceanography and Society
Elsevier Oceanography Series, 63

P. BOCOTTI
Wave Mechanics for Ocean Engineering
Elsevier Oceanography Series, 64

Cover illustrations

Centre image: Trajectory of surface drifter 15371 in the vicinity of the Kuril Islands, western Pacific Ocean, for the period September 4 to December 3, 1993. The triangle denotes the start of the 90-day track. Squares denote 10-day positions along the trajectory. Depths are in meters.

Top left Image: Time series of the east-west component of surface current speed (u-component) as a function of time and water depth as the drifter advanced from the coast of Urup Island, through Friz Strait, and into the Sea of Okhotsk. The current variability was alternatively dominated by eddy-like features, wind-generated inertial motions, and regionally enhanced diurnal tidal currents.

Bottom left image: Wavelet (frequency-time domain) analysis for the clockwise rotary component of the current spectra. The graph highlights the timing and duration of episodic current events in the inertial, semidiurnal, and diurnal frequency bands as the drifter progressed from the open ocean through Friz Strait.

Illustrations taken from Thomson, R.E., P.H. LeBlond, and A.B. Rabinovich. 1997. Oceanic odyssey of a satellite-tracked drifter: North Pacific variability delineated by a single drifter trajectory. Journal of Oceanography, 53, 81-87.

DATA ANALYSIS METHODS IN PHYSICAL OCEANOGRAPHY

Second and Revised Edition

WILLIAM J. EMERY

CCAR Box 431
University of Colorado
Boulder, CO 80309
USA

and

RICHARD E. THOMSON

Institute of Ocean Sciences
9860 West Saanich Road
Sidney, British Columbia V8L 4B2
Canada

ELSEVIER
AMSTERDAM - LONDON - NEW YORK - OXFORD - PARIS - SHANNON - TOKYO

ELSEVIER SCIENCE B.V.
Sara Burgerhartstraat 25
P.O. Box 211, 1000 AE Amsterdam, The Netherlands

© 2001 Elsevier Science B.V. All rights reserved.

This work is protected under copyright by Elsevier Science, and the following terms and conditions apply to its use:

Photocopying
Single photocopies of single chapters may be made for personal use as allowed by national copyright laws. Permission of the Publisher and payment of a fee is required for all other photocopying, including multiple or systematic copying, copying for advertising or promotional purposes, resale, and all forms of document delivery. Special rates are available for educational institutions that wish to make photocopies for non-profit educational classroom use.

Permissions may be sought directly from Elsevier Science Global Rights Department, PO Box 800, Oxford OX5 1DX, UK; phone: (+44) 1865 843830, fax: (+44) 1865 853333, e-mail: permissions@elsevier.co.uk. You may also contact Global Rights directly through Elsevier's home page (http://www.elsevier.nl), by selecting 'Obtaining Permissions'.

In the USA, users may clear permissions and make payments through the Copyright Clearance Center, Inc., 222 Rosewood Drive, Danvers, MA 01923, USA; phone: (+1) (978) 7508400, fax: (+1) (978) 7504744, and in the UK through the Copyright Licensing Agency Rapid Clearance Service (CLARCS), 90 Tottenham Court Road, London W1P 0LP, UK; phone: (+44) 207 631 5555; fax: (+44) 207 631 5500. Other countries may have a local reprographic rights agency for payments.

Derivative Works
Tables of contents may be reproduced for internal circulation, but permission of Elsevier Science is required for external resale or distribution of such material.
Permission of the Publisher is required for all other derivative works, including compilations and translations.

Electronic Storage or Usage
Permission of the Publisher is required to store or use electronically any material contained in this work, including any chapter or part of a chapter.

Except as outlined above, no part of this work may be reproduced, stored in a retrieval system or transmitted in any form or by any means, electronic, mechanical, photocopying, recording or otherwise, without prior written permission of the Publisher.
Address permissions requests to: Elsevier Science Global Rights Department, at the mail, fax and e-mail addresses noted above.

Notice
No responsibility is assumed by the Publisher for any injury and/or damage to persons or property as a matter of products liability, negligence or otherwise, or from any use or operation of any methods, products, instructions or ideas contained in the material herein. Because of rapid advances in the medical sciences, in particular, independent verification of diagnoses and drug dosages should be made.

First printing 1997 (Hardbound)

Second and revised edition 2001
Hardbound: isbn 0-444-50756-6
Paperback: isbn 0-444-50757-4

Library of Congress Cataloging in Publication Data
A catalog record from the Library of Congress has been applied for.

∞ The paper used in this publication meets the requirements of ANSI/NISO Z39.48-1992 (Permanence of Paper).
Printed in The Netherlands.

Contents

Preface — xi
Acknowledgments — xv

Chapter 1 Data Acquisition and Recording — 1

1.1 Introduction — 1
1.2 Basic sampling requirements — 2
   1.2.1 Sampling interval — 3
   1.2.2 Sampling duration — 4
   1.2.3 Sampling accuracy — 6
   1.2.4 Burst sampling versus continuous sampling — 6
   1.2.5 Regularly versus irregularly sampled data — 7
   1.2.6 Independent realizations — 8
1.3 Temperature — 8
   1.3.1 Mercury thermometers — 9
   1.3.2 The mechanical bathythermograph (MBT) — 12
   1.3.3 Resistance thermometers (expendable bathythermograph: XBT) — 14
   1.3.4 Salinity/conductivity–temperature–depth profilers — 18
   1.3.5 Dynamic response of temperature sensors — 19
   1.3.6 Response times of CTD systems — 22
   1.3.7 Temperature calibration of STD/CTD profilers — 23
   1.3.8 Sea surface temperature — 24
   1.3.9 The modern digital thermometer — 30
   1.3.10 Potential temperature and density — 32
1.4 Salinity — 33
   1.4.1 Salinity and electrical conductivity — 34
   1.4.2 The practical salinity scale — 39
   1.4.3 Nonconductive methods — 42
1.5 Depth or pressure — 42
   1.5.1 Hydrostatic pressure — 42
   1.5.2 Free-fall velocity — 43
   1.5.3 Echo sounding — 48
   1.5.4 Other depth sounding methods — 54
1.6 Sea-level measurement — 55
   1.6.1 Tide and pressure gauges — 58
   1.6.2 Satellite altimetry — 62
   1.6.3 Inverted echo sounder (IES) — 63

vi *Data Analysis Methods in Physical Oceanography*

	1.6.4 Wave height and direction	67
1.7	Eulerian currents	68
	1.7.1 Early current meter technology	70
	1.7.2 Rotor-type current meters	70
	1.7.3 Nonmechanical current meters	78
	1.7.4 Profiling acoustic Doppler current meters (ADCM)	83
	1.7.5 Comparisons of current meters	94
	1.7.6 Electromagnetic methods	95
	1.7.7 Other methods of current measurement	96
	1.7.8 Mooring logistics	97
	1.7.9 Acoustic releases	99
1.8	Lagrangian current measurements	102
	1.8.1 Drift cards and bottles	103
	1.8.2 Modern drifters	104
	1.8.3 Processing satellite-tracked drifter data	107
	1.8.4 Drifter response	109
	1.8.5 Other types of surface drifters	115
	1.8.6 Subsurface floats	116
	1.8.7 Surface displacements in satellite imagery	119
1.9	Wind	119
1.10	Precipitation	125
1.11	Chemical tracers	127
	1.11.1 Conventional tracers	128
	1.11.2 Light attenuation and scattering	138
	1.11.3 Oxygen isotope: $\delta^{18}O$	142
	1.11.4 Helium-3; helium/heat ratio	143
1.12	Transient chemical tracers	145
	1.12.1 Tritium	146
	1.12.2 Radiocarbon	149
	1.12.3 Chlorofluorocarbons	153
	1.12.4 Radon-222	155
	1.12.5 Sulfur hexafluoride	157
	1.12.6 Strontium-90	158

Chapter 2 Data Processing and Presentation 159

2.1	Introduction	159
2.2	Calibration	160
2.3	Interpolation	161
2.4	Data presentation	162
	2.4.1 Introduction	162
	2.4.2 Vertical profiles	167
	2.4.3 Vertical sections	170
	2.4.4 Horizontal maps	172
	2.4.5 Map projections	177
	2.4.6 Characteristic or property versus property diagrams	181
	2.4.7 Time-series presentation	185
	2.4.8 Histograms	187
	2.4.9 New directions in graphical presentation	187

Chapter 3 Statistical Methods and Error Handling — 193

3.1	Introduction	193
3.2	Sample distributions	194
3.3	Probability	197
	3.3.1 Cumulative probability functions	200
3.4	Moments and expected values	201
	3.4.1 Unbiased estimators and moments	203
	3.4.2 Moment generating functions	204
3.5	Common probability density functions	207
3.6	Central limit theorem	211
3.7	Estimation	214
3.8	Confidence intervals	216
	3.8.1 Confidence interval for μ (σ known)	217
	3.8.2 Confidence interval for μ (σ unknown)	218
	3.8.3 Confidence interval for σ^2	219
	3.8.4 Goodness-of-fit test	220
3.9	Selecting the sample size	224
3.10	Confidence intervals for altimeter bias estimates	225
3.11	Estimation methods	227
	3.11.1 Minimum variance unbiased estimation	228
	3.11.2 Method of moments	229
	3.11.3 Maximum likelihood	230
3.12	Linear estimation (regression)	233
	3.12.1 Method of least squares	234
	3.12.2 Standard error of the estimate	238
	3.12.3 Multivariate regression	239
	3.12.4 A computational example of matrix regression	240
	3.12.5 Polynomial curve fitting with least squares	242
	3.12.6 Relationship between least-squares and maximum likelihood	242
3.13	Relationship between regression and correlation	243
	3.13.1 The effects of random errors on correlation	244
	3.13.2 The maximum likelihood correlation estimator	245
	3.13.3 Correlation and regression: cause and effect	246
3.14	Hypothesis testing	249
	3.14.1 Significance levels and confidence intervals for correlation	253
	3.14.2 Analysis of variance and the F-distribution	254
3.15	Effective degrees of freedom	257
	3.15.1 Trend estimates and the integral time scale	261
3.16	Editing and despiking techniques: the nature of errors	266
	3.16.1 Identifying and removing errors	266
	3.16.2 Propagation of error	273
	3.16.3 Dealing with numbers: the statistics of roundoff	274
	3.16.4 Gauss–Markov theorem	277
3.17	Interpolation: filling the data gaps	277
	3.17.1 Equally and unequally spaced data	277
	3.17.2 Interpolation methods	279
	3.17.3 Interpolating gappy records: practical examples	286
3.18	Covariance and the covariance matrix	290

viii *Data Analysis Methods in Physical Oceanography*

	3.18.1 Covariance and structure functions	291
	3.18.2 A computational example	291
	3.18.3 Multivariate distributions	293
3.19	Bootstrap and jackknife methods	294
	3.19.1 Bootstrap method	295
	3.19.2 Jackknife method	301

Chapter 4   The Spatial Analyses of Data Fields 305

4.1	Traditional block and bulk averaging	305
4.2	Objective analysis	309
	4.2.1 Objective mapping: examples	314
4.3	Empirical orthogonal functions	319
	4.3.1 Principal axes of a single vector time series (scatter plot)	325
	4.3.2 EOF computation using the scatter matrix method	328
	4.3.3 EOF computation using singular value decomposition	332
	4.3.4 An example: deep currents near a mid-ocean ridge	334
	4.3.5 Interpretation of EOFs	336
	4.3.6 Variations on conventional EOF analysis	340
4.4	Normal mode analysis	344
	4.4.1 Vertical normal modes	344
	4.4.2 An example: normal modes of semidiurnal frequency	347
	4.4.3 Coastal-trapped waves (CTWs)	350
4.5	Inverse methods	356
	4.5.1 General inverse theory	356
	4.5.2 Inverse theory and absolute currents	361
	4.5.3 The IWEX internal wave problem	366
	4.4.4 Summary of inverse methods	370

Chapter 5   Time-series Analysis Methods 371

5.1	Basic concepts	371
5.2	Stochastic processes and stationarity	373
5.3	Correlation functions	374
5.4	Fourier analysis	380
	5.4.1 Mathematical formulation	381
	5.4.2 Discrete time series	384
	5.4.3 A computational example	387
	5.4.4 Fourier analysis for specified frequencies	388
	5.4.5 The fast Fourier transform	390
5.5	Harmonic analysis	392
	5.5.1 A least-squares method	392
	5.5.2 A computational example	395
	5.5.3 Harmonic analysis of tides	397
	5.5.4 Choice of constituents	398
	5.5.5 A computational example for tides	399
	5.5.6 Complex demodulation	402
5.6	Spectral analysis	404
	5.6.1 Spectra of deterministic and stochastic processes	409

	5.6.2	Spectra of discrete series	413
	5.6.3	Conventional spectral methods	417
	5.6.4	Spectra of vector series	425
	5.6.5	Effect of sampling on spectral estimates	432
	5.6.6	Smoothing spectral estimates (windowing)	441
	5.6.7	Smoothing spectra in the frequency domain	450
	5.6.8	Confidence intervals on spectra	454
	5.6.9	Zero-padding and prewhitening	455
	5.6.10	Spectral analysis of unevenly spaced time series	460
	5.6.11	General spectral bandwidth and Q of the system	461
	5.6.12	Summary of the standard spectral analysis approach	461
5.7	Spectral analysis (parametric methods)		464
	5.7.1	Some basic concepts	467
	5.7.2	Autoregressive power spectral estimation	468
	5.7.3	Maximum likelihood spectral estimation	478
5.8	Cross-spectral analysis		480
	5.8.1	Cross-correlation functions	480
	5.8.2	Cross-covariance method	482
	5.8.3	Fourier transform method	482
	5.8.4	Phase and cross-amplitude functions	484
	5.8.5	Coincident and quadrature spectra	485
	5.8.6	Coherence spectrum (coherency)	486
	5.8.7	Frequency response of a linear system	490
	5.8.8	Rotary cross-spectral analysis	495
5.9	Wavelet analysis		501
	5.9.1	The wavelet transform	502
	5.9.2	Wavelet algorithms	504
	5.9.3	Oceanographic examples	505
	5.9.4	The S-transformation	508
	5.9.5	The multiple filter technique	511
5.10	Digital filters		516
	5.10.1	Introduction	518
	5.10.2	Basic concepts	517
	5.10.3	Ideal filters	519
	5.10.4	Design of oceanographic filters	527
	5.10.5	Running-mean filters	532
	5.10.6	Godin-type filters	535
	5.10.7	Lanczos-window cosine filters	536
	5.10.8	Butterworth filters	543
	5.10.9	Frequency-domain (transform) filtering	551
5.11	Fractals		557
	5.11.1	The scaling exponent method	561
	5.11.2	The yardstick method	562
	5.11.3	Box counting method	563
	5.11.4	Correlation dimension	564
	5.11.5	Dimensions of multifractal functions	564
	5.11.6	Predictability	567

Appendices 569

Appendix A Units in physical oceanography — 570
Appendix B Glossary of statistical terminology — 572
Appendix C Means, variances and moment-generating functions for some common continuous variables — 576
Appendix D Statistical tables — 577
Appendix E Correlation coefficents at the 5% and 1% levels of significance for various degrees of freedom ν — 585
Appendix F Approximations and nondimensional numbers in physical oceanography — 586
Appendix G Convolution — 593

References 597

Index 621

Preface

Numerous books have been written on data analysis methods in the physical sciences over the past several decades. Most of these books lean heavily toward the theoretical aspects of data processing and few have been updated to include more modern techniques such as fractal analysis and rotary spectral decomposition. In writing this book we saw a clear need for a practical reference volume for earth and ocean sciences that brings established and modern techniques together under a single cover. The text is intended for students and established scientists alike. For the most part, graduate programs in oceanography have some form of methods course in which students learn about the measurement, calibration, processing and interpretation of geophysical data. The classes are intended to give the students needed experience in both the logistics of data collection and the practical problems of data processing and analysis. Because the class material generally is based on the experience of the faculty members giving the course, each class emphasizes different aspects of data collection and analysis. Formalism and presentation can differ widely. While it is valuable to learn from the first-hand experiences of the class instructor, it seemed to us important to have available a central reference text that could be used to provide some uniformity in the material being covered within the oceanographic community.

Many of the data analysis techniques most useful to oceanographers can be found in books and journals covering a wide variety of topics ranging from elementary statistics to wavelet transforms. Much of the technical information on these techniques is detailed in texts on numerical methods, time series analysis, and statistical techniques. In this book, we attempt to bring together many of the key data processing methods found in the literature, as well as add new information on data analysis techniques not readily available in older texts. We also provide, in Chapter 1, a description of most of the instruments used today in physical oceanography. Our hope is that the book will provide instructional material for students in the oceanographic sciences and serve as a general reference volume for those directly involved with oceanographic research.

The broad scope and rapidly evolving nature of oceanographic sciences has meant that it has not been possible for us to cover all existing or emerging data analysis methods. However, we trust that many of the methods and procedures outlined in the book will provide a basic understanding of the kinds of options available to the user for interpretation of data sets. Our intention is to describe general statistical and analytical methods that will be sufficiently fundamental to maintain a high level of utility over the years.

Finally, we believe that the analysis procedures discussed in this book apply to a wide readership in the geophysical sciences. As with oceanographers, this wider community of scientists would likely benefit from a central source of information that encompasses not only a description of the mathematical methods but also considers some of the practical aspects of data analyses. It is this synthesis between theoretical insight and the logistical limitations of real data measurement that is a primarily goal of this text.

William J. Emery and Richard E. Thomson
Boulder, Colorado and Sidney, BC

To our wives Dora Emery and Irma Thomson

Acknowledgments

Many people have contributed to this book over the years that it has taken to write. The support and encouragement we received from our colleagues while we were drafting the manuscript is most gratefully acknowledged. Bill Emery began work on the book during a sabbatical year at the Institut für Meereskunde (IFM) in Kiel and is grateful for the hospitality of Drs Gerold Siedler, Wolfgang Kraus, Walter Zenk, Rolf Kasë, Juergen Willebrand, Juergen Kielman, and others. Additional progress was made at the University of British Columbia where Mrs Hiltrud Heckel aided in typing the handwritten text. Colleagues at the University of Colorado who helped with the book include Dr Robert Leben, who reviewed the text, contributed advice, text and figures, and Mr Dan Leatzow, who converted all of the initial text and figures to the Macintosh. The author would like to thank Mr Tom Kelecy for serving as the teaching assistant the first time that the material was presented in a course on "Engineering Data Analysis". Bill Emery also thanks the students who helped improve the book during the course on data analysis and the National Space and Aeronautics Administration (NASA) whose grant funding provided the laptop computer used in preparation of the manuscript. The author also thanks his family for tolerating the ever growing number of files labelled "dampo" on the family computer.

Rick Thomson independently began work on the book through the frustration of too much time spent looking for information on data analysis methods in the literature. There was clearly a need for a reference-type book that covers the wide range of analysis techniques commonly used by oceanographers and other geoscientists. Many of the ideas for the book originated with the author's studies as a research scientist within Fisheries and Oceans Canada, but work on the book was done strictly at home during evenings and weekends. Numerous conversations with Drs Dudley Chelton and Alexander Rabinovich helped maintain the author's enthusiasm for the project. The author wishes to thank his wife and two daughters (Justine and Karen) for enduring the constant tapping of the keyboard and hours of dark despair when it looked as if the book would never come to an end, and his parents (John and Irene) for encouraging an interest in science.

The authors would like to thank the colleagues and friends who took time from their research to review sections of the text or provide figures. There were others, far too numerous to mention, whose comments and words of advice added to the usefulness of the text. We are most grateful to Dudley Chelton of Oregon State University and Alexander Rabinovich of Moscow State University who spent considerable time criticizing the more mathematical chapters of the book. Dudley proved to be a most impressive reviewer and Sasha contributed several figures that significantly improved the section on time series analysis. George Pickard, Professor Emeritus of the University of British Columbia (UBC), and Susumu Tabata, Research Scientist Emeritus of the Institute of Ocean Sciences (IOS), provided thorough, and much appreciated, reviews of Chapters 1 and 2. We thank Andrew Bennett of Oregon State University for his comments on inverse methods in Chapter 4, Brenda Burd of Ecostat Research for reviewing the bootstrap method in Chapter 3, and Steve Mihaly for reviewing Appendix A. Contributions to the text were provided by Libe Washburn (University of California, Santa Barbara), Peter Schlussel (IFM), Patrick Cummins (IOS), and Mike Woodward (IOS). Figures or data were generously provided by Mark E. Geneau (Inter-Ocean Systems, Inc.), Gail Gabel (G.S. Gabel Associates), R. Lee Gordon (RD Instruments, Inc.), Diane Masson, Humfrey Melling, George Chase, John Love, and Tom Juhász (IOS), Jason Middleton and Greg Nippard (University of New South Wales), Doug Bennett (Sea-Bird Electronics, Inc.), Dan Schaas and Marcia Gracia (General Oceanics), Mayra Pazos (National Oceanic and Atmospheric Administration), Chris Garrett (University of Victoria), David Halpern (Jet Propulsion Laboratory), Phillip Richardson (Woods Hole Oceanographic Institution), Daniel Jamous (Massachusetts Institute of Technology), Paul Dragos (Battelle Ocean Sciences), Thomas Rossby (University of Rhode Island), Lynne Talley (Scripps), Adrian Dolling and Jane Eert (Channel Consulting), and Gary Hamilton (Intelex Research).

The authors would also like to thank Anna Allen for continued interest in this project from the time she took over until now. There were a great many delays and postponements, but through it all she remained firm in her support of the project. This continued support made it easier to work through these delays and gave us courage to believe that the project would one day be completed.

Lastly, we would like to thank our colleagues who found and reported errors and omissions in the first printing of the book. Although the inevitable typos and mistakes are discouraging for the authors (and frustrating for the reader), it is better that we know about them so that they can be corrected in future printings and revisions. Our thanks to Brian Blanton (University of North Carolina), Mike Foreman (Institute of Ocean Sciences), Denis Gilbert (Institute Maurice-Lamontagne), Jack Harlan (NOAA, Boulder, Colorado), Clive Holden (Oceanographic Field Services, Pymbla, New South Wales), Frank Janssen (University of Hamburg), Masahisa Kubota (Tokai University), Robert Leben (University of Colorado), Rolf Lueck (University of Victoria), Andrew Slater (University of Colorado), and Roy Hourston (WeatherWorks Consulting, Victoria).

CHAPTER 1

Data Acquisition and Recording

1.1 INTRODUCTION

Physical oceanography is an evolving science in which the instruments, types of observations and methods of analysis have undergone considerable change over the last few decades. With most advances in oceanographic theory, instrumentation, and software, there have been significant advances in marine science. The advent of digital computers has revolutionized data collection procedures and the way that data are reduced and analyzed. No longer is the individual scientist personally familiar with each data point and its contribution to his or her study. Instrumentation and data collection are moving out of direct application by the scientist and into the hands of skilled technicians who are becoming increasingly more specialized in the operation and maintenance of equipment. New electronic instruments operate at data rates not possible with earlier mechanical devices and produce volumes of information that can only be handled by high-speed computers. Most modern data collection systems transmit sensor data directly to computer-based data acquisition systems where they are stored in digital format on some type of electronic medium such as a tape, harddrive, or optical disk. High-speed analog-to-digital (AD) converters and digital-signalprocessors (DSPs) are now used to convert voltage or current signals from sensors to digital values.

With the many technological advances taking place, it is important for oceanographers to be aware of both the capabilities and limitations of their sampling equipment. This requires a basic understanding of the sensors, the recording systems and the data-processing tools. If these are known and the experiment carefully planned, many problems commonly encountered during the processing stage can be avoided. We cannot overemphasize the need for thoughtful experimental planning and proper calibration of all oceanographic sensors. If instruments are not in nearoptimal locations or the researcher is unsure of the values coming out of the machines, then it will be difficult to believe the results gathered in the field. To be truly reliable, instruments should be calibrated on a regular basis at intervals determined by use and the susceptibility of the sensor to drift. More specifically, the output from some instruments such as the piezoelectric pressure sensors and fixed pathlength transmissometers drift with time and need to be calibrated before and after each field deployment. For example, the zero point for the Paroscientific Digiquartz (0–10,000 psi) pressure sensors used in the Hawaii Ocean Time-series (HOT) at station "Aloha" 100 km north of Honolulu drifts about 4 dbar in three years. As a consequence, the sensors are calibrated about every six months against a Paroscientific laboratory standard, which is recalibrated periodically at special calibration facilities in the United States (Lukas, 1994). Our experience also shows that over-the-side field calibrations during oceanic surveys can be highly valuable. As we discuss in the following chapters, there are a number of fundamental requirements to be considered when planning the collection of field records, including such basic considerations as the sampling interval, sampling duration and sampling location.

It is the purpose of this chapter to review many of the standard instruments and measurement techniques used in physical oceanography in order to provide the reader with a common understanding of both the utility and limitations of the resulting measurements. The discussion is not intended to serve as a detailed "user's manual" nor as an "observer's handbook". Rather, our purpose is to describe the fundamentals of the instruments in order to give some insight into the data they collect. An understanding of the basic observational concepts, and their limitations, is a prerequisite for the development of methods, techniques and procedures used to analyze and interpret the data that are collected.

Rather than treat each measurement tool individually, we have attempted to group them into generic classes and to limit our discussion to common features of the particular instruments and associated techniques. Specific references to particular company products and the quotation of manufacturer's engineering specifications have been avoided whenever possible. Instead, we refer to published material addressing the measurement systems or the data recorded by them. Those studies which compare measurements made by similar instruments are particularly valuable.

The emphasis of the instrument review section is to give the reader a background in the collection of data in physical oceanography. For those readers interested in more complete information regarding a specific instrument or measurement technique, we refer to the references at the end of the book where we list the sources of the material quoted. We realize that, in terms of specific measurement systems, and their review, this text will be quickly dated as new and better systems evolve. Still, we hope that the general outline we present for accuracy, precision and data coverage will serve as a useful guide to the employment of newer instruments and methods.

1.2 BASIC SAMPLING REQUIREMENTS

A primary concern in most observational work is the accuracy of the measurement device, a common performance statistic for the instrument. Absolute accuracy requires frequent instrument calibration to detect and correct for any shifts in behavior. The inconvenience of frequent calibration often causes the scientist to substitute instrument precision as the measurement capability of an instrument. Unlike absolute accuracy, precision is a relative term and simply represents the ability of the instrument to repeat the observation without deviation. Absolute accuracy further requires that the observation be consistent in magnitude with some absolute reference standard. In most cases, the user must be satisfied with having good precision and repeatability of the measurement rather than having absolute measurement accuracy. Any instrument that fails to maintain its precision, fails to provide data that can be handled in any meaningful statistical fashion. The best instruments are those that provide both high precision and defensible absolute accuracy.

Digital instrument resolution is measured in bits, where a resolution of N bits means that the full range of the sensor is partitioned into 2^N equal segments ($N = 1, 2,$ ...). For example, eight-bit resolution means that the specified full-scale range of the sensor, say $V = 10$ volts, is divided into $2^8 = 256$ increments, with a bit-resolution of $V/256 = 0.039$ volts. Whether the instrument can actually measure to a resolution or accuracy of $V/2^N$ units is another matter. The sensor range can always be divided into an increasing number of smaller increments but eventually one reaches a point where the value of each bit is buried in the noise level of the sensor.

1.2.1 Sampling interval

Assuming the instrument selected can produce reliable and useful data, the next highest priority sampling requirement is that the measurements be collected often enough in space and time to resolve the phenomena of interest. For example, in the days when oceanographers were only interested in the mean stratification of the world ocean, water property profiles from discrete-level hydrographic (bottle) casts were adequate to resolve the general vertical density structure. On the other hand, these same discrete-level profiles failed to resolve the detailed structure associated with interleaving and mixing processes that now are resolved by the rapid vertical sampling of modern conductivity–temperature–depth (CTD) profilers. The need for higher resolution assumes that the oceanographer has some prior knowledge of the process of interest. Often this prior knowledge has been collected with instruments incapable of resolving the true variability and may only be suggested by highly aliased (distorted) data collected using earlier techniques. In addition, theoretical studies may provide information on the scales that must be resolved by the measurement system.

For discrete digital data $x(t_i)$ measured at times t_i, the choice of the sampling increment Δt (or Δx in the case of spatial measurements) is the quantity of importance. In essence, we want to sample often enough that we can pick out the highest frequency component of interest in the time-series but not oversample so that we fill up the data storage file, use up all the battery power, or become swamped with a lot of unnecessary data. We might also want to sample at irregular intervals to avoid built-in bias in our sampling scheme. If the sampling interval is too large to resolve higher frequency components, it becomes necessary to suppress these components during sampling using a sensor whose response is limited to frequencies equal to that of the sampling frequency. As we discuss in our section on processing satellite-tracked drifter data, these lessons are often learned too late—after the buoys have been cast adrift in the sea.

The important aspect to keep in mind is that, for a given sampling interval Δt, the highest frequency we can hope to resolve is the *Nyquist* (or *folding*) *frequency*, f_N, defined as

$$f_N = 1/(2\Delta t) \tag{1.2.1}$$

We cannot resolve any higher frequencies than this. For example, if we sample every 10 h, the highest frequency we can hope to see in the data is $f_N = 0.05$ cph (cycles per hour). Equation (1.2.1) states the obvious—that it takes at least two sampling intervals (or three data points) to resolve a sinusoidal-type oscillation with period $1/f_N$ (Figure

1.2.1). In practice, we need to contend with noise and sampling errors so that it takes something like three or more sampling increments (i.e. ≥four data points) to accurately determine the highest observable frequency. Thus, f_N is an upper limit. The highest frequency we can resolve for a sampling of $\Delta t = 10$ h in Figure 1.2.1 is closer to $1/3\Delta t \approx 0.033$ cph.

An important consequence of (1.2.1) is the problem of *aliasing*. In particular, if there is considerable energy at frequencies $f > f_N$—which we obviously cannot resolve because of the Δt we picked—this energy gets folded back into the range of frequencies, $f < f_N$, which we are attempting to resolve. This unresolved energy doesn't disappear but gets redistributed within the frequency range of interest. What is worse is that the folded-back energy is disguised (or aliased) within frequency components different from those of its origin. We cannot distinguish this folded-back energy from that which actually belongs to the lower frequencies. Thus, we end up with erroneous (aliased) estimates of the spectral energy variance over the resolvable range of frequencies. An example of highly aliased data would be 13-h sampling of currents in a region having strong semidiurnal tidal currents. More will be said on this topic in Chapter 5.

As a general rule, one should plan a measurement program based on the frequencies and wavenumbers (estimated from the corresponding periods and wavelengths) of the parameters of interest over the study domain. This requirement then dictates the selection of the measurement tool or technique. If the instrument cannot sample rapidly enough to resolve the frequencies of concern it should not be used. It should be emphasized that the Nyquist frequency concept applies to both time and space and the Nyquist wavenumber is a valid means of determining the fundamental wavelength that must be sampled.

1.2.2 Sampling duration

The next concern is that one samples long enough to establish a statistically significant picture of the process being studied. For time-series measurements, this amounts to a requirement that the data be collected over a period sufficiently long that

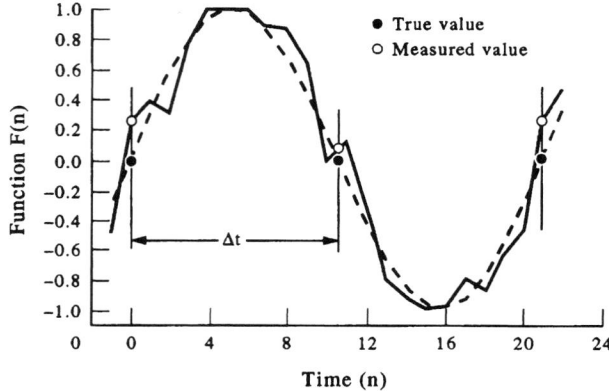

Figure 1.2.1. Plot of the function $F(n) = \sin(2\pi n/20 + \phi)$ *where time is given by the integer* $n = -1, 0, ..., 24$. *The period* $2\Delta t = 1/f_N$ *is 20 units and* ϕ *is a random phase with a small magnitude in the range* ± 0.1. *Open circles denote measured points and solid points the curve* $F(n)$. *Noise makes it necessary to use more than three data values to accurately define the oscillation period.*

repeated cycles of the phenomenon are observed. This also applies to spatial sampling where statistical considerations require a large enough sample to define multiple cycles of the process being studied. Again, the requirement places basic limitations on the instrument selected for use. If the equipment cannot continuously collect the data needed for the length of time required to resolve repeated cycles of the process, it is not well suited to the measurement required.

Consider the duration of the sampling at time step Δt. The longer we make the record the better we are to resolve different frequency components in the data. In the case of spatially separated data, Δx, resolution increases with increased spatial coverage of the data. It is the total record length $T = N\Delta t$ obtained for N data samples that: (1) determines the lowest frequency (the *fundamental frequency*)

$$f_o = 1/(N\Delta t) = 1/T \qquad (1.2.2)$$

that can be extracted from the time-series record; (2) determines the frequency resolution or minimum difference in frequency $\Delta f = |f_2 - f_1| = 1/N\Delta t$ that can be resolved between adjoining frequency components, f_1 and f_2 (Figure 1.2.2); and (3) determines the amount of band averaging (averaging of adjacent frequency bands) that can be applied to enhance the statistical significance of individual spectral estimates. In Figure 1.2.2, the two separate waveforms of equal amplitude but different frequency produce a single spectrum. The two frequencies are well resolved for $\Delta f = 2/N\Delta t$ and $3/2N\Delta t$, just resolved for $\Delta f = 1/N\Delta t$, and not resolved for $\Delta f = 1/2N\Delta t$.

In theory, we should be able to resolve all frequency components, f, in the frequency range $f_o \leq f \leq f_N$, where f_N and f_o are defined by (1.2.1) and (1.2.2), respectively. Herein lies a classic sampling problem. In order to resolve the frequencies of interest in a time-series, we need to sample for a long time (T large) so that f_o covers the low end of the frequency spectrum and Δf is small (frequency resolution is high). At the same time, we would like to sample sufficiently rapidly (Δt small) so that f_N extends beyond all frequency components with significant spectral energy. Unfortunately, the longer and more rapidly we want to sample the more data we need to collect and store, the more time, effort and money we need to put into the sampling and the better resolution we require from our sensors.

Our ability to resolve frequency components follows from Rayleigh's criterion for the resolution of adjacent spectral peaks in light shone onto a diffraction grating. It states that two adjacent frequency components are just resolved when the peaks of the spectra are separated by frequency difference $\Delta f = f_o = 1/N\Delta t$ (Figure 1.2.2). For example, to separate the spectral peak associated with the lunar–solar semidiurnal tidal component M_2 (frequency = 0.08051 cph) from that of the solar semidiurnal tidal component S_2 (0.08333 cph), for which $\Delta f = 0.00282$ cph, requires $N = 355$ data points at a sampling interval $\Delta t = 1$ h or $N = 71$ data points at $\Delta t = 5$ h. Similarly, a total of 328 data values at 1-h sampling are needed to separate the two main diurnal constituents K_1 and O_1 ($\Delta f = 0.00305$ cph). Note that if f_N is the highest frequency we can measure and f_o is the limit of frequency resolution, then

$$f_N/f_o = (1/2\Delta t)/(1/N\Delta t) = N/2 \qquad (1.2.3)$$

is the maximum number of Fourier components we can hope to estimate in any analysis.

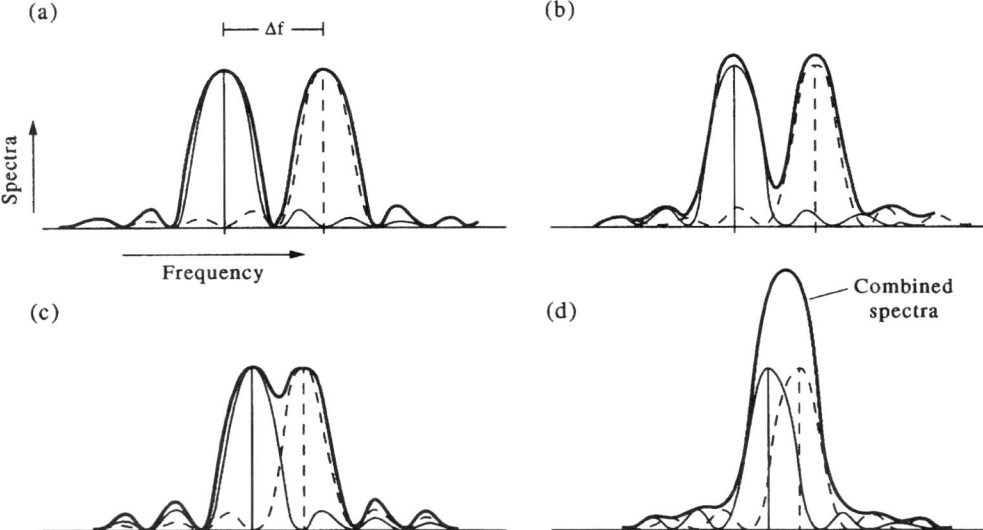

Figure 1.2.2. *Spectral peaks of two separate waveforms of equal amplitude and frequencies f_1 and f_2 (dashed and thin line) together with the calculated spectrum (solid line). (a) and (b) are well-resolved spectra; (c) just resolved spectra; and (d) not resolved. Thick solid line is total spectrum for two underlying signals with slightly different peak frequencies.*

1.2.3 Sampling accuracy

According to the two previous sections, we need to sample long and often if we hope to resolve the range of scales of interest in the variables we are measuring. It is intuitively obvious that we also need to sample as accurately as possible—with the degree of recording accuracy determined by the response characteristics of the sensors, the number of bits per data record (or parameter value) needed to raise measurement values above background noise, and the volume of data we can live with. There is no use attempting to sample the high or low ends of the spectrum if the instrument cannot respond rapidly or accurately enough to resolve changes in the parameter being measured. In addition, there are several approaches to this aspect of data sampling including the brute-force approach in which we measure as often as we can at the degree of accuracy available and then improve the statistical reliability of each data record through post-survey averaging, smoothing, and other manipulation.

1.2.4 Burst sampling versus continuous sampling

Regularly-spaced, digital time-series can be obtained in two different ways. The most common approach is to use a *continuous sampling mode*, in which the data are sampled at equally spaced intervals $t_k = t_o + k\Delta t$ from the start time t_o. Here, k is a positive integer. Regardless of whether the equally spaced data have undergone internal averaging or decimation using algorithms built into the machine, the output to the data storage file is a series of individual samples at times t_k. (Here, "decimation" is used in the loose sense of removing every nth data point, where n is any positive integer, and not in the sense of the ancient Roman technique of putting to death one in ten soldiers in a legion guilty of mutiny or other crime.) Alternatively, we can use a *burst sampling mode*, in which rapid sampling is undertaken over a relatively short time interval Δt_B or "burst" embedded within each regularly spaced time interval, Δt. That is, the data are sampled at high frequency for a short duration starting (or ending) at times t_k for which the burst duration $\Delta t_B \ll \Delta t$. The instrument "rests" between bursts.

There are advantages to the burst sampling scheme, especially in noisy (high frequency) environments where it may be necessary to average-out the noise to get at the frequencies of interest. Burst sampling works especially well when there is a "spectral gap" between fluctuations at the high and low ends of the spectrum. As an example, there is typically a spectral gap between surface gravity waves in the open ocean (periods of 1–20 s) and the 12-hourly motions that characterize semidiurnal tidal currents. Thus, if we wanted to measure surface tidal currents using the burst-mode option for our current meter, we could set the sampling to a 2-min burst every hour; this option would smooth out the high-frequency wave effects but provide sufficient numbers of velocity measurements to resolve the tidal motions. Burst sampling enables us to filter out the high-frequency noise and obtain an improved estimate of the variability hidden underneath the high-frequency fluctuations. In addition, we can examine the high-frequency variability by scrutinizing the burst sampled data. If we were to sample rapidly enough, we could estimate the surface gravity wave energy spectrum. Many oceanographic instruments use (or have provision for) a burst-sampling data collection mode. The "duty cycle" often used to collect positional data from satellite-tracked drifters is a cost-saving form of burst sampling in which all positional data within a 24-h period (about 10 satellite fixes) are collected only every third day. Tracking costs paid to Service Argos are reduced by a factor of three using the duty cycle. Problems arise when the length of each burst is too short to resolve energetic motions with periods comparable to the burst sample length. In the case of satellite-tracked drifters poleward of tropical latitudes, these problems are associated with highly energetic inertial motions whose periods $T = 1/(2\Omega \sin \theta)$ are comparable to the 24-h duration of the burst sample (here, $\Omega = 0.1161 \times 10^{-4}$ cycles per second is the earth's rate of rotation and $\theta \equiv$ latitude). Since 1992, it has been possible to improve resolution of high-frequency motions using a 1/3 duty cycle of 8 h "on" followed by 16 h "off". According to Bograd *et al.* (1999), even better resolution of high-frequency mid-latitude motions could be obtained using a duty cycle of 16 h "on" followed by 32 h "off".

1.2.5 Regularly versus irregularly sampled data

In certain respects, an irregular sampling in time or nonequidistant placement of instruments can be more effective than a more esthetically appealing uniform sampling. For example, unequal spacing permits a more statistically reliable resolution of oceanic spatial variability by increasing the number of quasi-independent estimates of the dominant wavelengths (wavenumbers). Since oceanographers are almost always faced with having fewer instruments than they require to resolve oceanic features, irregular spacing can also be used to increase the overall spatial coverage (fundamental wavenumber) while maintaining the small-scale instrument separation for Nyquist wavenumber estimates. The main concern is the lack of redundancy should certain key instruments fail, as often seems to happen. In this case, a quasi-regular spacing between locations is better. Prior knowledge of the scales of variability to expect is a definite plus in any experimental array design.

In a sense, the quasi-logarithmic vertical spacing adopted by oceanographers for bottle cast (hydrographic) sampling of 0, 10, 20, 30, 50, 75, 100, 125, 150 m, etc. represents a "spectral window" adaptation to the known physical–chemical structure of the ocean. Highest resolution is required near the surface where vertical changes are most rapid. Similarly, an uneven spatial arrangement of observations increases the number of quasi-independent estimates of the wavenumber spectrum. Digital data are most often sampled (or subsampled) at regularly-spaced time increments. Aside from the usual human propensity for order, the need for regularly-spaced data derives from the fact that most analysis methods have been developed for regular-spaced data. However, digital data do not necessarily need to be sampled at regularly-spaced time increments to give meaningful results, although some form of interpolation between values may eventually be required.

1.2.6 Independent realizations

As we review the different instruments and methods, the reader should keep in mind the three basic concerns of accuracy/precision, resolution (spatial and temporal), and statistical significance (statistical sampling theory). A fundamental consideration in ensuring the statistical significance of a set of measurements is the need for independent realizations. If repeat measurements of a process are strongly correlated, they provide no new information and do not contribute to the statistical significance of the measurements. Often a subjective decision must be made on the question of statistical independence. While this concept has a formal definition, in practice it is often difficult to judge. A simple guide suggested here is that any suite of measurements that is highly correlated (in time or space) cannot be independent. At the same time, a group of measurements that is totally uncorrelated, must be independent. In the case of no correlation, the number of "degrees of freedom" is defined by the total number of measurements; for the case of perfect correlation, the redundancy of the data values reduces the degrees of freedom to one for scalar quantity and to two for a vector quantity. The degree of correlation in the data set provides a way of roughly estimating the number of degrees of freedom within a given suite of observations. While more precise methods will be presented later in this text, a simple linear relation between degrees of freedom and correlation often gives the practitioner a way to proceed without developing complex mathematical constructs.

As will be discussed in detail later, all of these sampling recommendations have statistical foundations and the guiding rules of probability and estimation can be carefully applied to determine the sampling requirements and dictate the appropriate measurement system. At the same time, these same statistical methods can be applied to existing data in order to better evaluate their ability to measure phenomena of interest. These comments are made to assist the reader in evaluating the potential of a particular instrument (or method) for the measurement of some desired variable.

1.3 TEMPERATURE

The measurement of temperature in the ocean uses conventional techniques except for deep observations where hydrostatic pressures are high and there is a need to protect the sensing system from ambient depth/temperature changes higher in the water column as the sensor is returned to the ship. Temperature is the ocean property that is easiest to measure accurately. Some of the ways in which ocean temperature can be measured are:

(a) Expansion of a liquid or a metal.
(b) Differential expansion of two metals (bimetallic strip).
(c) Vapor pressure of a liquid.
(d) Thermocouples.
(e) Change in electrical resistance.
(f) Infrared radiation from the sea surface.

In most of these sensing techniques, the temperature effect is very small and some form of amplification is necessary to make the temperature measurement detectable. Usually, the response is nearly linear with temperature so that only the first-order term is needed when converting the sensor measurement to temperature. However, in order to achieve high precision over large temperature ranges, second, third and even fourth order terms must sometimes be used to convert the measured variable to temperature.

1.3.1 Mercury thermometers

Of the above methods, (a), (e), and (f) have been the most widely used in physical oceanography. The most common type of the liquid expansion sensor is the mercury-in-glass thermometer. In their earliest oceanographic application, simple mercury thermometers were lowered into the ocean with hopes of measuring the temperature at great depths in the ocean. Two effects were soon noticed. First, thermometer housings with insufficient strength succumbed to the greater pressure in the ocean and were crushed. Second, the process of bringing an active thermometer through the oceanic vertical temperature gradient sufficiently altered the deeper readings that it was not possible to accurately measure the deeper temperatures. An early solution to this problem was the development of min–max thermometers that were capable of retaining the minimum and maximum temperatures encountered over the descent and ascent of the thermometer. This type of thermometer was widely used on the Challenger expedition of 1873–1876.

The real breakthrough in thermometry was the development of reversing thermometers, first introduced in London by Negretti and Zambra in 1874 (Sverdrup et al., 1942, p. 349). The reversing thermometer contains a mechanism such that, when the thermometer is inverted, the mercury in the thermometer stem separates from the bulb reservoir and captures the temperature at the time of inversion. Subsequent temperature changes experienced by the thermometer have limited effects on the amount of mercury in the thermometer stem and can be accounted for when the temperature is read on board the observing ship. This "break-off" mechanism is based on the fact that more energy is required to create a gas-mercury interface (i.e. to break the mercury) than is needed to expand an interface that already exists. Thus, within the "pigtail" section of the reversing thermometer is a narrow region called the "break-off point", located near appendix C in Figure 1.3.1, where the mercury will break when the thermometer is inverted.

The accuracy of the reversing thermometer depends on the precision with which this break occurs. In good reversing thermometers this precision is better than 0.01°C. In standard mercury-in-glass thermometers, as well as in reversing thermometers, there are concerns other than the break point which affect the precision of the temperature measurement. These are:

(a) Linearity in the expansion coefficient of the liquid.
(b) The constancy of the bulb volume.

10 *Data Analysis Methods in Physical Oceanography*

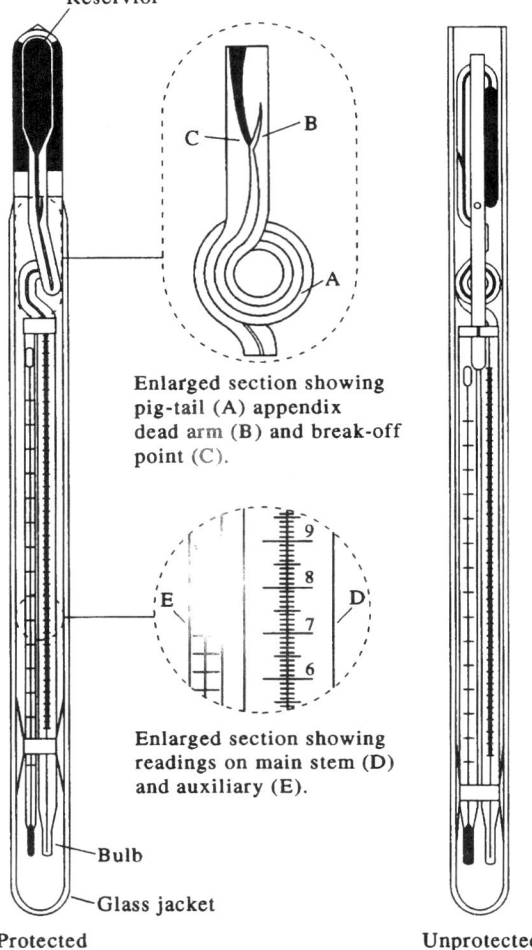

Figure 1.3.1. Details of a reversing mercury thermometer showing the "pigtail appendix".

(c) The uniformity of the capillary bore.
(d) The exposure of the thermometer stem to temperatures other than the bulb temperature.

Mercury expands in a near-linear manner with temperature. As a consequence, it has been the liquid used in most high precision, liquid–glass thermometers. Other liquids such as alcohol and toluene are used in precision thermometers only for very low temperature applications where the higher viscosity of mercury is a limitation. Expansion linearity is critical in the construction of the thermometer scale which would be difficult to engrave precisely if expansion were nonlinear.

In a mercury thermometer, the volume of the bulb is equivalent to about 6000 stem-degrees Celsius. This is known as the "degree volume" and usually is considered to comprise the bulb plus the portion of the stem below the mark. If the thermometer is to retain its calibration, this volume must remain constant with a precision not commonly realized by the casual user. For a thermometer precision within $\pm 0.01°C$, the bulb volume must remain constant within one part in 600,000. Glass does not have ideal mechanical properties and it is known to exhibit some plastic behavior and deform under sustained stress. Repeated exposure to high pressures may produce permanent deformation and a consequent shift in bulb volume. Therefore, precision can only be maintained by frequent laboratory calibration. Such shifts in bulb volume can be detected and corrected by the determination of the "ice point" (a slurry of water plus ice) which should be checked frequently if high accuracy is required. The procedure is more or less obvious but a few points should be considered. First the ice should be made from distilled water and the water–ice mixture should also be made from distilled water. The container should be insulated and at least 70% of the bath in contact with the thermometer should be chopped ice. The thermometer should be immersed for five or more minutes during which time the ice–water mixture should be stirred continuously. The control temperature of the bath can be taken by an accurate thermometer of known reliability. Comparison with the temperature of the reversing thermometer, after the known calibration characteristics have been accounted for, will give an estimate of any offsets inherent in the use of the reversing thermometer in question.

The uniformity of the capillary bore is critical to the accuracy of the mercury thermometer. In order to maintain the linearity of the temperature scale it is necessary to have a uniform capillary as well as a linear response liquid element. Small variations in the capillary can occur as a result of small differences in cooling during its construction or to inhomogeneities in the glass. Errors resulting from the variations in capillary bore can be corrected through calibration at known temperatures. The resulting corrections, including any effect of the change in bulb volume, are known as "index corrections". These remain constant relative to the ice point and, once determined, can be corrected for a shift in the ice point by addition or subtraction of a constant amount. With proper calibration and maintenance, most of the mechanical defects in the thermometer can be accounted for. Reversing thermometers are then capable of accuracies of $\pm 0.01°C$, as given earlier for the precision of the mercury break-point. This accuracy, of course, depends on the resolution of the temperature scale etched on the thermometer. For high accuracy in the typically weak vertical temperature gradients of the deep ocean, thermometers are etched with scale intervals between 0.1 and 0.2°C. Most reversing thermometers have scale intervals of 0.1°C.

The reliability and calibrated absolute accuracy of reversing thermometers continue to provide a standard temperature measurement against which all forms of electronic sensors are compared and evaluated. In this role as a calibration standard, reversing thermometers continue to be widely used. In addition, many oceanographers still believe that standard hydrographic stations made with sample bottles and reversing thermometers, provide the only reliable data. For these reasons, we briefly describe some of the fundamental problems that occur when using reversing thermometers. An understanding of these errors may also prove helpful in evaluating the accuracy of reversing thermometer data that are archived in the historical data file. The primary malfunction that occurs with a reversing thermometer is a failure of the mercury to break at the correct position. This failure is caused by the presence of gas (a bubble) somewhere within the mercury column. Normally all thermometers contain some gas within the mercury. As long as the gas bubble has sufficient mercury compressing it, the bubble volume is negligible, but if the bubble gets into the upper part of the capillary tube it expands and causes the mercury to break at the bubble rather than at the break-off point. The proper place for this resident gas is at the bulb end of the mercury; for this reason it is recommended that reversing thermometers always be stored and transported in the bulb-up (reservoir-down) position. Rough handling can be the cause of bubble formation higher up in the capillary tube. Bubbles lead to consistently offset temperatures and a record of the thermometer history can clearly indicate when such a malfunction has occurred. Again the practice of renewing, or at least checking, the thermometer calibration is essential to ensuring accurate temperature measurements. As with most oceanographic equipment, a thermometer with a detailed history is much more valuable than a new one without some prior use.

There are two basic types of reversing thermometers: (1) protected thermometers which are encased completely in a glass jacket and not exposed to the pressure of the water column; and (2) unprotected thermometers for which the glass jacket is open at one end so that the reservoir experiences the increase of pressure with ocean depth, leading to an apparent increase in the measured temperature. The increase in temperature with depth is due to the compression of the glass bulb, so that if the compressibility of the glass is known from the manufacturer, the pressure and hence the depth can be inferred from the temperature difference, $\Delta T = T_{\text{Unprotected}} - T_{\text{Protected}}$. The difference in thermometer readings, collected at the same depth, can be used to compute the depth to an accuracy of about 1% of the depth. This subject will be treated more completely in the section on depth/pressure measurement. We note here that the 1% accuracy for reversing thermometers exceeds the accuracy of 2–3% one normally expects from modern depth sounders.

Unless collected for a specific observational program or taken as calibrations for electronic measurement systems, reversing thermometer data are most commonly found in historical data archives. In such cases, the user is often unfamiliar with the precise history of the temperature data and thus cannot reconstruct the conditions under which the data were collected and edited. Under these conditions one generally assumes that the errors are of two types; either they are large offsets (such as errors in reading the thermometer) which are readily identifiable by comparison with other regional historical data, or they are small random errors due to a variety of sources and difficult to identify or separate from real physical oceanic variability. Parallax errors, which are one of the main causes of reading errors, are greatly reduced through use of an eye-piece magnifier. Identification and editing of these errors depends on the problem being studied and will be discussed in a later section on data processing.

1.3.2. The mechanical bathythermograph (MBT)

The MBT uses a liquid-in-metal thermometer to register temperature and a Bourdon tube sensor to measure pressure. The temperature sensing element is a fine copper tube nearly 17 m long filled with toluene (Figure 1.3.2). Temperature readings are recorded by a mechanical stylus which scratches a thin line on a coated glass slide. Although this instrument has largely been replaced by the expendable bathythermograph (XBT), the historical archives contain numerous temperature profiles collected using this device. It is, therefore, worthwhile to describe the instrument and the data it measures. Only the temperature measurement aspect of this device will be considered; the pressure/depth recording capability will be addressed in a latter section.

There are numerous limitations to the MBT. To begin with, it is restricted to depths less than 300 m. While the MBT was intended to be used with the ship underway, it is only really possible to use it successfully when the ship is traveling at no more than a

Data Acquisition and Recording 13

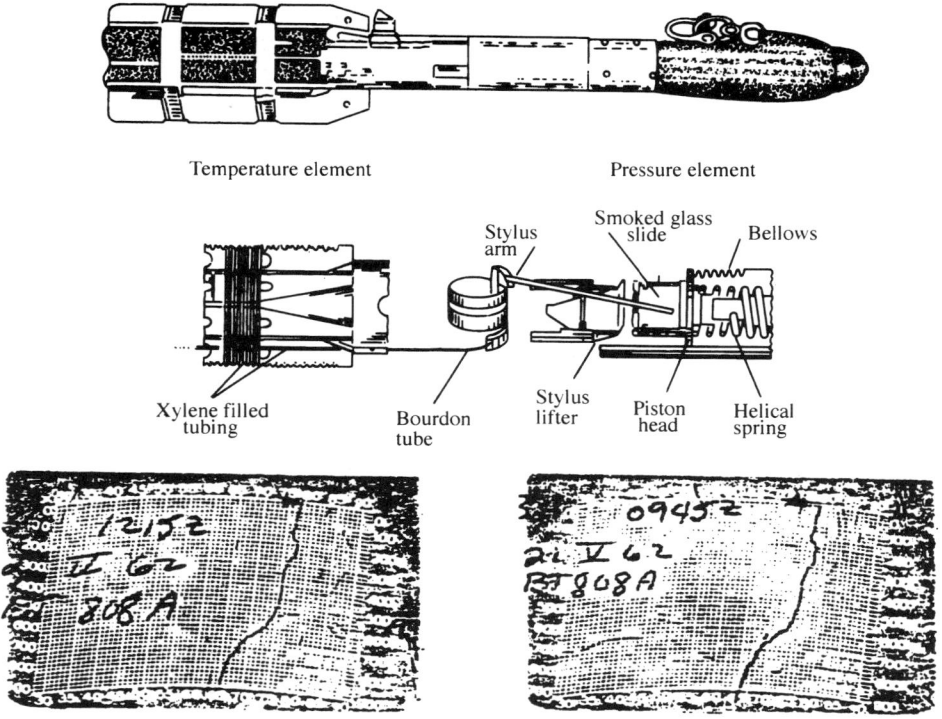

Figure 1.3.2. A bathythermograph showing its internal construction and sample BT slides.

few knots. At higher speeds, it becomes impossible to retrieve the MBT without the risk of hitting the instrument against the ship. Higher speeds also make it difficult to properly estimate the depth of the probe from the amount of wire out. The temperature accuracy of the MBT is restricted by the inherent lower accuracy of the liquid-in-metal thermometer. Metal thermometers are also subject to permanent deformation. Since metal is more subject to changes at high temperatures than is glass it is possible to alter the performance of the MBT by continued exposure to higher temperatures (i.e. by leaving the probe out in the sun). The metal return spring of the temperature stylus is also a source of potential problems in that it is subject to hysteresis and creep. Hysteresis, in which the up-trace does not coincide with the down-trace, is especially prevalent when the temperature differences are small. Creep occurs when the metal is subjected to a constant loading for long periods. Thus, an MBT continuously used in the heat of the tropics may be found later to have a slight positive temperature error.

Most of the above errors can be detected and corrected for by frequent calibration of the MBT. Even with regular calibration it is doubtful that the stated precision of 0.1°F (0.06°C) can be attained. Here, the value is given in °F since most of the MBTs were produced with these temperature scales. When considering MBT data from the historical data files, it should be realized that these data were entered into the files by hand. The usual method was to produce an enlarged black-and-white photograph of the temperature trace using the nonlinear calibration grid unique to each instrument. Temperature values were then read off of these photographs and entered into the data file at the corresponding depths. The usual procedure was to record temperatures for a fixed depth interval (i.e. 5 or 10 m) rather than to select out inflection points that best described the temperature profile. The primary weakness of this procedure is the ease with which incorrect values can enter the data file through misreading the temperature trace or incorrectly entering the measured value. Usually these types of errors result in large differences with the neighboring values and can be easily identified. Care should be taken, however, to remove such values before applying objective methods to search for smaller random errors. It is also possible that data entry errors can occur in the entry of date, time and position of the temperature profile and tests should be made to detect these errors.

1.3.3. Resistance thermometers (expendable bathythermograph: XBT)

Since the electrical resistance of metals, and other materials, changes with temperature, these materials can be used as temperature sensors. The resistance (R) of most metals depends on temperature (T) and can be expressed as a polynomial

$$R = R(1 + aT + bT^2 + cT^3 + ...) \tag{1.2.4}$$

where a, b, and c are constants. In practice, it is usually assumed that the response is linear over some limited temperature range and the proportionality can be given by the value of the coefficient a (called the temperature resistance coefficient). The most commonly used metals are copper, platinum, and nickel which have temperature coefficients of 0.0043, 0.0039, and 0.0066 (°C)$^{-1}$, respectively. Of these, copper has the most linear response but its resistance is low so that a thermal element would require many turns of fine wire and would consequently be expensive to produce. Nickel has a very high resistance but deviates sharply from linearity. Platinum has a relatively high resistance level, is very stable and has a relatively linear behavior. For these reasons, platinum resistance thermometers have become a standard by which the international scale of temperature is defined. Platinum thermometers are also widely used as laboratory calibration standards and have accuracies of ±0.001°C.

The semiconductors form another class of resistive materials used for temperature measurements. These are mixtures of oxides of metals such as nickel, cobalt, and manganese which are molded at high pressure followed by sintering (i.e. heating to incipient fusion). The types of semiconductors used for oceanographic measurements are commonly called thermistors. These thermistors have the advantages that: (1) the temperature resistance coefficient of -0.05(°C)$^{-1}$ is about ten times as great as that for copper; and (2) the thermistors may be made with high resistance for a very small physical size.

The temperature coefficient of thermistors is negative which means that the resistance decreases as temperature increases. This temperature coefficient is not a constant except over very small temperature ranges; hence the change of resistance with temperature is not linear. Instead, the relationship between resistance and temperature is given by

$$R(T) = R_o \exp[\beta(T^{-1} - T_o^{-1})] \tag{1.2.5}$$

where $R_o = \beta/T^2$ is the conventional temperature coefficient of resistance, and T and T_o are two absolute temperatures (K) with the respective resistance values of $R(T)$ and

R_o. Thus, we have a relationship whereby temperature T can be computed from the measurement of resistance $R(T)$.

One of the most common uses of thermistors in oceanography is in expendable bathythermographs (XBTs). The XBT was developed to provide an upper ocean temperature profiling device that operated while the ship was underway. The crucial development was the concept of depth measurement using the elapsed time for the known fall rate of a "freely-falling" probe. To achieve "free-fall", independent of the ship's motion, the data transfer cable is constructed from fine copper wire with feed-spools in both the sensor probe and in the launching canister (Figure 1.3.3). The details of the depth measurement capability of the XBT will be discussed and evaluated in the section on depth/pressure measurements.

The XBT probes employ a thermistor placed in the nose of the probe as the temperature sensing element. According to the manufacturer (Sippican Corp.; Marion, Massachusetts, U.S.A.), the accuracy of this system is $\pm 0.1°C$. This figure is determined from the characteristics of a batch of semiconductor material which has known resistance–temperature $(R-T)$ properties. To yield a given resistance at a standard temperature, the individual thermistors are precision-ground, with the XBT probe thermistors ground to yield 5000 Ω (Ω is the symbol for the unit of ohms) at 25°C (Georgi et al., 1980). If the major source of XBT probe-to-probe variability can be attributed to imprecise grinding, then a single-point calibration should suffice to reduce this variability in the resultant temperatures. Such a calibration was carried out by Georgi et al. (1980) both at sea and in the laboratory.

To evaluate the effects of random errors on the calibration procedure, twelve probes were calibrated repeatedly. The mean differences between the measured and bath temperatures was $\pm 0.045°C$ with a standard deviation of $0.01°C$. For the overall calibration comparison, 18 cases of probes (12 probes per case) were examined. Six cases of T7s (good to 800 m and up to 30 knots) and two cases of T6s (good to 500 m and at less than 15 knots) were purchased new from Sippican while the remaining 10 cases of T4s (good to 500 m up to 30 knots) were acquired from a large pool of XBT probes manufactured in 1970 for the U.S. Navy. The overall average standard deviation for the probes was $0.023°C$ which then reduces to $0.021°C$ when consideration is made for the inherent variability of the calibration procedure.

A separate investigation was made of the $R-T$ relationship by studying the response characteristics for nine probes. The conclusion was that the $R-T$ differences ranged from $+0.011°C$ to $-0.014°C$ which then means that the measured relationships were within $\pm 0.014°C$ of the published relationship and that the calculation of new coefficients, following Steinhart and Hart (1968), is not warranted. Moreover the final conclusions of Georgi et al. (1980) suggest an overall accuracy for XBT thermistors of $\pm 0.06°C$ at the 95% confidence level and that the consistency between thermistors is sufficiently high that individual probe calibration is not needed for this accuracy level.

Another method of evaluating the performance of the XBT system is to compare XBT temperature profiles with those taken at the same time with an higher accuracy profiler such as a CTD system. Such comparisons are discussed by Heinmiller et al. (1983) for data collected in both the Atlantic and the Pacific using calibrated CTD systems. In these comparisons, it is always a problem to achieve true synopticity in the data collection since the XBT probe falls much faster than the recommended drop rate for a CTD probe. Most of the earlier comparisons between XBT and CTD profiles (Flierl and Robinson, 1977; Seaver and Kuleshov, 1982) were carried out using XBT temperature profiles collected between CTD stations separated by 30 km. For the

16 *Data Analysis Methods in Physical Oceanography*

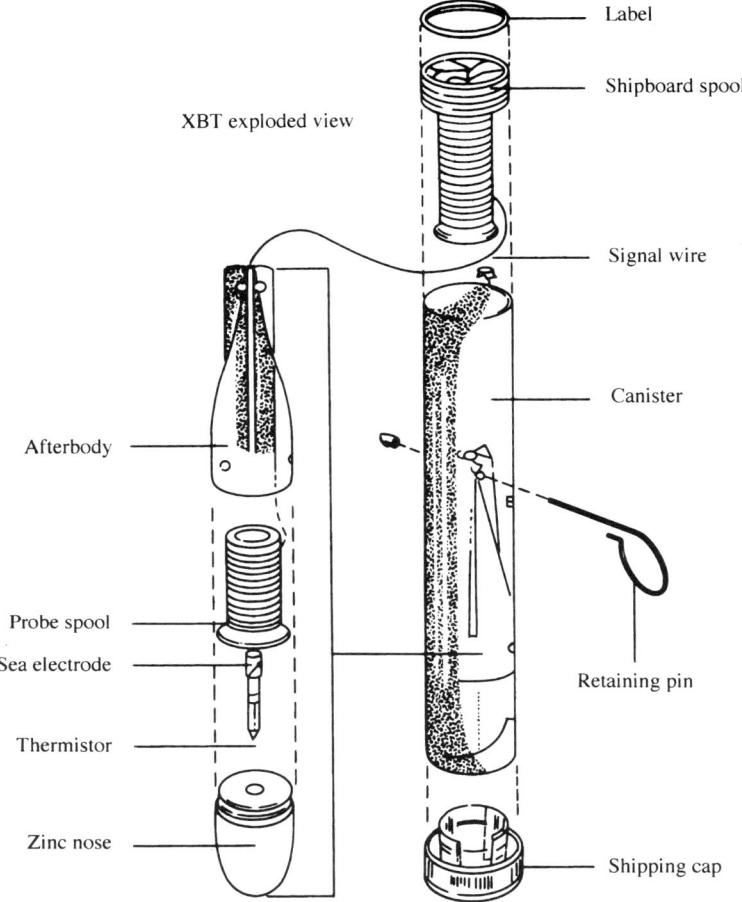

Figure 1.3.3. Exploded view of a Sippican Oceanographic Inc. XBT showing spool and canister.

purposes of intercomparison, it is better for the XBT and CTD profiles to be collected as simultaneously as possible.

The primary error discussed by Heinmiller *et al.* (1983) is that in the measurement of depth rather than temperature. There were, however, significant differences between temperatures measured at depths where the vertical temperature gradient was small and the depth error should make little or no contribution. Here, the XBT temperatures were found to be systematically higher than those recorded by the CTD. Sample comparisons were divided by probe type and experiment. The T4 probes (as defined above) yielded a mean XBT-CTD difference of about 0.19°C while the T7s (defined above) had a lower mean temperature difference of 0.13°C. Corresponding standard deviations of the temperature differences were 0.23°C, for the T4s, and 0.11°C for the T7s. Taken together, these statistics suggest an XBT accuracy less than the ±0.1°C given by the manufacturer and far less than the 0.06°C reported by Georgi *et al.* (1980) from their calibrations.

From these divergent results, it is difficult to decide where the true XBT temperature accuracy lies. Since the Heinmiller *et al.* (1983) comparisons were made *in situ* there are many sources of error that could contribute to the larger temperature differences. Even though most of the CTD casts were made with calibrated instruments, errors in operational procedures during collection and archival could add significant errors to the resultant data. Also, it is not easy to find segments of temperature profiles with no vertical temperature gradient and therefore it is difficult to ignore the effect of the depth measurement error on the temperature trace. It seems fair to conclude that the laboratory calibrations represent the ideal accuracy possible with the XBT system (i.e. better than ±0.1°C). In the field, however, one must expect other influences that will reduce the accuracy of the XBT measurements and an overall accuracy slightly more than ±0.1°C is perhaps realistic. Some of the sources of these errors can be easily detected, such as an insulation failure in the copper wire which results in single step offsets in the resulting temperature profile. Other possible temperature error sources are interference due to shipboard radio transmission (which shows up as high frequency noise in the vertical temperature profile) or problems with the recording system. Hopefully, these problems are detected before the data are archived in historical data files.

In closing this section we comment that, until recently, most XBT data were digitized by hand. The disadvantage of this procedure is that chart paper recording doesn't fully realize the potential digital accuracy of the sensing system and that the opportunities for operator recording errors are considerable. Again, some care should be exercised in editing out these large errors which usually result from the incorrect hand recording of temperature, date, time or position. It is becoming increasingly popular to use digital XBT recording systems which improve the accuracy of the recording and eliminate the possibility of incorrectly entering the temperature trace. Such systems are described, for example, in Stegen *et al.* (1975) and Emery *et al.* (1986). Today, essentially all research XBT data are collected with digital systems, while the analog systems are predominantly used by various international navies.

1.3.4 Salinity/conductivity–temperature–depth profilers

Resistance thermometers are widely used on continuous profilers designed to replace the earlier hydrographic profiles collected using a series of sampling bottles. The new *in situ* electronic instruments continuously sample the water temperature, providing much higher resolution information on the ocean's vertical and horizontal temperature structure. Since density also depends on salinity, electronic sensors were developed to measure salinity *in situ* and were incorporated into the profiling system. As discussed by Baker (1981), an early electronic profiling system for temperature and salinity was described by Jacobsen (1948). The system was limited to 400 m and used separate supporting and data transfer cables. Next, a system called the STD (salinity–temperature–depth) profiler was developed by Hamon and Brown in the mid-1950s (Hamon, 1955; Hamon and Brown, 1958). The evolution of the conductivity measurement, used to derive salinity, will be discussed in the section on salinity. This evolution led to the introduction of the conductivity–temperature–depth (CTD) profiling system (Brown, 1974). This name change identified improvements not only in the conductivity sensor but also in the temperature sensing system designed to overcome the mismatch in the response times of the temperature and conductivity sensors. This mismatch often resulted in erroneous salinity spikes in the earlier STD systems (Dantzler, 1974).

Most STD/CTD systems use a platinum resistance thermometer as one leg of an impedance bridge from which the temperature is determined. An important development was made by Hamon and Brown (1958) where the sensing elements were all connected to oscillators that converted the measured variables to audio frequencies which could then be sent to the surface via a single conducting element in the profiler support cable. The outer cable sheath acted as both mechanical support and the return conductor. This data transfer method has subsequently been used on most electronic profiling systems. The early STDs were designed to operate to 1000 m and had a temperature range of 0–30°C with an accuracy of ±0.15°C. Later STDs, such as the widely used Plessey Model 9040, had accuracies of ±0.05°C with temperature ranges of −2 to +18°C or +15 to +35°C (range was switched automatically during a cast). Modern CTDs, such as the Sea Bird Electronics SBE 25 and the General Oceanics MK3C (modified after the EG&G Mark V) (Figure 1.3.4) have accuracies of ±0.002°C over a range of −3 to +32°C and a stability of 0.001°C/month (Brown and Morrison, 1978; Hendry, 1993). To avoid the problem of sensor response mismatch the MK3C CTD combines the accuracy and stability of a platinum resistance thermometer with the speed of a thermistor. The response time of the platinum thermometer is typically 250 ms while the response time of the conductivity cell (for a fall rate of 1 m/s) is typically 25 ms. The miniature thermistor probe matches the faster response of the conductivity cell with a response time of 25 ms. These two temperature measurements are combined to yield a rapid and highly accurate temperature. The response of the combined system to a step input is shown in Figure 1.3.5 taken from Brown and Morrison (1978). Later modifications have sent the platinum resistance temperature up the cable along with the fast response thermistor temperature for later combination. It is also possible to separate the thermometer from the conductivity cell so that the spatial separation acts as a time delay as the unit falls through the water (Topham and Perkins, 1988).

1.3.5 Dynamic response of temperature sensors

Before considering more closely the problem of sensor response time for STD/CTD systems, it is worthwhile to review the general dynamic characteristics of temperature measuring systems. For example, no temperature sensor responds instantaneously to changes in the environment which it is measuring. If the environment temperature is changing, the sensor element lags in its response. A simple example is a reversing thermometer which, lowered through the water column, would at no time read the correct environment temperature until it had been stopped and allowed to equilibrate for some time. The time (K) that it takes the thermometer to respond to the temperature of a new environment is known as the response time or "time constant" of the sensor.

The time constant K is best defined by writing the heat transfer equation for our temperature sensor as

$$-\frac{dT}{dt} = \frac{1}{k}(T - T_w) \qquad (1.3.5.1)$$

where T_w and T are the temperatures of the medium (water) and thermometer and t refers to the elapsed time. If we assume that the temperature change occurs rapidly as the sensor descends, the temperature response can be described by the integration of equation (1.3.5.1) from which:

(a)

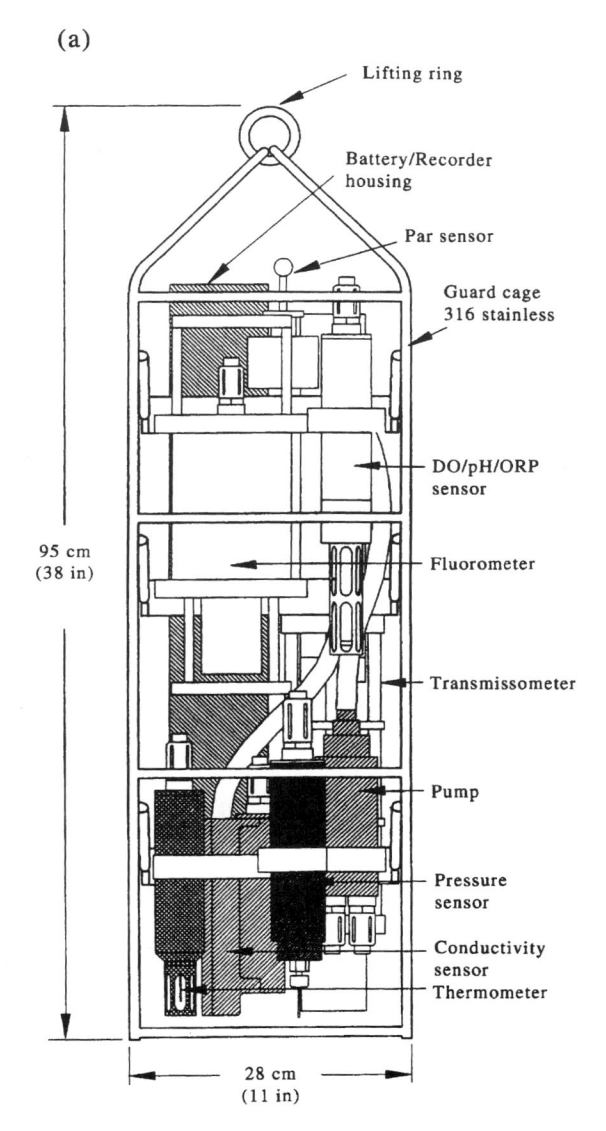

Figure 1.3.4. (a) Schematic of the Sea-Bird SBE 25 CTD and optional sensor modules (courtesy, Doug Bennett, Sea-Bird Electronics).

$$(T - T_w)/(T_o - T_w) = \Delta T/\Delta T_o = e^{-t/K} \qquad (1.3.5.2)$$

In this solution, T_o refers to the temperature of the sensor before the temperature change and K is defined so that the ratio $\Delta T/\Delta T_o$ becomes e^{-1} (= 0.368) when 63% of the temperature change, ΔT, has taken place. The time for the temperature sensor to reach 90% of the final temperature value can be calculated using $e^{-t/k} = 0.1$. A more complex case is when the temperature of the environment is changing at a constant rate; i.e.

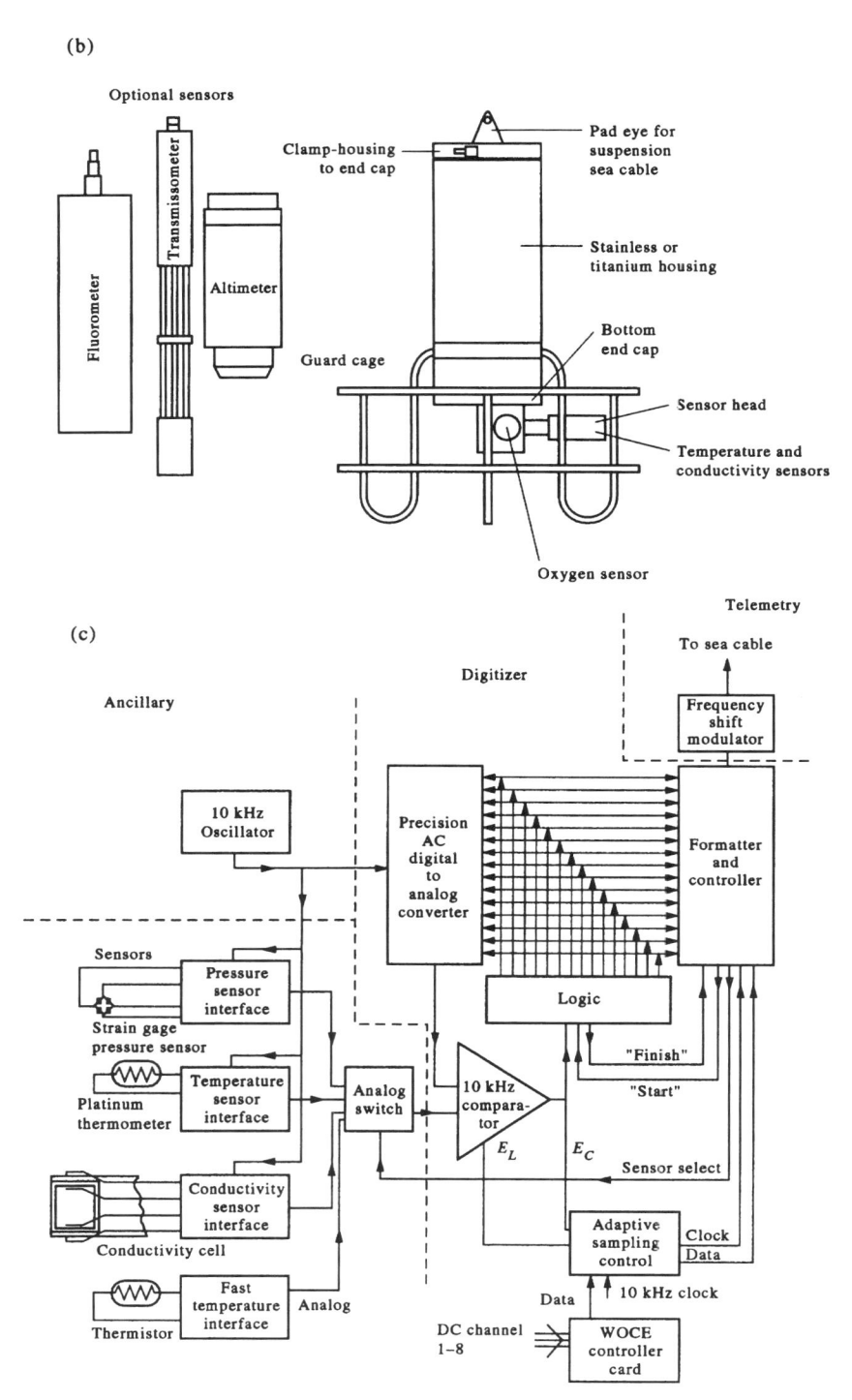

Figure 1.3.4. (b) schematic of General Oceanics MK3C/WOCE CTD and optional sensors; (c) schematic of electronics and sensors of General Oceanics MK3C/WOCE CTD (courtesy, Dan Schaas and Mabel Gracia, General Oceanics).

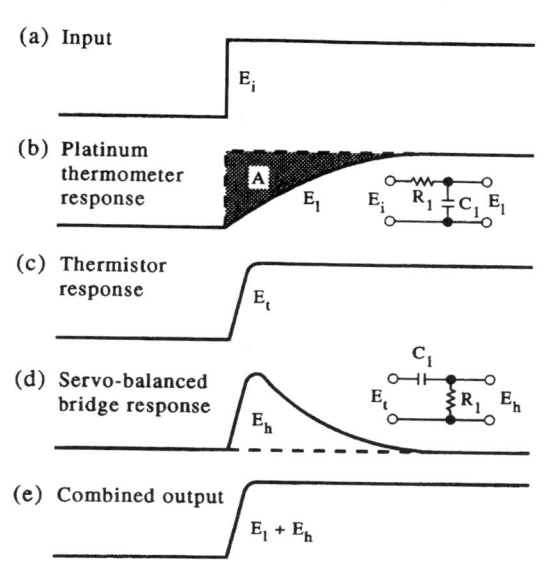

Figure 1.3.5. Combined output and response times of the resistance thermometer of a CTD.

$$T_w = T_1 + ct \quad (1.3.5.3)$$

where T_1 and c are constants. The temperature sensor then follows the same temperature change but lags behind so that

$$T - T_w = -cK \quad (1.3.5.4)$$

The response times, as defined above, are given in Table 1.3.1 for various temperature sensing systems. Values refer to the time in seconds for the sensor to reach the specified percentage of its final value.

The ability of the sensor to attain its response level depends strongly on the speed at which the sensor moves through the medium. An example of the application of these response times is an estimate for the period of time a reversing thermometer is allowed to "soak" to register the appropriate temperature. Suppose we desired an accuracy of ±0.01°C and that our reversing thermometer is initially 10°C warmer than the water. From equation (1.3.5.2), $0.01/10.0 = \exp(-t/K)$, so that $t = 550$ s or 9.2 min. Thus, the usually recommended soak period of 5 min (for a hydrographic cast) is set by thermometer limitations rather than by the imperfect flushing of the water sample bottles. Another application is the estimation of the descent rate for a STD/CTD.

Table 1.3.1 Response times (in s) for various temperature sensors

Device	$K_{63\%}$	$K_{90\%}$	$K_{99\%}$
Mechanical bathythermograph	0.13	0.30	0.60
STD	0.30	0.60	1.20
Thermistor	0.04	0.08	0.16
Reversing thermometer	17.40	40.00	80.00

Assuming that the critical temperature change is in the thermocline where the temperature change is about 2°C/m then to sense this change, with an accuracy of 0.1°C, the STD/CTD response requires that exp($-t/0.6$) = 0.1/2.0 from which t = 1.8 s. Thus, we have the usual recommendation for a lowering rate of about 1 m/s. Today sensors, such as those used in the SBE 25 CTD (Figure 1.3.4a), General Oceanics MK3C CTD (Figure 1.3.4b, c) and the Guildline 8737 CTD have response times closer to 1.0 s.

1.3.6 Response times of CTD systems

As with any thermometer, the temperature sensor on the CTD profiler does not respond instantaneously to a change in temperature. Heat must first diffuse through the viscous boundary layer set up around the probe and through the protective coatings of the sensor (Lueck et al., 1977). In addition, the actual temperature head must respond before the temperature is recorded. These effects lead to the finite response time of the profiler and introduce noise into the observed temperature data (Horne and Toole, 1980). A correction is needed to achieve accurate temperature, salinity and density data. Fofonoff et al. (1974) discuss how the single-pole filter model (1.3.5.2) may be used to correct the temperature data. In this lag-correction procedure, the true temperature at a point is estimated from (1.3.5.1) by calculating the time rate-of-change of temperature from the measured record using a least-square linear estimation over several neighboring points.

Horne and Toole (1980) argue that data corrected with this method may still be in error due to errors arising in the estimation of terms in the differential equation or the approximation of the equation to the actual response of the sensor. As an alternative, they suggest using the measured data to estimate a correction filter directly. This procedure assumes that the observed temperature data may be written as a convolution of the true temperature with the response function of the sensor such that

$$T(t) = H[T^*(t)] \qquad (1.3.6.1)$$

where T is the observed temperature at time t, T^* is the true temperature and H is the transfer or response function of the sensor. The filter g is sought so that

$$g \cdot H = \delta(t) \qquad (1.3.6.2)$$

where δ is the Dirac delta function. The filter g can be found by fitting, in a least-squares sense, its Fourier transform to the known Fourier transform of the function H. This method is fully described in the appendix to Horne and Toole (1980). The major advantage of this filter technique is only realized in the computation of salinity from conductivity and temperature.

In addition to the physical response time problem of the temperature sensor there is the problem of the nonuniform descent of the CTD probe due to the effects of a ship's roll or pitch (Trump, 1983). From a study of profiles collected with a Neil-Brown CTD, the effects of ship's roll were clearly evident at the 5-s period when the data were treated as a time-series and spectra were computed. High coherence between temperature and conductivity effects suggest that the mechanisms leading to these roll-induced features are not related to the sensors themselves but rather reflect an interaction between the environment and the sensor motion. Two likely candidates are: (a) the modification of the temperature by water carried along in the wake of the CTD probe from another depth; and (b) the effects of a turbulent wake overtaking the probe as it decelerates bringing with it a false temperature.

Trump (1983) concludes by saying that, while some editing procedure may yet be discovered to remove roll-related temperature variations, none is presently available. He, therefore, recommends that CTD data taken from a rolling ship not be used to compute statistics on vertical fine-structure and suggests that the best way to remove such contamination is to employ a roll-compensation support system for the CTD probe. Trump also recommends a series of editing procedures to remove roll effects from present CTD data and argues that of the 30,000 raw input data points in a 300 m cast that up to one-half will be removed by these procedures. A standard procedure is to remove any data for which there is a negative depth change between successive values on the way down and *vice versa* on the way up.

1.3.7 Temperature calibration of STD/CTD profilers

Although STD/CTD profilers were supposed to replace bottle casts, it was soon found that, due to problems with electronic drift and other subtle instrument changes, it was necessary to conduct *in situ* calibrations using one or more bottle samples. For this reason, and also to collect water samples for chemical analyses, most CTDs are used in conjunction with Rosette bottle samplers which can be commanded to sample at desired depths. A Rosette sampler consists of an aluminum cage at the end of the CTD conducting cable to which are fixed six, 12, or more water bottles that can be individually triggered electronically from the surface. Larger cages can accommodate larger volume bottles, typically up to 30 litres. While such *in situ* calibrations are more important for conductivity measurements, it is good practice to compare temperatures from the reversing thermometers with the CTD values. This comparison must be done in waters with near-uniform temperature and salinity profiles so that the errors between the CTD values and water sample are minimized. One must pick the time of the CTD values that coincide exactly with the tripping of the bottle. As reported by Scarlet (1975), *in situ* calibration usually confirms the manufacturer's laboratory calibration of the profiling instrument. Generally, this *in situ* calibration consists of comparisons between four and six temperature profiles, collected with reversing thermometers. Taken together with the laboratory calibration data these data are used to construct a "correction" curve for each temperature sensor as a function of pressure. Fofonoff *et al.* (1974) present a laboratory calibration curve obtained over an 18-month period for an early Neil-Brown CTD (the Niel-Brown CTD was the forerunner of the General Oceanics MK3C CTD). A comparison of 175 temperatures measured *in situ* with this profiler and those measured by reversing mercury thermometers is presented in Figure 1.3.6. In the work reported by Scarlet (1975), these calibration curves were used in tabular, rather than functional, form and intermediate values were derived using linear interpolation. This procedure was likely adequate for the study region (Scarlet, 1975) but may not be generally applicable. Other calibration procedures fit a polynomial to the reference temperature data to define a calibration curve.

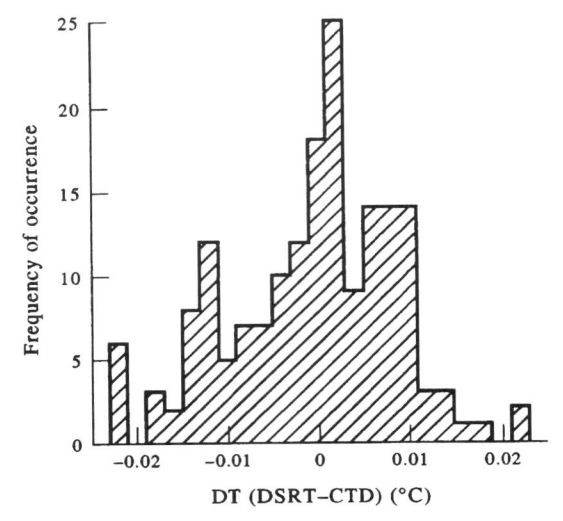

Figure 1.3.6. Histogram of temperature differences. Values used are the differences in temperature between deep-sea reversing mercury thermometers (DSRT) and the temperature recorded by an early Niel-Brown CTD. (From Fofonoff et al., 1974.)

1.3.8 Sea surface temperature

Sea surface temperature (SST) was one of the first oceanographic variables to be measured and continues to be one of the most widely made routine oceanographic measurements. Benjamin Franklin mapped the position of the Gulf Stream by suspending a simple mercury-in-glass thermometer from his ship while traveling between the U.S. and Europe.

1.3.8.1 Ship measurements

The introduction of routine SST measurements did away with the technique of suspending a thermometer from the ship. In its place, SST was measured in a sample of surface water collected in a bucket. When SST measurements were fairly approximate, this method was adequate. However, as the temperature sensors improved, problems with on-deck heating/cooling, conduction between bucket and thermometer, spillage, and other sources of error, led to modifications of the bucket system. New buckets were built that contained the thermometer and captured only a small volume of near-surface water. Due to its accessibility and location at the thermal boundary between the ocean and the atmosphere, the SST has become the most widely observed oceanic parameter. As in the past, the measurement of SST continues to be part of the routine marine weather observations made by ships at sea.

There are many possible sources of error with the bucket method including changes of the water sample temperature on the ship's deck, heat conduction through contact of the container with the thermometer, and the temperature change of the thermometer while it is being read (Tabata, 1978a). In order to avoid these difficulties, special sample buckets have been designed (Crawford, 1969) which shield both the container and the thermometer mounted in it from the heating/cooling effects of sun and wind. Comparisons between temperature samples collected with these special bucket samplers and reversing thermometers near the sea surface have yielded temperature differences of ±0.1°C (Tauber, 1969; Tabata, 1978a).

Seawater cooling for ship's engines makes it possible to measure SST from the temperature of the engine intake cooling system sensed by some type of thermometer imbedded in the cooling stream. Called "injection temperatures" these temperature values are reported by Saur (1963) to be on the average 0.7±0.9°C higher than corresponding bucket temperatures. For his study, Saur used SST data from 12 military ships transiting the North Pacific. Earlier similar studies by Roll (1951) and by Brooks (1926) found smaller temperature differences, with the intake-temperatures being only 0.1°C higher than the bucket values. Brooks found, however, that the engine-room crew usually recorded values that were 0.3°C too high. More recent studies by Walden (1966), James and Fox (1972) and Collins et al. (1975) have given temperature differences of 0.3, 0.3±1.3, and 0.3±0.7°C, respectively. Tabata (1978b) compared three-day average SSTs from the Canadian weather ships at Station "P" (which used a special bucket sampler) with ship injection temperatures from merchant ships passing close by. He found an average difference of 0.2±1.5°C (Figure 1.3.7). Again, the mean differences were all positive suggesting the heating effect of the engine room environment on the injection temperature.

The above comparisons between ship injection and ship bucket SSTs were made with carefully recorded values on ships that were collecting both types of measurements. Most routine ship-injection temperature reports are sent via radio by the ship's officers and have no corresponding bucket sample. As might be expected, the largest errors in these SST values are usually caused by errors in the radio transmission or in the incorrect reporting or receiving of ship's position and/or temperature value (Miyakoda and Rosati, 1982). The resulting large deviations in SST can normally be detected by using a comparison with monthly climatological means and applying some range of allowable variation such as 5°C.

This brings us to the problem of selecting the appropriate SST climatology—the characteristic SST temperature structure to be used for the global ocean. Until recently, there was little agreement as to which SST climatology was most appropriate. In an effort to establish a guide as to which climatology was the best, Reynolds (1983) compared the available SST climatologies with one he had produced (Reynolds, 1982). It was this work that led to the selection of Reynolds (1982) climatology for use in the Tropical Ocean Global Atmosphere (TOGA) research program.

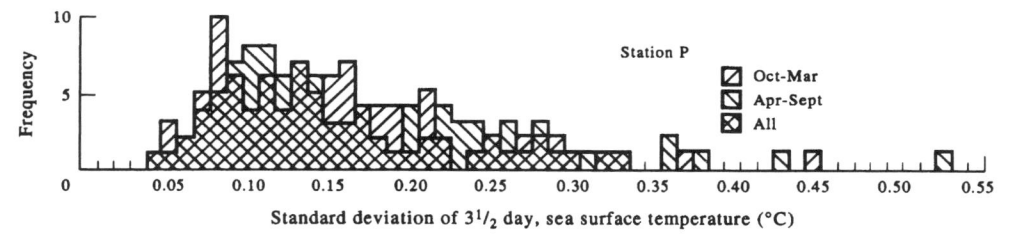

Figure 1.3.7. Frequency of occurrence of standard deviations associated with the 3.5-day mean sea surface temperature at Ocean Station "P"(50°N, 145°W). Difference between bucket temperature and ship intake surface temperature. (Modified after Tabata, 1978a.)

1.3.8.2 Satellite-sensed sea surface temperature (radiation temperature)

In contrast to ship and buoy measurements that sample localized areas on a quasi-synoptic basis, earth-orbiting satellites offer an opportunity to uniformly sample the surface of the globe on a nearly synoptic basis. Infrared sensors flown on satellites retrieve information that can be used to estimate SST, with certain limitations. Clouds are one of the major limitations in that they prevent long-wave, sea-surface radiation from reaching the satellite sensor. Fortunately, clouds are rarely stationary and all areas of the earth's surface are eventually observed by the satellite. In general, researchers wishing to produce "cloud-free" images of SST are required to construct composites over time using repeat satellite infrared data. These composites may require images spanning a couple of days to a whole week. Even with the need for composite images, satellites provide almost global coverage on a fairly short time scale compared with a collection of ship SST observations. The main problem with satellite SST estimates is that their level of accuracy is a function of satellite sensor calibration and specified corrections for the effects of the intervening atmosphere, even under apparently cloud-free conditions (Figure 1.3.8).

One of the first satellites capable of viewing the whole earth with an infrared sensor was ITOS 1 launched January 23, 1970. This satellite carried a scanning radiometer (SR) with an infrared channel in the 10.5–12.5 μm spectral band. The scanner viewed the earth in a swath with a nadir (sub-satellite) spatial resolution of 7.4 km as the satellite travelled in its near-polar orbit. A method that uses these data to map global SST is described in Rao *et al.* (1972). This program uses the histogram method of Smith *et al.* (1970) to determine the mean SST over a number of pixels (picture elements), including those with some cloud. A polar-stereographic grid of 2.5° of latitude by 2.5° of longitude, encompassing about 1024 pixels per grid point per day, was selected. In order to evaluate the calculated SST retrievals from the infrared measurements, the calibrated temperature values were compared with SST maps made from ship radio reports. The resulting root-mean-square (RMS) difference for the northern hemisphere was 2.6°C for three days in September 1970. When only the northern hemisphere ocean weather ship SST observations were used, this RMS value dropped to 1.98°C, a reflection of the improved ship SST observations. A comparison for all ships from the southern hemisphere, for the same three days, resulted in an RMS difference of 2.45°C. As has been discussed earlier in this chapter, one of the

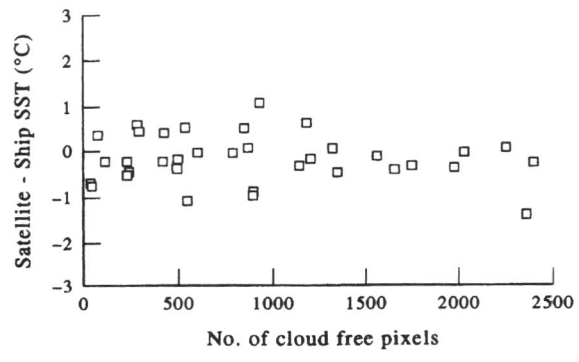

Figure 1.3.8. Sea surface temperature differences, satellite minus ship, plotted as a function of the number of cloud-free pixels in a 50 × 50 pixel array. (From Llewellyn-Jones et al., 1984.)

reasons for the magnitude of this difference is the uncertainty of the ship-sensed SST measurements.

The histogram method of Smith et al. (1970) was the basis for an operational SST product known as GOSSTCOMP (Brower et al., 1976). Barnett et al. (1979) examined the usefulness of these data in the tropical Pacific and found that the satellite SST values were negatively biased by 1–4°C and so were, by themselves, not very useful. The authors concluded that the systematically cooler satellite temperatures were due to the effects of undetected cloud and atmospheric water vapor. As reported in Miyakoda and Rosati (1982), the satellite SST retrieval procedure was evolving at that time and retrieval techniques had improved (Strong and Pritchard, 1980). These changes included new methods to detect clouds and remove the effects of atmospheric water vapor contamination.

More recently, a new satellite sensor system, the Advanced Very High Resolution Radiometer (AVHRR) has become the standard for infrared SST retrievals. In terms of SST the main benefits of the new sensor system are: (a) improved spatial resolution of about 1 km at nadir; and (b) the addition of other spectral channels which improve the detection of water vapor by computing the differences between various channel radiometric properties. The AVHRR is a four- or five-channel radiometer with channels in the visible (0.6–0.7 μm), near-infrared (0.7–1.1 μm), and thermal infrared (3.5–3.9, 10.5–11.5, and 11.5–12.5 μm). The channel centered at 3.7 μm was intended as a nighttime cloud discriminator but is more useful when combined with the 11 and 12 μm channels to correct for the variable amounts of atmospheric water vapor (Bernstein, 1982). While there are many versions of this "two-channel" or "dual-channel" correction procedure (also called the "split-window" method), the most widely used was developed by McClain (1981).

The above channel correction methods have been developed in an empirical manner using some sets of *in situ* SST measurements as a reference to which the AVHRR radiance data are adjusted to yield SST. This requires selecting a set of satellite and *in situ* SST data collected over coincident intervals of time and space. Bernstein (1982) chose intervals of several tens of kilometers and several days (Figure 1.3.9) while McClain et al. (1983) used a period of one day and a spatial grid of 25 km. In a recent evaluation of both of these methods, Lynn and Svejkovsky (1984) found the Bernstein method yielded a mean bias of +0.5°C and the McClain equation a bias of −0.4°C, relative to *in situ* SST measurements. In each case, the difference from *in situ* values was smaller than the RMS errors suggested by the authors of the two methods. Bernstein (1982) compared mean maps made from 10 days of AVHRR retrievals with similar maps made from routine ship reports. He found the maps to agree to within ±0.8°C (one standard deviation) and concluded that this level of agreement was limited by the poor accuracy of the ship reports. He suggested that properly handled, the radiometer data can be used to study climate variations with an accuracy of 0.5–1.0°C. This is consistent with the results of Lynn and Svejkovsky (1984) for a similar type of data analysis.

Another possible source of satellite SST estimates is the Visible Infrared Spin Scan Radiometer (VISSR) carried by the Geostationary Orbiting Earth Satellite (GOES). Unfortunately, the VISSR has a spatial resolution of about 8 km at the sub-satellite point for the infrared channel. Another disadvantage of the VISSR is the lack of onboard infrared calibration similar to that available from the AVHRR (Maul and Bravo, 1983). While the VISSR does provide a hemispherical scan every half hour its shortcomings have discouraged its application to the general estimation of SST. In

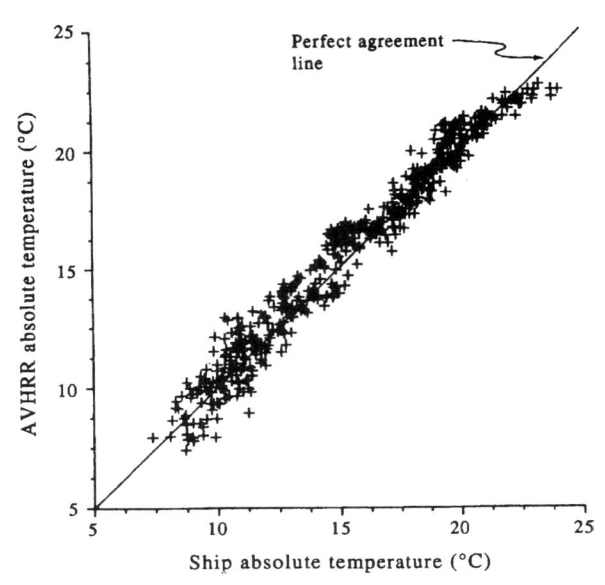

28 *Data Analysis Methods in Physical Oceanography*

Figure 1.3.9. Grid point by grid point cross plot of the mapped values of sea surface temperature from ship-based and AVHRR-based maps. (Bernstein, 1982.)

some cases, VISSR data have been examined where there is a lack of suitable AVHRR data. In one such study, Maul and Bravo (1983) found that a regression between VISSR infrared and *in situ* SST data, using the radiative transfer equations, yielded satellite SST estimates that were no better than ±0.9°C. The conclusion was that, in general, GOES VISSR SST estimates are accurate to within ±1.3°C only. The primary problem with improving this accuracy is the presence of sub-pixel size clouds which contaminate the SST regression.

Efforts are underway to improve the accuracy of retrievals from AVHRR through a better understanding of the on-board satellite calibration of the radiometer and by the development of regional and seasonal "dual-channel" atmospheric correction procedures. Evaluation of these correction procedures, compared with collections of atmospheric radiosonde measurements, has demonstrated the robust character of the "dual-channel" correction and improvements only require a better estimate of the local versus global effects in deriving the appropriate algorithm (Llewellyn-Jones *et al.*, 1984). Thus, it appears safe to suggest that AVHRR SST estimates can be made with accuracies of about 0.5°C, assuming that appropriate atmospheric corrections are performed.

Three workshops were held at the Jet Propulsion Laboratory (JPL) to compare the many different techniques of SST retrievals from existing satellite systems. The first workshop (January 27–28, 1983) examined only the microwave data from the scanning multichannel microwave radiometer (SMMR) while the second workshop (June 22–24, 1983) considered SMMR, HIRS (high resolution infrared sounder) and AVHRR for two time periods, November 1979 and December, 1981. The third workshop (February 22–24, 1984) examined SST products derived from SMMR, HIRS, AVHRR, and VAS (atmospheric sounder on the GOES satellites) for an additional two months (March and July, 1982). A series of workshop reports is available from JPL and the results are summarized in journal articles in the November 1985 issue of the *Journal of Geophysical Research*.

In their review of workshop-3 results, Hilland et al. (1985) reported that the overall RMS satellite SST errors range from 0.5 to 1.0°C. In a discussion of the same workshop results, Bernstein and Chelton (1985) were more specific, reporting RMS differences between satellite and SST anomalies ranging from 0.58 to 1.37°C. Mean differences for this same comparison ranged from −0.48 to 0.72°C and varied substantially from month to month and season to season. They also reported that the SMMR SSTs had the largest RMS differences and time-dependent biases. Differences for the AVHRR and HIRS computed SSTs were smaller. When the monthly ship SST data were smoothed spatially to represent 600 km averages, the standard deviations of the monthly ship averages from climatology varied from 0.35 to 0.63°C. Using these smoothed ship SST anomalies as a reference, the signal-to-noise variance ratios were 0.25 for SMMR and 1.0 for both the AVHRR and HIRS.

The workshop review by McClain et al. (1985) of the AVHRR based multichannel SST (MCSST) retrieval method found biases of 0.3–0.4°C (with MCSST lower than ship), standard deviations of 0.5–0.6°C, and correlations of +0.3 to +0.7 (see also Bates and Diaz, 1991). They also discussed a refined MCSST technique being used with more recent NOAA-9 AVHRR data which yielded consistent biases of −0.1°C and RMS differences (from ship SSTs) of 0.5°C. In an application of AVHRR data to the study of warm Gulf Stream rings, Brown et al. (1985) discuss a calibration procedure which provides SST estimates accurate to ±0.2°C. This new calibration method is the result of thermal vacuum tests which revealed instrument specific changes in the relative emittance between internal (to the satellite) and external (deep space) calibration targets. By reviewing satellite prelaunch calibration data they found that there was an instrument-specific, nonlinear departure from a two-point linear calibration for higher temperatures. In addition, it was found that the calibration relationship between the reference PRT (platinum resistance thermocouples) and the sensor systems changed in the thermal vacuum tests; hence a limited instrument retest, as part of the calibration cycle, is recommended as a way to improve AVHRR SST accuracy. Such higher accuracy absolute SST values are of importance for future climate studies where small, long-term temperature changes are significant.

The workshop results for the geostationary sounder unit (VAS) (Bates and Smith, 1985) revealed a warm bias of 0.5°C with an RMS scatter of 0.8–1.0°C. The positive bias was attributed to a diurnal sampling bias and a bias in the original set of empirical VAS/buoy matchups. Use of a second set of VAS/buoy matches reduced this warm bias making VAS SSTs more attractive due to the increased temporal coverage (every half hour) over that of the AVHRR (one to four images per day).

All of these satellite SST intercomparisons were evaluated against either ship or buoy measurements of the near surface bulk temperature. As is often acknowledged in these evaluations, the bulk temperature is not generally equal to the sea surface skin temperature measured by the satellite. Studies directed at a comparison between skin and bulk temperatures by Grassl (1976), as well as Paulson and Simpson (1981), demonstrate marked (about 0.5°C) differences between the surface skin and subsurface temperatures. In an effort to better evaluate the atmospheric attenuation of infrared radiance, Schluessel et al. (1987) compare precision radiometric measurements made from a ship with SST calculated using a variety of techniques from coincident NOAA-7 AVHRR imagery. In addition, subsurface temperatures were continuously monitored with thermistors in the upper 10 m for comparison with the ship and satellite radiometric SST estimates.

As part of this study, Schluessel *et al.* (1987) examined the effects of radiometer scan angle on the AVHRR attenuation and concluded that differences in scan angle, resulting in different atmospheric paths, resulted in significant changes in the computed SST. To correct for atmospheric water vapor attenuation, HIRS radiances were used to correct the multichannel computation of SST from the AVHRR. The correspondence between the HIRS radiances and atmospheric water vapor content was found by numerical simulation of 182 different atmospheres. With the HIRS correction, AVHRR-derived SST was found to have a bias of +0.03°C and an RMS error of 0.42°C when compared with the ship radiometer measurements. Comparison between ship radiometric and *in situ* temperatures yielded a mean offset of 0.2°C and a range of −0.5 to +0.9°C about this value (Figure 1.3.10). According to Weinreb *et al.* (1990), even if the nonlinearity corrections for the sensor remained valid after satellite launch, the error in AVHRR data is still 0.55°C of which 0.35°C is traceable to calibration of the laboratory blackbody.

From all these various studies, it appears that the infrared satellite SST, when computed using a multichannel algorithm corrected by HIRS, is capable of yielding reliable estimates of SST in the absence of visible cloud. Microwave satellite SST sensing shows promise for future all weather sensing but present systems have a poor a signal-to-noise ratio and a consequent low spatial resolution. Future microwave systems will be designed to specifically measure atmospheric water vapor (in a separate channel) thus making the correction of the infrared SST estimates more straightforward. The frequent global coverage of satellite systems makes them attractive for the global, long-term studies required for an understanding of the world's climate.

1.3.9 The modern digital thermometer

In many oceanographic institutions, the mercury thermometer has been replaced by a digital deep-sea reversing thermometer built by Sensoren-Instrutmente-System (SIS) of Kiel that uses a highly stable platinum thermistor to measure temperature. The SIS RTM 4002 digital reversing thermometer has the outer dimensions of mercury instruments so that it fits into existing thermometer racks. Instead of the lighted magnifying glass needed to read most mercury thermometers, the user simply touches a small permanent magnet to the sensor "trigger" spot on the glass face of the thermometer to obtain a bright digital readout of the temperature to three decimal places. Since the instrument displays the actual temperature at the sample depth, there is no need to read an auxillary thermometer to correct the main reading, as is the case for reversing mercury thermometers. This makes life much more pleasant on a rolling ship in the middle of the night. Because the response time of the platinum thermometer is rapid compared with the "soaking" time of several minutes required for mercury thermometers, less ship time is wasted at oceanographic stations.

The RTM 4002 has a range of −2 to 40°C and a stability of 0.00025°C per month. According to the manufacturer, the instrument has a resolution of 0.001°C and an accuracy of 0.005°C over the temperature range −2 to 20°C. Both resolution and accuracy are considerably lower for temperatures in the range 20–40°C. A magnet is used to reset the instrument and to activate the light-emitting diode (LED). Sampling can be performed in the three sequential modes. The "Hold" mode displays the last

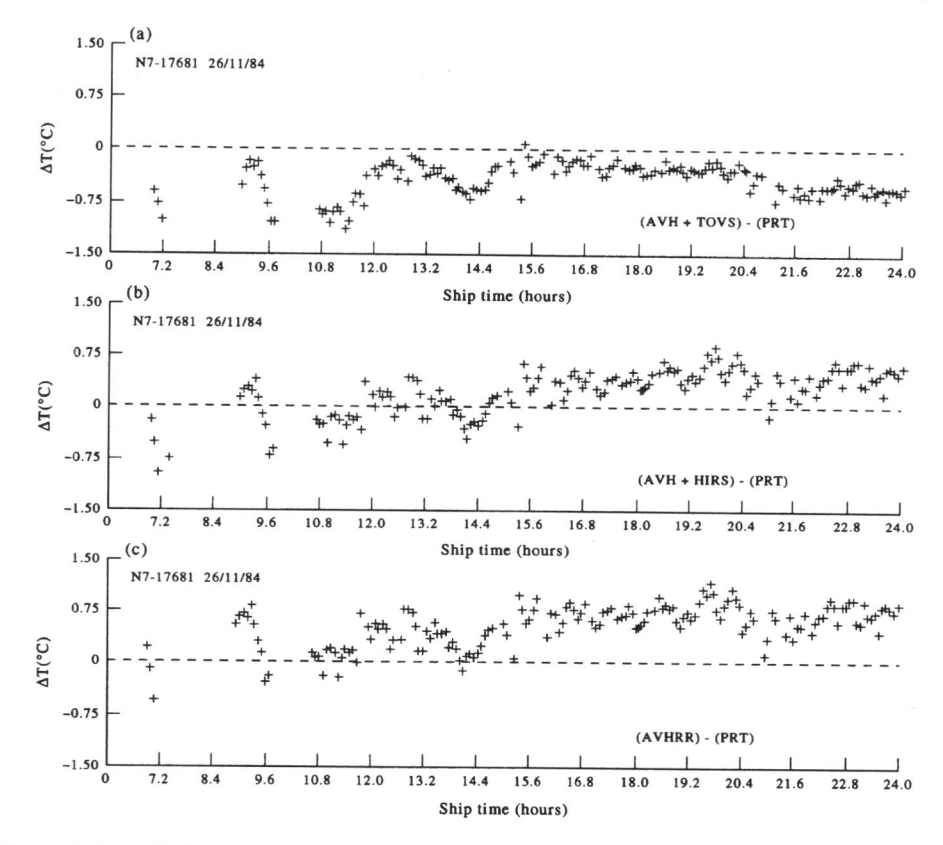

Figure 1.3.10. *Difference between uncorrected (a) and corrected (b, c) satellite-sensed sea surface temperatures and ship radiometric in situ temperatures. HIRS = high-resolution infrared sounder; PRT = platinum resistance thermocouples.*

temperature stored in memory; the "Cont" mode allows for continuous sampling for use in the laboratory while the "Samp" mode is used for reversing thermometer applications. The instrument allows for a minimum of 2700 samples on two small lithium batteries.

1.3.10 Potential temperature and density

The deeper one goes into the ocean, the greater the heating of the water caused by the compressive effect of hydrostatic pressure. The ambient temperature for a parcel of water at depth is significantly higher than it would be in the absence of pressure effects. Potential temperature is the *in situ* temperature corrected for this internal heating caused by adiabatic compression as the parcel is transported to depth in the ocean. To a high degree of approximation, the potential temperature defined as $\theta(p)$, or T_θ, is given in terms of the measured *in situ* temperature $T(p)$ as $\theta(p) = T(p) - F(R)$, where $F(R) \approx 0.1°C$ km^{-1} is a function of the adiabatic temperature gradient R. The results can have important consequences for oceanographers studying water mass characteristics in the deep ocean. The difference between the ambient temperature and θ increases slowly from zero at the ocean surface to about 0.5°C at 5000 m depth. At a depth of approximately 100 m and temperatures less than 5°C, the difference between the two forms of temperature reaches the absolute resolution (±0.01°C) of most thermistors (Figure 1.3.11). Such differences are significant in studies of deep ocean heating from hydrothermal venting or other heat sources where temperature anomalies of ten millidegrees (0.010°C) are considered large. In fact, if the observed temperatures are not converted to potential temperature, it is impossible to calculate the anomalies correctly.

We remark here that the use of potential temperature in the calculation of density leads to the definition of potential density, $\rho_\theta = \rho(S, \theta, 0)$ in kg m^{-3}, as the value of ρ for a given salinity and potential temperature at surface pressure, p = 0. The corresponding counterpart to $\sigma_t (= 10^3[\rho(S, T, 0) - 1000])$, is then called "sigma–theta" where $\sigma_\theta = 10^3[\rho(S, \theta, 0) - 1000]$. Since density surfaces (as well as isotherms) can be displaced vertically hundreds of meters by internal oscillations in the deep ocean, it is crucial that we compare temperatures correctly by taking into consideration the thermal compression effect. Readers familiar with the oceanographic literature will also note the use of σ_2, σ_4, and similar sigma expressions for density surfaces in the deep ocean. These expressions are used as reference levels for the calculation of density at depths where the effect of hydrostatic compression on density becomes important. For example, $\sigma_4 = 10^3[\rho(S, \theta, 4) - 1000]$ refers to density at the observed salinity and potential temperature referred to a pressure of 4000 dbar (40,000 kPa) or about 4000 m depth. Use of σ_θ in the deep Atlantic suggests a vertically unstable water mass below 4000 m whereas the profiles of σ_4 correctly increase toward the bottom (Pickard and Emery, 1992). As indicated by Table 1.3.2, the different sigma values differ significantly.

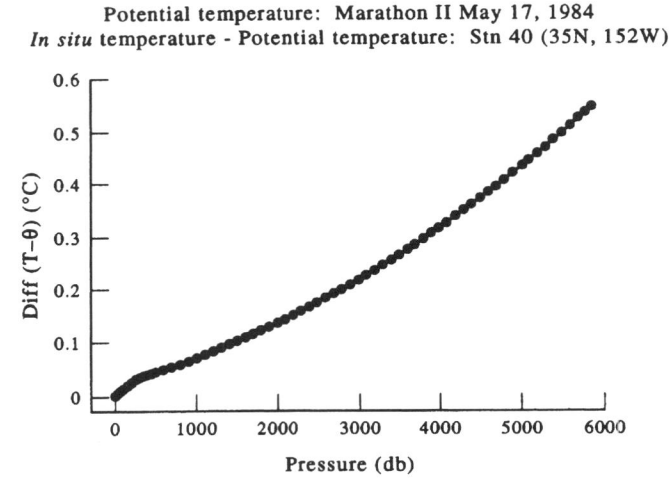

Figure 1.3.11. Difference between in situ temperature (T) recorded by a CTD versus the calculated potential temperature (θ) for a deep station in the North Pacific Ocean (35°N, 152°W). Below about 500 m, this curve is applicable to any region of the world ocean. (Data from Martin et al., 1987.)

Table 1.3.2 Comparison of different forms of sigma for the western Pacific Ocean near Japan (39°41'N, 147°56'E). (From Talley et al., 1988.) Columns 2 and 3 give the in situ and potential temperatures, respectively. Sigma units are kg m^{-3}

Depth (m)	In situ T (°C)	Pot. T (θ) (°C)	Salinity (psu)	σ_θ	σ_2	σ_4
0	18.909	18.909	32.574	23.192	31.706	39.852
100	1.160	1.156	33.158	26.555	35.830	44.689
500	3.338	3.305	34.108	27.145	36.286	45.020
1000	2.697	2.632	34.410	27.447	36.619	45.382
2000	1.868	1.734	34.600	27.672	36.890	45.696
3000	1.528	1.311	34.661	27.752	36.993	45.820
4000	1.456	1.138	34.679	27.778	37.029	45.865
5000	1.503	1.069	34.686	27.788	37.043	45.883
5460	1.547	1.054	34.688	27.791	37.046	45.886

1.4 SALINITY

It is the salt in the ocean that separates physical oceanography from other branches of fluid dynamics. Most oceanographers are familiar with the term "salinity" but many are not aware of its precise definition. Physical oceanographers often forget that salinity is a nonobservable quantity and was traditionally defined by its relationship to another measured parameter, "chlorinity". For the first half of this century, chlorinity was measured by the chemical titration of a seawater sample. In 1899 the International Council for the Exploration of the Sea (ICES) established a commission, presided over by Professor M. Knudsen, to study the problems of determining salinity and density from seawater samples. In its report (Forch et al., 1902), the commission recommended that salinity be defined as follows: "The total amount of solid material in grams contained in one kilogram of seawater when all the carbonate has been converted to oxide, all the bromine and iodine replaced by chlorine and all the organic material oxidized."

Using this definition, and available measurements of salinity, chlorinity and density for a relatively small number of samples (a few hundred), the commission produced the empirical relationship

$$S(‰) = 1.805 Cl(‰) + 0.03 \qquad (1.4.1)$$

known as Knudsen's equation and a set of tables referred to as Knudsen's tables. The symbol ‰ indicates "parts per thousand" (ppt) in analogy to percent (%) which is parts per hundred. In the more modern Practical Salinity Scale, salinity is a unitless quantity written as "psu" for *practical salinity units*. It is interesting to note that Knudsen himself considered using electrical conductivity (Knudsen, 1901) to measure salinity. However, due to the inadequacy of the apparatus available, or similar problems at the time, he decided that the chemical method was superior.

There are many different titration methods used to determine salinity but that most widely applied is the colorimetric titration of halides with silver nitrate (AgNO$_3$) using the visual end-point provided by potassium chromate (K$_2$CrO$_4$), as described in Strickland and Parsons (1972). With a trained operator, this method is capable of an accuracy of 0.02‰ in salinity using the empirical Knudsen relationship. For precise laboratory work Cox (1963) reported on more sensitive techniques for determining the titration end-point which yield a precision of 0.002‰ in chlorinity. Cox also describes an even more complex technique, used by the Standard Sea-water Service, which is capable of a precision of about 0.0005‰ in chlorinity. It is fairly safe to say that these levels of precision are not typically obtained by the traditional titration method and that preconductivity salinities are generally no better than ±0.02‰.

1.4.1 Salinity and electrical conductivity

In the early 1950s, technical improvements in the measurement of the electrical conductivity of seawater turned attention to using conductivity as a measure of salinity rather than the titration of chlorinity. Seawater conductivity depends on the ion content of the water and is therefore directly proportional to the salt content. The primary reason for getting away from titration methods was the development of reliable methods of making routine, accurate measurements of conductivity. As noted earlier, the potential for using seawater conductivity as a measurement of salinity was first recognized by Knudsen (1901). Later papers explored further the relationship between conductivity, chlorinity and salinity. A paper by Wenner *et al.* (1930) suggested that electrical conductivity was a more accurate measure of total salt content than of chlorinity alone. The authors' conclusion was based on data from the first conductivity salinometer developed for the International Ice Patrol. This instrument used a set of six conductivity cells, controlled the sample temperature thermostatically and was capable of measurements with a precision of better than 0.01‰. With an experienced operator the precision may be as high as 0.003‰ (Cox, 1963). The latter is a typical value for most modern conductive and inductive laboratory salinometers and is an order of magnitude improvement in the precision of salinity measurements over the older titration methods.

It is worth noting that the conductivity measured by either inductive or conductive laboratory salinometers, such as the widely-used Guildline 8410 Portable Salinometer, are relative measurements which are standardized by comparison with "standard seawater". As an outgrowth of the ICES commission on salinity, the reference or standard seawater was referred to as "Copenhagen Water" due to its earliest production by a group in Denmark. This standard water is produced by diluting a large sample of seawater until it has a precise salinity of 35‰ (Cox, 1963). Standard UNESCO seawater is now being produced by the "Standard Seawater Service" in Wormley, England as well as at other locations in the U.S.A. (i.e. Woods Hole Oceanographic Institution).

Standard seawater is used as a comparison standard for each "run" of a set of salinity samples. To conserve standard water, it is customary to prepare a "secondary standard" with a constant salinity measured in reference to the standard seawater. A common procedure is to check the salinometer every 10–20 samples with the secondary standard and to use the primary standard every 50 or 100 samples. In all of these operations, it is essential to use proper procedures in "drawing the salinity sample" from the hydrographic water bottle into the sample bottle. Assuming that the hydrographic bottle remains well sealed on the upcast, two effects must be avoided: first, contamination by previous salinity samples (that have since evaporated leaving a salt residue that will increase salinity in the sample bottle); and second, the possibility of evaporation of the present sample. The first problem is avoided by "rinsing" the salinity bottle and its cap two to three times with the sample water. Evaporation is avoided by using a screw cap with a gasket seal. A leaky bottle will give sample values that are distorted by upper ocean values. For example, if salinity increases and dissolved oxygen decreases with depth, deep samples drawn from a leaky bottle will have anomalously low salinities and high oxygens.

Salinity samples are usually allowed to come to room temperature before being run on a laboratory bench salinometer. In running the salinity samples one must be careful to avoid air bubbles and insure the proper flushing of the salinity sample through the conductivity cell. Some bench salinometers correct for the marked influence of ambient temperature on the conductivity of the sample by controlling the sample temperature while other salinometers merely measure the sample temperature in order to be able to compute the salinity from the conductivity and coincident temperature.

Another reason for the shift to conductivity measurements was the potential for *in situ* profiling of salinity. The development history of the salinity/conductivity–temperature–depth (STD/CTD) profilers has been sketched out in Section 1.3.4 in terms of the development of continuous temperature profilers. The salinity sensing aspects of the instrument played an important role in the evolution of these profilers. The first STD (Hamon, 1955) used an electrode-type conductivity cell in which the resistance or conductivity of the seawater sample is measured and compared with that of a sample of standard seawater in the same cell. Fouling of the electrodes can be a problem with this type of sensor. Later designs (Hamon and Brown, 1958) used an inductive cell to sense conductivity. The inductive cell salinometer consists of two coaxial torodoial coils immersed in the seawater sample in a cell of fixed dimensions. An alternating current is passed through the primary coil which then induces an EMF (electromagnetic force) and hence a current within the secondary coil. The EMF and current in the secondary coil are proportional to the conductivity (salinity) of the seawater sample. Again, the instrument is calibrated by measuring the conductivity of standard seawater in the same cell. The advantage of this type of cell is that there are no electrodes to become fouled. A widely used inductive type STD was the Plessey model 9040 which claimed an accuracy of ±0.03‰. Precision was somewhat better being between 0.01 and 0.02‰ depending on the resolution selected. Modern electrode-type cells measure the difference in voltage between conductivity elements at each end of the seawater passageway. With the conducting elements potted into the same material this type of salinity sensor is less prone to contamination by biological fouling. At the same time, the response time of the conductive cell is much greater than that of an inductive sensor leading to the problem of salinity spiking due to a mismatch with temperature response.

The mismatch between the response times of the temperature and conductivity sensors is the primary problem with STD profilers. Spiking in the salinity record occurs because the salinity is computed from a temperature measured at a slightly different time than the conductivity measurement. Modern CTD systems record conductivity directly, rather than the salinity computed by the system's hardware, and have faster response thermal sensors. In addition, most modern CTD systems use electrodes rather than inductive salinity sensors. As shown in Figure 1.4.1, this sensor has a set of four parallel conductive elements that constitute a bridge circuit for the measurement of the current passed by the connecting seawater in the glass tube containing the conductivity elements. The voltage difference is measured between the conducting elements in the bridge circuit of the conductivity cell. The primary advantage of the conductive sensor is its greater accuracy and faster time response. In their discussion of the predecessor of the modern CTD, Fofonoff *et al.* (1974) give an

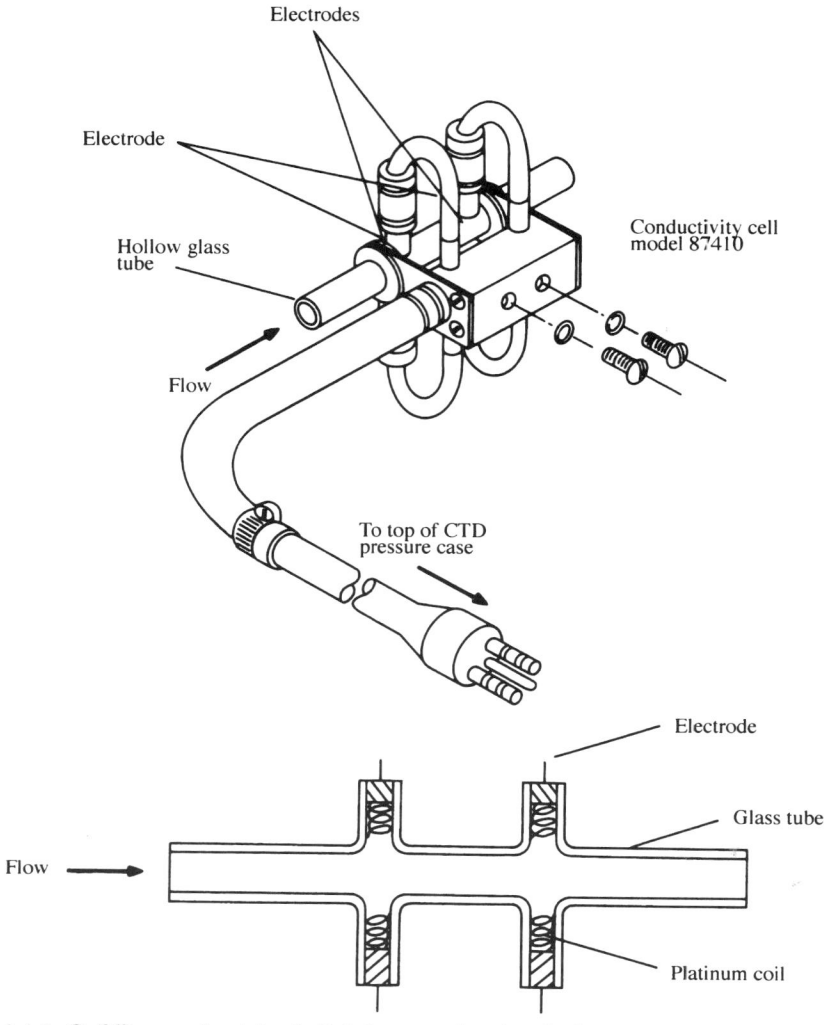

Figure 1.4.1. Guildline conductivity (salinity) sensor showing the location of four parallel conductive elements inserted into the hollow glass tube. Conductivity is measured as the water flows through the glass tube. Cable plugs into the top of the CTD end plate on the pressure case.

overall salinity accuracy for this instrument of ±0.003‰. This accuracy estimate was based on comparisons with *in situ* reference samples whose salinities were determined with a laboratory salinometer also accurate to this level (Figure 1.4.2). This accuracy value is the same as the standard deviation of duplicate salinity samples run in the lab, demonstrating the high level of accuracy of CTD profilers.

1.4.1.1 A comparison of two modern CTDs

In recent sea trials in the North Atlantic, scientists at the Bedford Institute of Oceanography (Bedford, Nova Scotia) examined *in situ* temperature and salinity records from a EG&G Mark V CTD and a Sea-Bird Electronics SBE 9 CTD (Hendry,

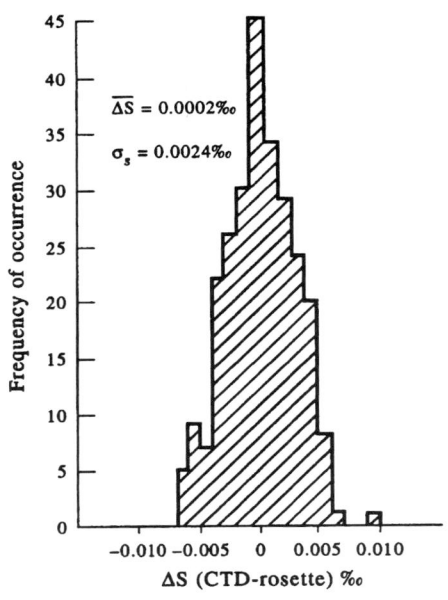

Figure 1.4.2. Histogram of salinity differences (in parts per thousand, ‰). Values used are the differences in salinity between salinity recorded by an early Niel-Brown CTD and deep-sea bottle samples taken from a Rosette sampler. $\overline{\Delta S}$ is the mean salinity difference and σ_s is the standard deviation. (Modified after Fofonoff et al., 1974.)

1993). The standards used for the comparisons were temperatures measured by Sensoren Instrumente Systeme (SIS) digital-reading reversing thermometers and salinity samples drawn from 10-litre bottles on a Rosette sampler and analyzed using a Guildline Instruments Ltd Autosal 8400A salinometer standardized with IAPSO Standard Water.

The Mark V samples at 15.625 Hz and used two thermometers and a standard inductive salinity cell. The fast response (250 ms time-constant) platinum thermometer is used to record the water temperature while the slower resistance thermometer, whose response time is more closely tuned to that of the conductivity cell, is used in the conversion of conductivity to salinity. Plots of the differences in temperature and salinity between the bottle samples and the CTD are presented in Figure 1.4.3. Using only the manufacturer's calibrations for all instruments, the Mark V CTD temperatures were lower than the reversing thermometer values by 0.0034±0.0023°C with no obvious dependence on depth (Figure 1.4.3a). In contrast, the Mark V salinity differences (Figure 1.4.3c) showed a significant trend with pressure, which may be related to the instrument used or a peculiarity of the cell. With pressure in dbar, regression of the data yields

$$\text{Salinity Diff (bottle} - \text{CTD)} = 0.00483 + 6.25910^{-4} \text{Pressure (CTD)} \quad (1.4.2.1)$$

with a correlation coefficient $r^2 = 0.84$. Removal of the trend gives salinity values accurate to about ±0.003 psu. Pressure errors of several dbars (several meters in depth) were noted.

The SBE 9 and SBE 25 sample at 24 Hz and use a high-capacity pumping system and TC-duct to flush the conductivity cell at a known rate (e.g. 2.5 m/s pumping

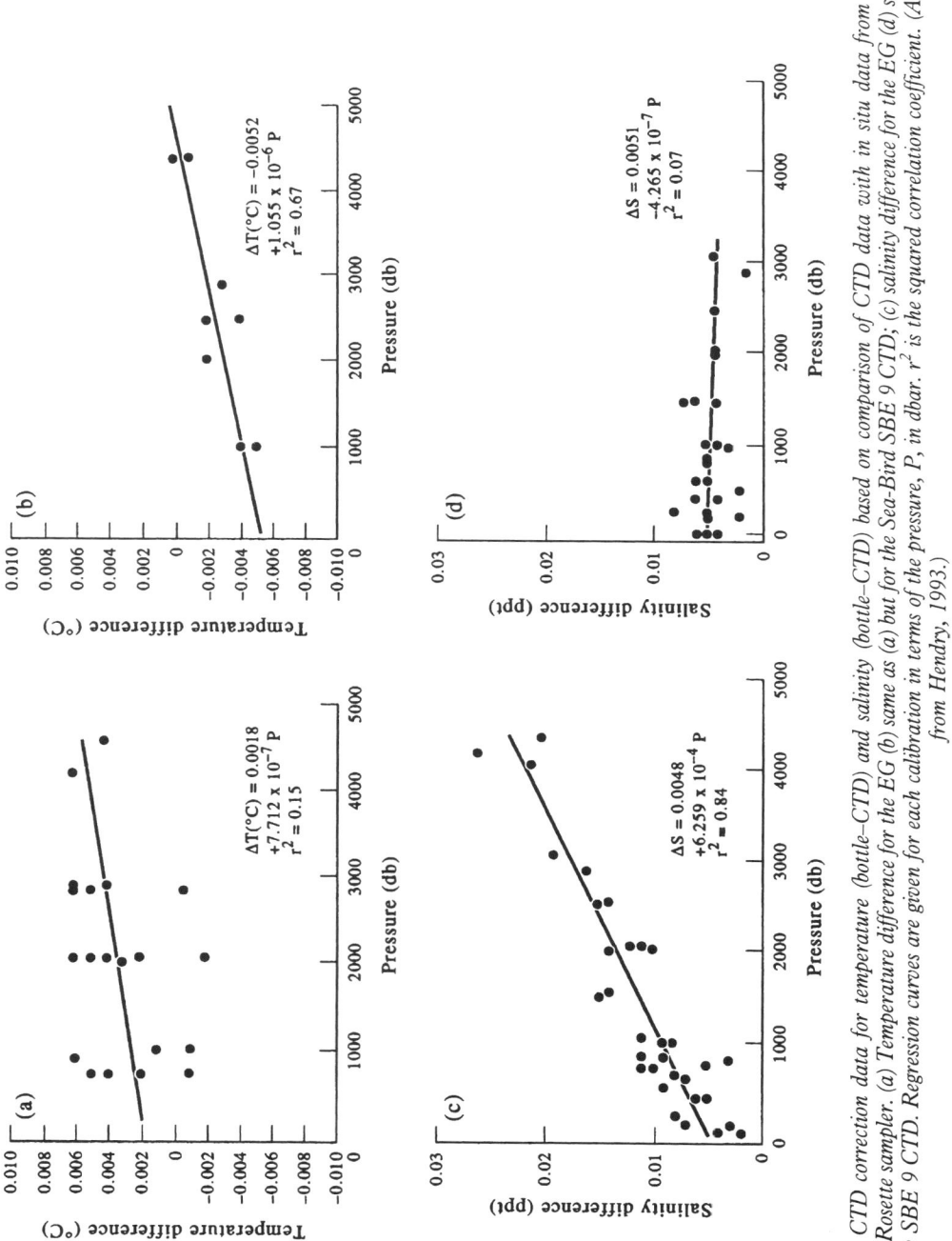

Figure 1.4.3. *CTD correction data for temperature (bottle–CTD) and salinity (bottle–CTD) based on comparison of CTD data with in situ data from bottles attached to a Rosette sampler. (a) Temperature difference for the EG (b) same as (a) but for the Sea-Bird SBE 9 CTD; (c) salinity difference for the EG (d) same as (c) but for the SBE 9 CTD. Regression curves are given for each calibration in terms of the pressure, P, in dbar. r^2 is the squared correlation coefficient. (Adapted from Hendry, 1993.)*

speed for a rate of 0.6–1.2 l/s). When on deck, the conductivity cell must be kept filled with distilled water. To allow for the proper alignment of the temperature and conductivity records (so that the computed salinity is related to the same parcel of water as the temperature), the instrument allows for a time shift of the conductivity channel relative to the temperature channel in the deck unit or in the system software SEASOFT (module AlignCTD). In the study, the conductivity was shifted by 0.072 s earlier to align with temperature. (The deck unit was programed to shift conductivity by one integral scan of 0.042 s and the software the remaining 0.030 s.) Using the manufacturer's calibrations for all instruments, the SBE 9 CTD temperatures for the nine samples were higher than the reversing thermometer values by $0.0002 \pm 0.0024°C$ with only a moderate dependence on depth (Figure 1.4.3b). Salinity data from 30 samples collected over a 3000 db depth range (Figure 1.4.3d) gave CTD salinities that were lower (fresher) than Autosal salinities by 0.005 ± 0.002 psu, with no depth dependence. By comparison, the precision of a single bottle salinity measurement is 0.0007 psu. Pressure errors were less than 1 dbar. Due to geometry changes and the slow degradation of the platinum black on the electrode surfaces, the thermometer calibration is expected to drift by 2 m°C/year and the electronic circuitry by 3 m°C/year.

Based on the Bedford report, modern CTDs are accurate to approximately 0.002°C in temperature, 0.005 psu in salinity and <0.5% of full-scale pressure in depth. The report provides some additional interesting reading on oceanic technology. To begin with, the investigators had considerable difficulty with erroneous triggering (misfiring) of bottles on the Rosette. Those of us who have endured this notorious "grounding" problem appreciate the difficulty of trying to decide if the bottle did or didn't misfire and if the misfire registered on the shipboard deck unit. If the operator triggers the unit again after a misfire, the question arises as whether the new pulse fired the correct bottle or the next bottle in the sequence. Several of these misfires can lead to confusing data, especially in well-mixed regions of the ocean. It is good policy to keep track of the misfires for sorting out the data later.

Another interesting observation was that variations in lowering speed had a noticeable influence on the temperature and conductivity measurements. Since most modern CTDs dissipate 5–10 watts, the CTD slightly heats the water through which it passes. At a 1 m/s nominal fall rate, surface swell can cause the actual fall rate to oscillate over an approximate range of ± 1 m/s with periodic reversals in fall direction at the swell period. (Heave compensation is needed to prevent the CTD from being pulled up and down.) As a result, the CTD sensors are momentarily yanked up through an approximately 1 m °C thermal wake which is shed from boundary layers of the package as it decelerates. Hendry (1993) claims that conditional editing based on package speed and acceleration is reasonably successful in removing these artifacts. Since turbulent drag varies with speed squared, mechanical turbulence was found to cause package vibration that affected the electrical connection from the platinum thermometer to the Mark V CTD. Mixing, entrainment and thermal contamination caused differences in down versus up casts in both instruments. The correction for thermal inertia of the conductivity cell in the SBE CTD resulted in salinity changes of 0.005 psu with negative downcast corrections when the cell was cooling and positive upcast corrections when the cell was warming (Lueck, 1990).

1.4.2 The practical salinity scale

In using either chlorinity titration or the measurement of conductivity to compute salinity, one employs an empirical definition relating the observed variable to salinity. In light of the increased use of conductivity to measure salinity, and its more direct relationship to total salt content, a new definition of salinity has been developed. As a first step in establishing the relationship between conductivity and salinity Cox et al. (1967) examined this relationship in a variety of water samples from various geographical regions. These results were used in formulating new salinity tables (UNESCO, 1966) from which Wooster et al. (1969) derived a polynomial fit giving a new formula for salinity in terms of the conductivity ratio (R) at 15°C. The RMS deviation of this fit from the tabulated values was 0.002‰ in chlorinity for values greater than 15‰ and 0.005‰ for smaller values. It is worth noting that Cox et al. (1967) found that deep samples (>2000 m depth) had a mean salinity, computed from chlorinity, that was 0.003‰ lower than that for conductivity. This was not true for the surface samples.

As noted earlier, a new salinity definition has been adopted called the "practical salinity scale" or PSS 78 (Lewis, 1980). This scale has been accepted by major oceanographic organizations and has been recommended as the scale in which to report future salinity data (Lewis and Perkin, 1981). The primary objections to the earlier salinity definition of Wooster et al. (1969) were:

(a) With salinity defined in terms of chlorinity it was independent of the different ionic ratios of seawater.
(b) The mixtures of reference seawaters used to derive the relationship between chlorinity and conductivity ratio were nonreproducible.
(c) The corresponding International Tables do not go below 10°C which makes them unsuitable for many *in situ* salinity measurements.

In the practical salinity scale, it is suggested that standard seawater should be a conductivity standard corresponding to, and having the same ionic content as, Copenhagen Water. The salinity of all other waters will be defined in terms of the conductivity ratio (R_{15} or C_{15} in the nomenclature of Lewis and Perkins) derived from a study of dilutions of standard seawater. This becomes then a practical salinity scale as distinct from an absolute salinity defined in terms of the total mass of salts per kilogram of solution.

A major problem in applying this new salinity scale is its application to archived hydrographic data. As discussed by Lewis and Perkins (1981), the correction procedure for such data depends not only on the reduction formula used but also on the calibration procedure used previously for the salinity instrument. Essentially, the correction procedure amounts to performing this calibration a second time using the differences, provided by Lewis and Perkins (1981) between the older salinity scale and PSS 78. Another alternative would be to return to the original raw data, if they have been saved, and to recompute the salinity according to PSS 78. From the discussion of Lewis and Perkins it is clear, however, that for salinities in the range of 33–37‰ differences of about ±0.01‰ can be anticipated between archived salinities and the corresponding values computed using PSS 78. This is about the same overall accuracy of modern CTD profilers.

It is interesting that the primary motivation for the development of PSS 78 came from people working in low salinity polar waters where the UNESCO tables did not apply. In areas such as estuarine environments which have very low salinities, and mid-ocean ridge regions with strong hydrothermal fluid venting, even PSS 78 is not adequate and there are still serious limitations to computing accurate salinities from conductivity measurements. There are several reasons for these limitations. First of all, the approach of relating specific conductance to salinity of total dissolved solids

Data Acquisition and Recording 41

requires that the proportions of all the major ions in the natural electolyte remain constant in time and space, and second, that the salinity expression represents all the dissolved solids in the fluid. Another factor to keep in mind is that the density calculated from the conductivity values is based on conducting ions in the fluid. If there are chemical components that contribute to the density of the fluid but not to the conductivity (i.e. nonconductive ions) then the density will be wrong. This is exactly the problem faced by McMannus *et al.* (1992) for deep CTD data from Crater Lake in Oregon. Here, silicic acid from hydrothermal venting in the south basin of the lake below 450 m depth did not contribute to the conductance but did alter the density. Without accounting for silicic acid the water column was weakly stratified; after accounting for it, the bottom waters became stratified. The different combinations of ions in hydrothermal fluids from mid-ocean ridges can seriously alter the salinity structure observed at the source during a normal deep CTD measurement. However, because of rapid entrainment of ambient bottom water, the ion mix becomes similar to that of normal seawater a few meters above the vent orifice. Typical ratios of ambient to hydrothermal fluid volumes are 7000:1. Vertical structure such as that in Figure 1.4.4 taken a few meters above venting fields on Juan de Fuca Ridge in the northeast Pacific, presumably results from unstable conditions arising from turbulent mixing in the rising plume, not from sensor response problems.

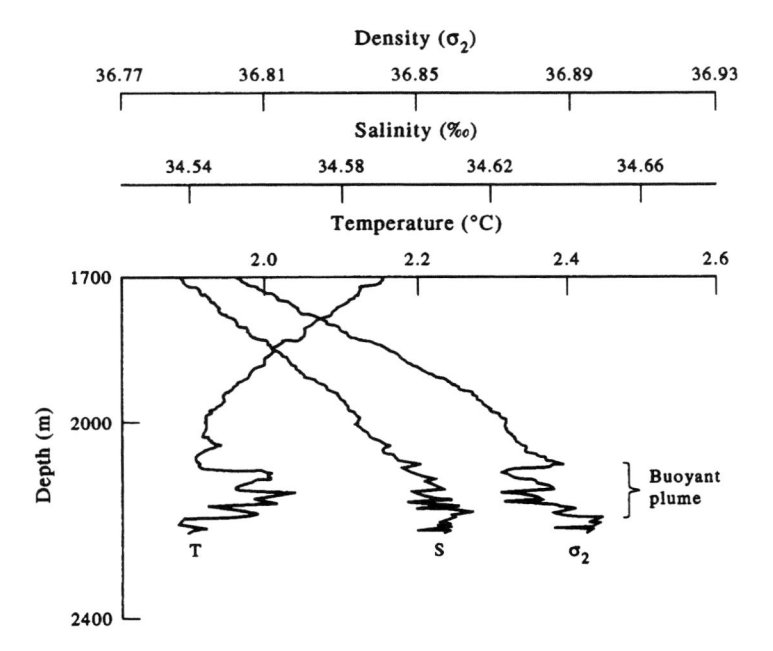

Figure 1.4.4. Vertical profiles of temperature, salinity (‰) and potential density collected from a CTD mounted on the submersible Alvin. Data collected during ascent away from the main hydrothemal vent field at Endeavour Ridge in the northeast Pacific (48°N, 129°W). Density is unstable over the depth range of the buoyant portion of the plume. (From Lupton et al., 1985.)

42 Data Analysis Methods in Physical Oceanography

1.4.3 Nonconductive methods

Efforts have been made to infer salinity directly from measurements of refractive index and density. Since the refractive index (n) varies with the temperature (T) and salinity (S) of a water sample (and with the wavelength of the illumination), measurements of n and T can be used to obtain *in situ* estimate of salinity. In order to achieve a salinity accuracy of 0.01 psu, it is necessary to measure n to within 20×10^{-7} and to control temperature to within 0.005°C. Some refractometers are capable of measuring n to 100×10^{-7}, leading to a salinity precision of 0.06 psu. Handheld refractometers are simple and easy to use but yield salinity measurements no better than ±0.2 psu. For higher sensitivity, interference methods can be used giving a precision in n of 5×10^{-7} corresponding to a salinity precision of 0.003 psu. This is a comparative interference technique and requires a reference seawater sample. Since it is a comparative method, knowledge of the exact temperature is not critical as long as both samples are observed at the same temperature. Direct measurements of seawater density can yield a precision of ±0.008 in sigma-t (Kremling, 1972) which can be used to calculate salinities to within 0.02 psu. Since the measurement of density is much more complicated than those of temperature and electrical conductivity, these latter quantities are usually observed and used to compute the *in situ* density.

1.5. DEPTH OR PRESSURE

1.5.1 Hydrostatic pressure

The depths of profiling instruments are mainly derived from measured hydrostatic pressure, p. This is possible because of the almost linear relationship between hydrostatic pressure, $p = p(z)$, and geometric depth, z. The relationship is such that the "pressure expressed in decibars is nearly the same as the numerical value of the depth expressed in meters" (Sverdrup *et al.*, 1942). The validity of this approximation can be seen in Table 1.5.1 in which we have compared values of hydrostatic pressure

Table 1.5.1 Comparison of pressure (dbar) and depth (m) at standard oceanographic depths using the UNESCO algorithms. Percent difference = (pressure − depth)/pressure × 100%

Pressure (dbar)	Depth (m)	Difference (%)
0	0	0
100	99	1
200	198	1
300	297	1
500	495	1
1000	990	1
1500	1483	1.1
2000	1975	1.3
3000	2956	1.5
4000	3932	1.7
5000	4904	1.9
6000	5872	2.1

and geometric depth for a standard ocean. At depths shallower than 4000 m, the difference is within 2%. For most applications, this error is sufficiently small that it can be neglected and hydrostatic pressure values can be converted directly into geometric depth. The cause for the slight difference between pressure in decibars and depth in meters is found in the familiar hydrostatic relation,

$$p(z) = -g \int_z^0 \rho(z)\,dz$$

where $g = 9.81$ m s^{-2} is the acceleration due to gravity and $\rho \approx 1.025 \times 10^3$ kg/m^3 is the mean water density. Units of p are [kg/m^3][m/s^2][m] = [kg/m][1/s^2]. Also, p = force/area has units [N/m^2] = [Pa] = [10^{-5} bar]. A newton (N) = kg m/s^2, so that $p \approx 1.025 \times 10^3 (9.81)z = 1.005525z$ (m). A different value of density gives a slightly different p versus z relation.

Certain techniques allow for continuous measurement of hydrostatic pressure while others can be carried out at discrete depths only. An example of the latter is the computation of "thermometric depth" using a combination of protected and unprotected reversing thermometers to sense the effects of pressure on the temperature reading. This is still considered one of the most accurate methods of determining hydrostatic pressure and is often used as an *in situ* calibration procedure for CTD profilers. Specifically, the pressure, p (in decibars), obtained from a CTD is related to the temperature difference, ΔT, between the protected and unprotected thermometers by $p \approx g\Delta T/k$, where $k \approx 0.1°$C/(kg cm^2). The details of this procedure are well described in Sverdrup *et al.* (1942, p. 350) but with a significant printing error (a missing plus sign in the second bracket) which is not corrected in Defant (1961, Vol. 1, p. 35). When correctly applied (see LaFond, 1951; Keyte, 1965), the thermometer technique is capable of yielding pressure measurements accurate to ±0.5% (Sverdrup, 1947). Most modern CTD systems claim a similar accuracy using strain-gauge sensors to directly measure pressure. The accuracy of early CTD pressure sensors was a function of the depth (pressure) itself and varied from 1.5 db in the upper 1500 db to over 3.5 db below 3500 db (Brown and Morrison, 1978). A recent test of a Sea Bird SBE 9 CTD (Hendry, 1993) found pre- and postcruise pressure calibration offsets of less than 1 dbar. Nonlinearity and hysteresis were less than 0.5 dbar over the full range of the sensor.

The mechanical bathythermograph introduced earlier in this chapter measures pressure with a Bourdon tube sensor. The problem with this sensor is that the response of the tube to volume change is nonlinear and any alteration in tube shape or diameter will lead to abrupt changes in the pressure response. As a result of the nonlinear scaling of the MBT pressure readout and special optical reader needed to read the scales, this reading error, added to the inaccuracies of the Bourdon tube, results in limited accuracy of the MBT.

1.5.2 Free-fall velocity

Unlike the MBT, the more commonly used XBT, does not measure depth directly but rather infers it from the elapsed time of a "freely-falling" probe. While this is a key element that makes such an expendable system feasible it is also a possible source of error. In their study, Heinmiller *et al.* (1983) first corrected XBT profiles for systematic temperature errors and then compared the XBT profiles with corresponding CTD temperature profiles. In all cases, the XBT isotherm depths were less than the corresponding CTD isotherm depths for observations deeper than an intermediate depth (150 m for T4s and 400 m for T7s) with the largest differences at the bottom of each trace. Near the bottom of the XBT temperature profile, the difference errors exceeded the accepted limit of 2% error, with the deviation being far greater for the shallower T4 probes (Figure 1.5.1a). Added to this systematic error is an RMS depth error of approximately 10 m regardless of probe type (Figure 1.5.1b). Based on the data they analyzed, Heinmiller *et al.* (1983) provide a formula to correct for the systematic depth error. There are two primary sources of this depth error: first, the falling probe loses weight as the wire runs out of the probe supply spool, thus changing the fall rate and second, frictional forces increase as the probe enters more dense waters.

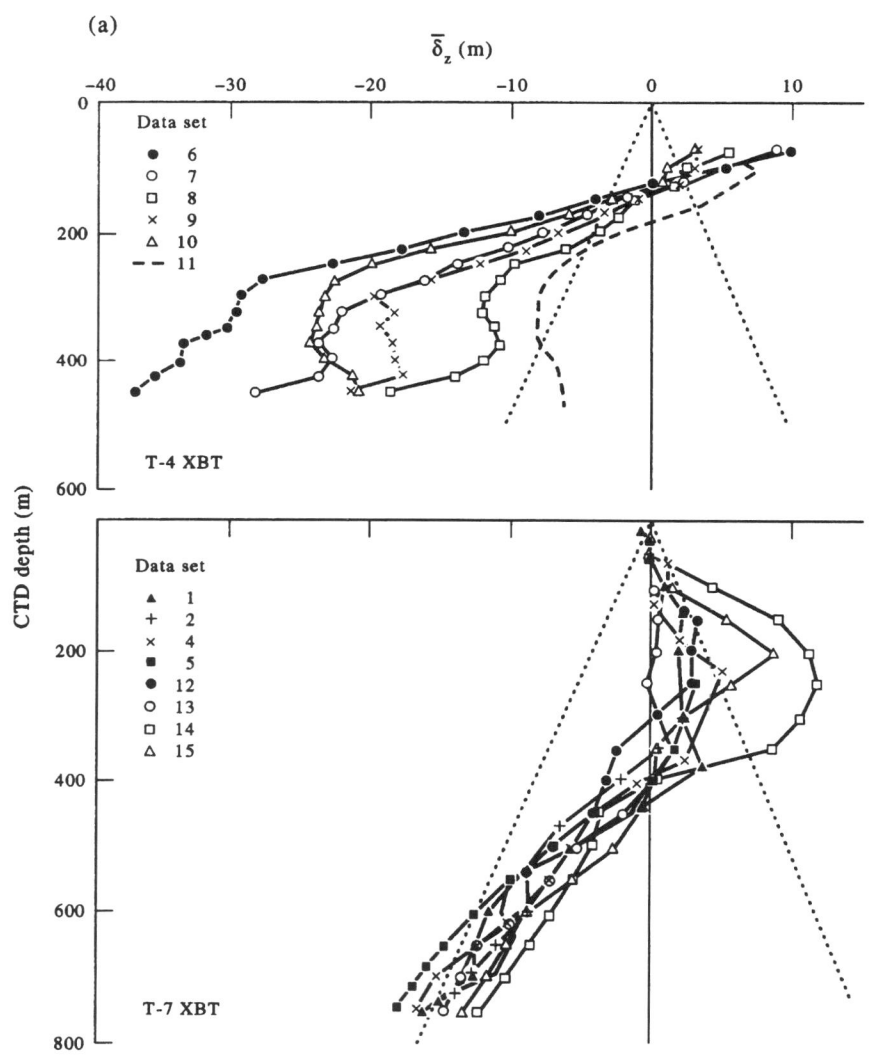

Figure 1.5.1. Vertical profiles of XBT-CTD depth differences for T-4 and T-7 XBTs for different data sets. (a) Mean values, $\overline{\delta}_z$ (m).

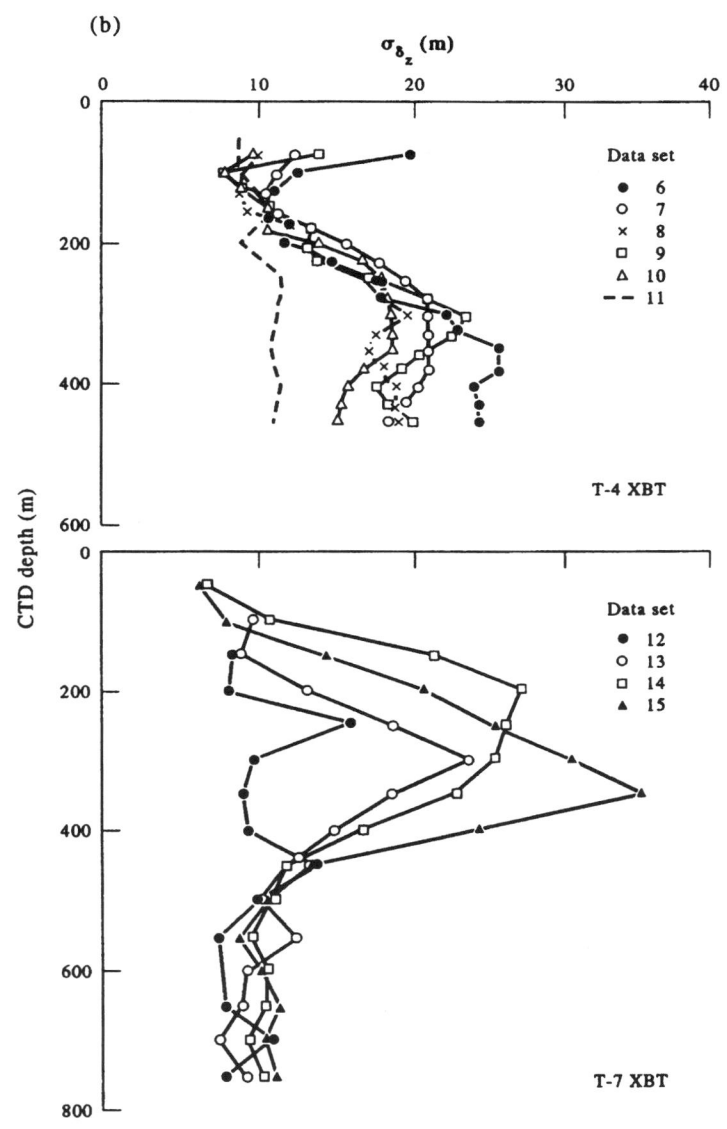

Figure 1.5.1. Vertical profiles of XBT-CTD depth differences for T-4 and T-7 XBTs for different data sets. (b) Standard deviations $\overline{\sigma_{\delta_z}}$ (m).

The issue of XBT depth error, first reported by Flierl and Robinson (1977) has been extensively investigated by many groups (Georgi et al., 1980; Seaver and Kuleshov, 1982; Heinmiller et al., 1983; Green, 1984; Hanawa and Yoritaka, 1987; Roemmich and Cornuelle, 1987; Hanawa and Yoshikawa, 1991; Hanawa and Yasuda, 1991; Rual, 1991) with varying results There is general agreement that the XBT probes fall faster than specified by the manufacturers and that some corrections are needed. Most of these assessments have been performed as a comparison with nearly coincident CTD profiles. The concentration assessments in the western Pacific shows the interest in this problem in Japan, Australia and Nouméa, New Caledonia.

46 *Data Analysis Methods in Physical Oceanography*

A sample comparison between XBT and CTD temperature profiles (Figure 1.5.2a) shows the differences between the XBT and CTD temperature profiles as a function of depth. These profiles have not been corrected using the standard depth equation. The sample in Figure 1.5.2(b) has been depth corrected using the formulation given by Hanawa and Yoritaka (1987). Note the substantial changes in the shape of the difference profile and how the depth correction eliminates apparent minima in the differences. The overall magnitude of the differences has also been sharply reduced, demonstrating that many of the apparent temperature errors are, in reality, depth errors.

These XBT depth errors are known to be functions of depth since they depend on an incorrect fall-rate equation. This is clearly demonstrated in Figure 1.5.3 which gives the mean depth difference of a collection of 126 simultaneous temperature profiles, along with the standard deviation (shown as bars that represent the standard deviation on either side of the mean line) at the various depths. Also shown are the ±2% or ±5 m limits which are given as depth error bounds by the manufacturer. From this figure it is clear that there is a bias with the XBT falling faster than specified by the fall-rate equation, resulting in negative differences with the CTD profiles. The mean depth error of 26 m at 750 m depth translates into 3.5%, which obviously exceeds the manufacturer's specification.

When various investigators reduced these comparisons to new fall rate equations, where the depth (z) is given by

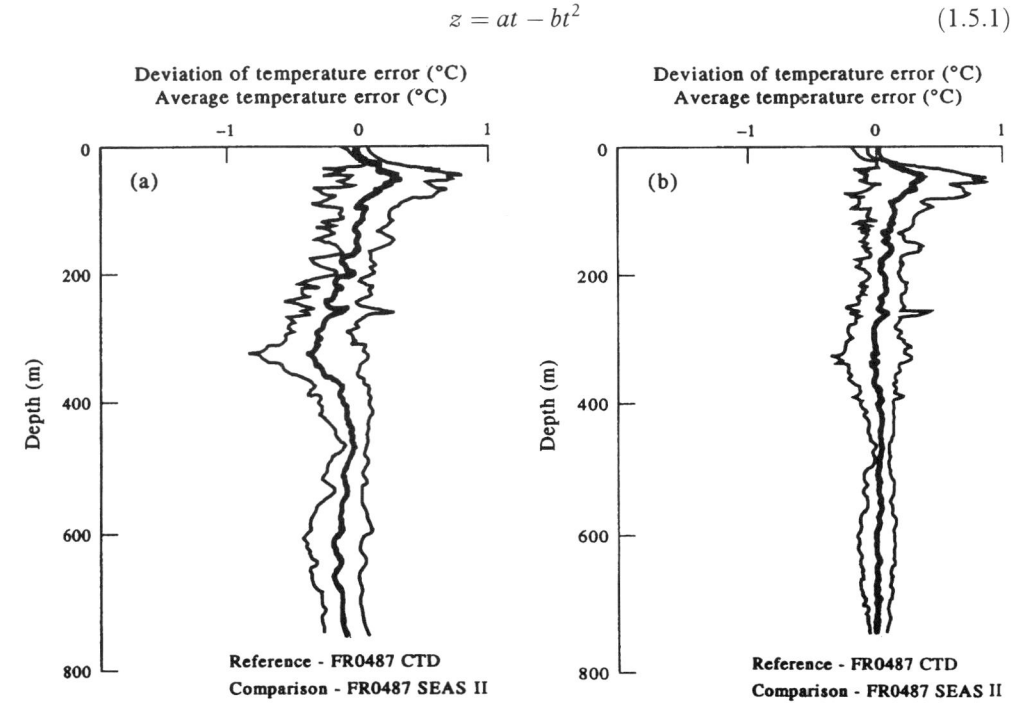

$$z = at - bt^2 \qquad (1.5.1)$$

Figure 1.5.2. Average temperature error profiles (TXBT-TCTD) for XBT/CTD comparisons on FR0487 using the SEAS II XBT system; center line gives the mean value. (a) Depth uncorrected; (b) depth corrected using the formulation given by Hanawa and Yoritaka (1987). (From IOC-888, Annex IV, p. 6.)

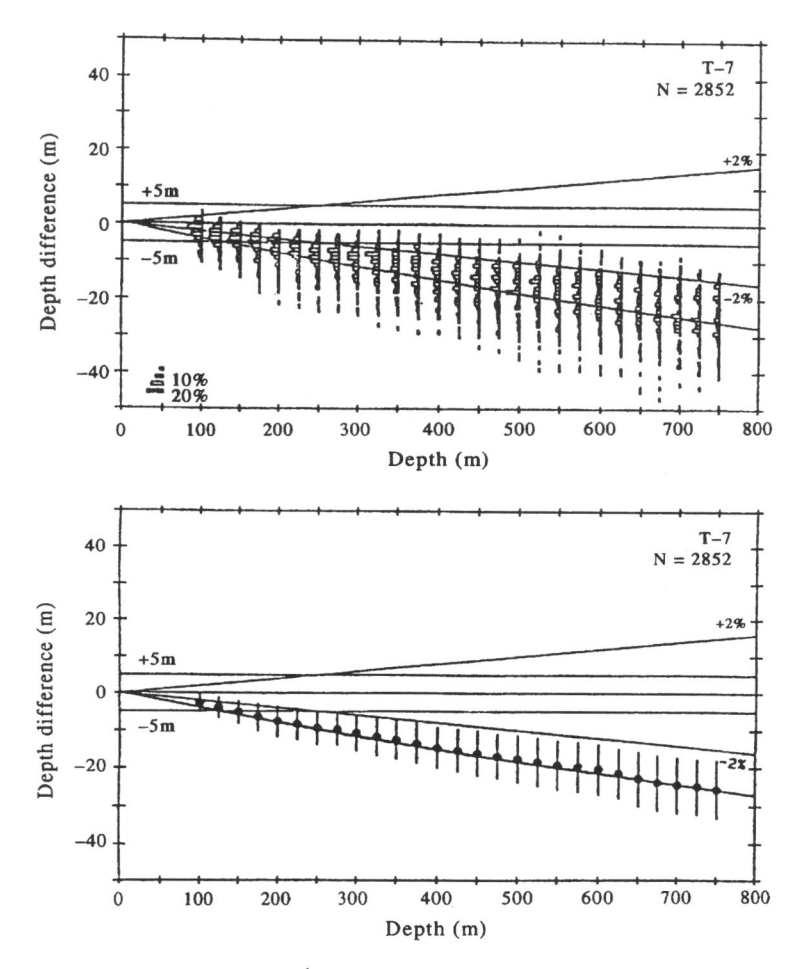

Figure 1.5.3. The mean depth difference of a collection of 126 simultaneous temperature profiles, along with the standard deviation (shown as bars that represent the standard deviation on either side of the mean line) at the various depths. (From IOC/INF-888, p. 13).

the coefficients were not very different (Figure 1.5.4). Along with the coefficients, this figure also shows contours of maximum deviations in depth relative to the revised equation of Hanawa and Yoshikawa (1991) for these different combinations of constants a and b in the fall rate equation (1.5.1). Most of the errors lie within the 110 m envelope of depth deviations, suggesting that it might be possible to develop a new fall-rate equation (i.e. new coefficients) that represents a universal solution to the fall-rate problem for XBT probes. An effort was made to develop this universal equation be reanalyzing existing XBT–CTD comparison profiles. This revised equation is

$$z = 6.733t - 0.00254t^2 \qquad (1.5.2)$$

This revised fall-rate equation only applies to the T7 (roughly 700 m depth) XBT

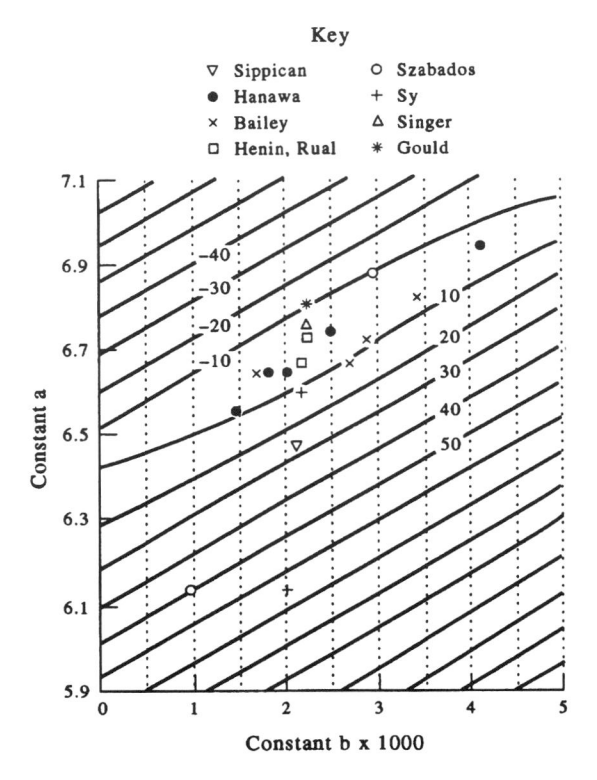

48 *Data Analysis Methods in Physical Oceanography*

Figure 1.5.4. Fall-rate equation coefficients for different XBT studies listed in the key. (From IOC/INF-888.)

probes that were used in the comparisons. It was concluded that similar comparisons must be carried out for the other types of XBT probes.

1.5.3 Echo sounding

Acoustic depth sounders are now standard equipment on all classes and sizes of vessels. Marketed under a variety of names including echo sounder, fish finder, or depth indicator, the instruments all work on the same basic principle: The time it takes for an acoustic signal to make the round-trip from a source to an acoustic reflector such as the seafloor is directly proportional to the distance traveled. Water supports the propagation of acoustic pressure waves because it is an elastic medium. The acoustic waves radiate spherically and travel with a speed $c(E, \rho)$ which depends on the elasticity (E) and density $[\rho = \rho(S, T, P)]$ of the water. If the speed of the sound is known at each time t along the sound path, then the distance d from the sound source to the seafloor is given in terms of the two-way travel time by

$$d = \frac{1}{2} \int_{t_i}^{t_r} c(t)\, dt \qquad (1.5.3.1)$$

where

$$\Delta t = t_r - t_t = 2 \int_{t_t}^{t_r} [1/c(S,T,P;t)] \, dz(t) \tag{1.5.3.2}$$

is the time between transmission time (t_t) of the sound pulse and reception time (t_r) of the reflected pulse or echo. In practice, the values of c along the sound paths are not known and equation (1.5.3.1) must be approximated by

$$d = \frac{1}{2} \langle c \rangle \Delta t \tag{1.5.3.3}$$

where $\langle c \rangle$ is a mean sound speed over the path length, a value normally entered into the echo sounder during its calibration. The depth determined using the time delay is called a "sounding". In hydrography, a "reduced" sounding is one that is referenced to a particular datum. As noted by Watts and Rossby (1977), equation (1.5.3.2) is similar in form to the equation for dynamic height (geopotential anomaly), suggesting that travel time measurements from an inverted echo sounder can be used to measure geostrophic currents (cf. Section 1.6.3).

Since the bulk properties of the medium depend on temperature, salinity and pressure, sound speed also depends on these parameters through the relation

$$c = c_{0,35,0} + \Delta c_T + \Delta c_S + \Delta c_P + \Delta c_{S,T,P} \tag{1.5.3.4}$$

in which $c_{0,35,0} = 1449.22$ m/s (= the speed of sound at 0°C, 35 psu, and pressure $p = 0$, at depth $z = 0$). The remaining terms are the first-order Taylor expansion corrections for temperature (T), salinity (S), and hydrostatic pressure (P); the final term, $\Delta c_{S,T,P}$, is a nonlinear corrective term for simultaneous variation of all three properties. A well known set of values for this equation, having stated experimental standard deviation of 0.29 m/s, is attributed to W. Wilson (Hill, 1962; p. 478). To a good approximation (Calder, 1975; MacPhee, 1976)

$$\begin{aligned} c \, (\text{m/s}) = {} & 1449.2 + 4.6T - 0.055T^2 + 0.00029T^3 \\ & + (1.34 - 0.010T)(S - 35) + 0.016z \end{aligned} \tag{1.5.3.5}$$

or (Mackenzie, 1981)

$$\begin{aligned} c \, (\text{m/s}) = {} & 1448.96 + 4.591T - 5.304 \times 10^{-2} T^2 + 2.374 \times 10^{-4} T^3 \\ & + 1.340(S - 35) + 1.630 \times 10^{-2} z + 1.675 \times 10^{-7} z^2 \\ & - 1.025 \times 10^{-2} T(S - 35) - 7.139 \times 10^{-13} T z^3 \end{aligned} \tag{1.5.3.6}$$

where T is the temperature (°C) and S is the salinity (psu) measured at depth z (m). Accurate profiles of sound speed clearly require accurate measurement of temperature which may not be available in advance. A commonly used oceanic approximation is the mean calibration speed $\langle c \rangle = 1490$ m/s that is generally applied to ship's sounders. Note that the speed of sound increases with increasing temperature, salinity and pressure, with temperature having by far the greatest effect (Figure 1.5.5). For exam-

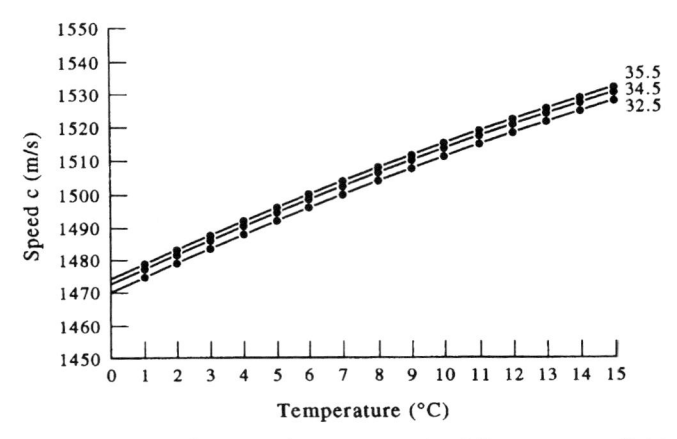

Figure 1.5.5. Speed of sound as a function of temperature for different mean salinities (32.5, 34.5, and 35.5 psu) and fixed depth, z = 1500 m.

ple, c increases by 1.3 m/s per 1 psu in salinity (range 34–35 psu); increases by 4.5 m/s per 1.0°C for temperature (range 0–10°C); and increases by 1.6 m/s per 100 m depth.

The depth capability of any sounder is limited by the power output of the transducer transmitting the sound pulses, by the sensitivity of the receiver listening for the echo returns and by the capability of the instrument electronics and software to resolve signal from noise. In modern sounders, pulse lengths typically range from 0.1 to 50 ms and a single transducer with a transmit/receive (T/R) switching arrangement is used to both generate and receive the acoustic signals. The depth capability of an echo sounder is limited also by a number of important environmental factors. Sound waves are attenuated rapidly in water according to the relation

$$\text{propagation loss (dB)} = 20 \log z + \alpha z/1000$$

where the propagation loss is measured in decibels (dB), z is the depth range in meters and α (Figure 1.5.6) is the attenuation coefficient (dB/km) in seawater as a function of frequency, temperature and salinity (Urick, 1967). The first term in the equation accounts for geometrical spreading of the transmitted and received signal while the second encompasses scattering and absorption. Diffraction and refraction arising from density gradients have a minor effect on the attenuation compared with these other factors. The higher the frequency of the source the greater the attenuation due to absorption and the more limited the depth range (Figure 1.5.7). It is for this reason that most deep-sea sounders operate in the 1.0–50 kHz range. Even though high-frequency sounders can provide more precise depth resolution through shorter wavelengths and narrower beam widths, they cannot penetrate deeply enough to be of use for general soundings. They do have other important applications in bioacoustical studies of zooplankton and fish.

The output transducer converts electrical energy to sound energy and the receiving transducer converts sound vibrations to electrical energy. Loss of the acoustic signal through geometrical spreading is independent of frequency and results from the spherical spreading of wave fronts, while frequency-dependent absorption leads to the conversion of sound into heat through viscosity, thermal conductivity, and

Data Acquisition and Recording   51

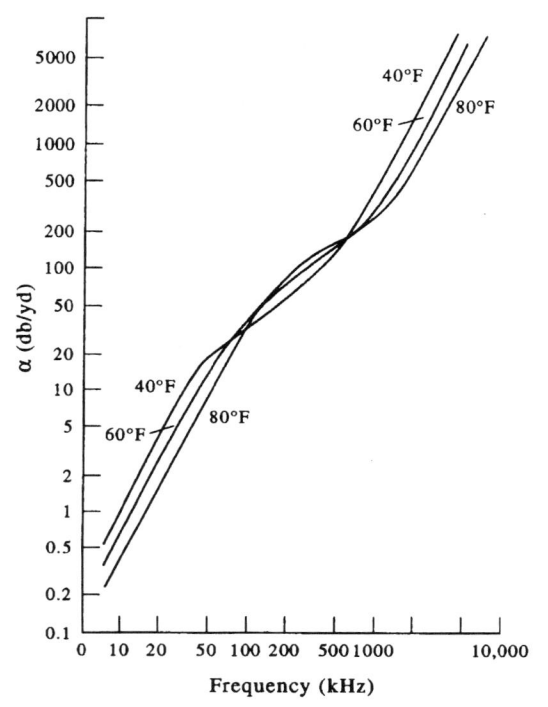

Figure 1.5.6. Absorption coefficient in seawater at salinity 35 psu as a function of frequency for three temperatures ($40°F = 3.6°C$; $60°F = 12.4°C$; $80°F = 21.3°C$). Conversion factor: 1 db/km = 1.0936 db/kyard; 1 kHz = 1000 cps. db ≡ dB (After Urick, 1967.)

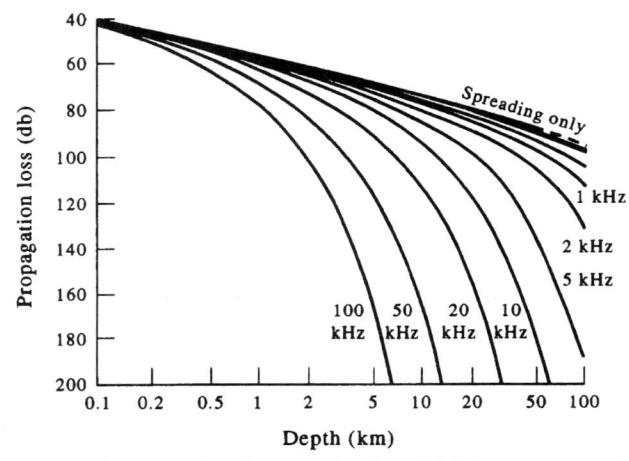

Figure 1.5.7. One-way sound attenuation (propagation loss, PL) in water as a function of sounder frequency (1 kHz = 1000 cps). Curves are derived using $PL = 20\log R + \alpha R/1000$, where R is in km and α is taken from Figure 1.5.6 using the conversion to db/km.

52 Data Analysis Methods in Physical Oceanography

inframolecular processes. Scattering is caused by suspended particles and living organisms. In the upper 25 m of the water column, air bubbles from breaking waves and gas exchange processes are major acoustic scatterers. If I_o is the intensity (e.g. power in watts) of the transducer and I_a is some reference intensity (nominally the output intensity in watts measured at 1-m distance from the transducer head) then the measured backscatter I_r is given by

$$I_r/I_a = b\exp(-\gamma z)/z^2 + A_n \qquad (1.5.3.8)$$

where $b = I_o/I_a$ is the gain of the transducer, γ is the inverse scale length for absorption of sound in water, $1/z^2$ gives the effect of geometric spreading over the distance z from which the sound is being returned, and A_n is a noise level. A return signal intensity reduction by a factor of 2 corresponds to a loss of -3 dB [$=10\log(1/2)$] while a reduction by a factor of 100 corresponds to a loss of -20 dB [$= 10\log(1/100)$]. Echo sounders are generally limited to depths of 10 km. The low (1–15 kHz) frequencies needed for these depths result in poor resolutions of only tens of meters. Since it takes roughly 13 s for a transducer "ping" to travel to 10 km and back, the recorded depth in deep water is not always an accurate measure of the depth beneath the ship. Better resolution is provided for depths less than 5 km using frequencies of 20–50 kHz. High resolution sounders operate in a few hundred meters of water using frequencies of 30–300 kHz. Transducer beam width, side-lobe contamination and side echos ultimately reduce the resolution of any sounder.

1.5.3.1 Height above the bottom

In many oceanographic applications, the investigator is interested in real-time, highly accurate measurements of the altitude of his or her instrument package above the bottom. A real-time reporting echo sounder on the package will serve as an altimeter. Alternatively, the investigator can choose a cheaper route and attach a high-power omnidirectional "pinger" to the package. Rather than trying to measure the total depth of water, one uses the pinger on the package together with ship's transducer (or hydrophone lowered over the side of the ship) to obtain the difference in depth (i.e. difference in time) between the signal which has taken a direct route from the pinger and one which has been reflected from the bottom at an angle θ, where θ is roughly the angle the tow line makes with the bottom (Figure 1.5.8). The direct path PC takes a time t_d while the path reflected from the bottom (PBC) takes a total time t_r. The height, d, of the package above the bottom is

$$d = \frac{1}{2}(c\Delta t)\sin\theta = \frac{1}{2}c(t_r - t_d)\sin\theta \qquad (1.5.3.9)$$

where Δt is the time delay between the direct and reflected pings and c is the speed of sound in water. As the instrument package approaches the bottom, the two strongest analog traces on the depth recorder can be seen converging toward a "crossover" point. When the time taken to cover the direct and reflected paths are equal ($\Delta t = 0$), the package has hit the bottom. The novice operator will be confused by the number of "false" bottom crossovers or wrap-around points when working in water that is deeper than integer multiples of the chosen sounder range. To avoid high levels of stress as each crossover is approached, the operator must know in advance how many false

Data Acquisition and Recording

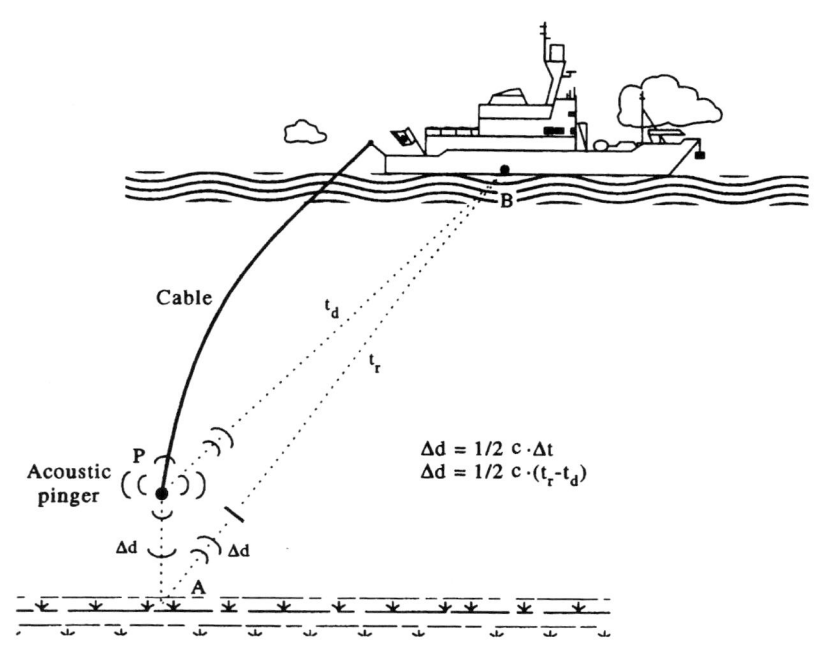

Figure 1.5.8. Schematic of how acoustic pinger and ship's sounder in the receive mode can be used to accurately determine proximity of a probe to the bottom. c is the speed of sound and t is time.

crossovers or wrap-arounds to expect before the instrument is truly in proximity to the bottom. For example, in water of depth 3230 m, a recorder chart set for a full-scale depth range of 0–750 m will register false bottoms when the pinger reaches water depths of 230, 980, and 1730 m (i.e. $3230 - n \times 750$ m, where $n = 4, 3, 2$). To the depth recorder, $3230 - 4 \times 750$ m is the same as 230 m.

Analog devices such as the PDR are the only real choice for this application since the analog trace provides a continuous visual record of how many crossovers have passed and how rapidly the final crossover is being approached. (The depth is first obtained from the ship's sounder which can then turned to "receive mode" only.) The two traces give a history of what has been happening so that it is easier to project in one's mind what to expect as the instrument nears the bottom. Problems arise if the package gets too far behind the vessel and the return echos become lost in the ambient noise or if the bottom topography is very rugged and numerous spurious side echos and shadows begin to appear. We recommend omnidirectional rather than strongly directional pingers so that if the package streams away from the ship or twists with the current and cable there is still some acoustic energy making its way to the ships hull. To help avoid hitting the bottom, it is best to have the ship's sounder output turned off so that only receive mode is working and the background noise is reduced; the operator can check on the total depth every once in a while by reconnecting the "transmit pulse". The ships echo sounder correctly measures the height above bottom since it is programmed to divide any measured time delay by a factor of 2 to account for two-way travel times. The depth accuracy of the method improves as the package approaches the bottom because both the direct and reflected paths experience the same sound speed c. A value of c more closely tuned to deep water is applicable here since all that really counts is c near the seafloor in the region of study. With a little experience and a good clean signal, one can get accuracies of several meters above the bottom using 8–12 kHz sounders in several kilometers of water.

Note that the depth errors using this method are negligible while the actual sounding depths from a depth sounder can be quite large. For example, if spatial differences in sound speed vary from 1470 to 1520 m/s over the sounding depth, then the percentage depth errors are $\Delta c/c = 50/1500 = 0.033 = 3.3\%$. In 4000 m of water, this amounts to an error of 0.033×4000 m ≈ 133 m.

1.5.4 Other depth sounding methods

For sake of completeness, several lesser-known, remote-sensing, depth-sounding methods are introduced in the following sections.

1.5.4.1 Laser induced detection and ranging (LIDAR)

Lidar is an active electro-optical (LASER) remote-sounding method using a pulsed laser system as a radiation source flown from an aircraft. An airborne sensor measures the distance to the surface of the ocean and to the seafloor along the appropriate light path by measuring the time interval between emission of the pulse and the reception of its reflection from the surface and the bottom. A typical LIDAR unit consists of a pulsed laser transmitter, a receiver, and a signal analyzer-recorder. The technique is good to depths of a few tens of meters in coastal waters where extinction coefficients are typically around 0.4–1.6 m^{-1}. The rapid spatial sampling capability of this technique makes it highly useful to hydrographers wanting to map shoals, rocks and other navigational hazards.

1.5.4.2 Synthetic aperture radar (SAR)

One of the surprising aspects of SAR is its ability to "see" shallow banks, ridges and shoals in the coastal ocean. In this case, SAR does not measure the bottom topography directly but, instead, detects the distortion of the wave ripplet field over the feature caused by deflection and/or acceleration of the ocean currents. For a discussion of this effect, the reader is referred to Robinson (1985).

1.5.4.3 Satellite altimetry

The suggestion that there is a close connection between oceanic gravity anomalies and water depth was postulated as early as 1859 (Pratt, 1859, 1871). The idea of using measured gravity anomalies to estimate water depth began with Siemens (1876) who designed a gravity meter that obviated the need to spend hours for a single sounding (Vogt and Jung, 1991). It was not until the launching of the Seasat radar altimeter in 1978 that this concept could be used as alternative to large-scale sounding line measurements. Indeed, the first Seasat-derived gravity map of the world oceans (Haxby, 1985) closely resembles a bathymetric map with a resolution of about 50 km (Vogt and Jung, 1991). The idea of using satellite radar altimetry as a "bathymeter" is based on the good correlation observed between gravity anomalies and bathymetry in the 25–150 km wavelength radar band. Satellite bathymetry is especially valuable for sparsely sounded regions of the world ocean such as the South Pacific, and in regions where depths are based on older soundings in which navigation errors are 10 km or more. As pointed out by Vogt and Jung, however, one-dimensional predictions cannot be accurately ground-truthed with a single ship survey track since the geoid measured along the track is partly a function of the off-track density distribution. (Geoid refers to a constant geopotential reference surface, such as long-term mean sea-level, which is everywhere normal to the earth's graviational field.) A broad swath of shipborne data is needed to overlap the satellite track swath.

Satellites presently carrying radar altimeters to map sea-surface topography include GEOS-3, SEASAT, GEOSAT, ERS-1, and TOPEX/POSEIDON. Of these, only data from the U.S. Navy's GEOSAT have been processed to the accuracy and density of coverage needed to clearly resolve tectonic features in the marine gravity field on a global basis (Marks et al., 1993; EOS). This mission was designed to map the marine geoid to a spatial resolution of 15 km. Detailed maps of the seafloor topography south of 30°S from the GEOSAT Geodetic Mission were declassified in 1992. These maps have a vertical RMS resolution of about 10–20 cm and, together with Seasat data, have been used to delineate fracture zones, active and extinct mid-ocean ridges, and propagating rifts. Satellite altimetry from ERS-1 has been used to map the marine gravity field over the permanently ice-covered Arctic Ocean (Laxon and McAdoo, 1994). Future declassification of military satellite data will lead to further analysis of the seafloor structure over the remaining portions of the world ocean.

As we discuss later in more detail, satellite altimetry shows much promise for interpreting large-scale circulation previously deduced from dynamic height data. Satellite altimetry is particularly well suited to examination of temporal variability of meso-scale features in the ocean, such as the propagation of Rossby waves.

1.6 SEA-LEVEL MEASUREMENT

The measurement of sea-level is one of the oldest forms of oceanic observation. Pytheus of Marseilles, who is reported to have circumnavigated Britain around 320BC, was one of the first to actually record the existence of tides and to note the close relationship between the time of high water and the transit of the moon. Nineteenth-century sea-level studies were related to vertical movements of the coastal boundaries in the belief that, averaged over time, the height of the mean sea-level was related to movements of the land. More recent applications of sea-level measurements include the resolution of tidal constituents for coastal tide height predictions and assisting in the prediction of El Niño/La Niña events in the Pacific. Tide gauge data are essential to studies of wind-generated storm surges which can lead to devastating flooding of highly populated low-lying areas such as Bangladesh. Tide gauges located along the perimeter of the Pacific Ocean and on Pacific islands form an important component of the tsunami warning system that alerts coastal residents to possible seismically-generated waves associated with major underwater earthquakes and crustal displacements.

In addition to measuring the vertical movement of the coastal land mass, long-term sea-level observations reflect variations in large-scale ocean circulation, surface wind stress, and oceanic volume. Because they provide a global-scale integrated measure of oceanic variability, long-term (>50 years) sea-level records from the global tide-gauge network provide some of the best information available on global climate change. Long-term trends in mean sea-level are called *secular* changes while changes in mean sea-level that occur throughout the world ocean are known as *eustatic* changes. As described later in this section, eustasy is associated with land-based glaciation, the accumulation of oceanic sediments, and tectonic activity, such as the change in ocean volume and the shape of the ocean basins. Since coastal stations really measure the movement of the ocean relative to the land, land-based sea-level measurements are referred to as relative sea-level (RSL) measurements. Mean sea-level is the long-term average sea-level taken over a periods of months or years. Datum levels used in hydrographic charts can be defined in many ways and generally differ from country to country (Thomson, 1981; Woodworth, 1991). For geodetic purposes, mean sea-level needs to be measured over several years.

Specifics of sea-level variability

As the previous discussion indicates, observed sea-level variations about some mean equilibrium "datum" level can arise from four principal components: (1) *short-term* temporal fluctuations in the height of the sea surface include those associated with wind waves and oceanic tides forced by the changing alignment of the sun, moon, and earth. In addition, there are changes due to atmospheric pressure (the inverse barometer effect), to wind-induced current set-up/set-down along the coast, to changes in river runoff, and to changes in the large-scale ocean circulation caused by fluctuations in the oceanic wind field. (2) *Long-term* temporal changes result from changes in the mass of the ocean due to melting or accumulation of land-based ice in the major ice sheets in Antarctica and Greenland, and in the smaller ice sheets and mountain glaciers. *Steric sea-level* changes arising from changes in ocean volume— without a change in mass—involve the heating (expansion) and cooling (contraction) of the ocean. Addition or removal of salt has the same steric effect as do cooling and heating. (3) *Coastal subsidence* involves the lowering of the land brought about by reduction in the thickness of unconsolidated coastal sediments, erosion, sediment deposition and with the withdrawal of fluids (water, oil, etc.) from the sediments. (4) Large-scale crustal movements produce sea-level change through *tectonic processes* (mountain building) and *glacio-isostatic rebound* (continued viscoelastic response of the earth to melting of glaciers during the last ice age).

The principal semidiurnal (M_2) and diurnal (K_1) tidal constituents with respective periods of 12.42 and 23.93 h can be accurately resolved using a 15.3-day tidal record of hourly values. Further resolution of the important spring-neap cycle of the tides (the 15-day fortnightly cycle) requires a record length of 29 days (0.98 lunar months; 1 lunar day = 25 h). For most practical purposes, this is the minimum length of record that is acceptable for construction of local tide tables. In fact, many countries maintain primary tide-gauge stations as reference locations for secondary (short-term) tide-gauge stations. Differences in the tide heights and times of high/low water are tabulated relative to the primary location. Accurate resolution of all 56 principal tidal constituents requires a record length of 365 days while an accurate measure of all components for long-term tidal applications requires a record of 18.6 years. The 18.6-year "Metonic" cycle or nuation is linked to the 5° tilt of the plane of the moon's orbit with respect to the plane of the earth's orbit and is the time it takes the line of intersection of these two planes to make one complete revolution (Thomson, 1981). Other tidal constituents include: The centimeter-scale Pole Tide (Chandler Effect) with a period of 14.3 months that arises from the Chandler Wobble in the instantaneous axis of the earth's rotation; an 8.8-year cycle associated with alterations in the eccentricity of the moon's orbit about the earth; and a 20,940-year cycle due to a wobble in the earth's orbit about the sun (precession of the equinox). Meteorological tides are caused by local atmospheric forcing and include large (~1 m) sea-level changes associated with storm surges that often flood low-lying areas (Murty, 1984). Their periodicities are related to changes in wind and atmospheric pressure.

In addition to the relatively small changes in sea-level associated with the Metonic cycle and other orbital factors, there are a variety of major variations due to geological processes. Geological techniques such as coring of Greenland glaciers show a relatively rapid rise of sea-level from 18,000 years ago, when sea-levels were roughly 130 m lower than today. The rate of sea-level rise, which averaged 10 mm/year during the glacial–interglacial transition period, slowed dramatically about 8000 years ago when the levels were 15 m below those of today. Present levels were reached roughly 4000 years ago. Since that time, mean sea-level changes have consisted of oscillations of small amplitude (Barnett, 1983). However, there is now concern that global sea-levels are rising at about 1–2 mm/year due to global warming through build up of CO_2 in the atmosphere. Many long-term stations show definite long-term trends (Figure 1.6.1) that may be related to global climate change. There is a general increase of about 0.1–0.15 m per century for gauges located in geologically "stable" regions of the world. One possible source of long-term variability is melting from global warming of water locked into the polar ice caps and northern hemisphere glaciers. Changes in the land-based ice cover alter both the ocean volume and the geodetic loading. However, it should be noted that the rate of rise varies considerably from station to station and is strongly dependent on regional tectonic activity and the lingering effects of the continuing glacio-isostatic response (Peltier, 1990; Tushingham and Peltier, 1992). Investigators attempting to extract a possible climate change component from global sea-level records spend considerable effort generating spatially-smoothed data sets consisting of a relatively small subset of the total tide gauge data available from the world archives.

Mean sea-levels are usually computed from long series of hourly observations. Generally, a simple arithmetic average of hourly values is computed, but other methods, including the application of low-pass numerical filters to eliminate tides and surges, may be used before the means are computed. The average of all high and low water levels is called the mean tide height; it is close to, but not identical with, mean sea-level. Monthly and annual mean sea-level series for a global network of stations are collected and published by the Permanent Service for Mean Sea-level (PSMSL) in England, together with details of gauge location and the definitions of the datums to which the measurements are referred. Data are held for 1446 "Metric" file stations of which 889 have had their data adjusted to a tide gauge benchmark datum to form the Revised Local Reference (RLR) data set. This datum is approximately 7.000 meters below mean sea-level, with the arbitrary choice made to avoid negative numbers in the resulting RLR monthly and annual mean values. Of these stations, 112 have data from before 1900. Most of these stations are in the northern hemisphere so that careful analysis is necessary to avoid geographic bias in their interpretation. Only one year of monthly data are needed for the Metric data to be converted to the common datum (Woodworth, 1991). Amsterdam has the longtest tide-gauge record in world but the oldest data which satisfy the selection criteria of the PSMSl are from Brest, starting in 1807.

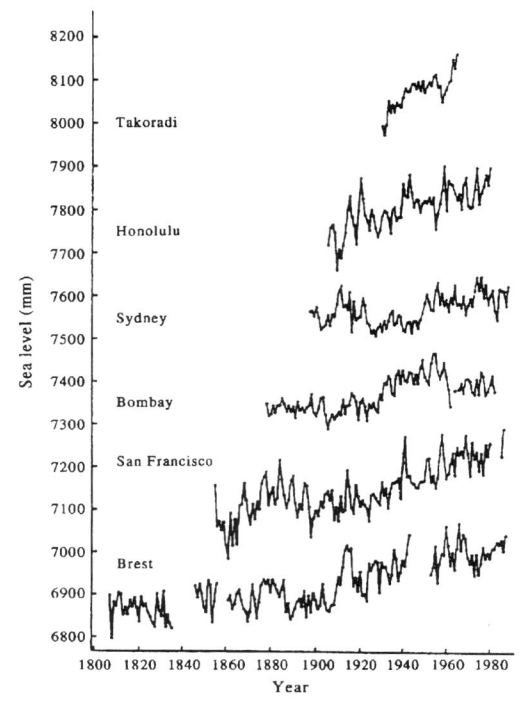

58   *Data Analysis Methods in Physical Oceanography*

Figure 1.6.1. Annual mean sea-level values for the longest records for each continent. Data are Revised Local Reference (RLR) records from the Permanent Service for Mean Sea-level at the Bidston Observatory in Merseyside. Each record has been given an arbitrary offset for presentation purposes. The Takoradi record was truncated in 1965 when major problems with the gauge were reported. (From Woodworth, 1991.)

1.6.1 Tide and pressure gauges

Although pressure and acoustic gauges are becoming increasingly more popular around the world, most sea-level measurements are still made using a float gauge, in which the float rises and falls with the water level (Figure 1.6.2a). Modern recording systems replace the analog pen with a digital recording system that records on punched paper, magnetic tape, or hard-drive. Many of the digital recording systems have been equipped with telemetering systems that send sea-level heights and time via satellite, direct radio link or meteor-burst communication. The most important aspect of this type of sea-level measurement is the installation of the float stilling well. The nature of this installation will determine the frequency response of the float system and will help damp out unwanted high-frequency oscillations due to surface gravity waves. Stilling wells can have their inlet at the bottom of the well or use a pipe inlet connected to the lower part of the well. Both designs damp out the high-frequency sea-level changes. Maintenance of the sea-level gauge insures that the water inlet orifice is kept clear of obstructions from silt, sand, or marine organisms. Also, in areas of strong stratification such as rivers or estuaries, the water in the stilling well can be of a different density than the water surrounding it. When installing such a measurement

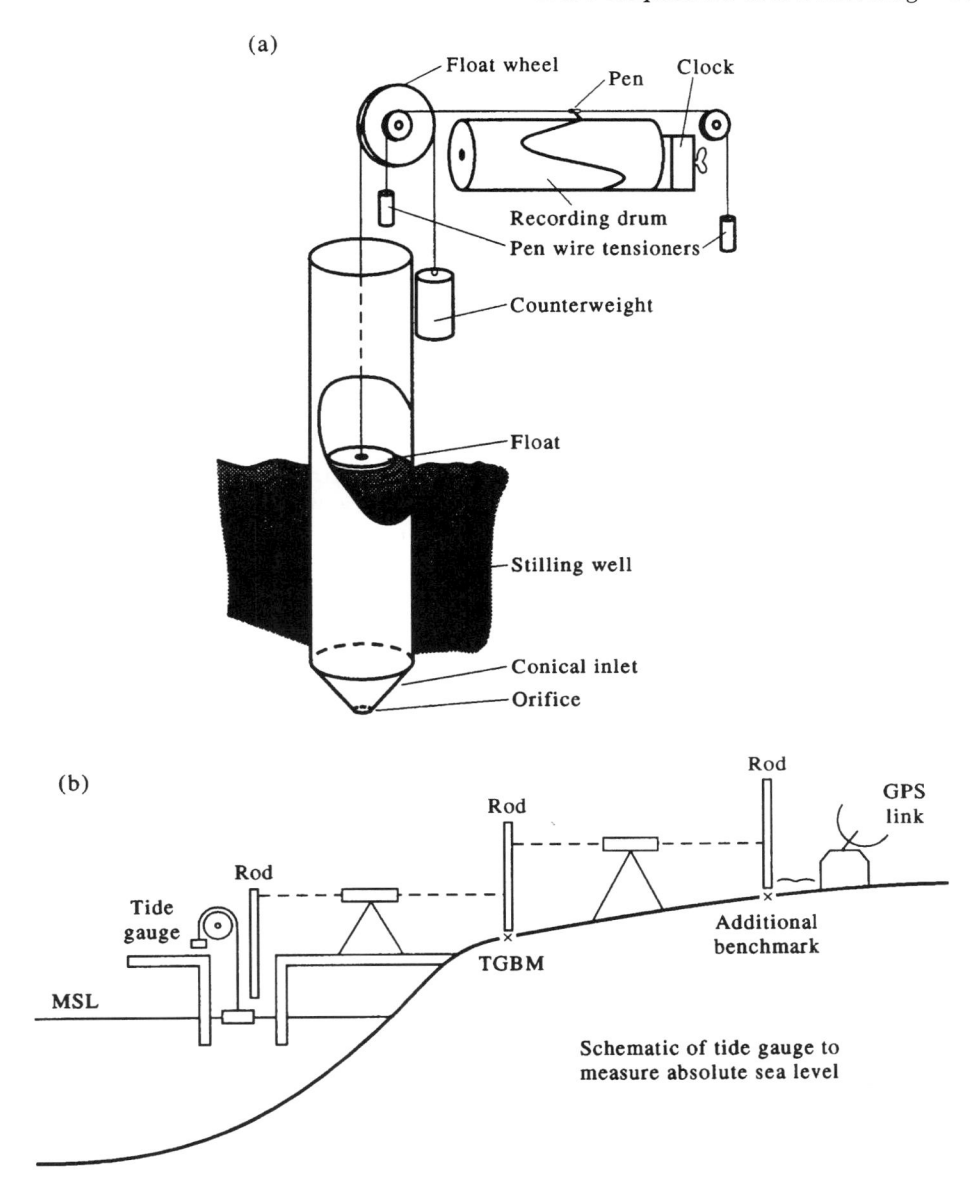

Figure 1.6.2. (a) A basic float stilling-well gauge used to measure water levels on the coast; (b) Schematic of tide-gauge station with the gauge, network of benchmarks and advanced geodetic link. TGBM = tide gauge bench mark. (After Woodworth, 1991.)

system, it is important to provide adequate protection from contamination of the stilling well and from damage to the recorder. A potential hazard in all harbor gauge installations is damage from ship traffic or contamination of the float response from ship wakes.

Proper installation of coastal sea-level gauges requires that they be surveyed into a legal bench mark so that measured changes will be known relative to a known land

60 Data Analysis Methods in Physical Oceanography

elevation (Figure 1.6.2b). When properly tied to a bench mark height, changes in gauge height relative to land can be taken into account when computing the mean sea-level. This tie-in with the local benchmark datum is done by running the level back to the nearest available geodetic datum. Modern three-dimensional satellite-based global positioning systems (GPS) and long-baseline telemetry systems will soon make it possible to accurately determine the vertical movement of a tide gauge relative to the geoid. Future accurate satellite altimeter measurements may also provide global measurements of the relative sea surface which can then be compared with the conventional sea-level measurements.

The usual test of a sea-level instrument (called the Van de Casteele test) involves operating the instrument over a full tidal cycle and comparing the results against simultaneous measurements made with a manual procedure. This procedure only shows if the recording device is operating properly so that a separate test of the stilling well is needed to accurately measure the response of the float. Other than mechanical problems, timing errors are one of the major sources of error in sea-level records. Sea-level gauges have either mechanical or electronic clocks which must be periodically checked to insure that there is no significant drift in the timing of the mechanism. When possible these checks should be made weekly. Depending on the specific instrument, well-maintained sea-level recorders are capable of measurements accurate to within a several millimeters.

Another type of sea-level measuring device is the pneumatic or bubbler gauge (Figure 1.6.3). This system links changes in the hydrostatic pressure at the outlet

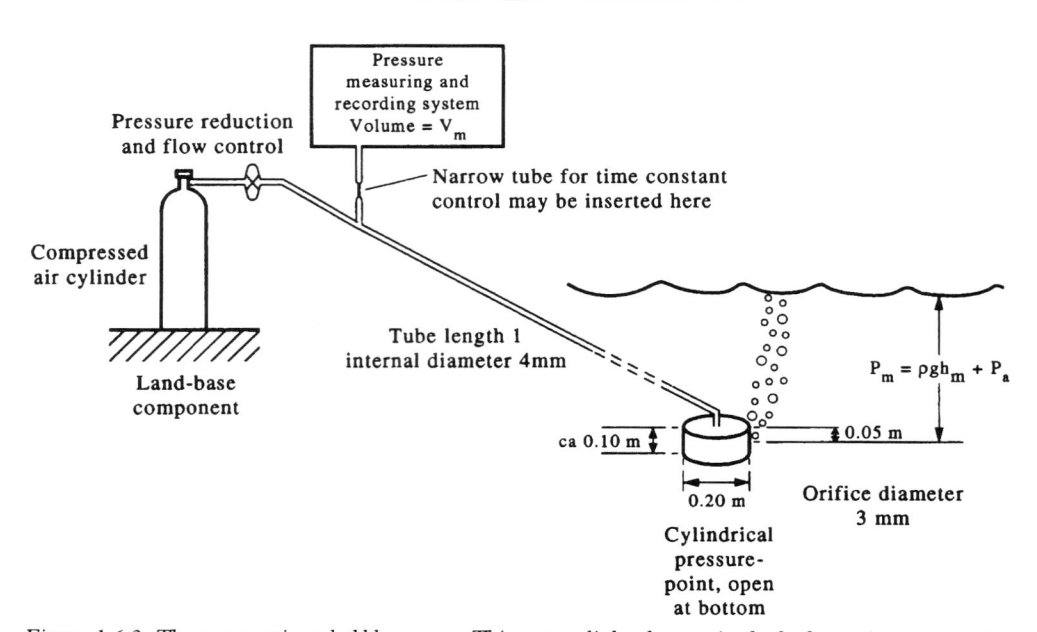

Figure 1.6.3. The pneumatic or bubbler gauge. This system links changes in the hydrostatic pressure, P_m, at the outlet point of the bubbles to variations in sea-level, h_m, water density and atmospheric pressure, P_a.

point of the bubbles to variations in sea-level. Like other pressure sensing gauges, this gauge measures the combined sea-level height and atmospheric pressure. As a consequence, most bubbler gauges operate in the differential mode whereby the recorded value is the difference between the measured pressure and the atmospheric pressure. While these instruments are somewhat less accurate than float gauges, they are useful in installations where a float gauge would be subject to either damage from ship traffic or strongly influenced by wave motion. In a study in Tasmania in the 1970s, the plastic pressure tubes leading from the electronics package to the ocean were constantly being destroyed by curious wombats.

Sea-level heights are recorded in a variety of formats. Graphical records must be digitized and care taken to record only values properly resolved by the instrument. Differences in recording scale will lead to variations in the resolution of the gauge thereby limiting the accuracy of the digitized data. Modern gauges eliminate this possible problem by recording digital data. During digitization of the data, it is important to edit out any of the obvious errors due to pen-ink problems or mechanical failures in the advance mechanism. Also, when long-term sea-level variations are of interest, one must be careful to filter out high-frequency fluctuations due to waves and seiches. The choice of time reference is important in creating a sea-level time series. The usual convention is to use the local time at the location of the tide gauge which can then be referenced to Greenwich Mean Time (GMT, now called Universal Temps Coordinée, UTC). The sea-level record should also contain some information about the reference height datum on which the sea-level heights are based. Digital recording systems are subject to clock errors and care should be taken to correct for these errors when the digital records are examined.

In recent years, sensitive and accurate pressure sensors have been developed for measurement of deep sea tides where fluctuations of the order of 1 mm need to be detected in depths of thousands of meters. At first, these sensors were largely based on the "Vibraton" built by United Control Corporation which measured pressure by changes in the frequency of oscillation of a wire under tension. This frequency change was measured to an accuracy of 6×10^{-4} Hz and led to a sea-level accuracy of 0.8 mm (Snodgrass, 1968). To maintain this high level of accuracy, it was necessary to correct for temperature effects to a resolution of 0.001°C. When Vibraton sensors ceased to be commercially available they were replaced by resonating quartz crystal transducers (Wimbush, 1977), which are the standard for measurement in both deep sea and coastal pressure gauge recorders. Produced by Parosscientific (Paros, 1976), these sensors have a sensitivity of 1×10^{-4} dbars for shallow applications (<500 m depth) and a sensitivity of 1×10^{-3} dbars for deep sea measurements. Most modern pressure gauges used in coastal and deep sea tidal measurements make use of these sensors. (Pressure gauges used in coastal waters are often known as water level gauges.) Temperature correction is required to maintain accurate depth measurements. Wearn and Baker (1980) report measurements made by such quartz sensors from year-long moorings in the Southern Ocean. Unfortunately, instabilities in the quartz sensors lead to sensor drifts which limit the use of the sensors in long-term, deep pressure measurements. The use of dual pressure sensors helps to correct for drift since each pressure sensor will have somewhat different drift characteristics but will produce similar responses to higher frequency oceanic variability.

1.6.2 Satellite altimetry

Conventional sea-level measurement systems are limited by the need for a fixed platform installation. As a result, they are only possible from coastal or island stations where they can be referenced to the land boundary. Unfortunately, there are large segments of the world's ocean without islands, so that the only hope for long-term sea-level measurements lies with satellite-borne radar altimetry. Early studies (Huang et al., 1978) using GEOS-3 altimeter data with its fairly low precision of 20–30 cm, demonstrated the value of such data for estimating the variability of the sea surface from repeated passes of the satellite radar. In this case, the difference between repeated collinear satellite passes eliminates the unknown contribution of the earth's geoid to the radar altimeter measurement. This same technique was employed by Cheney et al. (1983) using 1000 orbits of high-quality SEASAT radar altimeter data.

With a known precision of 5–8 cm, the early radar data provided some of the first large-scale maps of mesoscale variability in the world's ocean. Satellite altimetry is one of the future monitoring capabilities now being actively pursued by physical oceanographers interested in ocean circulation problems. While the experience with GEOS-3 and SEASAT altimeter data demonstrated the great potential of these systems, it is not yet entirely clear if planned altimetry satellites will provide sufficient accuracy to allow the specification of the mean ocean circulation related to the ocean surface topography. The primary concern is for the contribution of the earth's geoid to the satellite altimeter measurements. The geoid is known to have variations with space scales similar to the scales of sea-level fluctuations associated with the mean and mesoscale ocean circulation. In addition, satellite altimetry data must be corrected for atmospheric effects which requires knowledge of the intervening atmospheric temperature and water vapor profiles. It is hoped that present research efforts will better specify these constituents of the altimeter signal so that future satellite altimeter measurements can be used to monitor both the mean and time variable (using crossovers or colinear satellite tracks) ocean circulation.

Considerable headway has been made recently in the area of satellite altimetry due to the successful deployment of a number of space-borne altimeters. The first to generate a lot of new data was the GEOSAT satellite first launched in 1985 by the U.S. Navy in an effort to more precisely map the influences of the geoid on missile tracks. After an 18-month "geodetic mission" the Navy was convinced by Dr Jim Mitchell and others to put the satellite into an "exact repeat orbit" in November, 1986, using the same orbit as the previous SEASAT satellite (Tapley et al., 1982). The altimeter data from this orbit had already been made public data and thus the classified altimeter data from the geodetic mapping mission were already compromised for this orbit. By having the satellite operate in this orbit, scientists would be able to collect and analyze data on the ocean's height variability. Fortunately, the GEOSAT altimeter continued to function into 1989 providing almost three full years of repeat altimetry measurements. In addition, the navy has released the "crossover" data from the geodetic mission. In this mission, the track did not repeat but the crossovers between ascending and descending tracks provided valuable information on ocean height variability. Thus, it is possible to combine data from the earlier cross overs and repeat orbits from the "exact repeat mission" (ERM) to form a nearly five-year time-series of sea surface height variations. It should be stressed that without a detailed knowledge of the earth's geoid it is not possible to compute absolute currents and the main area of investigation provided by the GEOSAT data was in studying the ocean's height variability.

Considerable experience was gained in computing the various corrections that are needed to correct satellite altimeter data (Chelton, 1988). These include the ionospheric correction, the dry tropospheric correction, the wet tropospheric correction. Added to these are the errors due to EM (electromagnetic) bias, antenna mispointing, antenna gain calibration, the inverse barometer effect, ocean/earth tides and precise orbit determination. Since the GEOSAT satellite did not carry a radiometer to compute tropospheric water vapor other operational satellite sensors were used to compute the atmospheric moisture to correct the altimeter path length (Emery et al., 1989a). Many experiments were conducted to better understand the EM-bias correction (Born et al., 1982; Hayne and Hancock, 1982). Other corrections can be routinely computed from available sources, including the dry troposphere correction which requires knowledge of the atmospheric pressure (Chelton, 1988).

GEOSAT data have been used to map both the large-scale and smaller scale regional circulations of the ocean. Miller and Cheney (1990) used GEOSAT data to monitor the meridional transport of warm surface water in the tropical Pacific during an El Niño event. Combining crossover and colinear data, the authors constructed a continuous time-series of sea-level changes on a $2° \times 1°$ grid in the Pacific between 20°N and 20°S for the four-year period from 1985 to 1989. They concluded that the 1986–1987 El Niño was a low-frequency modulation of the normal seasonal sea-level cycle and that a build-up of sea-level in the western Pacific was not required as a precursor to an El Niño event. A similar analysis of colinear GEOSAT data for the tropical Atlantic (Arnault et al., 1990) showed good agreement between the satellite sensed sea-level changes and those measured *in situ* using dynamic height methods. Using GEOSAT sea-level residuals computed from a two-year mean Vazquez et al. (1990) examined the behavior of the Gulf Stream downstream of Cape Hatteras. Comparisons with NOAA infrared satellite imagery show a fair agreement in Gulf Stream path with some sea-level deviation maps not showing a clear location of the main stream. In the same geographic region Born et al. (1987) used a combination of GEOSAT altimetry and airborne expendable bathythermograph data, to map geoid profiles as the difference between the altimetric sea-level and the baroclinic dynamic height. Many other oceanographers have used GEOSAT data to study a great variety of oceanographic circulation systems (e.g. Figure 1.6.4).

In the summer of 1992, the long awaited TOPEX-POSEIDON altimetric satellite was launched. Carrying two altimeters (one French and one U.S.) with a single antenna, TOPEX/POSEIDON marked a significant step forward in altimetric remote sensing. The NASA altimeter is a dual-frequency altimeter which will be able to compensate for the influence of ionospheric changes. The French altimeter is the first solid-state instrument to be deployed in space. In addition, there is a boresight microwave radiometer (TOPEX Microwave Radiometer, TMR) to provide real time atmospheric water vapor measurements for the computation of wet troposphere corrections for the onboard altimeters. The resulting combination of data provides altimeter heights accurate to ± 10 cm. A truly joint project, the satellite was built in the U.S. and launched by the French *Ariane* launch vehicle. A large group of French and U.S. science teams have been working to prepare for this mission and will continue to work on the analysis of these important data.

1.6.3 Inverted echo sounder (IES)

As noted in Section 1.5, accurate depth measurements using acoustic sounders requires corrections on the order of $\pm 1\%$ for variations in sound speed introduced by

64 *Data Analysis Methods in Physical Oceanography*

TOPEX Eddy Kinetic Energy

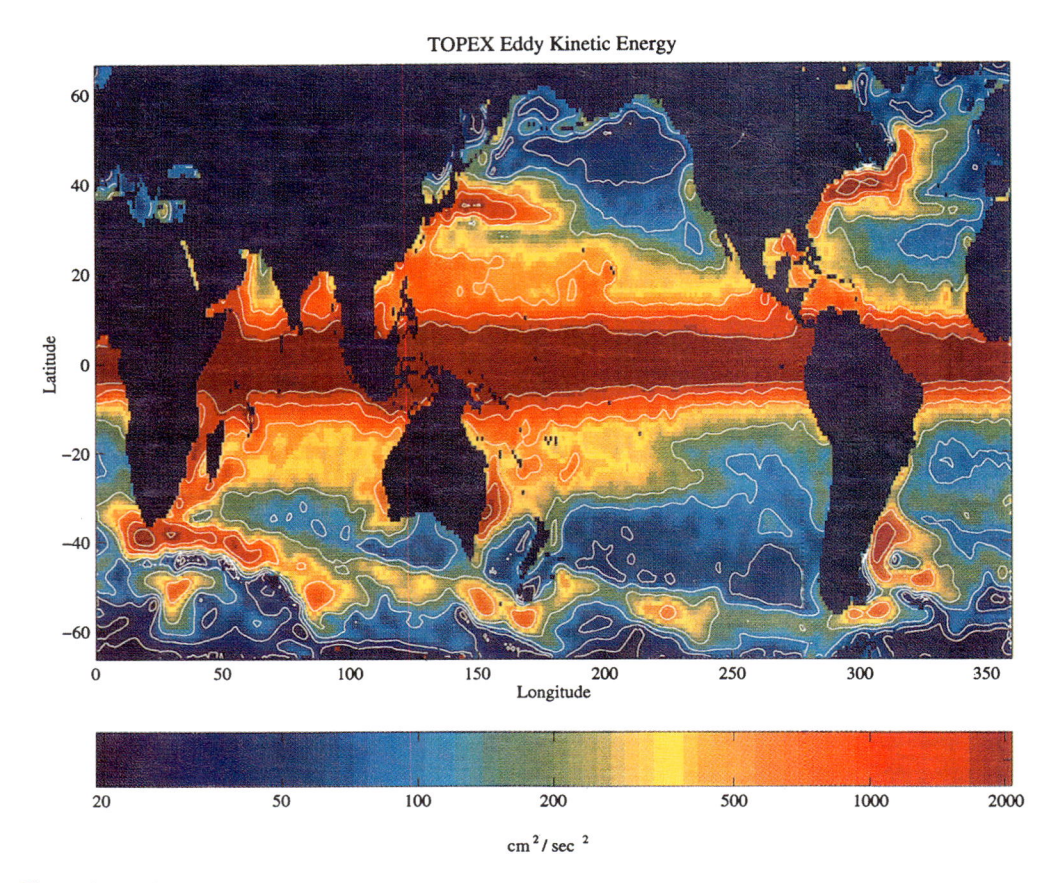

Figure 1.6.4. Average eddy kinetic energy maps from two years of Geosat altimeter data (November 1986 to November 1988) for ascending and descending passes (after Shum et al., 1990).

changes in oceanic density. Rossby (1969) suggested that this effect could be used to advantage since it provided a way to measure variations in travel times of acoustic pulses from the sea floor due to changes in the depth of the thermocline. Moreover, the fact that travel times are integrated measurements means that they effectively filter out all but the fundamental mode of any vertical oscillations. This idea led to the development of the inverted echo sounder (IES) in which the round-trip travel time of regularly-spaced 10 kHz acoustic pulses from the seafloor are now used to determine temporal variability in the integrated density structure of the ocean. The IES has been widely used in studies of the Gulf Stream where its records are interpreted in terms of thermocline depth, heat content and dynamic height (Rossby, 1969; Watts and Rossby, 1977). It has also been used in the equatorial Pacific and Atlantic although interpretation of the data is more uncertain because of a lack of repeated deep CTD casts to determine density variability (Chiswell *et al.*, 1988).

Tidal period variability and large changes caused by El Niño–Southern Oscillation (ENSO) events are potentially serious problems in the interpretation of echo sounding data. In particular, the CTD data are needed to convert time-series of acoustic travel-time Δt between two depth levels (z_1, z_2) to a time-series of dynamic height ΔD

integrated over the pressure range p_1 to p_2 with an accuracy of 0.01–0.04 dynamic meters. The obvious similarity between these two parameters (Watts and Rossby, 1977) can be seen from the relations

$$\Delta t_{z1/z2} = 2 \int_{z_1}^{z_2} [1/c(S, T, p)] \, dz \qquad (1.6.3.1)$$

and

$$\Delta D_{p2/p1} = \int_{p_1}^{p_2} [1/\rho(S, T, p) - 1/\rho(35, 0, p)] \, dp$$

$$= 10^3 \int_{p_1}^{p_2} \delta \, dp \qquad (1.6.3.2)$$

where c is the speed of sound (at specified salinity S, temperature T, and pressure p), z = depth (positive downward), and ρ is density. Finally, δ, defined as

$$\delta = \alpha(S, T, p) - \alpha(35, 0, p) = 1/\rho(S, T, p) - 1/\rho(35, 0, p) \qquad (1.6.3.3)$$

is the specific volume anomaly. In these expressions, we use SI units with depth in meters, density in kg/m^3, pressure in decibars, and dynamic height in dynamic meters (1 dyn m = 10 m^2 s^{-2}).

Chiswell *et al.* (1988) compare time-series of dynamic height from an inverted echo sounder with sea-level height ($z_1 = -\eta$) from a pressure sensor located 70 km away on Palmyra Island in the central equatorial Pacific. The spectra for the dynamic height variations determined from the ISE closely resembled those from the pressure gauge. Significant coherences were found between the two signals at the 99.9% level of significance. Although, in principle, varying mixtures of vertical internal modes could produce a frequency dependence in the conversion of ISE to dynamic height, the effect was not significant over the year-long data series. Wimbush *et al.* (1990) discussed moorings in 4325 m of water 72 km west of the subsurface pressure gauge in the Palmyra Lagoon (5°53'N, 162°05'W). The IES was set to 1/2-hour sampling, each sample consisting of 20 pulses 10 s apart. Outliers are eliminated and the median value taken as representative of the acoustic travel time. According to Wimbush *et al.*, a conventional IES without a pressure sensor adequately records synoptic-scale dynamic height oscillations with 20–100-day periods. Chiswell (1992) discussed 14-month records from five inverted echo sounders deployed in February 1991 in a 50-km array in 4780 m of water near 23°N, 158'W north of Hawaii. The CTD and ADCP (acoustic Doppler current profiler) data collected during monthly surveys at the array site provided sufficient density data to calibrate the IES data in terms of dynamic height and geostrophic currents.

Wimbush *et al.* (1990) used the response method of Munk and Cartwright (1966) to get the daily and semidiurnal tides and filtered with a 40-h Gaussian low-pass filter while Chiswell has attempted to resolve the tidal motions through 36-h burst sampling of the density structure from three-hourly CTD profiles. IES deployments show that there is a linear relationship between dynamic height and travel time (Figure 1.6.5), with the calibration slope dependent on the particular T–S properties of the region. In this case, we can link variations in ΔD (for depths shallower than the reference level p_{ref} used in the dynamic height calculation) to the acoustic travel time Δt_{ref}

$$\Delta D = m\Delta t_{\text{ref}} \qquad (1.6.3.4)$$

where total acoustic travel time to the bottom is

$$\Delta t = \Delta t_{\text{ref}} + \gamma H_2 \qquad (1.6.3.5)$$

in which $\gamma = 2/c_b$ and c_b is the average sound speed (assumed constant) between p_{ref} and the bottom. The depth H_2 is the depth range between the seafloor (pressure $= p_b$) and the reference pressure level, p_{ref}. Solving yields,

$$\Delta D = m[\Delta t - (\gamma/\rho_b' g') p_b] \qquad (1.6.3.6)$$

where gravity and bottom density are scaled as $g' = 0.1g$ and $\rho_b' = 10^{-3}\rho_b$, respectively. For oscillations in density having periods longer than about 20 days, the second term on the right-hand side of equation (1.6.3.6) may be dropped, whereby

$$\Delta D = m\Delta t \qquad (1.6.3.7)$$

Wimbush *et al.* find $m = -70$ dyn m s^{-1} to convert Δt to ΔD, while Chiswell (1992) finds $m = -57.8$ dyn m s^{-1} for Δt defined for $z = 0$–4500 m and ΔD at 100 m referenced to 1000 m. The high correlation coefficient, $r^2 = 0.93$, is based on 186 shallow (1000 m) and 17 deep (4500 m) CTD casts. The error in the slope using the deep casts is 4 dyn m s^{-1} with an RMS deviation of 0.017 dyn m; for the shallow casts the mean is 0.1 dyn m s^{-1} with a deviation of 0.029 dyn m. The travel times for the subtropical Pacific moorings of Chiswell correlate better with dynamic height

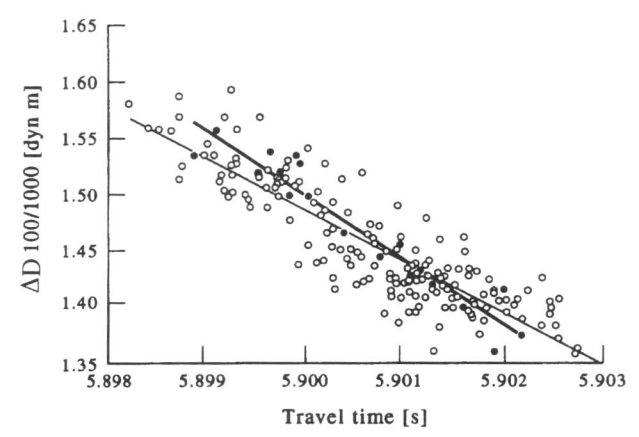

Figure 1.6.5. Dynamic height at 100 m relative to 1000 m ($\Delta D_{100/1000}$) from 186 shallow CTD casts plotted against corresponding travel time measured by IESs (open circles). Thin line is the least squares fit. Thick line and solid circles give $\Delta D_{100/1000}$ calculated from 17 deep casts plotted against the corresponding travel time from 4500 m to the surface, $T_{0/4500}$ ($r^2 = 0.93$ and slope $= -57.8$ dyn m/s). (From Chiswell, 1992.)

measured below 100 m than with surface dynamic heights. This is because large variations in the temperature and salinity relation in the upper 100 m affect dynamic height more than they affect acoustic travel time (Chiswell et al., 1988). The tidal range of 8 dyn cm is relatively large compared with the seasonal range of 25 dyn cm and illustrates the need for detailed CTD sampling. Geostrophic currents have been derived from the array using the time-series of dynamic height created from the multiple IES moorings. Aliasing of the records by high frequency motions and a lack of CTD data to the depth of the IES remain problems for this method.

1.6.4. Wave height and direction

Any discussion of sea-level would be incomplete without some mention of surface gravity wave measurement. Methods include: a capacitance staff which measures the change in capacitance of a conductor as the air–water interface moves up and down with passage of the waves; an upward-looking, high-frequency acoustic sounder or acoustic Doppler current profiler with a vertical-pointing transducer which can be used to examine both the surface elevation and the associated orbital currents; a fixed graduated-staff attached to a drill platform, stuck in the sand or otherwise attached to the seafloor; satellite altimetry; a bottom-mounted pressure gauge with rapid sampling time; a shipborne Tucker wave-recorder system; and the waverider and directional waverider buoys. For brevity, we limit our presentation to the directional waverider since it represents reliable off-the-shelf technology.

Built by the Datawell company of The Netherlands, the directional waverider is a spherical, 0.9-m diameter buoy for measuring wave height and wave direction. The buoy contains a heave–pitch–roll sensor, a three-axis fluxgate compass and two fixed "x" and "y" accelerometers. The directional (x, y, z) displacements in the buoy frame of reference are based on digitial integration of the horizontal (x, y) and vertical (z) accelerations. Horizontal motions rather than wave slope are measured by this system. Vertical motions are measured by an accelerometer placed on a gravity-stabilized platform. The platform consists of a disk which is suspended in a fluid within a plastic sphere placed at the bottom of the buoy. Accelerations are derived from the electrical coupling between a fixed coil on the sphere and a coil on the platform. A fluxgate compass is used to convert displacements from the buoy frame of reference to true earth coordinates.

Displacement records are internally filtered at a high-frequency cut-off of 0.6 Hz. Onboard data reduction computes energy density, the prevailing wave direction, and the directional spread of the waves. Frequency resolution is around 0.01 Hz for waves in the range 0.025–0.59 Hz. Transmission of data is to Argos satellite or through standard 27–40 MHz radio-link to shore. The buoy will measure heave in the range ± 20 m with 1 cm resolution for wave periods of 1.6–20 s in the moored configuration. The direction range is 0–360° with a resolution of 1.5°.

A crucial aspect of collecting reliable, long-term wave data is the mooring configuration. If not designed and moored correctly, there is little chance the mooring will survive the constant stresses of the wave motions. As illustrated in Figure 1.6.6, the recommended configuration consists of a single-point vertical mooring with two standard rubber shock cords and heavy bottom chain. This arrangement ensures sufficient symmetrical horizontal buoy response for small motions at low frequencies while the low stiffness of the rubber cords allows the waverider to follow waves up to

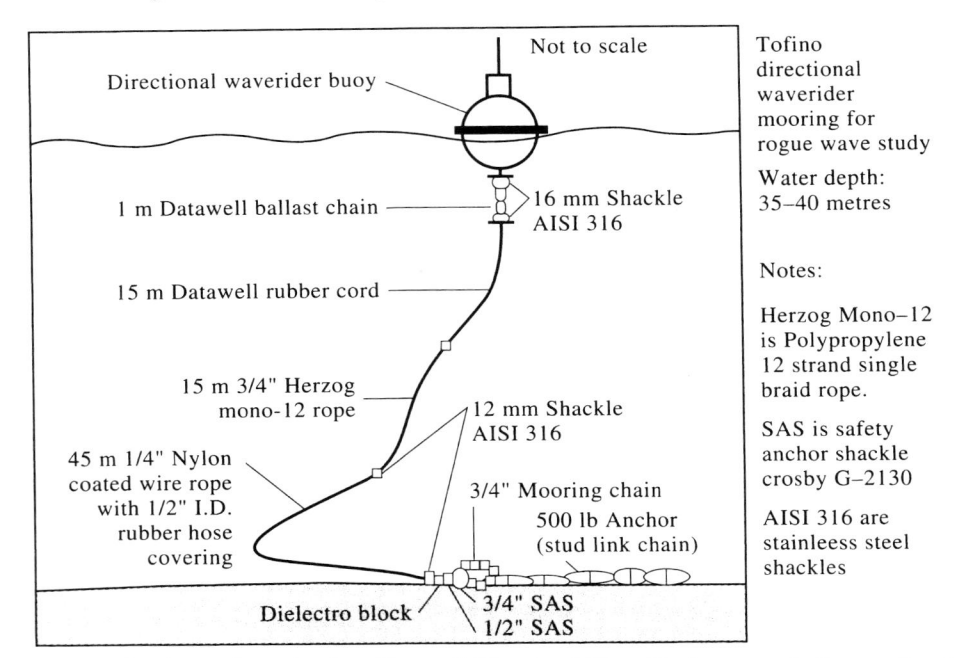

Figure 1.6.6. Mooring configuration for a Datawell Directional Waverider buoy on a shallow continental shelf. (Modified after Datawell bv, 1992; Courtesy T. Juhász and R. Kashino.)

40 m high. Current velocities can be up to 2.5 m/s, depending on water depth (Datawell bv, 1992).

1.7 EULERIAN CURRENTS

The development of reliable, self-recording current meters is one of the major technological advances of modern oceanography. These sturdy, comparatively lightweight instruments are, in part, a byproduct of the rapid improvement in electronic recording systems which make it possible to record large volumes of digital data at high sampling rates (Baker, 1981). Although they can be used in either moored or profiling modes, most current meters are used in time-series measurement of current speed and direction at fixed locations. [Such fixed-location measurements are called Eulerian measurements after the Swiss mathematician Leonhard Euler (1707–1783) who first formulated the equations for fluid motion in a fixed frame of reference.] The development of reliable mooring technology and procedures also has played a major role in advancing the use of moored current meters and associated instrumentation. Acoustic release technology, which has proven so critical to oceanic research, will be discussed at the end of this section.

Most commercially available current meters have sufficient internal power and data storage to be moored for several months to a year. The instrument's duration depends on the selected sampling rate, the data storage capacity, the battery life and the ambient water temperature. Greater power can be obtained from lithium batteries than from more conventional batteries but the user sometimes faces numerous transportation regulations and operational concerns with lithium batteries. Operating time for all types of batteries decreases with water temperature. Despite their sophistication, most current meters are made to withstand a fair amount of abuse during deployment and recovery operations. Typical "off-the-shelf" current meters (and releases) can be deployed to depths of 1000–2000 m, and many manufacturers fabricate deep versions of their products with heavy-duty pressure cases and connectors for deployments to depths of 6000 m. Most modern current meters also allow for the addition of ancillary sensors for concurrent measurement of temperature, conductivity (salinity), water clarity (light attenuation and turbidity), pressure, and other scalars.

Current meters differ in their type of speed and direction sensors, and in the way they internally process and record data. Although most oceanographers would prefer to work with the scalar components u, v of the horizontal current velocity vector, $\mathbf{u} = (u, v)$, it is a fact of life that current meters can directly measure only the speed ($|\mathbf{u}|$) and direction (θ) of the horizontal flow. (For now, we ignore the vertical velocity component, w.) It is because of this constraint that most current meter editing and analysis programs historically work with speed and direction. From a practical point of view, both the (u, v) and the ($|\mathbf{u}|, \theta$) representations have their advantages, despite the difficulties with the discontinuity in direction at the ends of the interval of 0–360°.

Speed sensors can be of two types: *mechanical sensors* which measure the current-induced spin of a rotor or paddle wheel; and *nonmechanical sensors* which measure the current-induced change in a known electromagnetic field or the differences in acoustic transmission times along an acoustic path. Despite these fundamental differences, all current meters have certain basic components that include speed sensors, a compass to determine orientation relative to the earth, built-in data averaging algorithms, and a digital storage device. Possible speed sensors include:

(a) Propellers (with or without ducts).
(b) Savonious rotors.
(c) Acoustic detectors (sound propagation or Doppler shift).
(d) Electromagnetic sensors (induced magnetic field).
(e) Platinum resistors (flow-induced cooling).

Flow direction relative to the axes of the current meter is usually sensed using a separate vane or by configuring the speed sensors along two or three orthogonal axes. In all current meters, the absolute orientation of the instrument relative to the earth's magnetic field is determined by an internal compass. At polar latitudes where the horizontal component of the earth's magnetic field is weak, measurement of absolute current requires that the meter be positioned rigidly in a known orientation. Direction resolution depends on the type of compass used in the measurement; e.g. clamped potentiometer for Aanderaa RCMs, optical disk for Marsh–McBirney electromagnetic current meters, and flux gate (Hall effect) compass for the EG&G Vector Measuring Current Meter, InterOcean S4 current meter and SimTronix Ultrasonic Current Meter UCM 40. For each deployment, compass direction must be corrected for the local deviation of the earth's magnetic field before the velocity data are converted to north–south and east–west components. The accuracy, precision, and reliability of a particular current meter are functions of the specific sensor configuration and the kind of processing applied to the data. Rather than comment on all the many possible variations, we will discuss a few of the more generic and successful configurations.

1.7.1 Early current meter technology

One of the earliest forms of current measurement was the tilt of a weighted line lowered from a ship. The time it took an object to travel the length of the ship also provided a measure of the surface flow. [The term "knot" is from the use by Dutch sailors of a knotted line to measure the speed of their sailing vessel.] Although we like to think of the current meter as a recent innovation, the Ekman current meter was in use as early as the 1930s (Ekman, 1932). Although many different mechanical current meters built in those days (see Sverdrup et al., 1942), few worked and most scientists went back to the Ekman meter. To measure the current, the instrument was lowered over the side of the ship to a specific depth, started by a messenger, and then allowed several minutes before being stopped by a second messenger. The current speed for each time increment was determined by reading a dial that recorded the number of revolutions of an impeller turned by the current. A table was used to convert impeller revolutions to current speed. Current direction was determined from the distribution of copper balls that fell into a compass box below the meter. A profile from 10 to 100 m typically took about 30 min. Obvious problems with this instrument included low accuracy in speed and direction, limited endurance, and the need to work from a ship or other stationary platform. One of the first commercial current meters was the self-contained Geodyne 850 current meter built in the United States in the 1960s. The Geodyne was a large and bulky, vertically-standing unit with a small direction vane and four-cup Savonious rotor. Burst sampling was permissible in the range of 60–660 s. The Nerpic CMDR current meter built in France in the 1960s was a torpedo-like device that oriented itself with the current flow and used an impellor-type rotor to measure current speed. In the original versions, data were recorded on punched paper tape. The Kaijo Denki current meter built in Japan in the 1970s was one of the earliest types of acoustic current meter.

Although some oceanographers might disagree, the age of the modern current meter appears to have started with the Aanderaa Recording Current Meter (RCM) developed by Ivar Aanderaa in Norway in the early 1960s under sponsorship of the North Atlantic Treaty Organizaton (NATO). The fact that many thousands of these internally recording current meters remain in operation attests to the instrument's durability. Many oceanographers still consider the Aanderaa RCM4 (Figure 1.7.1) and its deep (>2000 m) counterpart, the RCM5, the "workhorses" of physical oceanography. It certainly is the most common and reliable current meter used to measure ocean currents. For this reason, there have been more studies, intercomparisons and soul-searching with this instrument than with any other type of meter.

1.7.2 Rotor-type current meters

The RCM series of current meters

The Geodyne and RCM4 current meters were the first current meters to use a Savonious rotor to measure current speed. This rotor consists of six axisymmetric, curved blades enclosed in a vertical housing which is oriented normally to the direction of flow (Figure 1.7.1). Data in the RCM4 are recorded on a small 1/4 inch reel-to-reel magnetic tape. Allowable sampling rate settings are 3.75×2^N min (e.g. 3.75, 7.5, 15, 30, 60 min) where N ($= 0, 1, 2, ...$) is an integer. Although shorter sampling periods are possible, they are not practical given the mechanical limitations of the rotor. Speed is obtained from the number of rotor revolutions for the entire sample interval while direction is the single direction recorded at the end of the sample period. Thus, speed is based on the average value for the recording interval while direction involves a single measurement. In the past, the number of revolutions per recorded data "count" was varied by changing the entire rotor counter module. More recent RCMs allow the investigator to set the number of revolutions per count (e.g. 2^M revolutions per count, where $M = 1, 2, 3, ...$) so that the speed range of the instrument can be adjusted for the flow conditions. For example, in the coastal tidal passes of British Columbia and Alaska, the common upper range of 3 m/s for standard rotor settings is not always sufficient to measure peak tidal speeds; the peak speed of 7.5 m/s that occurs in Nakwakto Rapids in coastal British Columbia is beyond the design of most modern current meters. The direction vane of the RCM4 is rigidly affixed to the pressure case containing the data logger. The unit is then inserted in the mooring line and the entire current meter allowed to orient in the direction of the current. Although the RCM does not average internally, vector-averaged currents can be obtained through post-processing of the data (Thomson et al., 1985). A meteorological package for surface applications also is available with the same data logging system (Pillsbury et al., 1974).

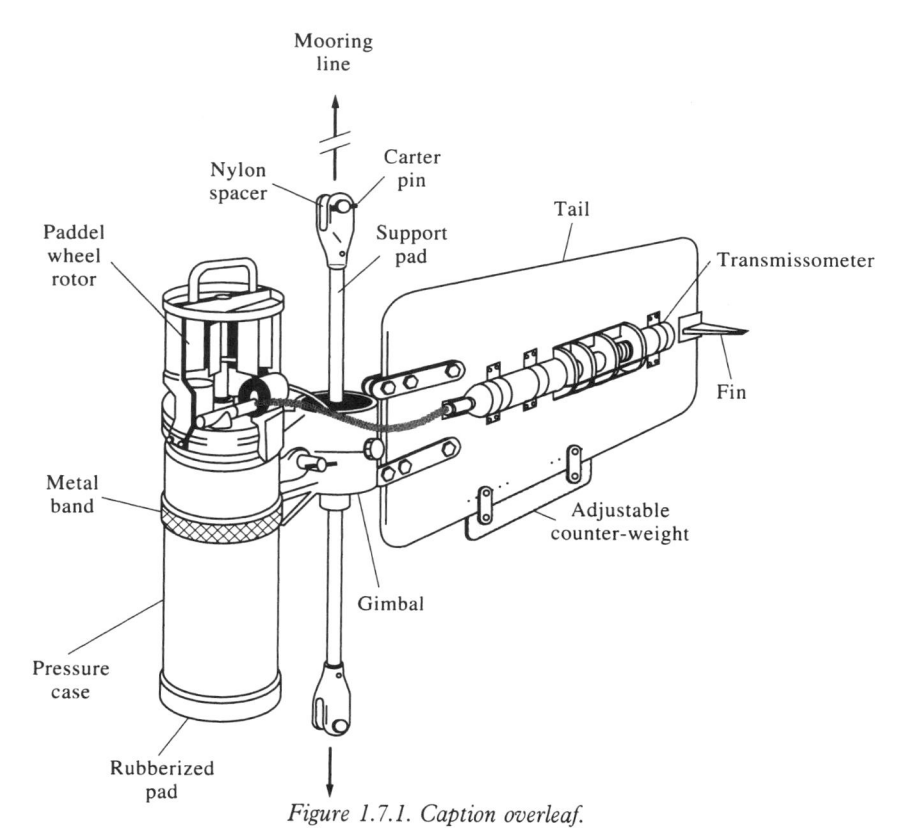

Figure 1.7.1. Caption overleaf.

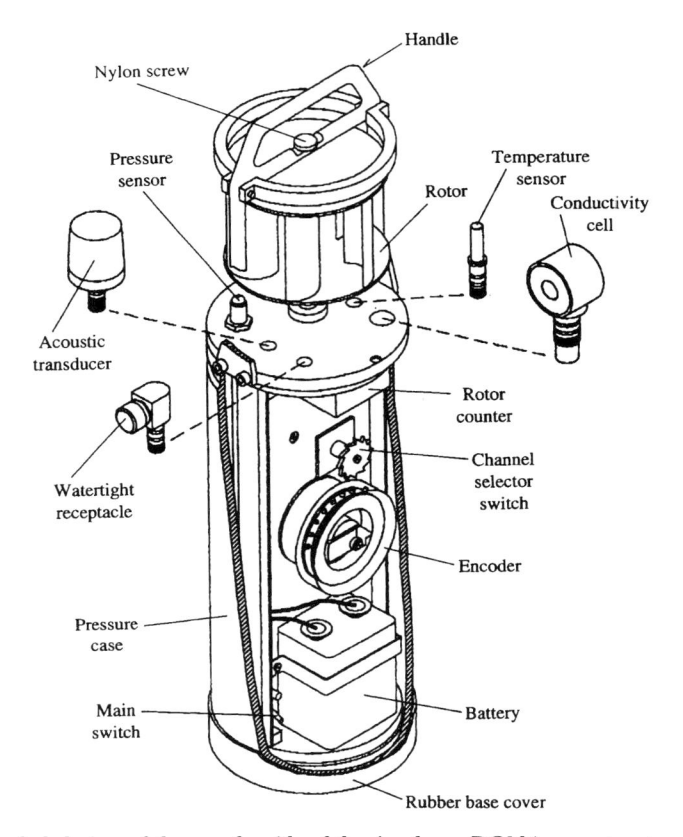

72 *Data Analysis Methods in Physical Oceanography*

Figure 1.7.1. Exploded view of the encoder side of the Aanderaa RCM4 current meter. The reverse side contains a reel-to-reel 1/4-in tape system for recording the data from the different channels. The recorder unit is attached to a directional vane. (Courtesy G. Gabel, G. S. Gabel Corp.)

Part of the reason for the popularity of the RCM series of current meters has been their reliability, comparatively low cost, and relatively simple operation. Both calibration and maintenance of the instruments can be performed by individuals with fairly limited electronics expertise. In more recent years, many of the other types of current meters, such as electromagnetic current meter (ECM) and the acoustic current meter (ACM), have advanced to the point that they require electronics expertise due to the advanced computer diagnostics available from the manufacturers of these instruments. Another attractive feature of the RCM is the easy addition of sensors for measuring temperature, conductivity and pressure (depth) on the same data logger. The Aanderaa RCM7 introduced in the late 1980s can be purchased with standard temperature (-2 to $+35°C$), expanded temperature (i.e. over a narrower range such as $0-10°C$), conductivity (for salinity), and total pressure. The 0–5 volt output from a Sea Tech transmissometer for measuring water clarity can be readily incorporated in the instrument package. Thus, there is the potential to collect a wide range of parameters other than just currents alone.

The profiling Cyclesonde (van Leer *et al.*, 1974; Baker, 1981) consists of a RCM4 current meter affixed to a buoyancy-driven platform which makes repeated automatic round trips between the surface and some specified depth (<500 m) along a taut-wire mooring. The vertical cycling of the instrument is controlled by changing the density of the instrument package by a few percent using an inflatable bladder. Depending on the prescribed sampling interval and the duration of each round-trip (or depth of water sampled), the instrument can provide time-series of currents, temperature and salinity over periods of weeks to months at depths of every 10 m or so through the water column.

Processing of RCM4 data includes four major steps: (1) tape transcription (quarter-inch tape to computer format); (2) calibration, or conversion to physical units; (3) error detection, spike removal and interpolation; and (4) data analysis. The last point will be discussed in detail in later chapters of this book. The first three steps provide an example of the procedure required in producing useful data from moored current meters. The data in an RCM4 are recorded as 10-bit binary words (numbers from 0 to 1023) on 1/4-inch magnetic tape. For each cycle, six binary words are written on the tape. For the temperature channnel, a near-linear calibration curve is applied to the measured value to convert it to temperature. Current speed for earlier versions is handled somewhat differently since speeds must be calculated as the difference between consecutive integers recorded on the appropriate data channel. The relationship between speed in physical units (e.g. cm/s) and rotor count (X) is nearly linear so that speed also can be calculated from a linear calibration.

Tape translation is carried out by connecting a 1/4-inch tape recorder to a digital computer. With this set-up, the digital data are transferred from the 1/4-inch tape to computer compatible format for further editing and analysis. To these raw character data must be added "header" information such as the start and stop times of the particular mooring. As a check one calculates the number of instrument cycles that should have occurred during the mooring period and this should equal the number of records in the raw data file. If this is not the case, then the data have timing errors which must be corrected before processing can continue. Timing errors will be more closely addressed in a later section.

To convert the dated raw data to physical units (i.e. speed, direction) calibration constants are needed for the individual sensors. For most parameters the calibration values are found for each meter separately as quadratic fits to the calibration data. As has been mentioned above, this is not the case for the speed parameter for which a general curve can be used for all rotors if currents are typically greater than 10 cm/s. Directions also are handled somewhat differently in that no formula is derived from the calibration data but rather a simple look-up table is developed for the calibration data from which the compass readings can be converted directly into degrees from true or magnetic north.

1.7.2.1 The vector averaging current meter (VACM)

As discussed by Baker (1981), one of the important data reduction techniques in oceanography was the introduction of the "burst sampling" scheme of Richardson *et al.* (1963) whereby short samples of densely packed data are interspersed with longer periods of no data. In continuous mode, the average current speed and instantaneous direction are recorded once per sampling interval. In burst mode, a rapid series of speed and direction measurements are averaged over a short segment of the sampling interval. In vector-average mode, the instrument uses speed and direction to calculate the horizontal and vertical components of the absolute velocity during the burst. It

74 *Data Analysis Methods in Physical Oceanography*

then separately averages each component internally to provide a single value of velocity vector for each burst. If enough is known about the spectra of the flow variability, the burst samples can be used to adequately estimate the total energy in the various frequency bands. This procedure greatly reduces the amount of recording space needed to sample the currents. The vector-averaging current meter (VACM) introduced in the 1970s uses both burst sampling and internal processing to compute the vector-average components of the current for each sampling period. Current speed is obtained using a Savonious rotor similar to that on the RCMs but direction is from a small vane that is free to rotate relative to the chassis of the current meter. Vectors are computed for every eight revolutions of the rotor and averaged over periods of from 4 to 15 min, depending on the selected sampling interval.

Problems with the Savonious rotor

Because of its widespread use in oceanography, the Savonious rotor sensor needs to be covered in some detail. We begin by noting that a principal shortcoming of the RCM4/ 5 is its inability to record currents accurately in regions affected by surface wave motions. The problem with the Savonius rotor response is that it is omnidirectional and therefore responds excessively to oscillatory wave action. An intercomparison experiment using a mooring array shown schematically in Figure 1.7.2(a) demonstrated the differences between Savonius rotor measurements and those made with an EM current meter (Woodward *et al.*, 1990). Even under moderate wave conditions, the near-surface moored RCM4 can have its speeds increased by a factor of two through wave pumping (Figure 1.7.2b). The effect of wave pumping on the Savonius rotor

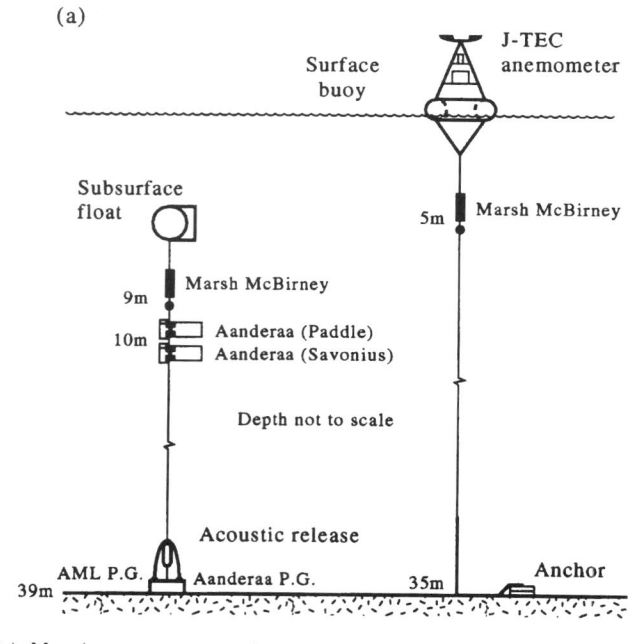

Figure 1.7.2. (a) Mooring arrangement for comparison of current speed and direction from Aanderaa RCM4 (Savonious rotor) and RCM7 (paddle wheel) current meters and Marsh-McBirney (Electromagnetic) current meters moored at 10 m depth during September 1983 in an oceanic wave zone (Hecate Strait, British Columbia).

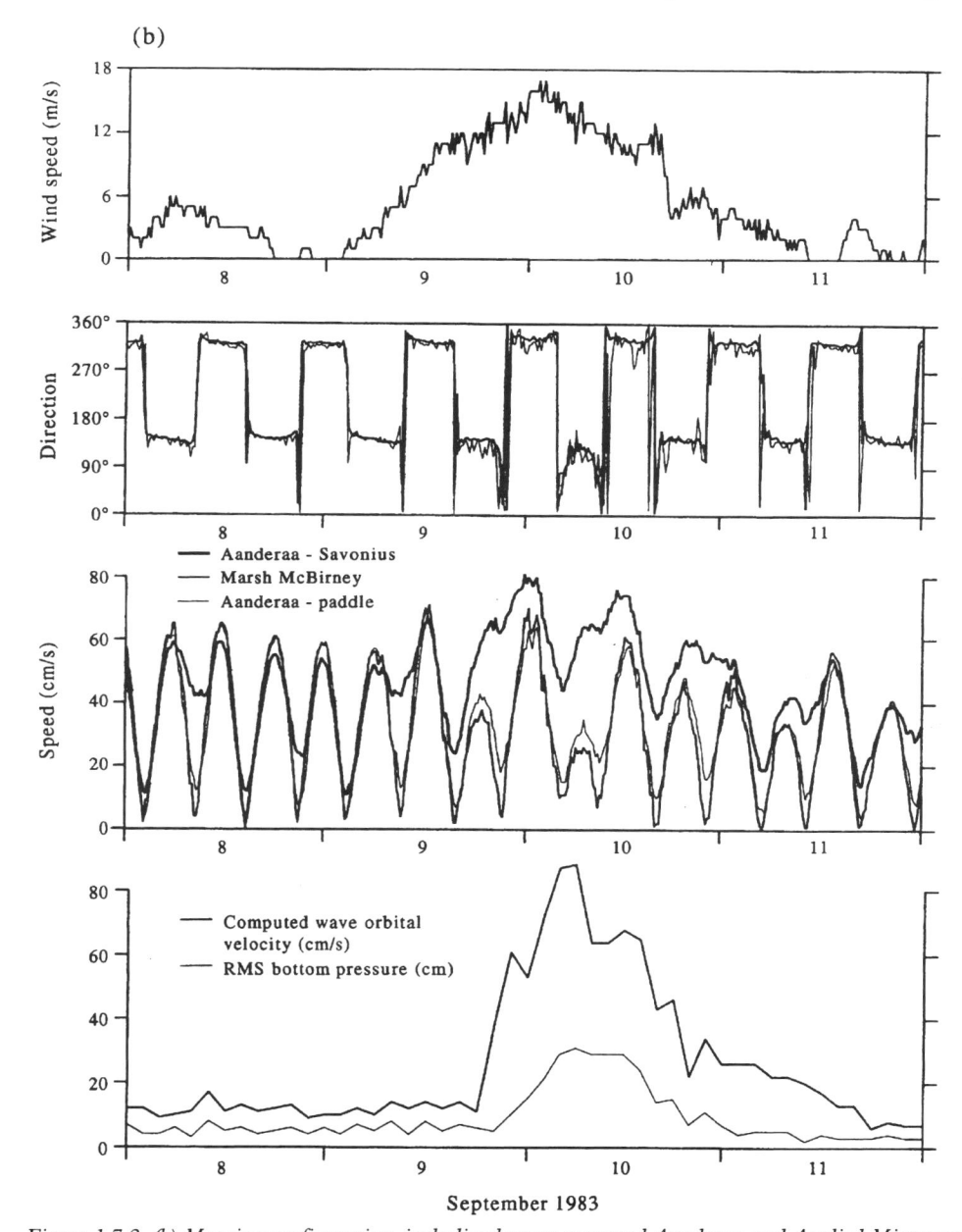

Figure 1.7.2. (b) Mooring configuration, including bottom-mounted Aanderaa and Applied Microsystems pressure gauges. Winds were measured using a J-Tec vortex-shedding anemometer. In moderate wind-wave conditions, a surface or near-surface moored RCM4 with Savonius rotor can have its speeds increased by a factor of 2 through wave pumping. The paddle wheel RCM7 behaves somewhat better.

significantly increases the spectral energy at both low and high frequencies (Figure 1.7.2c). Hence, the instrument is best suited to moorings supported with subsurface floats but is not suitable for mooring beneath surface buoys or in the upper ocean wave regime. Unlike the earlier Aanderaa current meters, VACMs provide accurate measurements when deployed in near-surface wave fields and from surface-following

76   *Data Analysis Methods in Physical Oceanography*

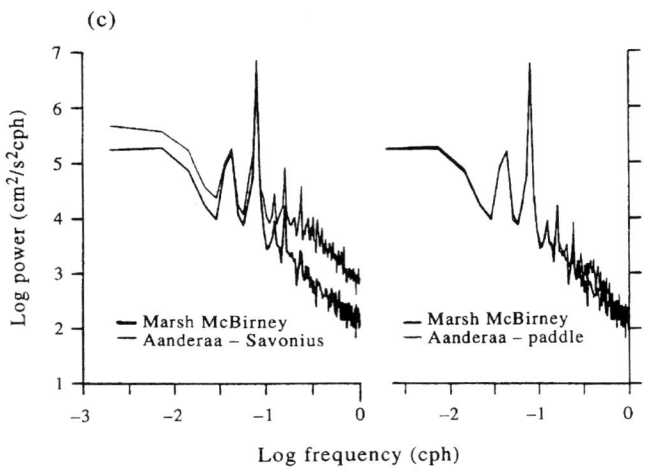

Figure 1.7.2. (c) Power spectra for current measurements in (a). (Adapted from Woodward et al., 1990.)

moorings (Halpern, 1978). In a comparison between Aanderaa and VACM measurements, Saunders (1976) concluded that "the Aanderaa instrument, excellent though it is on subsurface moorings, is not designed, nor should it be used, where wave frequency fluctuations are a significant fraction of the signal." In this, and a later paper, Saunders (1980) pointed out that the contamination of the Aanderaa measurements in near-surface applications is due also to a lag in the response of the direction vane to oscillatory flow.

Since 1991, Aanderaa has gone to a vector-averaging RCM7 current meter with a paddle-wheel rotor (Fig. 1.7.1a) and internal solid state, E-prom modular memory. In earlier versions of the RCM7, the paddle-wheel rotor was partially shielded by a semicircle baffle which was intended to reduce wave induced "pumping". This has now been abandoned since the baffle sheds small-scale eddies which interfere with the response of the paddle-wheel in other operations. Field tests indicate that the vector-averaging RCM7 has only slightly better wave-region performance than the earlier RCMs (Figure 1.7.2c) and the overall improvements are marginal for most applications. During the selected recording interval, the number of rotor revolutions and compass direction are sampled 50 times per recording interval; e.g. every 12 s for a 10-min sampling interval. As with other vector-averaging current meters, the speed and direction are then resolved internally into east–west and north–south components and successive components are added and temporarily stored. When the selected recording interval has elapsed, the resulting average vector and its angle are calculated and stored. A problem with the electronic memory is that data are lost if the instrument floods, as it often does when the instrument is hit by fishnet or tug lines. This was not the case for the 1/4-inch magnetic tapes used in the RCM4. Thomson (1977) reports finding an old RCM that had lain on the bottom of Johnstone Strait for over three years. Although the metal components and circuit boards had turned to mush, the salt-encrusted tape contained a full record of error-free data.

Another problem common to all Savonius rotor current meters is that bearing friction results in fairly high threshold of the rotor and an improper response of the rotor to low current speeds. For the Aanderaa RCM4/5, this threshold level is about

2 cm/s and current measurements taken in quiescent portions of the ocean will have many missing values where the currents were too slow to turn the rotors during the sampling interval. According to manufacturer specifications, the response is linear for current speeds between 2.5 and 250 cm/s so that once the rotor is turning it has acceptable response characteristics. In this range, accuracy is given as ±1 cm/s or 2% of the speed, whichever is greater. Accuracies for the other associated sensors are 1% for pressure, ±0.3°C for temperature and ±0.05 psu for salinity. All of these accuracies are really "relative" values and regular calibration is required to insure reliable measurements. Such a calibration procedure is discussed in detail in Pillsbury et al. (1974) for RCM4s. We will only highlight some of the more important aspects of this calibration in order to suggest problem areas where data from Aanderra current meters may be subject to error.

As described by Pillsbury et al. (1974), calibration of the RCM4 compass is important because more compass failures occurred for a set of instruments than all other sensor failures combined. Careful calibration will reveal the several different kinds of compass failure. The compass calibration is performed for selected compass bearings by rotating the instrument through 360° on a pivoted stand. This operation is repeated 10 times. A reliable compass is one which repeats its calibration curve within 3°. From calibration work reported by Gould (1973), it is clear that there is a significant departure from linearity in most RCM4 compasses. The magnitude of the nonlinearity errors (approximately 1% of the scalar mean speed per degree of compass nonlinearity) means that many of the residual velocity values observed in the ocean could be introduced by a nonlinearity of 1° or 2° in the direction sensor. If such residual values are to be trusted, care must be taken to "calibrate out" instrument nonlinearities in the data analysis procedure. Such precautions are particularly important if the current meter records are to be used to deduce shears from pairs of instruments or circulation patterns from horizontal current meter arrays.

Turning to the rotor, it was found that for speeds several centimeters per second above the threshold, the calibration of all rotors of a given type can be considered as equal. For calibration, this threshold was found to be roughly 10 cm/s, below which each rotor should be calibrated with its corresponding current meter. For mean speeds greater than 10 cm/s, a general calibration curve can be used for all instruments (Figure 1.7.3). This calibration curve is fitted by a line and used for all calibrations. Deviations from this line varied from 19% at 2 cm/s to less than 1% at 30 cm/s, with a mean value of 4%.

1.7.2.2 Vector measuring current meter (VMCM)

To circumvent the nonlinear response problems of the RCM4, Weller and Davis (1980) developed the vector measuring current meter (VMCM) which uses two orthogonal propeller current sensors with an accurate cosine response. This instrument produces negligible rectification and therefore should accurately measure mean flow in the presence of unsteady oscillating flow. In laboratory tests, the VMCM performed well in the presence of combined mean plus oscillatory flow as compared with poorer performances by Savonius rotor/vane systems and by electromagnetic and acoustic sensors. The open fan-type rotors of the VMCM are highly susceptable to fouling by small filaments of weed and other debris.

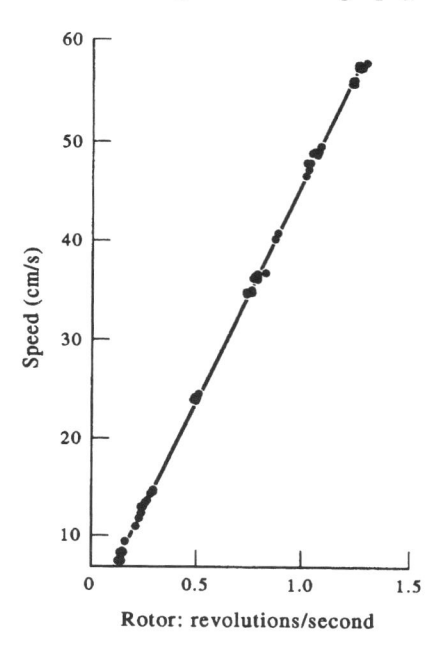

Figure 1.7.3. A general calibration curve of current speed (cm/s) versus rotor counts in revolutions/second for mean speeds greater than 10 cm/s. This particular calibration curve has a linear relation for all calibrations.

1.7.3 Nonmechanical current meters

1.7.3.1 Acoustic current meters (ACM)

Nonmechanical current meters determine current speed and direction by measuring speed along two or three orthogonal sensor axes. Once the flow direction relative to the current meter is determined, absolute direction is found using a built in magnetic compass. Acoustic current meters (ACMs) measure the difference in the time delay of short, high frequency (megahertz) sound pulses transmitted between an acoustic source and receiver separated by a fixed distance, L. In all cases, the transducer and receiver are combined into one source–receiver unit. The greater the speed of the current component in the direction of sound propagation, the shorter the pulse travel time and vice versa. For instance, suppose that the speed of sound in the absence of any current has a value c. The times for sound to travel simultaneously in opposite directions from two combined transducer-receiver pairs in the presence of an along-axis current of speed v is: $t_1 = L/(c + v)$ for transducer–receiver pair No. 1 and $t_2 = L/(c - v)$ for transducer–receiver pair No. 2. The velocity component along the transducer axis is therefore

$$v = L(t_2 - t_1)/(2t_1 t_2) \qquad (1.7.1)$$

A three-axis current meter determines the three-dimensional velocity by simultaneously measuring time differences along three orthogonal axes.

Examples of commercial acoustic current meters include the SimTronix UCM 40 and the Niel Brown ACM current meters. Because of the rapid ($\approx$1500 m/s) propagation of sound in water, these current meters are capable of high frequency sampling and processing, with typical data rates of 10–20 Hz. The instruments also can provide estimates of the sound velocity along the two paths of length L between the sensors. More specifically, $c = 2L/t$, where $t = t_1 t_2/(t_1 + t_2)$ is the effective time of propagation. Manufacturer specifications vary but may be characterized as follows:

- *Speed accuracy*: on the order ±1 cm/s at flow speeds of 10 cm/s.
- *Speed resolution*: approximately ±1 mm/s
- *Threshold speed*: 1 mm/s.
- *Speed range*: 0–5 m/s
- *Compass direction*: accuracy ±2°; resolution 0.1% of the range.
- *Sampling rate*: 20 Hz.
- *Acoustic frequency*: 4 mHz.
- *Allowable tilt*: a true cosine tilt response up to ±20°.
- *Sound speed*: range of 1350–1600 m/s and accuracy of ±5 m/s.

Because of their sophisticated technology, acoustic current meters are often difficult to operate and maintain without dedicated technical support. For example, biofouling of the transducers can be a problem on any long-term mooring in the euphotic (near-surface light influenced) zone. The instruments also must undergo frequent recalibration due to problems with sensor misalignment and changes in the physical dimensions of the transducer-receiver pairs. As discussed by Weller and Davis (1980), this is a particular weakness of the ACM which has proved difficult to calibrate due to drifts in the zero level and in the amplifier gain. In one comparison, they found that the background electrical noise of the ACM had the same level as the signal. As they point out, these problems are with the system electronics and should be solvable. Similar problems were encountered by Kuhn et al. (1980) in their intercomparison test but they were quick to point out that their ACM was an early prototype model and many of the problems that they encountered have since been solved by the manufacturer (Gytre, Norway).

1.7.3.2 Electromagnetic current meters (ECM)

Electromagnetic current meters such as the Marsh-McBirney 512 and the Inter-Ocean S4 use the fact that an oceanic current behaves as a moving electrical conductor. As a result, when an ocean current moves through a magnetic field generated within the instrument, an electromotive force is induced which is directly proportional to the speed of the ocean current and at right angles to both the magnetic field and the direction of the current (Faraday's law of electromagnetic induction). In general, the magnetic field may be that of the earth or the one generated by an electric current flowing through appropriately shaped coils (Figure 1.7.4). Faraday tried to measure the flow of the Thames River using electrodes on either side but was unsuccessful because his galvanometers were not sensitive enough. Following the Second World War, the principle was used successfully to estimate the flow along the English Channel by measuring the potential difference between electrodes on either side using a telegraph cable for the distant electrode and the vertical component of the earth's magnetic field.

A two-axis electromagnetic current meter with an internal compass is used to produce horizontal components referenced to earth coordinates. The induced elect-

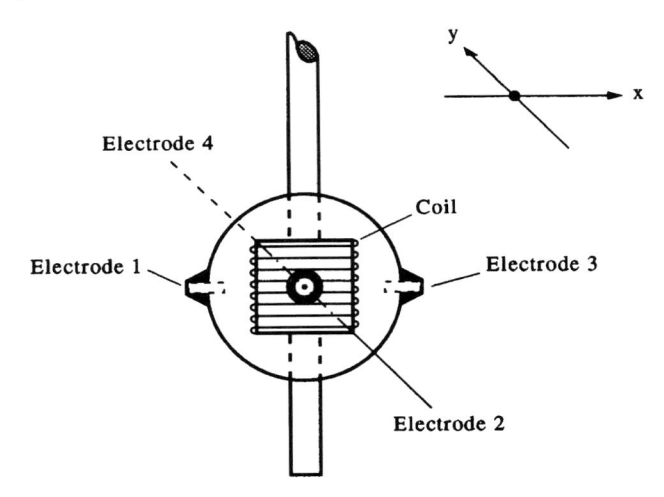

80 *Data Analysis Methods in Physical Oceanography*

Figure 1.7.4. Principle of the electromagnetic current meter. Instrument measures the electromotive force (EMF) on an electric charge (the oceanic flow) moving through the magnetic field generated by the coil. This produces a voltage potential at right angles to both the magnetic induction field and the direction of flow.

rical current gives the oceanic flow components relative to the instrument axes while the internal compass determines the orientation of the axes relative to the horizontal component of the earth's magnetic field. Electromagnetic current meters such as the S4 measure the electrical potential generated across two pairs of exposed metal (titanium) electrodes located on opposite sides of the equatorial plane on the surface of a plastic sphere (Figure 1.7.5). The electrodes form orthogonal (x, y) axes that detect changes in the induced electrical potential associated with the ocean current. The induced voltage potential (or electromagnetic force, EMF) **E** is found by Faraday's Law through the cross-product

$$\mathbf{E} = \int_0^\infty \mathbf{v} \times \mathbf{B} \, dL \qquad (1.7.2)$$

where **v** is the velocity of the flow past the electrodes, **B** is the strength of the applied magnetic field supplied by a battery-driven coil oriented along the vertical axis of the instrument, and L is the distance from the center of the coil. The magnetic field is directed vertically past the electrodes so that current flow parallel to the x-axis generates a voltage along the y-axis that is directly proportional to the strength of the current. The electric current induced by the voltage potential can be measured directly and converted to components of the flow velocity using laboratory calibration factors. Alternatively, a gain-controlled amplifier can be used to maintain a constant DC voltage at the logical output. The feedback current needed to maintain that electric current is directly proportional to the flow speed. As with the acoustic current meters, manufacturer's specifications vary but may be characterized as follows:

- *Speed accuracy*: roughly ±2% of reading (with a minimum of 1 cm/s).
- *Speed resolution*: about ±1–2 mm/s for a standard velocity range of 0–3.5 m/s (higher accuracies for narrower ranges).

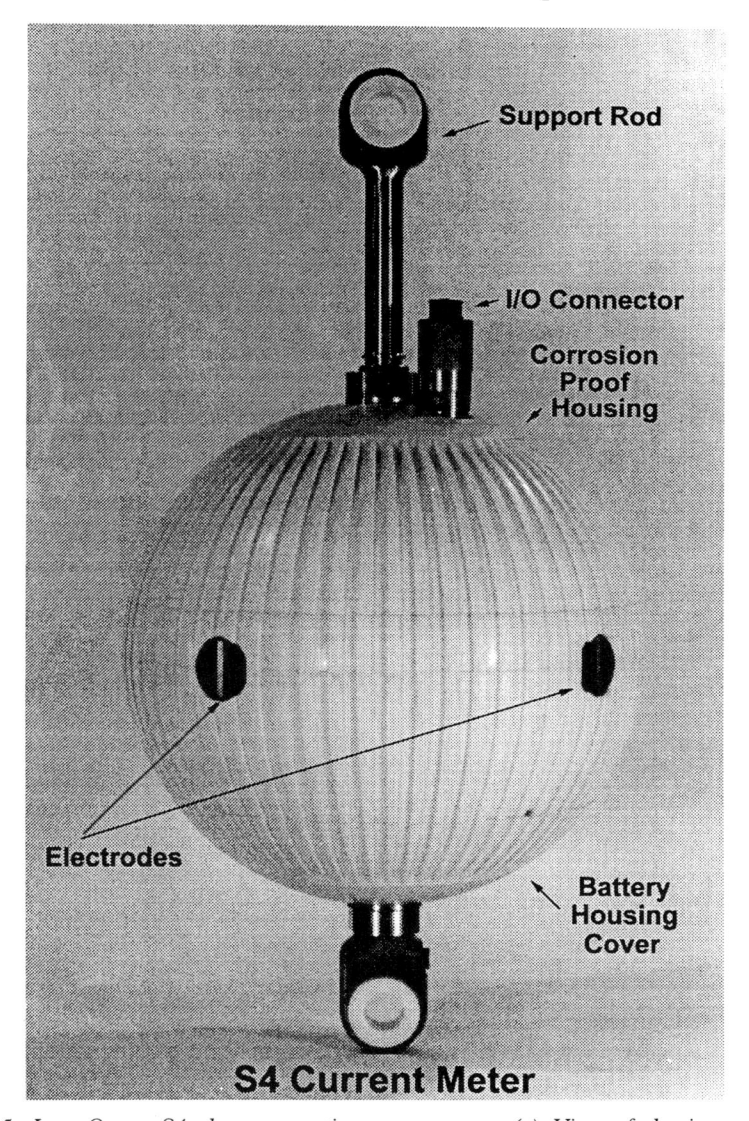

Figure 1.7.5. Inter-Ocean S4 electromagnetic current meter. (a) View of the instrument showing electrodes. (Courtesy, Mark Geneau, Inter-Ocean.)

- *Threshold speed*: 1 mm/s; limited by noise.
- *Speed range*: 0–3.5 m/s, but can expanded to 0–5 m/s or reduced for higher resolution.
- *Compass direction*: accuracy of ±2° and resolution of 0.5°.
- *Allowable tilt*: cosine tilt response to up to ±25°.

Most electromagnetic current meters allow for measurement of temperature, conductivity and pressure. Data can be averaged over regular intervals of a few seconds to tens of minutes, or set to burst sampling with a specified number of samples per burst at a given sampling interval. For example, one can set the number

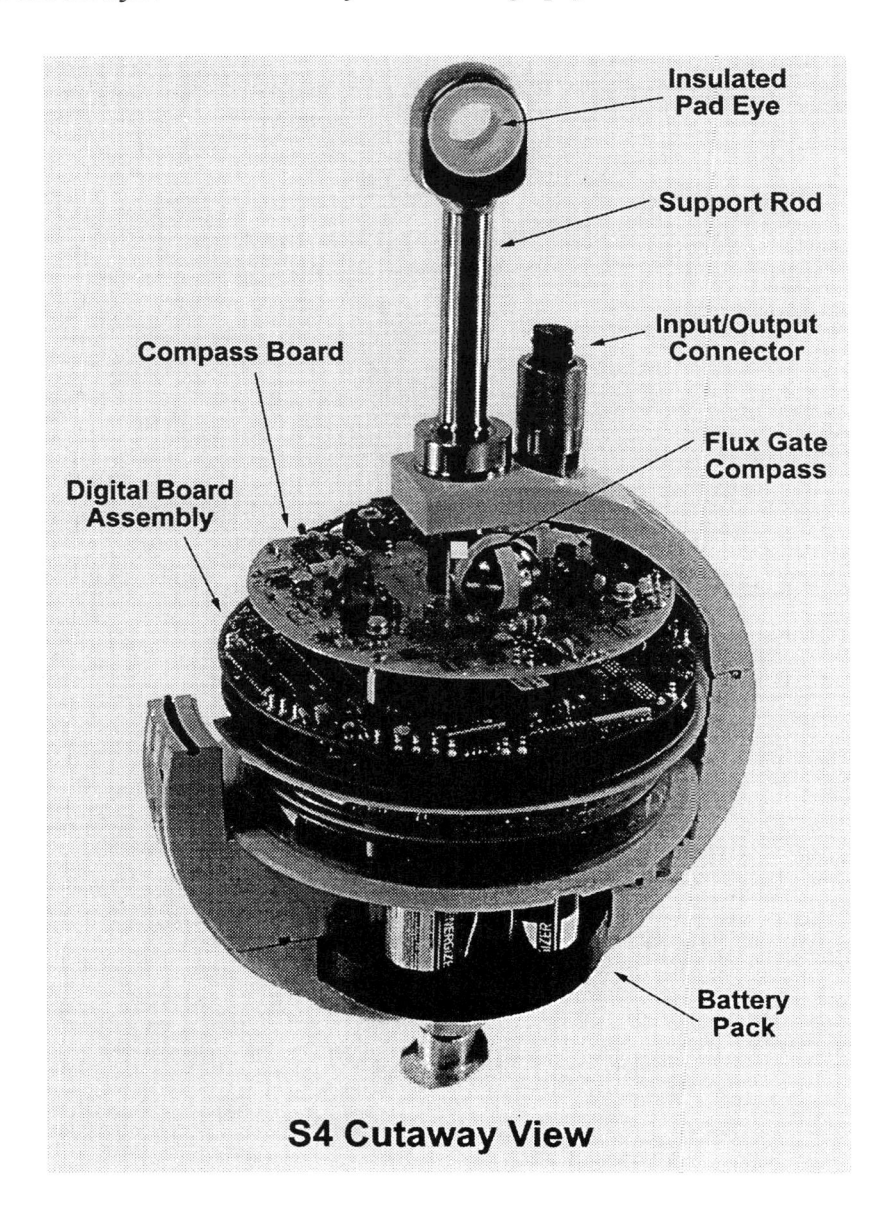

Figure 1.7.5. Inter-Ocean S4 electromagnetic current meter. (b) Cut-away view of the electronics. The spherical hull has a diameter of 25 cm and the instrument weighs 1.5 kg in water. (Courtesy, Mark Geneau, Inter-Ocean.)

of samples per burst (say continuous sampling for 2 min every hour) and set the number of times velocity is sampled compared with conductivity and temperature. The limitations are the storage capacity of the instrument (thousands of kilobytes) and the amount of power consumption. In the case of the S4, the surface of the housing is grooved to maintain a turbulent boundary layer and prevent flow separation at higher speeds.

1.7.4 Profiling acoustic Doppler current meters (ADCM)

Acoustic Doppler current meters (ADCMs) measure current speed and direction by transmitting high frequency sound waves and then determining the Doppler frequency shift of the return signal scattered from assemblages of "drifters" in the water column. In a sense, the instrument "whistles" at a known frequency and listens for changes in the frequency of the echo. The technique relies on the fact that: (1) sound is reflected and/or scattered when it encounters marked changes in density; and (2) the frequency of the reflected sound is increased (decreased) in direct proportion to the rate at which the reflectors are approaching (or receding from) the instrument. Principle (2) is used by astronomers to measure the rate at which stars and galaxies are moving relative to the earth. The commonly observed "red shift" of starlight suggests that most distant objects in the universe are receding from the earth. Reflectors ensonified by ADCMs include "clouds" of planktonic organisms such as euphausiids, copepods and gellies, fish (with and without swim bladders), suspended particles, and discontinuities in water density. Buoyant wastewater plumes from coastal sewage outfalls and hydrothermal plumes from seafloor spreading regions are two common examples of density discontinuities that can be detected acoustically.

Unlike the current meters discussed in the previous sections, which measure current time-series at a fixed depth, ADCMs provide time-series profiles of the flow averaged over a suite of depth bins. The ADCM is like having a stack of current meters. Commercial acoustic Doppler current meters are built by Amatak-Straza, Aanderaa Instruments and RD Instruments. Of the instruments available, the recently-commercialized Aanderaa Doppler current meter (DCM 12) and the better-known RD Instruments acoustic Doppler current profiler (ADCP) are specifically designed for oceanographic research. The ADCP, in particular, has been the focus of numerous comparisons and analyses (e.g. Pettigrew and Irish, 1983; Pettigrew et al., 1986; Flagg and Smith, 1989; Schott and Leaman, 1991). RD Instruments makes a self-contained (SC) internally-recording unit (Figure 1.7.6), a direct reading (DR) unit, and a vessel-mounted (VM) unit. The standard instruments are available at frequencies of 75, 150, 300, 600, and 1200 kHz; the more newly developed Broadband^{TM} ADCP also includes a 2400 kHz unit. The choice of frequency is dependent on the particular application. Because the ADCP is geared to oceanographic applications, we will consider this instrument in some detail. The standard ADCP measures current by first estimating the relative frequency change, Δf, of back-scattered echos from a single transmit pulse (Gordon, 1996). The newer Broadband ADCP measures the current by determining the phase shifts ("time dilation") $\Delta \phi$ of backscattered echos from a series of multiple transmitted pulses. The Aanderaa DCM, which was in development at the time this book was being written, operates at 607 kHz, has fewer vertical bins than the ADCP and considerably greater speed uncertainty (>3 cm s^{-1}). A report on an intercomparison between a 614 kHz Broadband ADCP and two 607 kHz DCMs moored in 11.5 m of water in Øresund, Denmark has been prepared by the Danish Hydraulic Institute (Rørbaek, 1994).

Aside from some custom-built units, the standard narrow-band ADCPs employ four separate transducers oriented in a Janus configuration with beams pointing at an angle of 30° to the plane of the transducers (Janus was the Roman god who looked both forward and backward at the same time). In the newer broadband unit, the angle has been reduced to 20°. We assume that the small drifters reflecting the transmitted sound pulse are being carried passively by the current and that their drift velocity has

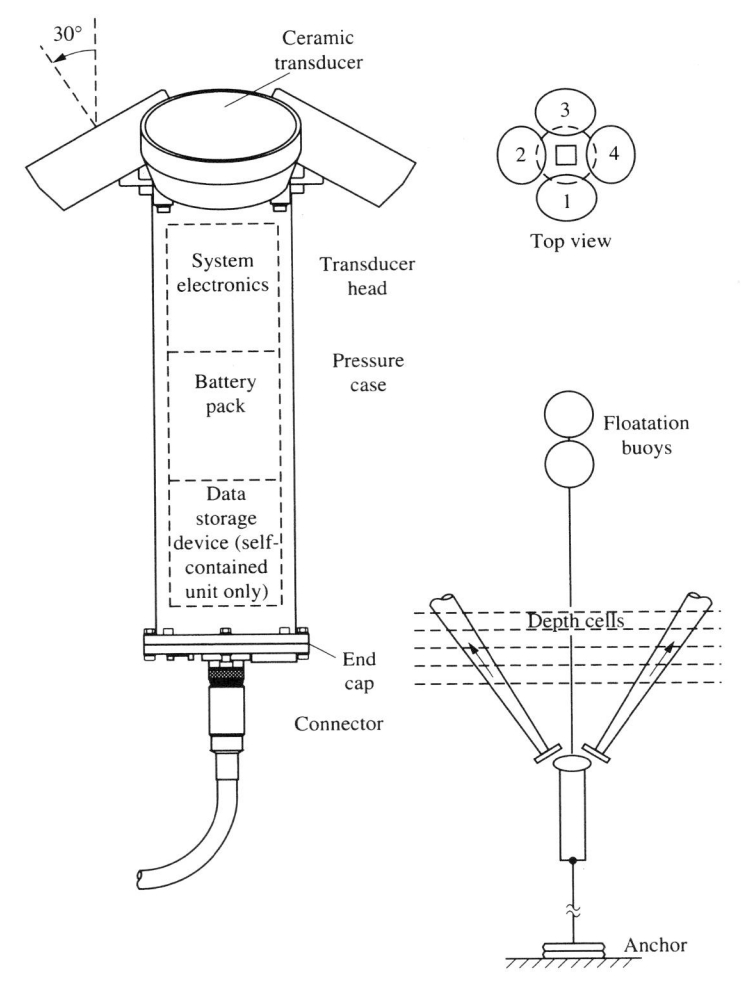

84 *Data Analysis Methods in Physical Oceanography*

Figure 1.7.6. A direct reading 150 kHz acoustic Doppler current meter with external RS-232 link manufactured by RD Instruments. Side view shows three of the four ceramic transducers. Each transducer is oriented at 30° to the axis of the instrument. The pressure case holds the system electronics and echo sounder power boards.

a near-uniform distribution over the horizontal area ensonified by the ADCP. For a narrow-band ADCP with a transmit pulse having a fixed length of a few milliseconds, the frequency shift, Δf, of the backscattered signal is proportional to the component of relative velocity, $v \cos \theta$, along the axis of the acoustic beam between the backscatterers and the transducer head (Figure 1.7.7a). For a given source frequency, f, and bin k (depth range $= D_k$) we find

$$v_k = \frac{\frac{1}{2}(\Delta f_k/f)c}{\cos \theta_k} \quad (1.7.3)$$

where v_k is the relative current velocity for *bin k* at depth D_k, θ_k is the angle between the relative velocity vector and the line between the scatters and the ADCP beam, c is

(a)

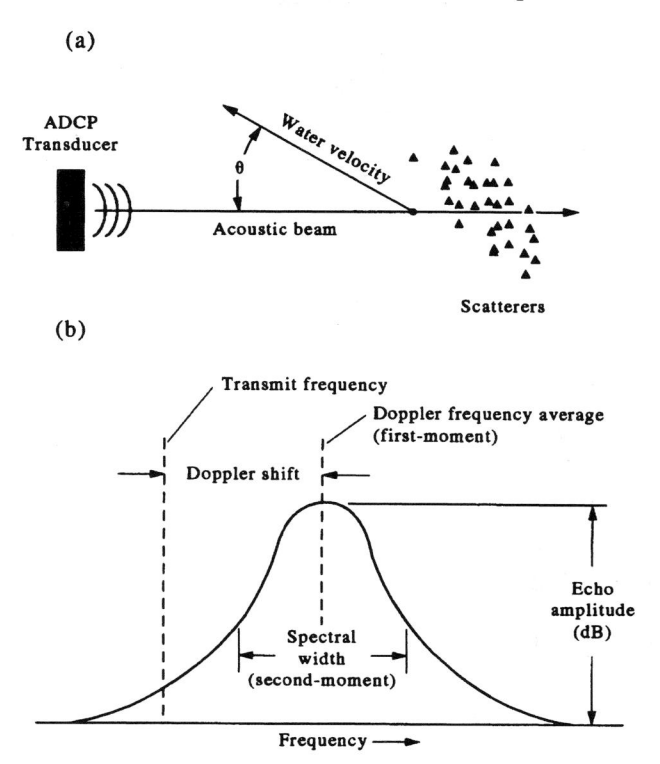

[Diagram: ADCP Transducer with acoustic beam at angle θ to water velocity, showing scatterers]

(b)

[Spectral diagram showing transmit frequency, Doppler frequency average (first-moment), Doppler shift, Spectral width (second-moment), and Echo amplitude (dB) vs Frequency]

Figure 1.7.7. Principles of ADCP measurement. (a) Relative velocity, $v \cdot \cos\theta$, along the axis of the acoustic beam between the backscatterers and the transducer head; (b) auto-spectrum of returned acoustic signal showing the Doppler frequency shift for a given bin. (RD Instruments, 1989.)

the speed of sound at the transducer and Δf is the frequency shift measured by the instrument. The ADCP first determines current velocity relative to the instrument by combining the observed values of frequency change along the axes of each of the acoustic beams (the instrument can only "see" along the axis of a given transducer, not across it). Absolute velocity components in east-west and north–south coordinates, called "earth" coordinates, are obtained using measurements from an internal magnetic compass.

The relative frequency shift, $\Delta f_k/f$ for bin D_k is derived from the observed frequency of the returning echo (Figure 1.7.7b). To calculate the Doppler frequency shift, the ADCP first estimates the autocovariance function, $C(\tau)$, of the echo using an internal hardware processing module. The slope of $C(\tau)$ as a function of time lag, τ, is then related to the frequency change due to the movement of the scatterer during the time that it was ensonified by the transmit pulse. Because of inherent noise in the instrumentation and the environment, as well as distortion of the backscattered signal due to differences in acoustic responses of the possible targets, the returned signal will have a finite spectral shape centered about the mean Doppler shifted frequency (Figure 1.7.7b). The spectral width SW of this signal has the form $SW = 500/D$, where D is the bin thickness in meters, and is a direct measure of the uncertainty of the velocity estimate due to the finite pulse length, turbulence and nonuniformity in scattering velocity. In the case of the standard RD Instruments ADCP, depth cell lengths, D, can range from 1 to 32 m but are usually set at 4–8 m. For the Broadband ADCP, depth cell size ranges from 0.12 m for the higher frequencies to 32 m for the lower frequencies. Each acoustic beam of the ADCP has a width of 2–4° (at the −3 dB or half-power point of the transducer beam pattern) so that the "footprint" over which the acoustic averaging is performed is fairly small. At a distance of 300 m, the footprint has a radius of 5–10 m. However, the horizontal separation between beams is roughly equal to the distance to the depth-cell so that the assumption of horizontal uniformity of the current velocity is not always valid, especially for those cells furthest from the transducers.

Sidelobes of the transducer acoustic pattern can limit the reliability of the data. For the standard 30° ADCP, measurements taken over the last 15% [$\approx (1 - \cos 30°)$] of the full-scale depth range are not valid if the ocean surface (or seafloor) are within the range of an upward (or downward) looking instrument. In general, the range R_{max} of acceptable data for a vertically-oriented ADCP within proximity to a "hard" reflecting surface such as the sea surface or sea floor is given by $R_{max} \approx H\cos\varphi$, where H is the distance from the ADCP to the reflecting surface and φ is the angle the transducers make with the instrument axis (for a 20° instrument, only 6% of the range is lost near the sea surface or seafloor). For vessel-mounted systems working in areas of rough or rapidly sloping bottom topography, a more practical estimate is $R_{max} \approx H(\cos\varphi - \alpha)$, where $\alpha \approx 0.05$ is a correction factor that accounts for differences in water depth during short (<10 min) ensemble averaging periods.

The higher the frequency, the shorter the distance an acoustic sounder can penetrate the water, but the greater the instruments ability to resolve velocity structure (Table 1.7.1). The 75 and 150 kHz units are mainly used for surveys over depth ranges of 0–500 m while higher frequencies such as the 600 and 1200 kHz units are favored for examining flow velocity in shallow water of 25–50 m depth. As noted above, ADCPs employ four separate transducers each pointing at an angle of 20° or 30° to the plane of the transducers. Since only the current speeds along each of the beam axes can be estimated, trigonometric functions must be applied to the velocities to transform them into horizontal and vertical velocity components. The instrument provides one estimate of the horizontal velocity and two independent estimates of the

Table 1.7.1. RD Instruments acoustic wavelengths (λ) and depth ranges (m) for different transducer frequencies for the low power and high power settings (low power is for self-contained units while high power is for either self-contained or externally powered units). Standard deviation for velocity of given frequency are for ensemble averages of N pings per ensemble, a depth cell size (bin length and length of transmit pulse) of 8 m and 30° beam angle orientation. For 20° angle multiply values by 1.5; for other depth cell sizes D (m) multiply values by 8/D. The values e_1 and e_2 are different published estimates of the absorption of sound at 4°C, 35psu and atmospheric pressure. At high frequencies, the range of transducers is limited by nonlinear dynamics (cavitation) and heat dissipation so that the ranges at high and low power output are the same. (From RD Instruments, 1989)

Freq. (kHz)	λ (mm)	Depth range (m) Low	High	Standard deviation (cm/s) $N=15$	$N=30$	$N=60$	e_1 (dB/m)	e_2 (dB/m)
76.8	20	400	700	6.72	4.75	3.36	0.025	0.0221
153.6	10	240	400	3.36	2.38	1.68	0.039	0.0395
307.2	5	120	240	1.68	1.19	0.84	0.062	0.0726
614.4	2.5	60	60	0.84	0.59	0.42	0.139	0.1884
1228.8	1.25	25	25	0.42	0.30	0.21	0.440	0.6466

vertical velocity. The ADCP senses the Doppler frequency shift in each 1-s acoustic "ping" by looking at the time-delayed gated signal returning from distinct "bins" (depth or distance ranges) from the transducer along each of the four-beam axes. The resultant speed estimates are then converted within the instrument to common bin positions centered at $p2^N$ meters ($N = 1, 2, ..., 8$ to a maximum of 128 bins) along the central axes normal to the plane of the transducers. Since the different time delays t_k of each pulse correspond to different distances D_k from the transducers, the instrument provides estimates of the horizontal (u, v) and vertical (w) components of velocity averaged over adjoining depth ranges (or depth bins). As illustrated by Figure 1.7.8, the averaging consists of a linear weighting over twice the bin length, $D = z_{k+1} - z_k = c(t_{k+1} - t_k)$, where c is the sound speed. For the 4-m bin length selected in Figure 1.7.8, the triangular weighted average is over 8 m. The depth range of a particular bin covers the distance:

from: blank depth + (bin number) × (bin length) − (bin length)/2

to: blank depth + (bin number) × (bin length) + (bin length)/2

A 4-m blanking is applied to the beginning of the beam to eliminate nonlinear effects near the transducer. The minimum length of the blank is frequency dependent but a larger value can be selected by the user. For the particular setup shown, there are 15 1-s pings for each 20-s ensemble; bottom-tracking is turned on every four pings. This option, together with machine processing "overhead" and time for transmission up the tow cable, uses up a segment of the total time available for each ensemble averaging period.

The maximum range of the standard (single transmit pulse) ADCP depends on the depth at which the strength of the return signal drops to the noise level. Depending on the rate of energy loss and heat dissipation, the instrument is generally capable of measuring current velocity to a range $R(m) = 250(300/f)$, where f is the frequency in kilohertz. The velocities (and backscatter intensity which we discuss later in the section) from a series of pings are averaged to form an "ensemble" record. This saves on storage space in memory, reduces the amount of processing and improves the error estimate for the velocity record. Each acoustic ping lasts about 1–10 ms and 10 or more separate pings, together with an equal number of compass readings, are typically used to calculate an ensemble-averaged velocity estimate for each recorded increment of time in the time-series. The random error of the horizontal velocity for each ensemble is given as

$$\sigma(\text{m/s}) = (1.6 \times 10^2)/(fDN^{1/2})$$

where N is the number of individual 1-s pings per ensemble and D is the bin length ($= 2^m$, $m = 1, 2, ...$) in meters. For example, a 30-s ensemble averaging period chosen during the instrument set-up procedure, generally allows for about 20 pings plus 10 s of processing time. This overhead time is inherent to the system and must be taken into account when determining the error estimates. As indicated in Table 1.7.2, the standard deviation of the vertical and horizontal velocity estimates for this case is about 3 cm/s for $D = 8$ m and a 150 kHz transducer. The greater the number of pings used in a given ensemble, the greater the accuracy of the velocity estimate, with $\sigma \approx N^{-1/2}$. Tilt sensors are used to calculate changes in the orientation of the transducer axis and to ensure that data are binned into correct depth ranges. These sensors

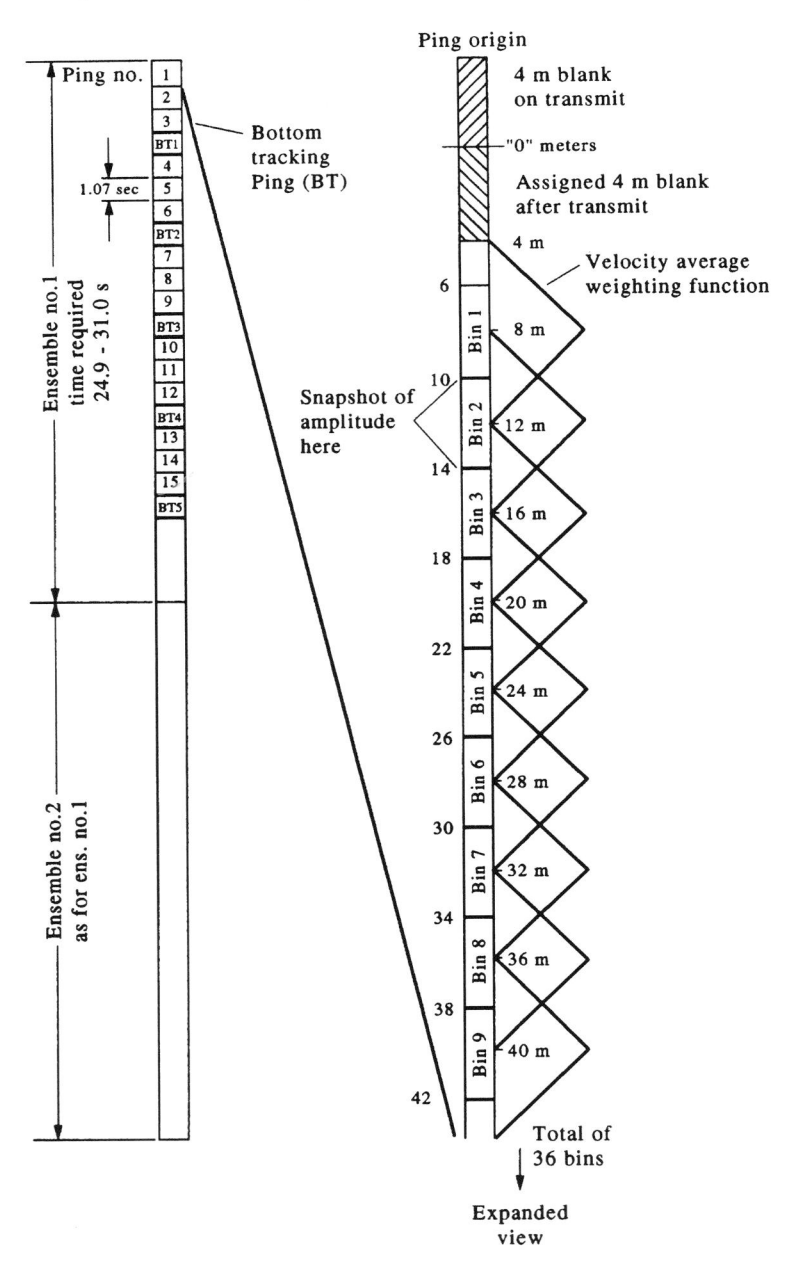

88 *Data Analysis Methods in Physical Oceanography*

Figure 1.7.8. Allocation of depth bins and machine overhead for a narrow band (standard) 150 kH. ADCP having a bin length of 4 m, a blanking range of 4 m and a depth range of 36 bins. The instrumen obtained 15 1-s pings for each 20-s ensemble and used the remaining time for internal processing and dat transmission up an electrical cable. The information on the right is an expansion of the bin allocation fo the first ping. A triangular weighting is used to determine the velocity for each bin. Similar results appl to the remaining pings for each of ensemble. A 4-m blanking is applied to the beginning of the beam t eliminate nonlinear effects near the transducer. (Courtesy, George Chase.)

Table 1.7.2. *Comparison of hourly time-series of longshore currents over 90 day period from 308 kHz ADCP and conventional current meters off northern California (adapted from Pettigrew and Irish, 1986). Results are found using the two-beam solution for the ADCP*

Depth (m)	Moored current meter	Correlation coefficient, r	Speed difference (cm/s) Mean	RMS
10	VACM	0.94	−3.7	8.1
20	VMCM	0.97	0.8	4.6
35	VMCM	0.98	0.2	2.7
55	VMCM	0.98	0.0	2.4
70	VMCM	0.98	0.3	2.2
90	VMCM	0.98	1.0	2.2
110	VMCM	0.98	0.5	1.9
120	VACM	0.97	−0.1	2.0

are limited to ±20° so that for greater tilts, the velocity components can not be determined accurately. Only three of the beams are needed for each three-dimensional velocity calculation. The built-in redundancy provides for an "error velocity" estimate for each ensemble velocity which involves subtracting the two independent estimates of the vertical velocity component for each ping. When the two vertical velocity estimates agree closely, the horizontal velocity components are most likely correct. In addition to the reliability check, the fourth beam serves as a backup should one of the transducers fails. Another measure provided by the ADCP is the "percent good" which is the percentage of pings that exceed the signal-to-noise threshold. Normally, the percent good rapidly falls below 50% at some depth and stays below that level. In practical terms, there usually is little difference in the data for assigned values of 25, 50, or 75 percent-good.

Since Doppler current meters were originally designed for measuring currents from a moving platform, the ADCP records instrument heading, pitch, roll and yaw. These data are then used to correct the measured velocities. In order to determine the true current velocity in "earth coordinates" from a moving vessel, the ADCP is capable of measuring the velocity of the instrument over the seafloor, providing the bottom is within range of the transducer and the bottom reflection exceeds the background noise level. A separate bin is used for this bottom-tracking. The bottom tracking mode is usually turned on for a fraction of the total sampling time, uses a longer pulse length and provides a more accurate estimate of relative velocity than other bins. Unfortunately, modern shipboard GPS systems in the standard (nondifferential) mode are only accurate to about ±100 m. This means that estimates of the ship speed taken at time increments of seconds to minutes will have errors of the order of 10–100 cm s^{-1} which usually are comparable to the kinds of current speeds we are trying to measure. Differential GPS, which relies on error corrections transmitted from a land-based reference station for which satellite positioning and timing errors have been calculated, is accurate to better than 10 m. Shipboard systems working in this mode can be used to determine absolute currents by subtracting the accurately determined ship's velocity over-the-ground from relative currents measured by the ADCP (see note at the end of this section).

There are several factors that limit the accuracy of the ADCP: (1) The accuracy of the frequency shift measurement used to obtain the relative velocity. This estimate is conducted by software within the instrument and strongly depends on the signal/noise ratio and the velocity distribution among the scatters; (2) the size of the foot-print and the homogeneity of the flow field. At a distance of 300 m from the transducer, the spatial separation between sampling volumes for opposite beams is 300 m so that they are seeing different parts of the water column, which may have different velocities; (3) The actual passiveness of the drifters (i.e. how representative are they of the *in situ* current?). In the shipboard system, the ADCP can track the bottom and obtain absolute velocity, provided the acoustic beam ranges to the bottom. Once out of range of the bottom, only the velocity relative to the ship or some level-of-no-motion can be measured. As noted above, standard GPS positioning without the highly accurate (±10 m) differential mode cannot be used to obtain ship velocity since the accuracy of the standard mode (±100 m) yields ship speed accuracies that are, at best, comparable to the current we are trying to measure. Erroneous velocity and backscatter data are commonly obtained from shipboard ADCP measurements due to vessel motions in moderate to heavy seas. In addition to exposure of the transducer head, the acoustic signal is strongly attenuated by air bubbles under the ship's hull or through the upper portion of the water column. Much better data are collected from a ship "running" with the seas than one lying in the trough or hove-to in heavy seas. Our experience is that data collected in moderate to heavy seas is often unreliable and needs to be carefully scrutinized. In deep water, zooplankton aggregations can lead to the formation of "false bottoms" in which the instrument mistakes the high reflectivity from the scattering layer as the seafloor.

The only way to improve velocity measurement accuracy with the standard single-pulse narrow-band ADCP is to lengthen the transmit pulse. A longer transmit pulse extends the length of the autocorrelation function and increases the number of lag values that can be used in the calculation of velocity. Since bin length is proportional to pulse length, this results in improved uncertainty in the velocity estimates. The tradeoff is reduced depth resolution. By transmitting a series of short pulses, the newer Broadband^{TM} ADCP circumvents these problems. Because of the multiple transmit pulses, the Broadband ADCP is capable of much better velocity resolution and higher vertical resolution. The time between pulses sets the correlation lags available for velocity computation while pulse length governs the size of the depth cells, as in the standard unit. Moreover, velocity is determined from differences in the arrival times of successive pulses. By increasing the effective bandwidth of the received signal by two orders of magnitude, the Broadband ADCP can reduce the variance of the velocity measurement by as much as two orders of magnitude. Figure 1.7.9 is an example of alongshore currents obtained using a shipboard Broadband ADCP during a cross-shelf transect of the inner continental shelf east of Sydney, Australia. The new ADCP system offers "real-time" computer screen display for at-sea operations. The standard narrow-band ADCP uses a data acquisition system, that is no longer supported by RD Instruments, to display output of velocity components, beam-averaged backscatter intensity, percent good, and other ship related parameters such as heading and pitch and roll.

A further note on GPS measurements. There are currently a variety of chart datums that are used in the setup menu of a GPS and one must be sure to select that datum which matches the chart being used for navigation. The general default datum is WGS-84 (World Geodetic Survey 1984) which applies to any region of the world. A commonly used datum in the eastern North Pacific and western North Atlantic is NAD-27 (North American Datum 1927) which has recently been replaced by NAD-83

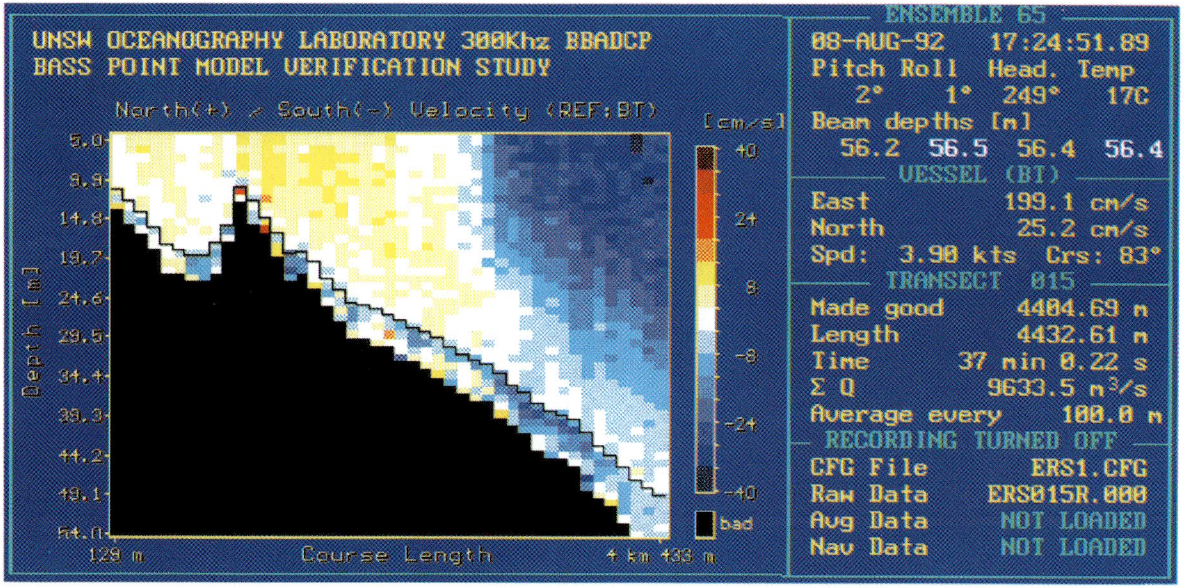

Figure 1.7.9. Alongshore currents measured by a Broadband RD Instruments Vessel-Mounted 300 kHz system along an eastward transect to the south of the Bass Point headland near Sydney, Australia (see Middleton et al., 1993). (a) Cross-section of the flow for all depth bins, with red corresponding to northward flow and blue to southward flow. (Courtesy, Jason Middleton and Greg Nippard.)

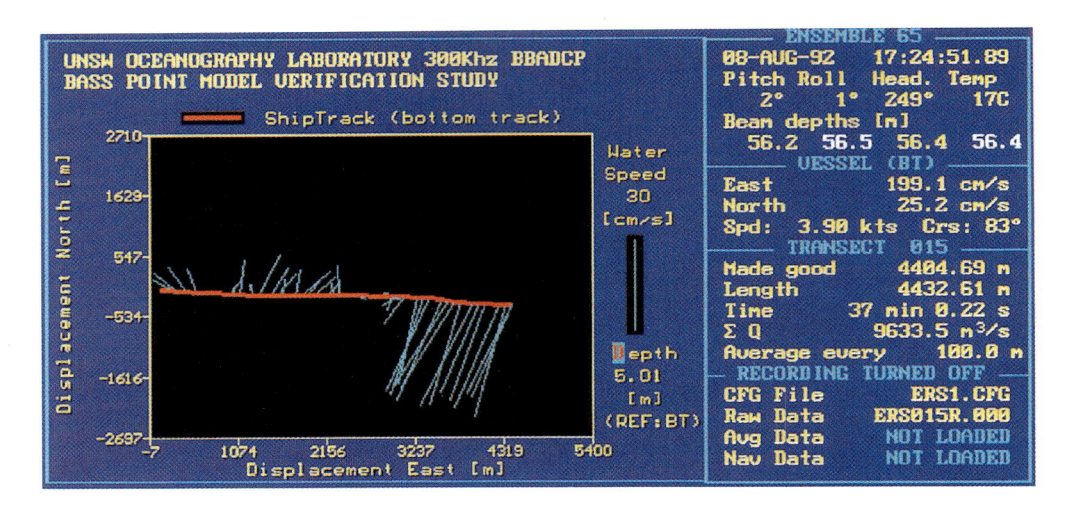

Figure 1.7.9. Alongshore currents measured by a Broadband RD Instruments Vessel-Mounted 300 kHz system along an eastward transect to the south of the Bass Point headland near Sydney, Australia (see Middleton et al., 1993). (b) Ensemble-average velocity at 5 m depth for the cross-section in (a). Flow was northward in the wake of the headland and southward seaward of the point. (Courtesy, Jason Middleton and Greg Nippard.)

(North American Datum 1983). Other datums are WGS-72, Australian, Tokyo, European and Alaska/Canada. *Selective Availability* is the name given by the United States Department of Defense for degradation of the GPS satellite constellation accuracy for civilian use. When disabled (as it was during the Gulf War with Iraq), GPS accuracy increases by about a factor of 10.

1.7.4.1 Acoustic backscatter

Although it was originally designed to measure currents, the ADCP has become a highly useful tool for investigating the distribution and abundance of zooplankton in the ocean. In particular, the intensity of backscattered sound waves for each depth bin—actually a "snapshot" of the intensity at a distance of two-thirds the way along the bin (Figure 1.7.8)—can be used to estimate the integrated mass of the backscatters over the "footprint" volume (width and thickness) of the original acoustic beams (Flagg and Smith, 1989). As with velocity, the instrument compensates for apparent changes in bin depth due to instrument tilt and roll. Calculation of the backscatter anomaly caused by plankton or other elements in the water column requires an understanding of the various factors causing dispersion and attenuation of the sound waves in water. Proper calibration of the acoustic signal as a function of acoustic range is essential for correct interpretation of the ADCP backscatter data. The measured backscatter intensity (also energy or amplitude squared) I_r is given by

$$I_r/I_a = b \exp(-2e_i z)/z^2 + A_n \tag{1.7.5}$$

where $b = I_o/I_a$ is the transducer gain, I_o is the intensity of the ADCP transducer output, I_a is a reference intensity, e_i is the absorption coefficient for water (cf. Table 1.7.1; $i = 1, 2$), $1/z^2$ is the effect of geometric beam spreading over the range z, and A_n is the relative noise level. The factor b arises because the ADCP does not record output intensity from the transducers, only relative intensity. The target strength TS of the ADCP is then given by the logarithm of (1.7.5) as

$$TS = 10 \log(I_r/I_o) - 10 \log(b) \tag{1.7.6}$$

where the first term is the absolute target strength of the ADCP and the second term is an unknown additive constant. Since the later term is unknown, a relative measure of the target strength TS_c to some standard calibration region can be determined as $TS' = TS - TS_c$ (Thomson et al., 1991, 1992; Burd and Thomson, 1994). Thomson et al. (1992) use a vertically towed vehicle and are therefore able to calibrate their data relative to the near-uniform backscatter layer at intermediate depths (1000–1500 m) in the northeast Pacific.

The ADCP does not measure directly the input or output of the acoustic backscatter intensity but rather the voltage from the so-called Automatic Gain Control (AGC) which is an internal adjustment, positive feedback circuit in the output device which attempts to keep the transducer output power constant. The average compensation voltage in the AGC is recorded and can be used to estimate the relative backscatter intensity. By incorporating a user exit program, the ensemble average AGC for each of the four beams for each bin can also be recorded. As we will discuss later, this is proportional to the biomass (density × cross-section) of the scatterers. The instrument also measures temperature—which it needs to calculate response correctly—and percent-good, which is a measure of the number of reasonable pings per ensemble.

The speed of sound in water varies with temperature, salinity and depth but is generally around 1500 m/s. Therefore, sound oscillations of 153 kHz (a common frequency used on shipboard systems and moored systems) have a wavelength of about 1 cm. Using the standard rule of thumb that the acoustic wave detects objects of about one-quarter wavelength, objects greater than 2.5 mm will reflect sound while objects less than this scatter the sound. The proportion of the sound beam transmitted, reflected or scattered by the object is influenced by small contrasts in compressibility and density between the water and the features of the object. Organisms with a bony skeleton, scaly integument and air bladder reflect/scatter more sound than an organism made up mostly of protoplasm such as salps and jellyfish (Flagg and Smith, 1989). Similarly, organisms which are aggregated into patches or layers return more scattered sound energy per unit volume (i.e. have a greater effective scattering volume) than uniform distributions of the same organisms.

A major problem with using the ADCP for plankton studies is common to all bioacoustical measurements; namely, determining the species composition and size distribution of the animals contributing to the acoustic backscatter. Invariably, *in situ* sampling using net tows is needed to calibrate the acoustic signal. If the ADCP is incorporated in the net system, the package has the advantage that the volume flow through each net can be determined accurately using the ADCP-measured velocity

(Burd and Thomson, 1993). An attempt to calibrate the ADCP against net samples was conducted by Flagg and Smith (1989) who also pointed out problems with the response of the shipboard system to temperature fluctuations in the ADCP electronics.

1.7.5 Comparisons of current meters

As noted earlier, a major problem with the Savonius rotor is contamination of speed measurements by mooring motions (Gould and Sambuco, 1975). The contamination of the rotor speed is caused primarily by vertical motion or "rotor pumping" as the mooring moves up and down under wave action. In effect, the speed overestimates of the rotor result from its ability to accelerate about three times faster than it decelerates. Pettigrew *et al.* (1986) summarize studies on the ability of VMCMs and VACMs in laboratory tests to accurately measure horizontal flow in the presence of surface waves. For wave orbital velocities, W, of the same magnitude as the steady towing speed, U, of the current meter through the water (i.e. $W/U \approx 1$), the accuracy of the VACM depends on the ratio W/U. The percentage error increases as the ratio W/U increases and substantial over-estimation of the true speed occurs for $W/U > 0.5$. The results for the VMCM differ significantly from those of the VACM. In particular, the VMCM underestimates the true velocity by as much as 30% for $W/U \approx 1$, while for $W/U > 2$, speed errors do not appear to be strongly dependent on either W/U or on the relative orientation of the mean and wave current motions. For $W/U < 1/3$, the VMCM was within 2% of the actual speed. While vector averaging can reduce the effect of vertical motion on the recorded currents by smoothing out the short-term oscillatory flow, the basic sensor response is not well tuned to conditions in the wave zone or those for surface moorings. Intercomparisons of conventional current meters (Quadfasel and Schott, 1979; Halpern *et al.*, 1981; Beardsley *et al.*, 1981) have shown that VACM speeds are only slightly higher on surface moorings than on subsurface moorings and that contamination by mooring motion was only important for higher frequencies (>1 cph). At frequencies above 3-4 cph ocean current spectra computed from VACM current meters did not flatten (i.e. not decrease with frequency) as much as spectra from other rotor equipped current meters. Near the surface this is due to horizontal motion of the mooring (Zenk *et al.*, 1980) which is rectified by the Savonius rotor while at greater depths the surface float motion translates into vertical motion which aliases the rotor speed due to rotor pumping. Further details can be found in Weller and Davis (1980), Mero *et al.* (1983), and Beardsely (1987).

Another problem with the Savonius rotor is that it does not have a cosine response to variations in the angle of attack of the flow due to interference of the support posts. In a study of rotor contamination, Pearson *et al.* (1981) conclude that Savonius rotor measurements, made from a mooring with a float 18 m below the sea surface, were not seriously contaminated by surface wave-induced mooring motion. In sharp contrast, Woodward *et al.* (1984) compared a standard Savonius rotor with a paddle-wheel (PW) rotor designed for wave-field applications, and an electromagnetic (EM) current meter. The EM speed sensors appeared to perform well in the near-surface wave field while the standard Savonius rotor was severely contaminated by wave induced currents (Figure 1.7.2).

Field comparisons (Halpern *et al.*, 1981) demonstrated that above the thermocline (5–27 m depth) the VMCM, the VACM and acoustic current meters (ACMs) all produced similar results for frequencies below 0.3 cph, regardless of mooring type. Above 4 cph, it was recommended that the VACM be used with a spar buoy surface float while both the VMCM and the ACM could be used with surface following floats such as a donut buoy. In general, better quality measurements were made at depths from subsurface moorings than from surface moorings, indicating that even the VMCM data were contaminated somewhat by mooring motion.

The processing of current meter data is specific to the type of meter being used. It is interesting to read in current meter comparisons such as Beardsley *et al.* (1981) or Kuhn *et al.* (1980) the variety of processing procedures required to produce compatible data for the intercomparison of observations from different current meters. An important part of the data processing is the application of the instrument specific calibration values to render measurements in terms of engineering units. In this regard, it is also important to have both a pre- and post-experiment calibration of the instrument to detect any serious changes in the equipment that might have occurred during the measurement period.

One of the earliest comparisons between a bottom-mounted ADCP and conventional mechanical current meters was conducted in 133 m of water near the shelf-break off northern California in 1982 (Pettigrew and Irish, 1983; Pettigrew *et al.*, 1986). The 90-day time-series of horizontal currents from a prototype upward-looking 308 kHz ADCP with 4 m bin length was compared with currents from a nearby ($\approx$300 m) string of VACMs and VMCMs. Despite the fact that only two of the beams could be used and the instrument had a 10° list, results show striking agreement between the two sets of data (Table 1.7.2). Mean differences between corresponding acoustic and mechanical current meters were typically less than 0.5 cm/s while RMS differences were about 2 cm/s. Since acoustic currents were based on two beams tilted at 20° to the vertical, the relatively poor correlation at 10 m depth probably resulted from rotor-pumping and over-speeding of the VACM rather than side-lobe contamination of the ADCP which would occur in the upper 6% of the depth range. Similar results were obtained by Schott (1986).

1.7.6 Electromagnetic methods

The dynamo interaction of moving, conducting seawater with the earth's stationary magnetic field induces electric currents in the ocean. These "motional" electric fields, whose existence in the ocean was first postulated by Faraday in 1832, produce a spatially-smoothed measure of the water velocity at subinertial periods [periods longer than $1/f = 11.964$ h/sin(latitude)]. For a given point on the seafloor, the electric fields are proportional to the vertically-averaged, seawater-conductivity weighted water velocity averaged over a horizontal radius of a few water depths (Chave and Luther, 1990). Technologies that measure the horizontal electric field (HEF) yield direct observations of the barotropic transport in the overlying water column. Electric field measurements of transport are obtained from abandoned submarine communication cables or from self-contained bottom recorders. For a submarine cable, the motional HEF is integrated along the cable length.

According to theory (Sanford, 1971; Chave and Luther, 1990; Chave *et al.*, 1992), the horizontal velocity vector field $\mathbf{v}^*$ is related to the horizontal electric field $\mathbf{E}_h$, by

$$\mathbf{E}_h = F_z \mathbf{k} \times \mathbf{v}^* \qquad (1.7.7a)$$

(sensor in a reference frame fixed to the seafloor)

96 Data Analysis Methods in Physical Oceanography

$$= -F_z \mathbf{k} \times (\mathbf{V} - \mathbf{v}^*) \qquad (1.7.7b)$$

(sensor moving relative to seafloor), where F_z is the local vertical component of the geomagnetic field, $\mathbf{k}$ is a unit vector in the upward vertical direction, $\mathbf{V}$ is the vector sum of the horizontal velocities of the ocean relative to the earth and the sensor relative to the ocean, and

$$\mathbf{v}^* = C \int_{-H}^{0} \sigma(z') \mathbf{v}_h(z') \, dz' \Big/ \int_{-H}^{0} \sigma(z') \, dz' \qquad (1.7.8)$$

is the scaled (by the constant C), horizontal water velocity. The water velocity is averaged vertically over the water column of thickness H and weighted by the seawater conductivity, $\sigma(z)$. Equation (1.7.8) reduces to the scaled barotropic velocity $C\mathbf{v}$ when either the conductivity profile or the horizontal velocity is depth-independent. In the northern hemisphere, where $\mathbf{F}$ points into the earth, the north electric field is proportional to the west component of velocity while the east electric field is proportional to its north component. Neglecting the noise, we can solve (1.7.7a) to obtain

$$\mathbf{v}^* = -\mathbf{k} \times \mathbf{E}_h / F_z \qquad (1.7.9)$$

Since $\mathbf{F}$ is known to one part in 10^4 for the entire globe, measurement of $\mathbf{E}_h$ yields the horizontal flow field.

Measurement of the HEF is entirely passive, being based on naturally occurring fields, and hence has low power requirements and is nonintrusive. Motional EM may be used in an Eulerian configuration (bottom recorders or submarine cables) or a Lagrangian configuration (surface drifter, subsurface float, or towed fish). Equation (1.7.7b) shows that a relative velocity estimate is possible by measuring the HEF from a moving platform. On many instances, lack of a specific knowledge of $\mathbf{v}^*$ is not a critical limitation since it is independent of depth by (1.7.8). The moving frame of reference equation (1.7.7b) is exploited by vertical profilers such as the electromagnetic velocity profiler (EMVP) and the expendable current profiler (XCP) produced by Sippican. Horizontal profiles of the HEF can be obtained from a towed instrument and used with precise navigation to yield estimates of $\mathbf{v}^*$ and the surface water velocity. The original form of such a towed instrument is the geomagnetic electrokinetograph (GEK) of von Arx (1950).

1.7.7 Other methods of current measurement

There are numerous other ways to measure currents though not all have been successfully commercialized. For example, prior to the ADCP, scientists in Japan used towed electrodes at the ocean surface (the GEK) to routinely monitor the currents off the east coast of Japan. Coastal Ocean Doppler Radar (CODAR) determines surface current velocity by using shore-based microwave radar with frequencies of around 12 MHz to sense the backscatter from wind-generated capillary waves. These waves ride on the ocean currents so that the Doppler shift of the radar signal can be used to estimate the current speed in the direction of the shore-based radar illumination. Two independent radar transmitters provide maps of the two-dimensional flow over the area covered by the radar signals. Goldstein *et al.* (1989) report on the use of synthetic aperture radar (SAR) to measure surface currents from the phase-delay maps of aircraft-borne radar. Other recent techniques, such as the correlation sonar and acoustic "scintillation" flow measurements use pattern recognition and cross-correlation methods, respectively, to determine the current over a volume of ensonified water (Farmer *et al.*, 1987; Lemon and Farmer, 1990). The acoustic scintillation method determines the flow in a turbulent medium by comparing the combined spatial and temporal variability of forward-scattered sound along two closely-spaced parallel acoustic paths separated by a distance, Δx. Assuming that the turbulent field does not change significantly during the time it takes the fluid to travel between the two paths, the pattern of amplitude and phase fluctuations at the downstream receiver will, for some time lag Δt, closely resemble that of the upstream receiver. Examination of the time delay in the peak of the covariance function for the two signals gives Δt which then determines the mean velocity $v = \Delta x/\Delta t$ normal to the two acoustic paths. The technique has been used successfully to measure the horizontal flow in tidal channels and rivers, as well as the vertical velocity of a buoyant-plume rising from a deep-sea hydrothermal vent in the northeast Pacific (Lemon *et al.*, 1996).

Numerous papers have discussed the computation of surface currents from the displacements of patterns of sea surface temperature (SST) in thermal AVHRR imagery. In the maximum cross-correlation (MCC) method, the cross-correlation between successive satellite images is used to map the displacements due to the advection of the SST pattern (Emery *et al.*, 1986). More recently, Wu (1991, 1993) has advanced a "relaxation labeling method" for computing sea surface velocity from sequential time-lapsed images. The method attempts to address two major deficiencies with the maximum cross-correlation method, namely: (1) the MCC approach is strictly statistical and does not exploit *a priori* knowledge of the physical problem; and (2) pattern deformation and rotation, as well as image noise, can introduce significant error into MCC vector estimates. The latter problem was addressed by Emery *et al.* (1992) who showed that rotation can be resolved using large search windows.

1.7.8 Mooring logistics

In terms of accuracy and reliability, current meter data from surface and subsurface moorings cannot be divorced from the mooring itself. While many common mooring procedures are available, there is no single accepted technique nor is there agreement on the subsequent behavior of the mooring while in the water. Surface moorings with their flotation on the wavy surface of the ocean will behave differently than subsurface moorings in which the buoyancy is distributed vertically along the mooring line. For the case of subsurface moorings, the addition of pressure sensors to most current meters has helped to characterize mooring motion and determine its effect on the measured currents. Variations in the depth of the sensor can be calculated from the pressure fluctuations and used to estimate the depth and position of the moored instruments as a function of time. Also, models of mooring behavior have been developed which enable the user to predetermine line tensions and mooring motions based on the cross-sectional areas of the mooring components and estimates of the horizontal current profile. For example, the program SSMOOR distributed by Cable Dynamics and Mooring Systems in Woods Hole (Berteaux, 1990), uses a finite element technique to integrate the differential equilibrium equations for cables subjected to steady state currents. Factors taken into consideration include: the mooring wire (or rope) diameter, weight in water, and modulus of elasticity; and the shapes, cross-sections, drag coefficients, weights, and centers of buoyancy of the recording instruments. Up to 10 current speeds can be specified for the current profile and as many as 20 instruments inserted in the anchoring line.

Mooring motions are largest when surface floats are used. For surface moorings in deep water, the length of the mooring line creates a relatively large "watch circle" that the surface float can occupy. This will add apparent horizontal motion to the attached current meters while, at depth, the surface wave and wind driven fluctuations translate into mainly vertical oscillations of the mooring elements. Some inter-comparison experiments have tried to use a variety of mooring types to test the effects of moorings alone. Zenk *et al.* (1980) compare VACM measurements from a taut-line surface mooring with a single line spar buoy float and a more rigid two-line, H-shaped mooring. As expected, the H-shaped mooring was more stable and the other two exhibited much stronger oscillations. The current meters on the rigid H-mooring registered the greater current oscillations since the meters on the other, less restricted, moorings moved with the flow rather than measuring it.

In their current meter comparison, Halpern *et al.* (1981) discuss four different types of mooring buoyancy; three surface and one subsurface. The surface floats were: a toroid, a spar-buoy and a torpedo-shaped float. They found that rotor-pumping was much greater under the toroid than under the spar buoy and that the effect of rotor-pumping on the resulting current spectra was significant at frequencies above 4 cph. While this was true for near-surface current meters, they also found that for deeper instruments the spar buoy float transmitted larger variations to the deeper meters making it a poor candidate for flotation in deep water current measurements. They found that both the VMCM and the ACM are less affected by the surface motions of a toroidal buoy. In a different comparison, Beardsley *et al.* (1981) tested an Aanderaa current meter suspended from a surface spar buoy, and found a significant reduction in the contamination of the measured signal by wave effects due to both currents and orbital motion with the spar buoy. Even with this flotation system, however, the Aanderaa continued to register high current speeds compared with other sensors.

In an overall review of the recent history of current meter measurement, Boicourt (1982) makes the interesting observation that "results from current measurement studies are independent of the quality of the data". In making this claim, he remarks that often the required results are only qualitative, placing less rigorous demands on the accuracy of the measurements. He also points out that present knowledge of the high-frequency performance of most flow sensors is inadequate to allow definitive analysis of the current measuring system. In this regard, he states that acoustic and electromagnetic current meters, with their fast velocity response sensors, hold great promise for overcoming the fundamental problems with mechanical current sensing systems. Finally, he calls for added research in defining the high-frequency behavior of common current meters.

Field work by the Bedford Institute of Oceanography on Georges Bank in the western Atlantic has revealed another unwelcome problem with moored rotor-type current meters. Comparisons between currents measured by a sub-surface array of Aanderaa current meters on the bank and a ship-board acoustic Doppler current profiler indicated that current speeds from the moored array were 20–30% lower than concurrent speeds from the profiler. To test the notion that the under-speeding was due to high-frequency mooring vibration caused by vortex shedding from the spherical floatation elements, an accelerometer was built into one of the sub-surface moorings. Accelerations measured by this device confirmed that the current meters were being subjected to high-frequency side-to-side motions. Under certain flow conditions, the amplitudes of the horizontal excursions were as large as 0.5 m at periods of 3 s. Tests confirmed that the spherical buoyancy packages were the source of the motions. By enclosing the spherically-shaped buoyancy elements in more streamlined torpedo-shaped packages, the mooring line displacements were reduced to about 10% of what they were for the original configuration. Excellent agreement was found between the current meter and vessel-mounted ADCP current records.

In certain areas of the world (e.g. Georges Bank), the survivability of a mooring can have more to do with fishing activity than to environmental conditions. Also, in the early days of deep-sea moorings, the Scripps Institution of Oceanography lost equipment on surface moorings to theft and vandalism. Preventing mooring and data loss in such regions can be difficult and expensive. For fishery oceanography studies the dilemma is that, to be of use, the measurements must be obtained in areas where they are most vulnerable to fish-net fouling and fish-line entanglement. Damage to nets equates to lost fishing time and damaged or lost instrumentation. Aside from providing detailed information on the mooring locations in printed material handed out to commercial fishermen, fish processing companies and coastguard, the scientist may need to resort to closely-spaced "guard buoys" in an attempt to keep fishermen and shipping traffic from subsurface moorings. Our experience is that a limited array of only three or so coastguard-approved buoys more than 0.5 km from the mooring is inadequate, and that certain operators will even use the buoys to guide their operations, thereby increasing the chance of damage.

1.7.9 Acoustic releases

An acoustic release is a remotely-controlled motorized linkage device that connects the expendable bottom anchor (often a set of train wheels) to the recoverable elements of a mooring (Figure 1.7.10). Modern acoustic releases are critical for free-fall deployment of moorings from ships and for reliable recovery of equipment on acoustic demand (Heinmiller, 1968). Operation of the release requires a deck unit specially built for the particular type of release and a transducer for acoustic interrogation of the release. A ship's sounder can be used in place of the hand-held transducer provided it is of compatible frequency and has a wide beam. This is useful since it allows the technician to talk to the release from the ship's laboratory rather than by lowering a transducer over the side of the ship. However, for "acoustically noisy" ships, lowering a transducer over the side is often the only way to talk to the release. More advanced releases enable the user to measure the *slant range* from the ship to the release based on the two-way time delay. By taking into account the slant of the acoustic path, the user can determine the coordinates of the release. Long-life (three-year) acoustic "pingers" built into the releases also are used to locate the depth and position of moorings using triangulation procedures. This is particulary useful for those moorings that fail to surface on command and must be dredged from the ship using a long line and hook. In some acoustic releases, a rough estimate of the orientation of the release can be obtained remotely through changes in ping-rate. For example, in the case of the Inter-Ocean release, a doubled ping-rate means that it is lying on its side rather being upright in the water column.

Figure 1.7.10. (a) Inter-Ocean acoustic release and attached anchor being lowered over the side of a ship. (Courtesy, Mark Geneau, Inter-Ocean.)

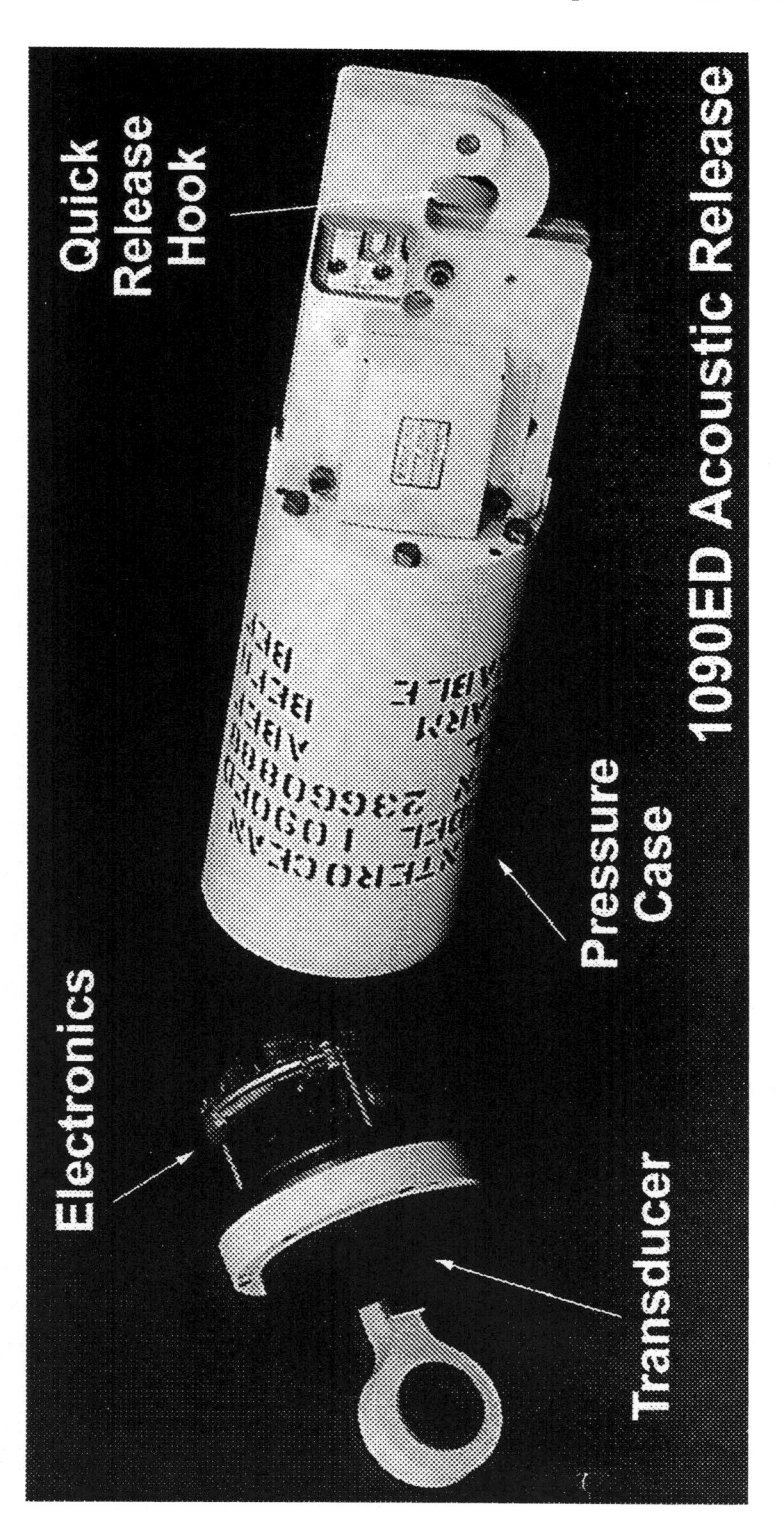

Figure 1.7.10. (b) Exploded view of the Inter-Ocean acoustic release. (Courtesy, Mark Geneau, Inter-Ocean.)

Most modern releases use separate "load" and "release" codes so that the release can be remotely opened and closed. Some releases also provide a release code that signals the oceanographer on the ship that the mooring has released and should be expected to surface in a time appropriate for the depth and net buoyancy of the mooring elements. This is always a tense time in mooring operations as it is sometimes difficult to predict precisely where the mooring will surface. Spotting the mooring from the ship can be a real challenge, especially in rough weather. Attachment of a pressure-rated radio beacon and flashing light to the top float of the mooring can aide considerably with the recovery operation. If the mooring fails to surface after an appropriate time, a search can initiated assuming that the mooring has surfaced and has not been spotted.

Past experience has demonstrated the wisdom of having a dual release system with two acoustic releases side-by-side in a parallel harness. A triangular bridle at the top connects the releases to a single point in the mooring line while a spreader bar connects the package to a single attachment point on the anchor chain. The extra cost can help avoid the need to dredge for the mooring if one of the releases should fail. Dredging is a last resort since in can be extremely harmful to the mooring hardware, leading to severe damage to the current meters and other instruments on the mooring line. In addition, there is a correct dredging procedure and oceanographers new to the field should talk to more experienced colleagues for guidance.

1.8 LAGRANGIAN CURRENT MEASUREMENTS

A fundamental goal of physical oceanography is to provide a first-order description of the global ocean circulation. The idea of following individual parcels of water (the Lagrangian perspective) is attractive since it permits investigation of a range of processes taking place within a tagged volume of water. Named after Joseph L. Lagrange (1736–1811), the French mathematician noted for his early work on fluid dynamics and tides, Lagrangian descriptions of flow can be used to investigate a range of processes from the dispersion of substances discharged into the ocean from a point source to the productivity of a semi-enclosed biological ecosystem as it drifts across the ocean. Early Lagrangian measurements consisted of tracking some form of tracer such as a surface float or dye patch. While giving vivid displays of water motions over short periods of time, these techniques demanded considerable onsite effort on the part of the investigator. Initial technical advances were made more rapidly in the development of moored current meters which yielded a strictly Eulerian picture of the current. However, improvements in tracking systems and buoy technology since the 1970s have made it possible to follow unattended surface and subsurface drifters for periods of many months to several years. Satellite-tracked surface buoys and acoustically-tracked, neutrally-buoyant SOFAR (SOund Fixing And Ranging or "Swallow") floats have been able to provide reliable, long-term, quasi-Lagrangian trajectories for many different parts of the world. (The trajectories are called quasi-Lagrangian since the drifters have a small "slip" of the order of 1–3 cm/s relative to the advective flow and because they do not move on true density surfaces. Surface drifters, for example, move on a two-dimensional plane rather than a three-dimensional density surface.)

Remotely-tracked drifters provide a convenient and relatively inexpensive tool for investigating ocean variability without continued direct involvement by the investigator. In the case of the satellite-tracked buoys, the scientist can now dial-up the position of drifters or collect data from ancillary sensors on the buoys. The number of possible satellite positional fixes varies with latitude (Table 1.8.1). Time delays between the time the data are collected by the spacecraft and the time they are available to the user is typically less than a few hours (Table 1.8.2). This feature makes the drifters useful for tracking floating objects or oil spills. Oceanic platforms and satellite data transmission systems have become so reliable that both moored and drifting platforms are now used for the collection of a variety of oceanic and meteorological data, including sea surface temperature, sea surface pressure, wind velocity and mixed layer temperature. A new era of oceanographic data collection is in progress with less direct dependence on ships and more emphasis on data collection from autonomous platforms.

Table 1.8.1. The mean number of satellite passes over a 24-h period for the Service Argos two-satellite system

Latitude (°)	Mean number of passes
0	7
15	8
30	9
45	11
55	16
65	22
75	28
90	28

Table 1.8.2. Data availability (global throughput times in hours) for June 1994. Percentage of the time that the delay between the satellite observation and the time that the data are processed and available from Service Argos. Service Argos Bulletin, July, 1994

Data availability	June (%)	December (%)
DA < 1 h	19.72	23.98
DA < 2 h	41.15	47.61
DA < 3 h	61.99	67.78
DA < 4 h	72.59	71.89
DA < 5 h	80.45	78.55
DA < 6 h	85.71	83.75
DA < 8 h	100.00	100.00

1.8.1 Drift cards and bottles

Until the advent of modern tracking techniques, estimates of Lagrangian currents were obtained by seeding the ocean surface with marked waterproof cards or sealed bottles and determining where these "drifters" came ashore. The card or bottle contained a note requesting that the finder notify the appropriate addressee of the time and location of recovery. To improve the chances of notification, a small token reward was usually offered (one Australian group gave out boomerangs). Although drift cards and bottles provide a relatively low-cost approach to Lagrangian measurements, they have major limitations. Because they float near the surface of the ocean, the movements of the cards and bottles are strongly affected by wind drag and wave-induced motions. In fact, much of what these type of drifters measure is wave-induced drift rather than underlying ocean currents. Moreover, even if the recovery rate was fairly high (1% is considered excellent for most drift card studies), the drifters provide, at best, an estimate of the lower bound of the mean current averaged over the time from deployment to recovery. Unless the card/bottle was recovered at sea, the scientist could never know if the drifter had recently washed ashore or had been laying on the beach for some time. In addition, the drifter provided no information on the current patterns between the deployment and recovery points.

1.8.2 Modern drifters

Quasi-Lagrangian drifters can be separated into two basic types: (1) Surface drifters having a surface buoy which is tethered to a subsurface drogue at some specified depth (typically less than 300 m); and (2) Subsurface, neutrally-buoyant floats which are designed to remain on fixed subsurface density surfaces. Modern surface drifters have a radio frequency transmitter (called a platform transmit terminal or PTT) for communication to a listening device while subsurface drifters may act either as a source or receiver of acoustic signals. Examples of the possible drogue configurations for modern satellite-tracked drifters are presented in Figure 1.8.1(a) along with the design for the standard holey sock WOCE/TOGA near-surface velocity drifter (Figure 1.8.1b). The purpose of the drogue is to reduce "slippage" between the drifter package

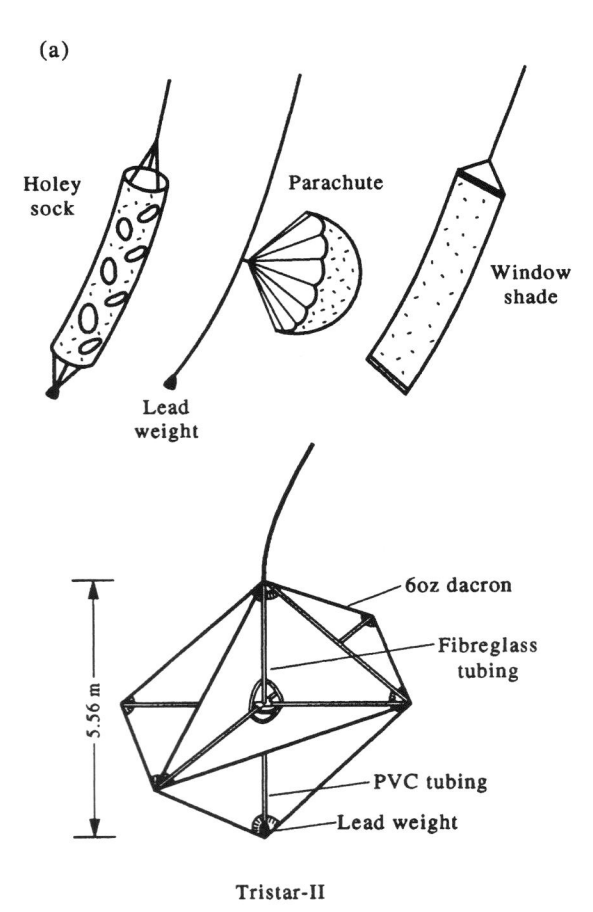

Figure 1.8.1. (a) Examples of the basic drogue designs for satellite-tracked drifters: holey-sock drogue; parachute drogue; window-shade drogue; and Tristar drogue (Niiler et al., 1987).

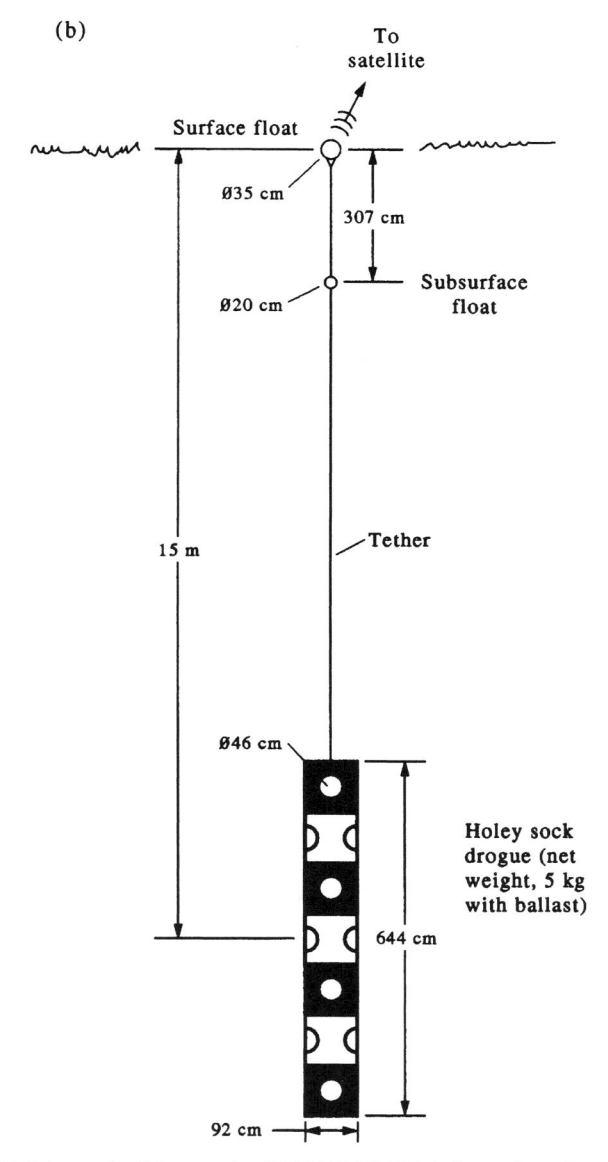

Figure 1.8.1. (b) Schematic of the standard WOCE/TOGA holey-sock surface-drifter showing pattern of holes in cloth panels. (From Sybrandy and Niiler, 1990.)

and the water. The surface float contains the PTT, temperature, and pressure sensors and other electronics (Figure 1.8.1c); the purpose of the subsurface buoy is reduce the "snap loading" on the drogue and cable by absorbing some of the shock from surface wave motion. In the case of the WOCE/TOGA holey sock drifter, the ratio of the drogue cross-sectional area to the cross-sectional area of the other drifter components (such as the wire tether and subsurface float) is about 45:1, a relatively high drag-area ratio for typical drogues.

Trajectories from an early type of drifter deployed in the NORPAX experiment are found in McNally et al. (1983). Similar tracks from more modern drifters are

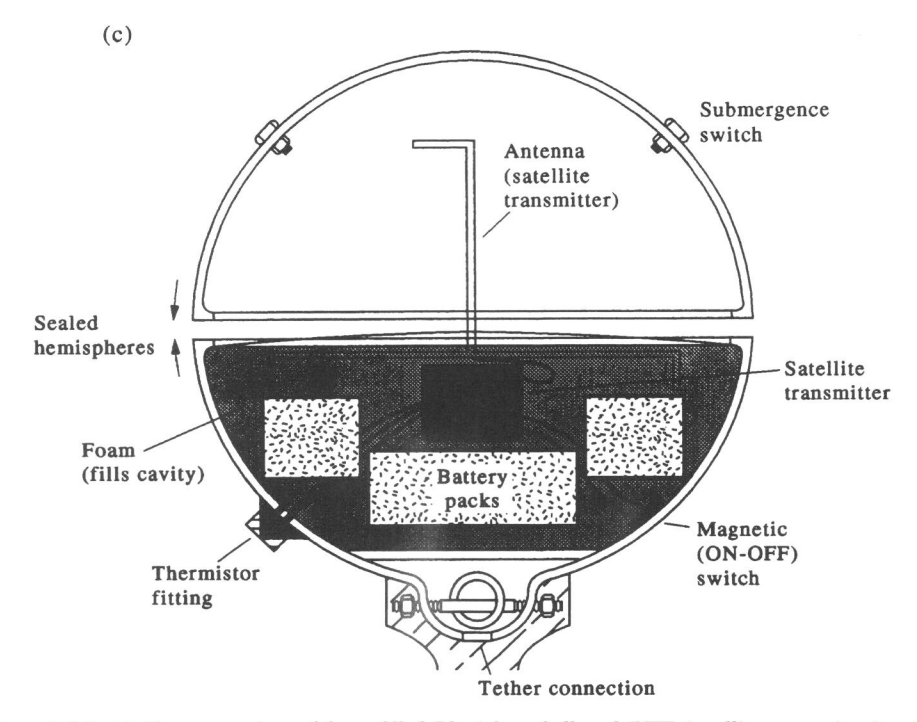

Figure 1.8.1. (c) Cut-away view of foam-filled Plexiglass shell and PTT (satellite transmitter) used for the surface buoy. When complete, the surface float has an excess buoyancy greater than 7 kg. (From Sybrandy and Niiler, 1990.)

presented in Figure 1.8.2. As examples, we have chosen trajectories from the North Atlantic near the Azores convergence zone (Figure 1.8.2a) and from the TOGA/ WOCE equatorial Pacific (Figure 1.8.2b). Note that, despite the extensive buoy coverages in these two cases, there are still regions unvisited by the drifters. Our final example (Figure 1.8.2c) is a unique point-source deployment from the 106-mile site southeast of New York city. This site was the only ocean disposal site in the U.S.A. designated for dumping sewage sludge during the 1980s.

The essential technology for the above type of tracking was the development of a random access positioning system for polar orbiting satellites that could simultaneously fix the positions of many platforms using the Doppler-shift of the radio signals transmitted at regular intervals from the buoy. Early versions of the satellite tracking system were flown on the NIMBUS 6 (Kirwan et al., 1975) and the French EOLE (Cresswell, 1976) satellites. Cresswell (1976) tested the accuracy of this system by examining the time-series from a moored buoy and from an antenna mounted on top of a laboratory. For both sites, the uncertainties were less than 1.0 km. Similar RMS position fix errors were reported for the NIMBUS 6 systems by Kirwan et al. (1975) and by Richardson et al. (1981).

The early satellite tracking systems have been replaced by the French ARGOS system (Collecte Localisation Satellit, CLS) carried onboard U.S. NOAA polar-orbiting weather satellites. As reported by Krauss and Käse (1984), this twin satellite system is capable of positional accuracies better than 0.2 km. Location quality depends on a number of factors such as the quality of the ephemeris data (orbital

Drifters in the atlantic ocean

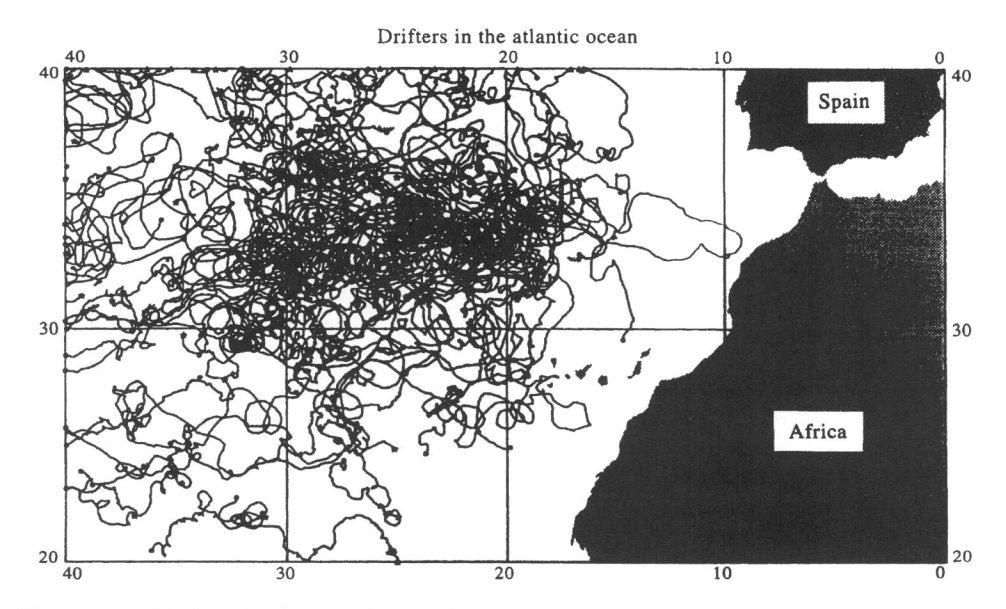

Figure 1.8.2. Trajectories from modern surface drifters with shallow (10–15 m) drogue depths. (a) Trajectories of 103 WOCE holey-sock drifters deployed near the Azores in the eastern North Atlantic from July 6, 1991 to October 25, 1993 (courtesy, Mayra Pazos, NOAA).

parameters), the stability of the receiver oscillator and temperature control, the duration of the satellite pass and the number of messages it receives from the drifter. Statistical information processed by Service ARGOS from thousands of fixed or slow-drifting platforms (Service ARGOS, 1992) indicates that quality locational fixes have 68% (= $\pm 1\sigma$) accuracies of 150 m while standard fixes have 68% accuracies of 350 m. As indicated by Table 1.8.1, the number of fixes per day is a function of latitude and higher accuracy is possible when the platform is fixed over periods longer than 7 min during two successive satellite passes. The drifters themselves cost a few thousand dollars and are considered expendable. Typical tracking costs are of the order of $5000 per year for full tracking (no positional fixes omitted) and one-third of this for the one-third duty cycle permitted by Service ARGOS (i.e. full-time tracking for 8 or 24 h followed by no tracking for 16 or 48 h, respectively).

1.8.3 Processing satellite-tracked drifter data

Position data obtained through satellite tracking need to be carefully examined for erroneous locations and loss of drogue. In fact, one of the main problems with surface drifters, aside from the need for accurate positioning, is knowing if and when the drogue has fallen off. Strain sensors are often installed to sense drogue attachment, but they have proven unreliable. The tether linkage between the surface buoy and the drogue is the major engineering problem in designing robust and long-life drifters. Because of this problem, drifters often have a subsurface float to help absorb the snap loading on the drogue caused by surface waves and also ensure that the surface element isn't constantly submerged in rough weather. An abrupt and sustained order-of-magnitude increase in the velocity variance derived from first differences of the

108   *Data Analysis Methods in Physical Oceanography*

Buoy trajectories from January 1990 thru July 1993

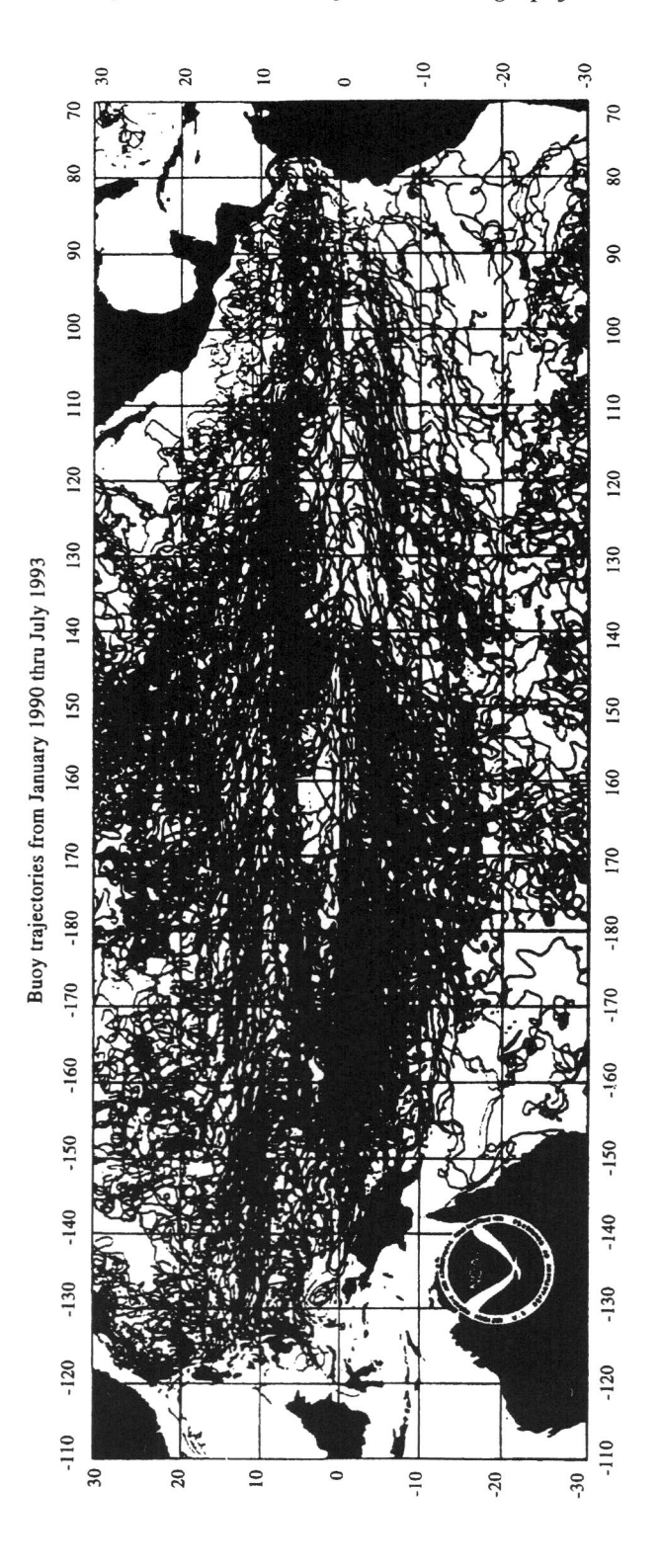

Figure 1.8.2. Trajectories from modern surface drifters with shallow (10–15 m) drogue depths. (b) Trajectories of TOGA/WOCE holey-sock drifters for the Equatorial Pacific from January 1990 to July 1993 (courtesy, Mayra Pazos, NOAA).

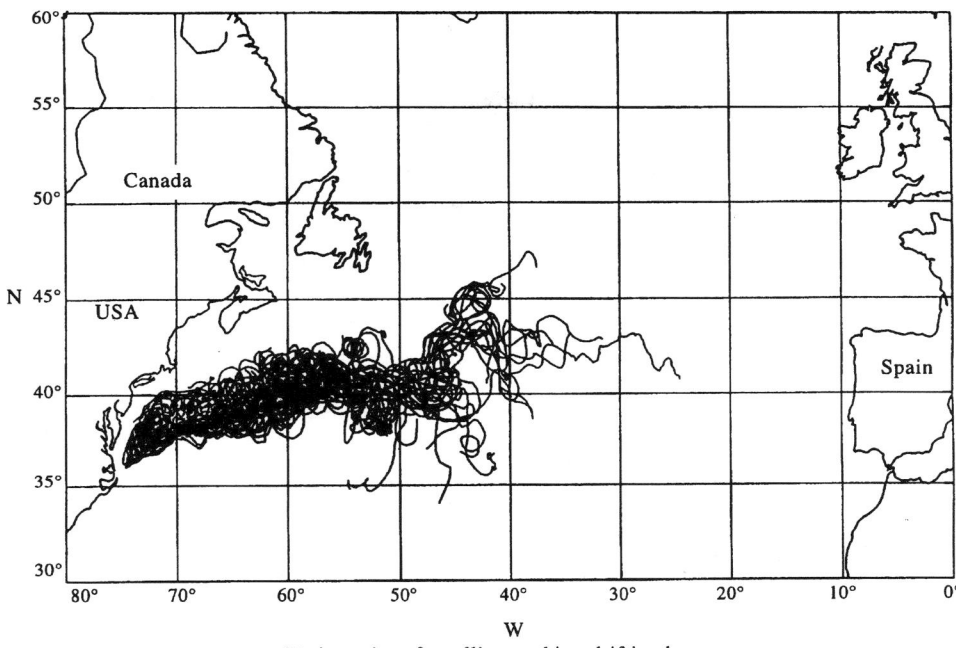

Trajectories of satellite-tracking drifting buoys
released at the 106-mile site October 1989 to June 1991

Figure 1.8.2. Trajectories from modern surface drifters with shallow (10–15 m) drogue depths. (c) Tracks of 66 holey sock drifters centered at 10 m depth released from 106-mile site southeast of New York between October 1989 and June 1991 (courtesy Paul Dragos, Battelle Ocean Sciences; Service Argos Newsletter 46, May 1993).

edited positional data can be considered as evidence for drogue loss. The cubic spline routine in most software analysis packages works well for positional data provided the sampling interval is only a few hours. Although it is not recommended, the user can obtain the velocity components (u, v) directly from spline coefficients for the positional data.

1.8.4 Drifter response

As with all Lagrangian tracers, it is difficult to know how accurately a drifter is coupled to the water and what effects external forces on the drifter's hull might have on its performance. In most applications, the coupling between the buoy and the water is greatly improved by the drogue. For shallow drifters with drogue depth centers less than 30 m, typical drogue-to-tether drag ratios are around 40:1. For deeper drogues (>100 m) the ratio decreases due to the added length of the tether. A smaller diameter wire can help offset the increased drag but at the expense of durability. There are as many different drogue designs as there are buoy hull shapes and it is difficult to get a consensus on the efficiencies of these drifter system elements. In a theoretical and experimental study, Kirwan *et al.* (1975) examined the effects of wind and currents on various hull and drogue types. They found that parachute drogues were more efficient than the common window-shade drogues, in strong contrast to the finding by Vachon (1973) that a bottom-ballasted window blind drogue was the most effective.

Subsequent studies by Dahlen and Chhabra (1983) have determined that a holy-sock drogue is more efficient than either the window shade or the parachute. This shape is easy to deploy and was selected for the standard drifter used in the WOCE program (Sybrandy and Niiler, 1990). Another innovative drogue called the Tristar developed by Niiler et al. (1987) uses a cross-pattern of window-shades with an additional horizontal plane (Figure 1.8.1a). The idea behind this drogue design is to reduce "sailing" of the drogue as is often occurs with a single window shade. Although easy to deploy (it goes into the water in a soluble box), this type of drifter is difficult to recover. For compatibility reasons, the standard drifter used in the WOCE Surface Velocity Program (WOCE-SVP) uses a holey-sock drogue.

In addition to the disagreements about which type of drogue is best, the field studies of Kirwan et al. (1978) reported that the wind drag correction formula, given by Kirwan et al. (1975), is much too large for periods of high wind. The subsequent conclusion was that drifter velocities uncorrected for wind drag are better indicators of the true prevailing surface currents than are those corrected for the influence of wind drag on the buoy hull. In this context, it should be recognized that Lagrangian drifting buoys respond to the integrated drag forces including the forces on the drogue and the direct forcing on the hull. The driving forces in the water column consist of a superposition of geostrophic currents plus wind- and tide-generated currents. To evaluate the role of wind forcing on drifter trajectories, McNally (1981) compared monthly mean drifter trajectories with the flow lines for mean monthly winds computed from Fleet Numerical Weather Central's (now the Fleet Numerical Ocean Center, FNOC) synoptic wind analysis. He found that the large-scale, coherent surface flow followed isobars of sea-level pressure and was 20–30° to the right of the surface wind in the North Pacific (Figure 1.8.3). Overall buoy speeds were 1.5 of the geostrophic wind speed during periods of strong atmospheric forcing (fall, winter, and spring). In the summer, mesoscale ocean circulation features, unrelated to the local wind, tended to determine the buoy trajectories.

McNally (1981) also compared trajectories among buoys with drogues at 30 m depth, buoys with drogues at 120 m, and buoys without drogues. He found that drifters drogued below 100 m depth behaved very differently from ones drogued at 30 m, but that those drogued at 30 m and those without drogues behaved similarly. Using the record from a drogue tension sensor (drogue on–off sensor), McNally found that it was not possible to detect from the trajectory alone when a buoy had lost its drogue. This result suggests a lack of vertical current shear in the upper 30 m where the flow apparently responds more directly to wind-driven currents than to baroclinic geostrophic flow. The result was supported by the poor correlation between mean seasonal dynamic height maps and the tracks of near-surface drifters reported by McNally et al. (1983). McNally (1981) also described an annual increase by a factor of 5 of the wind speed in the North Pacific while the drifter speeds increased by a factor of 3.5, somewhat surprising considering that the wind stress that drives the currents is proportional to the square of the wind speed. During this same time the mean seasonal dynamic height amplitude changed only slightly.

Emery et al. (1985) confirmed the lack of agreement between drifter tracks and the synoptic geostrophic current estimates, as well as the high correlation between drifter displacements and the geostrophic wind speed and direction (Figure 1.8.4). In a rather complex analysis of the wind driven current derived from drifter trajectories, Kirwan et al. (1979) concluded that, while the drifter response is best described by a two-parameter linear system (consistent with the driving of the buoy by wave-driven

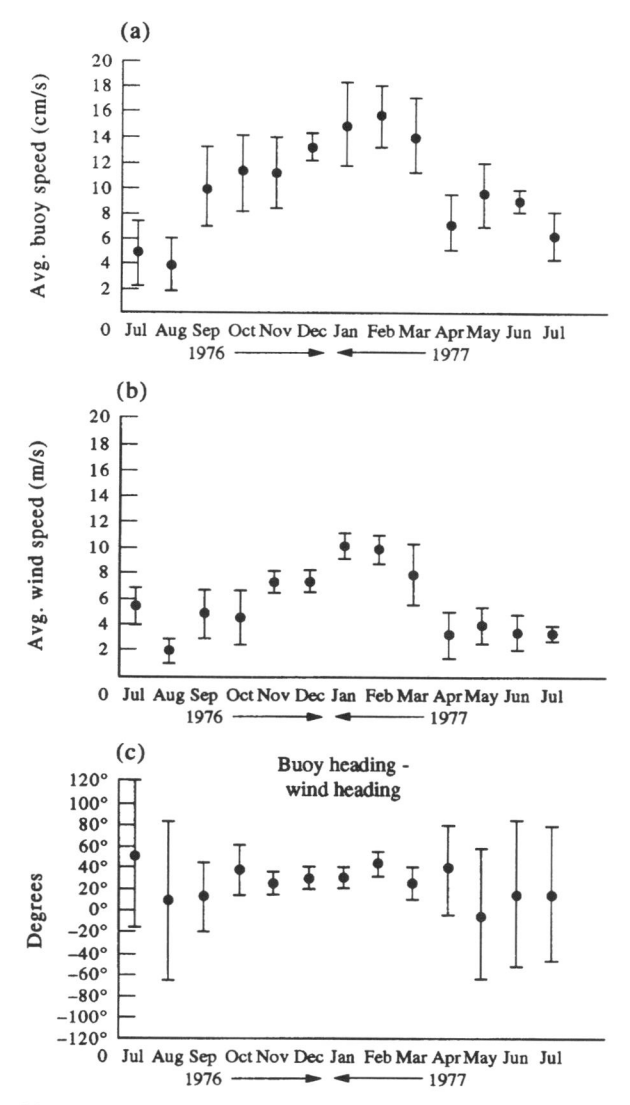

Figure 1.8.3. *Monthly average wind and buoy speeds over the ADS North Pacific region from June 1976 through July 1977. (a) Monthly average drifter speeds; (b) monthly average wind speeds; (c) monthly average difference angle between wind direction and drifter direction. Vertical bars denote ±1 standard deviation.*

Stokes drift), a combination of Ekman current plus Stokes drift also adequately described the resulting trajectories. Calculations by Emery *et al.* (1985), based on the nominal hull size, suggest that the Stokes drift component is relatively small and that the current in the surface Ekman layer is the primary driving mechanism for the mean drifter motion. That the angle this current makes to the wind is less than the 45° predicted by Ekman (1905), is expected since in the real ocean conditions never seem to meet the conditions for Ekman's derivation. McNally (1981) found an average angle

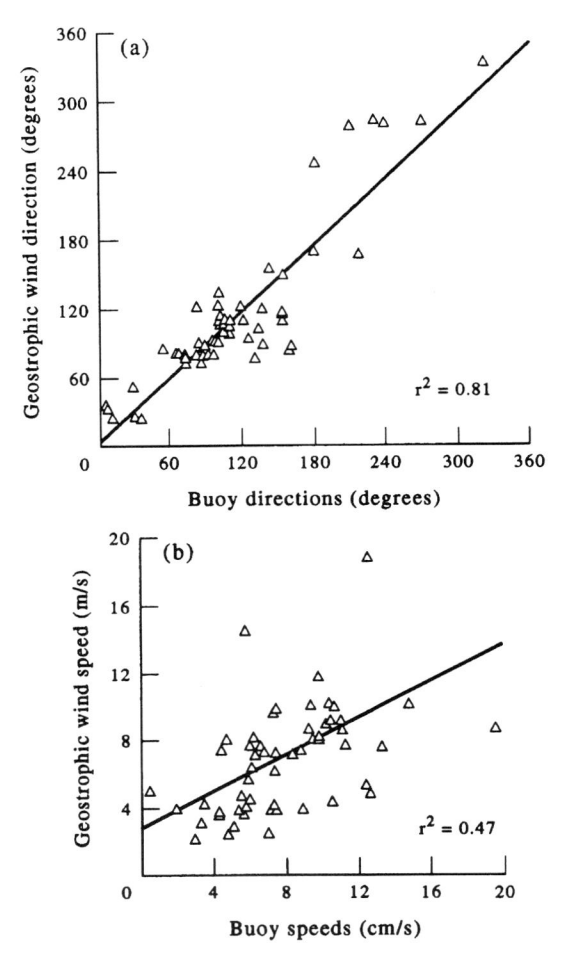

112 *Data Analysis Methods in Physical Oceanography*

Figure 1.8.4. (a) Comparison between monthly mean buoy and geostrophic wind directions. (b) Comparison between monthly mean buoy and geostrophic wind speeds. (From Emery et al., 1985.)

of 30° while Kirwan *et al.* (1979) reported an angle of 15°, both to the right of the wind for the northern hemisphere.

A search for the elusive Ekman spiral was conducted from November 20, 1991 to February 29, 1992 by Krauss (1993) using ten satellite drifters drogued at five different levels within well-mixed homogeneous water of 80 m depth in the North Sea midway between England and Norway. The holey-sock drogues used in the study were 10 m long and centered at 5 m depth intervals from 7.5 to 27.5 m (Figure 1.8.5a). Results for the first four weeks of drift when the drifters were relatively close together revealed a clockwise turning and decay of the apparent wind stress with depth as required by Ekman-layer theory (Figure 1.8.5b). Here, the apparent wind stress is derived from the fluctuations in current velocity shear measured by the satellite-tracked drifters. Sea surface slopes needed to complete the calculations are from a numerical model. The observed amplitude decay of 0.90 and deflection of 10° near the surface are in close agreement with theory (apparent wind increases from 0° at the surface and is associated with an Ekman current that should be 45° to the right of the

(a)

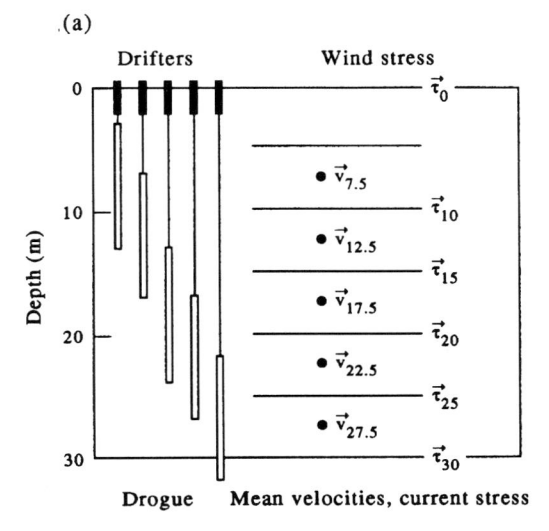

Figure 1.8.5. Test of Ekman's theory. The clockwise turning and decay of the apparent wind stress τ_D at depth D (m) relative to the observed surface wind stress. The apparent wind stress is derived from the current velocity shear dv_D/dz measured by satellite-tracked drifters (a) drogued at different depths during homogeneous winter conditions. (Courtesy W. Krauss, 1994.)

apparent wind). The angle increases to 41.6° in 25 m depth. The total current field is a superposition of barotropic currents due to sea-level variations and Ekman currents. The classical Ekman theory is unable to fully describe the observed deflection of the apparent wind (and Ekman current) to the right of the wind and its decay with depth. To be consistent with Ekman's theory, an eddy viscosity of 10^3 cgs units would be needed, which is well beyond the norm. However, as noted by Krauss, "... the deflections are a strong indication that some type of Ekman spiral dominates within the upper 30 m."

In an older study McNally and White (1985) examined wind-driven flow in the upper 90 m using a set of buoys drogued at different depths. They found a sharp change in buoy behavior when the drogue entered the deepening surface mixed layer. This response was characterized by a sudden increase in the amplitude of near-inertial motions with a downwind drifter velocity component three times that of the crosswind component. They also found that 80–90% of the observed crosswind component could be explained by an Ekman slab model. The large downwind response leads to surface currents, calculated from the buoy displacements, that are greater than 0° but less than 45° (about 30°) to the right of the wind. This behavior was true for all buoys with drogues above the upper mixed layer; once in the mixed layer all buoys behaved the same regardless of drogue depth.

In summary, it seems that the question of the relative coupling of drogued and undrogued drifting buoys to the water is still not completely resolved. Drifters measure currents, but which components of the flow dominate the buoy trajectories is still a topic of debate. Based on the recent literature, it appears that shallow drifters with drogue depths less than about 50 m are driven mainly by the wind-forced surface frictional Ekman layer whereas deep drifters with drogue depths exceeding 100 m are more related to geostrophic currents. The likely percentage of contribution by these

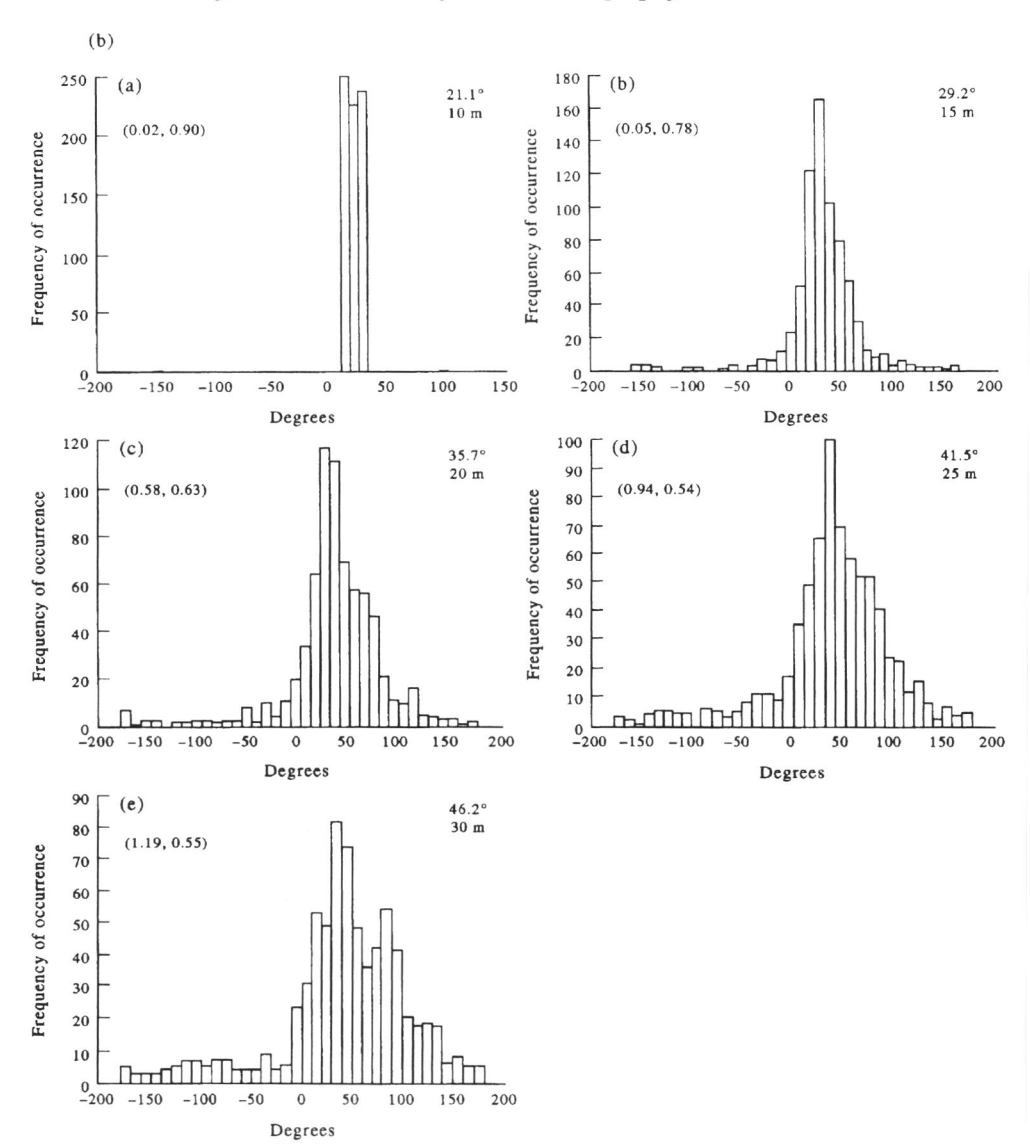

Figure 1.8.5. Test of Ekman's theory. The clockwise turning and decay of the apparent wind stress τ_D at depth D (m) relative to the observed surface wind stress. The apparent wind stress is derived from the current velocity shear dv_D/dz measured by satellite-tracked drifters (b) Histogram of the relative angle (in degrees) between the surface wind-stress vector and the calculated apparent wind-stress as a function of depth (surface wind minus apparent wind). Linear regression values (α, β) give apparent wind-stress as a function of surface wind-stress. Offset results from the different time scales of the winds and the currents. Mean values given in upper right corner of figure. (Courtesy W. Krauss, 1994.)

two current types depends on the type of drogue system that is used. A problem with trying to measure the deeper currents is that deeper drogue systems tend to fail sooner and it is difficult to access quantitatively the role of the drogue in the buoy trajectories. Drogue loss due to wave loading and mechanical decoupling of the surface buoy and the drogue is still the main technical problem to extending drifter life.

There are other problems with drifters worth noting. In addition to drogue loss and errors in positioning and data transmission, the transmitters submerge in heavy weather and loose contact with the passing satellite. Low drag ratios lead to poor flow response characteristics and, because of the time between satellite passes, there is generally inadequate sampling of tidal and near-inertial motions (especially at low latitudes or for the one day on–two days off duty cycles) leading to aliasing errors. Drifters also have an uncanny tendency to go aground and to concentrate in areas of surface convergence.

1.8.5 Other types of surface drifters

Before leaving this topic, it is appropriate to mention that while the satellite-tracked buoys are perhaps the most widely used type of surface follower for open waters, there are other buoy tracking methods being used in more confined coastal waters. A common method is to follow the surface buoy using ship's radar or radar from a nearby land-based station. More expensive buoys are instrumented with both radar reflectors and transponders to improve the tracking. The accuracies of such systems all depend on the ability of the radar to locate the platform and also on the navigational accuracy of the ship. Fixes at several near-simultaneous locations are needed to triangulate the position of the drifter accurately. Data recording techniques vary from hand plotting on the radar screen to photographing the screen continuously for subsequent digital analysis. These techniques are manpower intensive when compared with the satellite data transmission which provides direct digital data output.

In addition to radar, several other types of buoy tracking systems have been developed. Most rely on existing radio-wave navigation techniques such as LORAN or NAVSTAR (satellite navigation). For example, the subsurface drogued NAVocean and Candel Industries Sea Rover-3 Loran-C drifters have built-in Loran-C tracking systems that can both store and transmit the positional data to a nearby ship within a range of 25–50 km. Absolute positional accuracy in coastal regions is around 200 m but diminishes offshore with decreased Loran-C accuracy. However, relative positional errors are considerably smaller. Based on time-delay transmission data from three regional Loran-C transmitters, Woodward and Crawford (1992) estimated relative position errors of a few tens of meters and drift speed uncertainty of 2 cm s^{-1} for drifters deployed off the west coast of Canada. Once it is out of range of the ship, the Loran-C drifter can be lost unless it is also equipped with a satellite transmission system. Meteor-burst communication is a well-known technique that makes use of the high degree of ionization of the troposphere by the continuous meteor bombardment of the earth. A signal sent from a coastal master station skips from the ionosphere and is received and then retransmitted by the buoy up to several thousand kilometers from the source. Since the return signal is highly directional, it gives the distance and direction of the buoy from the master station. Buoys can also be positioned using VHF via direction and range. The introduction of small, low-cost GPS receivers makes it possible for buoy platforms to position themselves continuously to within 110 m. Provision for differential GPS (using a surveyed land-based shore station) has improved the accuracy to order of 10 m. Data are then relayed via satellite to provide a higher resolution buoy trajectory than is presently possible with ARGOS tracking buoys. Given the high positioning rate possible for GPS systems, it is the spatial accuracy of the fixes that limits the accuracy of the velocity measurements.

1.8.6 Subsurface floats

New technological advances in subsurface, neutrally-buoyant float design have improved interpretation and understanding of deep ocean circulation in the same way that surface drifters have improved research in the shallow ocean. In their earliest form (Swallow, 1955), subsurface quasi-Lagrangian drifters took advantage of the small absorption of low-frequency sound emitted in the sound channel (the sound velocity minimum layer) located at intermediate depths in the ocean. The sound-emitting drifters were tracked acoustically over a relatively short range from an attending ship. The development of the autonomous SOFAR (SOund Fixing And Ranging) float, which is tracked from listening stations moored in the sound channel (Rossby and Webb, 1970) has removed the burden of ship tracking and made the SOFAR float a practical tool for the tracking of subsurface water movements. Although positional accuracies of SOFAR floats depend on both the tracking and float–transponder systems, the location accuracy of 1 km given by Rossby and Webb (1970) is a representative value. In this case, neutrally-buoyant SOFAR floats have a positional accuracy that is comparable in magnitude to the satellite-tracked drifting buoys. Using high-power 250-Hz sound sources, the early SOFAR floats are credited with the discovery of mesoscale variability in the ocean and for pioneering our understanding of Lagrangian eddy statistics (Freeland et al., 1975). The now familiar "spaghetti-diagram" (Figure 1.8.6) is characteristic of the type of eddy-like variability measured by SOFAR floats deployed in the upper ocean sound channel (Richardson, 1993).

SOFAR floats transmit low-frequency sound pulses which are tracked from shore listening positions or from specially moored "autonomous" listening stations. The need to generate low-frequency sound means that the floats are long (8 m) and heavy (430 kg), making them expensive to build and difficult to handle. Since greater expense is involved in sending sound signals than receiving them, a new type of float called the RAFOS (SOFAR spelled backwards) float has been developed in which the buoys listen for, rather than transmit, the sound pulses (Figure 1.8.7). In this configuration, the float acts as a drifting acoustic listening station that senses signals emanating from moored sound sources (Rossby et al., 1986). The positions of RAFOS floats in a particular area are then determined through triangulation from the known positions of the moored source stations. A typical moored sound source, which broadcasts for 80 s every two days at a frequency of 260 Hz, has a range of 2000 km and an average lifetime of three years (WOCE Notes, June 3, 1991)

Since RAFOS floats are much less expensive to construct than SOFAR floats and more difficult to locate (since they are not a sound source), RAFOS floats are considered expendable. The data processed and stored by each RAFOS buoy as it drifts within the moored listening array must eventually be transmitted to shore via the ARGOS satellite link. To do this, the RAFOS float must come to the surface periodically to transmit its trajectory information. After "uplinking" its data, the buoy again descends to its programmed depth and continues to collect trajectory data. The cycle is repeated until the batteries run out.

The need for deep ocean drifters that are independent of acoustic tracking networks has led to the development of the "pop-up" float. The float is primarily a satellite PTT and a ballast device that periodically comes to the surface and transmits its location data and "health" status (an update on its battery voltage and other parameters) to the ARGOS system (Davis et al., 1992). The only known points on the buoy trajectory are

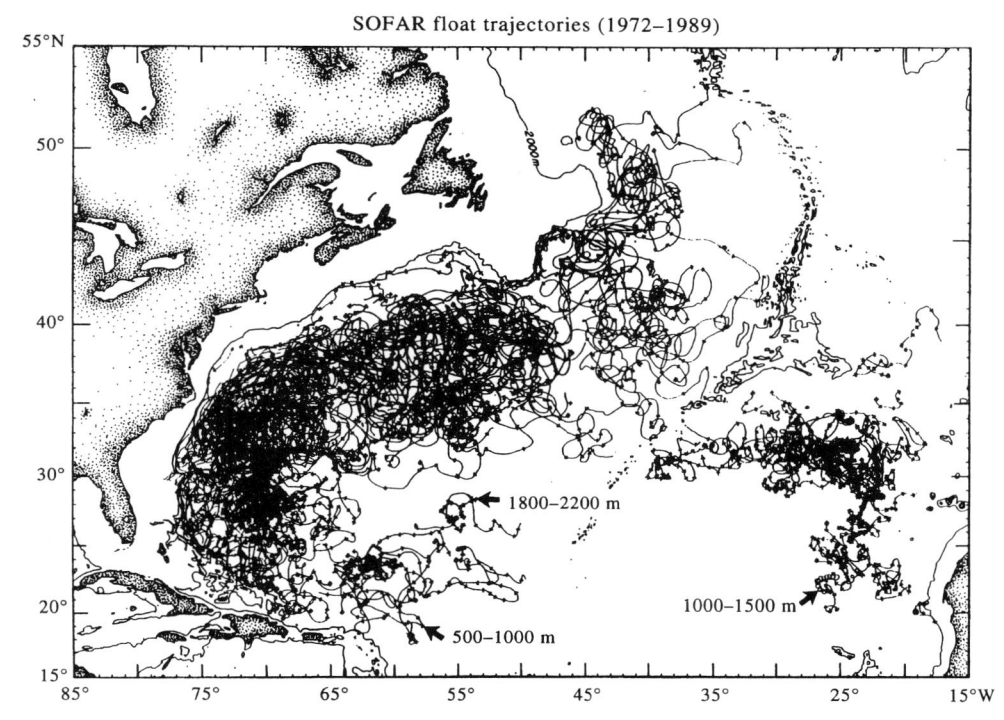

Figure 1.8.6. "Spaghetti-diagram" of all SOFAR float tracks from 1972 to 1989, excluding data from the POLYMODE Local Dynamics Experiment. Ticks on tracks denote daily fixes. Short gaps have been filled by linear interpretation. Plots are characteristic of the type of eddy-like variability measured by SOFAR floats deployed in the upper ocean sound channel. (Courtesy, Phillip Richardson, 1994.)

those obtained when the buoy is on the surface. As with the RAFOS buoys, the pop-up float sinks to its prescribed depth level after transmitting its data to the ARGOS system and continues its advection with the deep currents. The advantage of such a system is that it can be designed to survive for a considerable time using limited power consumption. Assuming that deep mean currents are relatively weak, the pop-up float is an effective tool for delineating the spatial pattern of the deep flow, which up to now has not been possible over large areas. The Autonomous Lagrangian Circulation Explorer (ALACE) described by Davis et al. (1992) drifts at a preset depth (typically less than 1000 m) for a set period of 25 days, then rises to the surface for about a day to transmit it position to the satellite. The drifter then returns to a prescribed depth which is maintained by pumping fluid to an external bladder which changes its volume and hence its buoyancy. Modern ALACE floats built by Webb Research Corp. are capable of making about 100 round-trips to depths of less than 1 km over a lifetime of about five years. Errors are introduced by surface currents when the device is on the surface. The floats also provide temperature and salinity profiles during ascent or descent.

As with the surface drifter data, the real problem in interpreting SOFAR float data is their fundamental "quasi-Lagrangian" nature (Riser, 1982). From a comparison of the theoretical displacements of true Lagrangian particles in simple periodic ocean current regimes with the displacements of real quasi-Lagrangian floats, Riser concludes that the planetary scale (Rossby wave) flows in his model contribute

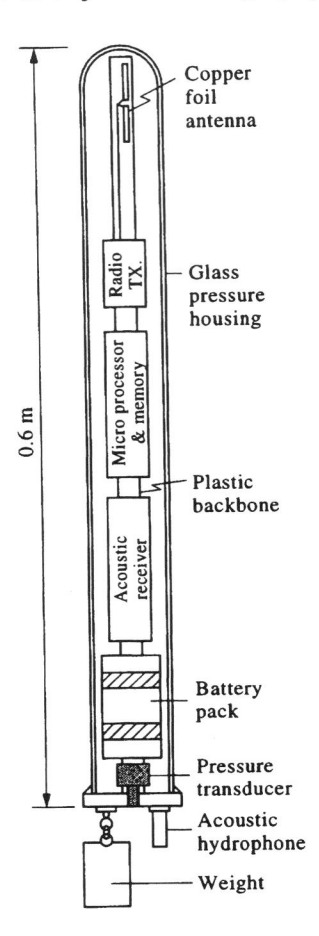

Figure 1.8.7. Schematic of a RAFOS float. (Courtesy, Thomas Rossby.)

more significantly to the dispersion of 700 m depth SOFAR floats than do motions associated with near-inertial oscillations or internal waves of tidal period. Based on these model speculations, he suggests that while a quasi-Lagrangian drifter will not always behave as a Lagrangian particle it nevertheless will provide a representative trajectory for periods of weeks to months. For his Rossby wave plus internal wave model, Riser derived a correlation time scale of about 100 days. He also suggests that the residence of some floats in the small scale (25 km) features, in which they were deployed, provides some justification for his conclusions.

For pop-up floats, problems in the interpretation of the positional data arise from: (1) interruptions in the deep trajectory every time the drifter surfaces; (2) uncertainty in the actual float position between satellite fixes; and (3) contamination of the deep velocity record by motions of the float on the surface or during ascent and descent. An essential requirement in the accurate determination of the subsurface drift is to find the exact latitude/longitude coordinates of the buoy when it first breaks the ocean surface and when first begins to re-sink. The ability to interpolate ARGOS fixes to these times is determined by the nature of the surface flow and the number satellite fixes. ALACE ascends more rapidly than it descends and spends little time at the surface. In a trial to 1 km, the drifter spent 0.3 h in the upper 150 m, and 4 h between 150 and 950 m depth. Thus, according to Davis *et al.* (1992), most of the error comes from vertical velocity shear at depths of 150 m and deeper, below the surface wind-driven layer (see Thomson and Freeland (1999) for further details).

1.8.7 Surface displacements in satellite imagery

As noted briefly at the end of Section 1.7.6, well-navigated (geographically-located) sequential satellite images can be used as "pseudo-drifters" to infer surface currents. The assumption is that the entire displacement of surface features seen in the imagery is caused by surface current advection. This displacement estimate method (called the maximum cross-correlation or MCC) was applied successfully to sea ice displacements by Ninnis *et al.* (1986). Later, the same approach was applied to infrared images of sea-surface temperature (SST) by Emery *et al.* (1986). The patterns and velocities of the SST-inferred currents were confirmed by the drifts of shallow (5 m drogue) drifters and by a CTD survey. Later studies (Tokamamkian *et al.*, 1990; Kelly and Strub, 1992) have confirmed the utility of this method in tracking the surface displacements in different current regimes. When applied to the Gulf Stream (Emery *et al.*, 1992), the MCC method reveals both the prevailing flow and meanders. A numerical model of the Gulf Stream, used to evaluate the reliability of the MCC currents found that, for images more than 24 h apart, noise in this strong flow regime begins to severely distort the surface advection pattern.

The MCC method can also be applied to other surface features such as chlorophyll and sediment patterns mapped by ocean color sensors. In the future, it may be possible to combine ocean color tracking with infrared image tracking. Infrared features are influenced by heating and cooling, in addition to surface advection, while surface chlorophyll patterns respond to *in situ* biological activity. Since these two features should reflect the same advective patterns (assuming similar advective characteristics for temperature and color), the differences in calculated surface vectors should reflect differences in surface responses. Thus, by combining both color and SST it should be possible to produce a unique surface flow pattern that corrects for heating/cooling and primary biological production.

1.9 WIND

Although it might be surprising to find a section on wind data in an oceanographic text, we can state with some confidence that most of the scientific assessment of wind data over the ocean has been done by oceanographers searching for the best way to define the meteorological forcing field for oceanic processes. This is especially true of observationalists working on upper ocean dynamics and numerical modelers who require climatological winds to drive their circulation models. It is not the intent of this book to discuss in detail the many types of available wind sensors and to evaluate their performance, as is done with the oceanographic sensors. Instead, we will briefly review the types of wind data available for ocean regions and make some general statements about the usefulness and reliability of these data.

Open-ocean wind data are of three types: (1) six-hourly geostrophic wind data computed from measured distributions of atmospheric sea surface pressure over the ocean; (2) directly measured wind data from ships and moored platforms (typically at hourly intervals); and (3) inferred six-hourly wind data derived from satellite sensors. Atmospheric pressure maps are prepared from combinations of data recorded by ships at sea, from moored or drifting platforms such as buoys, and from ocean island stations. Analysis procedures have changed over the years with early efforts depending on the subjective hand contouring of the available data. More recently, there has been a shift to computer-generated "objective analysis" of the atmospheric pressure data. Since they are derived from synoptic weather networks, the pressure data are originally computed at six-hourly intervals (00, 06, 12, and 18 UTC). While some work has been done to correct barometer readings from ships to compensate for installation position relative to sea-level, no systematic study has been undertaken to test or edit these data or analyses. However, in general, sea-level pressure patterns appear to be quite smooth, suggesting that the data are generally reliable. Objective analysis smooths the data and suppresses any noise that might be present.

It is not a simple process to conformally map a given atmospheric pressure distribution into a surface wind field. While the computation of the geostrophic wind velocity from the spatial gradients of atmospheric pressure is fairly straightforward, it is more difficult to extrapolate the geostrophic wind field through the sea-surface boundary layer. The primary problem is our imperfect knowledge of the oceanic boundary layer and the manner in which it transfers momentum from the wind to the ocean surface. While most scientists have agreed on the drag coefficient for low wind speeds (<5 m/s), there continues to be some disagreement on the appropriate coefficient for higher wind speeds. Added to this is a lack of understanding of boundary layer dynamics and how planetary vorticity affects this layer. This leads to a lack of agreement on the backing effect and the resulting angle one needs to apply between the geostrophic wind vector and the surface wind vector. Thus, wind stress computations have required the *a priori* selection of the wind stress formulae for the transformation of geostrophic winds into surface wind stresses. The application of these stress calculations will therefore always depend on the selected wind stress relation and any derived oceanographic inferences are always subject to this limitation.

Anemometers installed on ships, buoys or island stations provide another source of open-ocean wind data. The ship and buoy records are subject to problems arising from measuring the wind around structures and relative to a moving platform, which is itself being affected by the wind. These effects are difficult to estimate and even more difficult to detect once the data have been recorded or transmitted. Many of the earlier ship-wind data in climatological archives are based on wind estimates made by the ship's officers from their evaluation of the local sea state. (The Beaufort Scale was designed for the days of sailing vessels and uses the observed wave field to estimate the wind speed.) Analysis of the ship-reported winds from the Pacific (Wyrtki and Meyers, 1975a, b) has demonstrated that, with some editing and smoothing, these subjective data can yield useful estimates of the distribution of wind over the equatorial Pacific. Barnett (1983) has used objective analysis on these same data to produce an even more filtered set of wind observations for this region. Following a slightly different approach, Busalacchi and O'Brien (1981) reanalyzed the ship wind-data to fill in spatial gaps before applying the wind fields to oceanographic model studies.

Included in other widely used sets of wind data are the synoptic wind fields produced by the U.S. Fleet Numerical Ocean Center (FNOC) in Monterey, California. These analyses use not only ship, buoy and island reports but also winds inferred from the tracking of clouds in sequences of visible and infrared satellite imagery. In this technique, one uses the infrared image to estimate the temperature and, therefore, infer the elevation of the cloud mass being followed. By examining sequences of satellite images, specific cloud forms can be followed and the corresponding wind speed and direction computed for the altitude of the cloud temperature. Naturally, this procedure is dependent not only on the accuracies of the satellite sensors but also on the interpretive skills of the operator. As a consequence, no real quantitative levels of accuracy can be attached to these data. Comparison between the FNOC winds and coincident winds measured from an open-ocean buoy (Friehe and Pazan, 1978) showed excellent agreement in speed and direction over a period of 60 days. Although this single-point comparison is too limited to establish any uncertainty values for the FNOC wind fields, the comparison provides some confirmation of the validity of techniques used to derive the FNOC winds.

A wind product for the Pacific Ocean similar to the FNOC winds is generated by the National Marine Fisheries Service (NMFS) in Monterey, California (Holl and Mendenhall, 1972; Bakun, 1973). In this product, the geostrophic "gradient" winds are first computed at a 3° × 3° latitude–longitude grid spacing from spatial gradients in the six-hourly synoptic atmospheric pressure fields at the 500 or 800 mb surfaces. To obtain the surface-wind vectors in the frictional atmospheric boundary layer, the magnitudes of the calculated geostrophic wind vectors are reduced by a factor of 0.7 and the wind vectors rotated (backed) by 15°; here, "backed" refers to a counterclockwise motion in the northern hemisphere and a clockwise rotation in the southern hemisphere. (Some of the original work on this method can be traced to Fofonoff, 1960.)

Thomson (1983) compared winds computed by the NMFS with winds measured from moored buoys off the coast of British Columbia during the summers of 1979 and 1980 (Figure 1.9.1). In this comparison, it was concluded that winds computed from atmospheric pressure provided an accurate representation of the oceanic winds for time scales longer than several days but failed to accurately resolve short-term wind reversals associated with transient weather systems. Computed winds also tended to underestimate percentages of low and high wind speed. Similar results were reported by Marsden (1987) for the northeast Pacific (including Ocean Weather Station P) and by Macklin et al. (1993) for the rugged coast of western Alaska. The poor correlation of observed and computed winds at short time scales is thought to be due to the large (3° × 3°) spacing, the coarse six-hour sampling of the pressure field and the strong influence of orographic effects in mountainous coastal regimes. In Thomson's study, peak computed winds were roughly 20° to the right of the observed peak inner-shelf winds, suggesting that the computed winds were representative of more offshore conditions or that the 15° correction for frictional effects was too small. Spectra of observed winds were found to be dominated by motions at much larger wavelengths than were found in the computed values. The NMFS winds were found to contain a significant 24-h sea-breeze component in the inner shelf observed winds but not in the records farther offshore. Based on spectral comparisons it was concluded that the NMFS winds closely represented the actual winds for periods longer than two days (frequencies less than 0.02 cph) and only marginally matched actual winds for periods shorter than two days.

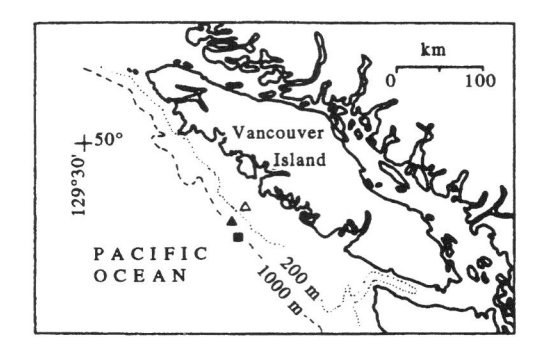

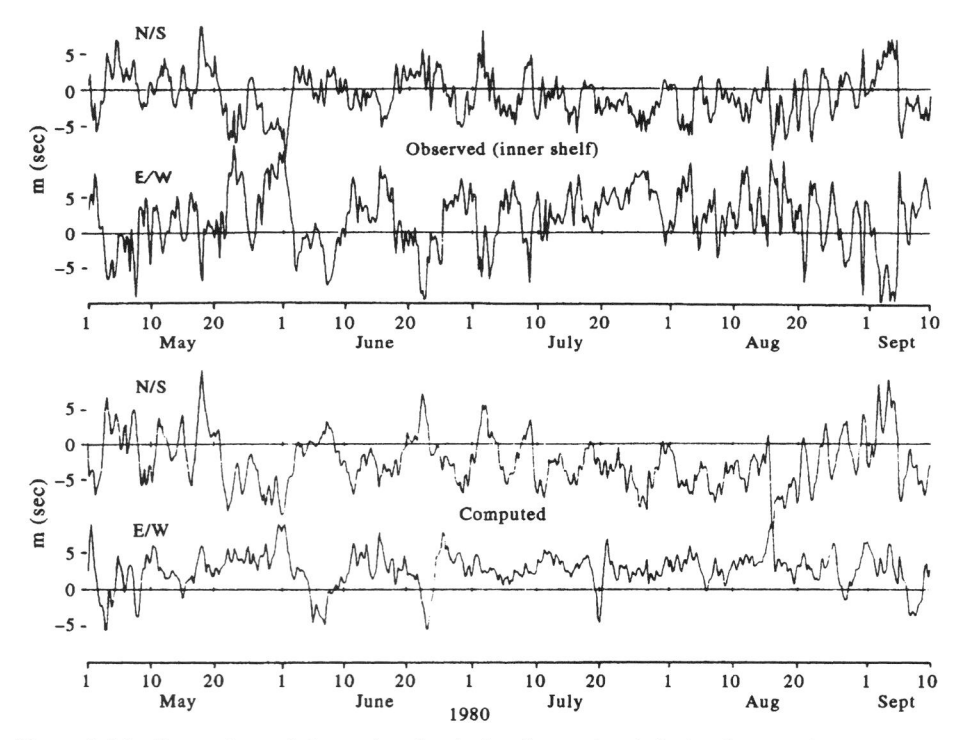

Figure 1.9.1. Comparison of observed and calculated oceanic winds for the period May to September 1980 on the west coast of Vancouver Island. Insert shows location of the moored bouys for 1979 and 1980 (triangles) and location of grid point (49°N, 127°W) for the geostrophic winds. Observed winds are from anemometers on moored buoys; calculated winds are the six-hourly geostrophic winds provided by the National Marine Fisheries Service (NMFS) in Monterey, California. (From Thomson, 1983.)

Oceanographers often use winds measured at coastal stations when studying problems of the nearshore marine environment. The primary caution with these data is that winds should be corrected for local orographic effects especially along mountainous coasts (Macklin et al., 1993). If the wind data are to be considered representative of the coastal ocean region, the wind sensor must be unobstructed along the direction of the wind. If not ideally situated, the measured wind data can still be used if the directional data are weighted to account for the bias due to local wind-channeling by the topography. Marsden (1987) found good agreement between measured and calculated winds at the rugged but exposed anemometer site at Cape St. James on the British Columbia coast, but relatively poor agreement for these winds at the protected coastal station at Tofino Airport 300 km to the south of the Cape.

Finally some comments about future wind measurements are appropriate. As new *in situ* sensing methods evolve, emphasis is being placed on the ability to measure wind over the ocean. An attractive new method is to detect changes in the ambient acoustic noise level due to wind-driven surface effects. The exact mechanisms causing these acoustic noise variations are still being investigated but empirical data clearly suggest a linear relationship to wind-stress fluctuations. Even more attractive is the future possibility of monitoring the wind over the ocean from polar orbiting satellites using a microwave scatterometer. The usefulness of such data was clearly demonstrated during the SEASAT mission (Brown, 1983) which confirmed scatterometer accuracies of 1–2 m/s (speed) and 1–20° (direction). A study of SEASAT data (Thompson *et al.*, 1983) has shown that radar backscatter from the ocean depends on surface wind stress for a wide range of transmitted wavelengths. These authors found that SEASAT synthetic aperture radar (SAR) data, combined with simultaneous SEASAT scatterometer data, provided a good estimate of the coefficient of wind speed to wind stress. Hence, in the future, it may be possible to measure wind stress directly rather than infer it from wind or pressure measurements. In the past decade, a number of new systems have been deployed that are capable of measuring wind speed over the ocean. The GEOSAT altimeter discussed earlier is able to observe wind speed from the change in the shape of the altimeter waveform. While direction sensing is not possible, the altimeter is able to provide relatively accurate wind speeds along the satellite subtracks every few kilometers (Witter and Chelton, 1991). This capability has been used by the U.S. Navy to routinely map the global wind field over the ocean. Comparisons of these winds with moored buoys and operational numerical model analyses have demonstrated the relative accuracy of these satellite winds.

In addition, the passive microwave sensor on the Defense Meteorological Satellite Program (DMSP) satellites, called the Special Sensor Microwave Imager (SSM/I), is able to sense wind speed but not direction (Figure 1.9.2; color plate). The SSMI is a seven-channel four-frequency, linearly-polarized microwave radar operating in a sun-synchronous orbit at an altitude of 860 km. Three of the four channels (19.3, 37.0, and 85.5 GHz) are dual-polarized while the 22.2 GHz channel is only vertically polarized, for a total of seven channels. The nearly 1400 km swathe of the conically scanned SSM/I produces complete coverage between 87°36′S to 87°36′N every three days per satellite (Halpern *et al.*, 1993). There are now at least two SSM/I operating. While the spatial resolution is poor due to the sensing capabilities at the microwave frequencies, algorithms have been developed that appear to produce reliable estimates of wind speed over the open ocean (Wentz *et al.*, 1986; Gooberlet *et al.*, 1990; Halpern *et al.*, 1993). Wind speed accuracies are about ±2 m/s for the range of speeds between 3 and 25 m/s under rain-free conditions. Since the emissivity of land is very different from that of water, the SSM/I cannot be used to estimate wind speed within 100 km of land. Similarly, surface wind speed within 200 km of the ice edge cannot be computed from SSM/I data. However, wind speeds computed from the SSM/I compare reasonably well with open-ocean winds (Emery *et al.*, 1994). Waliser and Gautier (1993) find that in the central and eastern equatorial Pacific, SSM/I wind-speed comparisons were well

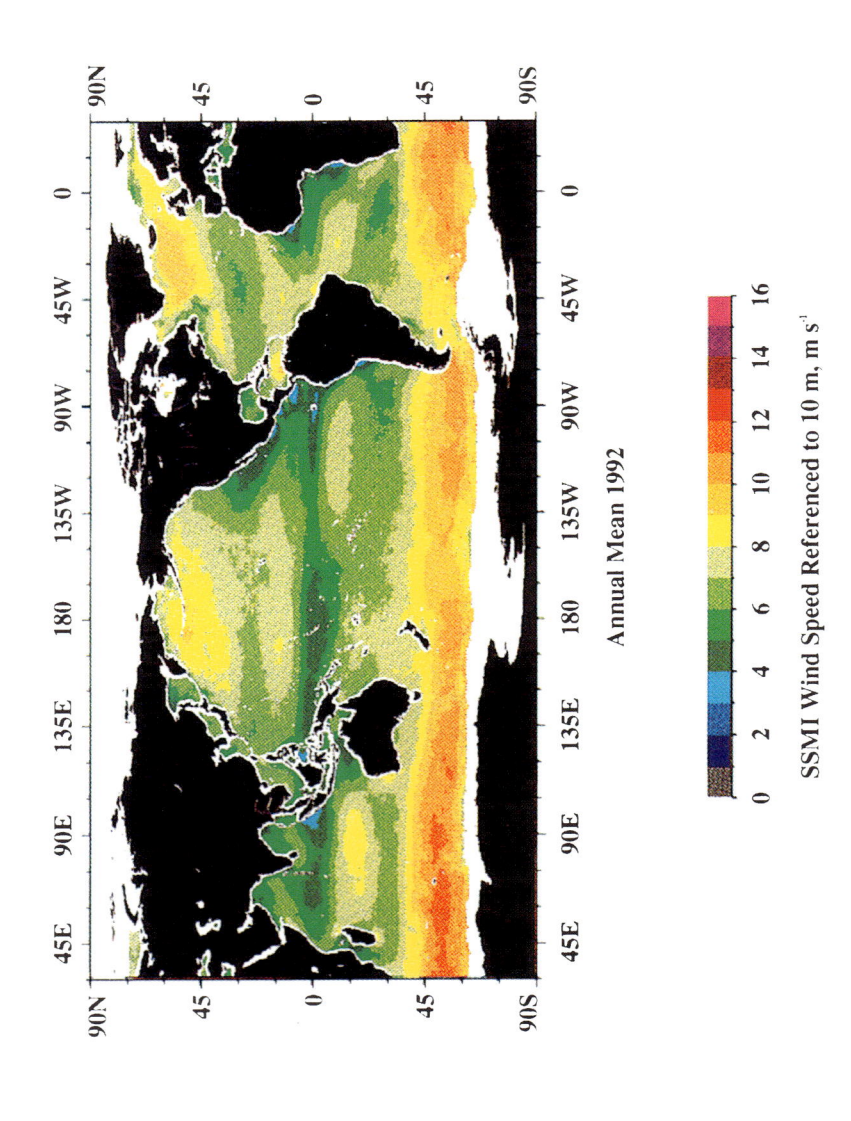

124 *Data Analysis Methods in Physical Oceanography*

Figure 1.9.2. *Global annual mean of the SSMI (Special Sensor Microwave Imager) surface wind speed for 1991. Courtesy of David Halpern (from Halpern et al., July 1993). JPL Publication 93-10.*

within the accuracies specified for the SSM/I. Biases (buoy-SSM/I) were generally less than 1 m/s and RMS differences were less than 2 m/s. However, in the western equatorial Pacific, biases were generally greater than 1–3 m/s and RMS differences closer to 2–3 m/s. According to Waliser and Gautier, "... there are still some difficulties to overcome in understanding the influences that local synoptic conditions (e.g. clouds/rainfall), and even background atmospheric and oceanic climatology effects, have on the retrieval of ocean-surface wind speeds from spaceborne sensors."

The most comprehensive space-borne measurement of the wind field is made using a microwave scatterometer which measures the radar scattering cross-section of the sea surface at different incidence and azimuthal angles. The SEASAT scatterometer demonstrated the applicability of this instrument for the measurement of open-ocean wind speed and direction. Using a combination of fan-beam antennas, the scatterometer is able to compute both the wind speed and direction. As with many other satellite borne systems, the scatterometer uses the Doppler shift of the received signal to compute the speed component while multiple fan-beam antennas (called sticks) are required to unambiguously resolve the wind direction. Since scattering cross-section at radar frequencies is mostly related to the small wavelets that form when the wind acts on the sea surface, the scatterometer signal is actually related to the wind stress rather than to the wind speed. Unlike anemometers and other like instruments, no additional conversion from wind speed to wind stress is needed. The problem is that all historical calibration information is based on wind speed and direction, rather than wind stress. As a consequence, all present algorithms still convert the scatterometer measurements to wind speed and direction. Studies of SEASAT scatterometer data (Pierson, 1981; Guymer et al., 1981) have demonstrated the ability of the satellite scatterometer to reliably measure wind speed and direction relative to ship and buoy observations. New scatterometers are flying on the European ERS-1 satellite and on the NASA–Japan ADEOS missions and other ERS satellites.

Trenworth and Olson (1988) consider the surface wind field computed by the European Centre for Medium-range Weather Forecasting (ECMWF) to be the best winds for general operational global analyses. ECMWF forecast-analyses of surface wind components at 10 m height are issued twice a day at 00 and 12 UTC. Numerical modelers examining large-scale circulation in the Pacific Ocean typically make use of the monthly mean and annual wind stress climatology provided by the Hellerman and Rosenstein (1983) wind fields. These data have problems near the equator where they tend to underestimate wind strength.

1.10 PRECIPITATION

Precipitation is one of the most difficult and challenging measurements to make over the ocean. Simple rain gauges installed on ships are invariably affected by salt spray and wind flow over the ship's hull and superstructure, and the short space and time scales of precipitation make it difficult to interpret point measurements. Rain gauges have two conflicting requirements that make use on shipboard difficult. First the gauge needs to be installed away from the ship influences, such as salt spray, which calls for positioning as high as possible on a mast. However, this conflicts directly with the second requirement, which calls for the regular maintenance of the gauge by ship's personnel. Few systematic studies have been made of precipitation measurements taken from ships, and little effort is made today to instrument ships to routinely observe rainfall over the ocean. A 25-year time-series from Ocean Station P (Figure 1.10.1) in the northeast Pacific is one of a few in the open ocean (most others were taken at Ocean Weather Stations similar to Station P). Unless some entirely new sampling procedure is developed, this situation is unlikely to change for the foreseeable future.

One new technique is to infer rainfall from variations in the upper-ocean acoustic noise. While it may seem a bit confusing to interpret ocean upper-layer acoustic noise both in terms of rainfall and wind, the frequency signatures of the two noise-generating mechanisms are sufficiently different to be distinguishable. Research is needed to define the accuracy of such a procedure and make at-sea precipitation measurements possible from moorings. These will still only provide point measurements and will not yield an improved spatial picture of the rainfall distribution. There are indications that future microwave satellite sensors, both active and passive, might be used to infer spatial variations in precipitation activity. SEASAT results were again encouraging in this regard but the validity of such measurements was not yet established. At any rate, it is presently impossible to attach any accuracy limits to what few ocean rain data are available. These data are so lacking that their relatively large uncertainty is not important. Any improvement in the analysis of precipitation measurements awaits the technical progress required to produce reliable at-sea rain data.

The 1987 launch of the Special Sensor Microwave Imager (SSM/I) on one of the Defense Meteorological Satellite Program satellites provided a new opportunity to infer precipitation from microwave satellite measurements. While a precipitation algorithm was developed prior to the launch (Hollinger, 1989) later studies have improved upon this algorithm to formulate better retrievals of precipitation over both

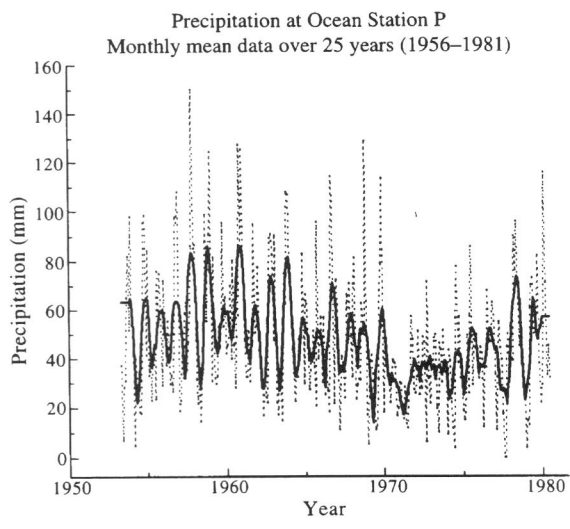

Figure 1.10.1. A 25-year time-series (1956–1981) of precipitation collected from Canadian Weather Ships at Ocean Station PAPA (50°N, 145°W) in the northeast Pacific. Solid line is from use of a Savitsky-Golay smoother (order = 13 months). (Data courtesy, Sus Tabata.)

land and ocean. In the list of "environmental products" for the SSM/I the "precipitation over water" field shows a 25 km resolution, a range of 0–80 mm/h, an absolute accuracy of 5 mm/h for quantization levels of 0, 5, 10, 15, 20, and ≥ 25 mm/h. This algorithm utilized both the 85.5 and 37 GHz SSM/I channels, thus limiting the spatial resolution to the 25 km spot sizes of the 37 GHz channel.

A study by Spencer *et al.* (1989) employed only the two different polarizations of the 85.5 GHz channel, thus allowing the resolution to improve to the 12.5 km per spot size of this channel. This algorithm was compared with 15-min rain gauge data from a squall system in the southeast United States (Spencer *et al.*, 1989). The 0.01 inch (0.039 mm) rain gauge data were found to correlate well ($r^2 = 0.7$) with the SSM/I polarization corrected 85.5 GHz brightness temperatures. This correlation is surprisingly high considering the difference in the sampling characteristics of the SSM/I versus the rain gauge data. Portions of a rain system adjacent to the squall line were found to have little or no scattering signature in either the 85.5 or the 37 GHz SSM/I data due likely to the lack of an ice phase presence in the target area. This appears to be a limitation of the passive microwave methods to discern warm rain over land.

1.11 CHEMICAL TRACERS

Oceanographers use a variety of chemical substances to track diffusive and advective processes in the ocean. These chemical tracers can be divided into two primary categories: *conservative* tracers such as salt and helium whose concentrations are affected only by mixing and diffusion processes in the marine environment; and *nonconservative* tracers such as dissolved oxygen, silicate, iron, and manganese whose concentrations are modified by chemical and biological processes, as well as by mixing and diffusion. The *conventional* tracers, temperature, salinity, dissolved oxygen and nutrients (nitrate, phosphate and silicate), have been used since the days of Wüst (1935) and Defant (1936) to study ocean circulation. More recently, *radioactive* tracers such as radiocarbon (^{14}C) and tritium (^{3}H) are being used to study oceanic motions. The observed concentrations of those substances which enter from the atmosphere must first be corrected for natural radioactive decay and estimates made of these substance's atmospheric distribution prior to their entering the ocean. If these radioactive materials decay to a stable daughter isotope, the ratio of the radioactive element to the stable product can be used to determine the time that the tracer was last exposed to the atmosphere. *Transient tracers*, which we will consider separately, are chemicals added to the ocean by anthropogenic sources in a short time span over a limited spatial region. Most transient tracers presently in use are radioactive. What is important to the physical oceanographer is that chemical substances that enter the ocean from the atmosphere or through the seafloor provide valuable information on a wide spectrum of oceanographic processes ranging from the ventilation of the bottom water masses, to the rate of isopycnal and diapycnal (cross-isopycnal) mixing and diffusion, to the downstream evolution of effluent plumes emanating from hydrothermal vent sites.

Until recently, many of these parameters required the collection and post-cast analysis of water bottle samples using some which are then subsampled and analyzed by various types of chemical procedures. There are excellent reference books presently available that describe in detail these methods and their associated problems (e.g. Grasshoff *et al.*, 1983; Parsons *et al.*, 1984). The book by Grasshof *et al.* (1983) also contains an excellent section on water samples and their application to chemical analyses. There are important concerns for the reliability of the chemical measurements regarding contamination of the sampling bottle or the subsampling procedure. Also, the volumes required for different chemical analyses vary greatly. A list of sample volumes for chemical observations as part of the World Ocean Circulation Experiment (WOCE) can be found in Volume I of the WOCE Implementation Plan (WOCE, 1988). It is certain that the collection of these volumes will include both presently available "off-the-shelf" samplers, sampling systems newly developed by private companies and sampling units designed and built by scientists. In any case, the precision and accuracy of these measurements depends, in part, on the sampling technique used.

Modern chemical "sniffers" (or chemical pumps) are being developed that allow for *in situ* analysis of samples (Lupton *et al.*, 1993). The requirement for *in situ* chemical sampling of hydrothermal vents lead to the development of the submersible chemical analyzer (SCANNER) for analyses of Mn and Fe, the SUAVE (submersible system used to assess vented emissions) for Mn, Fe, Si, H_2S, and one of PO_4 or Cl, and the ZAPS (zero angle photon spectrophotometer) for detecting dissolved Mn to ambient seawater concentrations (≤ 1 nmol l^{-1}) (Lilley *et al.*, 1995). The SCANNER and SUAVE systems comprise online colorimetric chemical detectors while ZAPS is a fiber-optic spectrometer which combines solid-state chemistry with PMT detection to make flow-through *in situ* chemical measurements.

For many chemical measurements, no single set of procedures applies so that groups, or individual scientists, must be responsible for their own data quality. It is impossible to evaluate after-the-fact the influences of sampling technique, sample history (storage, etc.) and analysis technique. It is therefore more difficult to attach levels of accuracy to these diverse methods. In this text, we will make some general comments regarding potential problems for each of the important parameters. For a more extensive discussion of chemical tracers, the reader is referred to Broecker and Peng (1982) and Charnock *et al.* (1988).

1.11.1 Conventional tracers

1.11.1.1 Temperature and salinity

If it were not for large-scale geographical differences in heat and buoyancy fluxes through the ocean surface from the overlying atmosphere, ocean temperatures and salinity would be nearly homogeneous, disrupted only by input from geothermal heating through the seafloor (Warren, 1970; Jenkins *et al.*, 1978; Reid, 1982). In fact, below 1500 m depth the salinity range throughout the world ocean is only about 0.5 psu despite the regular deep-water formation at high latitudes (Warren, 1983). Temperature, salinity, and density distributions enable us to identify different water masses and track the movement of these water masses in the world oceans.

Atlases of temperature and salinity for the Atlantic Ocean were produced by Wüst (1935) and Defant (1936) using data from the 1925–1927 *Meteor Expedition*. These maps help define the depths of vertical mixing and upwelling in the upper ocean and reveal the extent of ventilation of deep and intermediate waters by sinking of cold, high salinity, high density water from the Southern Ocean and the Labrador Sea.

Updated atlases for the Atlantic were presented in Fuglister (1960) and Worthington (1976). Similar maps for the Pacific Ocean were produced by Reid (1965) and Barkley (1968). Reid's atlas included distributions of dissolved oxygen and inorganic phosphate/phosphorous. An atlas of water properties for the North Pacific was presented by Dodimead et al. (1963) and Favorite et al. (1976). Wyrtki (1971) provided conventional tracer data for the Indian Ocean obtained from the International Indian Ocean Expedition. An atlas of the Bering Sea is provided by Sayles et al. (1979). A summary of the global water mass distribution can be found in Emery and Meincke (1985). Surveys conducted during the World Ocean Circulation Experiment (1991–1997) will provide updated maps of conventional tracer distributions in the global ocean.

1.11.1.2 Dissolved oxygen

Along with temperature and salinity, dissolved oxygen concentration is considered one of the primary scalar properties needed to characterize the physical attributes of marine and freshwater environments. Although it is not usually a conservative quantity, dissolved oxygen serves as a valuable tracer for mixing and ventilation throughout the water column and is a key index of water quality in regions of strong biological oxygen demand (BOD). This demand may arise from animal respiration, bacteria-driven decay, or nonorganic chemical processes (the discharge of pulp-mill effluent into the marine environment places a heavy burden on oxygen levels). Dissolved oxygen is widely used by physical oceanographers to delineate water-mass distributions, to estimate the timing and intensity of coastal upwelling processes and to establish the occurrence of deep water renewal events in coastal fjords. In a study of the North Pacific, Reid and Mantyla (1978) found that dissolved oxygen gives the clearest signal of the subarctic cyclonic gyre in the deep ocean.

The apparent oxygen utilization (AOU) is the difference between the possible saturated oxygen content at a given pressure and temperature, and the actually observed oxygen content (Figure 1.11.1). Below the euphotic zone, this parameter provides an approximate measure of biological demand due to respiration and decay. It also is commonly used to trace water-mass movement and to determine the "age" (defined as the time away from exposure to the surface source) of oceanic water masses. Use of AOU suggests that the intermediate waters of the northeast Pacific have an age of several thousand years and are among the oldest (last to be ventilated) waters of the world ocean. Mantyla and Reid (1983) arrived at similar conclusions based on global distributions of potential temperature, salinity, oxygen and silicate. A more complete discussion of this parameter can found in Chapter 3 of Broecker and Peng (1982).

The "core-layer" method introduced by Wüst (1935) identified water masses, and their boundaries, on the basis of maxima or minima in temperature, salinity and dissolved oxygen content. In the ocean, dissolved oxygen levels are high near the surface where they contact the atmosphere but rapidly diminish to a minimum near 500–1000 m due to the decay of upper-ocean detritus. Oxygen values again increase with depth toward the bottom. Wyrtki (1962) discusses the relationship between the observed subsurface oxygen minimum in the North Pacific and the general circulation of the ocean, suggesting that it is to be a balance between upward advection, downward diffusion and *in situ* biological/chemical consumption. Miyake and Saruhashi (1967) argued that the effect of horizontal advection has a much greater effect on

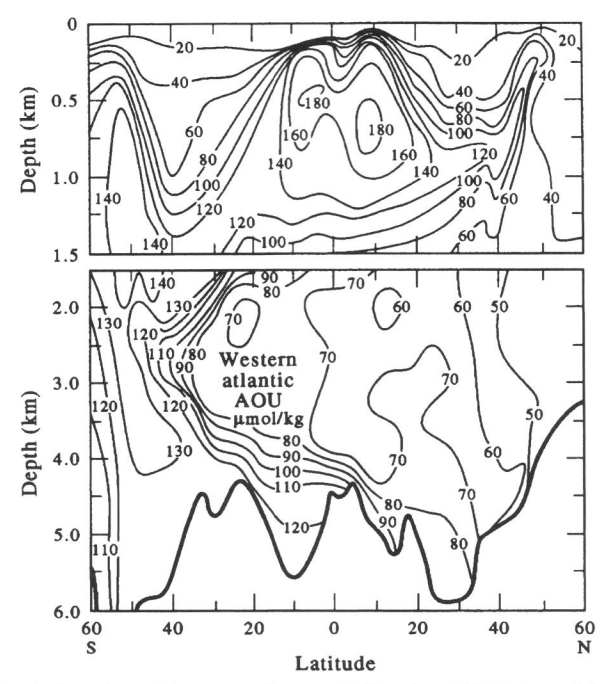

Figure 1.11.1. Vertical section of Apparent Oxygen Utilization (AOU) in mol/kg for the western basin of the Atlantic Ocean. (Figure 3.9 from GEOSECS program, Broecker and Peng, 1982.) The section is broken at 1500 m depth.

dissolved oxygen distributions than horizontal diffusion and biological consumption. In certain deep regions of the ocean, such as the Weddell Enderby Basin off Antarctica, the consumption of oxygen is below the detectable limit of the data so that oxygen may serve as a conservative chemical tracer (Edmond *et al.*, 1979).

When water bottle sampling was the only method for oceanographic profiling, the measurement of dissolved oxygen was only slightly more cumbersome and time-consuming than the measurement of temperature and salinity. The advent of the modern CTD with its rapid temperature and conductivity responses has left oxygen sampling behind. Thus, despite the importance of dissolved oxygen distributions to our understanding of chemical processes and biological consumption in the ocean, dissolved oxygen is far less widely observed than temperature or salinity. At present, there are two principal methods for measurement of dissolved oxygen: (1) water bottle sampling followed by chemical "pickling" and endpoint titration using the Winkler method (Strickland and Parsons, 1968; Hichman, 1978); and (2) electronic sampling using a membrane covered polarographic "Clark" cell (Langdon, 1984). The primary problems with standard water-bottle sampling of dissolved oxygen are the potential for sample contamination by the ambient air when the subsampling is carried out on deck, poor sampling procedure (such as inadequate rinsing of the sample bottles), and the oxidization effects caused by sunlight on the sample. Thus, laboratory procedures call for the immediate fixing of the solution after it is drawn from the water bottle by the addition of manganese chloride and alkaline iodide. During the pickling stage of the Winkler method, the dissolved oxygen in the sample oxidizes Mn(II) to Mn(III) in alkaline solution to form a precipitate MnO_2. This is followed by oxidation of added I^- by the Mn(III) in acidic solution. The resultant I_2 is titrated with thiosulfate solution using starch as an endpoint indicator. After the sample is chemically "fixed", the precipitate that forms can be allowed to settle for 10–20 min. At this stage, samples may be stored in a dark environment for up to 12 h before they need to be titrated. Parsons *et al.* (1984) give the precision of their recommended spectrophotometric method as $\pm 0.064/N$ (mg/l), where N is the number of replicate subsamples processed. The Winkler method is accurate to 1% provided the chemical analysis methods are rigorously applied. Another measure is the percentage saturation, which is the ratio of dissolved oxygen in the water to the amount of oxygen the water could hold at that temperature, salinity and pressure. Saturation curves closely follow those for dissolved oxygen.

In situ electronic dissolved oxygen sensors have been developed for use with profiling systems such as the CTD. All existing sensors use a version of the Clark cell which operates on the basis of electro-reduction of molecular oxygen at a cathode. When used in a polarographic mode, the electric current supplied by the cathode is proportional to the oxygen concentration in the surrounding fluid. To lessen the sensitivity of the device to turbulent fluctuations in the fluid, the electrode is covered with an electrolyte and membrane. Oxygen must diffuse down-gradient through the membrane into the electrolyte before it can be reduced at the surface of the cathode. There are a number of drawbacks with the present systems. First of all, the diffusion of oxygen through the boundary layer near the surface of the probe is slow, limiting the response time of the cell to several minutes. Also, the electrochemical reaction within the cell consumes oxygen and stirring may be required to maintain the correct external oxygen concentration. Changes in the structure of the cell—due to alterations in the diffusion characteristics of the membrane as a result of temperature, mechanical stress and biofouling and to deterioration of the electrolyte and surfaces— require that the cell be recalibrated every several hours. The need for frequent recalibration limits the use of the polarographic technique for profiling and mooring applications. Langdon (1984) uses a pulse technique to reduce the calibration drift. This improves long-term stability but time constants are still the order of minutes.

The YSI (Yellow Springs Instruments) and Beckman (Beckman part No. 147737) polarographic dissolved oxygen sensors (Brown and Morrison, 1978) sense the oxygen content by the current in an electrode membrane combined with a thermistor for membrane temperature correction. The current through this membrane depends on the dissolved oxygen in the water and the temperature of the membrane. Samples of both membrane current and temperature are averaged every 1.024s giving a resolution of 0.5 μA (microamps) with an accuracy of ± 2 μA over a range of 0–25 μA. These *in situ* sensors have yet to be critically evaluated with reference to well tested and approved methods. There are concerns with changes in the membrane over the period of operations and problems with calibration. Nevertheless, as measurement technology improves, an *in situ* oxygen sensor will be a high priority in that it saves considerable processing time and avoids errors possible with shipboard processing.

Fluorescence quenching is a promising technique that may make it possible to couple the modern CTD with a rapid and stable dissolved oxygen sensor. Although the use of fluorescence quenching for oxygen determination has been known since the 1930s (Kautsky, 1939) and widely used for *in vivo* measurement of the partial pressure of oxygen in blood (Peterson *et al.*, 1984), the first application in oceanography was not reported until 1988 (Thomson *et al.*, 1988). This fluorescence-based dissolved oxygen

132   *Data Analysis Methods in Physical Oceanography*

sensor operates on the principle that the fluorescence intensity of an externally light-excited fluorophore will be attenuated or "quenched" in direct relation to the concentration of dissolved oxygen in an ambient fluid (Figure 1.11.2a). Optimum results are obtained using high-intensity blue-light source (wavelength of 450–500 nm) since this is the wavelength that most readily excites the known fluorophores. Results from a six-day time-series record of dissolved oxygen concentration from a moored instrument in Saanich Inlet in 1987 suggests that the technique can be used to

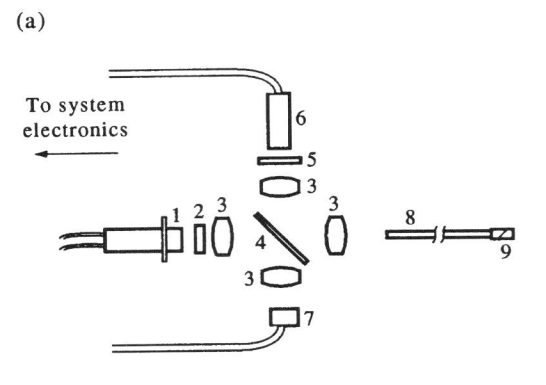

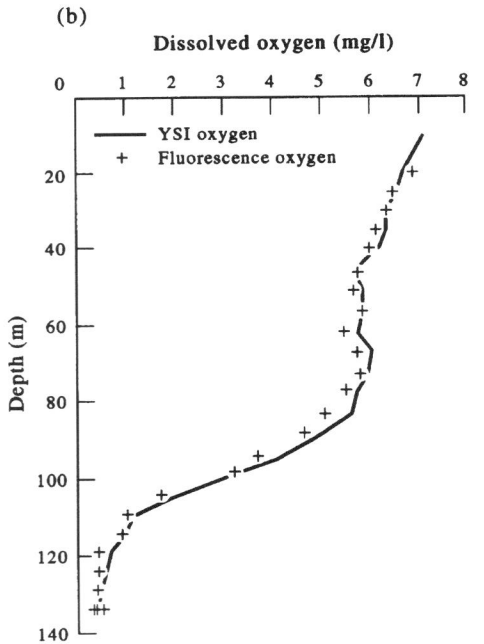

Figure 1.11.2. (a) Schematic of the first solid-state dissolved oxygen sensor. System uses blue-light from (1) to excite a fluorophore in the sensor tip (9). The concentration of dissolved oxygen in the ambient fluid sensed by (6) is proportional to the degree of quenching of blue light fluoresed by the chemical-dopped sensor. (b) Simultaneous profiles of oxygen in Saanich Inlet. YSI = YSI dissolved oxygen sensor. (From Thomson et al., 1988.)

build a rapid (<1 s) response profiling sensor (Figure 1.11.2b) with long-term stability and high (<0.1 ml/l) sensitivity. The fact that the oxygen spectra closely resemble the temperature spectra for the entire frequency band up to a period of 2 h suggests that the oxygen data are at least as stable as the thermistor on the Aanderaa RCM4 current meter that was used in the moored study. Since no blue-light source was available, the prototype device relied on a high-power white-light source and a car battery to drive the system. The present technological problem is to fabricate a blue-light source and a chemically-stable, fiber-optic, fluorescence-quenching probe capable of withstanding the rigors of shipboard operations and high hydrostatic pressures. Until now, lack of a commercial blue-light source with sufficient power (≈ 1 mW) to produce a strong fluorescence response appears to be the main impediment the development of the new dissolved oxygen sensor. The recent fabrication of blue-light light-emitting diodes (LED) and lasers will make it possible to rapidly sample dissolved oxygen as well as other dissolved gases such as carbon dioxide.

1.11.1.3 Nutrients

Nutrients such as nitrate, nitrite, phosphate and silicate are among the "old guard" of oceanic properties obtained on standard oceanographic cruises. One need only examine the early technical reports published by oceanographic institutions to appreciate the considerable effort that went into collection of these data on a routine basis. Oceanographers are again beginning to use these data on a routine basis to understand the distribution and evolution of water masses. However, there are a number of problems with nutrient collection that need to be heeded. To begin with, the data must be collected in duplicate (preferably triplicate) in small 10 mm vials and frozen immediately after the samples are drawn using a "quick freeze" device or alcohol bath. This is to prevent chemical and biological transformations of the sample while it is waiting to be processed. Careful rinsing of the nutrient vials is required as the samples are being drawn. Silicate must be collected using plastic rather than glass vials to prevent contamination by the glass silicate. Plastic caps must not be placed on too tightly and some space must be left in the vials for expansion of the fluid during freezing. Nutrient sample analysis is labour-intensive, time-consuming work. Although storage time can be extended to several weeks, we strongly recommend that nutrients be processed as soon as possible after collection, preferably on board the research ship using an autoanalyzer. With individual parameter techniques this is less likely to be possible than with more recent automated methods which have been developed to handle most nutrients (Grasshof et al., 1983). These automated systems, which use colorimetric detection for the final measurement, need to be carefully standardized and maintained. Under these conditions, they are capable of providing high quality nutrient measurements on a rapid throughput basis.

Profiles of nutrients and dissolved oxygen for the North Pacific are presented in Figure 1.11.3. As first reported by Redfield (1958), the concentrations of nitrate, phosphate, and oxygen are closely linked except near source or sink regions of the water column. A weaker relationship exists between these variables and silicate. Nitrite only occurs in significant amounts near the sea surface where it is associated with phytoplankton activity in the photic zone and in the detritus layer just below the seasonal depth of the mixed layer. Although the linear relationships between these parameters varies from region to region, the reason for the strong correlations is readily explained. Within the photic zone, phytoplankton fix nitrogen, carbon and

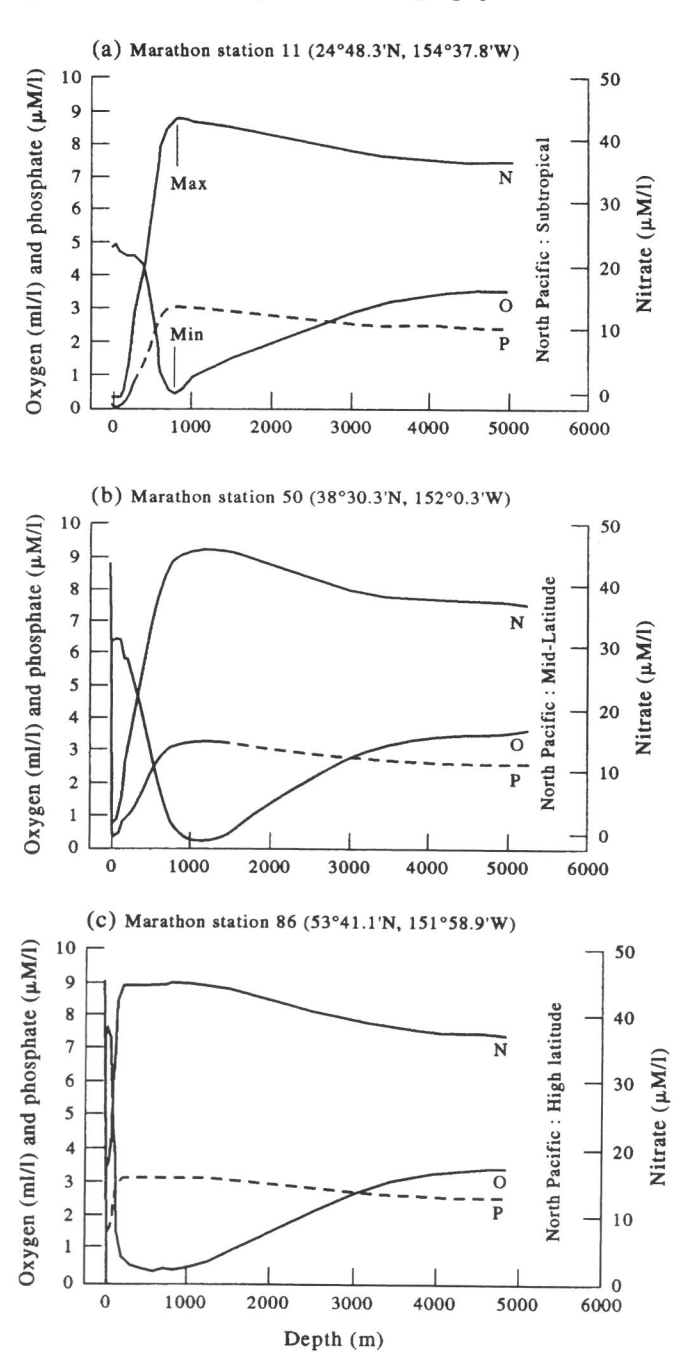

Figure 1.11.3. Plots of nitrate (N), phosphate (P) and dissolved oxygen (O) for the North Pacific. (a) Station 11 at 24°48.3'N; 154°37.8'W; (b) Station 50 at 38°30.3'N; 152°00.3'W; (c) Station 86 at 53°41.1'N; 151°58.9'W. (Data from Martin et al., 1987.)

other materials using sunlight as an energy source and chlorophyll as a catalyst. In regions of high phytoplankton activity such as mid-latitudes in summer, the upper layers of the ocean are supersaturated in oxygen and depleted in nutrients. That is, there are sources and sinks for oxygen and nutrients. However, below the photic zone, bacterial decay and dissolution of detritus raining downward from the upper ocean leads to chemical transformations of oxidized products. This, in turn, leads to a reduction of oxygen compounds and corresponding one-to-one release of nitrate, phosphate and silicate. This linear relation would prevail throughout the ocean below the photic zone if weren't for other sources and sinks for these chemicals. For example, we now know that silicate enters the ocean through resuspension of bottom sediments and from hydrothermal fluids vents from mid-ocean ridge systems (Talley and Joyce, 1992). Chemosynthetic production by bacteria in hydrothermal plumes is also a source/sink region as the analog to photosynthetic processes in the upper ocean.

It is generally thought that limitations in upper ocean nutrients, especially nitrates, combined with zooplankton predation (grazing) and turbulent mixing processes control primary (phytoplankton) productivity in the ocean. More recently, it has been proposed that other nutrients such as the aeolian supply of iron compounds might ultimately control productivity in areas such as the equatorial and subarctic Pacific and the Southern Ocean where nitrate concentrations are high year-round but spring and fall blooms do not occur (Chisholm and Morel, 1991). These high nutrient, low chlorophyll (HNLP) regions have become the focus of increasing numbers of multi-disciplinary studies.

In a classic paper, Redfield (1958) suggested that organisms both respond to and modify their external environments. His premise was that the nitrate of the ocean and the oxygen of the atmosphere are determined by the biochemical cycle and not conditions imposed on the organisms through factors beyond their control. Support for his thesis was derived from the fact that the well-defined nitrogen, phosphorous, carbon, and oxygen compositions of plankton in the upper ocean were almost identical to the concentrations of these elements regenerated from chemical processes in the 95% of the ocean that lies below the autotrophic zone. As pointed out by Redfield, the synthesis of organic material by phytoplankton leads to oceanic changes in concentration of phosphorous, nitrogen, and carbon in the ratio 1:15:106. During heterotrophic oxidation and remineralization of this biogenic material (i.e. decomposition of these organisms), the observed ratios are 1:15:105. Thus, for every phosphorous atom that is used by phytoplankton during photosynthesis in the euphotic zone, exactly 15 nitrogen atoms and 106 carbon atoms are used up. Alternatively, for every phosphorous atom that is liberated during decomposition in the deep ocean, exactly 15 nitrogen and 105 carbon atoms are liberated. The oxidation of these atoms during photosynthesis requires about 276 oxygen atoms while during decomposition 235 oxygen atoms are withdrawn from the water column for each atom of phosphorous that is added. If this process were simply one way, the primary nutrients would soon be completely depleted from the upper ocean. That is why life supporting replenishment of depleted nutrients to the upper ocean through upwelling and vertical diffusion of deeper nutrient rich waters is such an important process to the planet. [Bruland *et al.* (1991) give a modern version of the Redfield ratios based on phytoplankton collected under bloom conditions as: C:N:P:Fe:Zn:Cu,Mn,Ni,Cd = 106:16:1:0.005:0.002:0.0004 (see also Martin and Knauer, 1973).]

According to the above ratios, the formation of organic matter by phytoplankton in the surface autotrophic zone leads to the withdrawal of carbonate, nitrate, and phosphate from the water column. Oxygen is released as part of photosynthesis and the upper few meters of the ocean can be supersaturated in oxygen at highly productive times of the year. When the plants die and sink into the deeper ocean, decomposition by oxidation returns these compounds back to the seawater. Thus, increases in carbonate, nitrate, and phosphate concentrations below the euphotic zone are accompanied by a corresponding decrease in oxygen levels. This process leads to a rapid increase in nitrate and phosphate and a corresponding rapid decrease in oxygen within the upper kilometer or so of the ocean (Figure 1.11.3). Nitrate and phosphate reach subsurface maximums at intermediate depths and then begin to decrease slowly with depth to the seafloor. Oxygen, on the other hand, falls to a mid-depth minimum (the oxygen minimum layer) before starting to increase slowly with depth toward the seafloor. In the upper zone, the balance of chemicals is altered considerably by biological activity while near the coast, the balance is altered by runoff which provides a different ratio of nutrients. However, below the surface layer, the changes occur in the manner suggested by the Redfield ratios (Redfield *et al.*, 1963). Note that the concentration of silicate is almost like that of the other nutrients, except that it doesn't reach a maximum at mid depth and becomes more decoupled from the accompanying oxygen curve. This suggests a source function for silicate in the deep ocean. Indeed, there are two sources; resuspension and dissolution of siliceous material from rocks and other inorganic material on the seafloor and the injection of silicates into the ocean from hydrothermal venting along mid-ocean ridges and other magmatic source regions in the deep ocean.

The fact that carbon and oxygen concentrations greatly exceed the levels required by plankton while those of phosphorous and nitrogen were identical to those observed on average in the ocean (carbon is at least 10 times that needed for photosynthesis), prompted Redfield to suggest that phosphate and nitrate are limiting factors to oceanic primary productivity. It is thought that nitrate (NO_3) is the primary limiting factor although phosphorous limitation is still important in certain coastal areas. Airborne iron is also thought to be a limiting nutrient for primary productivity in the open ocean. Evidence for this is based on the year-round absence of phytoplankton blooms in the subarctic Pacific, equatorial Pacific and Southern Ocean despite the high near-surface concentrations of nitrate and phosphate. In these areas, autotrophic processes fail to exploit NO_3 and PO_4. The idea is that iron, or some other mineral, limits growth, which is not the case in areas served by aeolion transport from the land. Unfortunately, it not yet possible to sort out the effects of iron limitations from grazing by herbivorous zooplankton or from physical mixing in the surface layer which prevents stratification from confining the animals to a thin upper layer. A recent experiment conducted over an 8 km square area of the equatorial Pacific 500 km south of the Galápagos Islands showed that iron enrichment can dramatically increase surface productivity. Using sulfur hexafluoride to track the 480 kg of iron sulfate solution added to the ocean, scientists found that the rate of growth and total mass of phytoplankton doubled over a period of three days. However, the iron soon precipitated out of solution as ultra-fine particles and sank, causing a sharp decrease in productivity levels. The question of iron enrichment and ocean productivity remains unresolved.

1.11.1.4 Silicate

The oceanic distribution of many elements is determined by their involvement in the biochemical cycle. Nitrate and phosphate are associated with the labile tissue and protoplasm of surface plankton whereas silicate and alkalinity are linked to the refractory hard parts of the organisms. The term "silicate" applies to dissolved reactive silicate [monosilicic acid, $Si(OH)_4$; Iler, 1979] measured from water samples. Since most of the silicate undergoes dissolution in the water column rather than the seafloor (Edmond et al., 1979), its distribution serves as tracer for water-mass mixing and advection. The advantage of silica over carbonates or other compounds is that siliceous sediments are found only in well-defined areas associated with surface upwelling and their distribution is not particularly dependent on depth. According to Edmond et al. (1979) the average flux of dissolved silica from the sediments to the deep ocean is about 3 $\mu mol/cm^2/year$ which is sufficient to make it a useful tracer of deep sea flow. Large fluxes are observed in the Weddell-Enderby Basin off Antarctica and in the northern Indian Ocean.

In the extreme northeast Pacific (northeast of 45°N, 160°W), silicate concentration increases with depth to the bottom while in the equatorial Pacific no anomalies are observed despite the presence of opaline deposits. The increased silicate with depth in the northeast Pacific appears to be associated, in part, with westward advection of dissolved siliceous sediments deposited over the continental margin of the wind-induced upwelling domain that extends from British Columbia to Baja California along the west coast of North America. As noted by Edmond et al. (1979), the existence of silica sources at the seafloor makes it impossible to use global correlations with the extensive silicate distribution data to determine the distribution of other variables such as trace metals. Historical data, together with transect data collected along 47°N (Talley et al., 1988), further suggest that both the intermediate silica maximum in the depth range 2000–2400 m and the near-bottom silica maximum in Cascadia Basin to the east of the Juan de Fuca Ridge (Figure 1.11.4) may be due, in part, to hydrothermal venting of high silicate waters (Talley and Joyce, 1992). The silica in the hydrothermal plumes emanating from the vents originates as silicates stripped from the crustal rocks by the high-temperature hydrothermal fluids. Other factors include vertical flux divergence of settling silicate particles, dissolution from opaline bottom sediments, and up-slope injection from the bottom boundary layer.

Macdonald et al. (1986) point out that improper thawing of frozen silicate samples can result in a significant and variable negative bias in seawater determination of silicate. The problem arises from conversion of reactive silicate to a nonreative, polymetric form in the frozen sample. This polymerization need not affect accuracy for frozen samples provided that sufficient thawing time is allowed for depolymerization to the reactive form. To control bias, the analyst must adjust the length of time between thawing and analysis. The appropriate "waiting time" varies according to the salinity of the sample, the silicate concentration and the length of time the sample was frozen. Waiting time increases with the time that the samples were frozen and with silicate/salinity ratio. For example, deep silicate samples collected from the northeast Pacific (salinity $\approx$ 35 psu; silicate $\approx$ 180 $\mu mol/l$) and stored for one to two months must be thawed for about 8 h before processing. This increases to 24 h for samples stored for more than five to six months. Macdonald et al. conclude that "If the objectives of sampling can accept a 5% negative bias and a slight loss of precision,

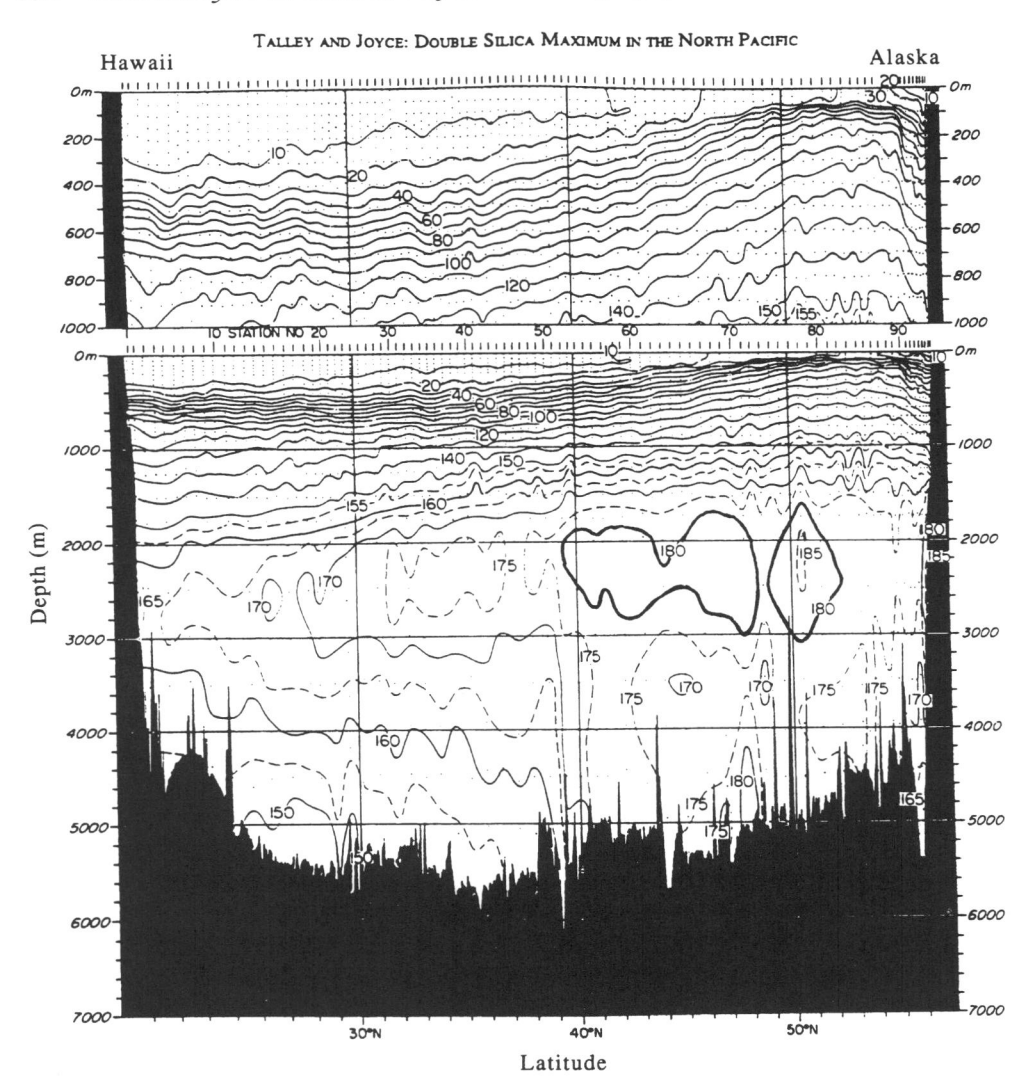

Figure 1.11.4. The meridional distribution of silica (micromoles per liter) in the North Pacific along approximately 152°W (Hawaiian region to Kodiak Island, Alaska). Mid-depth maximum values in excess of 180 μmol/l are emphasized. (From Talley et al., 1991; Talley and Joyce, 1992.)

then freezing is a simple method for storing a wide range of samples. However, samples should be analyzed as soon as practicable".

1.11.2 Light attenuation and scattering

The light energy in a fluid is attenuated by the combined effects of absorption and scattering. In the ocean, absorption involves a conversion of light into other forms of energy such as heat; scattering involves the redirection of light by water molecules, dissolved solids and suspended material without the loss of total energy. Transmissometers are optical instruments that measure the clarity of water by measuring the fraction of light energy lost from a collimated light beam as it passes along a known pathlength (Figure 1.11.5). Attenuation results from the combined effects of absorption and shallow-angle Rayleigh (forward) scattering of the light beam by impurities and fine particles in the water. Water that is completely free of impurities is optically pure. Nephelometers (or turbidity meters) measure scattered light and respond primarily to the first-order effects of particle concentrations and size. Depending on manufacturer, commercially available nephelometers examine scattered light in the range from 90° to 165° to the axis of the light beam. Most instruments use infrared light with a wavelength of 660 nm. Because light at this wavelength is rapidly absorbed in water (63% attenuation every 5 cm), there is little contamination of the source beam due to sunlight except within the top meter or so of the water column.

The intensity $I(r)$ of a light beam of wavelength λ traveling a distance r through a fluid suspension attenuates as

$$I(r) = I_o \exp(-cr) \qquad (1.11.1)$$

where I_o is the initial intensity at $r = 0$ and $c = c(\lambda)$ is the rate of attenuation per unit distance. Attenuation of the light source occurs through removal or redirection of light beam energy by scattering and absorption. In the ocean, visible long-wave radiation (red) is absorbed more than visible short-wave radiation (blue and green) and what energy is left at long wavelengths undergoes less scattering than at short wavelengths. As a consequence, the ocean appears blue to blue–green when viewed from above. The exact color response depends on the scattering and absorption characteristics of the materials in the water including the dissolved versus the suspended phase—factors that are used to advantage in remote sensing techniques. For a fixed monochromatic light source, the clarity of the water, measured relative to distilled uncontaminated water, provides a quantitative estimate of the mass or volume concentration of suspended particles. Such material can originate from a variety of sources including terrigenous sediment carried into the coastal ocean by runoff, from current-induced resuspension of material in the benthic layer, or from detectable concentrations of plankton.

The "Secchi disk" is one of the simplest and earliest methods for measuring light attenuation in the upper layer of the ocean. A typical Secchi disk consists of a flat, 30-

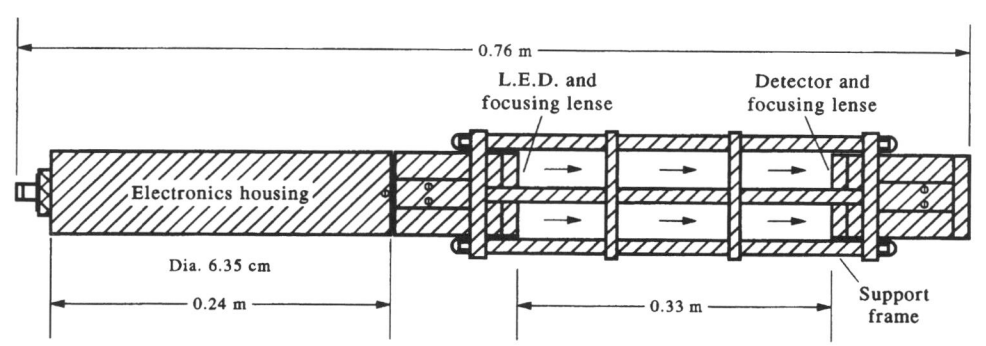

Figure 1.11.5. *Exploded view of a Sea-Tech transmissometer. Red light of wavelength 690 nm passes from the light-emitting diode (LED) to the sensor over a fixed path length of 0.25 m.*

cm diameter white plate that is lowered on a marked line (suspended from the disk center) over the side of the ship. The depth at which the disk can no longer be seen from the ship is a measure of the amount of surface light that reaches a given depth and can be used to obtain a single integrated estimate of the extinction coefficient, $c(\lambda)$. The disk is still in use today. For example, Dodson (1990) used Secchi disk data from a series of lakes in Europe and the U.S.A. that suggest a direct relationship between the depth of day–night (diel) migration of zooplankton and the amount of light penetrating the epihelion. In this case, the zooplankton minimize mortality from visually feeding fish and maximize grazing rate. Despite its simplicity, there are a number of problems with this technique, notwithstanding the fact that it fails to give a measure of the water clarity as a function of depth and is limited to near-surface waters. In addition, the visibility of the disk will depend on the amount of light at the ocean surface (and type of light through cloud cover), on the roughness of the ocean surface, and the eyesight of the observer. Today, oceanographers rely on transmissometers and nephelometers to determine the clarity of the water as a function of depth.

A typical transmissometer consists of a constant intensity, single frequency light source and receiving lens separated by a fixed pathlength, r_o. The attenuation coefficient in units of m^{-1} is then found from the relation

$$c = -(1/r_o) \ln (I/I_o) \qquad (1.11.2)$$

in which r_o is measured in meters, and I/I_o is the ratio of the light intensity at the receiver versus that transmitted by the red (660 nm) LED. This choice of light wavelength is useful because it eliminates attenuation from dissolved organic substances consisting mainly of humic acids or "yellow matter" (also called "gelbstoff"; Jerlov, 1976). The Sea Tech transmissometer (Bartz et al., 1978) has an accuracy of ±0.5% and a small (<1.03° or 0.018 radians) receiver acceptance angle that minimizes the complication of the collector receiving specious forward-scattering light. To obtain absolute values, the source and lens must be calibrated in distilled water and air since scatter can effect the results. As an example, a 0.25 m pathlength transmissometer which has a calibration value of $I_o = 94.6\%$ in clean water and reading of 89.1% in the ocean corresponds to a light attenuation coefficient

$$c = -4 \ln (I/I_o) = -4 \ln (0.891/0.946) = 0.240 \, \text{m}^{-1} \qquad (1.11.3)$$

Values of c in the ocean range from around 0.15 m^{-1} for relatively clear offshore water for concentrations of particles as low as 100 μg/l to around 21 m^{-1} for turbid coastal water with particle concentrations of 140 mg/l (Sea Tech user's manual). In studies of hydrothermal venting, measurement of water clarity is often one of the best methods to determine the location and intensity of the plume (Baker and Massoth, 1987; Thomson et al., 1992; Figure 1.11.6).

Problems with the transmissometer technique are: (1) drift in the intensity of the light source with time; (2) clouding of the lens by organic and inorganic material which affect the *in situ* calibration of the instrument; and (3) scattering, rather than absorption, of the light. If we ignore the influence of dissolved substances, the attenuation coefficient, c, depends on the concentration of the suspended material but also on the size, shape, and index of refraction of the material (Baker and Lavelle, 1984). Thus, a linear relationship between c and particle concentration C such that

Data Acquisition and Recording 141

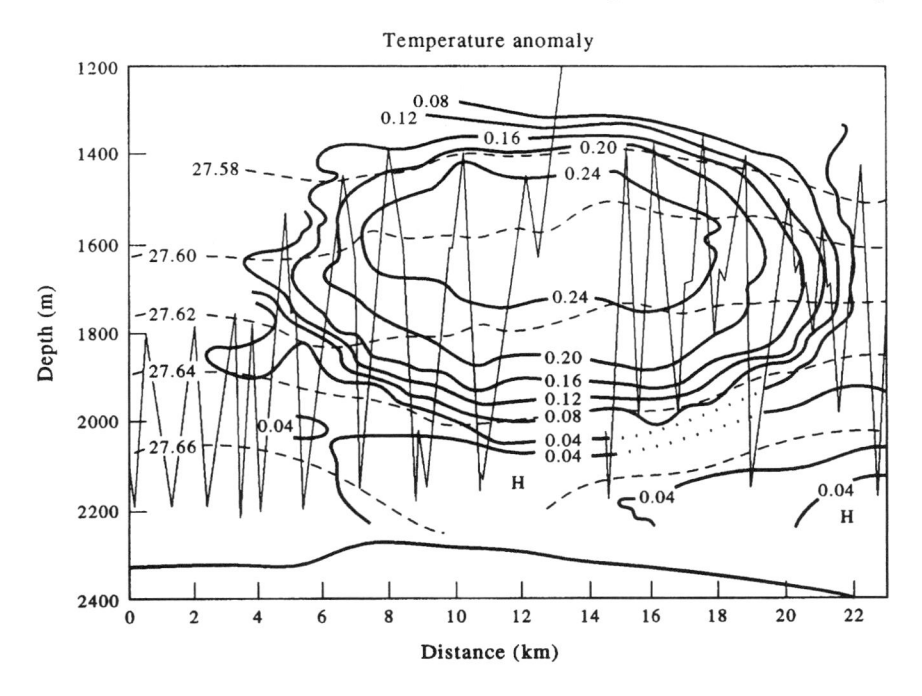

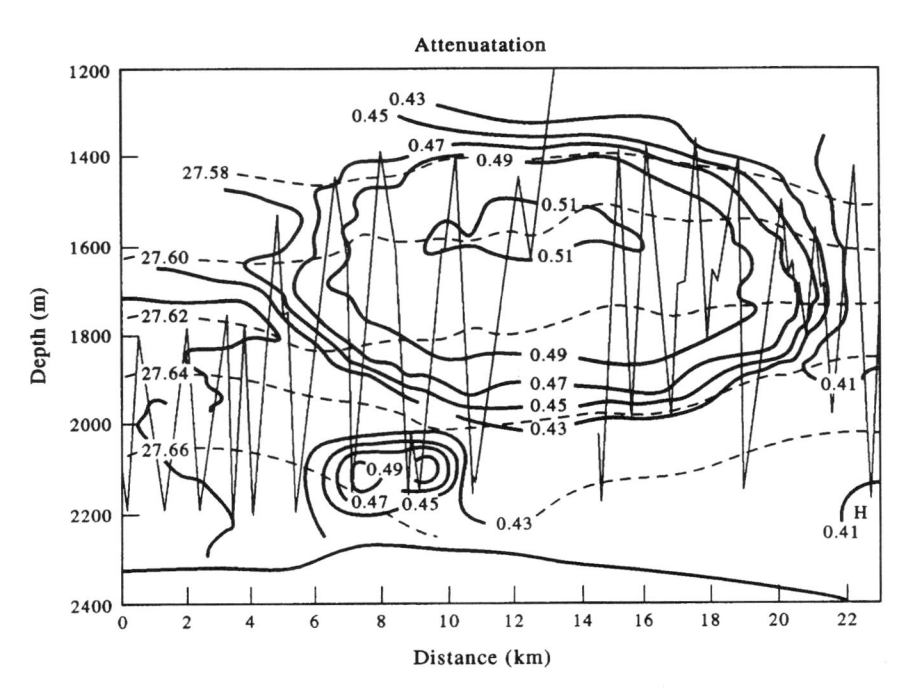

Figure 1.11.6. Cross-sections of temperature anomaly (°C) and light attenuation coefficient (m^{-1}) for the "megaplume" observed near the hydrothermal main site on the Cleft Segment of Juan de Fuca Ridge in the northeast Pacific in September 1986. Temperature anomaly gives temperature over the plume depth relative to the observed background temperature. Dotted line shows σ_θ surfaces and solid line the sawtooth track of the towed CTD path. (From Baker et al., 1989.)

142 *Data Analysis Methods in Physical Oceanography*

$$c = \alpha C + \alpha_o \qquad (1.11.4)$$

only occurs when the effects of size, shape and index of refraction are negligible or mutually compensating. After concentration, particle size is the next most important variable effecting clarity. Accurate estimates of concentration therefore require calibration in terms of the distribution of particle sizes and shapes in suspension as for example in Baker and Lavelle (1984). Laboratory results demonstrate that calibrations of beam transmissometer data in terms of particle mass or volume concentration are acutely sensitive to the size distribution of the particle population under study. There is also a trend of decreasing calibration slopes from environments where large particles are rare (deep ocean) to those where they are common (shallow estuaries and coastal waters). Theoretical attenuation curves agree more with observation when the natural particles are treated as disks rather than as spheres as in Mie scattering theory. The need for field calibration is stressed.

The results of Baker and Lavelle (1984) can be summarized as follows: (1) calibration of beam transmissometers is acutely sensitive to the size distribution of the particle population under study; (2) theoretical calculations based on Mie scattering theory and size distributions measured by a Coulter counter-agree when attenuation for glass spheres is observed but underestimate the attenuation of natural particles when these particles are assigned an effective optical diameter equal to their equivalent spherical diameter deduced from particle volume measurements; (3) treating particles as disks expands their effective optical diameter and increases the theoretical attenuation slope close to the observed values; (4) there is a need to collect samples along with the transmissometer measurements, especially where the particle environment is nonhomogeneous.

Transmissometers are best used for measuring the optical clarity of relatively clear water whereas nephelometers are most suitable for measuring suspended particles in highly turbid waters. In murky waters, nephelometers have superior linearity over transmissometers while transmissometers are more sensitive at low concentrations. "Turbidity" or cloudiness of the water is a relative, not an absolute term. It is an apparent optical property depending on characteristics of the scattering particles, external lighting conditions and the instrument used. Turbidity is measured in nephelometer units (NTUs) referenced to a turbidity standard or in Formazin Turbidity Units (FTUs) derived from diluted concentrations of 4000-FTU formazin, a murky white suspension that can be purchased commercially. Since turbidity is a relative measure, manufacturers recommend that calibration involve the use of suspended matter from the waters to be monitored. This is not an easy task if one is working in a deep or highly variable regime.

1.11.3 Oxygen isotope: $\delta^{18}O$

The ratio of oxygen isotope 18 to oxygen isotope 16 in water is fractionated by differences in weight. The lighter element ^{16}O is more easily evaporated than ^{18}O and is therefore a measure of temperature; the higher the temperature the greater the $H_2^{18}O/H_2^{16}O$ ratio. In contrast to the variability in the surface ocean, average $H_2^{18}O/H_2^{16}O$ ratios for the deep ocean (>500 m depth) vary by less than 1%. This ratio (in percent) is expressed in conventional delta "δ" notation as

$$\delta^{18}O(\%) = (R_{std}/R_{sample} - 1) \times 10^{10}$$

where $R = H_2^{18}O/H_2^{16}O$ is the ratio of the two main isotopes of oxygen and the subscript "std" refers to Standard Mean Ocean Water (SMOW). The low variability in $\delta^{18}O$ values in waters in the deep sea has led to widespread use of oxygen isotopes as a paleothermometric indicator. These methods assume relatively little variation (about 1%) in the $\delta^{18}O$ values of deep-ocean water over geological time. The $\delta^{18}O$ values of carbonate, silica, and phosphate precipitated by both living and fossil marine organisms, such as foramininferans, radiolarians, coccolithophorids, diatoms, and barnacles, have been used to estimate temperatures of the water in which the organism lived based on temperature-dependent equilibria between the oxygen in the water and the biomineralized phase of interest. The $\delta^{18}O$ values vary in space and time in different regions of the ocean. For example, shallow continental shelves are influenced by freshwater input, particularly at high latitudes. Thus, oxygen removed from seawater by organisms should reflect oceanic conditions at the time. Salinity and ^{18}O content are related in most ocean waters with similar processes influencing both in tandem.

According to Kipphut (1990), the $H_2^{18}O/H_2^{16}O$ ratio in seawater in the Gulf of Alaska shows only slight variation except near those coastal margins where there is significant input of freshwater from melting of large glaciers ($\delta^{18}O \approx -23\%$) and runoff from coastal precipitation ($\delta^{18}O \approx -10\%$). Precipitation is generally depleted in the heavier isotopes of oxygen because of isotopic fractionation processes which occur during evaporation and condensation. Since the fractionation processes are temperature dependent, precipitation at higher latitudes and elevations shows progressively lower $H_2^{18}O/H_2^{16}O$ ratios. The ratio is a conservative property of water and when combined with salinity may be useful in determining distinct components of water masses. The isotope data south of Alaska suggest that the coastal waters in southwestern Alaska are derived from a combination of glacier melt and runoff from as far east as south-central Alaska. If we add the freshwater added by runoff from the large rivers of northern British Columbia, the Alaska Coastal Current (Royer, 1981; Schumacher and Reed, 1986) is continuous feature flowing more than 1500 km from the southern Alaska Panhandle to Unimak Pass at the beginning of the Aleutian Island chain.

1.11.4 Helium-3; helium/heat ratio

Helium-3 (^{3}He) is an inert and stable isotope of helium whose residence time of about 4000 years in the ocean makes it a useful tracer for oceanic mixing times and deepsea circulation. There are two main sources in the ocean. In the upper mixed layer and thermocline, ^{3}He is produced by the β-decay of anthropogenic tritium; in the deep ocean, ^{3}He originates with mantle degassing of primordial helium from mid-ocean ridge hydrothermal vents. Anderson (1993) also argues that ^{3}He and neon from hotspot magmas and gases may reflect an extraterrestrial origin; specifically, subduction of ancient pelagic sediments rich in solar ^{3}He and neon originate with interplanetary dust particles now being recycled at oceanic hotspots. [For counterarguments see Hiyagon (1994) and Craig (1994)]. The distinct isotopic ratio of mantle helium (^{3}He/^{4}He = 10^{-5}) versus a ratio of 10^{-6} for atmospheric helium makes ^{3}He/^{4}He a useful tracer in the ocean. In a classic paper, Lupton and Craig (1981) showed that the ^{3}He/^{4}He ratio in the 2500 m deep core of the hydrothermal plume emanating from the East Pacific Rise at 15°S in the Pacific Ocean was 50 higher than the ratio of atmospheric helium. The helium plume could be traced more than 2000 km westward from the venting region on the crest of the mid-ocean ridge (Figure 1.11.7). To quote the authors, "In magnitude, scale, and striking asymmetry, this plume is one of the most remarkable features of the deep ocean, resembling a volcanic cloud injected into a steady east wind". Helium-3 is now used extensively as tracer for hydrothermal plumes in active spreading regions such as the Juan de Fuca Ridge in the northeast Pacific and the East Pacific Rise in the South Pacific.

Data collected during GEOSECS indicates that the deep Pacific is the oceanic region most enriched in ^{3}He with a mean ratio concentration δ^3He value of 17% compared with 10% in the Indian Ocean, 7% in the Southern Ocean and 2% in the Atlantic (Jamous et al., 1992). The core of the plume at the East Pacific Rise has a value of 50%. [Here, δ^3He(%) = $(R/R_a - 1) \times 100$, where $R = {}^3$He$/^2$He is the isotopic ratio of the sample and R_a is the atmospheric ratio.] The differences in concentration relate directly to the differences in hydrothermal input and inversely to the degree of deep-water ventilation. For example, there is a considerably greater hydrothermal activity in the Pacific than in the Atlantic while the Atlantic deep water is highly ventilated compared with the Pacific. Similarly, the values of δ^3He ($\approx$ 28) at the bottom of the Black Sea reflect the presence of a strong source at the seafloor. In contrast, the strong correlation between dissolved oxygen concentration and ^{3}He in the Southern Ocean (Figure 1.11.8) indicates that the distributions of these tracers in this region of the world ocean are mainly determined by ventilation processes.

Early vent-fluid samples taken from hydrothermal systems on the Galapagos Rift and at 21°N on the East Pacific Rise were found to have nearly equal ratios of ^{3}He to heat despite the considerable geographical separation of the sites and widely different fluid exit temperatures ($\approx$20 and 350°C, respectively). Here, "heat" is the excess

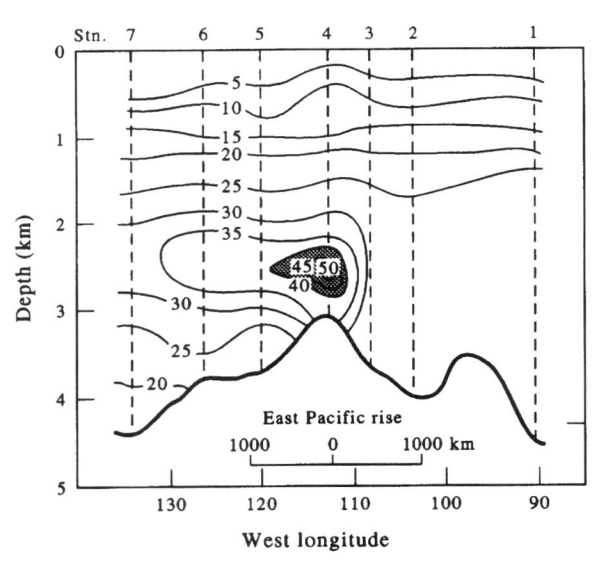

Figure 1.11.7. Cross-section of $\delta(^3He)$ over the East Pacific Rise at 15°S. The level of neutral plume bouyancy, as determined by the core depth of the ^3He plume, is about 400 m above the ridge crest. The ratio is defined as $\delta(^3He) = (R/R_{ATM} - 1) \times 100$ where $R = {}^3He/^4He$ and $R_{ATM} = 1.40 \times 10^{-6}$. (From Lupton and Craig, 1981.)

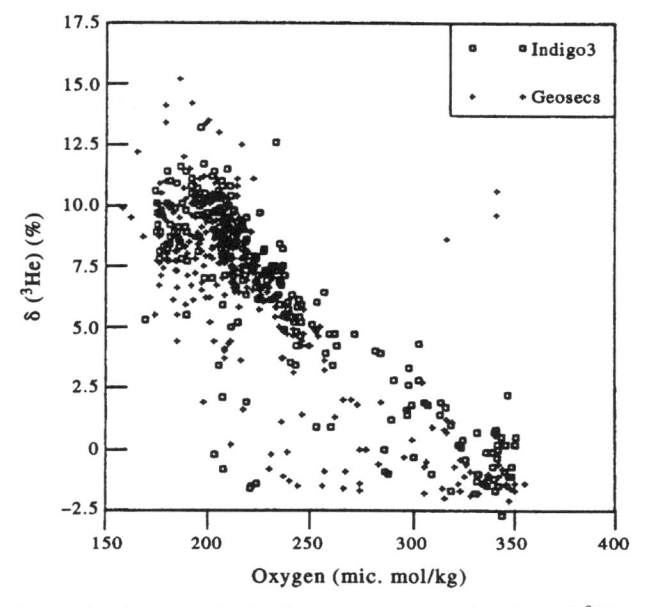

Figure 1.11.8. Correlation between dissolved oxygen concentration O_2 and ^3He in the Southern Ocean indicates that the distributions of these tracers in the region of the world ocean are mainly determined by ventilation processes. Combination of GEOSECS and INDIGO-3 data. (From Jamous et al., 1992.)

amount of heat (in calories or joules) added to the ambient water by geothermal processes. By combining independent estimates of the mantle flux of ^3He within the ocean with the observed ratio ^3He/Heat $\approx 0.5 \times 10^{-12}$ cm^3 STP cal^{-1}, Jenkins et al. (1978) calculated a global oceanic hydrothermal heat flux of 4.9×10^{19} cal/year. An examination of the ^3He/Heat ratios in the 20-km wide megaplume observed in August 1986 on Juan de Fuca Ridge (Lupton et al., 1989) has shown that the ratios can vary by as much as an order of magnitude and that heat fluxes based on ^3He measurements must be taken with caution. Specifically, the ratio ^3He/Heat was found to vary with height within the megaplume formed during the hydrothermal event. The megaplume had lower helium values and five times the temperature anomaly as the near-bottom chronic venting regime. Since helium is extracted from the magma by the circulating fluids in the hydrothermal system, the relatively low ratios of ^3He/Heat in the megaplume presumably resulted from relatively high water-to-rock ratios and the youth of the hydrothermal fluid prior to its injection into the overlying ocean. Lupton et al. (1989) suggest that a value of $\approx 2 \times 10^{-12}$ cm^3 STP He cal^{-1} may be a reasonable estimate for the average ^3He/Heat signature of fluids vented into the oceans by mid-ocean ridge hydrothermal systems.

1.12 TRANSIENT CHEMICAL TRACERS

"Transient tracers" are anthropogenic compounds that are injected into the ocean over spatially limited regions within well-defined periods of time. The time "window" makes these compounds especially well suited to studies of upper-ocean mixing and deep-sea ventilation. Transient tracers are commonly used to constrain solutions of global "box" models used to investigate climate-scale carbon dioxide fluxes within coupled atmosphere–ocean systems (Broecker and Peng, 1982; Sarmiento *et al.*, 1988), and in generalized inverse models incorporating both data and ocean dynamics to determine oceanic flow structure (see Bennett, 1992). The timed release into the ocean may take place over a few hours, as in the case of rhodamine dye, or last longer than a century, as in the case of chlorofluorocarbons (CFCs). Injection of certain tracers, such as radiocarbon (^{14}C) greatly augments the natural distributions of these chemicals while for others, such as CFCs and tritium (^3H), the tracer is superimposed on an almost nonexistent background concentration. Because of the slow advection and mixing processes in the ocean, as well as the extensive research needed to measure the tracer distributions, most tracers are used in the study of seasonal to decadal scale oceanic variability. For all transient tracers, studies are limited by imperfect knowledge of the surface boundary conditions during water-mass formation. This is especially true of tracers entering from the atmosphere. Tritium and radiocarbon are radioactive isotopes whose observed concentrations must first be corrected for natural radioactive decay. Both tracers have widespread use in descriptive studies and large-scale numerical modeling of ventilation of the deep ocean and the transformation of water masses over periods of decades. Our main purpose in this section is to provide a brief outline of the types of studies possible with transient tracers. Only results for the main tracers will be presented; secondary tracers such as krypton-85 and argon-39 are not discussed.

1.12.1 Tritium

During the late 1950s and early 1960s, large amounts of bomb-produced radiocarbon (^{14}C), strontium (^{90}Sr) and tritium (^3H) were released into the stratosphere during above-ground testing of thermonuclear weapons (Figure 1.12.1a, b). Of these, "bomb" tritium (the heaviest isotope of hydrogen) has an extensive database and is measurable to high precision and sensitivity. Tritium is incorporated directly in water molecules as HTO so that it is a true water-mass tracer. Most of the tritium was produced by tests conducted in the northern hemisphere and was eventually deposited onto the earth's surface north of 15°N (Weiss and Roether, 1980; Broecker *et al.*, 1986). Deposition into the oceans is through vapor diffusion and rainfall at a ratio of roughly 2:1 according to observational data. A study by Lipps and Hemler (1992) suggests that the ratio varies according to the type of rainfall. The large fronts across the Pacific and Atlantic oceans at subtropical latitudes impede lateral mixing and the southward transport of tritium. As a result, tritium with a half-life of 12.43 years serves as a useful tracer for water motions on time-scales of decades. It is most useful when combined with measurements of its stable, inert daughter product ^3He. This combination helps determine the age of tritium entering the ocean and provides additional information on the distribution of tritium in the atmosphere before it entered the ocean (Jenkins, 1988). Most large-scale studies are based on the extensive tritium data collected in the North Pacific during the Geochemical Ocean Sections Study (GEOSECS: 1972–1974) and Long Lines (1983–1985). Roughly 0.3 litres of seawater are required for the measurement of tritium by beta-decay counting.

Tritium in natural waters is expressed in "tritium units" (TU), which is the abundance ratio $^3H/^1H \times 10^{18}$. The ratio abundance corresponds to 7.09 disintegrations per min per kg of water. To remove the effect of normal radioactive decay from a data series, the tritium concentrations are corrected to a common

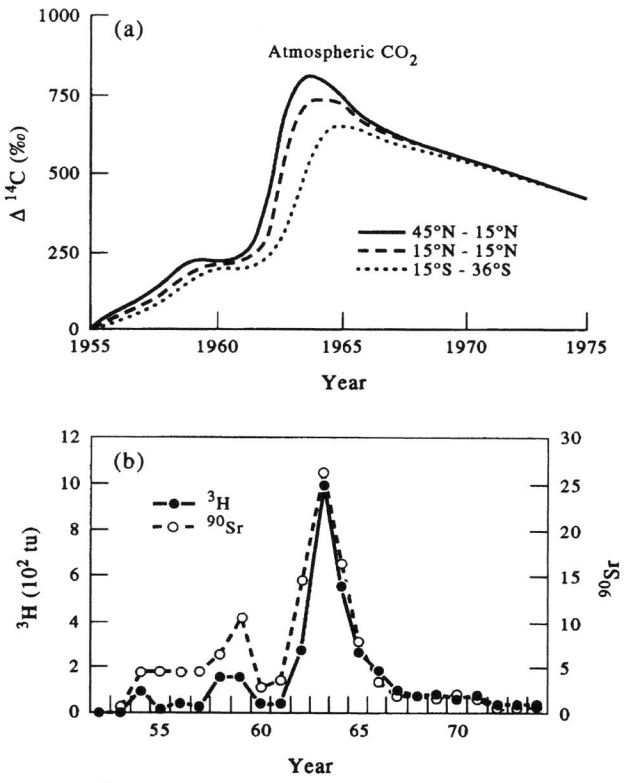

Figure 1.12.1. Time-series of bomb-produced elements released into the stratosphere during above-ground testing of thermonuclear weapons during the late 1950s and early 1960s: (a). radiocarbon (^{14}C) and (b). strontium (^{90}Sr) from measurements of atmospheric carbon dioxide and tritium (^{3}H) based on rain at Valencia Ireland. (Adapted from Quay et al., 1983; Broecker and Peng, 1982.)

reference of January 1, 1981. Thus, TU81N is the ratio of $^{3}H/^{1}H$ a sample would have as of 1981/01/01. The measurement error for "decay-corrected" data is 0.05TU or 3.5%, whichever is greater (Van Scoy et al., 1991). Water having values less than 0.2TU81 are considered to reflect cosmogenic background levels or arise from dilution by mixing of bomb tritium. The fact that decay-corrected tritium is a conservative quantity that was added to a selected area of the world ocean in a relatively short period of time (Figure 1.12.2) makes it attractive as an oceanic tracer. Changes in the spatial distribution of tritium with time provide a measure of horizontal advection while depth penetration on isopycnals that do not outcrop to the atmosphere are indicative of cross-isopycnal (diapycnal) mixing. Fine (1985) uses upper ocean tritium data from the GEOSECS program to show that there is a net transport of 5×10^{6} m^{3}/s in the upper 300 m from the Pacific to the Indian Ocean through the Indonesian Archipelago. This contrasts with values of 1.7×10^{6} m^{3}/s obtained using hydrographic data (Wyrtki, 1961) and 5–14 $\times 10^{6}$ m^{3}/s from salt and mass balances (Godfrey and Golding, 1981; Piola and Gordon, 1984; Gordon, 1986). Gargett et al. (1986) have examined the nine-year record of tritium from Ocean Station P (50°N, 145°W) in the northeast Pacific. Results suggest that the observed vertical distribution of tritium in this region is determined mainly through advection along isopycnals rather than by

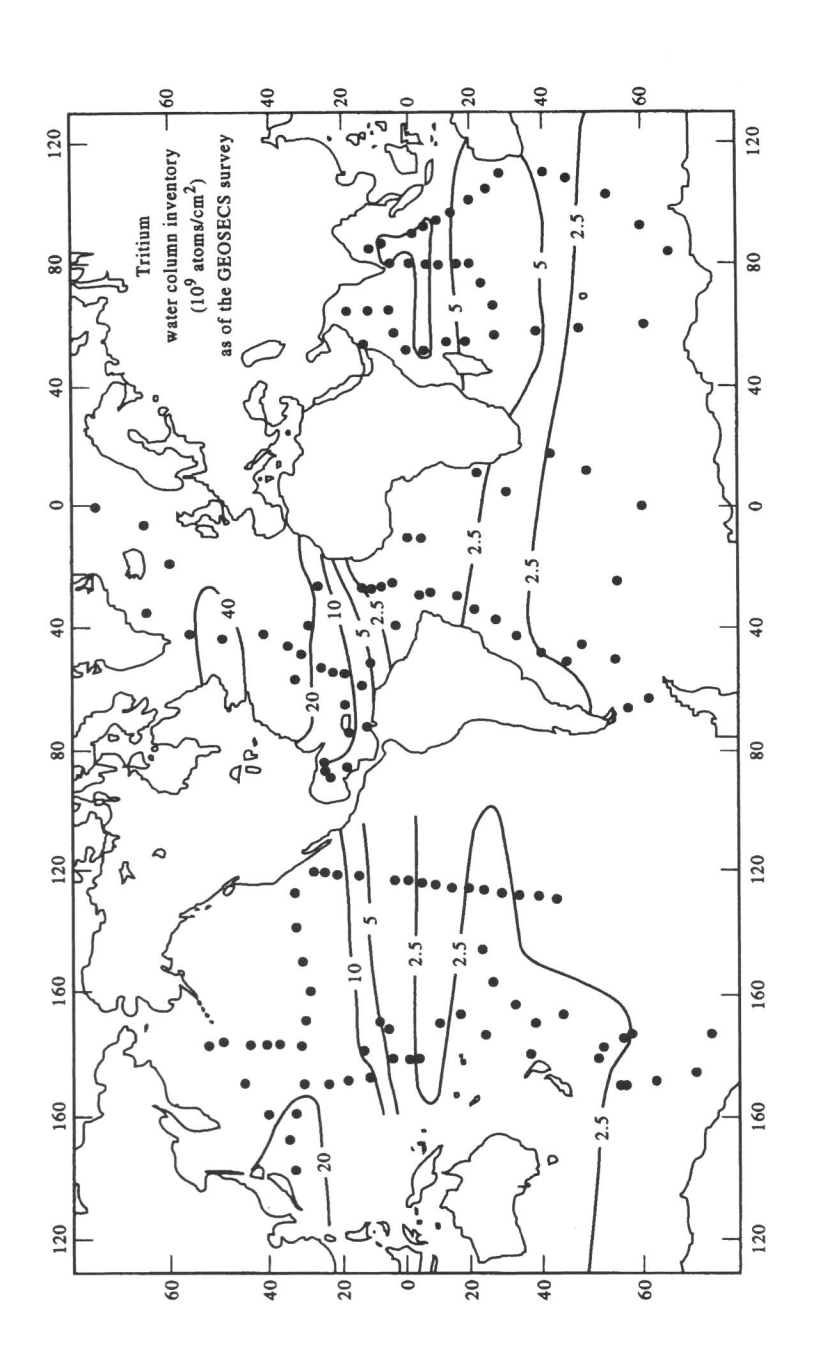

Figure 1.12.2. *Decay-corrected tritium (TU81) water column inventories over the world oceans based on results obtained as part of the GEOSECS program and NAGS expedition. (Adapted from Broecker et al., 1986.)*

isopycnal or diapycnal diffusion in the density range of maximum vertical tritium gradient. Tritium data studied by Van Scoy et al. (1991) show evidence for wind-driven circulation to the depth of the dissolved oxygen minimum near 1000 m depth ($\sigma_t =$ 27.40) in subpolar regions of the North Pacific. The authors conclude that, after two decades of mixing, advection along isopycnal surfaces appears to be the dominant process influencing the distribution of tritium in the North Pacific and that cross-isopycnal mixing in the subpolar region is important for ventilating the nonoutcropping isopycnals. According to Van Scoy et al. (1991), tritium has penetrated on average 100 m deeper into the ocean during the 10 years between the GEOSECS and Long Lines surveys. Depletions of tritium in the upper ocean are seen in the tropics and at high southern latitudes. Moreover, the above-background tritium levels observed on nonoutcropping isopycnals surfaces in the North Pacific indicate that ventilation is still taking place despite the absence of deep convective mixing in this region. In the Atlantic Ocean, deep convection is the dominant mechanism for the invasion of surface waters into the deep ocean (Figure 1.12.3).

Tritium data are used to constrain circulation models for the world ocean. For example, tritium records combined with a three-box model of the Japan Sea—a comparatively isolated oceanic region with a mean depth of 1350 m—have yielded overturn times for the deep water of 100 years and overall residence times of 1000 years (Watanabe et al., 1991). Similar estimates for this region based on the same box-model constrained by ^{226}Ra and ^{14}C data yielded a turnover time of 300–500 years for deep water and 600–1300 years for the residence time (Harada and Tsunogai, 1986). Applications to larger oceanic basins are generally less successful. Memery and Wunsch (1990) found that the tritium data did not strongly constrain their circulation model for the North Atlantic and that large errors ($\approx 20\%$) in the input of tritium at the surface can be accommodated by relatively minor changes in the model circulation. According to Wunsch (1988), "Any uncertainty in the transient tracer boundary conditions and sparse interior ocean temporal coverage greatly weakens the ability of such tracers to constrain the ocean circulation". Although the authors still believe in the usefulness of tritium records, they suggest that chlorofluorocarbons will improve modeling capability since the atmospheric concentration of these compounds remains relatively high despite the 1988 Montreal Accord and are better known than for tritium.

Jenkins (1988) describes the use of the tritium–^{3}He age, which takes advantage of the radioactive clock of ^{3}He and the long time-scale of tritium to measure the elapsed time since the Helium gas was in equilibrium with the atmosphere. Time scales for which this combined tracer is useful are 0.1–10 years.

1.12.2 Radiocarbon

Carbon-14 (^{14}C) dating requires prior knowledge of long-term variations in the ^{14}C/^{12}C ratio in the atmosphere. Because of the difficulties in separating radiocarbon produced from thermonuclear devices and cosmic rays, bomb-generated radiocarbon is a less useful tracer of upper ocean processes than is tritium. The problem of using radiocarbon data collected prior to 1958 together with tritium measurements to establish the prenuclear levels of radiocarbon is discussed by Broecker and Peng (1982). Once the prenuclear surface-water cosmic radiocarbon concentration is known for each locality, water column inventories for bomb-radiocarbon can be obtained from the depth profiles of ^{14}C/C, ^{3}H, and $\sum CO_2$ concentration obtained as part of

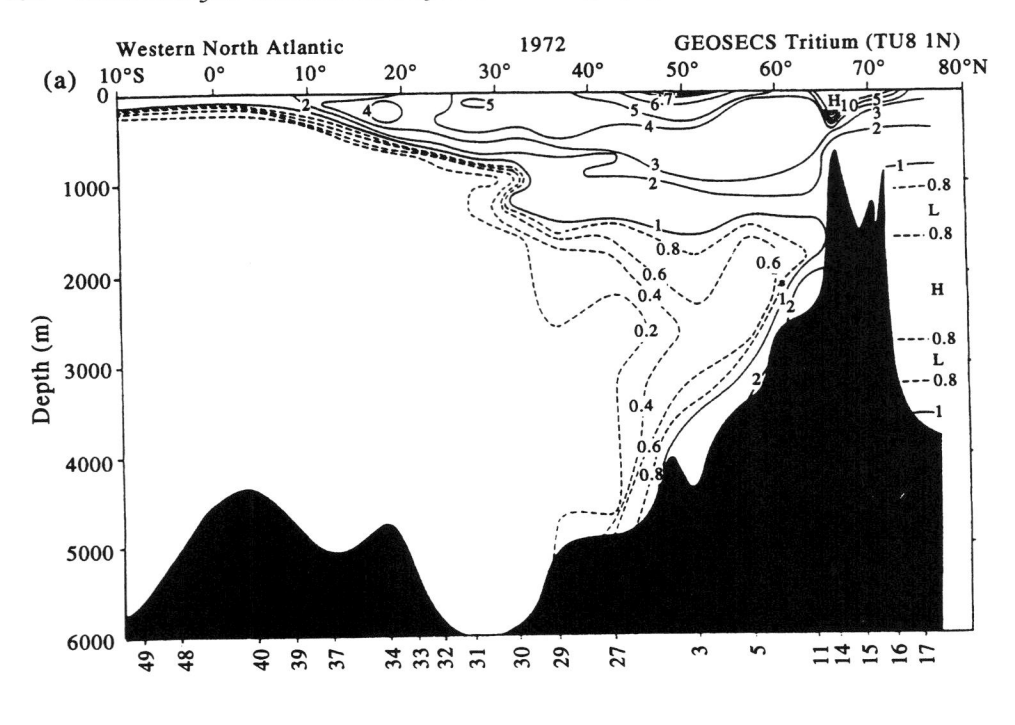

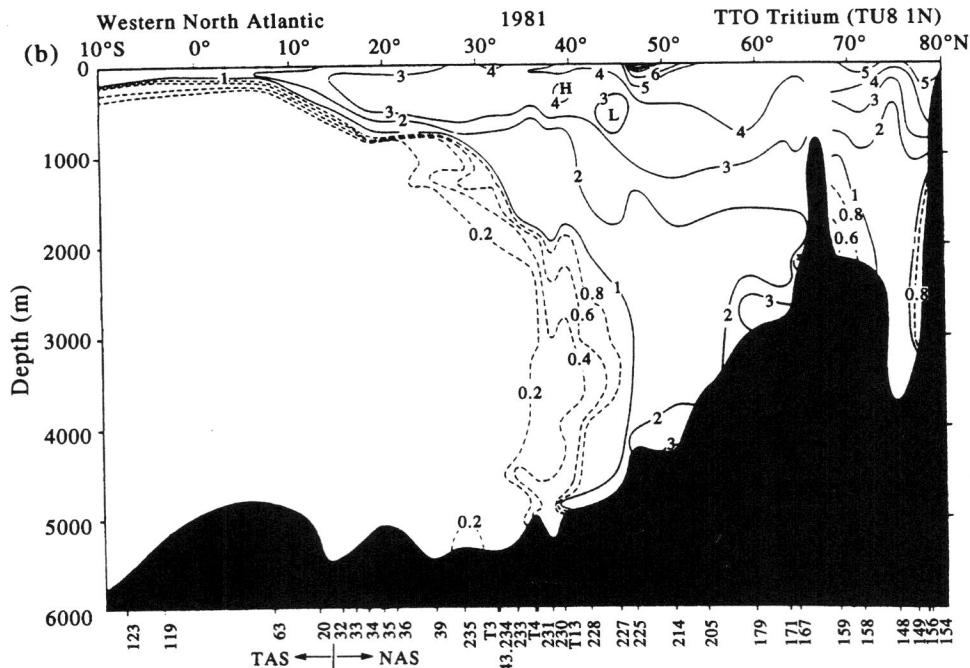

150 *Data Analysis Methods in Physical Oceanography*

Figure 1.12.3. Cross-section of tritium (TU81N) concentrations in the western Atlantic Ocean (a) GEOSECS 1972; (b) TTO 1981. (From Östlund and Rooth, 1990.) Results suggest that the observed vertical distribution of tritium in this region is determined mainly through advection along isopycals rather than by isopycnal or diapyncal diffusion in the density range of maximum vertical tritium gradient.

the GEOSECS, NORPAX and TTO programs (Broecker et al., 1985). Bomb-produced radiocarbon is delivered through a nearly irreversible process from the atmosphere to the ocean so that it is possible to estimate the amount of this isotope that has entered any given region of the ocean. As a result of this production, levels of $^{14}CO_2$ increased by about a factor of two in the northern hemisphere during the late 1950s and early 1960s. Measurement of radiocarbon by beta decay requires 200–250 litres of seawater to give the desired accuracy of 3–4 ppt. Age resolution is 25–30 years for abyssal oceanic conditions for which the introduction of bomb-radiocarbon effects remain negligible. Radiocarbon has a half-life of 5680 years and decays at a fixed rate of 1% every 83 years. A rapid onboard technique for measuring radiocarbon using an accelerator mass spectrometer is described by Bard et al. (1988). This technique decreases the sample size by 2000 compared with that using the standard β-counting method.

By convention, radiocarbon assays are expressed as $\Delta^{14}C$, which is the deviation in parts per thousand (ppt) of the $^{14}C/^{12}C$ ratio from that of a hypothetical wood standard with $\delta^{13}C = {^{13}C/^{12}C} = -25$ ppt and corrected from the actual $\delta^{13}C$ values (around 0 ppt for seawater to exactly -25 ppt to compare with the wood standard). The standard is a way to compare the observed ratio of carbon isotopes to the atmospheric value prior to the industrial revolution of about 1850. The quantity of ^{14}C in a sample of seawater is proportional to the actual uncorrected $^{14}C/^{12}C$ ratio (1 + $0.001\delta^{14}C$). More precisely

$$\Delta^{14}C = \delta^{14}C - 2(\delta^{13} + 25)(1 + \delta^{14}C/1000) \tag{1.12.1}$$

where

$$\delta^{14}C = 1000[(^{14}C/C) \text{ sample} - (^{14}C/C) \text{ standard}]/(^{14}C/C) \text{ standard}] \tag{1.12.2}$$

Pre-bomb $\Delta^{14}C$ values from corals collected in the early 1950s average around -50 (± 5) ppt (Druffel, 1989). Thus, any $\Delta^{14}C$ value above -50 ppt will indicate the presence of anthropogenic radiocarbon, mainly produced by the atmospheric nuclear testing in the early 1960s. The determination of inventories for bomb-produced radiocarbon in the ocean is much more complex than for bomb tritium. The reason is that the amount of natural tritium in the sea is negligible compared with the amount of bomb-produced tritium. In the case of radiocarbon, the delivery of isotopes to the ocean requires a better knowledge of wind speeds over the ocean and of the wind speed dependence of the CO_2 exchange rate.

The concentration of ^{14}C in the ocean is influenced by several processes. For example, bottom water formation in the Weddell Sea and the North Atlantic provides a direct input of surface water ^{14}C (Figure 1.12.4). Additional input of ^{14}C to the deep sea can occur by transport along isopycnals, by vertical mixing in the main oceanic thermocline, by lateral mixing of water masses and by upwelling in coastal and equatorial regions. Addition of CO_2 and ^{14}C comes from the dissolution of carbonate skeletons and the oxidation of organic materials from sinking particles. Stuiver et al. (1982) use radiocarbon data from GEOSECS to estimate abyssal (>1500 m) waters replacement times for the Pacific, Atlantic and Indian Oceans of 510, 275, and 250 years, respectively. The deep waters of the entire world ocean are replaced on average

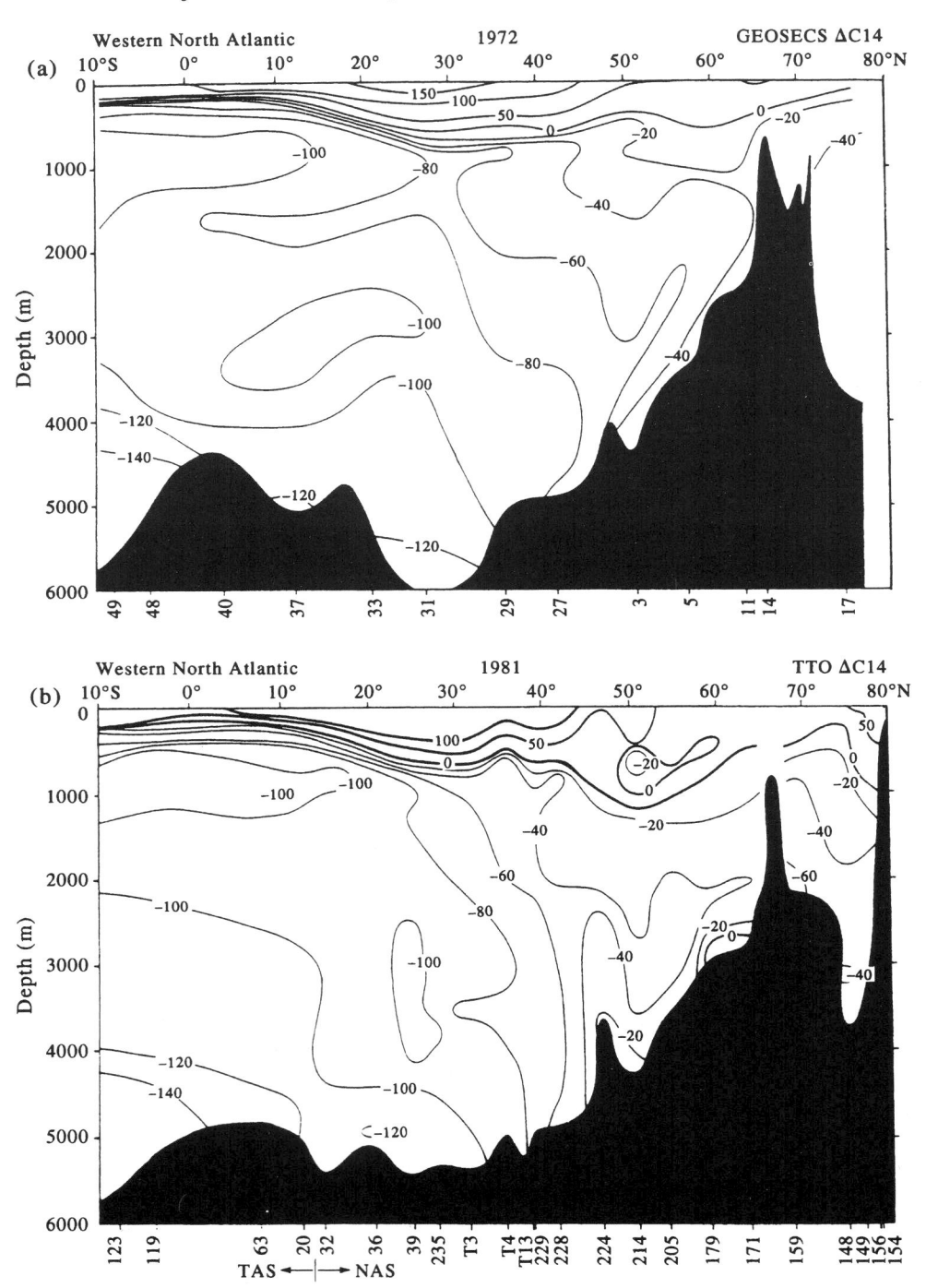

152 *Data Analysis Methods in Physical Oceanography*

Figure 1.12.4. Cross-section of radiocarbon concentrations in the western Atlantic Ocean (a) GEOSECS 1972; (b) TTO 1981 (cf. Figure 1.12.3). Note that significant changes occur mainly in the deep waters north of 40°N. (From Östlund and Rooth, 1990.)

every 500 years. Östlund and Rooth (1990) found a relative decrease in the difference in $\Delta^{14}C$ between the surface and the northerly abyssal layers of the North Atlantic of 25–30%. If this were due to vertical diffusivity a high value of 10 cm²/s would be required based on a scale depth of 1 km and 10 years between surveys. This is a factor of 10 too large so that high latitude injection processes must be responsible for the observed evolution below 1000 m depth. Measurements of $\Delta^{14}C$ from seawater and organisms from the Pacific coast of Baja California (Druffel and Williams, 1991) revealed the effects of coastal upwelling and bottom-feeding habits. Dilution of nearshore waters by upwelling accounts for reduced radioactive carbon levels observed near the coast while feeding on sediment-derived carbon explains the reduced levels of ^{14}C in sampled organisms relative to dissolved inorganic carbon in the water column. Broecker et al. (1991) have addressed the concerns about the accuracy of ventilation flux estimates for the deep Atlantic due temporal changes in the $^{14}C/C$ ratio for atmospheric CO_2. Despite the fact that $\Delta^{14}C$ values have declined from about 10 to -20 ppt over the past 300 years due to changes in the solar wind and the addition of $\Delta^{14}C$-free CO_2 to the atmosphere from fossil fuel burning, temporal effects have been considerably buffered in the ocean and errors in radiocarbon ages are too low by only 10–15%. The reason is that the northern and southern source waters for the Atlantic deep water have $\Delta^{14}C$ ratios, and hence relative time variabilities, considerably lower than the atmospheric ratio.

1.12.3 Chlorofluorocarbons

Chlorofluorocarbons (CFCs) are a group of volatile anthropogenic compounds that until the 1988 Montreal Protocol found increasingly widespread use in aerosol propellants, plastic foam blowing agents, refrigerants and solvents. Also known as chlorofluoromethanes (CFMs) and "Freons" (a Dupont tradename), most of these chemicals eventually find their way into the atmosphere where they play a primary role in the destruction of stratospheric ozone. The two primary compounds CFC-12 or F-12 (CF_2Cl_2) and CFC-11 or F-11 ($CFCl_3$) have respective lifetimes in the troposphere of 111 and 74 years. Although more than 90% of production and release of F-11 and F-12 takes place in the northern hemisphere, the meridional distributions of these compounds in the global troposphere are relatively uniform due to the high stability of the compounds and the rapid mixing that occurs in the lower atmosphere. The source function at the ocean surface differs by only about 7% from the northern hemisphere to the southern hemisphere (Bullister, 1989). During the period 1930–1975 the ratio F-11/F-12 in the atmosphere and ocean surface increased with increasing uses of these chemicals (Figure 1.12.5). The regulation of CFC use in spray cans in the U.S.A. during the late 1970s decreased the rate of CFC-11 increase so that the ratio F-11/F-12 ratio in the atmosphere has remained nearly constant. As a consequence, measurements of the ratio provide information on when a particular water mass was last in contact with the atmosphere. In shelf waters the CFCs concentration is determined by rates of mixed layer entrainment, gas exchange and mixing with source water. At a removal rate of about 1% per year from the atmosphere by stratospheric photolysis, CFCs will serve as ocean tracers well into the next century.

Since they are chemically inert in seawater, chlorofluorocarbons are used to examine gas exchange between the atmosphere and ocean, ocean ventilation and mixing on decadal scales. The limit of detection of F-11 and F-12 in seawater volumes

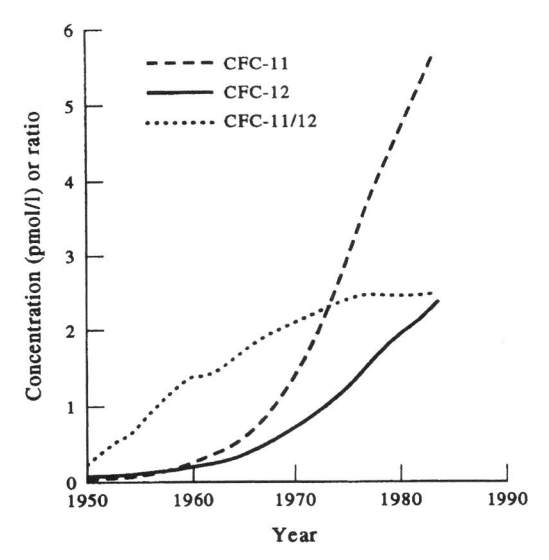

Figure 1.12.5. CFC-12 and CFC-11 concentrations in the upper ocean for T = −1°C and S = 34.3 psu as function of time. (From Trumbore et al., 1991.)

as small as 30 ml is better than 5×10^{-15} mole/kg seawater (Bullister and Weiss, 1988), or roughly three orders of magnitude higher than near-surface concentrations in the ocean. Modern techniques allow for processing of CFCs at sea with processing times of the order of hours. Gammon *et al.* (1982) examined the vertical distribution of CFCs at two offshore sites in the northeast Pacific. Using a one-dimensional vertical diffusion/advection model driven by an exponential surface source term, they obtained a characteristic depth penetration of 120–140 m. For the Gulf of Alaska station at 50°N, 140°W, vertical profiles of F-11 and F-12 gave consistent vertical diffusivities of order 1 cm^2/s and an upwelling velocity of 12–14 m/year. Woods (1985) used CFCs to estimate the transit time and mixing of Labrador seawater from its northern source region to the equator along the western Atlantic Ocean boundary. In a related study, Wallace and Lazier (1988) used CFCs and a simple convection model to examine recently renewed Labrador seawater formed by deep convection to depths greater than 1500 m following a severe winter in the North Atlantic. Their observed CFC levels of 60% saturation with respect to contemporary atmospheric concentrations suggest that deep convection took place too rapidly for air–sea gas exchange to bring CFC levels to equilibrium. Trumbore *et al.* (1991) have used CFCs collected in 1984 to examine recent deep water ventilation and bottom water formation near the continental shelf in the Ross Sea in the Antarctic Ocean. Using CFC data combined with conventional (temperature, salinity, dissolved oxygen and nutrient) tracer data in a time-dependent convection model they estimate shelf-water resident times of about three years for the Ross Sea. At the other end of the globe, Schlosser *et al.* (1991) use hydrographic and CFC data to suggest that formation of Greenland Sea Deep Water decreased in the 1980s. The dissolved F-12 concentration in Figure 1.12.6 illustrates several aspects of the circulation in the North Atlantic. In particular, we note the core of the Labrador Seawater mentioned earlier in this section, the presence of a lens of Mediterranean outflow water ("Meddy") at 22°N and the core of high CFC over the equator which is

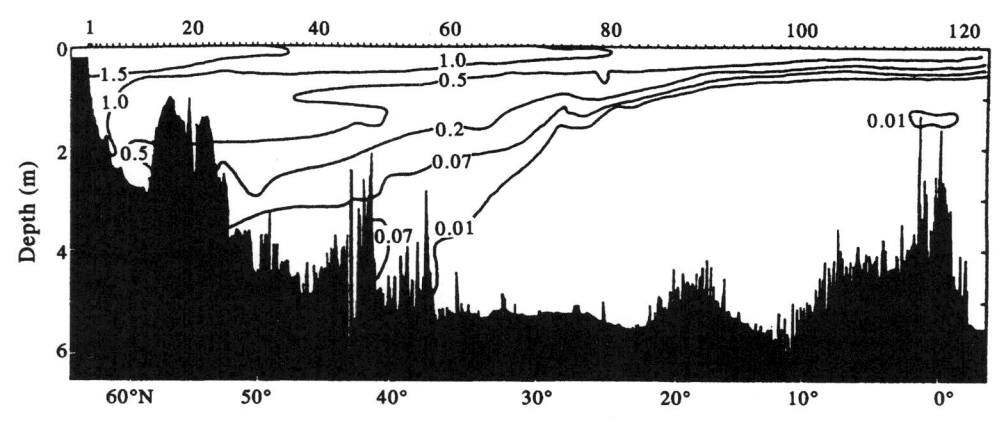

Figure 1.12.6. Dissolved CFC-12 concentrations ($\times 10^{-12}$ mole/kg) along a North Atlantic section. (From Bullister, 1989.)

thought to be a longitudinal extension of flow from the western boundary near Brazil (Bullister, 1989).

1.12.4 Radon-222

Radon (^{222}Rn) is a chemically inert gas with a radioactive half-life of 3.825 days. It occurs naturally as a radionuclide of the ^{238}U series and is injected into the atmosphere by volcanic eruptions. The gas has proven particularly useful at time-scales of a few days to weeks for examining the rate of gas transfer between the atmosphere and the ocean surface (Peng *et al.*, 1979), in studies of water column mixing rates (Sarmiento *et al.*, 1976), and for estimating the heat and chemical fluxes from hydrothermal venting at mid-ocean ridges (Rosenberg *et al.*, 1988; Kadko *et al.*, 1990).

The new application of ^{222}Rn studies to hydrothermal venting regions has been especially successful (Rosenberg *et al.*, 1988). In this case, it is assumed that there is a constant flux of radon into the effluent plume that typically rises several hundred meters above the venting region at depths of 2–3 km on the ridge axis. Typical venting regions have scales of 100 m and are spaced at several kilometers along the ridge axis. Waters exiting from black smokers can be up to 400°C. At steady state, the amount of radon lost to radioactive decay at some point in the laterally spreading nonbuoyant plume is balanced by a supply of radon from the venting region. To obtain the total heat (or chemical species) issuing from the venting region, the observer first uses a submersible or towed sensor package to measure the ratio of radon to heat (or species) anomaly, ^{222}Rn/ΔT, in the plume near the vent orifice—before the radon in the plume has a chance to disperse or age. The observer then uses a towed sensor package to map the total inventory of radon in the spreading plume (Figure 1.12.7). Taking into account the effect of cold water entrainment on the rising plume at Endeavour Ridge in the northeast Pacific (47°47'N, 129°06'W), Rosenberg *et al.* (1988) found an initial radon/ΔT value of 0.03 dpm (disintegrations per minute—the standard unit of measurement for radioactive materials) or 55 atoms per joule. They then used hydocast bottle data to estimate the standing crop of radon above 2100 m depth as ^{222}Rn(Total) = 8×10^{12} dpm. At steady-state, hydrothermal venting must be adding this much radon to the system so that the total heat emanating from the vents is

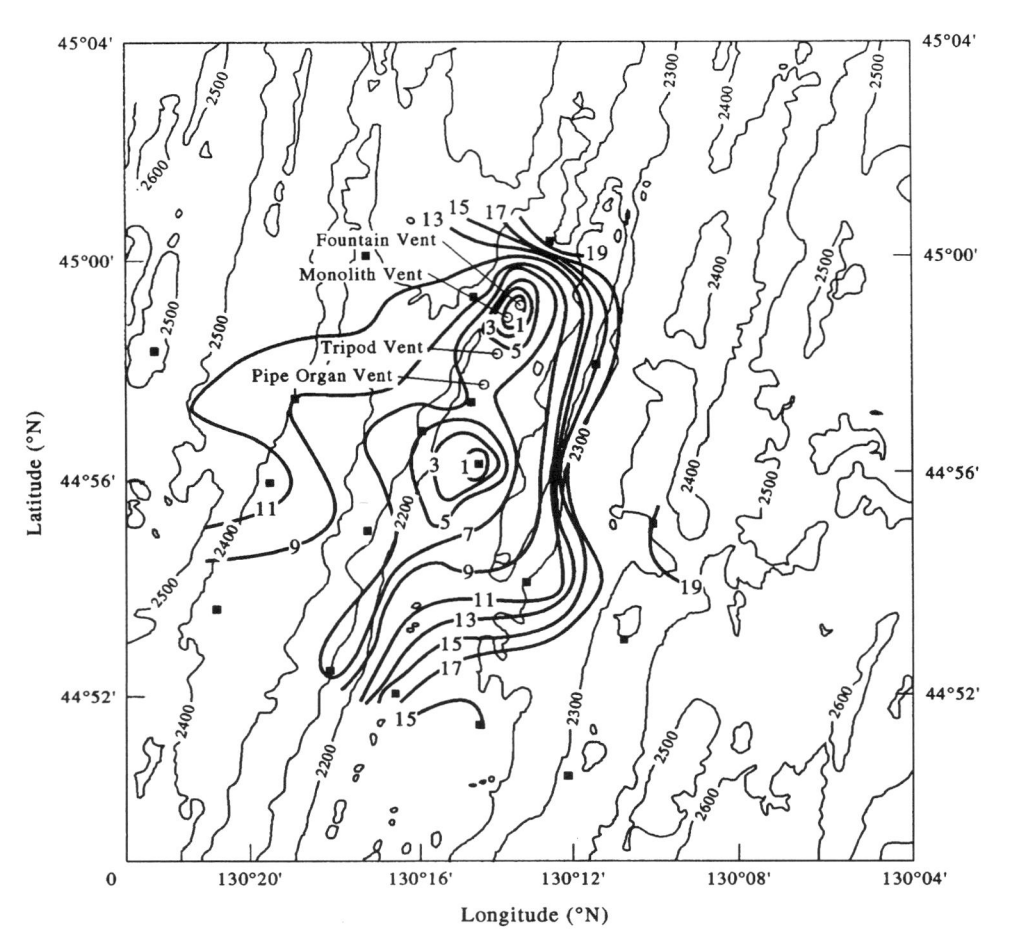

Figure 1.12.7. Apparent age of the neutrally-buoyant plume on the isopyncnal surface $\sigma_\theta = 27.68$ (roughly 2100 m depth) at the Cleft Segment of Juan de Fuca Ridge in the northeast Pacific. Distribution based on Radon-222 data for September 1990. Depths in meters. Plume rises from the hydrothermal vent depth of 2280 m to approximately 2100 m. (From Gendron et al., 1993.)

$$^{222}\text{Rn (Total)}/(^{222}\text{Rn}/\Delta T) = 3(\pm 2) \times 10^9 \text{ watts} \qquad (1.12.3)$$

which compares with estimates based on direct measurements of the total heat content anomaly of the plume in combination with local currents (Baker and Massoth, 1986, 1987; Baker *et al.*, 1995). Gendron *et al.* (1993) have used ^{222}Rn to examine time variability in hydrothermal venting on the Cleft Segment of Juan de Fuca Ridge and to estimate the age of the plume as a function of location relative to the known vent sites. They found that the hydrothermal flux decreased from 2.2±0.3 GW in 1990 to 1.2±0.2 GW in 1991 (1 GW = 10^9 watts).

The estimates using radon-222 in the ocean are complicated by the fact that radon concentration is a function of both radioactive decay and dilution with ambient seawater. Similar estimates can be made using ^{3}He to heat ratios combined with the total inventory of ^{3}He in the ocean. The result (Jenkins, 1978) is a global hydrothermal heat flux of 4.9×10^{19} cal/year. Baker and Lupton (1990) have used the ^{3}He/heat ratio as a possible indicator of magmatic/tectonic activity at ridge segments. The change from a ratio of 4.4×10^{-12} cm^3 STP cal^{-1} immediately following the megaplume eruption at Cleft segment to 1.3×10^{-12} cm^3 STP cal^{-1} two years later suggests that high ratios may be indicative of venting created or profoundly perturbed by a magmatic–tectonic event, while lower values may typify systems at equilibrium.

1.12.5 Sulfur hexafluoride

In certain instances, there is a distinct advantage to a controlled and localized release of a chemical into the environment. Prefluorinated tracers such as sulfur hexachloride (SF$_6$) and perfluorodeclin (PFD) are among the new generation of deliberately released tracers used to measure mixing and diffusion rates in the ocean. These substances are particularly good at examining vertical mixing. Their appeal is that they are a readily detectable conservative tracers that have no significant effect on the environment and no toxicity. A thorough description of the use of these tracers as well as rhodamine dyes can be found in Watson and Ledwell (1988). In the case of rhodamine dyes, the detection limit by fluorometers is set by the background fluorescence of natural substances in water which is about 1 part in 10^{12} in the deep ocean. For SF$_6$ the background limit is set by dissolution from the atmosphere where the compound is present at 1–2 parts in 10^{12} by volume. Surface values in the ocean are roughly 5×10^{-17} and diminish to zero in deep water. The instrumental detection for SF$_6$ is limited to about 1/10 of the near-surface value (Watson and Liddicoat, 1985). PFD has no measurable background level in the ocean and is limited by instrumental detection to about 1 part in 10^{16}. For a release of 1 metric ton ($\equiv$ 1 tonne) at a given density level in an experiment, these detection limits translate to maximum horizontal scales of 100 km for rhodamine dye, 1000 km for PFDs and basin scales for SF$_6$. Lifetimes for the tracers range from months to about a year. Despite their usefulness, the long-term prognosis for SF$_6$ and PFDs is limited as industrial injection of SF$_6$ into the atmosphere and medical use of PFDs will eventually increase background levels and take away from their ability to serve as tracers.

Rhodamine dye is used mainly in coastal studies. SF$_6$ has been used successfully in WOCE. The North Atlantic Tracer Release Experiment (NATRE) was a large-scale WOCE-related study using SF$_6$ to examine the stirring and diapycnal mixing in the pycnocline of the North Atlantic. In May 1992, 139 kg of sulfur hexafluoride was released on the isopycnal surface 26.75 kg/m^3 (310 dbar) along with eight SOFAR floats and six pop-up drifters in the eastern subtropical Atlantic near 25.7°N, 28.3°W. To sample the tracer, investigators towed a vertical array of 20 integrating sample at 0.5 m/s through the patch. A prototype 18-chamber sampler at the center of the array obtained a lateral resolution of about 360 m. The average profile increased from a RMS thickness of 6.8 m after 14 days to a RMS thickness of about 45 m by April 1993, yielding a diapycnal eddy diffusivity of 0.1–0.2 cm^2 s^{-1}. To be successful, experiments like NATRE require the tracer to be injected on a constant density surface rather than a constant depth. Internal wave oscillations and other vertical motions would broaden the tracer concentration more than necessary if it were released at a constant depth. Care must be taken during injection to ensure the tracer's buoyancy is correct and that the turbulent wake of the injection apparatus is not excessive.

1.12.6 Strontium-90

The distribution of bomb-produced ^{90}Sr in the ocean is quite similar to that of tritium. However as pointed out by Toggweiler and Trumbore (1985), ^{90}Sr has the virtue that the ratio ^{90}Sr/Ca incorporated into coral skeletons has the same value as this ratio in seawater. Corals average out seasonal variations in the ^{90}Sr content of seawater so that annual bands provide a time-averaged measure of the amount of strontium in the water. The results of Toggweiler and Trumbore (1985) suggest that waters move into the Indian Ocean via passages through the Indonesian Archipelago. In addition, the data suggest that there is a large-scale transport of water between the temperate and tropical North Pacific.

CHAPTER 2

Data Processing and Presentation

2.1 INTRODUCTION

Most instruments do not measure oceanographic properties directly nor do they store the related engineering or geophysical parameters that the investigator eventually wants from the recorded data. Added to this is the fact that all measurement systems alter their characteristics with time and therefore require repeated calibration to define the relationship between the measured and/or stored values and the geophysical quantities of interest. The usefulness of any observations depends strongly on the care with which the calibration and subsequent data processing are carried out. Data processing consists of using calibration information to convert instrument values to engineering units and then using specific formulae to produce the geophysical data. For example, calibration coefficients are used to convert voltages collected in the different channels of a CTD to temperature, pressure, and salinity (a function mainly of conductivity, temperature, and pressure). These can then be used to derive such quantities as potential temperature (the compression-corrected temperature) and steric height (the vertically integrated specific volume anomaly derived from the density structure).

Once the data are collected, further processing is required to check for errors and to remove erroneous values. In the case of temporal measurements, for example, a necessary first step is to check for timing errors. Such errors arise because of problems with the recorder's clock which cause changes in the sampling interval (Δt), or because digital samples are missed during the recording stage. If N is the number of samples collected, then $N\Delta t$ should equal the total length of the record, T. This points to the obvious need to keep accurate records of the exact start and end times of the data record. When $T \neq N\Delta t$, the investigator needs to conduct an initial search for possible missing records. Simultaneous, abrupt changes in recorded values on all channels often point to times of missing data. Changes in the clock sampling rate (clock "speed") are more of a problem and one has often to assume some sort of linear change in Δt over the recording period. When either the start or end time is in doubt, the investigator must rely on other techniques to determine the reliability of the sampling clock and sampling rate. For example, in regions with reasonable tidal motions, one can check that the amplitude ratios among the normally dominant K_1, O_1 (diurnal) and M_2, S_2 (semidiurnal) tidal constituents (Table 2.1) are consistent with previous observations. If they aren't, there may be problems with the clock (or calibration of amplitude). If the phases of the constituents are known from previous observations in the region, these can be compared with phases from the suspect instrument. For diurnal motions, each one hour error in timing corresponds to a phase

Table 2.1. *Frequencies (cycles per hour) for the major diurnal (O_1, K_1) and semidiurnal (M_2, S_2) tidal constitutents*

Tidal constituent	O_1	K_1	M_2	S_2
Frequency (cph)	0.03873065	0.04178075	0.08051140	0.08333333

change of 15°; for semidiurnal motions, the change is 30° per hour. Large discrepancies suggest timing problems with the data.

Two types of errors must be considered in the editing stage: (1) large "accidental" errors or "spikes" that result from equipment failure, power surges, or other major data flow disruptions (including some planktors such as salps and small jellyfish which squeeze through the conductivity cell of a CTD); and (2) small random errors or "noise" that arise from changes in the sensor configuration, electrical and environmental noise, and unresolved environmental variability. The noise can be treated using statistical methods while elimination of the larger errors generally requires the use of some subjective evaluation procedure. Data summary diagrams or distributions are useful in identifying the large errors as sharp deviations from the general population, while the treatment of the smaller random errors requires a knowledge of the population density function for the data. It is often assumed that random errors are statistically independent and have a normal (Gaussian) probability distribution. A summary diagram can help the investigator evaluate editing programs that "automatically" remove data points whose magnitudes exceed the record mean value by some integer multiple of the record standard deviation. For example, the editing procedure might be asked to eliminate data values $|x - X| > 3\sigma$, where X and σ are the mean and standard deviation of x, respectively. This is wrought with pitfalls, especially if one is dealing with highly variable or episodic systems. By not directly examining the data points in conjunction with adjacent values, one can never be certain that he/she is not throwing away reliable values. For example, during the strong 1983–1984 El Niño, water temperatures at intermediate depths along Line P in the northeast Pacific exceeded the mean temperature by 10 standard deviations (10σ). Had there not been other evidence for basin-wide oceanic heating during this period, there would have been a tendency to dispense with these "abnormal" values.

2.2 CALIBRATION

Before data records can be examined for errors and further reduced for analysis, they must first be converted to meaningful physical units. The integer format generally used to save storage space and to conduct onboard instrument data processing is not amenable to simple visual examination. Binary and ASCII formats are the two most common ways to store the raw data, with the storage space required for the more basic Binary format about 20% of that for the integer values of ASCII format. Conversion of the raw data requires the appropriate calibration coefficients for each sensor. These constants relate recorded values to known values of the measurement parameter. The accuracy of the data then depends on the reliability of the calibration procedure as well as on the performance of the instrument itself. Very precise instruments with poor calibrations will produce incorrect, error-prone data. Common practice is to fit the set of calibration values by least-squares quadratic expressions, yielding either functional (mathematical) or empirical relations between the recorded values and the appropriate physical values. This simplifies the post-processing since the raw data can readily be passed through the calibration formula to yield observations in the correct units. We emphasize that the editing and calibration work should always be performed on *copies* of the original data; never work directly on the raw, unedited data.

In some cases the calibration data do not lend themselves to description in terms of polynomial expressions. An example is the direction channel in Aanderaa current meter data for which the calibration data consists of a table relating the recorded direction in raw 10-byte integer format (0–1024) to the corresponding direction in degrees from the compass calibration (Pillsbury *et al.*, 1974). Some thought should be given to producing calibration "functions" that best represent the calibration data. With the availability of modern computing facilities, it is no more burdensome to build the calibration into a table than it is to convert it to a mathematical expression. Most important, however, is the need to ensure that the calibration accurately represents the performance range and characteristics of the instrument. Unquestioned acceptance of the manufacturer's calibration values is not recommended for the processing of newly collected data. Instead, individual laboratory and/or field calibration may be needed for each instrument. In some cases this is not possible (for example, in the case of XBT probes which come prepackaged and ready for deployment) and some overall average calibration relation must be developed for the measurement system regardless of individual sensor.

Some instruments are individually calibrated before and after each experiment to determine if changes in the sensor unit had occurred during its operation. The conversion to geophysical units must take both pre- and postcalibrations into account. Often the pre- and postcalibration are averaged together or used to define a calibration trend-line which can then be used to transform the instrument engineering units to the appropriate geophysical units. Sometimes a postcalibration reveals a serious instrument malfunction and the data record must be examined to find the place where the failure occurred. Data after this point are eliminated (or modified to account for the instrumental problems) and the postcalibration information is not used in the conversion to geophysical values. Even if the instrument continues to function in a reasonable manner, the calibration history of the instrument is important to producing accurate geophysical measurements from the instrument.

Since each instrument may use a somewhat different procedure to encode and record data it is not possible to discuss all of the techniques employed. We therefore have outlined a general procedure only. Appendix A provides a list of the many physical units used today in physical oceanography. Although there have been many efforts to standardize these units one must still be prepared to work with data in nonstandard units. This may be particularly true in the case of older historical data collected before the introduction of acceptable international units. These standard units also are included in Appendix A.

2.3 INTERPOLATION

Data gaps or "holes" are a problem fundamental to many geophysical data records. Gappy data are frequently the consequence of uneven or irregular sampling (in time and/or space), or they may result from the removal of erroneous values during editing and from sporadic recording system failures. Infrequent data gaps, having limited duration relative to strongly energetic periods of interest, are generally of minor concern, unless one is interested in short-term episodic events rather than stationary periodic phenomena. Major difficulties arise if the length of the holes exceeds a significant fraction (1/3–1/2) of the signal of interest and the overall data loss rises beyond 20–30% (Sturges, 1983). Gaps have a greater effect on weak signals than on strong signals and the adverse effects of the gaps increases most rapidly for the smallest percentages of data lost. While some useful computational techniques have been developed for unevenly spaced data (Meisel, 1978, 1979) and even some advantages to having a range of Nyquist frequencies within a given data set (Press *et al.*, 1992), most analysis methods require data values that are regularly spaced in time or space. As a consequence, it is generally necessary to use an interpolation procedure to create the required regular set of data values as part of the data processing. The problem of interpolation and smoothing is discussed in more detail in Chapter 3.

2.4 DATA PRESENTATION

2.4.1 Introduction

The analysis of most oceanographic records necessitates some form of "first-look" visual display. Even the editing and processing of data typically requires a display stage, as for example in the exact determination of the start and end of a time series, or in the interactive removal and interpolation of data spikes and other erroneous values. A useful axiom is, "when in doubt, look at the data". In order to look at the data, we need specific display procedures. A single set of display procedures for all applications is not possible since different oceanographic data sets require different displays. Often, the development of a new display method may be the substance of a particular research project. For instance, the advent of satellite oceanography has greatly increased the need for interactive graphics display and digital image analysis.

Our discussion begins with traditional types of data and analysis product presentations. These have been developed as oceanographers sought ways to depict the ocean they were observing. The earliest shipboard measurements consisted of temperatures taken at the sea surface and soundings of the ocean bottom. These data were most appropriately plotted on maps to represent their geographical variability. The data were then contoured by hand to provide a smooth picture of the variable's distribution over the survey region. Examples of historical interest are the meridional sections of salinity from the eastern and western basins of the North Atlantic based on data collected during the German *Meteor* Expedition of 1925–1927 (Figure 2.1; Spiess, 1928). The water property maps from this expedition were among the first to indicate the north–south movements of water masses in the Atlantic basin.

As long as measurements were limited to the sea surface or sea floor, the question of horizontal level for display was never raised. As oceanographic sampling became more sophisticated and the vertical profiling of water properties became possible, new data displays were required. Of immediate interest were simple vertical profiles of temperature and salinity such as those shown in Figure 2.2. These property profiles, based on a limited number of sample bottles suspended from the hydrographic wire at

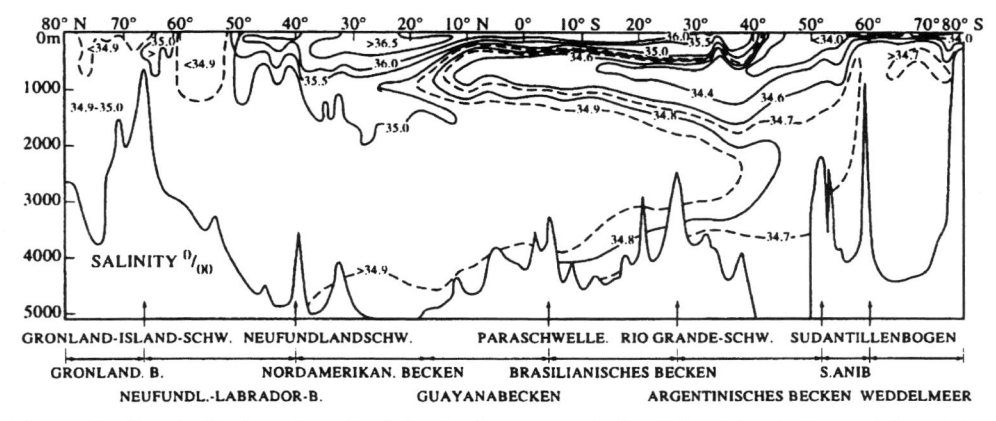

Figure 2.1. Longitudinal section of salinity in the western basin of the Atlantic Ocean (after Spiess, 1928).

standard hydrographic depths, originally served to both depict the vertical stratification of the measured parameter and to detect any sampling bottles that had not functioned properly. The data points could then either be corrected or discarded from the data set. Leakage of the watertight seals, failure of the bottle to trip, and damage against the side of the ship are the major causes of sample loss. Leakage problems can be especially difficult to detect.

The data collected from a research vessel at a series of hydrographic stations may be represented as vertical section plots. Here, the discretely sampled data are entered into a two-dimensional vertical section at the sample depths and then contoured to produce the vertical structure along the section (Figure 2.3). Two things need to be considered in this presentation. First, the depth of the ocean, relative to the horizontal distances, is very small and vertical exaggeration is required to form readable sections. Second, the stratification can be separated roughly into two near-uniform layers with a strong density-gradient layer (the pycnocline) sandwiched between. This two-layer system led early German oceanographers to introduce the terms "troposphere" and "stratosphere" (Wüst, 1935; Defant, 1936) which they described as the warm and cold water spheres of the ocean. Introduced by analogy to the atmospheric vertical structure, this nomenclature has not been widely used in oceanography. The consequence of this natural vertical stratification, however, is that vertical sections are often best displayed in two parts, a shallow upper layer, with an expanded scale, and a deep layer with a much more compressed vertical resolution.

Vertical profiling capability makes it possible to map quantities on different types of horizontal surface. Usually, specific depth levels are chosen to characterize spatial variability within certain layers. The near-vertical homogeneity of the deeper layers means that fewer surfaces need to be mapped to describe the lower part of the water column. Closer to the ocean surface, additional layers may be required to properly represent the strong horizontal gradients.

The realization by oceanographers of the importance of both along- and cross-isopycnal processes has led to the practice of displaying water properties on specific isopycnal surfaces. Since these surfaces do not usually coincide with constant depth levels, the depth of the isopycnal (equal density) surface also is sometimes plotted.

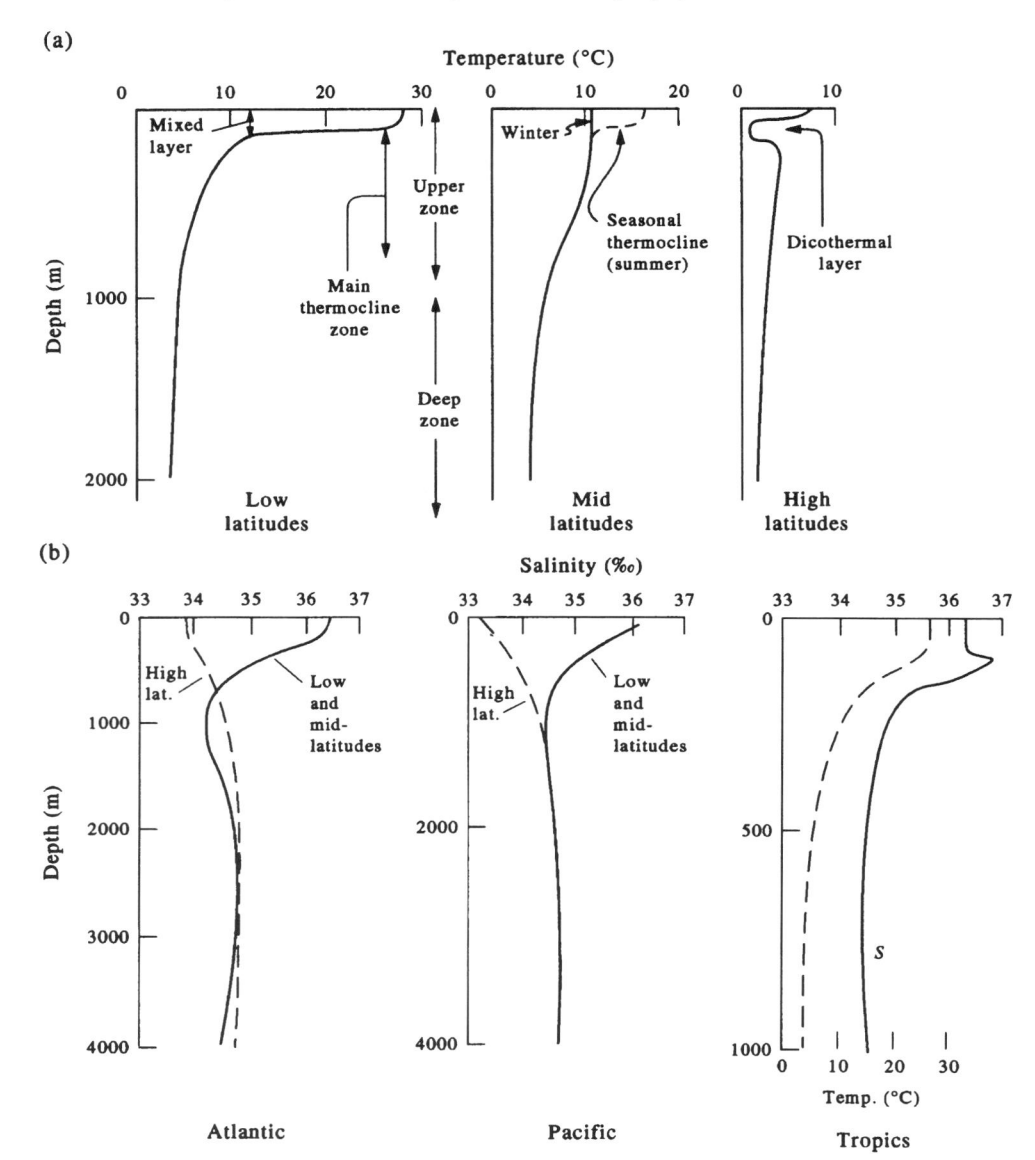

Figure 2.2. Vertical profiles. (a) Temperature profiles for tropical (low) latitudes, mid-latitudes, and polar (high) latitudes in the Pacific Ocean. (b) Salinity profiles for the Atlantic, Pacific and tropical oceans for different latitudes. The dicothermal layer in (a) is formed from intense winter cooling followed by summer warming to shallower depths. Both salinity (solid line) and temperature (dashed line) are plotted for the tropics in (b). (From Pickard and Emery, 1992.)

Isopycnal surfaces are chosen to characterize the upper and lower layers separately. Often, processes not obvious in a horizontal depth plot are clearly shown on selected isopycnal (sigma) surfaces. This practice is especially useful in tracking the lateral distribution of tracer properties such as the deep and intermediate depth silicate maxima in the North Pacific (Talley and Joyce, 1992) or the spreading of hydrothermal plumes that have risen to a density surface corresponding to their level of neutral buoyancy (Feely et al., 1994).

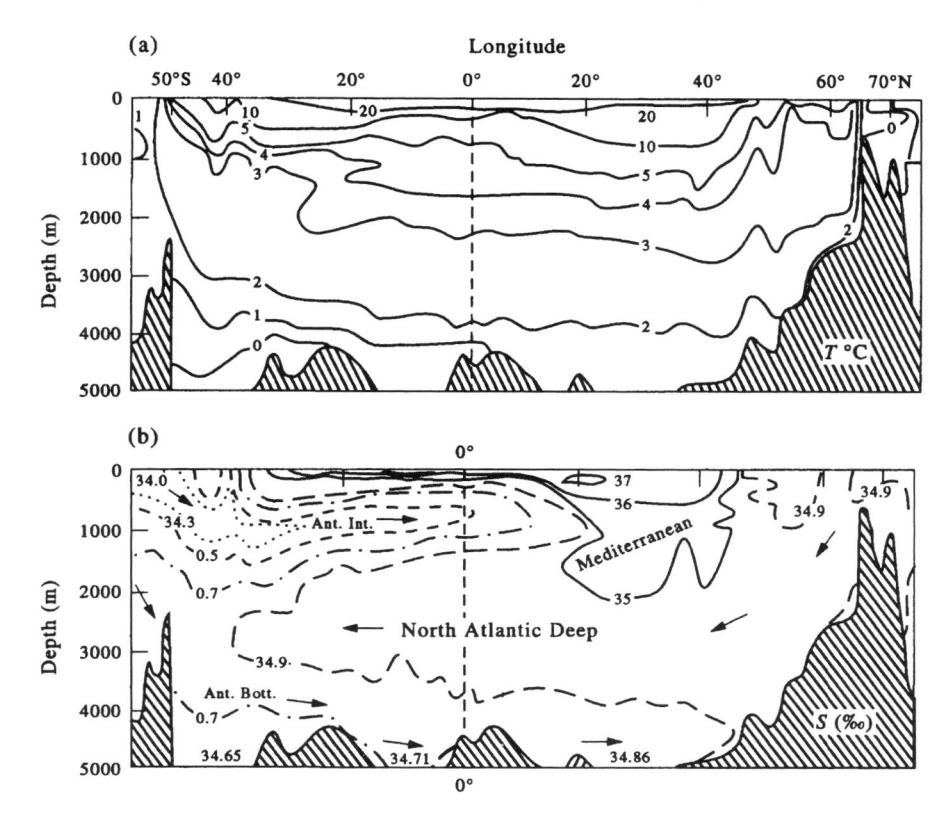

Figure 2.3. Longitudinal cross-sections of (a) in situ temperature and (b) salinity for the Atlantic Ocean. Arrows denote direction of water mass movement based on the distribution of properties. Ant. Bott. = Atlantic Bottom Water; Ant. Int. = Antarctic Intermediate Water. (From Pickard and Emery, 1992.)

Another challenge to the graphical presentation of oceanographic data is the generation of time series at specific locations. Initially, measured scalar quantities were simply displayed as time-series plots. Vector quantities, however, require a plot of two parameters against time. A common solution is the use of the "stick plot" (Figure 2.4) where each stick (vector) corresponds to a measured speed and direction at the specified time. The only caution here is that current vectors are plotted as the direction the current is toward (oceanographic convention) whereas winds are sometimes plotted as the direction the wind is from (meteorological convention). The progressive vector diagram (PVD) also is used to plot vector velocity time series (Figure 2.5). In this case, the time-integrated displacements along each of two orthogonal directions (x, y) are calculated from the corresponding velocity components $(x, y) = (x_o, y_o) + \sum(u_i, v_i)\Delta t_i$, $(i = 1, 2, ...)$ to give "pseudo" downstream displacements of a parcel of water from its origin (x_o, y_o).

A plot relating one property to another is of considerable value in oceanography. Known as a "characteristic diagram" the most common is that relating temperature and salinity called the *TS* diagram. Originally defined with temperature and salinity values obtained from the same sample bottles, the *TS* relationship was used to detect incorrect bottle samples and to define oceanic water masses. *TS* plots have been shown

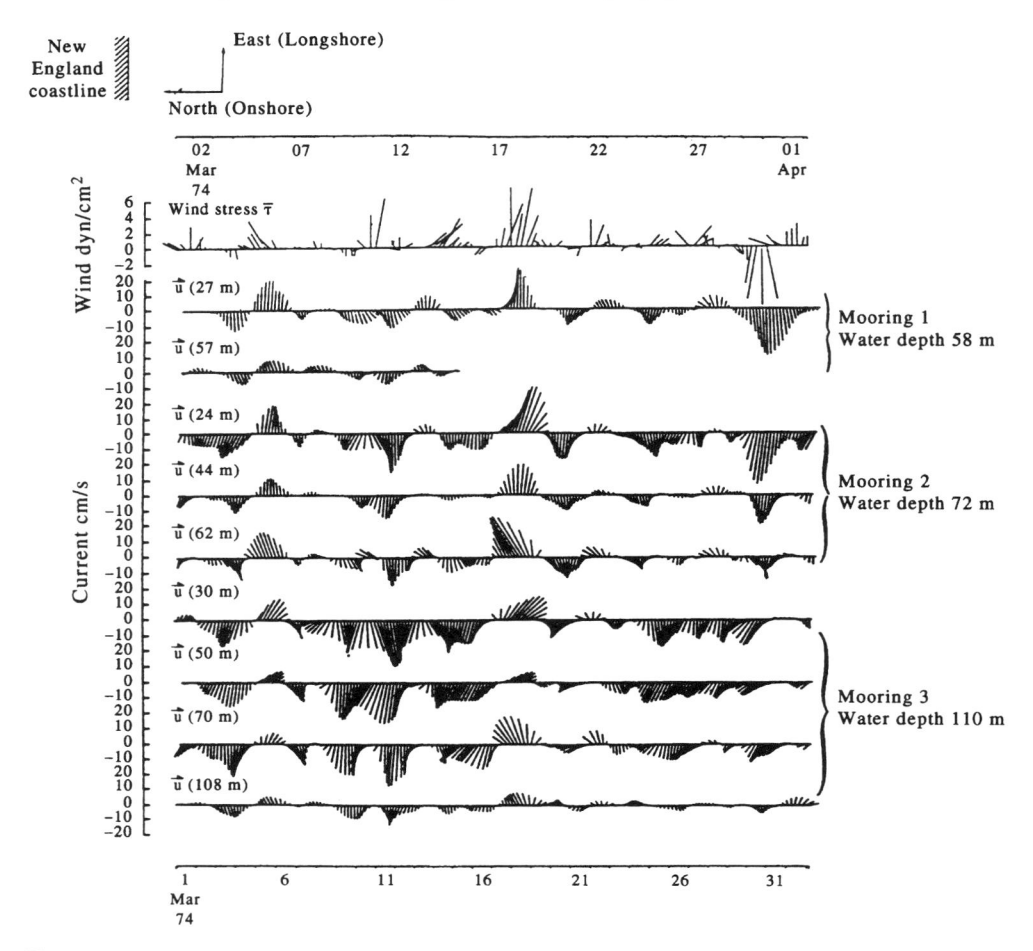

166 *Data Analysis Methods in Physical Oceanography*

Figure 2.4. Vector (stick) plots of low-pass filtered wind stress and subtidal currents at different depths measured along the East Coast of the United States about 100 km west of Nantucket Shoals. East (up) is alongshore and north is cross-shore. Brackets give the current meter depth (m). (Figure 7.11 from Beardsley and Boicourt, 1981.)

to provide consistent relationships over large horizontal areas (Helland-Hansen, 1918) and have recently been the focus of studies into the formation of water masses (McDougal, 1985a, b). Plots of potential temperature versus salinity (the θ–S relationship) or versus potential density (the θ–σ_θ relationship) have proven particularly useful in defining the maximum height of rise of hydrothermal plumes formed over venting sites along mid-ocean ridges (Figure 2.6; Thomson et al., 1992).

Except for some minor changes, vertical profiles, vertical sections, horizontal maps, and time series continue to serve as the primary display techniques for physical oceanographers. The development of electronic instruments, with their rapid sampling capabilities and the growing use of high-volume satellite data, may have changed how we display certain data but most of the basic display formats remain the same. Today, a computer is programmed to carry out both the required computations and to plot the results. Image formats, which are common with satellite data, require further sophisticated interactive processing to produce images with accurate geographical corres-

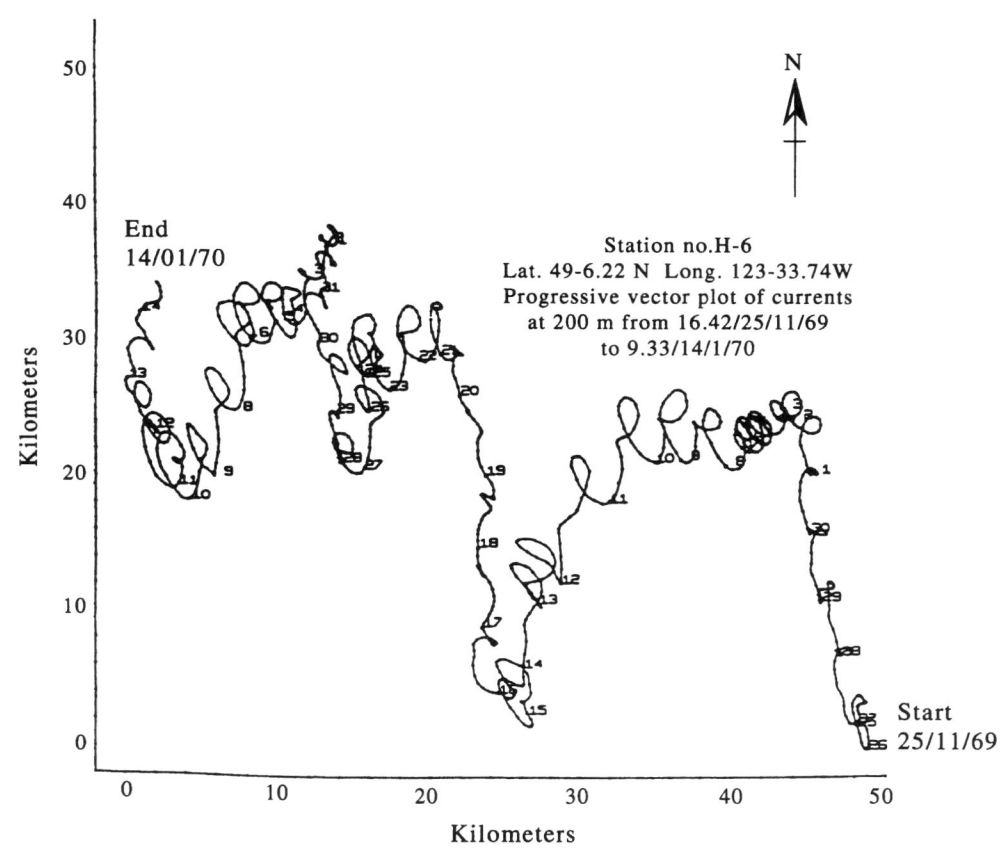

Figure 2.5. Progressive vector diagram (PVD) constructed from the east–west and north–south components of velocity for currents measured every 10 min for a period of 50 days at a depth of 200 m in the Strait of Georgia, British Columbia. Plotted positions correspond to horizontal displacements of the water that would occur if the flow near the mooring location was the same as that at this location. (From Tabata and Stickland, 1972.)

pondence. Despite this, the combination of vertical sections and horizontal maps continues to provide most investigators with the requisite geometrical display capability.

2.4.2 Vertical profiles

Vertical profiles obtained from ships, buoys, aircraft or other platforms provide a convenient way to display oceanic structure (Figure 2.2). One must be careful in selecting the appropriate scales for the vertical and the horizontal property axes. The vertical axis may change scale or vary nonlinearly to account for the marked changes in the upper ocean compared with the relative homogeneity of the lower layers. The property axis needs to have a fine enough scale so as to define the small vertical gradients in the deeper layer without the upper layer going off-scale. When considering a variety of different vertical profiles together (Figures 2.7 and 2.8), a common property scale is an advantage although consideration must be given to the strong dependence of vertical property profiles on latitude and season.

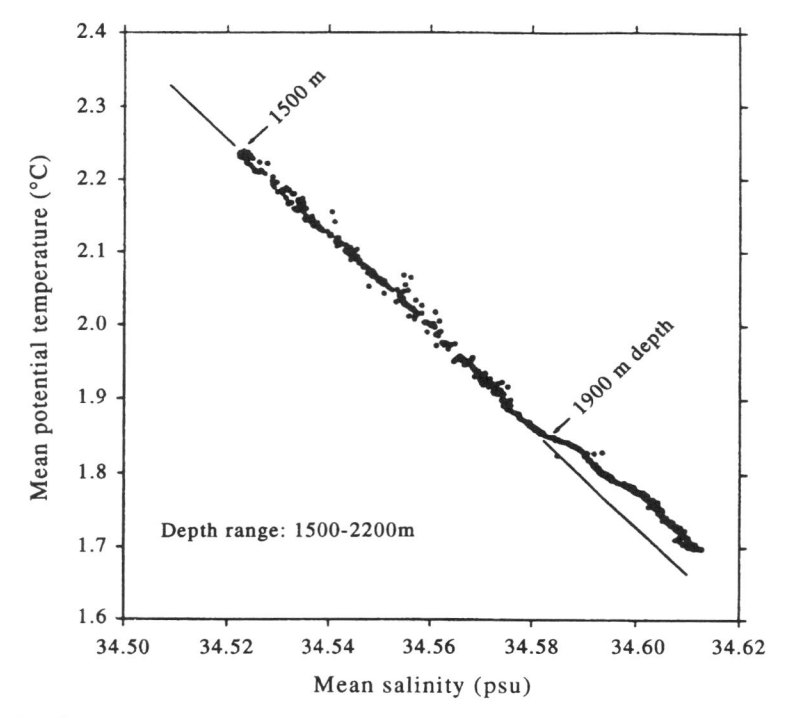

Figure 2.6. Plot of mean potential temperature (θ) versus mean salinity (S) for depths of 1500–2200 m over Endeavour Ridge in the northeast Pacific. The least squares linear fit covers the depth range 1500–1900 m, where $\overline{\theta} = -6.563\overline{S} + 228.795°C$. The abrupt change in the θ–S relationship at a depth of 1900 m marks the maximum height of rise of the hydrothermal plume. (From Thomson et al., 1992.)

A dramatic change has taken place recently in the detailed information contained in vertical profiles. The development and regular use of continuous, high-resolution, electronic profiling systems have provided fine-structure information previously not possible with standard hydrographic casts. Profiles from standard bottle casts required smooth interpolation between observed depths so that structures finer in scale than the smallest vertical sampling separation were missed. Vertical profiles from modern CTD systems are of such high resolution that they are generally either vertically averaged or subsampled to reduce the large volume of data to a manageable level for display. For example, with the rapid (≈10 Hz) sampling rates of modern CTD systems, parameters such as temperature and salinity, which are generated approximately every 0.01 m, are not presentable in a plot of reasonable size. Thus data are either averaged or subsampled to create files with sampling increments of 1 m or larger.

Studies of fine-scale (centimeter scale) variability require the display of full CTD resolution and will generally be limited to selected portions of the vertical profile. These portions are chosen to reflect that part of the water column of greatest concern for the study. Full-resolution CTD profiles reveal fine-scale structure in both T and S, and can be used to study mixing processes such as interleaving and double-diffusion. Expressions of these processes are also apparent in full-resolution TS diagrams using CTD data. One must be careful, however, not to confuse instrument noise (e.g. those due to vibrations or "strumming" of the support cable caused by vortex shedding)

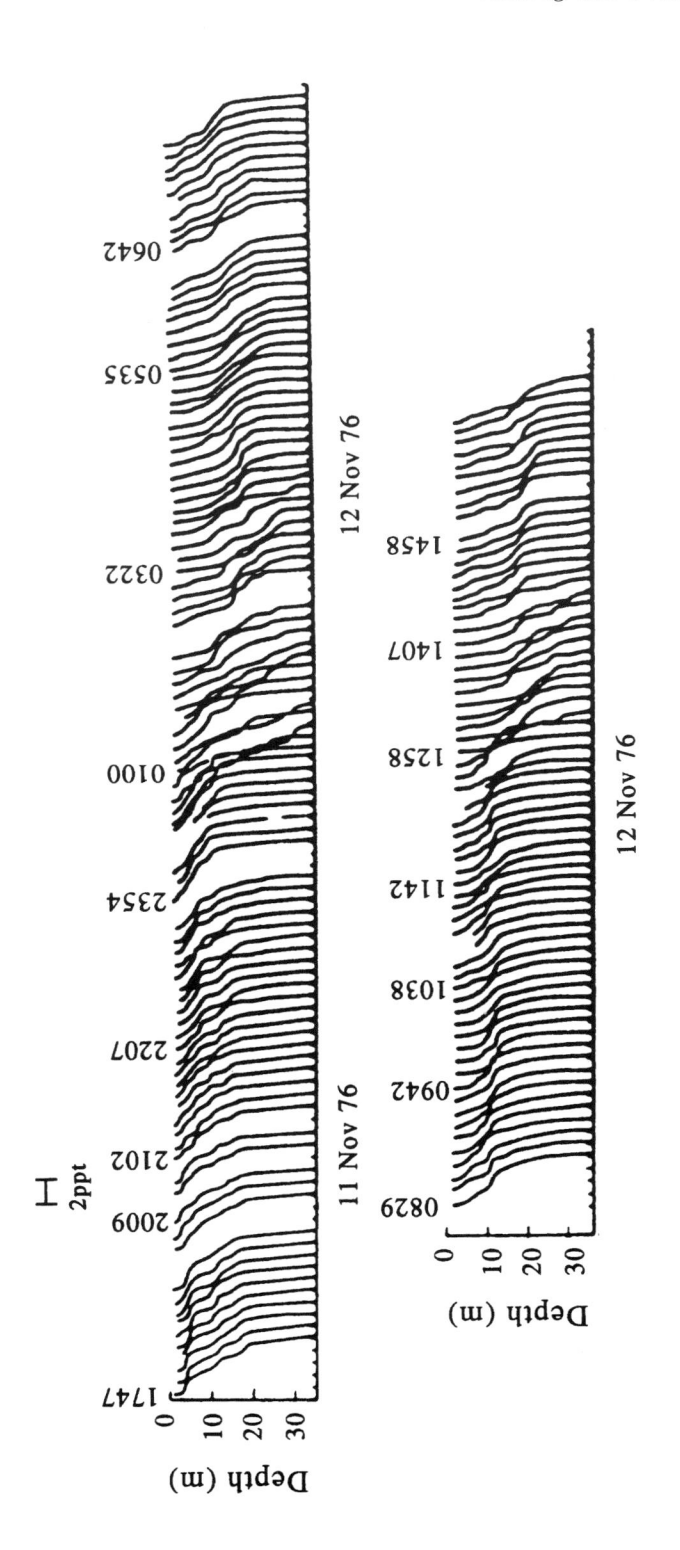

Figure 2.7. Time series of salinity profiles ("waterfall plot") taken in a highly stratified fjord. The effects of large internal waves can be seen around 0100 and 1300 on 12 November. (From Farmer and Smith, 1980.)

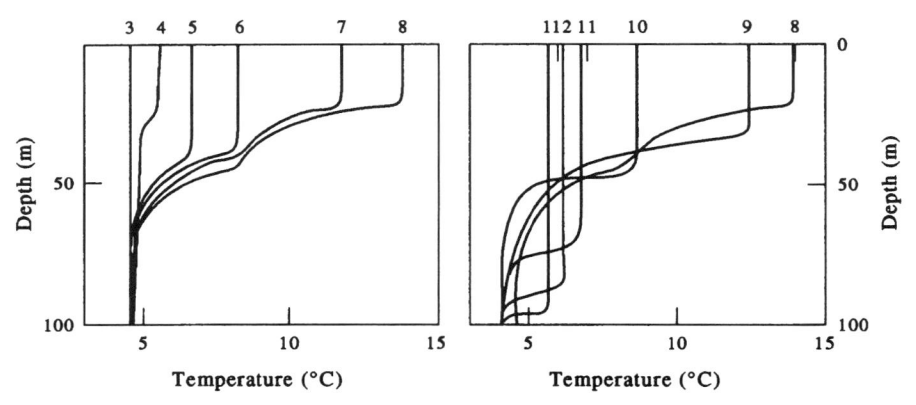

Figure 2.8. Time series of monthly mean profiles of upper ocean temperature at Ocean Weather Station "P", northeast Pacific (50°N, 145°W). Numbers denote the months of the year. (From Pickard and Emery, 1992.)

with fine-scale oceanic structure. Processing should be used where possible to separate the instrument noise from the wave number, band-limited signal of mixing processes.

Often, computer programs for processing CTD data contain a series of different display options that can be used to manipulate the stored high-resolution digital data. The abundance of raw CTD digital data, and the variety of *in situ* calibration procedures, make it difficult to interpret and analyze CTD records using a universal format. This is a fundamental problem in assembling a historical file of CTD observations. Hopefully, the statistics of CTD data that have been smoothed to a resolution comparable to that of traditional bottle casts are sufficiently homogeneous to be treated as updates to the hydrographic station data file. The increasingly wide use of combined CTD and rosette profiling systems has led to a dramatic decrease in the number of standard bottle casts. (A rosette system consists of a carrousel holding 12 or so hydro bottles that can be "tripped" from the ship by sending an electric pulse down the conducting CTD support cable. The CTD is generally placed in the center of the carrousel.)

2.4.3 Vertical sections

Vertical sections are a way to display vertically profiled data collected regionally along the track of a research vessel or taken from more extended crossings of an ocean basin (usually, meridionally or zonally). Marked vertical exaggeration is necessary to make oceanic structure visible in these sections. A basic assumption in any vertical section is that the structure being mapped has a persistence scale longer than the time required to collect the section data. Depending on the type of data collected at each station, and on the length of the section, shipboard collection times can run from a few days to a few weeks. Thus, only phenomena with time scales longer than these periods are properly resolved by the vertical sections. Recognizing this fact leads to a trade-off between spatial resolution (between-station spacing) and the time to complete the section. Sampling time decreases as the number of profiles decreases and the samples taken approach a true *synoptic* representation (samples collected at the same time). Airborne surveys using expendable probes such as AXBTs (airborne XBTs) from fixed-wing aircraft and helicopters yield much more synoptic information but are limited in the type of measurement that can be made and by the depth range of a given measurement. Although aircraft often have hourly charge-out rates that are similar to ships and generally are more cost-effective than ships on a per datum basis, operation of aircraft is usually the domain of the military or coastguard.

Fewer sample profiles means wider spacing between stations and reduced resolution of smaller, shorter-term variability. There is a real danger of short time-scale or space-scale variability aliasing quasi-synoptic, low-resolution vertical sections. Thus, the data collection scheme must be designed to either resolve or eliminate (by filtering) scales of oceanic variability shorter than those being studied. With the ever-increasing interest in ocean climate, and at a time when the importance of mesoscale oceanic circulation features has been recognized, investigators should give serious consideration to their intended sampling program to optimize the future usefulness of the data collected.

Traditional bottle hydrographic casts were intended to resolve the slowly changing background patterns of the property distributions associated with the mean "steady-state" circulation. As a result, station spacings were usually too large to adequately resolve mesoscale features. In addition, bottle casts require long station times leading to relatively long total elapsed times for each section. The fact that these data have provided a meaningful picture of the ocean suggests that there is a strong component of the oceanic property distributions related to the steady-state circulation. For these reasons, vertical sections based on traditional bottle-cast station data provide useful definitions of the meridional and zonal distributions of individual water masses (Figure 2.1).

The importance of mesoscale oceanic variability has prompted many oceanographers to decrease their sample spacing. Electronic profiling systems, such as the CTD and CTD-rosette, require less time per profile than standard bottle casts so that the total elapsed time per section has been reduced over the years despite the need for greater spatial resolution. Still, most oceanographic sections are far from being synoptic owing to the low speeds of ships and some consideration must be given to the definition of which time/space scales are actually being resolved by the measurements. For example, suppose we wish to survey a 1000 km oceanic section and collect a meagre 20 salinity–temperature profiles to 2000 m depth along the way. At an average speed of 12 knots, steaming time alone will amount to about two days. Each bottle cast would take about two hours and each CTD cast about one hour. Our survey time would range from three to four days, which is just marginally synoptic by most oceanographers' standards.

Expendable profiling systems such as the XBT make it possible to reduce sampling time by allowing profile collection from a moving ship. Ships also can be fitted with an acoustic current profiling system which allows for the measurement of ocean currents in the upper few hundred meters of the water column while the ship is underway. The depth of measurement is determined by frequency and is about 500 m for the commonly used 150 kHz transducers. Most modern oceanographic vessels also have SAIL (Shipboard ASCII Interrogation Loop) systems for rapid (≈ 1 min) sampling of the near-surface temperature and salinity at the intake for the ship's engine cooling system. SAIL data are typically collected a few meters below the ship's waterline. Oceanographic sensor arrays towed in a saw-tooth pattern behind the ship provide another technique for detailed sampling of the water column. This method has wide application in studying near-surface fronts, turbulent microstructure, and hydrothermal venting (Figure 1.11.6). These technological improvements have lowered the sample time and increased the vertical resolution.

As referred to earlier, it is common practice when plotting sections to divide the vertical axis into two parts, with the upper portion greatly expanded to display the larger changes of the upper layer. The contour interval used in the upper part may be larger than that used for the weaker vertical gradients of the deeper layer. It is important, however, to maintain a constant interval within each layer to faithfully represent the gradients. In regions with particularly weak vertical gradients, additional contours may be added but a change in line weight, or type, is customary to distinguish the added line from the other contours. All contours must be clearly labeled. Color is often very effective in distinguishing gradients represented by the contours. While it is common practice to use shades of red to indicate warm regions, and shades of blue for cold, there is no recommended color coding for properties such as salinity, dissolved oxygen or nutrients. The color atlas of water properties for the Pacific Ocean published by Reid (1965) provides a useful color scheme.

In sections derived from bottle samples, individual data points are usually indicated by a dot or by the actual data value. In addition, the station number is indicated in the margin above or below the profile. Stations collected with CTDs usually have the station position indicated but no longer have dots or sample values for individual data points. Because of the high vertical resolution, only the contours are plotted.

The horizontal axis usually represents distance along the section and many sections have a small inset map showing the section location. Alternatively, the reader is referred to another map which shows all section locations. Since many sections are taken along parallels of latitude or meridians of longitude, it is customary to include the appropriate latitude or longitude scale at the top or bottom of each section (Figure 2.3). Even when a section only approximates zonal or meridional lines, estimates of the latitude or longitude are frequently included in the axis label to help orient the reader. Station labels should also be added to the axis.

A unique problem encountered when plotting deep vertical sections of density is the need to have different pressure reference levels for the density determination to account for the dependence of sea-water compressibility on temperature. Since water temperature generally decreases with pressure (greater depths), artificially low densities will be calculated at the greatest depths when using the surface pressure as a reference (Lynn and Reid, 1968, Reid and Lynn, 1971). When one wants to resolve the deep density structure, and at the same time display the upper layer, different reference levels are used for different depth intervals. As shown in Figure 2.9, the resulting section has discontinuities in the density contours as the reference level changes.

A final comment about vertical sections concerns the representation of bottom topography. The required vertical exaggeration makes it necessary to represent the bottom topography on an exaggerated scale. This often produces steep-looking islands and bottom relief. There is a temptation to ignore bottom structure, but as oceanographers become more aware of the importance of bottom topography in dictating certain aspects of the circulation, it is useful to include some representation of the bottom structure in the sections.

2.4.4 Horizontal maps

In the introduction, we mentioned the early mapping of ocean surface properties and bottom depths. Following established traditions in map making, these early maps

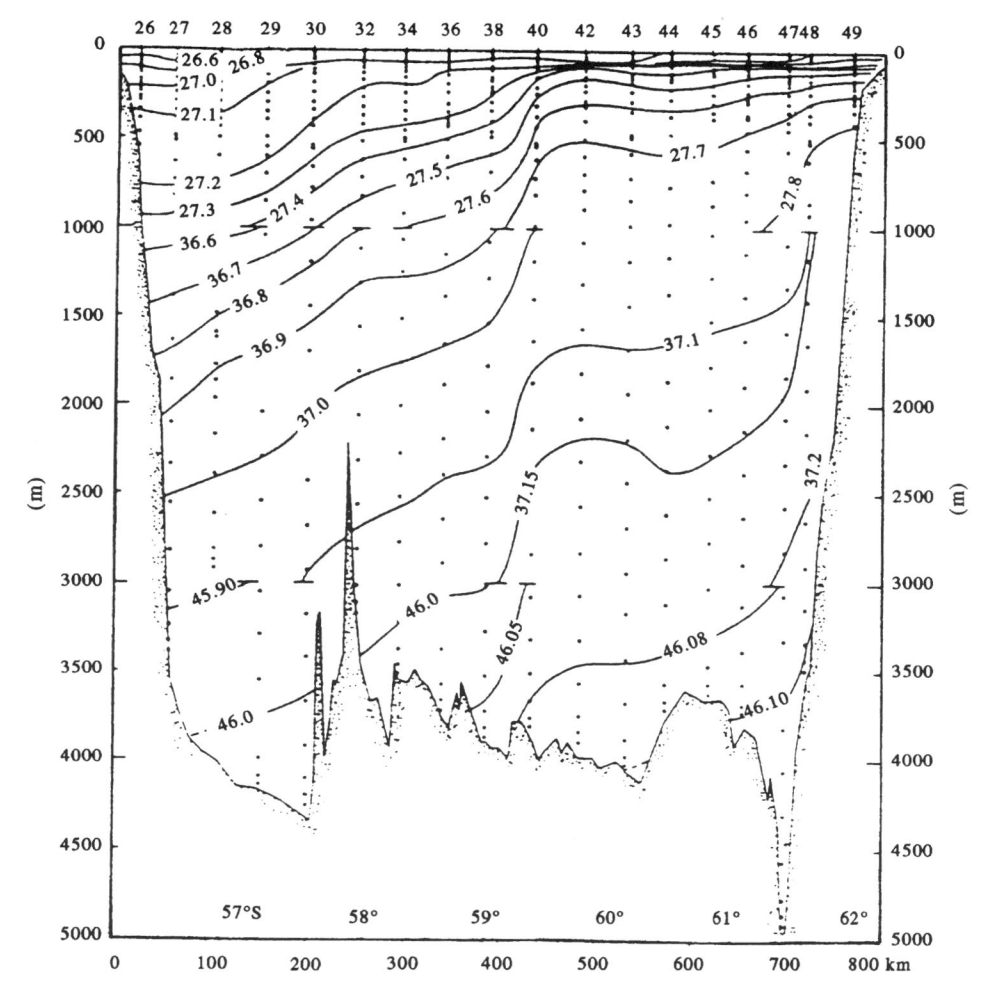

Figure 2.9. Cross-section of density (σ_t) (kg/m^3) across Drake Passage in 1976. (From Nowlin et al., 1986.)

were as much works of art as they were representations of oceanographic information. The collection of hydrographic profiles later made it possible to depict property distributions at different levels of the water column (Figure 2.10). As with vertical sections, the question of sample time versus horizontal resolution needs to be addressed, especially where maps cover large portions of an ocean basin. Instead of the days to weeks needed to collect data along a single short section, it may take weeks, months and even years to obtain the required data covering large geographical regions. Often, horizontal maps consist of a collection of sections designed to define either the zonal/meridional structure or cross-shore structure for near-coastal regions. In most cases, the data presented on a map are contoured with the assumption that the map corresponds to a stationary property distribution. For continental shelf regions, data used in a single map should cover a time period that is less than the approximately 10 day e-folding time scale of mesoscale eddies. In this context, the "e-folding time" is the time for the mesoscale currents to decay to $1/e^1 = 0.368$ of their peak values.

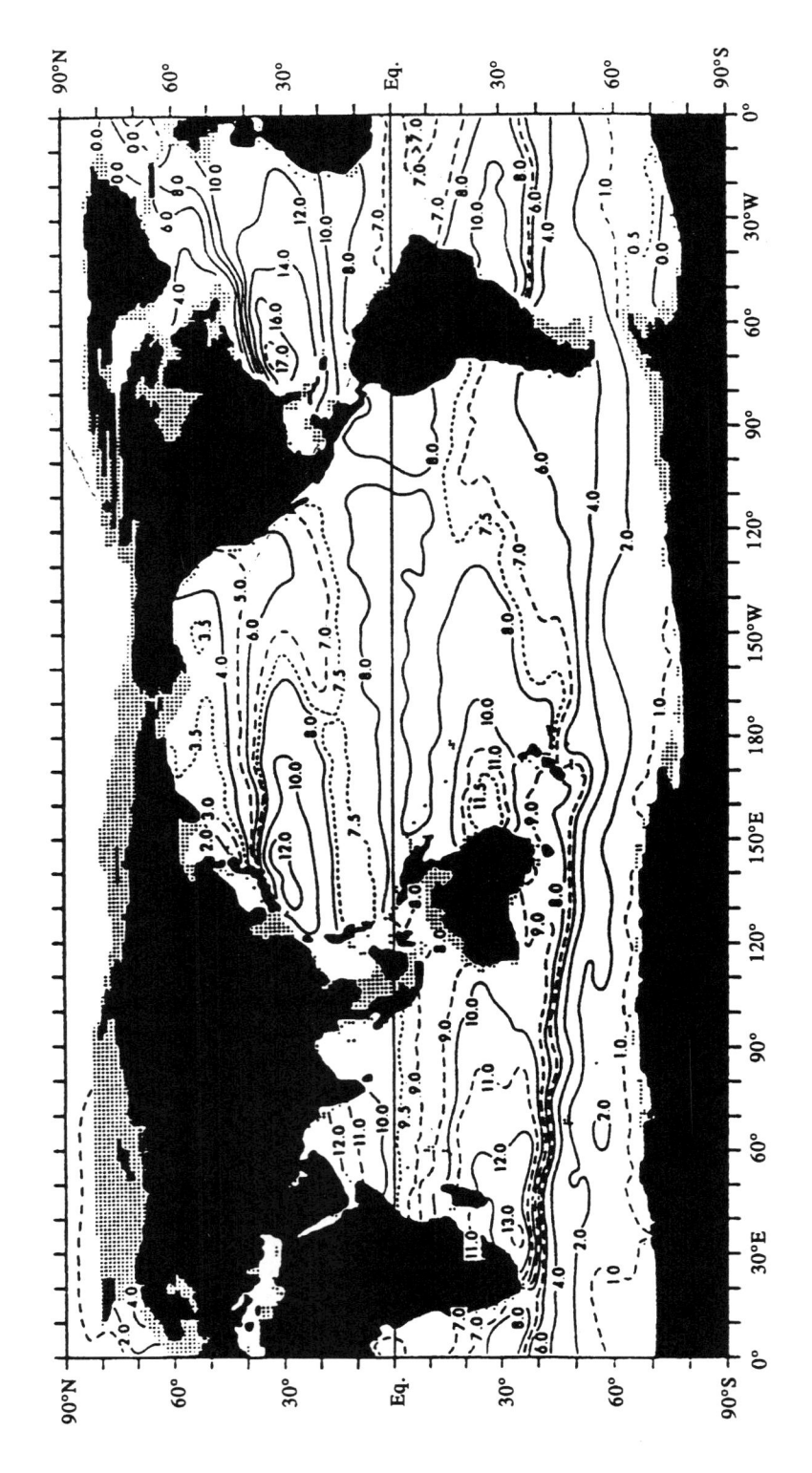

174 *Data Analysis Methods in Physical Oceanography*

Figure 2.10. Horizontal maps of annual mean potential temperature in the world ocean at (a) 500 m. (From Levitus, 1982.)

Data Processing and Presentation 175

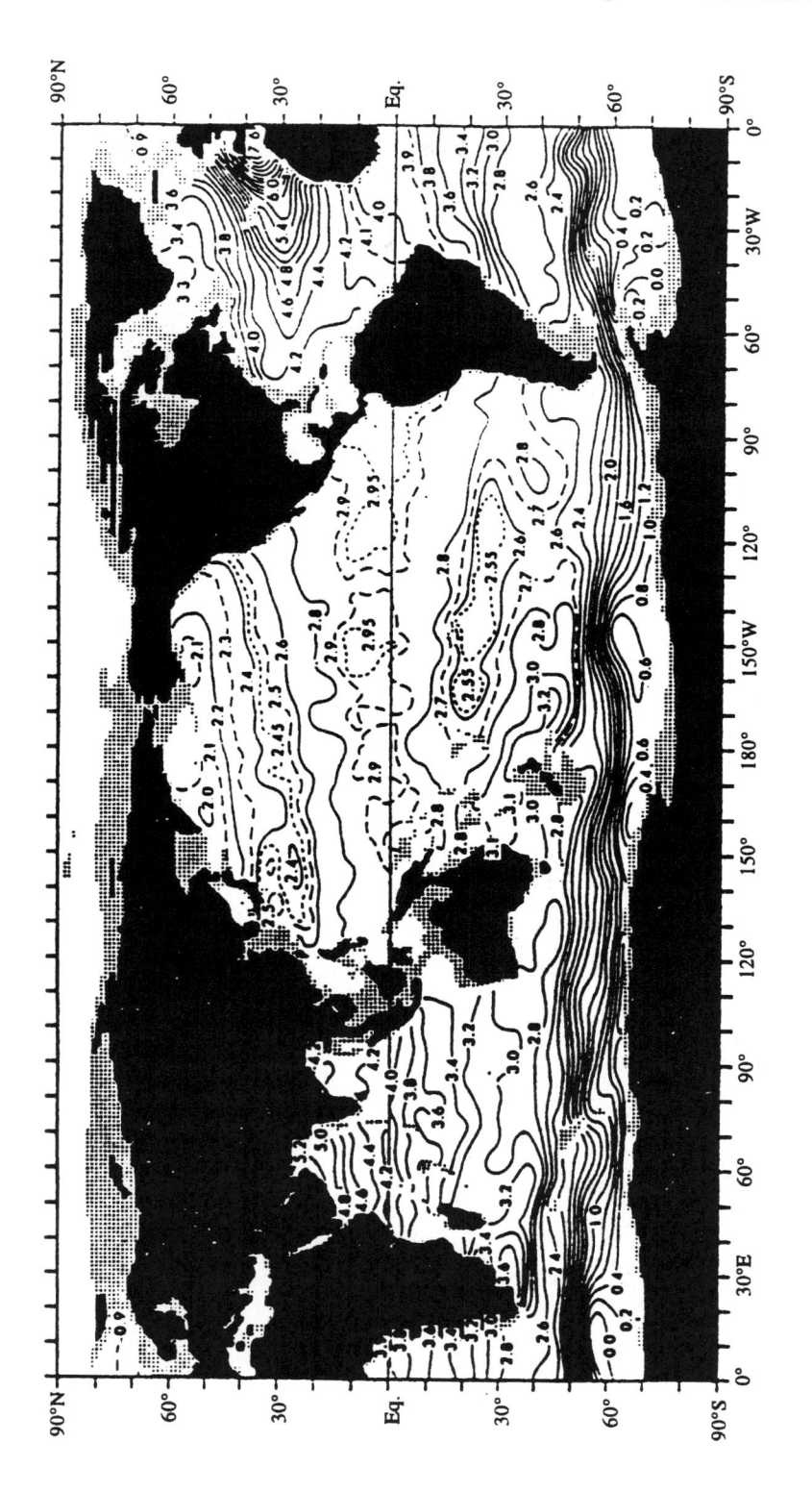

Figure 2.10. Horizontal maps of annual mean potential temperature in the world ocean at (b) 1000 m depth. (From Levitus, 1982.)

Much of what we know about the overall structure of the ocean, particularly the deep ocean, has been inferred from large-scale maps of water properties. A presentation developed by Wüst (1935) to better display the horizontal variations of particular water masses is based on the *core-layer* method. Using vertical property profiles, vertical sections, and characteristic (one property versus another property) diagrams, Wüst defined a core-layer as a property extremum and then traced the distribution of properties along the surface defined by this extremum. Since each core layer is not strictly horizontal, it is first necessary to present a map showing the depth of the core-layer in question. Properties such as temperature, salinity, oxygen, and nutrients also can be plotted along these layers in addition to the percentage of the appropriate water mass defined from the characteristic diagrams. A similar presentation is the plotting of properties on selected density surfaces. This practice originated with Montgomery (1938) who argued that advection and mixing would occur most easily along surfaces of constant entropy. Since these isentropic surfaces are difficult to determine, Montgomery suggested that surfaces of constant potential density would be close approximations in the lower layers and that sigma-t would be appropriate for the upper layers. Known as *isentropic analysis* because of its thermodynamic reasoning, this technique led to the practice of presenting horizontal maps on sigma-t or sigma-θ (potential density) surfaces. While it may be difficult to visualize the shape of the density surfaces, this type of format is often better at revealing property gradients. As with the core-layer method, preparing maps on density surfaces includes the plotting of characteristic property diagrams to identify the best set of density surfaces. Inherent in this type of presentation is the assumption that diapycnal (cross-isopycnal) mixing does not occur. Sometimes steric surfaces or surfaces of thermosteric anomaly are chosen for plotting rather than density.

The definition and construction of contour lines on horizontal maps has evolved in recent years from a subjective hand-drawn procedure to a more objective procedure carried out by a computer. Hand analyses usually appear quite smooth but it is impossible to adequately define the smoothing process applied to the data since it varies with user experience and prejudice. Only if the same person contoured all of the data, is it possible to compare map results directly. Differences produced by subjective contouring are less severe for many long-term and stationary processes, which are likely to be well represented regardless of subjective preference. Shorter-term and smaller space-scale variations, however, will be treated differently by each analyst and it will be impossible to compare results. In this regard, we note that weather maps used in six-hourly weather forecasts are still drawn by hand since this allows for needed subjective decisions based on the accumulated experience of the meteorologist.

Objective analysis and other computer-based mapping procedures have been developed to carry out the horizontal mapping and contouring. Some of these methods are presented individually in later sections of this text. Since there is such a wide selection of mapping methods, it is not possible to discuss each individually. However, the reader is cautioned in applying any specific mapping routine to ensure that any implicit assumptions are satisfied by the data being mapped. The character of the result needs to be anticipated so that the consequences of the mapping procedure can be evaluated. For example, the mapping procedure called objective analysis or optimum interpolation, is inherently a smoothing operation. As a consequence, the output gridded data may be smoothed over a horizontal length scale greater than the scale of interest in the study. One must decide how best to retain the variability of interest and still have a definable mapping procedure for irregularly spaced data.

2.4.5 Map projections

One neglected aspect of mapping oceanographic variables is the selection of an appropriate map projection. A wide variety of projections has been used in the past. The nature of the analysis, its scale and geographic region of interest dictate the type of map projection to use (Bowditch, 1977). Polar studies generally use a conic or other polar projection to avoid distortion of zonal variations near the poles. An example of a simple conic projection for the northern hemisphere is given in Figure 2.11. In this case, the cone is tangent at a single latitude (called a standard parallel) which can be selected by changing the angle of the cone (Figure 2.11a). The resulting latitude–longitude scales are different around each point and the projection is said to be nonconformal (Figure 2.11b). A conformal (=orthomorphic; conserves shape and angular relationships) conic projection is the Lambert conformal projection which cuts the earth at two latitudes. In this projection, the spacing of latitude lines is altered so that the distortion is the same as along meridians. This is the most widely used conic projection for navigation since straight lines nearly correspond to great circle routes. A variation of this mapping is called the "modified Lambert conformal projection". This projection amounts to selecting the top standard parallel very near the pole, thus closing off the top of the map. Such a conic projection is conformal over most of its domain. Mention should also be made of the "polar stereographic projection" that is favored by meteorologists. Presumably, the advantages of this projection is its ability to cover an entire hemisphere, and its low distortion at temperate latitudes.

At mid- and low-latitudes, it is common to use some form of Mercator projection which accounts for the meridional change in earth radius by a change in the length of the zonal axis. Mercator maps are conformal in the sense that distortions in latitude and longitude are similar. The most common of these is the transverse Mercator or cylindrical projection (Figure 2.12). As the name implies it amounts to projecting the earth's surface onto a cylinder which is tangent at the equator (equatorial cylindrical). This type of projection, by definition, cannot include the poles. A variant of this is called the oblique Mercator projection, corresponding to a cylinder which is tangent to the earth along a line tilted with respect to the equator. Unlike the equatorial

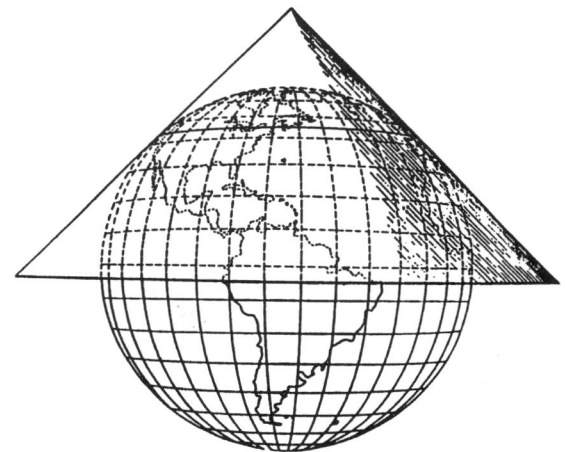

Figure 2.11. An example of a simple conic projection for the northern hemisphere. The single tangent cone in (a) is used to create the map in (b). (From Bowditch, 1977.)

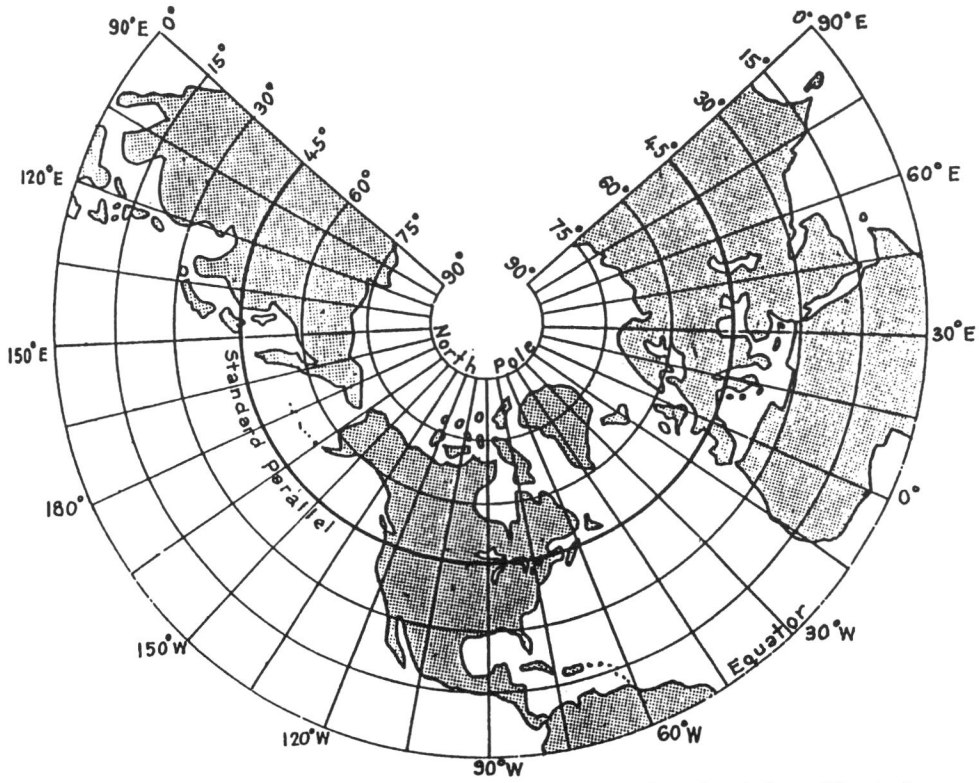

178   *Data Analysis Methods in Physical Oceanography*

Figure 2.11. An example of a simple conic projection for the northern hemisphere. The single tangent cone in (a) is used to create the map in (b). (From Bowditch, 1977.)

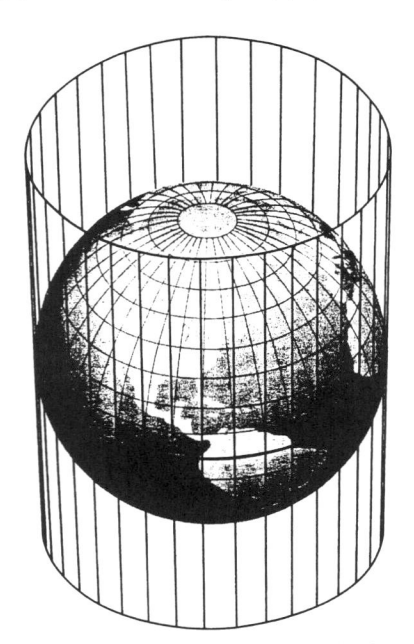

Figure 2.12. The transverse Mercator or cylindrical projection. (From Bowditch, 1977.)

(a)

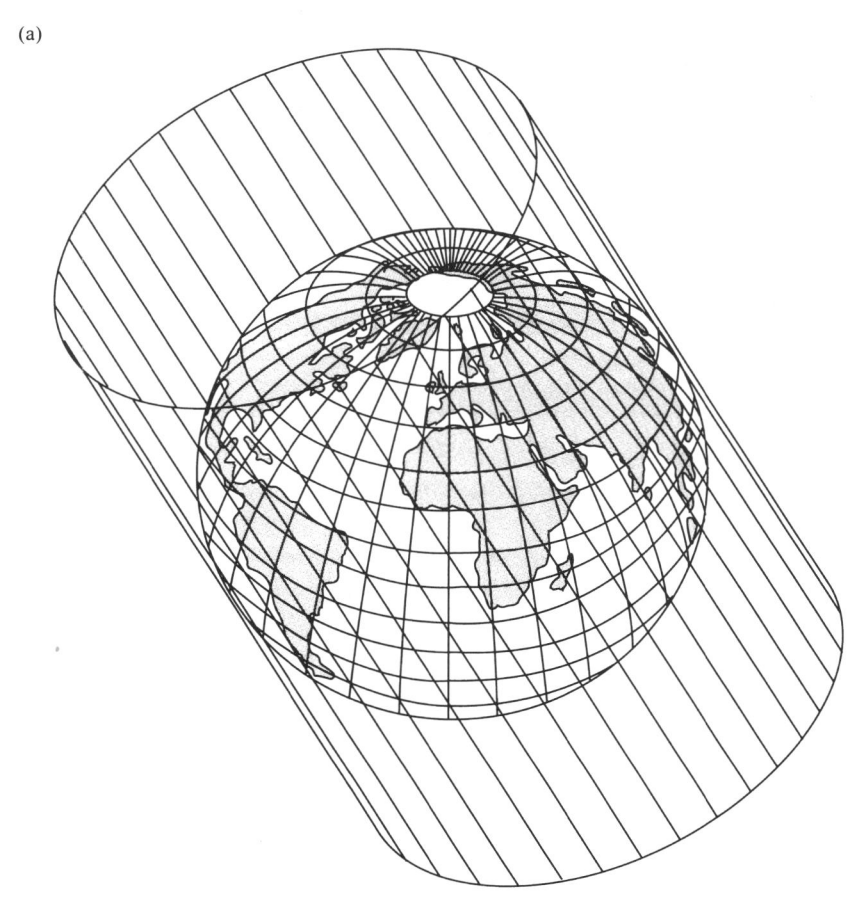

Figure 2.13. An oblique Mercator or oblique cylindrical projection that includes the poles. The cylinder in (a) is used to generate the transverse Mercator map of the western hemisphere in (b). (From Bowditch, 1977.)

cylindrical this oblique projection can represent the poles (Figure 2.13a). This form of Mercator projection also has a conformal character, with equal distortions in lines of latitude and longitude (Figure 2.13b). The most familiar Mercator mapping is the universal transverse Mercator (UTM) grid which is a military grid using the equatorial cylindrical projection. Another popular mid-latitude projection is the rectangular or equal-area projection which is a cylindrical projection with uniform spacing between lines of latitude and lines of longitude. In applications where actual earth distortion is not important, this type of equal area projection is often used. Whereas Mercator projections are useful for plotting vectors, equal-area projections are useful for representing scalar properties. For studies of limited areas, special projections may be developed such as the azimuthal projection, which consists of a projection onto a flat plane tangent to the earth at a single point. This is also called a gnomonic projection. Stereographic projects perform similar projections; however, where gnomonic projections use the center of the earth as the origin, stereographic projections use a point on the surface of the earth.

The effects of map projection on mapped oceanographic properties should always be considered. Often the distortion is unimportant since only the distribution relative

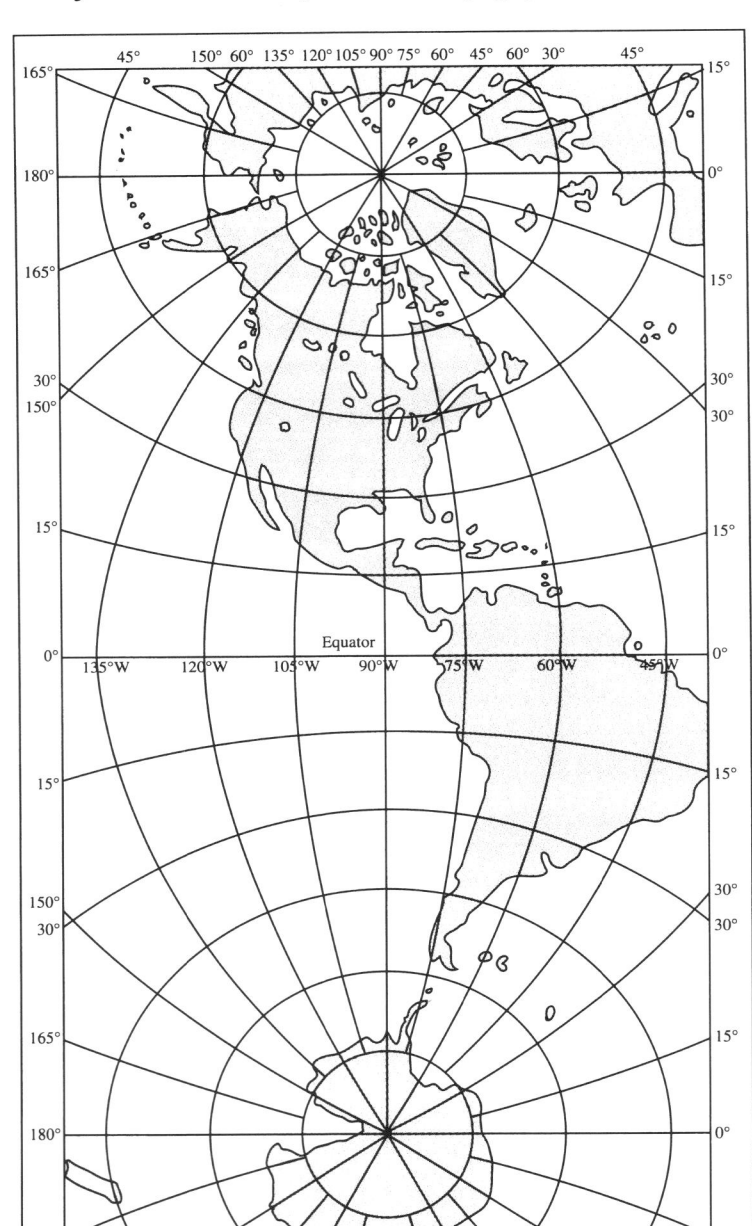

Figure 2.13. An oblique Mercator or oblique cylindrical projection that includes the poles. The cylinder in (a) is used to generate the transverse Mercator map of the western hemisphere in (b). (From Bowditch, 1977.)

2.4.6 Characteristic or property versus property diagrams

In many oceanographic applications, it is useful to relate two simultaneously observed variables. Helland-Hansen (1918) first suggested the utility of plotting temperature (T) against salinity (S). He found that TS diagrams were similar over large areas of the ocean and remained constant in time at many locations. An early application of the TS diagram was the testing and editing of newly acquired hydrographic bottle data. When compared with existing TS curves for a particular region, TS curves from newly collected data quickly highlighted erroneous samples which could then be corrected or eliminated. Similar characteristic diagrams were developed for other ocean properties. Many of these, however, were not conservative and could not be expected to exhibit the constancy of the TS relationship (we will use TS as representative of all characteristic diagrams.)

As originally conceived, characteristic diagrams such as the TS plots were straightforward to construct. Pairs of property values from the same water bottle sample constituted a point on the characteristic plot. The connected points formed the TS curve for the station (Figure 2.14). Each TS curve represented an individual oceanographic station and similarities between stations were judged by comparing their TS curves. These traditional TS curves exhibit a unique relationship between T, S, and Z (the depth of the sample). What stays constant is the TS relationship, not its correspondence with Z. As internal waves, eddies, and other unresolved dynamical

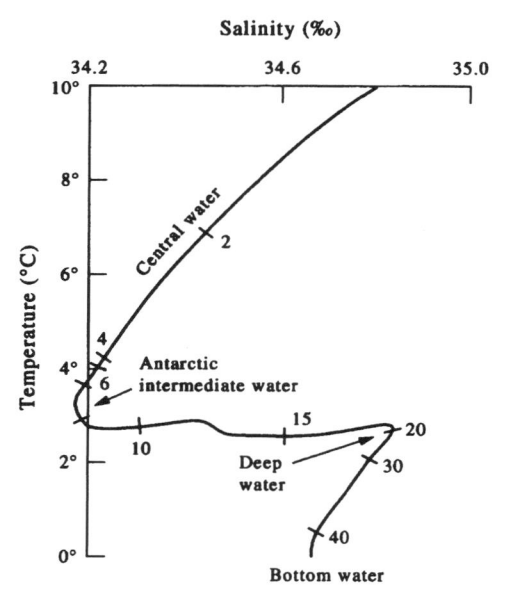

Figure 2.14. Temperature–salinity curve for the western basin of the South Atlantic at 41°S latitude. Depths are marked in hundreds of meters. (Adapted from Tchernia, 1980.)

182 *Data Analysis Methods in Physical Oceanography*

features move through a region, the depth of the density structure changes. In response, the paired *TS* value moves up and down along the *TS* curve, thus maintaining the water mass structure. This argument does not hold in frontal zones where the water mass itself is being modified by mixing and interleaving.

Temporal oceanic variability has important consequences for the calculation of mean *TS* diagrams where *TS* pairs, from a number of different bottle or CTD casts, are averaged together to define the *TS* relationship for a given area or lapsed time interval. Perhaps the easiest way to present this information is in the form of a scatter plot (Figure 2.15) where the dots represent individual *TS* pairs. The mean *TS* relationship is formulated as the average of S over intervals of T. Depth values have been included in Figure 2.15 and represent a range of Z values spanning the many possible depths at which a single *TS* pair is observed. Thus, it is not possible to define a unique mean T, S, Z relationship for a collection of different hydrographic profiles.

The traditional *TS* curve presented in Figure 2.15 is part of a family of curves relating measured variables such as temperature and salinity to density (sigma-*t*) or thermosteric anomaly ($\Delta_{S,T}$). The curvature of these lines is due to the nonlinear nature of the ocean's equation of state. In a traditional single-cast *TS* diagram, the stability of the water column, represented by the *TS* curve, can be easily evaluated. Unless one is in an unstable region, density should always increase with depth along the *TS* curve. Furthermore the analysis of *TS* curves can shed important light on the advective and mixing processes generating these characteristic diagrams. We note that the thermosteric anomaly, $\Delta_{S,T}$, is used for *TS* curves rather than specific volume

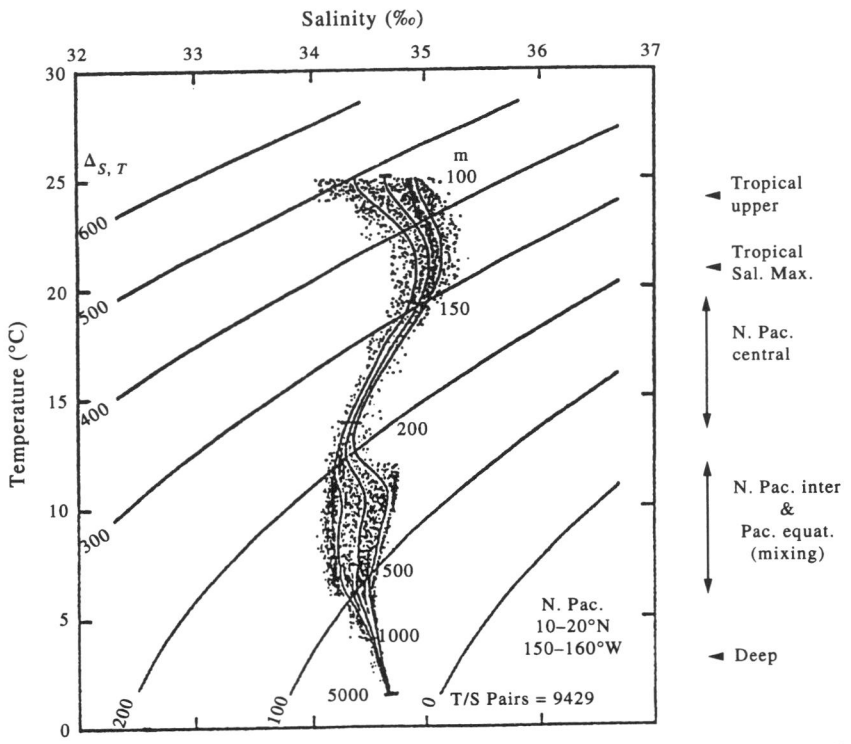

Figure 2.15. Mean temperature–salinity curves for the North Pacific (10–20°N; 150–160°W). Also shown is the density anomaly $\Delta_{S,T}$. (From Pickard and Emery, 1992.)

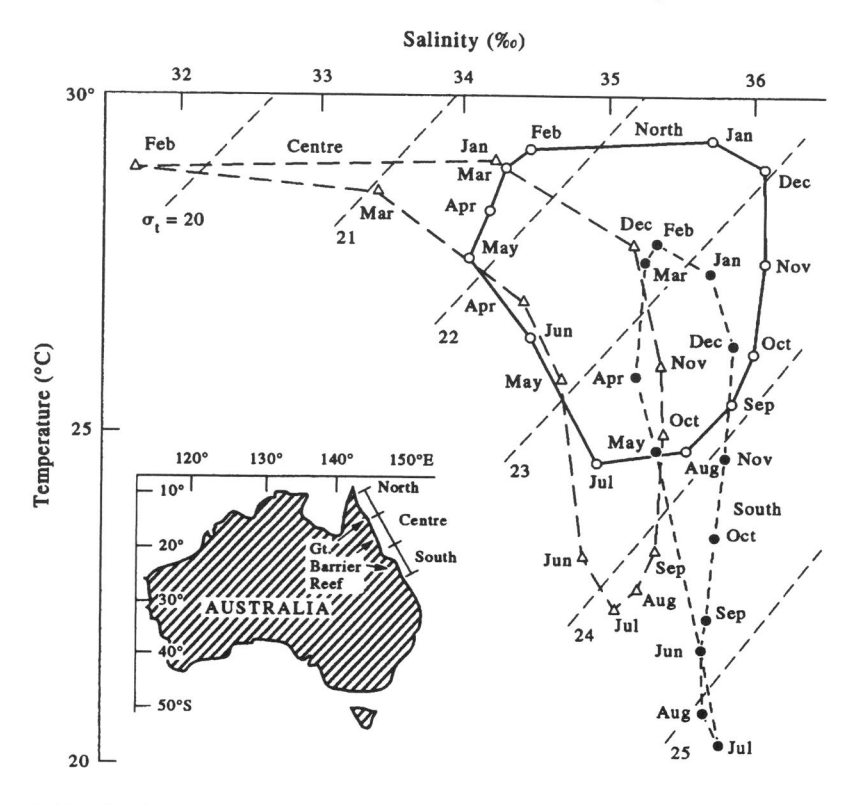

Figure 2.16. Monthly mean temperature–salinity pairs for surface water samples over a year in the lagoon waters of the Great Barrier Reef. (From Pickard and Emery, 1992.)

anomaly, $\delta_{S,T}$, since the pressure term in included in $\delta_{S,T}$ has been found to be negligible for hydrostatic computation and can be approximated by $\Delta_{S,T}$, which lacks the pressure term.

The time variability of the TS relation is also a useful quantity. A simple extension of this characteristic diagram shown in Figure 2.16 reveals the monthly mean TS pairs for surface water samples over a year in the vicinity of the Great Barrier Reef. The dominant seasonal cycle of the physical system is clearly displayed with this format.

Another more widely used variation of the TS diagram is known as the volumetric TS curve. Introduced by Montgomery (1958), this diagram presents a volumetric census of the water mass with the corresponding TS properties. The analyst must decide the vertical and horizontal extent of a given water mass and assign to it certain TS properties. From this information, the volume of the water mass can be estimated and entered on the TS diagram (Figure 2.17). The border values correspond to sums across T and S values. Worthington (1981) used this procedure, and a three-dimensional plotting routine, to produce a volumetric TS diagram for the deep waters of the world ocean (Figure 2.18). The distinct peak in Figure 2.18 corresponds to a common deep water which fills most of the deeper parts of the Pacific. Sayles *et al.* (1979) used the method to produce a good descriptive analysis of Bering Sea water. This type of diagram has been made possible with the development of computer graphics techniques which greatly enhance our ability to display and visualize data.

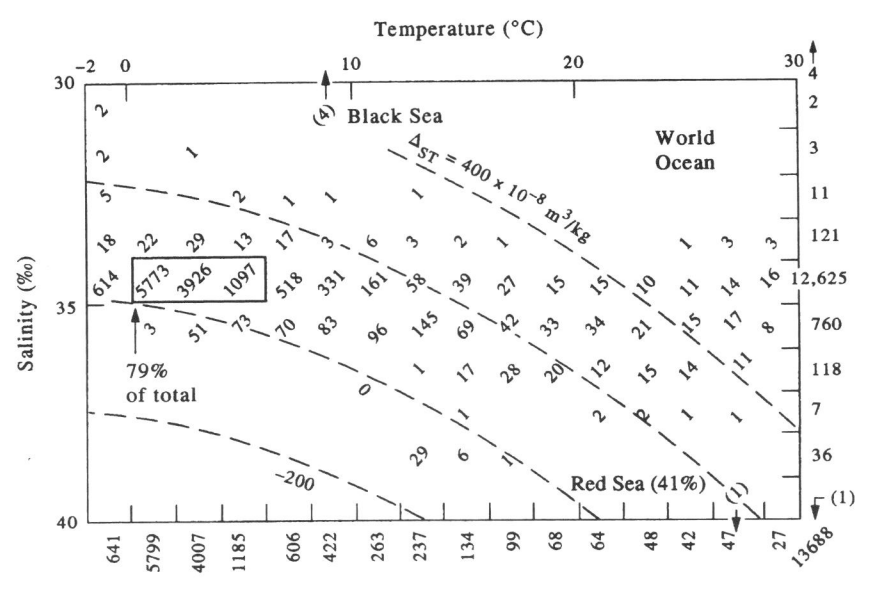

184 *Data Analysis Methods in Physical Oceanography*

Figure 2.17. Volumetric temperature–salinity (T–S) curve in which the number of T–S pairs in each segment of the plot can be calculated. (From Pickard and Emery, 1992.)

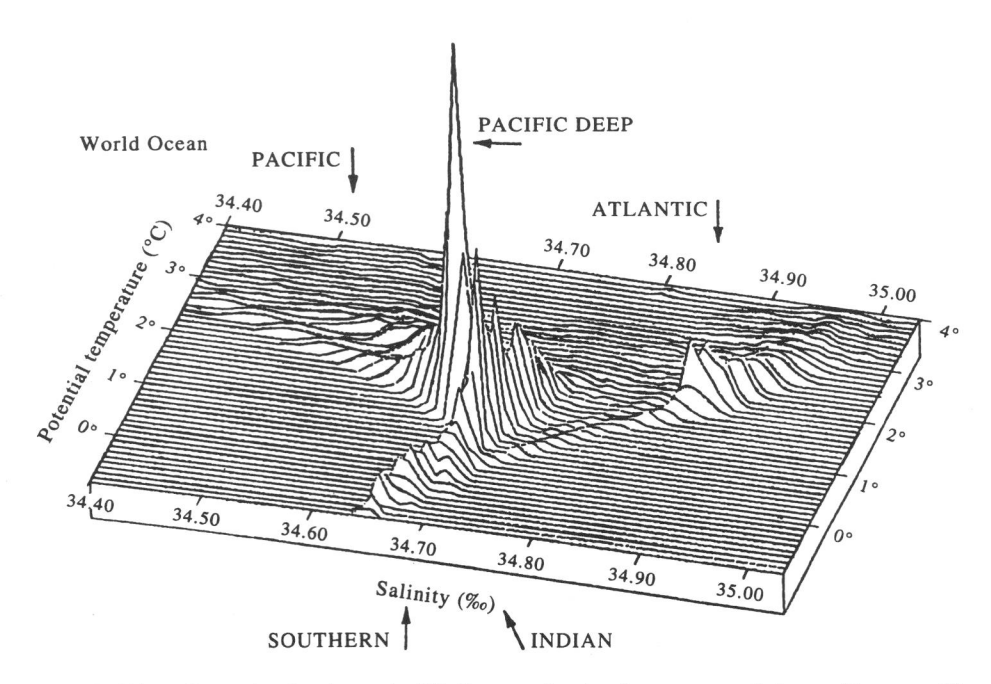

Figure 2.18. Three-dimensional volumetric TS diagram for the deep waters of the world ocean. The distinct peak corresponds to common deep water which fills most of the deeper parts of the Pacific. (From Pickard and Emery, 1992.)

In a highly site-specific application of *TS* curves, McDuff (1988) has examined the effects of different source salinities on the thermal anomalies produced by buoyant hydrothermal plumes rising from mid-ocean ridges. In potential temperature–salinity ($\theta - S$) space, the shapes of the $\theta - S$ curves are strongly dependent on the salinity of the source waters and lead to markedly different thermal anomalies as a function of height above the vent site.

2.4.7 Time-series presentation

In oceanography, as with other environmental sciences, there is a need to present time-series information. Early requirements were generated by shore-based measurements of sea-level heights, sea surface temperature and other relevant parameters. As ship traffic increased, the need for offshore beacons led to the establishment of light- or pilot-ships which also served as platforms for offshore data collection. Some of the early studies, made by geographers in the emerging field of physical oceanography, were carried out from light-ships. The time series of wind, waves, surface currents, and surface temperature collected from these vessels needed to be displayed as a function of time. Later, dedicated research vessels such as weather ships were used as "anchored" platforms to observe currents and water properties as time series. Today, many time-series data are collected by moored instruments which record internally or telemeter data back to a shore station. The need for real-time data acquisition for operational oceanography and meteorology has created an increased interest in new methods of telemetering data. The development of bottom-mounted acoustical modem systems and satellite data collection systems such as Service Argos have opened new

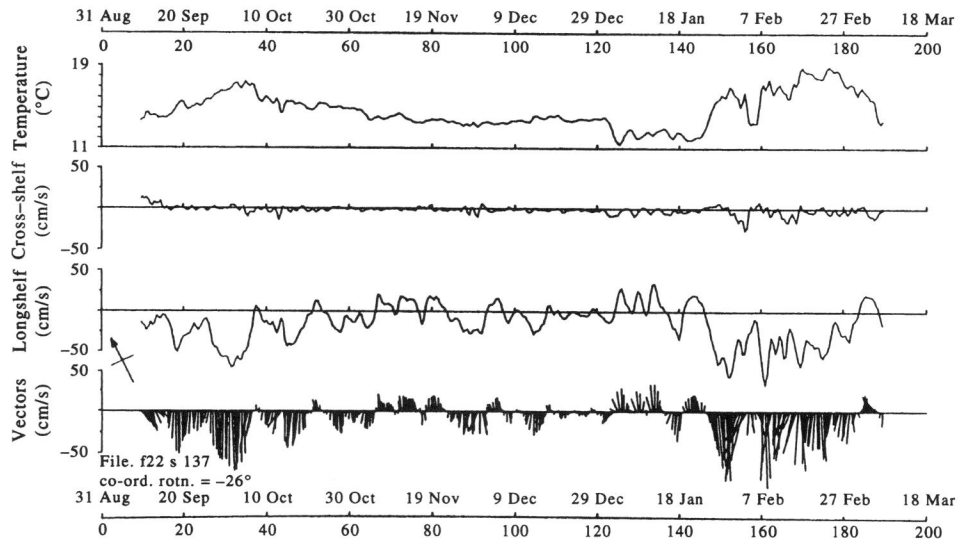

Figure 2.19. Time series of the low-pass filtered u (cross-shelf, x) and v (longshelf, y) components of velocity together with the simultaneously collected values of temperature (T) for the east coast of Australia immediately south of Sydney, 31 August, 1983 to 18 March, 1984. The axes for the stick vectors are rotated by −26° from North so that "up" is in the alongshore direction. The current meter was at 137 m depth in a total water depth of 212 m. Time in Julian days as well as calendar days. (Freeland et al., 1985.)

possibilities for the transmission of oceanographic data to shore stations and for the transmission of operational commands back to the offshore modules.

The simplest way to present time-series information is to plot a scalar variable (or scalar components of a vector series) against time. The time scale depends on the data series to be plotted and may range in intervals from seconds to years. Scalar time series of the u (cross-shore, x) and v (alongshore, y) components of velocity are presented in Figure 2.19 along with the simultaneously collected values of temperature. Note that it is common practice in oceanography to rotate the x, y velocity axes to align them with the dominant geographic or topographic orientation of the study region. The horizontal orthogonal axes can be along- and cross-shore or along- and across-isobath. Sometimes the terms cross-shelf and long-shelf are used in place of cross-shore and longshore. Since current meters and anemometers actually measure speed and direction, it is also customary to display time series of speed and direction as well as components of velocity. Keep in mind that oceanographic convention has vectors of current (and wind) pointing in the direction that the flow is *toward* whereas meteorological convention has wind vectors pointing in the direction the wind is *from*.

As noted in section 2.4.1, two common methods of displaying the actual vector character of velocity as a function of time are stick-plots and progressive vector diagrams (PVDs). A stick-plot (Figures 2.4 and 2.19) represents the current vector for a specific time interval with the length of the stick (or "vector") scaled to the current speed and the stick orientation representing the direction. Direction may be relative to true north (pointed upward on the page) or the coordinate system may be rotated to align the axes with the dominant geographic or topographic boundaries. The stick-plot presentation is ideal for displaying directional variations of the measured currents. Rotational oscillations, due to the tides and inertial currents, are clearly represented. The PVD (Figure 2.5) presents the vector sum of the individual current vectors plotting them head to tail for the period of interest. Residual or long-term vector-mean currents are readily apparent in the PVD and rotational behavior also is well represented. The signature of inertial and tidal currents can be easily distinguished in this type of diagram. The main problem with PVDs is that they have the appearance of a Lagrangian drift with time, as if measurements at one location could tell us the downstream trajectory of water parcels once they had crossed the recording location. Only if the flow is uniform in space and constant in time does the PVD give a true representation of the Lagrangian motion downstream. In that regard, we note that Lagrangian data are presented either as trajectories, in which the position of the drifting object is traced out on a chart or as time series of latitude $x(t)$ and longitude $y(t)$. Distance in kilometers may be used in place of earth coordinates although there are distinct advantages to sticking with Mercator projections.

Another type of time series plot consists of a series of vertical profiles at the same locations as functions of time (Figure 2.20a). The vertical time-series plot has a vertical axis much like a vertical section with time replacing the horizontal distance axis. Similarly, a time series of horizontal transects along a repeated survey line is like a horizontal map but with time replacing one of the spatial axes. Property values from different depth–time (z, t) or distance–time (x, t) pairs are then contoured to produce time-series plots (Figure 2.20b) which look very similar to vertical sections and horizontal maps, respectively. This type of presentation is useful in depicting temporal signals that have a pronounced vertical structure such as seasonal heating

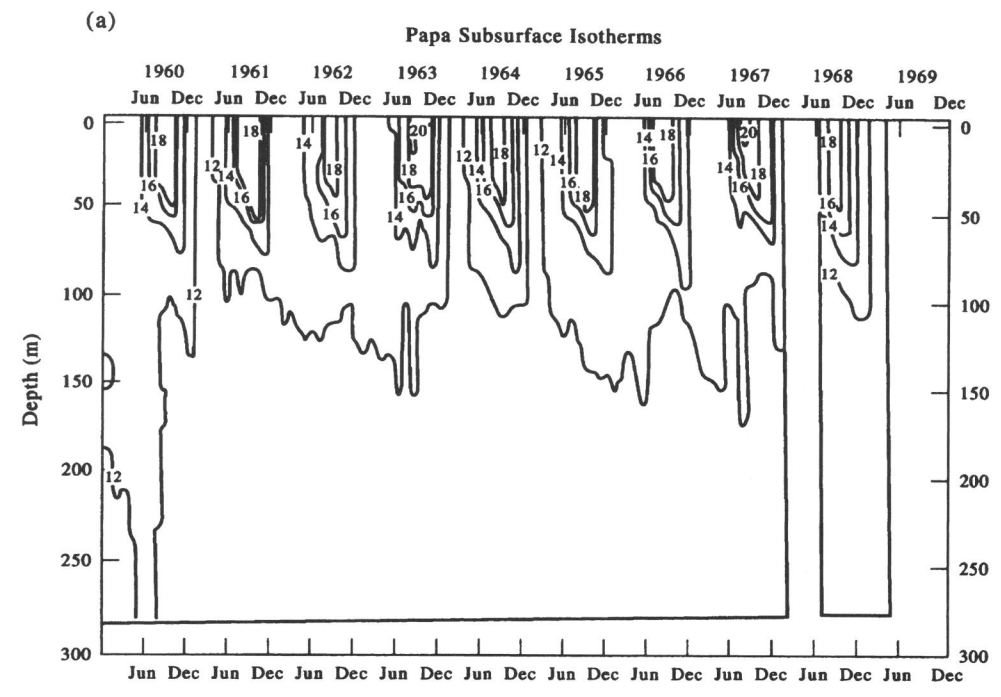

Figure 2.20. Time-series plots for: (a) Repeated vertical profiles at Ocean Weather Station "P" (50°N, 145°W in the northeast Pacific) for the period 1960–69. (From Fofonoff and Tabata, 1966.)

and cooling. Other temporal changes due to vertical layering (e.g. from river a plume) are well represented by this type of plot.

2.4.8 Histograms

As oceanographic sampling matured, the concept of a stationary ocean has given way to the notion of a highly variable system requiring repeated sampling. Data display has graduated from a purely pictorial presentation to statistical representations. A plot format, related to fundamental statistical concepts of sampling and probability, is the histogram or frequency-of-occurrence diagram. This diagram presents information on how often a certain value occurred in any set of sample values. As we discuss in the section on basic statistics, there is no set rule for the construction of histograms and the selection of a sample variable interval (called "bin size") is completely arbitrary. This choice of bin size will dictate the smoothness of the presentation but an appropriately wide enough interval must be used to generate statistically meaningful frequency-of-occurrence values.

2.4.9 New directions in graphical presentation

Plotting oceanographic data has gone from a manpower-intensive process to one primarily carried out by computers. Computer graphics have provided oceanographers with a variety of new presentation formats. For example, all of the data display formats previously discussed can now be carried out by computer systems. Much of the investigator's time is spent ensuring that computer programs are

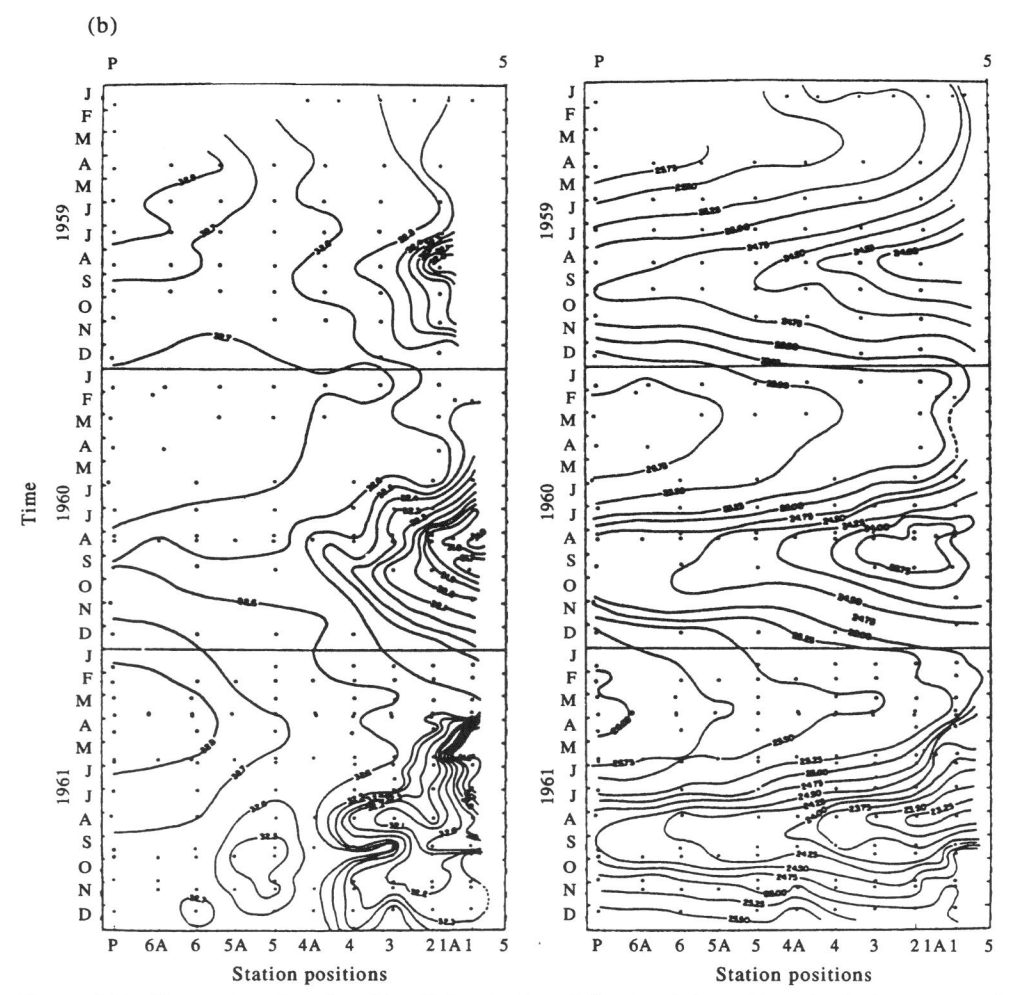

Figure 2.20. Time-series plots for: (b) salinity (psu) and density (σ_t) at 10 m depth from repeated transects along Line P between Station P and the coast of North America for the period January 1959 to December 1961. (From Fofonoff and Tabata, 1966.)

developed, not only for the analysis of the data, but also for the presentation of results. These steps are often combined, as in the case of objective mapping of irregularly spaced data. In this case, an objective interpolation scheme is used to map a horizontal flow or property field. Contouring of the output objective map is then done by the computer. Frequently, both the smoothing provided by objective analysis and the computer contouring can be performed by existing software routines. Sometimes problems with these programs arise, such as continuing to contour over land or the restriction to certain contour intervals. These problems must either be overcome in the computer routine or the data altered in some way to avoid the problems.

In addition to computer mapping, the computer makes it possible to explore other presentations not possible in hand analyses. Three-dimensional plotting is one of the more obvious examples of improved data display possible with computers. For

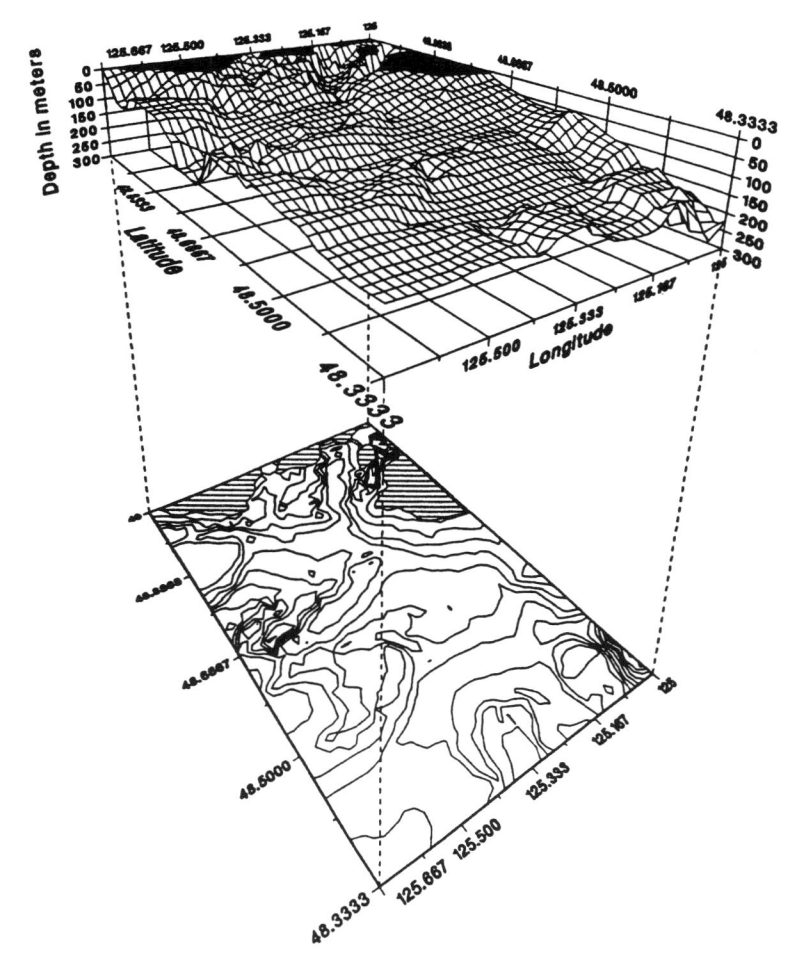

Figure 2.21. Three-dimensional plot of water depth at 20 m contour interval off the southwest coast of Vancouver Island. The bottom plot is the two-dimensional projection of the topography. (Courtesy Gary Hamilton, Intelex Research.)

example, Figure 2.21 shows a three-dimensional plot of coastal bottom topography and a two-dimensional projection (contour map) of the same field. One main advantage of the three-dimensional plot is the geometrical interpretation given to the plot. We can more clearly see both the sign and the relative magnitudes of the dominant features. A further benefit of this form of presentation is the ability to present views of the data display from different angles and perspectives. For example, the topography in Figure 2.21 can be rotated to emphasize the different canyons that cut across the continental slope. Any analysis which outputs a variable as a function of two others can benefit from a three-dimensional display. A well-known oceanic example is the Garrett-Munk spectrum for internal wave variability in the ocean (Figure 2.22) in which spectral amplitude based on observational data is plotted as a function of vertical wavenumber (m) and wave frequency (ω). The diagram tells the observer what kind of spectral shape to expect from a specific type of profiling method.

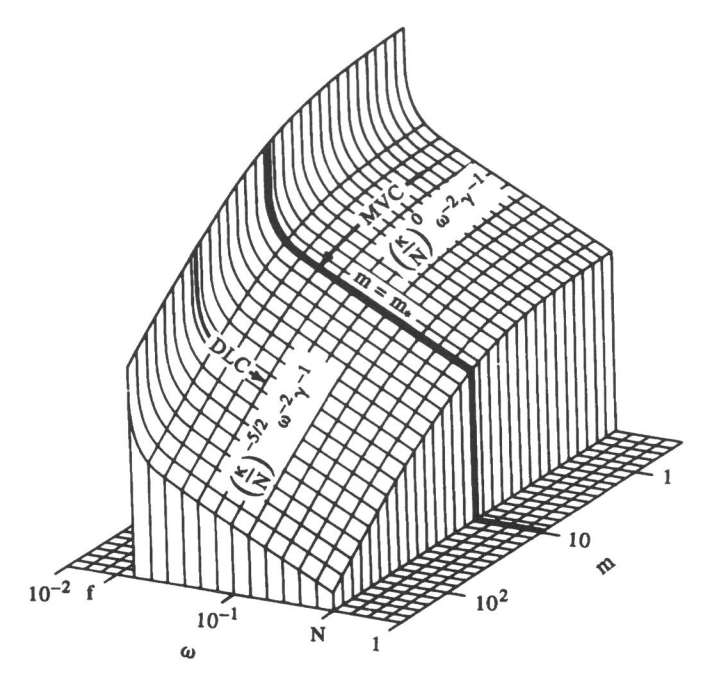

Figure 2.22. Garrett-Munk energy spectrum for oceanic internal waves based on different types of observations. Spectral amplitude (arbitrary units) is plotted against m (the vertical wavenumber in cycles per meter) and ω (the wave frequency in cycles per hour). Here, m^* is the wavenumber bandwidth, κ is the horizontal wavenumber, N the buoyancy frequency, f the Coriolis parameter, and $\gamma = (1 - f^2/\omega^2)^{1/2}$. MVC = moored vertical coherence and DLC = dropped lag coherence between vertically separated measurements. (From Garrett and Munk, 1979.)

The introduction of color into journal papers represents another important change in presentation method. As mentioned in the discussion of vertical sections, color shading has been used traditionally to better visually resolve horizontal and vertical gradients. Most of these color presentations have been restricted to atlas and report presentations and were not available in journal articles. New printing procedures have made color more affordable and much wider use is being made of color displays. One area of recent study where color display has played a major role, is in the presentation of satellite images. Here, the use of false color enables the investigator to expand the dynamic range of the usual gray shades so that they are more easily recognizable by eye. False color is also used to enhance certain features such as sea surface temperature patterns inferred from infrared satellite images. The enhancements, and pseudo-color images, may be produced using a strictly defined function or may be developed in the interactive mode in which the analyst can produce a pleasing display. One important consideration in any manipulation of satellite images is to have each image registered to a ground map which is generally called "image navigation" in oceanographic jargon. This navigation procedure (Emery *et al.*, 1989b) can be carried out using satellite ephemeris data (orbital parameters) to correct for earth curvature and rotation. Timing and spacecraft attitude errors often require the image to be "nudged" to fit the map projection exactly. An alternative method of image correction is to use a series of ground-control-points (GCPs) to navigate the image. GCPs are usually features such as bays or promontories that stand out in both the satellite image and the base map. In using GCP navigation a primary correction is made assuming a circular orbit and applying the mean satellite orbital parameters.

Access to digital image processing has greatly increased the investigator's capability to present and display data. Conventional data may be plotted in map form and overlain on a satellite image to show correspondence. This is possible since most image systems have one or more graphics overlay planes. Another form of presentation, partly motivated by satellite imagery, is the time-sequence presentation of maps or images. Called "scene animation", this format produces a movie-style output which can be conveniently recorded on video tape. With widespread home use of video recorder systems, this form of data visualization is readily accessible to most people. A problem with this type of display is the present inability to publish video tapes or film loops. This greatly restricts the communication of results which show the time evolution of a spatial field such as that shown by a series of geographically coincident satellite images.

Digital image manipulation also has changed the way oceanographers approach data display. Using an interactive system the scientist–operator can change not only the brightness scale assignment (enhancement) but can also alter the orientation, the size (zoom in, zoom out) and the overall location of the output scene using a joystick, trackball or mouse (digital tablet and cursor). With an interactive system, the three-dimensional display can be shifted and rotated to view all sides of the output. This allows the user to visualize areas hidden behind prominent features.

As more oceanographers become involved with digital image processing and pseudo-color displays, there should be an increase in the variety of data and results presentations. These will not only add new information to each plot but will also make the presentation of the information more interesting and "colorful". The old adage of a picture being worth a thousand words is often true in oceanography and the interests of the investigators are best served when their results can be displayed in some interesting graphical or image form.

CHAPTER 3

Statistical Methods and Error Handling

3.1 INTRODUCTION

This chapter provides a review of some of the basic statistical concepts and terminology used in processing data. We need this information if we are to deal properly with the specific techniques used to edit and analyze oceanographic data. Our review is intended to establish a common level of understanding by the readers, not to provide a summary of all available procedures.

In the past, all collected data were processed and reduced by hand so that the individual scientist had an opportunity to become personally familiar with each data value. During this manual reduction of data, the investigator took into account important information regarding the particular instrument used and was able to determine which data were "better" in the sense that they had been collected and processed correctly. Within the limits of the observing systems, an accurate description of the data could be achieved without the application of statistical procedures. Individual intuition and familiarity with shipboard procedures took precedence in this type of data processing and analyses were made on comparatively few data. In such investigations, the question of statistical reliability was seldom raised and it was assumed that individual data points were correct.

For the most part, the advent of the computer and electronic data collection methods has meant that a knowledge of statistical methods has become essential to any reliable interpretation of results. Circumstances still exist, however, for which physical oceanographers still assign considerable weight to the quality of individual measurements. This is certainly true of water sample data such as dissolved oxygen, nutrients, and chemical tracers collected from bottle casts. In these cases, the established methods of data reduction, including familiarity with the data and knowledge of previous work in a particular region, still produce valuable descriptions of oceanic features and phenomena with a spatial resolution not possible with statistical techniques. However, for those more accustomed to having data collected and/or delivered on high density storage media such as magnetic tape, CD-ROM, or floppy disk, statistical methods are essential to determining the value of the data and to decide how much of it can be considered useful for the intended analysis. This statistical approach arises from the fundamental complexity of the ocean, a multivariate system with many degrees of freedom in which nonlinear dynamics and sampling limitations make it difficult to separate scales of variability.

A fundamental problem with a statistical approach to data reduction is the fact that the ocean is not a stationary environment in which we can make repeated measurements. By "stationary" we mean a physical system whose statistical properties remain unchanged with time. In order to make sense of our observations, we are forced to make some rather strong assumptions about our data and the processes we are trying to investigate. Basic to these assumptions is the concept of randomness and the consequent laws of probability. Since each oceanographic measurement can be considered a superposition of the desired signal plus unwanted noise (due to measurement errors and unresolved geophysical variability), the assumption of random behavior often is applied to both the signal and the noise. We must consider not only the statistical character of the signal and noise contributions individually but also the fact that the signal and the noise can interact with each other. Only through the application of the concept of probability can we make the assumptions required to reduce this complex set of variables to a workable subset. Our brief summary of statistics will emphasize concepts pertinent to the analysis of random variables such as probability density functions and statistical moments (mean, variance, etc.). A brief glossary of statistical terms can be found in Appendix B.

3.2 SAMPLE DISTRIBUTIONS

Fundamental to any form of data analysis is the realization that we are usually working with a limited set (or sample) of random events drawn from a much larger population. We use our sample to make estimates of the true statistical properties of the population. Historically, studies in physical oceanography were dependent on too few data points to allow for statistical inference and individual samples were considered representative of the true ocean. Often, an estimate of the population distribution is made from the sample set by using the relative frequency distribution, or histogram, of the measured data points. There is no fixed rule on how such a histogram is constructed in terms of ideal bin interval or number of bins. Generally, the more data there are, the greater the number of bins used in the histogram. Bins should be selected so that the majority of the measurements do not fall on the bin boundaries. Since the area of a histogram bin is proportional to the fraction of the total number of measurements in that interval, it represents the probability that an individual sample value will lie within that interval (Figure 3.1).

The most basic descriptive parameter for any set of measurements is the sample mean. The mean is generally taken over the duration of a time series (time average) or over an ensemble of measurements (ensemble mean) collected under similar conditions (Table 3.1). If the sample has N data values, $x_1, x_2, \ldots, x_N$, the sample mean is calculated as

$$\bar{x} = \frac{1}{N}\sum_{i=1}^{N} x_i \qquad (3.2.1)$$

The sample mean is an unbiased estimate of the true population mean, μ. Here, an

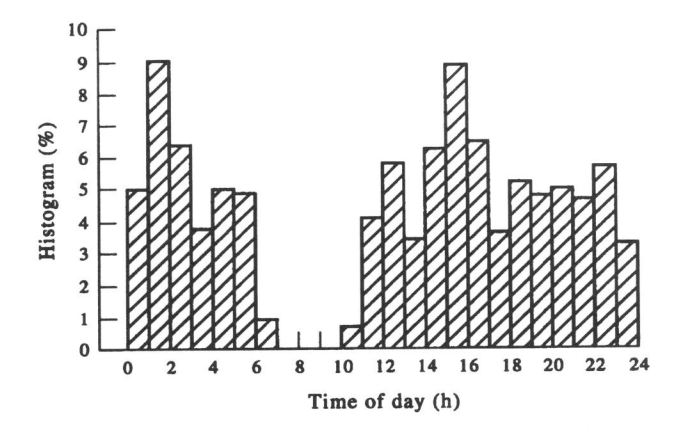

Figure 3.1. Histogram giving the percentage occurrences for the times of satellite position fixes during a 24-h day. Data are for satellite-tracked surface drifter #4851 deployed in the northeast Pacific Ocean from 10 December 1992 to 28 February 1993. During this 90-day period, the satellite receiver on the drifter was in the continuous receive mode.

Table 3.1. Statistical values for the data set $x = \{x_i, i = 1, \ldots, 9\} = \{-3, -1, 0, 2, 5, 7, 11, 12, 12\}$

Mean $\bar{x}$	Variance s'^2	Variance s^2	Standard deviation, s	Range	Median	Mode
5.00	30.22	34.00	5.83	15	5	12

"unbiased" estimator is one for which the expected value, $E[x]$, of the estimator is equal to the parameter being estimated. In this case, $E[x] = \mu$ for which $\bar{x}$ is an unbiased estimator. The sample mean locates the center of mass of the data distribution such that

$$\sum_{i=1}^{N} (x_i - \bar{x}) = 0$$

that is, the sample mean splits the data so that there is an equal weighting of negative and positive values of the fluctuation, $x' = x_i - \bar{x}$, about the mean value, $\bar{x}$. The weighted sample mean is the general case of (3.2.1) and is defined as

$$\bar{x} = \frac{1}{N} \sum_{i=1}^{N} f_i x_i \qquad (3.2.2)$$

where f_i/N is the relative frequency of occurrence of the ith value for the particular experiment or observational data set. In (3.2.1), $f_i = 1$ for all i.

The sample mean values give us the center of mass of a data distribution but not its width. To determine how the data are spread about the mean, we need a measure of the sample variability or *variance*. For the data used in (3.2.1), the *sample variance* is the average of the square of the sample deviations from the sample mean, expressed as

$$s'^2 = \frac{1}{N}\sum_{i=1}^{N}(x_i - \bar{x})^2 \qquad (3.2.3)$$

The *sample standard deviation* $s' = \sqrt{s'^2}$ the positive square root of (3.2.3), is a measure of the typical difference of a data value from the mean value of all the data points. In general, these differ from the corresponding true *population variance*, σ^2, and the *population standard deviation*, σ. As defined by (3.2.3), the sample variance is a biased estimate of the true population variance. An unbiased estimator of the population variance is obtained from

$$s^2 = \frac{1}{N-1}\sum_{i=1}^{N}(x_i - \bar{x})^2 \qquad (3.2.4a)$$

$$= \frac{1}{N-1}\left[\sum_{i=1}^{N}(x_i)^2 - \frac{1}{N}\left(\sum_{i=1}^{N}x_i\right)^2\right] \qquad (3.2.4b)$$

where the denominator $N-1$ expresses the fact that we need at least two values to define a sample variance and standard deviation, s. The use of the estimators s versus s' is often a matter of debate among oceanographers, although it should be noted that the difference between the two values decreases as the sample size increases. Only for relatively small samples ($N < 30$) is the difference significant. Because s' has a smaller mean square error than s and is an unbiased estimator when the population mean is known *a priori*, we recommend the use of (3.2.4). However, a word of caution: if your hypothesis depends on the difference between s and s', then you have ventured onto shaky statistical ground supported by questionable data. We further note that the expanded relation (3.2.4b) is a more efficient computational formulation than (3.2.4a) in that it allows one to obtain s^2 from a single pass through the data. If the sample mean must be calculated first, two passes through the same data set are required rather than one, which is computationally less efficient when dealing with large data sets.

Other statistical values of importance are the range, mode, and median of a data distribution (Table 3.1). The *range* is the spread or absolute difference between the end-point values of the data set while the *mode* is the value of the distribution that occurs most often. For example, the data sequence 2, 4, 4, 6, 4, 7 has a range of $|2 - 7| = 5$ and a mode of 4. The *median* is the middle value in a set of numbers arranged according to magnitude (the data sequence $-1, 0, 2, 3, 5, 6, 7$ has a median of 3). If there is an even number of data points, the median value is chosen mid-way between the two candidates for the central value. Two other measures, *skewness* (the third moment of the distribution and degree of asymmetry of the data about the mean) and *kurtosis* (a nondimensional number measuring the flatness or peakedness of a distribution) are less used in oceanography.

As we discuss more thoroughly later in this chapter, the shapes of many sample distributions can be approximated by a *normal* (also called a *bell* or *Gaussian*) distribution. A convenient aspect of a normal population distribution is that we can apply the following empirical "rule of thumb" to the data:

$\mu \pm \sigma$ spans approximately 68% of the measurements;
$\mu \pm 2\sigma$ spans approximately 95% of the measurements;
$\mu \pm 3\sigma$ spans most (99%) of the measurements.

The percentages are represented by the areas under the normal distribution curve spanned by each of the limits (Figure 3.2). We emphasize that the above limits apply only to normal distributions of random variables.

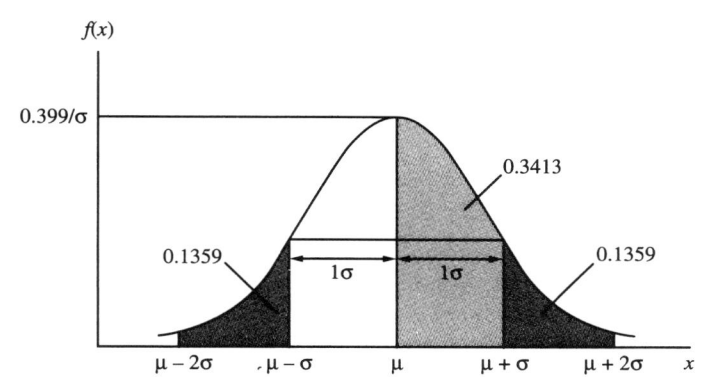

Figure 3.2. Normal distribution f(x) for mean μ and standard deviation σ of the random variable X. (From Harnett and Murphy, 1975.)

3.3 PROBABILITY

Most data collected by oceanographers are made up of samples taken from a larger unknown population. If we view these samples as random events of a statistical process, then we are faced with an element of uncertainty: "What are the chances that a certain event occurred or will occur based on our sample?" or "How likely is it that a given sample is truly representative of a certain population distribution? " (The last question might be asked of political pollsters who use small sample sizes to make sweeping statements about the opinions of the populace as a whole.) We need to find the best procedures for inferring the population distribution from the sample distribution and to have measures that specify the goodness of the inference. Probability theory provides the foundation for this type of analysis. In effect, it enables us to find a value between 0 and 1 which tells us just how likely is a particular event or sequence of events. A probability is a proportional measure of the occurrence of an event. If the event has a probability of zero, then it is impossible; if it has a probability of unity, then it is certain to occur. Probability theory as we know it today was initiated by Pascal and Fermat in the seventeenth century through their interest in games of chance. In the eighteenth century, Gauss and Laplace extended the theory to social sciences and actuarial mathematics. Well-known names like R. A. Fisher, J. Neyman, and E. S. Pearson are associated with the proliferation of statistical techniques developed in the twentieth century.

The *probability mass function*, $P(x)$, gives the relative frequency of occurrence of each possible value of a discrete random variable, X. Put another way, the function specifies the point probabilities $P(x_i) = P(X = x_i)$ and assumes nonzero values only at points $X = x_i$, $i = 1, 2, \ldots$. One of the most common examples of a probability mass function is the sum of the dots obtained from the roll of a pair of dice (Table 3.2). According to probability theory, the dice player is most likely to roll a 7 (highest probability mass function) and least likely to roll a 2 or 12 (lowest probability mass

198 *Data Analysis Methods in Physical Oceanography*

Table 3.2. *The discrete probability mass function and cumulative probability functions for the sum of the dots (variable X) obtained by tossing a pair of dice*

Sum of dots (X)	Frequency of occurrence	Relative frequency	Probability mass function, $P(x)$	Cumulative probability function $F(x) = P(X \leq x)$
2	1	1/36	$P(x = 2) = 1/36$	$F(2) = P(X \leq 2) = 1/36$
3	2	2/36	$P(x = 3) = 2/36$	$F(3) = P(X \leq 3) = 3/36$
4	3	3/36	$P(x = 4) = 3/36$	$F(4) = P(X \leq 4) = 6/36$
5	4	4/36	$P(x = 5) = 4/36$	$F(5) = P(X \leq 5) = 10/36$
6	5	5/36	$P(x = 6) = 5/36$	$F(6) = P(X \leq 6) = 15/36$
7	6	6/36	$P(x = 7) = 6/36$	$F(7) = P(X \leq 7) = 21/36$
8	5	5/36	$P(x = 8) = 5/36$	$F(8) = P(X \leq 8) = 26/36$
9	4	4/36	$P(x = 9) = 4/36$	$F(9) = P(X \leq 9) = 30/36$
10	3	3/36	$P(x = 10) = 3/36$	$F(10) = P(X \leq 10) = 33/36$
11	2	2/36	$P(x = 11) = 2/36$	$F11) = P(X \leq 11) = 35/36$
12	1	1/36	$P(x = 12) = 1/36$	$F(12) = P(X \leq 12) = 1$
SUM	36	1.00		1.00

function). The dice example reveals two of the fundamental properties of all discrete probability functions: (1) $0 \leq P(X = x)$; and (2) $\sum P(x) = 1$, where the summation is over all possible values of x. The counterpart to $P(x)$ for the case of a continuous random variable X is the *probability density function* (abbreviated, PDF), $f(x)$, which we discuss more fully later in the chapter. For the continuous case, the above fundamental properties become: (1) $0 \leq f(x)$; and (2) $\int f(x)\, dx = 1$ where the integration is over all x in the range $(-\infty, \infty)$.

To further illustrate the concept of probability, consider N independent trials, each of which has the same probability of "success" p and probability of "failure" $q = 1 - p$. The probability of success or failure is unity; $p + q = 1$. Such trials involve binomial distributions for which the outcomes can be only one of two events: for example, a tossed coin will produce a head or a tail; an XBT will work or it won't work. If X represents the number of successes that occur in the N trials, then X is said to be a discrete random variable having parameters (N, p). The term "Bernoulli trial" is sometimes used for X. The probability mass function which gives the relative frequency of occurrence of each value of the random variable X having parameters (N, p) is the binomial distribution

$$p(x) = \binom{N}{x} p^x (1-p)^{N-x}, \quad x = 0, 1, ..., N \qquad (3.3.1a)$$

where the expression

$$\binom{N}{x} = \binom{N}{N-x} = {_N}C_x \equiv N!/[(N-x)!x!] \qquad (3.3.1b)$$

is the number of different *combinations* of groups of x objects that can be chosen from a total set of N objects without regard to order. The number of different combinations of x objects is always fewer than the number of *permutations*, ${_N}P_x$, of x objects $[{_N}P_x \equiv N!/(N-x)!]$. In the case of permutations, different ordering of the same objects counts for a different permutation (i.e. ab is different than ba). As an example, the number of possible different batting orders (permutations) a coach can create among the first four hitters on a nine-person baseball team is $9!/(9 - 4)! = 9!/5! = 3024$. In contrast, the number of different groups of ball-players a coach can put in the first four lead-off batting positions without regard to batting order is $9!/[(9 - 4)!4!] = 9!/5!4! = 126$. The numbers

$$\binom{N}{x}$$

often are called *binomial coefficients* since they appear as coefficients in the expansion of the binomial expression $(a + b)^N$ given by the binomial theorem:

$$(a + b)^N = \sum_{k=0}^{N} \binom{N}{k} a^k b^{N-k} \qquad (3.3.2)$$

The summed probability mass function

$$P(a \leq x \leq b) = \sum_{a}^{b} P(x)$$

for variable X over a specified range of values (a, b) can be demonstrated by a simple oceanographic example. Suppose there is a probability $1 - p$ that a current meter will fail when moored in the ocean and that the failure is independent from current meter to current meter. Assume that a particular string of meters will successfully measure the expected flow structure if at least 50% of the meters on the string remain operative. For example, a two-instrument string used to measure the barotropic flow will be successful if one current meter remains operative while a four-instrument string used to resolve the baroclinic flow will be successful if at least two meters remain operative. We then ask: "For what values of p is a four-meter array preferable to a two-meter array?" Since each current meter is assumed to fail or function independently of the other meters, it follows that the number of functioning current meters is a binomial random variable. The probability that a four-meter mooring is successful is then

$$P(2 \leq x \leq 4) = \sum_{k=2}^{4} \binom{4}{k} p^k (1 - p)^{4-k}$$
$$= \binom{4}{2} p^2 (1 - p)^2 + \binom{4}{3} p^3 (1 - p)^1 + \binom{4}{4} p^4 (1 - p)^0$$
$$= 6p^2 (1 - p)^2 + 4p^3 (1 - p)^1 + p^4$$

Similarly, the probability that a two-meter array is successful is

$$P(1 \leq x \leq 2) = \sum_{k=1}^{2} \binom{2}{k} p^k (1 - p)^{2-k}$$
$$= 2p(1 - p) + p^2$$

From these two relations, we find that the four-meter string is more likely to succeed when

$$6p^2(1-p)^2 + 4p^3(1-p)^1 + p^4 \geq 2p(1-p) + p^2$$

or, after some factoring and simplification, when

$$(p-1)^2 + (3p-2) \geq 0$$

for which we find $3p - 2 \geq 0$, or $p \geq 2/3$. When compared to the two-meter array, the four-meter array is more likely to do its intended job when the probability, p, that the instrument works is $p \geq 2/3$. The two-meter array is more likely to succeed when $p \leq 2/3$.

As the previous example illustrates, we often make the fundamental assumption that each sample in our set of observations is an independent realization drawn from a random distribution. Individual events in this distribution cannot be predicted with certainty but their relative frequency of occurrence, for a long series of repeated trials (samples), is often remarkably stable. We further remark that the binomial distribution is only one type of probability density function. Other distribution functions will be discussed later in the chapter.

3.3.1 Cumulative probability functions

The probability mass function yields the probability of a specific event or probability of a range of events. From this function we can derive the *cumulative probability function*, $F(x)$—also called the cumulative distribution function, cumulative mass function, and probability distribution function—defined as that fraction of the total number of possible outcomes X (a random variable) which are less than a specific value x (a number). Thus, the distribution function is the probability that $X \leq x$, or

$$F(x) = P(X \leq x)$$
$$= \sum_{\text{all } X \leq x} P(x), \quad -\infty < x < \infty \text{ (discrete random variable, } X\text{)} \quad (3.3.3a)$$
$$= \int_{-\infty}^{x} f(x)\,dx \text{ (continuous random variable, } X\text{)} \quad (3.3.3b)$$

The discrete cumulative distribution function for tossing a pair of fair dice (Table 3.2) is plotted in Figure 3.3. Since the probabilities P and f are limited to the range 0 and 1, we have $F(-\infty) = 0$ and $F(\infty) = 1$. In addition, the distribution function $F(x)$ is a nondecreasing function of x, such that $F(x_1) \leq F(x_2)$ for $x_1 < x_2$, where $F(x)$ is continuous from the right (Table 3.2).

It follows that, for the case of a continuous function, the derivative of the distribution function F with respect to the sample parameter, x

$$f(x) = \frac{dF(x)}{dx} \quad (3.3.4)$$

recovers the probability density function (PDF), f. As noted earlier, the PDF has the property that its integral over all values is unity

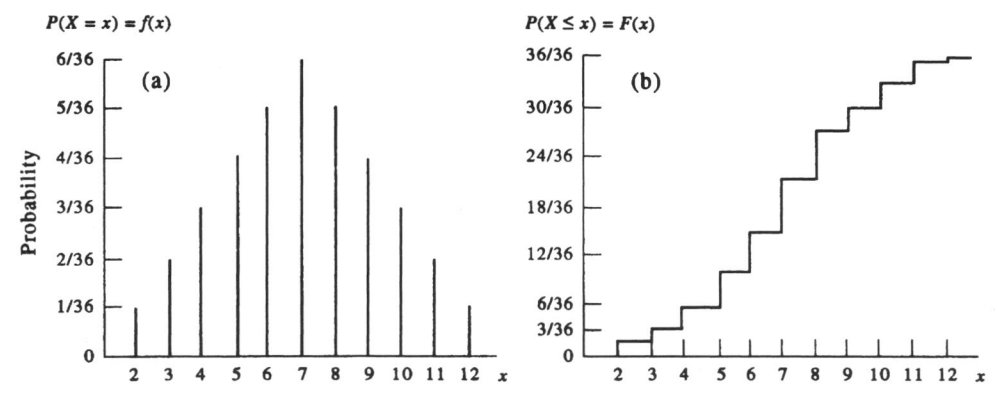

Figure 3.3. The discrete mass function P(x) and cumulative distribution function F(x) from tossing a pair of dice (see Table 3.2). (From Harnett and Murphy, 1975.)

$$\int_{-\infty}^{\infty} f(x)\,dx = F(\infty) - F(-\infty) = 1$$

In the limit $dx \to 0$, the fraction of outcomes for which x lies in the interval $x < x' < x + dx$ is equal to $f(x')\,dx$, the probability for this interval. The random variables being considered here are continuous so that the PDF can be defined by (3.3.4). Variables with distribution functions that contain discontinuities, such as the steps in Figure 3.3, are considered discrete variables. A random variable is considered discrete if it assumes only a countable number of values. In most oceanographic sampling, measurements can take on an infinity of values along a given scale and the measurements are best considered as continuous random variables. The function $F(x)$ for a continuous random variable X is itself continuous and appears as a smooth curve. Similarly, the PDF for a continuous random variable X is continuous and can be used to evaluate the probability that X falls within some interval $[a, b]$ as

$$P(a \leq X \leq b) = \int_{a}^{b} f(x)\,dx \qquad (3.3.5)$$

3.4 MOMENTS AND EXPECTED VALUES

The discussion in the previous section allows us to determine the probability of a single event or experiment, or describe the probability of a set of outcomes for a specific random variable. However, our discussion is not concise enough to describe fully the probability distributions of our data sets. The situation is similar to section 3.2 in which we started with a set of observed values. In addition to presenting the individual values, we seek properties of the data such as the sample mean and variance to help us characterize the structure of our observations. In the case of probability distributions, we speak not of *observed* mean and variance but of the *expected* mean and variance obtained from an infinite number of realizations of the random variable under consideration.

Before discussing some common PDFs, we need to review the computation of the parameters used to describe these functions. These parameters are, in general, called "moments" by analogy to mechanical systems where moments describe the distribution of forces relative to some reference point. The statistical concept of degrees-of-freedom is also inherited from the terminology of physical–mechanical systems where the number of degrees-of-freedom specifies the motion possible within the physical constraints of the mechanical system and its distribution of forces. As noted earlier, the population mean, μ, and standard deviation, σ, define the first and second moments which describe the center and spread of the probability function. In general, these parameters do not uniquely define the PDF since many different PDFs can have the same mean and standard deviation. However, in the case of the Gaussian distribution, the PDF is completely described by μ and σ. In defining moments we must be careful to distinguish between moments taken about the origin and moments taken about the mean (central moments).

When discussing moments it is useful to introduce the concept of expected value. This concept is analogous to the notion of weighted functions. For a discrete random variable, X, with a probability function $P(x)$ (the discrete analogue to the continuous PDF), the expected value of X is written as $E[X]$ and is equivalent to the arithmetic mean, μ, of the probability distribution. In particular, we can write the expected value for a discrete PDF as

$$E[x] = \sum_{i=1}^{N} x_i P(x_i) = \mu \tag{3.4.1}$$

where μ is the population mean introduced in Section 3.2. The probability function $P(x)$ serves as a weighting function similar to the function f_i/N in equation (3.2.2). The difference is that f_i/N is the relative frequency for a single set of experimental samples whereas $P(x)$ is the expected relative frequency for an infinite number of samples from repeated trials of the experiment. The expected value, $E[X]$, for the sample which includes X, is the sample mean, $\bar{x}$. Similarly, the variance of the random variable X is the expected value of $(X - \mu)^2$, or

$$V[X] = E[(X - \mu)^2] = \sum_{i=1}^{N} (x - \mu)^2 P(x_i) = \sigma^2 \tag{3.4.2}$$

In the case of a continuous random variable, X, with PDF $f(x)$, the expected value is

$$E[X] = \int_{-\infty}^{\infty} x f(x) \, dx \tag{3.4.3}$$

while for any function $g(X)$ with a PDF $f(x)$, the expected value can be written as

$$E[X] = \int_{-\infty}^{\infty} g(x) f(x) \, dx \quad \text{(continuous variable)} \tag{3.4.4a}$$

$$= \sum_{i=1}^{N} g(x_i) P(x_i) \quad \text{(discrete case)} \tag{3.4.4b}$$

Some useful properties of expected values for random variables are:

(1) For c = constant; $E[c] = c$, $V[c] = 0$;
(2) $E[cg(X)] = cE[g(X)]$, $V[cg(X)] = c^2 V[g(X)]$;
(3) $E[g_1(X) \pm g_2(X) \pm ...] = E[g_1(X)] \pm E[g_2(X)] \pm ...]$;
(4) $V[g(X)] = E[(g(X) - \mu)^2] = E[g(X)^2] - \mu^2$, (variance about the mean);
(5) $E[g_1 g_2] = E[g_1] E[g_2]$;
(6) $V[g_1 \pm g_2] = V[g_1] + V[g_2] \pm 2C[g_1, g_2]$.

Property (6) introduces the *covariance function* of two variables, C, defined as

$$C[g_1, g_2] = E[g_1 g_2] - E[g_1] E[g_2] \tag{3.4.5}$$

where $C = 0$ when g_1 and g_2 are independent random variances. Using properties (1) to (3), we find that $E[Y]$ for the linear relation $Y = a + bX$ can be expanded to

$$E[Y] = E[a + bX] = a + bE[X]$$

while from (1) and (6) we find

$$V[Y] = V[a + bX] = b^2 V[X]$$

At this point, we remark that averages, expressed as expected values, $E[X]$, apply to ensemble averages of many (read, infinite) repeated samples. This means that each sample is considered to be drawn from an infinite ensemble of identical statistical processes varying under exactly the same conditions. In practice, we do not have repeated samples taken under identical conditions but rather time (or space) records. In using time or space averages as representative of ensemble averages, we are assuming that our records are *ergodic*. This implies that averages over an infinite ensemble can be replaced by an average over a single, infinitely long time series. An ergodic process is not to be confused with a stationary process for which the PDF of $X(t)$ is independent of time. In reality, time/space series can be considered stationary if major shifts in the statistical characteristics of the series occur over intervals that are long compared to the averaging interval so that the space/time records remain homogeneous (exhibit the same general behavior) throughout the selected averaging interval. A data record that is quiescent during the first half of the record and then exhibits large irregular oscillations during the second half of the record is not stationary.

3.4.1 Unbiased estimators and moments

As we stated earlier, $\bar{x}$ and s^2 defined by (3.2.2) and (3.2.4) are unbiased estimators of the true population mean, μ, and variance, σ^2. That is, the expected values of $\bar{x}$ and $(x - \bar{x})^2$ are equal to μ and σ^2, respectively. To illustrate the nature of the expected value, we will first prove that $E(\bar{x}) = \mu$. We write the expected value as the normalized sum of all $\bar{x}$ values

$$E[\bar{x}] = E\left[\frac{1}{N}\sum_{i=1}^{N} x_i\right] = \frac{1}{N}\sum_{i=1}^{N} E[x_i] = \frac{1}{N}\sum_{i=1}^{N} \mu = \mu$$

as required. Next, we demonstrate that $E[s^2] = \sigma^2$. We again use the appropriate definitions and write

$$\begin{aligned}E[s^2] &= E\left[\frac{1}{N-1}\sum_{i=1}^{N}(x_i - \bar{x})^2\right]\\ &= E\left[\frac{1}{N-1}\left\{\sum_{i=1}^{N}\left[(x_i - \mu)^2 - N(\bar{x} - \mu)^2\right]\right\}\right]\\ &= \frac{1}{N-1}\left\{\sum_{i=1}^{N} E\left[(x_i - \mu)^2\right] - NE\left[(\bar{x} - \mu)^2\right]\right\}\\ &= \frac{1}{N-1}\left\{\sum_{i=1}^{N}(\sigma^2) - N\frac{\sigma^2}{N}\right\} = \frac{\sigma^2}{N-1}(N-1) = \sigma^2\end{aligned}$$

where we have used the relations $x_i - \bar{x} = (x_i - \mu) - (\bar{x} - \mu)$, $E[(x_i - \mu)^2] = V[x_i] = \sigma^2$ (the variance of an individual trial) and $E[(\bar{x} - \mu)^2] = V[\bar{x}] = \sigma^2/N$ (the variance of the sample mean relative to the population mean). The last expression derives from the central limit theorem discussed in Section 3.6.

Returning to our discussion of statistical moments, we define the ith moment of the random variable X, taken about the origin, as

$$E[X^i] = \mu_i \qquad (3.4.6)$$

Thus, the first moment about the origin ($i=1$) is the population mean, $\mu = \mu_1$. Similarly, we can define the ith moment of X taken about the mean (called the ith central moment of X) as

$$E[(X - \mu)^i] = \mu_i \qquad (3.4.7)$$

The population variance, σ^2, is the second ($i = 2$) central moment, μ_2.

3.4.2 Moment generating functions

Up to this point, we have computed the various characteristics of the random variable X using the probability functions directly. Now, suppose we look for a "generating" function that enables us to find all of the expected properties of the variable X using just this one function. For a discrete or continuous random variable X we define a *moment generating function* as $m(t) = E[e^{tX}]$ for the real variable, t. The moment generating function $m(t)$ serves two purposes. First, if we can find $E[e^{tX}]$, we can find any of the moments of X; second, if $m(t)$ exists it is unique and can be used to establish that both random variables have the same probability distributions. In other words, it is not possible for random variables with different probability distributions to have the same moment generating functions. Likewise, if the moment generating functions for two random variables are the same, then both variables must have the same

Statistical Methods and Error Handling 205

probability distribution. For a single-valued function X with a probability function, $P(X = x_k), k = 1, 2, \ldots$ in the discrete case, and $f(x)$ in the continuous case, the moment generating function, $m(t)$, is

$$m(t) = E[e^{tX}] = \sum_{i=1}^{\infty} e^{tx_i} P(x_i) \tag{3.4.7a}$$

$$= \int_{-\infty}^{\infty} e^{tx} f(x) \, dx \tag{3.4.7b}$$

The advantages of the moment generating function become more apparent if we expand e^{tx} in the usual way to get

$$e^{tX} = 1 + tX + (tX)^2/2! + \ldots + (tX)^n/n! + \ldots$$

and apply this to $m(t)$ so that

$$\begin{aligned} m(t) &= E[e^{tX}] = E[1 + tX + (tX)^2/2! + \ldots + (tX)^n/n! + \ldots] \\ &= 1 + tE[X] + t^2 E[X^2]/2! + \ldots + t^n E[X^n]/n! + \ldots \end{aligned} \tag{3.4.8}$$

Taking the derivatives of (3.4.8), we find

$$m'(t) = E[X] + tE[X^2] + \ldots + t^{n-1} E[X^n]/(n-1)! + \ldots \tag{3.4.9a}$$
$$m''(t) = E[X^2] + tE[X^3] + \ldots + t^{n-2} E[X^n]/(n-2)! + \ldots \tag{3.4.9b}$$

and so on (here, $m' \equiv dm/dt$). Setting $t = 0$ in (3.4.9) and continuing in the same way, we obtain

$$m'(0) = E[X]; m''(0) = E[X^2]; \ldots; m^{(n)}(0) = E[X^n] \tag{3.4.10}$$

In other words, we can easily obtain all the moments of the generating function $m(t)$ from the derivatives evaluated at $t = 0$. Specifically, we note that

$$E[X] = m'(0) \tag{3.4.11a}$$

$$V[X] = E[X^2] - (E[X])^2 = m''(0) - [m'(0)]^2 \tag{3.4.11b}$$

As a first example, suppose that the discrete variable X is binomially distributed with parameters N and p as in (3.3.1a). Then

$$m(t) = \sum_{k=0}^{N} e^{tk} \binom{N}{k} p^k (1-p)^{N-k}$$

$$= \sum_{k=0}^{N} \binom{N}{k} (p e^t)^k (1-p)^{N-k} = [p e^t + (1-p)]^N$$

where we have used the binomial expansion

$$\binom{N}{k}$$

from (3.3.2). Taking the derivatives of this function and evaluating the results at $t = 0$, as per (3.4.11), yields the mean and variance for the binomial probability function,

$$E[X] = m'(0) = Np$$
$$V[X] = m''(0) - \{m'(0)\}^2 = Np(1-p) = Npq$$

As a further example, consider the density of a continuous random variable x given by

$$f(x) = \alpha^2 x\, e^{-\alpha x}, \quad \text{if } x > 0$$
$$= 0, \quad \text{otherwise}$$

Using (3.4.7b), we first write the moment generating function $m(t)$ as

$$m(t) = \int_{-\infty}^{\infty} e^{tx} f(x) dx = \alpha^2 \int_0^{\infty} x\, e^{-(\alpha-t)x}\, dx$$

For $\alpha - t > 0$, and hence $t < \alpha$

$$m(t) = \alpha^2 \int_0^{\infty} x\, e^{-(\alpha-t)x} dx$$

$$= \alpha^2 \left[\frac{\alpha e^{-(\alpha-t)x}}{(\alpha-t)} \bigg|_0^{\infty} + \int_0^{\infty} \frac{e^{-(\alpha-t)x}}{(\alpha-t)} dx \right] = \alpha^2 \frac{[-e^{-(\alpha-t)}]}{(\alpha-t)^2} \bigg|_0^{\infty} = \frac{\alpha^2}{(\alpha-t)^2}, \quad \text{for } t < \alpha$$

For $t \geq \alpha$, $m(t)$ is not defined. Using (3.4.11), we find

$$E[X] = m'(t=0) = \mu = \frac{2\alpha^2}{(\alpha-t)^3}\bigg|_{t=0} = \frac{2}{\alpha}$$

Similarly, we find the second moment $V[X] = m''$ as

$$V[X] = m''(t=0) - \mu^2 = \frac{6\alpha^2}{(\alpha-t)^4}\bigg|_{t=0} - \mu^2 = 6/\alpha^2 - 4/\alpha^2 = 2/\alpha^2$$

Several properties of moment generating functions (MGFs) are worth mentioning since they may be used to simplify more complicated functions. These are: (1) if the random variable X has a moment generating function $m(t)$, then the MGF of the random variable $Y = aX + b$ is $m(t) = e^{bt} m(at)$; (2) if X and Y are random variables with respective MGFs $m(t; X)$ and $m(t: Y)$, and if $m(t: X) = m(t: Y)$ for all t, then X and Y have the same probability distribution; (3) If X_k, $k = 1, \ldots, n$, are independent random variables with MGFs defined by $m(t: X_k)$, then the MGF of the random variable $Y = X_1 + X_2 + \ldots + X_n$ is given by the product, $m(t: Y) = m(t: X_1) m(t: X_2) \ldots m(t: X_n)$.

For convenience, the probability density functions, means, variances, and moment generating functions for several common continuous variables are presented in Appendix C. Moments allow us to describe the data in terms of their PDFs. Comparisons between moments from two random variables will establish whether or not they have the same PDF.

3.5 COMMON PROBABILITY DENSITY FUNCTIONS

The purpose of this section is to provide examples of three common PDFs. The first is the uniform PDF given by

$$f(x) = \frac{1}{x_2 - x_1}, \quad x_1 \leq x \leq x_2 \qquad (3.5.1)$$
$$= 0, \quad \text{otherwise}$$

(Figure 3.4) which is the intended PDF of random numbers generated by most computers and handheld calculators. The function is usually scaled between 0 and 1. The cumulative density function $F(x)$ given by (3.3.3b) has the form

$$F(x) = 0, \quad x < x_1$$
$$= \frac{x - x_1}{x_2 - x_1}, \quad x_1 \leq x \leq x_2$$
$$= 1, \quad x \geq x_2$$

while the mean and standard deviation of (3.5.1) are given by $\mu = (x_2 + x_1)/2$ and $\sigma = (x_2 - x_1)/2\sqrt{3}$.

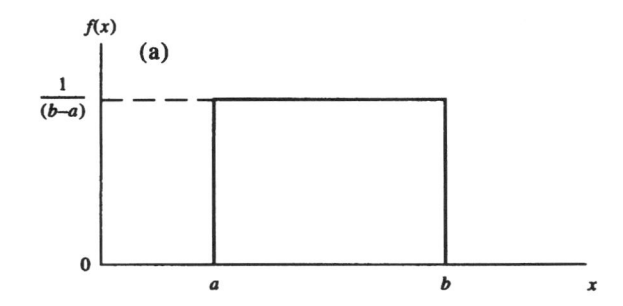

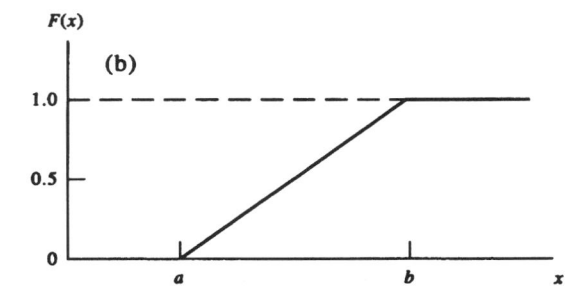

Figure 3.4. Uniform probability density distribution functions. (a) The probability density function, $f(x)$; and (b) the corresponding cumulative probability distribution function, $F(x)$. (From Bendat and Piersol, 1986.)

208 Data Analysis Methods in Physical Oceanography

Perhaps the most familiar and widely used PDF is the normal (or Gaussian) density function:

$$f(x) = \frac{e^{[-(x-\mu)^2/2\sigma^2]}}{\sigma\sqrt{(2\pi)}}, \quad \sigma > 0, \ -\infty < \mu < \infty, \ -\infty < x < \infty \quad (3.5.2)$$

where the parameter σ represents the standard deviation (or spread) of the random variable X about its mean value μ (Figure 3.2). For convenience, (3.5.2) is often written in shorthand notion as $N(\mu, \sigma^2)$. The height of the density function at $x = \mu$ is $0.399/\sigma$. The cumulative probability distribution of a normally distributed random variable, X, lying in the interval a to b is given by the integral (3.3.5)

$$P(a \leq X \leq b) = \int_a^b \frac{e^{[-(x-\mu)^2/2\sigma^2]}}{\sigma\sqrt{(2\pi)}} dx \quad (3.5.3)$$

which is the area under the normal curve between a and b. Since a closed form of this integral does not exist, it must be evaluated by approximate methods, often involving the use of tables of areas. We have included a table of curve areas in Appendix D (Table D.1). The normal distribution is symmetric with respect to μ so that areas need to be tabulated only on one side of the mean. For example, $P(\mu \leq x \leq \mu + 1\sigma) = 0.3413$ so by symmetry $P(\mu - 1\sigma \leq x \leq \mu + 1\sigma) = 2(0.3413) = 0.6826$. The latter is the value used in the rule of thumb estimates for the range of the standard deviation, σ. For the normal distribution, the tabulated values represent the area between the mean and a point z, where z is the distance from the mean measured in standard deviations. This leads to the familiar transform for a normal random variable X given by

$$Z = \frac{X - \mu}{\sigma} \quad (3.5.4)$$

called the standardized normal variable. The variable Z gives the distances of points measured from the mean of the normal random variable in terms of the standard deviation of the normal random variable, X (Figure 3.5). The standard normal variable Z is normally distributed with a mean of zero (0) and a standard deviation of unity (1). Thus, if X is described by the function $N(\mu, \sigma^2)$, then Z is described by the function $N(0, 1)$.

Our third continuous PDF is the gamma density function which applies to random variables which are always nonnegative thus producing distributions that are skewed to the right. The gamma PDF is given by

$$f(x) = \frac{x^{\alpha-1}e^{-x/\beta}}{\beta^\alpha \Gamma(\alpha)} \quad \alpha, \beta > 0; \ 0 \leq x \leq \infty \quad (3.5.5)$$
$$= 0, \text{elsewhere}$$

where α and β are parameters of the distribution and $\Gamma(\alpha)$ is the gamma function

$$\Gamma(\alpha) = \int_0^\infty x^{\alpha-1}e^{-x}dx \quad (3.5.6)$$

For any integer n

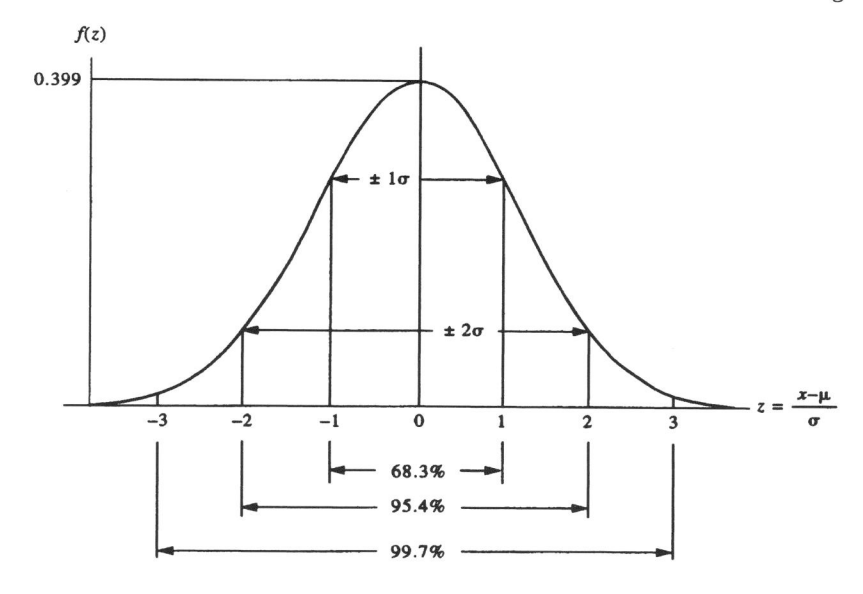

Figure 3.5. Distribution f(z) for the standardized normal random variable, $Z = (X - \mu)/\sigma$ (cf. Figure 3.2). (From Harnett and Murphy, 1975.)

$$\Gamma(n) = (n-1) \qquad (3.5.7)$$

while for a continuous variable α

$$\Gamma(\alpha) = (\alpha - 1)\Gamma(\alpha - 1), \text{ for } \alpha \geq 1 \qquad (3.5.8)$$

where $\Gamma(1) = 1$. Plots of the gamma PDF for $\beta = 1$ and three values of the parameter α are presented in Figure 3.6. Since it is again impossible to define a closed form of the integral of the PDF in (3.5.5), tables are used to evaluate probabilities from the PDF.

One particularly important gamma density function has a PDF with $\alpha = \nu/2$ and $\beta = 2$. This is the *chi-square random distribution* (written as χ_ν^2 and pronounced "ki square") with ν degrees of freedom (Appendix D, Table D.2). The chi-square distribution gets its name from the fact that it involves the square of normally distributed random variables, as we will explain shortly. Up to this point, we have dealt with a

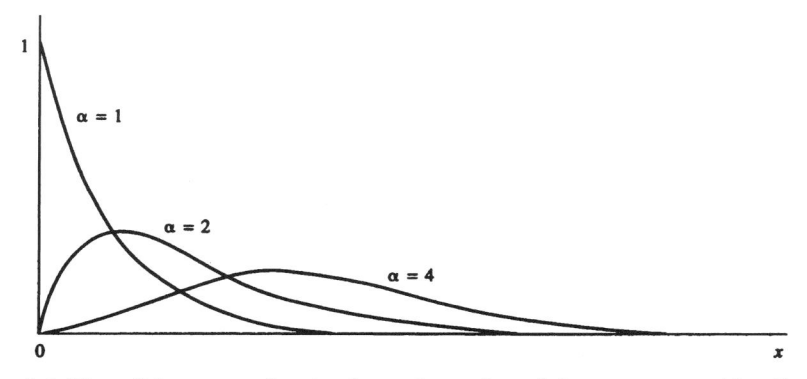

Figure 3.6. Plots of the gamma function for various values of the parameter α ($\beta = 1$).

single random variable X and its standard normalized equivalent, $Z = (X - \mu)/\sigma$. We now wish to investigate the combined properties of more than one standardized independent normal variable. For example, we might want to investigate the distributions of temperature differences between reversing thermometers and a CTD thermistor for a suite of CTD versus reversing thermometer intercomparisons taken at the same location. Each cast is considered to produce a temperature difference distribution x_k with a mean μ_k and a variance σ_k^2. The set of standardized independent normal variables Z_k formed from the casts is assumed to yield ν independent standardized normal variables $Z_1, Z_2, \ldots, Z_\nu$. The new random variable formed from the sum of the squares of the variables $Z_1, Z_2, \ldots, Z_\nu$ is the chi-square variable χ_ν^2 where

$$\chi_\nu^2 = Z_1^2 + Z_2^2 + \ldots + Z_\nu^2 \qquad (3.5.9)$$

has ν degrees of freedom. For the case of our temperature comparison, this represents the square of the deviations for each cast about the mean. Properties of the distribution are

$$\text{Mean} = E[\chi_\nu^2] = \nu \qquad (3.5.10a)$$

$$\text{Variance} = E[(\chi_\nu^2 - \nu)^2] = 2\nu \qquad (3.5.10b)$$

We will make considerable use of the function χ_ν^2 in our discussion concerning confidence intervals for spectral estimates.

It bears repeating that probability density functions are really just models for real populations whose distributions we do not know. In many applications, it is not important that our PDF be a precise description of the true population since we are mainly concerned with the statistics of the distributions as provided by the probability statements from the model. It is not, in general, a simple problem to select the right PDF for a given data set. Two suggestions are worth mentioning: (1) Use available theoretical considerations regarding the process that generated the data; and (2) use the data sample to compute a frequency histogram and select the PDF that best fits the histogram. Once the PDF is selected, it can be used to compute statistical estimates of the true population parameters.

We also keep in mind that our statistics are computed from, and thus are functions of, other random variables and are, therefore, themselves random variables. For example, consider sample variables $X_1, X_2, \ldots, X_N$ from a normal population with mean μ and variance σ^2, then

$$\overline{X} = \frac{1}{N} \sum_{i=1}^{N} X_i \qquad (3.5.11)$$

is normally distributed with mean μ and variance σ^2/N. From this it follows that

$$Z = \frac{\overline{X} - \mu}{\sigma_x} = \frac{\overline{X} - \mu}{\sigma/\sqrt{N}} = \sqrt{N}\frac{\overline{X} - \mu}{\sigma} \qquad (3.5.12)$$

has a standard normal distribution $N(0, 1)$ with zero mean and variance of unity. Using the same sample, $X_1, X_2, \ldots, X_N$, we find that

$$\frac{1}{\sigma^2}\sum_{i=1}^{N}(X_i-\overline{X})^2 = \frac{(N-1)s^2}{\sigma^2} = \chi_\nu^2 \qquad (3.5.13)$$

has a chi-square distribution (χ_ν^2) with $\nu = (N-1)$ degrees of freedom. (Only $N-1$ degrees of freedom are available since the estimator requires use of the mean which reduces the degrees of freedom by one.) Here, the sample standard deviation, s, is an unbiased estimate of σ. We also can use $(X-\overline{X})/(s/\sqrt{N})$ as an estimate of the standard normal statistic, $(X-\mu)/(\sigma/\sqrt{N})$. The continuous sample statistic $(X-\overline{X})/(s/\sqrt{N})$ has a PDF known as the *Student's t-distribution* (Appendix D, Table D.3) with $(N-1)$ degrees of freedom. The name derives from an Irish statistician named W. S. Gossett who was one of the first to work on the statistic. Because his employer would not allow employees to publish their research, Gossett published his results under the name "Student" in 1908. Mathematically, the random variable t is defined as a standardized normal variable divided by the square root of an independently distributed chi-square variable divided by its degrees of freedom; viz. $t = z/\sqrt{(\chi^2/\nu)}$, where z is the standard normal distribution. Thus, one can safely use the normal distribution for samples $\nu > 30$, but for smaller samples one must use the t-distribution. In other words, the normal distribution gives a good approximation to the t-distribution only for $\nu > 30$.

The above relations for statistics computed from a normal population are important for two reasons:

(a) often, the data or the measurement errors can be assumed to have population distributions with normal probability density functions;
(b) one is working with averages that themselves are normally distributed regardless of the probability density function of the original data. This statement is a version of the well-known *central limit theorem*.

3.6 CENTRAL LIMIT THEOREM

Let $X_1, X_2, \ldots, X_i, \ldots$ be a sequence of independent random variables with $E[X_i] = \mu_i$ and $V[X_i] = \sigma_i^2$. Define a new random variable $X = X_1 + X_2 + \ldots + X_N$. Then, as N becomes large, the standard normalized variable

$$Z_N = \frac{\left(X - \sum_{i=1}^{N}\mu_i\right)}{\left(\sum_{i=1}^{N}\sigma_i^2\right)^{1/2}} \qquad (3.6.1)$$

takes on a normal distribution regardless of the distribution of the original population variable from which the sample was drawn. The fact that the X_i values may have any kind of distribution, and yet the sum X may be approximated by a normally distributed random variable, is the basic reason for the importance of the normal distribution in probability theory. For example, X might represent the summation of fresh water added to an estuary from a large number of rivers and streams, each with its own particular form of variability. In this case, the sum of the rivers and stream input would result in a normal distribution of the input of fresh water. Alternatively, the variable X, representing the success or failure of an AXBT launch, may be represented as the sum of the following independent binomially-distributed random variables (a variable that can only take on one of two possible values)

$$X_i = 1 \text{ if the } i\text{th cast is a success}$$
$$= 0 \text{ if the } i\text{th cast is a failure}$$

with $X = X_1 + X_2 + ... + X_N$. For this random variable, $E[X] = Np$ and $V[X] = Np(1-p)$. For large N, it can be shown that the variable $(X - E[X])/\sqrt{V[X]}$ closely resembles the normal distribution, $N(0, 1)$.

A special form of the central limit theorem may be stated as: the distribution of mean values calculated from a suite of random samples X_i ($X_{i,1}$, $X_{i,2}$, ...) taken from a discrete or continuous population having the same mean μ and variance σ^2 approaches the normal distribution with mean μ and variance σ^2/N as N goes to infinity. Consequently, the distribution of the arithmetic mean

$$\overline{X} = \frac{1}{N} \sum_{i=1}^{N} X_i \tag{3.6.2}$$

is asymptotically normal with mean μ and variance σ^2/N when N is large. Ideally, we would like $N \to \infty$ but, for practical purposes, $N \geq 30$ will generally ensure that the population of X is normally distributed. When N is small, the shape of the sample distribution will depend mainly on the shape of the parent population. However, as N becomes larger, the shape of the sampling distribution becomes increasingly more like that of a normal distribution no matter what the shape of the parent population (Figure 3.7). In many instances, the normality assumption for the sampling distribution for $\overline{X}$ is reasonably accurate for $N > 4$ and quite accurate for $N > 10$ (Bendat and Piersol, 1986).

The central limit theorem has important implications for we often deal with average values in time or space. For example, current meter systems average over some time interval, allowing us to invoke the central limit theorem and assume normal statistics for the resulting data values. Similarly, data from high-resolution CTD systems are generally vertically averaged (or averaged over some set of cycles in time), thus approaching a normal PDF for the data averages, via the central limit theorem. An added benefit of this theorem is that the variance of the averages is reduced by the factor N, the number of samples averaged. The theorem essentially states that individual terms in the sum contribute a negligible amount to the variation of the sum, and that it is not likely that any one value makes a large contribution to the sum. Errors of measurements certainly have this characteristic. The final error is the sum of many small contributions none of which contributes very much to the total error. Note that the sample standard error is an unbiased estimate (again in the sense that the expected value is equal to the population parameter being estimated) even though the component sample standard deviation is not.

To further illustrate the use of the central limit theorem, consider a set of independent measurements of a process whose probability distribution is unknown. Through previous experimentation, the distribution of this process was estimated to have a mean of 7 and a variance of 120. If $\overline{x}$ denotes the mean of the sample measurements, we want to find the number of measurements, N, required to a give a probability

$$P(4 \leq \overline{x} \leq 10) = 0.866$$

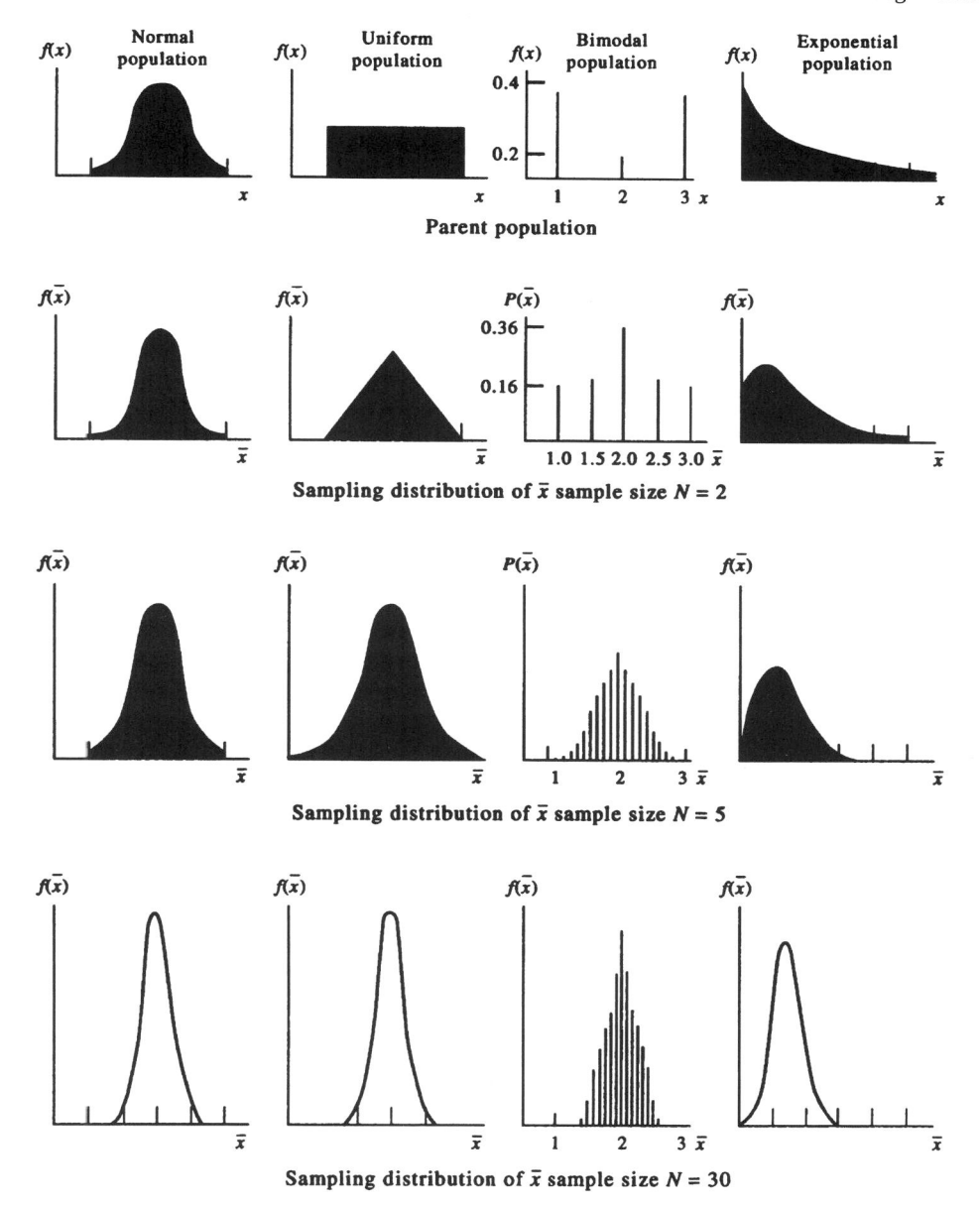

Figure 3.7. Sampling distribution of the mean $\bar{x}$ for different types of population distributions for increasing sample size, N = 2, 5, and 30. The shape of the sampling distribution becomes increasingly more like that of a normal distribution regardless of the shape of the parent population.

where 4 and 10 are the chosen problem limits. Here, we use the central limit theorem to argue that, while we don't know the exact distribution of our variable, we do know that means are normally distributed. Using the standard variable, $z = (x - \mu)/(\sigma/\sqrt{N})$, substituting $\bar{x}$ for x, and using the fact that $\sigma = \sqrt{120} = 2\sqrt{30}$, we can then write our probability function as

$$P(4 \leq \bar{x} \leq 10) = P\left[\frac{(4-\mu)\sqrt{N}}{\sigma} < z < \frac{(10-\mu)\sqrt{N}}{\sigma}\right]$$
$$= P\left[\frac{(4-7)\sqrt{N}}{2\sqrt{30}} < z < \frac{(10-7)\sqrt{N}}{2\sqrt{30}}\right]$$
$$= P\left[\frac{-3\sqrt{N}}{2\sqrt{30}} < z < \frac{3\sqrt{N}}{2\sqrt{30}}\right]$$
$$= 2P\left[z < \frac{3\sqrt{N}}{2\sqrt{30}}\right] - 1 = 0.866$$

from which we find

$$P\left[z < \frac{3\sqrt{N}}{2\sqrt{30}}\right] = 0.933$$

Assuming that we are dealing with a normal distribution, we can look up the value 0.933 in a table to find the value of the integrand to which this cumulative probability corresponds. In this case, $3/2\sqrt{(N/30)} = 1.5$, so that $N = 30$.

3.7 ESTIMATION

In most oceanographic applications, the population parameters are unknown and must be estimated from a sample. Faced with this estimation problem, the objective of statistical analysis is twofold: To present criteria that allow us to determine how well a given sample represents the population parameter; and to provide methods for estimating these parameters. An *estimator* is a random variable used to provide *estimates* of population parameters. "Good" estimators are those that satisfy a number of important criteria: (1) Have average values that equal the parameter being estimated (*unbiasedness* property); (2) have relatively small variance (*efficiency* property); and (3) approach asymptotically the value of the population parameter as the sample size increases (*consistency* property). We have already introduced the concept of estimator bias in discussing variance and standard deviation. Formally, an estimate $\hat{\theta}$ of a parameter θ (here, the hat symbol (^) indicates an estimate), is an unbiased estimate provided that $E[\hat{\theta}] = \theta$; otherwise, it is a biased estimate with a bias $B = E[\hat{\theta}] - \theta$. An unbiased estimator is any estimate whose average value over all possible random samples is equal to the population parameter being estimated. An example of an unbiased estimator is the mean of the noise in an acoustic current meter record created by turbulent fluctuations in the velocity of sound speed in water; an example of a biased estimator is the linear slope of a sea-level record in the presence of a long-term trend (a slow change in average value). Other examples of unbiased estimators are $\bar{x}$ for $\hat{\theta}$, μ for $E[\hat{\theta}]$, and σ^2/N for $\sigma_{\hat{\theta}}^2$. The mean square error of our estimate $\hat{\theta}$ is

$$E[(\hat{\theta} - \theta)^2] = V[\hat{\theta}] + B^2 \tag{3.7.1}$$

The most efficient estimator (property 2) is the estimator with the smallest mean square error. Since it is possible to obtain more than one unbiased estimator for the same target parameter, θ, we define the efficiency of an estimator as the ratio of the variances of the two estimators. For example, if we have two unbiased estimates $\hat{\theta}_1$ and $\hat{\theta}_2$, we can compute the relative efficiency of these two estimates as

$$\text{efficiency} = V[\hat{\theta}_2]/V[\hat{\theta}_1] \tag{3.7.2}$$

where $V[\hat{\theta}_1]$ and $V[\hat{\theta}_2]$ are the variances of the estimators. A low value of the ratio would suggest that $V[\hat{\theta}_2]$ is more efficient while a high value would indicate that $V[\hat{\theta}_1]$ is more efficient. As an example, consider the efficiency of two familiar estimators of the mean of a normal distribution. In particular, let $\hat{\theta}_1$ be the median value and $\hat{\theta}_2$ be the sample mean. The variance of the sample median is $V[\hat{\theta}_1] = (1.2533\sigma)^2/N$ while the sample mean has a variance $V[\hat{\theta}_2] = \sigma^2/N$. Thus, the efficiency is

$$\begin{aligned}\text{efficiency} &= V[\hat{\theta}_2]/V[\hat{\theta}_1] \\ &= (\sigma^2/N)/(1.2533^2\sigma^2/N) \\ &= 0.6366\end{aligned}$$

Therefore, the variability of the sample mean is 63% of the variability of the sample median, which indicates that the sample *mean* is a more efficient estimator than the sample *median*.

As a second example, consider the sample variances s'^2 and s^2 given by (3.2.3) and (3.2.4), respectively. The efficiency of these two sample variances is the ratio of s'^2 to s^2

$$\frac{1/N \sum_{i=1}^{N}(x_i - \bar{x})^2}{1/(N-1) \sum_{i=1}^{N}(x_i - \bar{x})^2} = \frac{N-1}{N} < 1$$

which indicates that s'^2 is a more efficient statistic than s^2.

We can view the difference $\hat{\theta} - \theta$ as the distance between the population "target" value and our estimate. Since this difference is also a random variable, we can ask probability-related questions, such as: "What is the probability

$$P(-b < (\hat{\theta} - \theta) < b)$$

for some range $(-b, b)$?" It is common practice to express b as some multiple of the sample standard deviation of θ (e.g. $b = k\sigma_\theta$, $k > 1$). A widely used value is $k = 2$, corresponding to two standard deviations. Here, we can apply an important result known as *Tshebysheff's theorem* which states that for any random variable Y, for $k > 0$:

$$P(|Y - \mu| < k\sigma) \geq 1 - \frac{1}{k^2} \tag{3.7.3a}$$

or

$$P(|Y - \mu| \geq k\sigma) \leq \frac{1}{k^2} \tag{3.7.3b}$$

where $\mu = E[\hat{Y}]$ and $\sigma^2 = V[\hat{Y}]$. Applying this to the problem at hand, we find that for $k = 2$, $P(|\hat{\theta} - \theta| < 2\sigma_\theta) \geq 1 - 1/(2)^2 = 0.75$. Therefore, most random variables occurring in nature can be found within two standard deviations ($\pm 2\sigma$) of the mean

216 Data Analysis Methods in Physical Oceanography

with a probability of 0.75. Note that the probability statement (3.7.3a) indicates that the probability is greater than or equal to the value of $1 - 1/k^2$ for any type of distribution. We can, therefore, expect somewhat more than 75% of the values to lie with the range $(-2\sigma, 2\sigma)$. In fact, this is generally a conservative estimate. If we assume that oceanographic measurements are typically normally distributed, we find $P(|Y - \mu| < 2\sigma) = 0.95$, so that 95% of the observations lie within $\pm 2\sigma$. This is an important conclusion in terms of editing methods which use criteria designed to select erroneous values from data samples based on probabilities.

3.8 CONFIDENCE INTERVALS

An important application of interval estimates for probability distribution functions is the formulation of *confidence intervals* for parameter estimates. These intervals define the degree of certainty that a given quantity θ will fall between specified lower and upper bounds θ_L, θ_U, respectively, of the parameter estimates. The confidence interval (θ_L, θ_U) associated with a particular confidence statement is usually written as

$$P(\theta_L < \theta < \theta_U) = 1 - \alpha, \quad 0 < \alpha < 1 \qquad (3.8.1)$$

where α is called the *level of significance* (or confidence coefficient) for the confidence statement and $(1 - \alpha)100$ is the percent significance level for the variable θ. (The terms confidence coefficient, significance level, confidence level and confidence are commonly used interchangeably). A typical value for α is 0.05, which means that 95% of the cumulative area under the probability curve (3.8.1) is contained between the points θ_L and θ_U (Figure 3.8). For both symmetric and nonsymmetric probability distributions, each of the two points cuts off $\alpha/2$ of the total area under the distribution curve, leaving a total area under the curve of $1 - \alpha$; θ_L cuts off the left-hand part of the distribution tail and θ_U cuts off the right-hand part of the tail.

If θ_L, θ_U are derived from the true value of the variable θ (such as the population mean, μ), then the probability interval is fixed. However, where we are using sample estimates (for example, the mean, $\overline{X}$) to determine the variable value, θ, the probability interval will vary from sample to sample because of changes in the sample mean and standard deviation. Thus, we must inquire about the probability that the true value of θ will fall within the intervals generated by each of the given sample estimates. The statement $P(\theta_L < \theta < \theta_U)$ does not mean that the population variable θ has a probability of $P = 1 - \alpha$ of falling in the sample interval (θ_L, θ_U), in the sense that θ was behaving like a sample. The population variable is a fixed quantity. Once

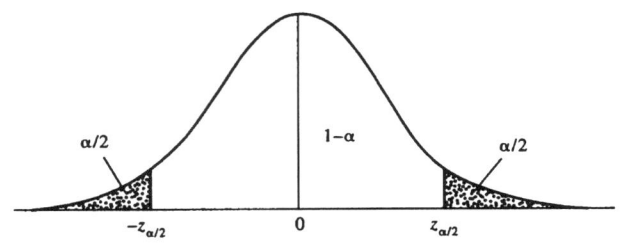

Figure 3.8. Location of the limits $(\theta_L, \theta_U) = (-z_{\alpha/2}, +z_{\alpha/2})$ for a normal probability distribution. For $\alpha = 0.05$, the cumulative area $1 - \alpha$ corresponds to the 95% interval for the distribution.

the interval is picked, the population variable θ is either in the interval or it isn't (probability 1 or 0). For the sample data, the interval may shift depending on the mean and variance of the particular sample we select from the population. We should, therefore, interpret (3.8.1) to mean that there is a probability P that the specified random sample interval (θ_L, θ_U) contains the true population variable θ a total of $(1 - \alpha)$ 100% of the time. That is, $(1 - \alpha)$ is the fraction of the time that the true variable value θ is contained by the sample interval (θ_L, θ_U).

In general, we need a quantity, called a *pivotal quantity*, that is a function of the sample estimator $\hat{\theta}$ and the unknown variable θ, where θ is the only unknown. The pivotal quantity must have a PDF that does not depend on θ. For large samples ($N \geq 30$) of unbiased point estimators, the standard normal distribution $Z = (\hat{\theta} - \theta)/\sigma_{\hat{\theta}}$ is a pivotal quantity. In fact, it is common to express the confidence interval in terms of Z. For example, consider the statistic $\hat{\theta}$ with $E[\hat{\theta}] = \theta$ and $V[\hat{\theta}] = \sigma_{\hat{\theta}}^2$; find the $100(1 - \alpha)$% confidence interval. To do this, we first define

$$P(-Z_{\alpha/2} < Z < Z_{\alpha/2}) = 1 - \alpha \qquad (3.8.2)$$

and then use the above relation $Z = (\hat{\theta} - \theta)/\sigma_{\hat{\theta}}$ to get

$$P(\hat{\theta} - Z_{\alpha/2}\sigma_{\hat{\theta}} < \theta < \hat{\theta} + Z_{\alpha/2}\sigma_{\hat{\theta}}) = 1 - \alpha \qquad (3.8.3)$$

This formula can be used for large samples to compute the confidence interval for θ once α is selected. Again, the significance level, $1 - \alpha$, refers to the probability that the population parameter θ will be bracketed by the given confidence interval. The meaning of these limits is shown graphically in Figure 3.8 for a normal population. We remark that if the population standard deviation σ is known it should be used in (3.8.3) so that $\sigma_{\hat{\theta}} = \sigma$; if not, the sample standard deviation s can be used with little loss in accuracy if the sample size is sufficiently large (i.e. $N > 30$).

The three types of confidence intervals commonly used in oceanography are listed below. Specific usage depends on whether we are interested in the mean or the variance of the quantity being estimated.

3.8.1 Confidence interval for μ (σ known)

When the population standard deviation, σ, is known and the parent population is normal (or $N > 30$), the $100(1 - \alpha)$ percent confidence interval for the population mean is given by the symmetrical distribution for the standardized normal distribution, z

$$\bar{x} - z_{\alpha/2}\frac{\sigma}{\sqrt{N}} < \mu < \bar{x} + z_{\alpha/2}\frac{\sigma}{\sqrt{N}} \qquad (3.8.4)$$

As an example, we wish to find the 95% confidence interval ($\alpha = 0.05$) for μ given the sample mean $\bar{x}$ and known normally distributed population variance, σ^2. Suppose that a thermister installed at the entrance to the ship's engine cooling water intake samples every second for $N = 20$ s and yields a mean ensemble temperature $\bar{x} = 12.7°C$ for the particular burst. Further, suppose that the water is isothermal and that the only source of variability is instrument noise, which we know from previous calibration in the lab has a known noise level $\sigma = 0.5°C$. Since we want the 95% confidence interval, the appropriate values of z for the normal distribution are $z_{\alpha/2} = 1.96$ and $-z_{\alpha/2} = -1.96$ (Appendix D, Table D.1). Substituting these values into (3.8.4) along with $N = 20$, $\sigma = 0.5°C$, and $\bar{x} = 12.7°C$ we find

$$[12.7 - (1.96)0.5/\sqrt{20}]°C < \mu < [12.7 + (1.96)0.5/\sqrt{20}]°C$$

$$12.48°C < \mu < 12.92°C$$

Based on our 20 data values, there is a 95% probability that the true mean temperature of the water will be bracketed by the interval (12.48°C, 12.92°C) derived from the random interval

$$(\bar{x} - z_{\alpha/2}\sigma/\sqrt{N}, \bar{x} + z_{\alpha/2}\sigma/\sqrt{N})$$

3.8.2 Confidence interval for μ (σ unknown)

In most real circumstances, σ is not known and we must resort to the use of the sample standard deviation, s. Similarly, for small samples ($N < 30$), we cannot use the above technique but must introduce a formalism that works for any sample size and distribution, as long as the departures from normality are not excessive. Under these conditions, we resort to the variable $t = (\bar{x} - \mu)/(s/\sqrt{N})$, which has a Student's t-distribution with $\nu = (N - 1)$ degrees of freedom. Derivation of the $100(1 - \alpha)\%$ confidence interval follows the same procedure used for the symmetrically distributed normal distribution, except that we must modify the limits. In this case

$$P\left[-t_{\alpha/2,\nu} < (\bar{x} - \mu)\bigg/\frac{s}{\sqrt{N}} < t_{\alpha/2,\nu}\right] = 1 - \alpha \qquad (3.8.5)$$

This is easily arranged to give the $100(1 - \alpha)\%$ confidence interval for μ

$$\bar{x} - t_{\alpha/2,\nu}\frac{s}{\sqrt{N}} < \mu < \bar{x} + t_{\alpha/2,\nu}\frac{s}{\sqrt{N}} \qquad (3.8.6)$$

Note the similarity between (3.8.6) and the form (3.8.3) obtained for μ when σ is known. We return to our previous example of ship injection temperature and this time assume that $s = 0.5°C$ is a measured quantity obtained by subtracting the mean value $\bar{x} = 12.7°C$ from the series of 20 measurements. Turning to Appendix D (Table D.3) for the cumulative t-distribution, we look for values of $F(t)$ under the column for the 95% confidence interval ($\alpha = 0.05$) for which $F(t) = 1 - \alpha/2 = 0.975$. Using the fact that $\nu = (N - 1) = 19$, we find $t_{\alpha/2,\nu} = t_{0.025,19} = 2.093$. Substituting these values into (3.8.6), yields

$$[12.7 - 2.093(0.5/\sqrt{20})]°C < \mu < [12.7 + 2.093(0.5/\sqrt{20})]°C$$

$$12.47°C < \mu < 12.93°C$$

There is a 95% chance that the interval (12.47°C, 12.93°C) will bracket the true mean temperature. Because of the large sample size, this result is only slightly different than the result obtained for the normal distribution in the previous example when σ was known *a priori*.

3.8.3 Confidence interval for σ^2

Under certain circumstances, we are more interested in the confidence interval for the signal variance than the signal mean. For example, to determine the reliability of a spectral peak in a spectral density distribution (or spectrum), we need to know the confidence intervals for the population variance, σ^2, based on our sample variance, s^2. To do this, we seek a new pivotal quantity. Recall from (3.5.13) that for N samples of a variable x_i from a normal population, the expression

$$\frac{1}{\sigma^2} \sum_{i=1}^{N} (x_i - \bar{x})^2 = \frac{(N-1)/s^2}{\sigma^2} \qquad (3.8.7)$$

is a χ^2 variable with $(N-1)$ degrees of freedom. Using this as a pivotal quantity, we can find upper and lower bounds χ_U^2 and χ_L^2 for which

$$P\left[\chi_L^2 < \frac{N-1}{\sigma^2/s^2} < \chi_U^2\right] = 1 - \alpha \qquad (3.8.8)$$

or, upon rearranging terms,

$$P\left[\frac{(N-1)s^2}{\chi_L^2} < \sigma^2 < \frac{(N-1)s^2}{\chi_U^2}\right] = 1 - \alpha \qquad (3.8.9)$$

Note that χ^2 is a skewed function (Figure 3.9), which means that the upper and lower bounds in (3.8.9) are asymmetric; the point $1 - \alpha/2$ rather than $-\alpha/2$ determines the point that cuts off $\alpha/2$ of the area at the lower end of the chi-square distribution.

From expression (3.8.9) we obtain the well-known $100(1 - \alpha)\%$ confidence interval for the variance σ^2 when sampled from a normal population

$$\frac{(N-1)s^2}{\chi_{\alpha/2,\nu}^2} < \sigma^2 < \frac{(N-1)s^2}{\chi_{1-\alpha/2,\nu}^2} \qquad (3.8.10)$$

where the subscripts $\alpha/2$ and $1 - \alpha/2$ characterize the endpoint values of the confidence interval and $\nu = (N-1)$ gives the degrees of freedom of the chi-square distribution. The larger value of $\chi^2 (= \chi_{\alpha/2,\nu}^2)$ appears in the denominator of the lower endpoint for σ^2 while the smaller value of $\chi^2 (= \chi_{1-\alpha/2,\nu}^2)$ is in the denominator of the

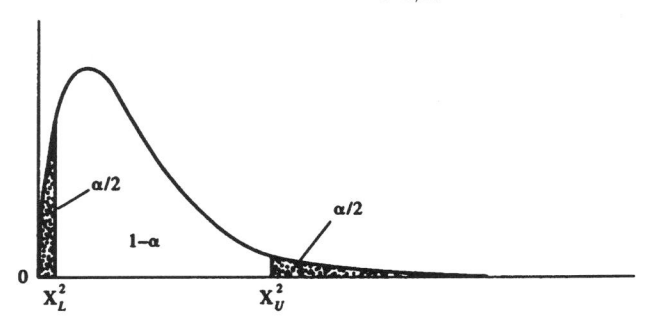

Figure 3.9. Location of the limts $(\theta_L, \theta_U) = (\chi_L^2, \chi_U^2)$ for a chi-square probability distribution. For $\alpha = 0.05$, the cumulative area $1 - \alpha$ corresponds to the 95% interval for the distribution. χ^2 is a skewed function so that the upper and lower bounds are asymmetric.

upper endpoint for σ^2. As an example, suppose that we have $\nu = 9$ in our spectral estimate of the eastward component of current velocity and that the background variance of our spectra near a distinct spectral peak is $s^2 = 10$ cm^2/s^2. What is the 95% confidence interval for the variance? How big would the peak have to be to stand out statistically from the background level? (Details on spectral estimation can be found in Chapter 5.) In this case, $\alpha/2 = 0.025$ and $1 - \alpha/2 = 0.975$. Turning to the cumulative distribution $F(\chi^2)$ for 9 degrees of freedom in Appendix D (Table D.2), we find that $\chi_9^2 = 2.70$ for a cumulative integral $F(\alpha/2 = 0.025)$ and that $\chi_9^2 = 19.02$ for a cumulative integral $F(1 - \alpha/2 = 0.975)$. Thus, $P(2.70 < \chi_{\nu=9}^2 < 19.02) = 1 - \alpha = 0.95$. Substituting $N - 1 = 9$, $s^2 = 10$ cm^2/s^2, $\chi_{\alpha/2,\nu}^2 = 19.02$ for the value that cuts off $\alpha/2$ of the upper end area under the curve and $\chi_{1-\alpha/2,\nu}^2 = 2.70$ for the value that cuts off $1 - \alpha/2$ of the lower end area of the curve, (3.8.10) yields

$$9(10 \text{ cm}^2/\text{s}^2)/19.02 < \sigma^2 < 9(10 \text{ cm}^2/\text{s}^2)/2.70$$
$$4.7 \text{ cm}^2/\text{s}^2 < \sigma^2 < 33.3 \text{ cm}^2/\text{s}^2$$

Thus, the true background variance will lie between 4.7 and 33.3 cm^2/s^2. If a spectral peak has a greater range than these limits then it represents a statistically significant departure from background energy levels.

In most instances, spectral densities are presented in terms of the log of the spectral density function versus frequency or log-frequency (see Chapter 5). Dividing through by s^2 in (3.8.10) and taking the log, yields

$$\log(N-1) - \log\left(\chi_{\alpha/2,\nu}^2\right) < \log(\sigma^2/s^2) < \log(N-1) - \log\left(\chi_{1-\alpha/2,\nu}^2\right)$$

or, upon subtracting $\log(N-1)$ and rearranging the inequality

$$\log\left(\chi_{1-\alpha/2,\nu}^2\right) < \log(N-1) - \log(\sigma^2/s^2) < \log\left(\chi_{\alpha/2,\nu}^2\right)$$

The range R of the variance is then

$$R = \log\left(\chi_{\alpha/2,\nu}^2\right) - \log\left(\chi_{1-\alpha/2,\nu}^2\right) \tag{3.8.11}$$

while the pivot point p_o of the interval is

$$p_o = \log(N-1) - \log(\sigma^2/s^2) \tag{3.8.12}$$

If we assume that the measured background value of s^2 is a good approximation to σ^2, so that $\sigma^2/s^2 = 1$, then $p_o = \log(N-1)$. The ranges between the maximum value and p_o, and the minimal value and p_o, are $\log(\chi_{\alpha/2,\nu}^2) - p_o$ and $p_o - \log(\chi_{1-\alpha/2,\nu}^2)$, respectively. Returning to our previous example for the 95% confidence interval, we find that

$$\log(2.70) < \log(9) < \log(19.02), \quad 0.43 < 0.95 < 1.28$$

giving a range $R = 0.848$ with the pivot point at $p_o = 0.95$.

3.8.4 Goodness-of-fit test

When the set of outcomes for an experiment is limited to two outcomes (such as success or failure, on or off, and so on), the appropriate test statistic for the distribution is the binomial variable. However, when more than two outcomes are possible, the preferred statistic is the chi-square variable. In addition to providing confidence intervals for spectral estimates and other measurement parameters, the chi-square variable is used to test how closely the observed frequency distribution of a given parameter corresponds to the expected frequency distribution for the parameter. The expected frequencies represent the average number of values expected to fall in each frequency interval based on some theoretical probability distribution, such as a normal distribution. The observed frequency distribution represents a sample of values drawn from some probability distribution. What we want to know is whether the observed and expected frequencies are similar enough for us to conclude that they are drawn from the same probability function (the "null hypothesis"). The test for this similarity using the chi-square variable is called a "goodness-of-fit" test.

Consider a sample of N observations from a random variable X with observed probability density function $p_o(X)$. Let the N observations be grouped into K intervals (or categories) called *class intervals*, whose graphical distribution forms a frequency histogram (Bendat and Piersol, 1986). The actual number of observed values that fall within the ith class interval is denoted by f_i, and is called the *observed frequency* in the ith class. The number of observed values that we would expect to fall within the ith class interval if the observations really followed the theoretical probability distribution, $p(X)$, is denoted F_i, and is called the *expected frequency* in the ith class interval. The difference between the observed frequency and the expected frequency for each class interval is given by $f_i - F_i$. The total discrepancy for all class intervals between the expected and observed distributions is measured by the sample statistic

$$\mathbf{X}^2 = \sum_{i=1}^{K} \frac{(f_i - F_i)^2}{F_i} \qquad (3.8.13)$$

where division by F_i transforms the sum of the squares into the chi-square-type variable, X^2.

The number of degrees of freedom, ν, for the variable X^2 is equal to K minus the number of different independent linear restrictions imposed on the observations. As discussed by Bendat and Piersol (1986), one degree of freedom is lost through the restriction that, if $K - 1$ class intervals are determined, the Kth class interval follows automatically. If the expected theoretical density function is normally distributed then the mean and variance must be computed to allow comparison of the observed and expected distributions. This results in the loss of two additional degrees of freedom. Consequently, if the chi-square goodness-of-fit test is used to test for normality of the data, the true number of degrees of freedom for X^2 is $\nu = K - 3$.

Formula (3.8.13) measures the goodness-of-fit between f_i and F_i as follows: when the fit is good (that is, f_i and F_i are generally close), then the numerator of (3.8.13) will be small and the hence the value of X^2 will be low. On the other hand, if f_i and F_i are not close, the numerator of (3.8.13) will be comparatively large and the value of X^2 will be large. Thus, the critical region for the test statistic X^2 will always be in the upper tail of the chi-square function because we wish to reject the null hypothesis whenever the difference between f_i and F_i is large. More specifically, the region of acceptance of the null hypothesis (see Section 3.14) is

$$X^2 \leq \chi^2_{\alpha;\nu} \qquad (3.8.14)$$

where the value of $\chi^2_{\alpha;\nu}$ is available from Appendix D (Table D.2). If the sample value X^2 is greater than $\chi^2_{\alpha;\nu}$, the hypothesis that $p(X) = p_o(X)$ is rejected at the level of significance. If X^2 is less than or equal to $\chi^2_{\alpha;\nu}$, the hypothesis is accepted at the α level of significance (i.e. there is a $100\alpha\%$ chance that we are wrong in accepting the null hypothesis that our data are drawn from a normal distribution). For example, suppose our analysis involves 15 class intervals and that the fit between the 15 estimates of f_i and F_i (where F_i is normally distributed) yields $X^2 = 23.1$. From tables for the cumulative chi-square distribution, $F(X) = p(x > \chi^2_{\alpha;\nu})$, we find that $p(X^2 > 21.03) = 0.05$ for $\nu = K - 3 = 12$ degrees of freedom. Thus, at the $\alpha = 0.05$ level of significance (95% certainty level) we cannot accept the null hypothesis that the observed values came from the same distribution as the expected values.

Chi-square tests for normality are typically performed using a constant interval width. Unless one is dealing with a uniform distribution, this will yield different expected frequency distributions from one class interval to the next. Bendat and Piersol recommend a class interval width of $\Delta x \approx 0.4s$, where s is the standard deviation of the sample data. A further requirement is that the expected frequencies in all class intervals be sufficiently large that X^2 in (3.8.13) is an acceptable approximation to $\chi^2_{\alpha;\nu}$. A common recommendation is that $F_i > 3$ in all class intervals. When testing for normality, where the expected frequencies diminish on the tails of the distribution, this requirement is attained by letting the first and last intervals extend to $-\infty$ to $+\infty$, respectively, so that $F_1, F_K > 3$.

As an example of a goodness-of-fit test, we consider a sample of $N = 200$ surface gravity wave heights measured every 0.78 s by a Datawell waverider buoy deployed off the west coast of Canada during the winter of 1993-1994 (Table 3.8.1). The wave record spans a period of 2.59 min and corresponds to a time of extreme (5 m high) storm-generated waves. According to one school of thought (e.g. Phillips et al., 1993),

Table 3.8.1. *Wave heights (mm) during a period of anomalously high waves as measured by a Datawell waverider buoy deployed in 30 m depth on the inner continental shelf of Vancouver Island, British Columbia. The original $N = 200$ time-series values have been rank ordered. The upper bounds of the K-class intervals have been underlined. (Courtesy, Diane Masson)*

4636	4840	4901	4950	4980	5014	5034	5060	5095	5130
4698	4842	4904	4954	4986	5014	5037	5066	5095	5135
4702	4848	4907	4955	4991	5015	5037	5066	5096	5135
4731	4854	4907	4956	4994	5017	5038	5069	5102	5145
4743	4856	4908	4956	4996	5020	5039	5069	5103	5155
4745	4867	4914	4956	4996	5020	5040	5071	5104	5157
4747	4867	4916	4959	4996	5021	5040	5072	5104	5164
4749	4870	4917	4960	4997	5023	5044	5073	5104	5165
4773	4870	4923	4961	4998	5024	5045	5074	5106	5166
4785	4874	4925	4963	5003	5025	5045	5074	5110	5171
4793	4876	4934	4964	5006	5025	5047	5074	5111	5175
4814	4877	4935	4964	5006	5025	5048	5078	5115	5176
4817	4883	4937	4966	5006	5025	5050	5079	5119	5177
4818	4885	4939	4966	5006	5028	5051	5080	5119	5181
4823	4886	4940	4970	5006	5029	5052	5081	5120	5196
4824	4892	4941	4971	5010	5029	5053	5086	5121	5198
4828	4896	4942	4972	5011	5029	5057	5089	5122	5201
4829	4897	4942	4974	5011	5030	5058	5091	5123	5210
4830	4898	4943	4977	5012	5031	5059	5093	5125	5252
4840	4899	4944	4979	5012	5032	5059	5094	5127	5299

Statistical Methods and Error Handling

extreme wave events in the ocean are part of a Gaussian process and the occurrence of maximum wave heights is related in a linear manner to the statistical distribution of the surrounding wave field. If this is true, then the heights of high-wave events relative to the background seas should follow a normal frequency distribution. To test this at the $\alpha = 0.05$ significance level, $K = 10$ class intervals for the observed wave heights were fitted to a Gaussian probability distribution. The steps in the goodness-of-fit test are as follows:

(1) Specify the class interval width Δx and list the upper limit of the standardized values, z_α, of the normal distribution that correspond to these values (as in Table 3.8.2). Commonly Δx is assumed to span 0.4 standard deviations, s, such that $\Delta x \approx 0.4s$; here we use $\Delta x \approx 0.5s$. For $\Delta x = 0.4s$, the values of z_α we want are (... , $-2.4, -2.0, ... , 2.0, 2.4, ...$) while for $\Delta x = 0.5s$, the values are (... , $-2.5, -2.0, ...$, $2.0, 2.5, ...$).

(2) Determine the finite upper and lower bounds for z_α from the requirement that $F_i > 3$. Since $F_i = NP_i$ (where $N = 200$ and P_i is the normal probability distribution for the ith interval), we require $P > 3/N = 0.015$. From tables of the standardized normal density function, we find that $P > 0.015$ implies a lower bound $z_\alpha = -2.0$, and an upper bound $z_\alpha = +2.0$. Note that for a larger sample, say $N = 2000$, we have $P > 3/2000 = 0.0015$ and the bounds become ± 2.8 for the interval $\Delta x = 0.4s$ and ± 2.5 for the interval $\Delta x = 0.5s$.

(3) Calculate the expected upper limit, $x = sz_\alpha + \bar{x}$ (mm), for the class intervals and mark this limit on the data table (Table 3.8.1). For each upper bound, z_α, in Table 3.8.2, find the corresponding probability density value. Note that these values apply to intervals so, for example, $P(-2.0 < x < -1.5) = 0.0668 - 0.0228 = 0.044$; $P(2.0 < x < \infty) = 0.0228$.

(4) Using the value of P, find the expected frequency values $F_i = NP_i$. The observed frequency f_i is found from Table 3.8.1 by counting the actual number of wave heights lying between the marks made in step 3. Complete the table and calculate X^2. Compare to $\chi^2_{\alpha;\nu}$.

Table 3.8.2. Calculation table for goodness-of-fit test for the data in Table 3.8.1. The number of intervals has been determined using an interval width $\Delta x = 0.5s$, with z_α in units of 0.5 and requiring that $F_i > 3$. $N = 200$, $\bar{x}$ (mean) $= 4997.6$ mm, s (standard deviation) $= 115.1$ mm, and ν (degrees of freedom) $= K - 3 = 7$

Class interval	z_α	Upper limit of data interval $x = sz_\alpha + \bar{x}$	P_i	$F_i = NP_i$	f_i	$F_i - f_i$	$\frac{(F_i - f_i)^2}{F_i}$
1	-2.0	4767.4	0.0228	4.6	8	3.4	2.51
2	-1.5	4825.0	0.0440	8.8	8	0.8	0.07
3	-1.0	4882.5	0.0919	18.4	16	2.4	0.31
4	-0.5	4940.1	0.1498	30.0	23	7.0	1.63
5	0	4997.6	0.1915	38.3	33	5.3	0.73
6	0.5	5055.2	0.1915	38.3	48	9.7	2.46
7	1.0	5112.7	0.1498	30.0	35	5.0	0.83
8	1.5	5170.2	0.0919	18.4	18	0.4	0.01
9	2.0	5227.8	0.0440	8.8	9	0.2	0.00
10	∞	∞	0.0228	4.6	2	2.6	1.47
Totals			1.0000	200	200		10.02

In the above example, $X^2 = 10.02$ and there are $\nu = 7$ degrees of freedom. From Table A.3, we find $P(X^2 > \chi^2_{\alpha;\nu}) = P(X^2 > \chi^2_{0.05;7}) = 14.07$. Thus, at the $\alpha = 0.05$ level of significance (95% significance level), we can accept the null hypothesis that the large wave heights measured by the waverider buoy had a Gaussian (normal) distribution in time and space.

3.9 SELECTING THE SAMPLE SIZE

It is not possible to determine the required sample size N for a given confidence interval until a measure of the data variability, the population standard deviation, σ, is known. This is because the variability of $\overline{X}$ depends on the variability of X. Since we do not usually know *a priori* the population standard deviation (the value for the true population), we use the best estimate available, the sample standard deviation, s. We also need to know the frequency content of the data variable so that we can ensure that the N values we use in our calculations are statistically independent samples. As a simple example, consider a normally distributed, continuous random variable, Y, with the units of meters. We wish to find the average of the sample and want it to be accurate to within ± 5 m. Since we know that approximately 95% of the sample means will lie within $\pm 2\sigma_Y$ of the true mean μ, we require that $2\sigma_Y = 5$ m. Using the central limit theorem for the mean, we can estimate σ_Y by

$$\hat{\sigma}_Y = \frac{\sigma}{\sqrt{N}}$$

so that $2\sigma/\sqrt{N} = 5$, or $N = 4\sigma^2/25$ (assuming that the N observations are statistically independent). If σ is known, we can easily find N.

When we don't know σ, we are forced to use an estimate from an earlier sample within the range of measurements. If we know the sample range, we can apply the empirical rule for normal distributions that the range is approximately 4σ and take one-fourth the range as our estimate of σ. Suppose our range in the above example is 84 m. Then, $\sigma = 21$ m and

$$N = 4\sigma^2/25 = (4)(21 \text{ m})^2/(25 \text{ m}^2) = 70.56 \approx 71$$

This means that, for a sample of $N = 71$ statistically independent values, we would be 95% sure (probability = 0.95) that our estimate of the mean value would lie within $\pm 2\sigma_Y = \pm 5$ m of the true mean.

One method for selecting the sample size for relatively large samples is based on Chebyshev's theorem known as the "weak law of large numbers". Let $f(x)$ be a density function with mean μ and variance σ^2, and let $\bar{x}_N$ be the sample mean of a random sample of size N from $f(x)$. Let ε and δ be any two specified numbers satisfying $\varepsilon > 0$ and $0 < \delta < 1$. If N is any integer greater than $(\sigma^2/\varepsilon^2)\delta$ then

$$P[-\varepsilon < \bar{x}_N - \mu < \varepsilon] \geq 1 - \delta \tag{3.9.1}$$

To show the validity of condition (3.9.1), we use Thebyshev's inequality

$$P[g(x) \geq k] \geq \frac{E[g(x)]}{k} \tag{3.9.2}$$

for every $k > 0$, random variable x, and nonnegative function $g(x)$. An equivalent formula is

$$P[g(x) < k] \geq 1 - \frac{E[g(x)]}{k} \quad (3.9.3)$$

Let $g(x) = (\bar{x}_N - \mu < \varepsilon)^2$ and $k = \varepsilon^2$, then

$$P[-\varepsilon < \bar{x}_N - \mu < \varepsilon] = P[|\bar{x}_N - \mu| < \varepsilon]$$

$$= P\left[|\bar{x}_N - \mu|^2 < \varepsilon^2\right] \geq 1 - \frac{E\left[(\bar{x}_N - \mu)^2\right]}{\varepsilon^2} \quad (3.9.4)$$

$$= 1 - \frac{\sigma^2}{N\varepsilon^2} \geq 1 - \delta$$

For $\delta > \sigma^2/N\varepsilon^2$ or $N > \sigma^2/\delta\varepsilon^2$, the latter expression becomes

$$P[|\bar{x}_N - \mu| < \varepsilon] \geq 1 - \delta \quad (3.9.5)$$

We illustrate the use of the above relations by considering a distribution with an unknown mean and variance $\sigma^2 = 1$. How large a sample must be taken in order that the probability will be at least 0.95 that the sample mean, $\bar{x}_N$, will lie within 0.5 of the true population mean? Given are: $\sigma^2 = 1$ and $\varepsilon = 0.5$. Rearranging the inequality (3.9.5)

$$\delta \geq 1 - P[|\bar{x}_N - \mu| < 0.5] = 1 - 0.95 = 0.05$$

Substituting into the relation $N > (\sigma^2/\delta\varepsilon^2) = 1/(0.05)(0.5)^2$ tells us that $N \geq 80$ independent samples.

3.10 CONFIDENCE INTERVALS FOR ALTIMETER BIAS ESTIMATES

As an example of how to estimate confidence limits and sample size, consider an oceanographic altimetric satellite where the altimeter is to be calibrated by repeated passes over a spot on the earth where surface-based measurements provide a precise, independent measure of the sea surface elevation. A typical reference site is an offshore oil platform having sea-level gauges and a location system, such as the multi-satellite global positioning system (GPS). For the TOPEX/POSEIDON satellite one reference site was an oil platform in the Santa Barbara channel off southern California (Christensen et al., 1994). Each pass over the reference site provides a measurement of the satellite altimeter bias which is used to compute an average bias after repeated calibration observations. This bias is just the difference between the height measured by the altimeter and the height measured independently by the *in situ* measurements at the reference site. If we assume that our measurement errors are normally distributed with a mean of zero, then the uncertainty of the true mean bias, σ_b, is

$$\sigma_b = z\, s_b/\sqrt{N}$$

where z is the standard normal distribution, s_b is the estimated standard deviation of the measurements, and N is the number of measurements (i.e. the number of calibration passes over the reference site).

Suppose we are required to know the true mean of the altimeter bias to within 2 cm, and that we estimate the uncertainty of the individual measurements to be 3 cm. We then ask: "What is the number of independent measurements required to give a bias of 2 cm at the 90%, 95%, and 99% confidence intervals?" Using the above formulation for the standard error we find

$$N = (z_{\alpha/2} s_b / \sigma_b)^2 \qquad (3.10.1)$$

from which we can compute the required sample size. As before, the parameter α refers to the chosen significance level. Now $\sigma_b = 2$ cm (required) and $s_b = 3$ cm (estimated), so that we can use the standard normal table for $z_{\alpha/2} = N(0, 1)$ in the appendix to obtain the values shown in Table 3.10.1. If we require the true mean to be 1.5 cm instead of 2.0 cm, the values in Table 3.10.1 become those in Table 3.10.2.

Finally, suppose the satellite is in a 10-day repeat orbit so that we can only collect a reference measurement every 10 days at our ground site; we are given 240 days to collect reference observations. What confidence intervals can be achieved for both of the above cases if we assume that only 50% of the calibration measurements are successful and that the 10-day observations are statistically independent? We can, in general, write the confidence intervals as

$$P(-c < z < c) = \alpha, \text{ and } P(z < c) = (\alpha + 1)/2$$

Now, since we have only one calibration measurement every 10 days for 50% of 240 days we have

$$c = (0.5)(240 \text{ days})(1 \text{ measurement}/10 \text{ days})$$
$$= 12 \text{ measurements/year}$$

Referring to the above tables, we see that for the first case (Table 3.10.1), where the mean bias was required to be 2.0 cm we can achieve the 95% interval; for the case where the mean bias is restricted to 1.5 cm (Table 3.10.2), only the 90% confidence interval is possible.

Table 3.10.1. Calculation of the number of satellite altimeter observations required to attain a given level of confidence in elevation using the relation (3.10.1) for $\sigma_b = 2$ cm and $s_b = 3$ cm

Confidence level (α)	Standard normal value (z_α)	Exact number of observations (N)	Actual number of observations
90%	1.645	6.089	7
95%	1.960	8.644	9
99%	2.576	14.931	15

Table 3.10.2. Calculation of the number of satellite altimeter observations needed for a given level of confidence in sea level elevation using the equation (3.10.1) for $\sigma_b = 1.5$ cm and $s_b = 3$ cm

Confidence level (α)	Standard normal value (z_α)	Exact number of observations (N)	Actual number of observations
90%	1.645	10.82	11
95%	1.960	15.37	16
99%	2.576	26.54	27

3.11 ESTIMATION METHODS

Now that we have introduced methods to calculate confidence intervals for our estimates of μ and σ^2, we need procedures to estimate these quantities themselves. There are many different methods we could use but space does not allow us to discuss them all. We first introduce a very general technique, known as minimum variance unbiased estimation (MVUE), and then later discuss a popular method called the maximum likelihood method which leads to MVUE estimators. We will also discuss one of the oldest methods for finding point estimates, the method of moments.

Before introducing the MVUE procedure, we need to define two terms: *sufficiency* and *likelihood*. Let $x_1, x_2, \ldots, x_N$ be a random sample from a probability distribution with an unknown statistical parameter, θ (mean, variance, etc.). The statistic $U = g(x_1, x_2, \ldots, x_N)$ is said to be sufficient for θ if the conditional distribution $x_1, x_2, \ldots, x_N$, given U, does not depend on θ. In other words, once U is known, no other combination of $x_1, x_2, \ldots, x_N$ provides additional information about θ. This tells us how to check if our statistic is sufficient but does not tell how to compute the statistic.

To define likelihood, let $y_1, y_2, \ldots, y_N$ be sample observations of random variables $Y_1, Y_2, \ldots, Y_N$. For continuous variables, the likelihood $L(y_1, y_2, \ldots, y_N)$ is the joint probability density $f(y_1, y_2, \ldots, y_N)$ evaluated at the observations, y_i. Assuming that the Y_i are statistically independent

$$L(y_1, y_2, \ldots, y_N) = f(y_1, y_2, \ldots, y_N) = f(y_1)f(y_2)\ldots f(y_N) \qquad (3.11.1)$$

where $f(y_i)$, $i = 1, 2, \ldots, N$, is the probability density function (PDF) for the random variable Y_i.

As an oceanographic example, consider a record of daily-average current velocities obtained using a single current meter moored for a period of one month ($N = 30$ days). Show that the monthly mean velocity, V, is a sufficient statistic for the population mean if the variance is known (in this case, estimated from the range of current values). Since the daily velocities are average values of shorter-term current velocity measurements (e.g. 30 min values), we can invoke the central limit theorem to conclude that the daily velocities are normally distributed. Hence the probability density function can be written as

$$f(V) = \frac{1}{\sigma(2\pi)^{1/2}} \exp\left[-\frac{1}{2\sigma^2}(V - \mu)^2\right]$$

We can write the likelihood L of our sample as

$$\begin{aligned} L = f(V_1, V_2, \ldots, V_{30}) &= f(V_1)f(V_2)\ldots f(V_{30}) \\ &= \frac{1}{\sigma(2\pi)^{1/2}} \exp\left[-\frac{1}{2\sigma^2}(V_1 - \mu)^2\right] \\ &\quad \times \frac{1}{\sigma(2\pi)^{1/2}} \exp\left[-\frac{1}{2\sigma^2}(V_2 - \mu)^2\right] \\ &\quad \ldots \frac{1}{\sigma(2\pi)^{1/2}} \exp\left[-\frac{1}{2\sigma^2}(V_{30} - \mu)^2\right] \\ &= \frac{1}{[\sigma(2\pi)]^{15}} \exp\left[-\frac{1}{2^{30}\sigma^{60}} \sum_{i=1}^{30}(V_i - \mu)^2\right] \end{aligned}$$

228 Data Analysis Methods in Physical Oceanography

Because σ is known from our range of current velocities then L is a function of V and μ only. Hence, V is a sufficient statistic for μ the population mean.

3.11.1 Minimum variance unbiased estimation

For random variables $Y_1, Y_2, \ldots, Y_N$, with probability density function, $f(y)$, and unknown parameter θ, let one set of sample observations be $(x_1, x_2, \ldots, x_N)$ and another be $(y_1, y_2, \ldots, y_N)$. The ratio of the likelihoods of these two sets of observations can be written as

$$\frac{L(x_1, x_2, \ldots, x_N)}{L(y_1, y_2, \ldots, y_N)} \tag{3.11.2}$$

In general, this ratio will not be a function of θ if, and only if, there is a function $g(x_1, x_2, \ldots, x_N)$ such that $g(x_1, x_2, \ldots, x_N) = g(y_1, y_2, \ldots, y_N)$ for all choices of x and y. If such a function can be found, it is the minimum sufficient statistic for θ. Any unbiased estimator that is a function of a minimal sufficient statistic will be an MVUE; this means that it will possess the smallest possible variance among the unbiased estimators.

We illustrate what we mean with an example. Let $x_1, x_2, \ldots, x_n$ be a random sample from a normal population with the unknown parameters μ and σ^2. We want to find the MVUE of μ and σ^2. Writing the likelihood ratio we have

$$\frac{L(x_1, x_2, \ldots, x_n)}{L(y_1, y_2, \ldots, y_n)} = \frac{f(x_1, x_2, \ldots, x_n)}{f(y_1, y_2, \ldots, y_n)}$$

$$= \frac{\frac{1}{\sigma\sqrt{2\pi}}\exp\left[-\frac{1}{2\sigma^2}\sum_{i=1}^{N}(x_i - \mu)^2\right]}{\frac{1}{\sigma\sqrt{2\pi}}\exp\left[-\frac{1}{2\sigma^2}\sum_{i=1}^{N}(y_i - \mu)^2\right]}$$

$$= \exp\left\{-\frac{1}{2\sigma^2}\left[\sum_{i=1}^{N}(x_i - \mu)^2 - \sum_{i=1}^{N}(y_i - \mu)^2\right]\right\} \tag{3.11.3}$$

$$= \exp\left\{-\frac{1}{2\sigma^2}\left[\left(\sum_{i=1}^{N}x_i^2 - \sum_{i=1}^{N}y_i^2\right) - 2\mu\left(\sum_{i=1}^{N}x_i - \sum_{i=1}^{N}y_i\right)\right]\right\}$$

For this ratio to be independent of μ, we must have

$$\sum_{i=1}^{N}x_i = \sum_{i=1}^{N}y_i \tag{3.11.4}$$

for the ratio to be independent of σ^2, requires both (3.11.4) as well as

$$\sum_{i=1}^{N}x_i^2 = \sum_{i=1}^{N}y_i^2 \tag{3.11.5}$$

Thus, both $\sum x_i$ and $\sum x_i^2$ are minimum sufficient statistics for μ and σ^2. Since $\bar{x}$ is an unbiased estimate of μ

$$s^2 = \frac{1}{N-1}\sum_{i=1}^{N}(x_i - \bar{x})^2 = \frac{1}{N-1}\left(\sum -N\bar{x}^2\right) \qquad (3.11.6)$$

is an unbiased estimate of σ^2. Since both $\bar{x}$ and s^2 are functions of the minimal sufficient statistics

$$\sum_{i=1}^{N} x_i \quad \text{and} \quad \sum_{i=1}^{N} x_i^2$$

as expressed by (3.11.4) and (3.11.5), they also are MVUEs for μ and σ^2.

3.11.2 Method of moments

As suggested earlier, the method of moments is one of the oldest methods for parameter estimation. It is simple and straightforward to apply. Recall that the kth moment of a random variable Y, taken about the origin, is

$$\mu'_k = E[Y^k] \qquad (3.11.7)$$

and the corresponding sample moment is

$$m'_k = \frac{1}{N}\sum_{i=1}^{N} y_i^k \qquad (3.11.8)$$

The method of moments is based on the assumption that sample moments should provide good estimates of the corresponding population moments (i.e. m'_k is a good estimate of μ'_k). Thus we choose our estimates as those parameter values which are solutions of the equations $\mu'_k = m'_k$, $k = 1, 2, \ldots, r$ where r is the number of parameters.

We again illustrate with an example. A random sample $y_1, y_2, \ldots, y_N$ is selected from a population with a uniform PDF over the interval $(0, \theta)$, where θ is unknown. We use the method of moments to estimate θ. The first moment of the population is $\mu'_1 = \mu = \theta/2$ (see Appendix C). The corresponding first sample moment is

$$m' = \frac{1}{N}\sum_{i=1}^{N} y_i = \bar{y}$$

If we equate the moments and solve for θ

$$\hat{\theta}/2 = \bar{y}, \text{ or } \hat{\theta} = 2\bar{y}$$

Thus, θ has a moment estimate of $2\bar{y}$.

We remark, that while the method of moments is straightforward to apply, the resulting estimates are not minimal sufficient statistics. In addition, these estimates may not even be unbiased. The primary advantage of this procedure is that it often yields results when others do not.

3.11.3 Maximum likelihood

The procedure introduced earlier to compute the MVUE is complicated by the fact that one must find some function of the minimal sufficient statistic that gives the sought-after target parameter. Finding this function is, in general, a matter of trial and error. We then introduced the method of moments which, while it is easy to apply, yields estimates which may not be optimal. A more sophisticated procedure, the maximum likelihood method, often leads to the MVUE.

The formal statement of this method is quite simple. Choose as estimates those parameter values that maximize the likelihood $L(y_1, y_2, \ldots, y_N)$. A simple example using discrete variables helps to illustrate the logic in the maximum likelihood method. Assume we have a bag containing three marbles. The marbles can be black or white. We randomly sample two of the three and find that they are both black. What is the best estimate of the total number of black marbles in the bag? If there are actually two black and one white in the bag, the probability of sampling two black marbles is

$$\frac{\binom{2}{2}\binom{1}{0}}{\binom{3}{2}} = 1/3$$

where, as in Section 3.3, the binomial expression is

$$\binom{N}{r} = {_NP_r}/r! = N!/[r!(N-r)!] \qquad (3.11.9)$$

and $_NP_r$ is the number of permutations of N discrete variables sampled r at a time. In the above expression

$$\binom{2}{2}$$

indicates the first sample of two marbles, with both being black. The next term is the remaining unsampled marble (hence the 0 in the denominator) if it were white. Now if there are three black marbles in the bag the probability of sampling two blacks is

$$\frac{\binom{3}{2}\binom{0}{0}}{\binom{3}{2}} = 1$$

On this basis, it seems reasonable to choose three as the estimate of the number of black marbles in the bag in order to maximize the probability of the observed sample.

A more complex example can be used to illustrate the application of this method to our estimates of the mean, μ, and variance, σ^2, for a normal population. Again, let $y_1, y_2, \ldots, y_N$ be a random sample from a normal population with parameters μ and σ^2. We want to find the maximum likelihood estimators of μ and σ^2. Note we used this same example for our discussion of the method of moments. To find the maximum likelihood, we need to write the joint PDF of the independent observations $y_1, y_2, \ldots, y_N$

$$L = f(y_1, y_2, \ldots, y_N) = f(y_1)f(y_2)\ldots f(y_N)$$

$$= \left\{ \frac{1}{\sigma\sqrt{(2\pi)}} \exp\left[\frac{-(y_1 - \mu)^2}{2\sigma^2}\right] \right\}$$

$$\times \left\{ \frac{1}{\sigma\sqrt{(2\pi)}} \exp\left[\frac{-(y_2 - \mu)^2}{2\sigma^2}\right] \right\} \quad (3.11.10)$$

$$\cdots \left\{ \frac{1}{\sigma\sqrt{(2\pi)}} \exp\left[\frac{-(y_N - \mu)^2}{2\sigma^2}\right] \right\}$$

$$= \left\{ \frac{1}{\sigma^N (2\pi)^{N/2}} \exp\left[-\sum_{i=1}^{N} \frac{-(y_i - \mu)^2}{2\sigma^2}\right] \right\}$$

We simplify this expression by taking $\log_N(L)$, which we then differentiate to find the maximum. Specifically

$$\log_N(L) = -\frac{N}{2}\log_N(\sigma^2) - \frac{N}{2}\log_N(2\pi) - \sum_{i=1}^{N} \frac{(y_i - \mu)^2}{2\sigma^2} \quad (3.11.11)$$

Taking derivatives of (3.11.11) with respect to μ and σ^2, we find

$$\frac{d[\log_N(L)]}{d\mu} = \sum_{i=1}^{N} \frac{(y_i - \mu)}{\sigma^2} \quad (3.11.12a)$$

$$\frac{d[\log_N(L)]}{d\sigma^2} = -\frac{N}{\sigma^2} + \sum_{i=1}^{N} \frac{(y_i - \mu)}{2\sigma^4} \quad (3.11.12b)$$

Setting (3.11.12a, b) to zero and solving yields the required estimates of μ and σ^2

$$\sum_{i=1}^{N} \frac{(y_i - s)}{\sigma^2} = 0 \quad \text{or} \quad \mu = \frac{1}{N}\sum_{i=1}^{N} y_i = \bar{y} \quad (3.11.13)$$

Substituting $\bar{y}$ into (3.11.12b)

$$-N/\hat{\sigma}^2 + \sum_{i=1}^{N} (y_i - \bar{y})^2/\hat{\sigma}^4 = 0 \quad (3.11.14a)$$

or

$$\hat{\sigma}^2 = \sum_{i=1}^{N} (y_i - \bar{y})^2/N = s'^2 \quad (3.11.14b)$$

Thus, $\bar{y}$ and s'^2 are the maximum likelihood estimators of μ and σ^2. Although, $\bar{y}$ is an unbiased estimate of μ, s'^2 is not unbiased for σ^2, as noted at the beginning of the chapter. However, s'^2 can easily be adjusted to the unbiased estimator

$$s^2 = \frac{1}{N-1}\sum_{i=1}^{N}(y_i - \bar{y})^2 \qquad (3.11.15)$$

Since the Maximum Likelihood Method has widespread application, we present another simple example to illustrate its use. Let $y_1, y_2, \ldots, y_N$, be a random sample taken from a uniform distribution with $f(y_i) = 1/\theta = $ constant, $0 \leq y_i \leq \theta$, and $i = 1, 2, \ldots, N$. We want to find the maximum likelihood estimate of θ. Again, we write the likelihood, L, as the joint probability function

$$\begin{aligned}L &= f(y_1, y_2, \ldots, y_N) = f(y_1)f(y_2)\ldots f(y_N) = (1/\theta)(1/\theta)\ldots(1/\theta) \\ &= (1/\theta)^N\end{aligned} \qquad (3.11.16)$$

In this case, L is a monotonically decreasing function of θ and nowhere is $dL/d\theta = 0$. Instead, L increases monotonically as θ decreases and must be greater than or equal to the largest sample value, y_N. L is, therefore, not an unbiased estimate of θ. It can be adjusted to

$$\theta = \frac{(N+1)}{N}y_N \qquad (3.11.17)$$

which is unbiased. We note that if any statistic U can be shown to be a sufficient statistic for estimating θ then the maximum likelihood estimator is always some function of U. If this maximum likelihood estimate can be found, and then adjusted to be unbiased, the result will generally be an MVUE.

To demonstrate the application of the maximum likelihood approach, assume that a random sample of size N is selected from the normal distribution (equation (3.5.2)) with μ and σ^2 as the mean and variance for each x_i (where we assume that the x_i values are independent). We ask: If $\bar{\theta} = (\theta_1, \theta_2) = (\mu, \sigma^2)$ is the parameter space for the probability density function $f(x_1, x_2, \ldots, x_N)$, then what is the likelihood function? Also, find the maximum likelihood estimator $\hat{\theta}_1$ of θ_1 which maximizes the likelihood function and find the maximum likelihood estimator $\hat{\theta}_2$ which maximizes the likelihood function θ_1. We first write the PDF as

$$\begin{aligned}f(\bar{x}, \bar{\theta}) &= (\tfrac{1}{2}\pi\sigma^2)^{N/2} = \exp\left[\frac{1}{2\sigma^2}\sum_{i=1}^{N}(x_i - \mu)^2\right] \\ &= \prod_{i=1}^{N}\left\{\frac{1}{\sqrt{(2\pi\sigma)}}\exp\left[-(x_i - \mu)^2/\sigma^2\right]\right\} = L(\bar{x}, \bar{\theta})\end{aligned}$$

which is the likelihood function written in terms of the product, Π, of the exponential. Taking the natural log of the above expression with respect to our estimated parameter, θ_1, and setting it equal to zero to find the maximum, we get

$$\ln(L) = -\frac{N}{2}\ln(2\pi) - \frac{N}{2}\ln(\sigma^2) - \frac{1}{2\sigma^2}\sum_{i=1}^{N}(x_i - \mu)^2$$

where $\sigma > 0$ and $-\infty < \mu < \infty$. The derivative of this function with respect to θ_1 (which is μ) is

$$\frac{\partial L}{\partial \mu} = -\frac{1}{2\sigma^2} \sum_{i=1}^{N} (x_i - \mu)(-2) = \frac{1}{\sigma^2} \sum_{i=1}^{N} (x_i - \mu) = 0$$

so that our estimate of μ is

$$\hat{\mu} = \bar{x} = \frac{1}{N} \sum_{i=1}^{N} x_i$$

Furthermore, the maximum likelihood estimator of θ_2 (which is σ^2) is given by

$$\frac{\partial L}{\partial \sigma^2} = -\frac{N}{2\sigma^2} - \frac{(-1)}{2\sigma^2} \sum_{i=1}^{N} (x_i - \mu)^2 = \frac{1}{2\sigma^2} \left[\frac{1}{\sigma^2} \sum_{i=1}^{N} (x_i - \mu)^2 - N \right] = 0$$

which yields the estimator

$$\hat{\sigma}^2 = \frac{1}{N} \sum_{i=1}^{N} (x_i - \hat{\mu})^2$$

For a normally distributed oceanographic data set, we can readily obtain maximum likelihood estimates of the mean and variance of the data. However, the real value of this technique is for variables that are not normally distributed. For example, if we examine spectral energy computed from current velocities, the spectral values have a chi-square distribution rather than a normal distribution. If we follow the maximum likelihood procedure, we find that the spectral values have a mean of ν, the number of degrees of freedom, and a variance of 2ν. These are the maximum likelihood estimators for the mean and variance. This example can be used as a pattern for applying the maximum likelihood method to a particular sample. In particular, we first determine the appropriate PDF for the sample values. We then find the joint likelihood function, take the natural logs and then differentiate with respect to the parameter of interest. Setting this derivative equal to zero to find the maximum subsequently yields the value of the parameter being sought.

3.12 LINEAR ESTIMATION (REGRESSION)

Linear regression is one of a number of statistical procedures that fall under the general heading of linear estimation. Since linear regression is widely treated in the literature and is available in many software packages, our primary purpose here is to establish a common vocabulary for all readers. In our previous discussion and examples, we assumed that the random variables $Y_1, Y_2, \ldots, Y_N$ were independent (in a probabilistic sense) and identically distributed, which implies that $E[Y_i] = \mu$ is a constant. Often this is not the case and the expected value $E[Y_i]$ of the variable is a function of some other parameter. We now consider the values y of a random variable, Y, called the dependent variable, whose values are a function of one or more *nonrandom* variables $x_1, x_2, \ldots, x_N$, called independent variables (in a mathematical, rather than probabilistic sense).

If we model our random variable as

$$y = E[y] + \varepsilon = b_0 + b_1 x + \varepsilon \tag{3.12.1}$$

we invariably find that the points y are scattered about the regression line $E[y] = b_0 + b_1 x$. The random variable ε in the right-hand term of (3.12.1) gives the departure from linearity and has a specific PDF with a mean value $\mu_\varepsilon = 0$. In other words, we can think of y as having a deterministic part, $E[y]$, and a random part, ε, that is randomly distributed about the regression line. By definition, simple linear regression is limited to finding the coefficients b_0 and b_1. If N independent variables $(x_1, x_2, \ldots, x_N)$ are involved in the variability of each value y, we must deal with *multiple linear regression*. In this case, (3.12.1) becomes

$$y = b_0 + b_1 x_1 + b_2 x_2 + \ldots + b_N x_N + \varepsilon \tag{3.12.2}$$

3.12.1 Method of least squares

One of the most powerful techniques for fitting a dependent model parameter y to independent (observed input) variables x_i ($i = 1, 2, \ldots, N$) is the *method of least squares*. We apply the method in terms of linear estimation and will later readdress the topic in terms of more general statistical models. (Note: by "linear" we mean linear in the parameters $b_0, b_1, \ldots, b_N$. Thus, $y = b_0 + b_1 x_1^2 + \varepsilon$ is linear but $y = b_0 + \sin(b_1 x_1) + \varepsilon$ is not.) We begin with the simplest case, that of fitting a straight line to a set of points using the "best" coefficients b_0, b_1 (Figure 3.10). In a sense, the least squares procedure does what we do by eye—it minimizes the vertical deviations (residuals) of data points from the fitted line. Let

$$y_i = \hat{y}_i + \varepsilon_i \tag{3.12.3}$$

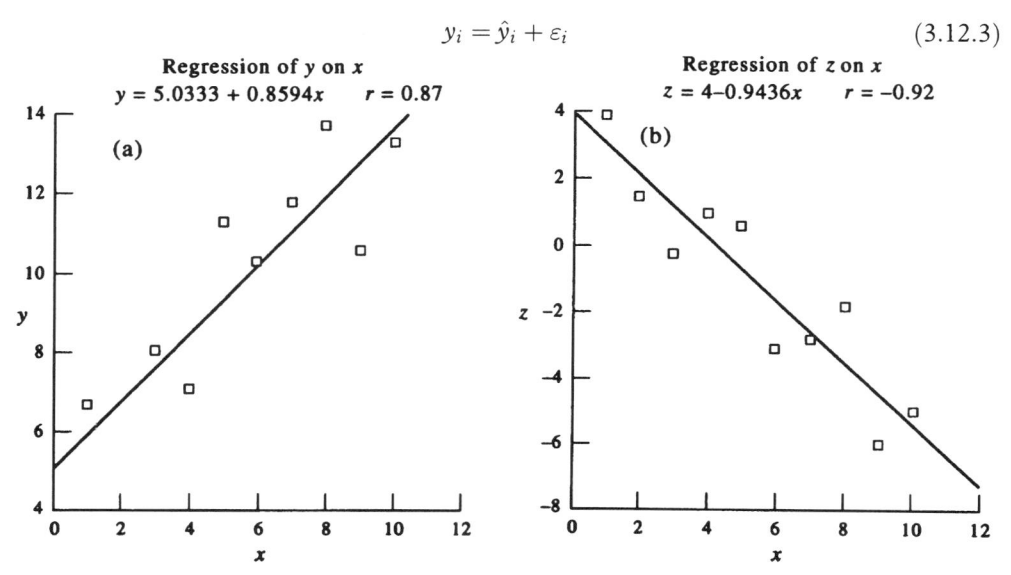

Figure 3.10. Straight line (linear regression) fits to the sets of points in Table 3.12.1 using the "best" coefficients b_0, b_1. (a) Regression of y on x, for which $(b_0, b_1) = (5.0333, 0.8594)$; and (b) regression of z on x, for which $(b_0, b_1) = (4.0, -0.9436)$. r is the correlation coefficient.

where

$$\hat{y}_i = b_0 + b_1 x_i \tag{3.12.4}$$

is our estimator for the deterministic portion of the data and ε is the residual or error. To find the coefficients b_0, b_1 we need to minimize the sum of the squared errors (SSE) where SSE is the total variance that is not explained (accounted for) by our linear regression model given by (3.12.3) and (3.12.4)

$$\text{SSE} = \sum_{i=1}^{N} \varepsilon_i^2 = \sum_{i=1}^{N} (y_i - \hat{y}_i)^2 = \sum_{i=1}^{N} [y_i - (b_0 + b_1 x_i)]^2 \tag{3.12.5a}$$

$$= \text{SST} - \text{SSR} \tag{3.12.5b}$$

in which

$$\text{SST} = \sum_{i=1}^{N} (y_i - \bar{y})^2 \quad \text{and} \quad \text{SSR} = \sum_{i=1}^{N} (\hat{y}_i - \bar{y})^2 \tag{3.12.5c}$$

Here, SST (sum of squares total) is the variance in the data and SSR (sum of squares regression) is the amount of variance explained by our regression model. Minimization amounts to finding those coefficients that minimize the unexplained variance (SSE). Taking the partial derivatives of (3.12.5a) with respect to b_0 and b_1 and setting the resultant values equal to zero, the minimization conditions are

$$\frac{\partial \text{SSE}}{\partial b_0} = 0; \quad \frac{\partial \text{SSE}}{\partial b_1} = 0 \tag{3.12.6}$$

Substituting (3.12.5a) into (3.12.6), we have for b_0

$$\frac{\partial \text{SSE}}{\partial b_0} = \frac{\partial}{\partial b_0} \left\{ \sum_{i=1}^{N} [y_i - (b_0 + b_1 x_i)]^2 \right\} = -2 \sum_{i=1}^{N} [y_i - (b_0 + b_1 x_i)]$$
$$= -2 \left(\sum_{i=1}^{N} y_i - N b_0 - b_1 \sum_{i=1}^{N} x_i \right) = 0 \tag{3.12.6a}$$

Now for b_1

$$\frac{\partial \text{SSE}}{\partial b_1} = \frac{\partial}{\partial b_1} \left\{ \sum_{i=1}^{N} [y_i - (b_0 + b_1 x_i)]^2 \right\} = -2 \sum_{i=1}^{N} [y_i - (b_0 + b_1 x_i)]$$
$$= -2 \left(\sum_{i=1}^{N} x_i y_i - b_0 \sum_{i=1}^{N} x_i - b_1 \sum_{i=1}^{N} x_i^2 \right) = 0 \tag{3.12.6b}$$

Once the mean values of y and x are calculated, these least squares equations can be solved simultaneously to find an estimate of the coefficient b_1 (the slope of the regression line); this is then used to obtain an estimate of the second coefficient, b_0 (the intercept of the regression line). In particular

$$\hat{b}_1 = \frac{\sum_{i=1}^{N}(x_i - \bar{x})(y_i - \bar{y})}{\sum_{i=1}^{N}(x_i - \bar{x})^2} = \frac{\left[N\sum_{i=1}^{N} x_i y_i - \sum_{i=1}^{N} x_i \sum_{i=1}^{N} y_i\right]}{\left[N\sum_{i=1}^{N} x_i^2 - \left(\sum_{i=1}^{N} x_i\right)^2\right]} \quad (3.12.7a)$$

$$\hat{b}_0 = \bar{y} - \hat{b}_1 \bar{x} \quad (3.12.7b)$$

Several features of the regression values are worth noting. First, if we substitute the intercept $b_0 = \bar{y} - b_1\bar{x}$ into the line $\hat{y} = b_0 + b_1 x$, we obtain

$$\hat{y} = \bar{y} + b_1(x - \bar{x})$$

As a result, whenever $x = \bar{x}$, we have $\hat{y} = \bar{y}$. This means: (1) That the regression line always passes through the point $(\bar{x}, \bar{y})$; and (2) that because the operation $\partial \text{SSE}/\partial b_0 = 0$ minimizes the error $\sum \varepsilon_i = 0$, the regression line not only goes through the point of averages $(\bar{x}, \bar{y})$ but it also splits the scatter of the observed points so that the positive residuals (where the regression line passes below the true point) always cancel exactly the negative residuals (where the line passed above the true point). The sample regression line is therefore an unbiased estimate of the population regression line.

To summarize the linear regression procedure, we note that:

(1) For each selected x (independent variable) there is a distribution of y from which the sample (dependent variable) is drawn at random.
(2) The population of y corresponding to a selected x has a mean μ that lies on the straight line $\mu = b_0 + b_1 x$, where b_0 and b_1 are regression parameters.
(3) In each population, the standard deviation of y about its mean, $b_0 + b_1 x$, has the same value $(s_{xy} = s_\varepsilon, y = b_0 + b_1 x + \varepsilon)$. Note that ε is a random variable drawn from a normal population with $\mu = 0$ and $s = s_{xy}$.

Table 3.12.1. Values for dependent variables y_i, z_i as functions of x_i. The estimated values $\hat{y}$ and $\hat{z}$ are derived from the linear regression analysis. Formulae at the bottom of the table are the total sum of squares (SST), sum of squares for the regression (SSR) and the sum of squares of the errors (SSE) to be derived in our regression analysis for $N = 10$

x_i	y_i	$\hat{y}_i$	z_i	$\hat{y}_i$
1.0	6.7	5.9	3.9	3.1
2.0	4.7	6.8	1.5	2.1
3.0	8.1	7.6	−0.2	1.2
4.0	7.1	8.5	1.0	0.2
5.0	11.3	9.4	0.6	−0.7
6.0	10.5	10.2	−3.1	−1.7
7.0	11.8	11.1	−2.8	−2.6
8.0	13.7	11.9	−1.8	−3.6
9.0	10.6	12.8	−6.0	−4.5
10.0	13.3	13.7	−5.0	−5.4

SST(y) = 80.64; SSR(y) = 61.11; SSE(y) = 19.53
SST(z) = 86.39; SSR(z) = 73.46; SSE(z) = 12.93

Thus, y is the sum of a random part ε and a fixed part x; the fixed part determines the mean values of the y population samples, with one distribution of y for each x that we pick. The mean values of y lie on the straight line, $\mu = b_0 + b_1 x$, which is the population regression line. The regression parameter b_0 is the y mean for $x = 0$ and b_1 is the slope of the regression line. The random part, ε, is independent of x and y. To compute the regression parameters, we need values of $N, \bar{x}, \bar{y}, \sum x^2, \sum y^2$, and $\sum xy$. Earlier, we discussed the computational shortcuts to compute $\sum x^2$ and $\sum y^2$ without first computing the means of x and y. The same can be accomplished for xy using,

$$\sum (x - \bar{x})(y - \bar{y}) = \sum xy - \sum x \sum y / N$$

As examples of linear regression, consider the data sets in Table 3.12.1 for dependent variables y_i and z_i which are both functions of the same independent variable x_i (for example, y_i could be the eastward and z_i the northward component of velocity as functions of time x_i). We will compute the regression coefficients b_0, b_1 plus the sample variance s^2 and percent of explained variance (100 SSR/SST) for each data set.

To estimate the regression parameters, we must first compute the means of the three series

$$\bar{x} = 5.50; \quad \bar{y} = 9.78; \quad \bar{z} = -1.19$$

We then use the means to calculate the sums in (3.12.6)

$$\sum_{i=1}^{10} (x_i - \bar{x})^2 = 82.50; \quad \sum_{i=1}^{10} (x_i - \bar{x})(y_i - \bar{y}) = 71.00;$$

$$\sum_{i=1}^{10} (x_i - \bar{x})(z_i - \bar{z}) = -77.85$$

$$SST(y) = \sum_{i=1}^{10} (y_i - \bar{y})^2 = 80.64$$

$$SST(z) = \sum_{i=1}^{10} (z_i - \bar{z})^2 = 86.36$$

For the regression of y on x ($\hat{y} = b_0 + b_1 x$), we find

$$b_0 = 5.05; \; b_1 = 0.861; \; s^2 = 2.44$$
$$100 \cdot SSR(y)/SST(y) = (100) 61.11/80.64 = 75.8\%$$

while for the regression of z on x ($\hat{z} = b_0 + b_1 x$), we have

$$b_0 = 4.00; \quad b_1 = -0.94; \quad s^2 = 1.62$$
$$100 \cdot SSR(z)/SST(z) = (100) 73.46/86.36 = 85.0\%$$

The ratio SSR/SST (variance explained/total variance) is a meaure of the goodness of fit of the regression curves called the *correlation of determination*, r^2. If the regression line fits perfectly all the sample values, all residuals would be zero. In turn, SSE $= 0$ and SSR/SST $= r^2 = 1$. As the fit becomes increasingly less representative of the data points, r^2 decreases towards a possible minimum of zero.

3.12.2 Standard error of the estimate

The measure of the absolute magnitude of the goodness of fit is the standard error of the estimate, s_ε, defined as

$$s_\varepsilon = [SSE/(N-2)]^{1/2}$$
$$= \left[\frac{1}{N-2} \sum_{i=1}^{N} (y - \hat{y})^2 \right]^{1/2} \quad (3.12.8)$$

The number of degrees of freedom, $N-2$, for s_ε is based on the fact that two parameters, b_0 and b_1 are needed for any linear regression estimate. If s_ε is from a normal distribution and has a mean of zero, then, in analogy with our discussion of the standard deviation of values about their mean, approximately 68.3% of the observations will fall within $\pm 1 s_\varepsilon$ units of the regression line, 95.4% will fall within $\pm 2 s_\varepsilon$ units of the line and 99.7% will fall within $\pm 3 s_\varepsilon$ units of this line. For the examples of Table 3.12.1

Variable y: $s_\varepsilon = [SSE(y)/(N-2)]^{1/2} = (19.53/8)^{1/2} = 1.56$
Variable z: $s_\varepsilon = [SSE(z)/(N-2)]^{1/2} = (12.93/8)^{1/2} = 1.27$

As a result, the $\pm 2 s_\varepsilon$ ranges are $\pm 2(1.56)$ and $\pm 2(1.27)$, respectively.

We next turn to our estimate of the slope, b_1. Recalling that $b_1 = \sum x / \sum x^2$, we find

$$s_{b1}^2 = \frac{s_{xy}^2}{\sum x^2} = \frac{s_\varepsilon^2}{\sum x^2} \quad (3.12.9)$$

where s_{b1}^2 is the sample variance for our estimate, $\hat{b}_1$, for the slope of the regression line. For small samples ($N < 30$), we can write the 95% confidence interval as

$$\hat{b}_1 - t_{0.05} s_{b1} \leq b_1 \leq \hat{b}_1 + t_{0.05} s_{b1} \quad (3.12.10)$$

Turning to the regression line itself, we wish to say something about the standard deviation about $\bar{y}$ (i.e. the regression line). First we note that $\hat{\varepsilon}$ has variance σ_{xy}^2/N and $\hat{b}_1$ has variance $\sigma_{xy}^2/\sum x^2$. Since the errors, ε, are assumed independent, the variance of the sums is the sum of the variances

$$\sigma_y^2 = \sigma_{xy}^2 \left[1/(N-2) + x^2/\sum x^2 \right] \quad (3.12.11)$$

which leads to the standard error given above. These confidence limits would appear as hyperbolae in regression diagrams such as Figure 3.10. The hyperbolae are the confidence belts for the different significance levels. Note the increasing hazard of making predictions for values of x far removed from the mean value $\bar{x}$. Since the lines indicate that y must be within the confidence belt, higher signifance levels have narrower belts. Thus, estimates of $\bar{y}$ get worse as we move away from $\bar{x}, \bar{y}$. Remember that these confidence belts are for the regression line itself and not for the individual points. Hence, if repeated samples of y_i are taken of the same size and the same fixed value of x, the 95% of the confidence intervals, constructed for the mean value of y and x, will contain the true value of the mean of y and x. If only one prediction is made of x, then the probability that the calculated interval of this will contain the true value is 95%.

3.12.3 Multivariate regression

To extend the regression procedure to multivariate regression, we must formulate our linear estimation model in matrix terms. Suppose our model is of the form

$$Y = b_0 + b_1 X_1 + b_2 X_2 + \ldots + b_k X_k + \varepsilon \qquad (3.12.12)$$

and that we make N independent (probabilistic) observations $y_1, y_2, \ldots, y_N$ of Y. This means that we can write

$$y_i = b_0 + b_1 x_{i1} + b_2 x_{i2} + \ldots + b_k x_{ik} + \varepsilon_i \qquad (3.12.13)$$

where x_{ij} is the jth independent variable for the ith observation. Writing this in matrix form we have

$$\mathbf{Y} = \begin{pmatrix} y_1 \\ y_2 \\ \ldots \\ \ldots \\ y_N \end{pmatrix}, \quad \mathbf{X} = \begin{pmatrix} x_0 & x_{11} & \ldots & x_{1k} \\ x_0 & x_{21} & \ldots & x_{2k} \\ \ldots & \ldots & \ldots & \ldots \\ \ldots & \ldots & \ldots & \ldots \\ x_0 & x_{N1} & \ldots & x_{Nk} \end{pmatrix}$$

$$\mathbf{B} = \begin{pmatrix} b_0 \\ b_1 \\ \ldots \\ \ldots \\ b_k \end{pmatrix}, \quad \mathbf{E} = \begin{pmatrix} \varepsilon_1 \\ \varepsilon_2 \\ \ldots \\ \ldots \\ \varepsilon_N \end{pmatrix} \qquad (3.12.14)$$

where the boldface letters indicate matrices. Using (3.12.14), we can represent the N equations relating y_i to the independent variable x_{ij} as

$$\mathbf{Y} = \mathbf{B} \cdot \mathbf{X} + \mathbf{E} \qquad (3.12.15)$$

If we restrict our analysis to the first two coefficients, (3.12.15) reduces to the simple straight line fit model (3.12.4). In this case, the matrices for N observations become

$$\mathbf{Y} = \begin{pmatrix} y_1 \\ y_2 \\ \ldots \\ \ldots \\ y_N \end{pmatrix}, \quad \mathbf{X} = \begin{pmatrix} x_0 & \ldots & x_{1N} \\ x_0 & \ldots & x_{2N} \\ x_0 & \ldots & \ldots \\ x_0 & \ldots & \ldots \\ x_0 & \ldots & x_{NN} \end{pmatrix}$$

$$\mathbf{B} = \begin{pmatrix} b_o \\ b_1 \end{pmatrix}, \quad \mathbf{E} = \begin{pmatrix} \varepsilon_1 \\ \varepsilon_2 \\ \ldots \\ \ldots \\ \varepsilon_N \end{pmatrix} \qquad (3.12.16)$$

Using these N observations in (3.12.15), our least squares equations are

$$Nb_0 + b_1 \sum_{i=1}^{N} x_i = \sum_{i=1}^{N} y_i$$
$$b_0 \sum_{i=1}^{N} x_i + b_1 \sum_{i=1}^{N} x_i^2 = \sum_{i=1}^{N} x_i y_i \quad (3.12.17)$$

which we can solve for b_0 and b_1. We can generalize the procedure further by realizing that for $x_0 = 1$ we have

$$\mathbf{x}' \cdot \mathbf{x} = \begin{pmatrix} 1 \dots & \dots & \dots & 1 \\ \dots & \dots & \dots & \dots \\ x_1 & \dots & \dots & x_N \end{pmatrix} \begin{pmatrix} 1 & x_i \\ \dots & \dots \\ 1 & x_N \end{pmatrix} = \begin{pmatrix} N & \sum x_i \\ \dots & \dots \\ \sum x_i & \sum x_i^2 \end{pmatrix} \quad (3.12.18)$$

where $\mathbf{X}'$ is the transpose of the matrix $\mathbf{X}$ and, the sums are from 1 to N, and

$$\mathbf{X}' \cdot \mathbf{Y} = \begin{pmatrix} \sum_{i=1}^{N} y_i \\ \sum_{i=1}^{N} x_i y_i \end{pmatrix} \quad (3.12.19)$$

Our least squares equations can then be expressed as

$$(\mathbf{X}'\mathbf{X}) \cdot \mathbf{B} = \mathbf{X}' \cdot \mathbf{Y} \quad (3.12.20)$$

where

$$B = \begin{pmatrix} b_0 \\ b_1 \end{pmatrix} \quad (3.12.21)$$

Solving the above equations for **B**, we obtain

$$\mathbf{B} = (\mathbf{X}' \cdot \mathbf{X})^{-1} \mathbf{X}' \cdot \mathbf{Y} \quad (3.12.22)$$

3.12.4 A computational example of matrix regression

Since linear regression is widely used in oceanography, we will illustrate its use by a simple example. Suppose we want to fit a line to the data pairs consisting of the independent variable x_i and the dependent variable y_i given in Table 3.12.2.
From these we find

$$\sum_{i=1}^{N} x_i = 0, \quad \sum_{i=1}^{N} y_i = 5, \quad \sum_{i=1}^{N} x_i y_i = 7, \quad \sum_{i=1}^{N} x_i^2 = 10$$

Substituting into equation (3.12.14), we have

Statistical Methods and Error Handling 241

Table 3.12.2. *Data values used in least squares linear fit of a two-coefficient regression model,* $y_i = F(x_i)$

Data		Solution values	
x_i	y_i	$(x_i)(y_i)$	x_i^2
−2	0	0	4
−1	0	0	1
0	1	0	0
1	1	1	1
2	3	6	4

$$b_1 = \frac{\left[N\sum_{i=1}^{N} x_i y_i - \sum_{i=1}^{N} x_i \sum_{i=1}^{N} y_i\right]}{\left[N\sum_{i=1}^{N} x_i^2 - \left(\sum_{i=1}^{N} x_i\right)^2\right]}$$

$$= \frac{[(5)(7) - (0)(5)]}{[(5)(10) - 10^2]} = 0.7$$

$$b_0 = \bar{y} - b_1\bar{x} = 5/5 - (0.7)(0) = 1$$

This same problem can be put in matrix form

$$\mathbf{Y} = \begin{pmatrix} 0 \\ 0 \\ 1 \\ 1 \\ 3 \end{pmatrix}, \quad \mathbf{X} = \begin{pmatrix} 1 & -2 \\ 1 & -1 \\ 1 & 0 \\ 1 & 1 \\ 1 & 2 \end{pmatrix}$$

$$\mathbf{X}' \cdot \mathbf{X} = \begin{pmatrix} 5 & 0 \\ 0 & 10 \end{pmatrix}, \quad \mathbf{X}' \cdot \mathbf{Y} = \begin{pmatrix} 5 \\ 7 \end{pmatrix}, \quad (\mathbf{X}' \cdot \mathbf{X})^{-1} = \begin{pmatrix} 1/5 & 0 \\ 0 & 1/10 \end{pmatrix}$$

$$\mathbf{B} = (\mathbf{X}' \cdot \mathbf{X})^{-1} \mathbf{X}' \cdot \mathbf{Y} = \begin{pmatrix} 1/5 & 0 \\ 0 & 1/10 \end{pmatrix} \begin{pmatrix} 5 \\ 7 \end{pmatrix} \begin{pmatrix} 1 \\ 0.7 \end{pmatrix}$$

so that by (3.12.21), $b_0 = 1$ and $b_1 = 0.7$.

An important property of the simple straight line least-square estimators we have just derived is that b_0 and b_1 are unbiased estimates of their true parameter values. We have assumed that $E[\varepsilon] = 0$ and that $V[\varepsilon] = \sigma^2$; thus the error variance is independent of x and $V[Y] = V[\varepsilon] = \sigma^2$. Since σ^2 is usually unknown we estimate it using the sample variance (3.2.4) given by

$$s^2 = \frac{1}{N-1}\sum_{i=1}^{N}(y_i - \bar{y})^2 \quad (3.12.23)$$

However, if we use the output values, $\hat{y}_i$, from the least squares, to estimate $\varepsilon_i(Y) = y_i - \hat{y}_i$, we must write (3.12.23) as

$$s^2 = \frac{1}{N-2}\sum_{i=1}^{N}(y_i - \hat{y}_i)^2 = \frac{1}{N-2}\text{SSE} \qquad (3.12.24)$$

where SSE, given by (3.12.23), represents the sum of the squares of the errors and the $N - 2$ corresponds to the fact that two parameters, b_0 and b_1, are needed in the model.

In matrix notation we can write the SSE as

$$\mathbf{SSE} = \mathbf{Y}' \cdot \mathbf{Y} - (\mathbf{B}' \cdot \mathbf{X}') \cdot \mathbf{Y} \qquad (3.12.25)$$

Using this with our previous numerical example we write (3.12.25) as

$$(0 \; 0 \; 1 \; 1 \; 3) \begin{pmatrix} 0 \\ 0 \\ 1 \\ 1 \\ 3 \end{pmatrix} - (1 \; 0.7) \begin{pmatrix} 1 & 1 & 1 & 1 & 1 \\ -2 & -1 & 0 & 1 & 2 \end{pmatrix} \begin{pmatrix} 0 \\ 0 \\ 1 \\ 1 \\ 3 \end{pmatrix}$$

$$= 11 - (1 \; 0.7) \begin{pmatrix} 5 \\ 7 \end{pmatrix} = 11 - 9.9 = 1.1$$

Since $s^2 = \text{SSE}/(N - 2)$, we have $s^2 = 1.1/(3) = 0.367$ as our estimator of σ^2.

3.12.5 Polynomial curve fitting with least squares

The use of least-squares fitting is not limited to the straight line regression model discussed thus far. In general, we can write our linear model as any polynomial of the form

$$Y = b_0 + b_1 x + b_2 x^2 + \ldots + b_N x^N + \varepsilon \qquad (3.12.26)$$

The procedure is the same as with the straight line case except that now the **X** matrix has $N+1$ columns. Thus, our least-squares fit will have $N+1$ linear equations with $N+1$ unknowns, $b_0, b_1, \ldots, b_N$. These equations are called the *normal equations*.

3.12.6 Relationship between least-squares and maximum likelihood

As discussed earlier, the maximum likelihood estimator is one that maximizes the likelihood of sampling a given parameter. In general, if we have a sample x_i from a population with the PDF $f(X_i, \theta)$, where θ is the parameter of interest, the maximum likelihood estimator $L(\theta)$ is the product of the individual independent probabilities

$$L(\theta) = f(x_1, \theta) f(x_2, \theta) \ldots f(x_N, \theta) \qquad (3.12.27)$$

If the errors all come from a normal distribution, this becomes from equation (3.11.10)

$$L(\theta) = \frac{\exp\left[-\sum_{i=1}^{N}(x_i - \theta)^2/2\sigma^2\right]}{\sigma^N(2\pi)^{N/2}} \quad (3.12.28)$$

When this is maximized, it leads to the least-squares estimate

$$\hat{\theta} = \frac{1}{N}\sum_{i=1}^{N} x_i = \bar{x}$$

In other words, the least-squares estimate of the mean of θ can be derived from a normal distribution using the maximum likelihood criterion. This value is found to be the average of the independent variable x.

3.13 RELATIONSHIP BETWEEN REGRESSION AND CORRELATION

The subject of correlation will be considered in more detail when we examine time-series analysis methods. Our intension, here, is simply to introduce the concept in general statistical terms and relate it to the simple regression model just discussed. As with regression, correlation relates two variables but unlike regression it is measured without estimation of the population regression line.

The *correlation coefficient*, r, is a way of determining how well two (or more) variables co-vary in time or space. For two random variables x ($x_1, x_2, \ldots, x_N$) and y ($y_1, y_2, \ldots, y_N$) the correlation coefficient can be written

$$r = \frac{1}{N-1}\sum_{i=1}^{N} \frac{(x_i - \bar{x})(y_i - \bar{y})}{s_x s_y} \quad (3.13.1a)$$

$$= C_{xy}/s_x s_y$$

where

$$C_{xy} = \frac{1}{N-1}\sum_{i=1}^{N}(x_i - \bar{x})(y_i - \bar{y}) \quad (3.13.1b)$$

is the *covariance* of x and y, and s_x and s_y are the standard deviations for the two data records as defined by equation (3.2.4). We note two important properties of r:

(1) r is a dimensionless quantity since the units of the numerator and the denominator are the same;
(2) the value of r lies between -1 and $+1$ since it is normalized by the product of the standard deviations of both variables.

For $r = \pm 1$, the data points (x_i, y_i) cluster along a straight line and the samples are said to have a perfect correlation (+ for "in-phase" fluctuations and minus (−) for 180° "out-of-phase" fluctuations). For $r \approx 0$, the points are scattered randomly on the graph and there is little or no relationship between the variables. The variables x_i, y_i in equation (3.13.1a, b) could be samples from two different, independent random variables or they could represent the independent (input) and dependent (output) variables of an estimation model. Alternatively, they could be samples from the same variable. Known as an *auto-correlation*, the later is usually computed for increasing lag or shifts in the starting value for one of the time series. A lag of "m" means that the first m values of one of the series, say the x series, are removed so that x_{m+1} becomes the new x_1 and so on.

Some authors prefer to use r^2 (the coefficient of determination discussed in Section 3.12.1 in the context of straight-line regression) rather than r (the correlation coefficient) since the squared value can be used to construct a significance level for r^2 in terms of a hypothesis test that the true correlation squared is zero. Writing

$$C_{xy}^2/(s_x s_y)^2 = \text{SSR}/\text{SST} = r^2 \tag{3.13.2}$$

we see that r^2 = variance explained/total variance, as stated earlier. A value $r = 0.75$ means that a linear regression of y on x explains $r^2 = 56.25\%$ of the total sample variance. Our approach is to use r to get the sign of the correlation and r^2 to estimate the joint variances.

3.13.1 The effects of random errors on correlation

Before discussing the relationship between r and our simple regression model, it is important to realize that sampling errors in x_i and y_i can only cause r to decrease. This can be shown by writing our two variables as a combination of true values (α_i, β_i) and random errors (δ_i, ε_i). In particular

$$x_i = \alpha_i + \delta_i$$
$$y_i = \beta_i + \varepsilon_i \tag{3.13.3}$$

Using equations (3.13.2) and (3.13.3), we can write the correlation between x_i and y_i as

$$r_{xy} = \frac{s_\alpha s_\beta r_{\alpha\beta} + s_\beta s_\delta r_{\beta\delta} + s_\alpha s_\varepsilon r_{\alpha\varepsilon} + s_\delta s_\varepsilon r_{\delta\varepsilon}}{s_x s_y} \tag{3.13.4}$$

where for convenience we have dropped the index i. Since the random errors δ and ε are assumed to be independent of each other and of the variables α and, β we know that

$$r_{\beta\delta} = r_{\alpha\varepsilon} = r_{\delta\varepsilon} = 0$$

so that (3.13.4) becomes

$$r_{xy} = \frac{s_\alpha s_\beta}{s_x s_y} r_{\alpha\beta} \tag{3.13.5}$$

This result means that the ratio between the product of the true standard deviations (s_α, s_β) to the product of the measured variable (s_x, s_y) determines the magnitude of the computed correlation coefficient (r_{xy}) relative to the true value ($r_{\alpha\beta}$).

Statistical Methods and Error Handling 245

To determine (3.13.5), we expand the variances of x and y as

$$\left(s_x^2, s_y^2\right) = \frac{1}{N-1} \sum_{i=1}^{N} \left[(x_i - \bar{x})^2, (y_i - \bar{y})^2\right]$$

where, as usual, $\bar{x}, \bar{y}$ are the average values for samples x_i, y_i respectively. Expanding the numerator into its component terms through (3.13.3), and using the fact that the errors are independent of one another, and of x and y, yields

$$\sum_{i=1}^{N} (x_i - \bar{x})^2 = \sum_{i=1}^{N} \left[(\alpha_i - \bar{\alpha})^2 + \delta_i^2\right]$$

$$\sum_{i=1}^{N} (y_i - \bar{y})^2 = \sum_{i=1}^{N} \left[(\beta_i - \bar{\beta})^2 + \varepsilon_i^2\right]$$

Dividing through by $(N-1)$ and using the definitions for standard devation, we find

$$s_x^2 = s_\alpha^2 + \frac{\sum_{i=1}^{N} \delta_i^2}{N-1}; \quad s_y^2 = s_\beta^2 + \frac{\sum_{i=1}^{N} \varepsilon_i^2}{N-1} \quad (3.13.6)$$

Since the second terms in each of the above expressions can never be negative ($N > 1$), the observed variances s_x^2 and s_y^2 are always greater than the corresponding true variances. Applying this result to equation (3.13.5), we see that the calculated correlation, r_{xy}, derived from the observations is always smaller than the true correlation, $r_{\alpha\beta}$. Because of random errors, the correlation coefficient computed from the observations will be smaller than (or, at best, equal to) the true correlation coefficient.

3.13.2 The maximum likelihood correlation estimator

Returning to the relationship between correlation and regression, we note the maximum likelihood estimator of the correlation coefficient is, by (3.13.1a)

$$r = \left[\sum_{i=1}^{N} (x_i - \bar{x})(y_i - \bar{y})\right] / \left[\sum_{i=1}^{N} (x_i - \bar{x})^2 \sum_{i=1}^{N} (y_i - \bar{y})^2\right]^{1/2} \quad (3.13.7)$$

for a bivariate normal population (x_i, y_i). We can expand this using (3.13.1b) to get

$$r = \frac{\left[N \sum_{i=1}^{N} x_i y_i - \sum_{i=1}^{N} x_i \sum_{i=1}^{N} y_i\right]}{\left\{\left[N \sum_{i=1}^{N} x_i^2 - \left(\sum_{i=1}^{N} x_i\right)^2\right]\left[N \sum_{i=1}^{N} y_i^2 - \left(\sum_{i=1}^{N} y_i\right)^2\right]\right\}^{1/2}} \quad (3.13.8)$$

Note that the numerator in equation (3.13.8) is similar to the numerator of the estimator for b_1 in equation (3.12.7a). For the case where the regression line passes through the origin in (3.12.7b), we have $b_1 = 0$ and our model is

$$\hat{y}_i = \hat{b}_1 x_i$$

and we can rewrite (3.12.7a) as

$$\hat{b}_1 = \frac{\left[\sum_{i=1}^{N} x_i y_i\right]\left[\sum_{i=1}^{N} y_i^2\right]^{1/2}}{\left[\sum_{i=1}^{N} x_i^2 \sum_{i=1}^{N} y_i^2\right]^{1/2} \left[\sum_{i=1}^{N} x_i^2\right]^{1/2}} \qquad (3.13.9)$$

$$= r s_y / s_x; \text{ or, } r = \hat{b}_1 s_x / s_y$$

Thus, r can be computed from $\hat{b}_1$ and vice versa if the standard deviations of the sample values x and y are known. Also, using the relationship between $\hat{b}_1$ and r we can write the variance of the parameter estimate in equation (3.13.9) as

$$s^2 = \frac{1}{N-2} \sum_{i=1}^{N} (y_i - \bar{y})^2 = \frac{1}{N-2} \text{SSE} \qquad (3.13.10)$$

We can use this result to better understand the relationship between correlation and regression by writing the ratio of the regression variance in equation (3.13.10) to the sample variance in y alone; for large N, this becomes

$$\frac{s^2}{s_y^2} = \frac{(N-1)(1-r^2)}{N-2} \approx (1-r^2) \qquad (3.13.11)$$

Thus, for N large, r^2 is that portion of the variance of y that can be attributed to its regression on x while $(1-r^2)$ is that portion of y's variance that is independent of x.

Earlier it was noted that a computationally efficient way to calculate the variance was to use equation (3.2.4b) which required only a single pass through the data sample. A similar saving can be gained in computing the covariance by expanding the product

$$\sum_{i=1}^{N} (x_i - \bar{x})(y_i - \bar{y}) = \sum_{i=1}^{N} (x_i y_i) - \frac{\left(\sum_{i=1}^{N} x_i\right)\left(\sum_{i=1}^{N} y_i\right)}{N} \qquad (3.13.12)$$

3.13.3 Correlation and regression: cause and effect

A point worth stressing is that a high correlation coefficient or a "good" fit of a regression curve $y = y(x)$ to a set of observations x, does not imply that x is "causing" y. Nor does it imply that x will provide a good predictor for y in the future. For example, the number of sockeye salmon returning to the Fraser River of British Columbia each fall from the North Pacific Ocean is often highly correlated with the mean fall surface water temperature at Amphitrite Point on the southwest coast of Vancouver Island. No one believes that the fish are responding directly to the temperature, but rather that temperature is a proxy variable for the real factor (or combination of factors) influencing the homeward migration of the fish. Of course, we are not saying that one should not draw inferences or conclusions from correlation or regression analysis, but only that caution is advised when seeking cause-and-effect relationships between variables. We further remark that there is little point in drawing any type of line through the data unless the scatter about the line is appreciably less than the overall spread of the observations. There is a tendency to fit trend lines to data with large variability and scatter even if a trend is not justified on statistical grounds. If $|r| < 0.5$, it hardly seems reasonable to fit a line for predictive purposes.

There is another important aspect of regression–correlation analysis that is worth stressing: Although the value of the correlation coefficient or coefficient of determination does *not* depend on which variable (x or y) is designated as the independent variable and which is designated as the dependent variable, this distinction *is* very important when it comes to regression analysis. The regression coefficients a, b for the conditional distribution of y given x ($y = a_1 + b_1 x$) are different than those for the conditional distribution of x given y ($x = a_2 + b_2 y$). In general, $a_1 \neq -a_2/b_2$ and $b_1 \neq 1/b_2$ so that the regression lines are different. In the first case, we are solving for the line shown in Figure 3.11(a), while in the second case we are solving for the line in Figure 3.11(b).

As an example, consider the broken lines in Figure 3.11(c) which show the two different linear regression lines for the regression of the observed cross-channel sea-level differences $y = \Delta \eta_c$, as measured by coastal tide gauges, and the calculated cross-channel sea-level difference $x = \Delta \eta_m$ obtained using concurrent current meter data from cross-channel moorings. The term $\Delta \eta_c$ is simply the difference in the mean sea-level from one side of the 25-km-wide channel to the other, while $\Delta \eta_m$ is calculated from the current meter records assuming that the time-averaged along-channel flow is in geostrophic balance (Labrecque et al., 1994). The dotted line is the regression $\Delta \eta_c = a_1 + b_1 \Delta \eta_m$ while the dashed line is the regression $\Delta \eta_m = a_2 + b_2 \Delta \eta_c$, with $b_1 \neq b_2$. The correlation coefficient $r = 0.69$ is the same for the two regressions. The solid line in Figure 3.11(c) is the so-called *neutral* regression line for the two parameters (Garrett and Petrie, 1981) and might seem the line of choice since it is not obvious which parameter should be the independent parameter and which should be the dependent parameter. Neutral regression is equivalent to minimizing the sum of the square distances from the regression line (Figure 3.11d).

In fisheries research, neutral regression is known as *geometeric mean functional regression* (GMFR) and is commonly used to relate fish body proportions when there is no clear basis to select dependent and independent variables (Sprent and Dolby, 1980). For two variables with zero means, the slope estimator, b, is given by the square roots of the variance ratios

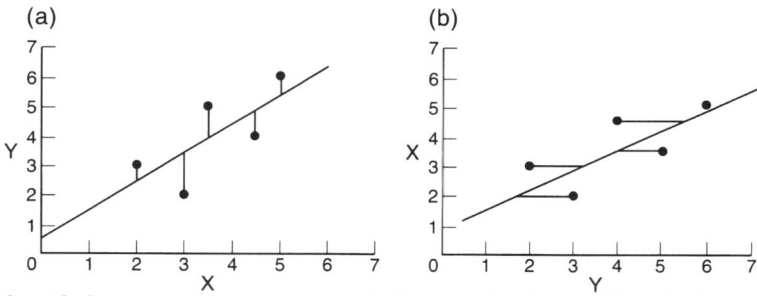

Figure 3.11. Straight line regressions (a) y on x, and (b) x on y showing the "direction" along which the variance is minimized.

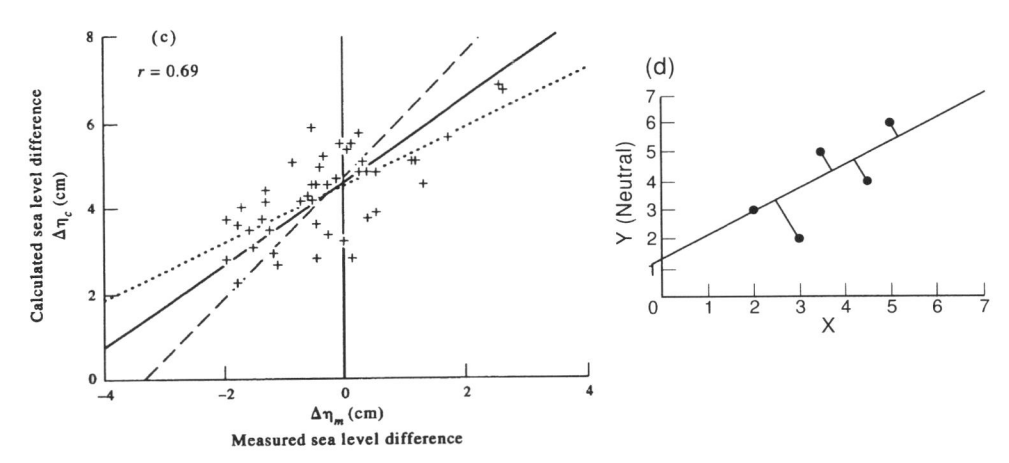

Figure 3.11. (c) *Scatter plot of $\Delta\eta_c$ versus $\Delta\eta_m$ for a cross-section of the 22-km-wide Juan de Fuca Strait separating Vancouver Island from Washington State. Plots give the regression of the observed cross-channel sea level differences $y = \Delta\eta_c$, as measured by coastal tide gauges, and the calculated cross-channel sea level difference, $x = \Delta\eta_m$, obtained using concurrent current meter data from cross-channel moorings. The solid line gives the bisector regression fit to the data (slope and 95% confidence level = 0.96 ± 0.37); the dotted line (slope = 0.66 ± 0.14) and the dashed line (slope = 1.40 ± 0.32) are the standard slopes for $\Delta\eta_c$ versus $\Delta\eta_m$ and $\Delta\eta_m$ versus $\Delta\eta_c$, respectively. $r = 0.69$. (From Labrecque et al., 1994.) (d) The "direction" along which the variance for the data points in (a) and (b) is minimized.*

$$b_{yx} = \text{sgn}\,(s_{xy}) \frac{\sum_{i=1}^{N}(y_i - \bar{y})^2}{\left[\sum_{i=1}^{N}(x_i - \bar{x})^2\right]^{1/2}}; \quad \text{regression } \hat{y}_i = \hat{b}_{yx}x_i$$

$$b_{xy} = \text{sgn}\,(s_{xy}) \frac{\sum_{i=1}^{N}(x_i - \bar{x})^2}{\left[\sum_{i=1}^{N}(y_i - \bar{y})^2\right]^{1/2}}; \quad \text{regression } \hat{x}_i = \hat{b}_{xy}y_i$$

(3.13.13)

where sgn (s_{xy}) is the sign of the covariance function $s_{xy} = \sum(x_i - \bar{x})(y_i - \bar{y})$ and $b_{yx} = 1/b_{xy}$, as required. Note that the slope b_{yx} lies midway between the slopes b_1 and b_2

$$b_1 = \frac{\sum_{i=1}^{N}(x_i - \bar{x})(y_i - \bar{y})}{\sum_{i=1}^{N}(x_i - \bar{x})^2}; \quad \text{regression line } \hat{y}_i = a_1 + \hat{b}_1 x_i$$

$$b_2 = \frac{\sum_{i=1}^{N}(x_i - \bar{x})(y_i - \bar{y})}{\sum_{i=1}^{N}(y_i - \bar{y})^2}; \quad \text{regression line } \hat{x}_i = a_2 + \hat{b}_2 y_i$$

(3.13.14)

given by (3.12.7a) for standard regression analyses (Figure 3.11a). The GMFR is then the geometric mean slope of the least-squares regression coefficient for the regression slope of y on x and the regression of x on y; $b_{yx} = [b_1(b_2)^{-1}]^{1/2}$. Since the slope from the GMFR is simply a ratio of variances, it is "transparent" to the determination of correlation coefficients or coefficients of determination. It is these correlations, not the slope of the line, that test the strength of the linear relationship between the two variables. Moreover, none of the standard linear regression models reduces to the GMFR slope estimate except under unlikely circumstances. According to Sprent and Dolby (1980), *ad hoc* use of the GMFR is not recommended when there are errors in both variables. The GMFR model, though appealing, rests on shaky statistical ground and its use remains controversial.

3.14 HYPOTHESIS TESTING

Statistical inference takes one of two forms. Either we make estimates of population variables, as we have done thus far, or we test hypotheses about the implications of these variables. Statistical inference in which we choose between two conflicting hypotheses about the value of a particular population variable is known as *hypothesis testing*.

Hypothesis testing follows scientific methodology from whose nomenclature the terms are borrowed. The investigator forms a "hypothesis," collects some sample data and uses a statistical construct to either reject or accept the original hypothesis. The basic elements of a statistical test are: (1) the *null hypothesis*, H_o (the hypothesis to be tested); (2) the alternate hypothesis, H_a; (3) the test statistic to be used; and (4) the region of rejection of the hypothesis. The active components of a statistical test are the test statistic and the associated rejection region, with the latter specifying the values of the test statistic for which the null hypothesis is rejected. We emphasize the point that "pure" hypothesis testing originated from early work in which the null hypothesis corresponded to an idea or theory about a population variable that the scientist hoped *would be rejected*. "Null" in this case means incorrect and invalid so that we could call it the "invalid hypothesis". In other words, the null hypotheis specified those values of the population variable which it was thought did *not* represent the true value of the variable. This is a form of negative thinking and is the reason that many of us would rather think in terms of the *alternate hypothesis* in which we specify those values of the variable that we hope will hold true (the "valid" hypothesis). Regardless of which hypothesis is chosen, it is important to remember that the true population value under consideration must either lie in the test set covered by H_o or in the set covered by H_a. There are no other choices.

We restrict consideration of hypothesis testing to large samples ($N > 30$). In hypothesis testing, two types of errors are possible. In a type-1 error, the null hypothesis H_o is rejected when it is true. The probability of this type of error is denoted by α. Type-2 errors occur when H_o is accepted when it is false (H_a is true). The probability of type-2 errors is written as β. In Table 13.14.1, the probability P (accept $H_o | H_o$ is true) $= 1 - \alpha$ corresponds to the $100(1 - \alpha)\%$ confidence interval. Alternatively, the probability $P(\text{reject } H_o | H_o \text{ is false}) = 1 - \beta$ is the power of the statistical test since it indicates the ability of the test to determine when the null hypothesis is false and H_o should be rejected.

Table 13.14.1. The four possible decision outcomes in hypothesis testing and the probability of each decision outcome in a test hypothesis

		Possible situation	
		H_o is true	H_o is false
Action	Accept H_o	Correct decision; confidence level $1 - \alpha$	Incorrect decision; (Type-2 error); β
	Reject H_o	Incorrect decision (Type-1 error); α	Correct decision; power of the test $1 - \beta$
	Sum	1.00	1.00

For a parameter θ based on a random sample $x_1, \ldots, x_N$, we want to test various values of θ using the estimate $\hat{\theta}$ as a test statistic. This estimator is assumed to have an approximately normal sampling distribution. For a specified value of $\hat{\theta}(= \theta_o)$, we want to test the hypothesis, H_o, that $\hat{\theta} = \theta_o$ (written H_o: $\theta = \theta_o$) with the alternate hypothesis, H_a, that $\hat{\theta} > \theta_o$ (written H_a: $\theta > \theta_o$). An efficient test statistic for our assumed normal distribution is the standard normal Z defined as

$$Z = \frac{(\hat{\theta} - \theta)}{\hat{\sigma}_{\hat{\theta}}} \qquad (3.14.1)$$

where $\hat{\sigma}_{\hat{\theta}}$ is the standard deviation of the approximately normal sampling distribution of $\hat{\theta}$, which can then be computed from the sample. For this test statistic, the null hypothesis (H_o: $\theta = \theta_o$) is rejected for $Z > Z_\alpha$ where α is the probability of a type-1 error. Graphically, this rejection region is depicted as the shaded portion in Figure 3.12(a), which is called an "upper-tail" test. Similarly, a "lower-tail" test would have the shaded rejection region starting at $-Z_\alpha$ and corresponds to $Z < -Z_\alpha$ and $\theta < \theta_o$ (Figure 3.12b). A two-tailed test (Figure 3.12c) is one for which the null hypothesis rejection region is $|Z| > Z_{\alpha/2}$ and $\theta \neq \theta_o$. The decision of which test alternative to use should be based on the form of the alternate hypothesis. If one is interested in parameter values greater than θ_o, an upper-tail test is used; for values less than θ_o, a lower-tail test is appropriate. If one is interested in any change from θ_o, it is best to use a two-tailed test. The following is an example for which a two-tailed test is appropriate.

Suppose that daily averaged currents for some mooring locations are available for the same month from two different years (e.g. January 1984 and January 1985). We wish to test the hypothesis that the monthly means of the longshore component of the flow, V, for these two different years are the same. If the daily averages are computed from hourly observations, we invoke the central limit theorem and conclude that our sampling distributions are normally distributed. Taking each month as having 31 days, we satisfy the condition of a large sample ($N > 30$) and can use the procedure outlined above. Suppose we observe that for January 1984 the mean and standard deviation of the observed current is $V_{84} = 23 \pm 3$ cm/s while for January 1985 we find a monthly mean speed $V_{85} = 20 \pm 2$ cm/s (here, the standard deviations are obtained from the signal variances). We now wish to test the null hypothesis that the true (as opposed to our sampled) monthly mean current speeds were statistically the same for the two separate years. We use the two-tailed test to detect any deviations from

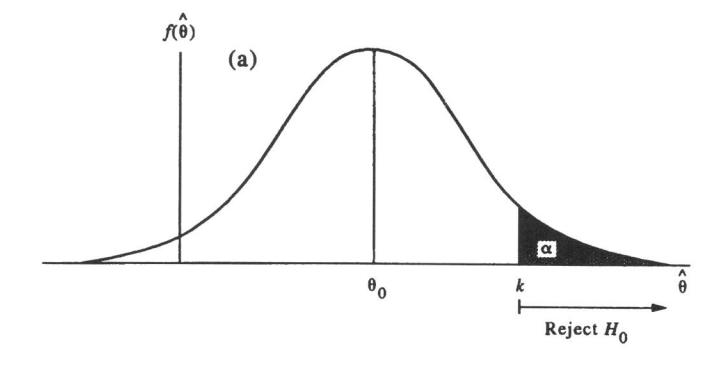

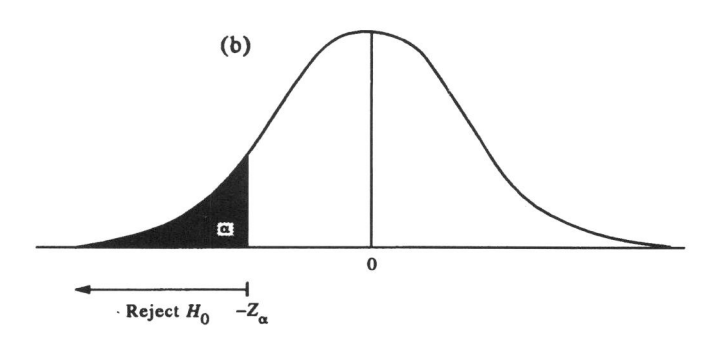

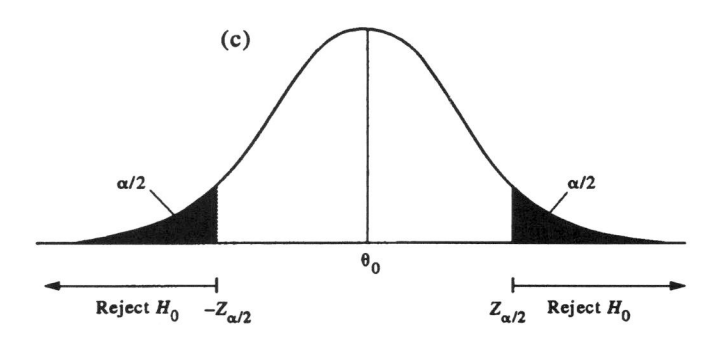

Figure 3.12. Large-sample rejection regions (shaded areas) for the null hypothesis $H_o : \theta = \theta_o$, for the normally distributed function $f(\theta)$. (a) Upper-tail test for $H_o : \theta = \theta_o$, $H_a : \theta > \theta_o$; (b) lower-tail test with the rejection region for $H_o : \theta = \theta_o$, $H_a : \theta < \theta_o$; and (c) two-tailed test for which the null hypothesis rejection region is $|z| > z_{\alpha/2}$ and $H_a : \theta \neq \theta_o$.

equality. In this example, the *point estimator* used to detect any difference between the monthly mean records calculated from daily observed values is the sample mean difference, $\hat{\theta} - \theta_o = V_{84} - V_{85}$. Our test statistic (3.14.1) is

$$Z = \frac{(V_{84} - V_{85})}{[s_{84}^2/N_{84} + s_{85}^2/N_{85}]^{1/2}}$$

which yields

$$Z = \frac{(23 - 20)}{[9/31 + 4/31]^{1/2}} = 4.63$$

To determine if the above result falls in the rejection region, $Z > Z_\alpha$, we need to select the significance level α for type-1 errors. For the 95% significance level, $\alpha = 0.05$ and $\alpha/2 = 0.025$. From the standard normal table (Appendix D, Table D.1) $Z_{0.025} = 1.2$. Our test value $Z = 4.63$ is greater than 1.2 so that it falls within the rejection region, and we must reject the hypothesis that the monthly mean current speeds are the same for both years. In most oceanographic applications hypothesis testing is limited to the null hypothesis and thus type-1 errors are most appropriate. We will not consider here the implementation of type-2 errors which lead to the acceptance of an alternate hypothesis as described in Table 3.14.1.

Turning again to satellite altimetry for an example, we note that the altimeter height bias discussed earlier is one of the error sources that contributes to the overall error "budget" of altimetric height measurements. Suppose that we wish to know if the *overall* height error H_T is less than some specified amount, H_ε. We first set up the null hypothesis (H_o: $H_T < H_\varepsilon$) that the true mean is less than H_ε. At this point, we must also select a significance level for our test. A significance level of $1 - \alpha$ means that we do not want to make a mistake and reject the null hypothesis more than $\alpha(100)\%$ of the time. We begin by defining our hypothesis limit, H_T, as

$$H_T = H_\varepsilon + \frac{z_\alpha s_b}{\sqrt{N}} \tag{3.14.2}$$

where the standard normal distribution Z_α is given by equation (3.14.1) and s_b is the standard error (uncertainty) in our measurements. If the mean of our measurements is greater than H_T, then we reject H_o and conclude that the mean is greater than H_ε with a probability α of being wrong.

Suppose we set $H_\varepsilon = 13$ cm and consider $N = 9$ consecutive statistically independent satellite measurements in which each measurement is assumed to have an uncertainty of $s_b = 3$ cm. If the mean height error is 15 cm, do we accept or reject the null hypothesis for the probability level $\alpha = 0.10$? What about the cases for $\alpha = 0.05$ and $\alpha = 0.01$? Given our hypothesis limit $H_\varepsilon = 13$ cm and the fact that $N = 9$ and $s_b = 3$ cm, we can write equation (3.14.2) as $H_T = 13 + Z_\alpha$ cm. According to the results of

Table 3.14.2. Testing the null hypothesis that the overall bias error H_T of satellite altimetry data is less than 13 cm. Assumes normal error distributions

Significance level, α	Standard normal distribution, Z_α	Total error height, H_T	Decision
0.10	2.326	15.326 cm	reject H_o
0.05	1.645	14.645 cm	accept H_o
0.01	1.280	14.280 cm	accept H_o

Table 3.14.2, this means that we accept the null hypothesis that the overall error is less than 13 cm at the 5 and 10% probability levels but not at the 1% probability level (these are referred to as the 95%, 90%, and 99% significance levels, respectively).

3.14.1 Significance levels and confidence intervals for correlation

One useful application of null hypothesis testing is the development of significance levels for the correlation coefficient, r. If we take the null hypothesis as $r = r_o$, where r_o is some estimate of the correlation coefficient, we can determine the rejection region in terms of r at a chosen significance level α for different degrees of freedom $(N-2)$. A list of such values is given in Appendix E. In that table, the correlation coefficient r for the 95 and 99% significance levels (also called the 5 and 1% levels depending on whether or not one is judging a population parameter or testing a hypothesis) are presented as functions of the number of degrees of freedom.

For example, a sample of 20 pairs of (x, y) values with a correlation coefficient less than 0.444 and $N - 2 = 18$ degrees of freedom would not be significantly different from zero at the 95% confidence level. It is interesting to note that, because of the close relationship between r and the regression coefficient b_1 of these pairs of values, we could have developed the table for r values using a test of the null hypothesis for b_1.

The procedure for finding confidence intervals for the correlation coefficient r is to first transform it into the standard normal variable Z_r as

$$Z_r = \frac{1}{2}[\ln(1+r) - \ln(1-r)] \quad (3.14.3)$$

which has the standard error

$$\sigma_z = \frac{1}{(N-3)^{1/2}} \quad (3.14.4)$$

independent of the value of the correlation. The appropriate confidence interval is then

$$Z_r - Z_{\alpha/2}\sigma_z < Z < Z_r + Z_{\alpha/2}\sigma_z \quad (3.14.5)$$

which can be transformed back into values of r using equation (3.14.3).

Before leaving the subject of correlations we want to stress that correlations are merely statistical constructs and, while we have some mathematical guidelines as to the statistical reliability of these values, we cannot replace common sense and physical insight with our statistical calculations. It is entirely possible that our statistics will deceive us if we do not apply them carefully. We again emphasize that a high correlation can reveal either a close relationship between two variables or their simultaneous dependence on a third variable. It is also possible that a high correlation may be due to complete coincidence and have no causal relationship behind it. A classic example (Snedecor and Cochran, 1967) is the high negative correlation (-0.98) between the annual birthrate in Great Britain and the annual production of pig iron in the United States for the years 1875–1920. This high correlation is statistically significant for the available $N - 2 = 43$ degrees of freedom, but the likelihood of a direct relationship between these two variables, is very low.

3.14.2 Analysis of variance and the *F*-distribution

Most of the statistical tests we have presented to this point are designed to test for differences between two populations. In certain circumstances, we may wish to investigate the differences among three or more populations simultaneously rather than attempt the arduous task of examining all possible pairs. For example, we might want to compare the mean lifetimes of drifters sold by several different manufacturers to see if there is a difference in survivability for similar environmental conditions; or, we might want to look for significant differences among temperature or salinity data measured simultaneously during an intercomparison of several different commercially available CTDs. The *analysis of variance* (ANOVA) is a method for performing simultaneous tests on data sets drawn from different populations. In essence, ANOVA is a test between the amount of variation in the data that can be attributed to chance and that which can be attributed to specific causes and effects. If the amount of variability *between* samples is small relative to the variability *within* samples, then the null hypothesis H_o—that the variability occurred by chance—cannot be rejected. If, on the other hand, the ratio of these variations is sufficiently large, we can reject H_o. "Sufficiently large", in this case, is determined by the ratio of two continuous χ^2 probability distributions. This ratio is known as the *F-distribution*.

To examine this subject further, we need several definitions. Suppose we have samples from a total of J populations and that a given sample consists of N_j values. In ANOVA, the J samples are called J "treatments", a term that stems from early applications of the method to agricultural problems where soils were "treated" with different kinds of fertilizer and the statistical results compared. In the one-factor ANOVA model, the values y_{ij} for a particular treatment (input), x_j, differ from some common background value, μ, because of random effects; that is

$$y_{ij} = \mu + x_j + \varepsilon_{ij}; \quad j = 1, 2, ..., J \\ \qquad\qquad\qquad\quad i = 1, 2, ..., N_j \tag{3.14.6}$$

where the outcome y_{ij} is made up of a common (grand average) effect (μ), plus a treatment effect (x_j) and a random effect, ε_{ij}. The grand mean, μ, and the treatment effects, x_j, are assumed to be constants while the errors, ε_{ij}, are independent, normally distributed, variables with zero mean and a common variance, σ^2, for all populations. The null hypothesis for this one-factor model is that the treatments have zero effect. That is, $H_o: x_j = 0$ ($j=1, 2, ..., J$) or, equivalently, $H_o: \mu_1 = \mu_2 = ... = \mu_J$ (i.e. there is no difference between the populations aside from that due to random errors). The alternative hypothesis is that some of the treatments have a nonzero effect. Note that "treatment" can refer to any basic parameter we wish to compare such as buoy design, power supply, or CTD manufacturer. To test the null hypotheses, we consider samples of size N_j from each of the J populations. For each of these samples, we calculate the mean value $\bar{y}_j$ ($j = 1, 2, ..., J$). The grand mean for all the data is denoted as $\bar{y}$.

As an example, suppose we want to intercompare the temperature records from three types of CTDs placed in the same temperature bath under identical sampling conditions. Four countries take part in the intercomparison and each brings the same three types of CTD. The results of the test are reproduced in Table 3.14.3.

If H_o is true, $\mu_1 = \mu_2 = \mu_3$, and the measured differences between $\bar{y}_1, \bar{y}_2$, and $\bar{y}_3$ in Table 3.14.3 can be attributed to random processes.

Statistical Methods and Error Handling   255

Table 3.14.3. Temperatures in °C measured by three makes of CTD in the same calibration tank. Four instruments of each type are used in the test. The grand mean for the data from all three instruments is $\bar{\bar{y}}$ = 15.002°C

Measurement (i)	CTD Type 1 Sample j = 1	CTD Type 2 Sample j = 2	CTD Type 3 Sample j = 3
1	15.001	15.004	15.002
2	14.999	15.002	15.003
3	15.000	15.001	15.000
4	14.998	15.004	15.002
Mean $\bar{y}$ °C	15.000 = $\bar{y}_1$	15.003 = $\bar{y}_2$	15.002 = $\bar{y}_3$

The treatment effects for the CTD example are given by

$$x_1 = \bar{y}_1 - \bar{\bar{y}} = -0.002°C$$
$$x_2 = \bar{y}_2 - \bar{\bar{y}} = +0.001°C$$
$$x_3 = \bar{y}_3 - \bar{\bar{y}} = 0.0°C$$

where $\bar{\bar{y}} = (\bar{y}_1 + \bar{y}_2 + \bar{y}_3)/3$. The ANOVA test involves determining whether the estimated values of x_j are large enough to convince us that H_o is not true. Whenever H_o is true, we can expect that the variability between the J means is the same as the variability within each sample (the only source of variability is the random effects, ε_{ij}). However, if the treatment effects are not all zero, then the variability between samples should be larger than the variability within the samples.

The variation within the J samples is found by first summing the squared deviations of y_{ij} about the mean value $\bar{y}_j$ for each sample, namely

$$\sum_{i=1}^{N_j} (y_{ij} - \bar{y}_j)^2$$

If we then sum this variation over all J samples, we obtain the *sum of squares within* (SSW)

$$\text{Sum of squares within: SSW} = \sum_{j=1}^{J} \sum_{i=1}^{N_j} (y_{ij} - \bar{y}_j)^2 \qquad (3.14.7)$$

Note that the sample lengths, N_j, need not be the same since the summation for each sample uses only the mean for that particular sample. Next, we will need the amount of variation between the samples (SSB). This is obtained by taking the squared deviation of the mean of the Jth sample, $\bar{y}_j$, and the grand mean, $\bar{\bar{y}}$. This deviation must then be weighted by the number of observations in the Jth sample. The overall sum is given by

$$\text{Sum of squares between: SSB} = \sum_{j=1}^{J} N_j (\bar{y}_j - \bar{\bar{y}})^2 \qquad (3.14.8)$$

To compare the variability within samples to the variability between samples, we need to divide each sum by its respective number of degrees of freedom, just as we did with

256 *Data Analysis Methods in Physical Oceanography*

other variance expressions such as s^2. For SSB, the degrees of freedom (DOF) = $\mathcal{J} - 1$ while for SSW

$$\text{DOF} = \left(\sum_{j=1}^{\mathcal{J}} N_j\right) - \mathcal{J}$$

The *mean square* values are then:

$$\text{Mean square between: MSB} = \frac{\text{SSB}}{\mathcal{J} - 1} \qquad (3.14.9\text{a})$$

$$\text{Mean square within: MSW} = \frac{\text{SSW}}{\left(\sum_{j=1}^{\mathcal{J}} N_j\right) - \mathcal{J}} \qquad (3.14.9\text{b})$$

In the above example, $\mathcal{J} - 1 = 2$ and $\sum_{j=1}^{\mathcal{J}} N_j - \mathcal{J} = 9$. The calculated values of MSB and MSW for our CTD example are given in Table 3.14.4. Specifically

$$\text{SSW} = \sum_{i=1}^{4}(y_{i1} - \bar{y}_1)^2 + \sum_{i=1}^{4}(y_{i2} - \bar{y}_2)^2 + \sum_{i=1}^{4}(y_{i3} - \bar{y}_3)^2$$

$$\text{SSB} = N_1(\bar{y}_1 - \bar{y})^2 + N_2(\bar{y}_2 - \bar{y})^2 + N_3(\bar{y}_3 - \bar{y})^2 + N_4(\bar{y}_4 - \bar{y})^2$$

where $N_j = 4$ ($j = 1, \ldots, 4$). To determine if the ratio of MSB to MSW is large enough to reject the null hypothesis, we use the *F*-distribution for $\mathcal{J} - 1$ and

$$\left(\sum_{j=1}^{\mathcal{J}} N_j\right) - \mathcal{J}$$

degrees of freedom.

Named after R. A. Fisher who first studied it in 1924, the *F*-distribution is defined in terms of the ratio of two independent χ^2 variables divided by their respective degrees of freedom. If X_1 is a χ^2 variable with ν_1 degrees of freedom and X_2 is another χ^2 variable with ν_2 degrees of freedom, then the random variable

$$F(\nu_1, \nu_2) = \frac{X_1/\nu_1}{X_2/\nu_2} \qquad (3.14.10)$$

is a nonnegative chi-square variable with ν_1 degrees of freedom in the numerator and ν_2 degrees of freedom in the denominator. If $\mathcal{J} = 2$, in the CTD example above, the *F*-

Table 3.14.4. *Calculated values of sum of squares and mean square values for the CTD temperature intercomparison. DOF = number of degrees of freedom*

Type of variation	Sum of squares (°C^2)	DOF	Mean square (°C^2)
Between samples (type of CTD)	20×10^{-6}	2	10×10^{-6}
Within samples (all CTDs)	18×10^{-6}	9	2×10^{-6}
Total	38×10^{-6}	11	(ratio = 5.0)

Statistical Methods and Error Handling 257

test is equivalent to a one-sided *t*-test. There is no upper limit to *F*, which like the χ^2 distribution is skewed to the right. Tables are used to list the critical values of $P(F > F_\alpha)$ for selected degrees of freedom ν_1 and ν_2 for the two most commonly used significance levels, $\alpha = 0.05$ and $\alpha = 0.01$. In ANOVA, the values of SSB and SSW follow χ^2-distributions. Therefore, if we let $X_1 = $ SSB and $X_2 = $ SSW, then

$$F\left(\mathcal{J} - 1, \sum N_j - \mathcal{J}\right) = \frac{[\text{SSB}/(\mathcal{J}-1)]}{\text{SSW}(\sum N_j - \mathcal{J})} = \frac{\text{MSB}}{\text{MSW}} \quad (3.14.11)$$

When MSB is large relative to MSW, *F* will be large and we can justifiably reject the null hypothesis that the different CTDs (different treatment effects) measure the same temperature within the accuracy of the instruments. For our CTD intercomparison (Table 3.14.4), we have MSB/MSW = 5.0, $\nu_1 = 2$ and $\nu_2 = 9$. Using the values for the *F*-distribution for 2 and 9 degrees of freedom from Appendix D, Table D.4a, we find $F_\alpha(2,9) = 4.26$ for $\alpha = 0.05$ (95% confidence level) and $F_\alpha(2,9) = 8.02$ for $\alpha = 0.01$ (99% confidence level). Since, *F* = 5.0 in our example, we conclude that a difference exists among the different makes of CTD at the 95% confidence level, but not at the 99% confidence level.

3.15 EFFECTIVE DEGREES OF FREEDOM

To this point, we have assumed that we are dealing with random variables and each of the *N* values in a given sample are statistically independent. For example, in calculating the unbiased standard deviation for *N* data points, we assume there are *N* − 1 degrees of freedom. (We use *N* − 1 rather than *N* since we need a minimum of two values to calculate the standard deviaton of a sample.) Similarly, in Sections 3.8 and 3.10, we specify confidence limits in terms of the number of samples rather than the "true" number of degrees of freedom of the sample. In reality, consecutive data values may not be independent. Contributions from low-frequency components and narrow band oscillations such as in inertial motions may lead to a high degree of correlation between values separated by large times or distances. The most common example of highly coherent narrow band signals are the tides and tidal currents which possess a strong temporal and spatial coherence. If we want our statistics to have any real meaning, we are forced to find the *effective number of degrees of freedom* using information on the coherence and autocorrelation of our data.

The effects of coherent nonrandom processes on data series lead us into the question of data redundancy in multivariate linear regression. Our general model is

$$\hat{y}(t_i) = \sum_{k=1}^{M} b_k x_k(t_i); \quad i = 1, \ldots, N \quad (3.15.1)$$

where the x_k represents *M* observed parameters or quantities at times t_i. The b_k are *M* linear-regression coefficients relating the independent variables $x_k(t_i)$ to the *N* model estimates, $\hat{y}(t_i)$. Here, the x_k observations can be measurements of different physical quantities or of the same quantity measured at different times or locations.

The estimate $\hat{y}$ differs from the true parameter by an error $\varepsilon_i = \hat{y}(t_i) - y(t_i) = \hat{y}_i - y_i$. Following our earlier discussion, we assume that this error is randomly distributed and is therefore uncorrelated with the input data $x_k(t_i)$. To find the best estimate, we apply the method of least squares to minimize the mean square error

$$\overline{\varepsilon^2} = \sum_{i=1}^{M}\sum_{j=1}^{M} b_i b_j \overline{x_i x_j} - 2\sum_{j=1}^{M} b_j \overline{x_j y} + \overline{y^2} \qquad (3.15.2)$$

In this case, the overbars represent ensemble averages. To assist us in our minimization, we invoke the Gauss–Markov theorem which says that the estimator, given by equation (3.15.1), with the smallest mean square error is that with coefficients

$$b_k = \sum_{j=1}^{M} \left[\{\overline{x_k x_j}\}^{-1} \overline{x_j y} \right] \qquad (3.15.3)$$

where $\{\overline{x_k x_j}\}^{-1}$ is the i,j element of the inverse of the $M \times M$ cross covariance matrix of the input variables (note: $\{\overline{x_k x_j}\}^{-1} \neq 1/\overline{x_k x_j}$). This mean-square product matrix is always positive definite unless one of the input variables x_k can be expressed as an exact combination of the other input values. The presence of random measurement errors in all input data make this "degeneracy" highly unlikely. It should be noted however, that it is the partial correlation between inputs that increases the uncertainty in our estimator by lowering the degrees of freedom through a reduction in the independence of our input parameters.

We can write the minimum least-square error ε_o^2 as

$$\overline{\varepsilon_o^2} = \overline{y^2} - \sum_{i=1}^{M}\sum_{j=1}^{M} \overline{y x_j} \{\overline{x_k x_j}\}^{-1} \overline{y x_j} \qquad (3.15.4)$$

At this point, we introduce a measure of the reliability of our estimate called the *skill* (S) of the model. This skill is defined as the fraction of the true parameter variance explained by our linear statistical estimator; thus

$$S = \{\overline{y^2}\}^{-1} \sum_{i=1}^{N}\sum_{j=1}^{N} \left[\overline{y x_j} \{\overline{x_k x_j}\}^{-1} \overline{y x_j} \right] \qquad (3.15.5)$$

The skill value ranges from no skill ($S = 0$) to perfect skill ($S = 1$). We note that for the case ($M = 1$), S is the square of the correlation between x_1 and y.

The fundamental trade-off for any linear estimation model is that, while one wants to use as many independent input variables as possible to avoid interdependence among the estimates of the dependent variable, each new input contributes random measurement errors that degrade the overall estimate. As pointed out by Davis (1977) the best criterion for selecting the input data parameters is to use *a priori* theoretical considerations. If this is not possible, some effort should be made to select those inputs which contribute most to the estimation skill.

The conflicting requirements of limiting M (the observed parameters) and including all candidate input parameters is a dilemma. In considering this dilemma Chelton (1983) concludes that the only way to reduce the error limits on the estimated regression coefficients is to increase what are called the "effective degrees of freedom N^*." This can be done only by increasing the sample size of the input variable (i.e using a longer time series) or by high-passing the data to eliminate contributions from unresolved, and generally coherent, low-frequency components. Since we are forced to deal with relatively short data records in which ensemble averages are replaced by sample averages over time or space, we need a procedure to evaluate N^*, the effective degrees of freedom.

In the case of real data, ensemble averages are generally replaced with sample averages over time or space so that the resultant values become estimates. Thus, the skill can be written as S given by (3.15.5). If we assume for a moment that the x_k input data are serially uncorrelated (i.e. we expand the data series into orthogonal functions), we can write the sample estimate of the skill as

$$\hat{S} = \sum_{i=1}^{M} \sum_{j=1}^{M} \frac{\overline{x_i x_j}^2}{\overline{x_j^2 y^2}}$$

Following Davis (1978) we can expand this skill estimate into a true skill plus an artificial skill

$$\hat{S} = S + S_A \qquad (3.15.7)$$

The artificial skill, S_A, arises from errors in the estimates and can be calculated by evaluating the skill in equation (3.15.6) at a very long time (or space lag) where no real skill is expected. At this point, there is no true estimation skill and $\hat{S} = S_A$.

Davis (1976) derived an appropriate expression for the expected (mean) value of this artificial skill which relates it to the effective degrees of freedom N^*

$$\overline{S}_A = \sum_{k=1}^{M} (N_k^*)^{-1} \qquad (3.15.8)$$

where N_k^* is the effective degrees of freedom associated with the sample estimate of the covariance between the output y and and input x_k of the model. Under the conditions that S (the true skill) is not large, that the record length N is long compared to the autocovariance scales of y and x, and that the N_k^* are the same for all N, we can write N^* as

$$N^* = \frac{N}{\left[\sum_{\tau=-\infty}^{\infty} C_{xx}(\tau) C_{yy}(\tau) \right] / [C_{xx}(0) C_{yy}(0)]} \qquad (3.15.9a)$$

$$= \frac{N}{\left[\sum_{\tau=-\infty}^{\infty} \rho_{xx}(\tau) \rho_{yy}(\tau) \right]} \qquad (3.15.9b)$$

where $\rho_{\zeta\zeta}(\tau) = C_{\zeta\zeta}(\tau)/C_{\zeta\zeta}(0) = C_{\zeta\zeta}(\tau)/s_\zeta^2$ is the normalized autocovariance function for any variable ζ (with variance s_ζ^2), and

$$C_{\zeta\zeta}(\tau) = E[(\zeta(t_i) - \bar{\zeta})(\zeta(\tau + t_i) - \bar{\zeta})]$$

$$= \frac{1}{N'} \sum_{i=1}^{N'} \{(\zeta(t_i) - \bar{\zeta})(\zeta(\tau + t_i) - \bar{\zeta})\} \qquad (3.15.10)$$

where N' is the number of data values used in the summation for the particular lag, τ. A more complete form of this expression was given by Chelton (1983) as

$$N^* = \frac{N}{\left[\sum_{\tau=-\infty}^{\infty} C_{xx}(\tau)C_{yy}(\tau) + C_{xy}(\tau)C_{yx}(\tau)\right]/[C_{xx}(0)C_{yy}(0)]} \quad (3.15.11a)$$

$$= \frac{N}{\left[\sum_{\tau=-\infty}^{\infty} \rho_{xx}(\tau)\rho_{yy}(\tau) + \rho_{xy}(\tau)\rho_{yx}(\tau)\right]} \quad (3.15.11b)$$

This expression now includes the cross-covariances between y and x [e.g. $C_{xy}(\tau)$ and $\rho_{xy}(\tau)$] and is not limited to cases where S is small.

In general, the true auto- and cross-covariances are not known and the computation of N^* requires the substitution of sample estimates over finite lags for the correlations in equation (3.15.11). The resulting effective degrees of freedom, N^*, can be used with standard tables to find the selected significance levels for $\hat{S}$. In the ideal case, when all input variables are neither cross- nor serially correlated (and therefore independent) the effective number of degrees of freedom is N, the sample size. In general, however input data series are serially correlated and $N^* \ll N$. The larger the time/space correlation scales in equation (3.15.11), the smaller the value of N^*. This means that it is the large scale, low-frequency components of the input data that lead to a decrease in the number of independent values in the data series.

The limitations of estimating regression characteristics for real data can be summarized as follows:

(1) Accurate statistical results require the use of the effective number of degrees of freedom, N^*, with N^* generally much less than the total number of observations N.
(2) The accuracy of the estimated regression coefficients increases as N^* increases.
(3) The accuracy of the regression coefficient decreases as the number of inputs M increases (measurement error is added).
(4) The accuracy increases as the model skill increases and decreases as the input parameters become more correlated.

The above considerations emphasize the need for careful selection of the input data and the careful evaluation of the characteristics of these data. As pointed out by Davis (1977), a fundamental part of this selection process is the determination of the space and time scales to be studied. The methods used to extract this fundamental scale information from the input data can range from cross-spectral analysis (see Chapter 5) to a filtering of the data using preselected windows. Performing this filtering in the time domain rather than the frequency domain is often less complicated. The filtering process has the goal of eliminating scales that are not expected to contribute to the true correlation but which will add artificial correlation due to instrument and sampling errors.

Once the space and/or time scales are determined, selected or set by filtering, the next step is the selection of the input series to use in the estimate. At this stage, the dilemma arises between limiting the effects of errors and at the same time including as many as possible uncorrelated input variables to increase the degrees of freedom. Davis (1977) recommends using dynamical considerations to make this selection and shows how the data required for proper statistical estimation are generally those required to make the dynamical system well posed. However, he also mentions that, in general, the dynamics of most processes are not well enough understood and that specification data are not known with certainty. Nevertheless, some quantitative understanding of the physical system can serve as a useful guide to the selection of estimation data.

3.15.1 Trend estimates and the integral time scale

Most oceanographic variability arises through a combination of random and non-random processes. The presence of tidal and low-frequency components means that data points in time or space series are not independent of one another. The data that we collect are not truly random samples drawn from random populations. There is invariably a nonzero correlation between values in the series which must be taken into account when we tally-up the true number of independent samples or degrees of freedom we think we have in our system. This number is important when it comes to determining the confidence limits of linear regression slopes and parameter estimates.

As an example, consider the confidence limits on the slope of the least squares linear regression $\hat{y} = b_0 + b_1 x$ (where, again, ˆ denotes an estimator for the function y). From equation (3.8.6), the limits are

$$\pm (s_\varepsilon t_{\alpha/2,\nu})/[(N-1)s_x]^{1/2} \quad (3.15.12a)$$

or, in terms of the estimator β_1 for b_1

$$b_1 - \frac{(s_\varepsilon t_{\alpha/2,\nu})}{[(N-1)s_x]^{1/2}} < \beta_1 < b_1 + \frac{(s_\varepsilon t_{\alpha/2,\nu})}{[(N-1)s_x]^{1/2}} \quad (3.15.12b)$$

where $\nu = N - 2$ is the number of degrees of freedom for the Student's t-distribution at the $(1-\alpha)100\%$ confidence level, and the standard error of the estimate, s_ε, is given by

$$s_\varepsilon = \left[\frac{1}{N-2} \sum_{i=1}^{N} (y_i - \hat{y}_i)^2\right]^{1/2} = \left[\frac{1}{N-2} \text{SSE}\right]^{1/2} \quad (3.15.13)$$

The standard deviation for the x variable, s_x, is given by

$$s_x = \left[\frac{1}{N-1} \sum_{i=1}^{N} (x_i - \bar{x})^2\right]^{1/2} \quad (3.15.14a)$$

or

$$\sqrt{N-1}\, s_x = \left[\sum_{i=1}^{N} (x_i - \bar{x})^2\right]^{1/2} \quad (3.15.14b)$$

The question is: what do we use for the number of degrees of freedom if the N samples in our series are not statistically independent? The reason we ask this question is that the characteristic amplitudes of the fluctuations s_ε and s_x are calculated using all N values in our data series when we really should be using some sort of *effective* number of degrees of freedom $N^*(< N)$ which takes into account the degree of correlation that exists between data points (as discussed in the previous section).

Suppose we decide to err on the conservative side by agreeing to work with that value of N^* which makes the confidence limits $\pm(s_\varepsilon t_{\alpha/2,\nu})/[(N-1)s_x]^{1/2}$ as small as justifiably possible. This means that when we estimate the confidence limits for a regression slope for a given confidence coefficient, α, we know that we have probably been too cautious and that the confidence limits on the slope probably bracket those that we derive.

We begin by keeping s_ε as it is. If there are high frequency (possibly random) fluctuations superimposed on coherent low-frequency motions, retaining the high-frequency variability adds to the magnitude of s_ε. Had we low-pass filtered the data first and recomputed s_ε based on the true number of data points in our low-pass filtered record, we would expect s_ε to be somewhat smaller. By using s_ε as it is we are assuming that it is a fixed quantity no matter how we subsample or filter the data (s_ε = constant). We do the same with s_x but now replace $N - 1$ with $N^* - 1$, where $N^* < N$. This increases the magnitude of the confidence limits. All that remains is to assume that the number of degrees of freedom for the t-distribution are given by the effective number of degrees of freedom $\nu = N^* - 2$. This statistic has a larger value than for $\nu = N - 2$ so that, again, we are overestimating the magnitude of the confidence interval. This confidence interval is then given by

$$\pm(s_\varepsilon t_{\alpha/2,\nu})/[(N^* - 1)s_x]^{1/2} \qquad (3.15.15a)$$

i.e.

$$b_1 - \frac{(s_\varepsilon t_{\alpha/2,\nu})}{[(N^* - 1)s_x]^{1/2}} < \beta_1 < b_1 + \frac{(s_\varepsilon t_{\alpha/2,\nu})}{[(N^* - 1)s_x]^{1/2}} \qquad (3.15.15b)$$

with $\nu = N^* - 2$.

Our final task is define the effective number of degrees of freedom, N^*, based on a knowledge of the autocovariance function $C(\tau)$ (3.15.10) as a function of lag τ. To do this, we must first find the integral time scale T for the data record

$$T = \frac{1}{C(0)} \sum_{k=0}^{m-1} \frac{\Delta\tau}{2} [C(\tau_k + \Delta\tau) + C(\tau_k)] \text{(discrete case)} \qquad (3.15.16a)$$

$$= \frac{1}{C(0)} \int_0^\infty C(\tau) d\tau \text{ continuous case} \qquad (3.15.16b)$$

where m is the number of lag values to be incorporated in the integral, $\Delta\tau$ is the time increment between data values and $\frac{1}{2}[C(\tau_k + \Delta\tau) + C(\tau_k)]$ is the mean value of C for the midpoint of the lag interval $(\tau_k, \tau_k + \Delta\tau)$. Once the integral time scale is known, the effective degrees of freedom are found by

$$N^* = \frac{N\Delta t}{T} \qquad (3.15.17)$$

where Δt is the sampling increment and $N\Delta t$ is the total length (duration or distance) of the record. If, for example, $N = 120$, $\Delta t = 1$ h, and $T = 10$ h, then $N^* = 12$ ($\ll N$).

To find the autocovariance function, we let $\tau_k = k\Delta\tau$ be the kth lag ($k = 0, 1, \ldots$), then

$$C(\tau_k) = \frac{1}{N-1-k}\sum_{i=1}^{N-k}(y_i - \bar{y_i})(y_{i+k} - \bar{y_{i+k}}); \quad k = 0, \ldots, N_{\max} \quad (3.15.18a)$$

$$C(0) = \frac{1}{N-1}\sum_{i=1}^{N}(y_i - \bar{y_i})^2 = s_y^2 \quad (3.15.18b)$$

where $C(0)$ is the just the variance s_y^2 of the full data series. In both equations (3.15.18a) and (3.15.18b), the data start with the first value for $i = 1$; $N_{\max}$ is the maximum number of reasonable lag values (starting at zero lag and going to $\ll N/2$) that can be calculated before the summation becomes erratic. In theory, we would like $C(\tau) \to 0$ as $\tau \to N$. In reality, however, the data series will contain low-frequency components which will cause the autocovariance function to oscillate about zero or asymptote towards a nonzero value. It should also be obvious that the statistical significance of the summation becomes meaningless at large lag due to the fact that the statistic is based on fewer and fewer values as the lag becomes large. For example, at a lag $k = (N - 3)$ there are only four values that go into the summation and these are derived from neighboring points that are likely highly correlated.

We can picture the integral time scale using equation (3.15.16b). Writing

$$T \cdot C(0) = \int_{\text{all } \tau} C(\tau)d\tau$$

we see that the area under the curve $C(\tau)$ has been equated to the rectangular region $T \cdot C(0)$ (Figure 13.13). In essence, we take a reasonable portion of the curve $C(\tau)$, obtain its area and divide the integral (sum) by its value at zero lag, $C(0)$. An example of the autocovariance function and the integral time scale derived from it are shown in Figure 13.14 for satellite-tracked drifter data in the North Pacific.

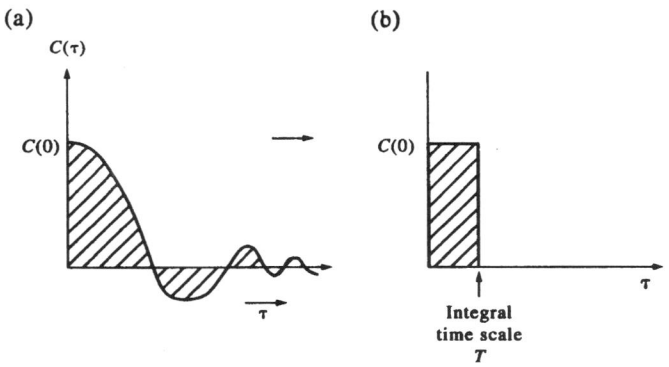

Figure 3.13. Definition of the integral time scale. The area under the curve $C(\tau)$ versus τ in (a) is equated to the rectangular region $TC(0)$ in (b). In practice, only a reasonable portion of the curve $C(\tau)$ is used to obtain the area in (a).

264 *Data Analysis Methods in Physical Oceanography*

3.16 EDITING AND DESPIKING TECHNIQUES: THE NATURE OF ERRORS

A major concern in processing oceanographic data is how to distinguish the true oceanic signal from measurement "errors" or other erroneous values. There are two very different types of measurement errors that can affect data. *Random errors*, usually equated with "noise", have random probability distributions and are generally small compared to the signal. Random errors are associated with inaccuracies in the measurement system or with real variability that is not resolved by the measurement system. The well-accepted statistical techniques for estimating the effects of such random errors are based largely on the statistics of a random population (see previous sections on statistics). Other errors which strongly influence data analysis are *accidental* errors. These errors are not representative of the true population and occur as a result of undetected instrument failures, misreading of scales, incorrect recording of data, and other human failings. In the following discussion, we will handle these two error types in reverse order since the large accidental errors must be removed first before techniques can be applied to treat the "statistical" (random) errors.

One example of a large accidental error would be assigning an incorrect geographic location to an oceanographic measurement which then transfers the observations to a

(a)

POLYCONIC PROJECTION

North Pacific

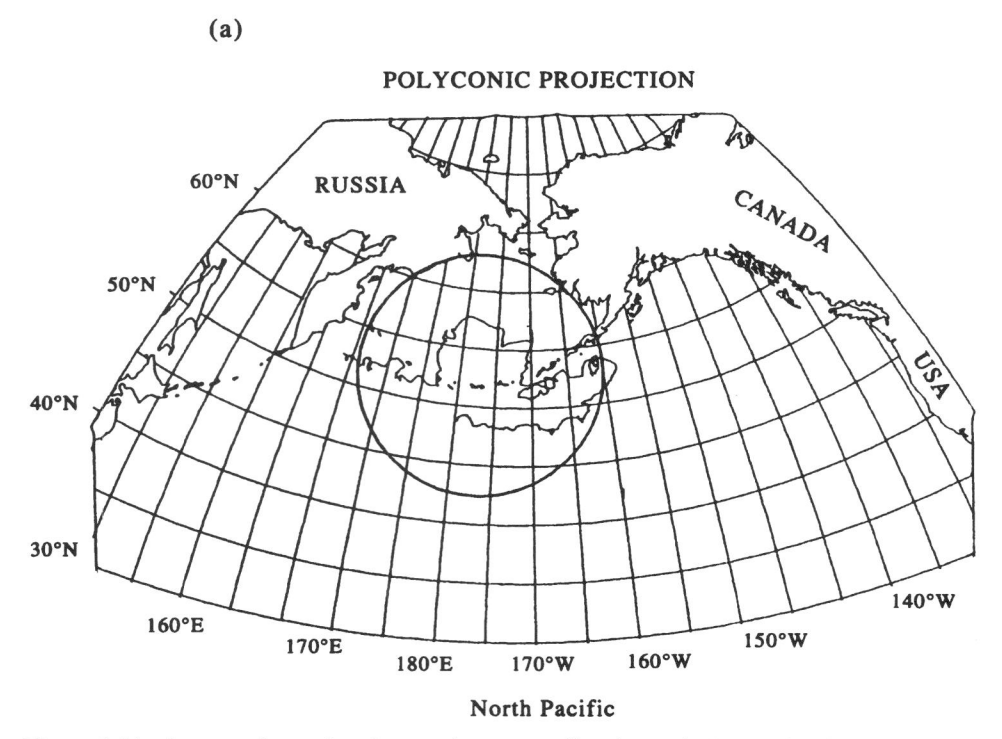

Figure 3.14. Autocovariance functions and corresponding integral time scales for zonal (u) and meridional (v) velocities of satellite-tracked drifter deployed to the south of the Aleutian Islands in the northeast Pacific (see insert) and covering the period 13 November 1991 to 30 July 1993 based on six-hourly sampling interval. (Courtesy of Adrian Dolling.)

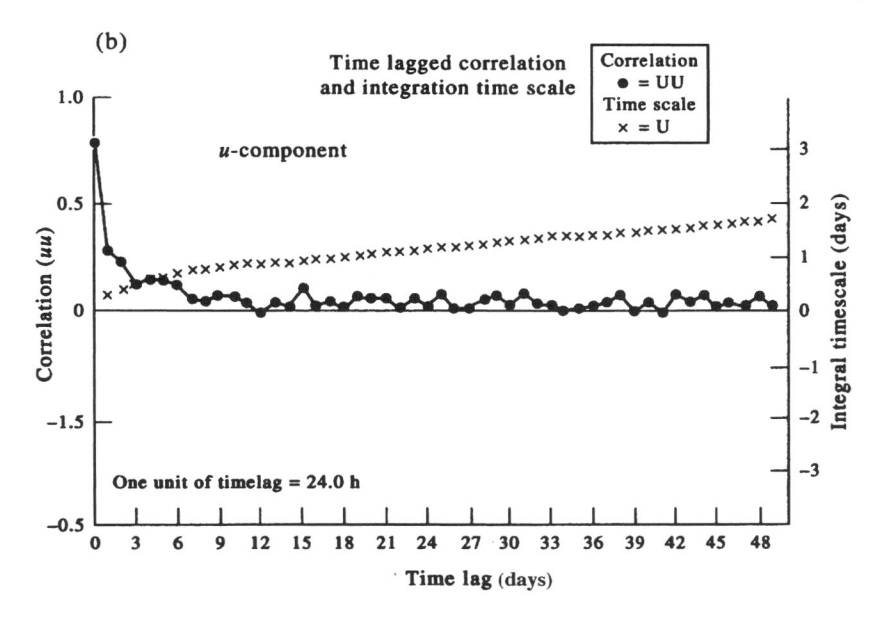

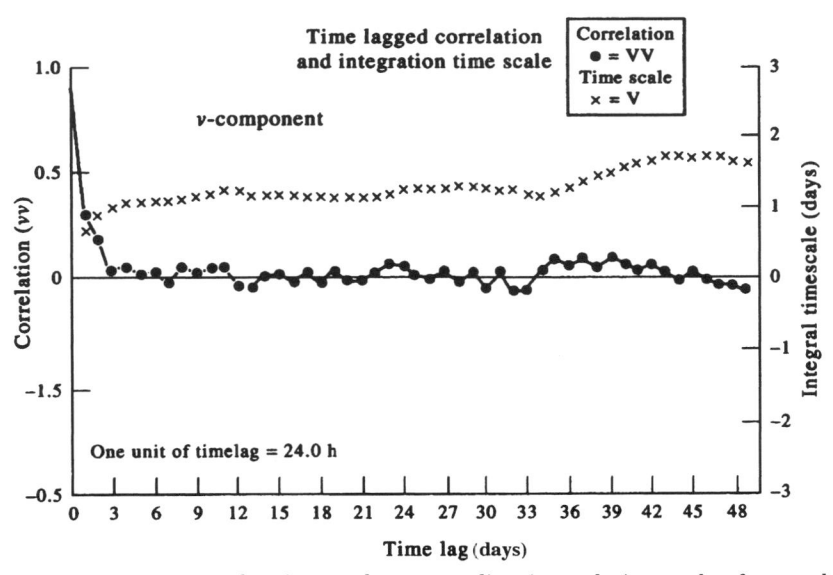

Figure 3.14. Autocovariance functions and corresponding integral time scales for zonal (u) and meridional (v) velocities of satellite-tracked drifter deployed to the south of the Aleutian Islands in the northeast Pacific (see insert) and covering the period 13 November 1991 to 30 July 1993 based on six-hourly sampling interval. (Courtesy of Adrian Dolling.)

region with which they have no direct relationship. Some of these errors, such as oceanographic stations on land, are easily detected, while others are less obvious. Another example of such errors would be biases in a group of measurements due to the application of incorrect sensor calibrations or undetected instrument malfunctions.

Again, the new data would be inconsistent with other existing measurements of the same phenomenon. Our goal is to remove or correct such errors in order to make the data set as self consistent as possible. If we know the history of the data, meaning the details of its collection and reduction, we may be in a better position to understand the sources of these errors. If we have received the data from another source, or are looking at archived data, we may not have available the necessary details on the "petigree" of the data and may have to come to some rather arbitrary decisions regarding its reliability. Considering the widespread use of computer-linked data banks, this is not a trivial problem. The question is how to ensure the necessary quality control yet ensure rapid dissemination and accessibility to data files.

3.16.1 Identifying and removing errors

There are two important axioms to follow when dealing with large erroneous values or "spikes":

(1) To identify the large errors, it is necessary to examine all of the data in visual form and to get a "feel" for the data;
(2) When large errors are encountered, it is usually best to eliminate them all together rather than try to "correct" them and incorporate them back into the data set.

Of course, care must be taken not to reject important data points just because they don't fit either the previous data structure or our preconceived notion of the process. A good example is the determination of heat transport in the South Atlantic. Bennett (1976) suggested that the oceanic heat transport in this ocean is directed toward the equator, contrary to the widely accepted notion that oceanic heat transports are generally poleward. Stommel (personal communication) noted that, in his tabulation of property fluxes for the South Atlantic, Wüst (1957) conspicuously left out the flux of heat while treating other less easily computed transports such as those of nutrients and oxygen. Through an exchange of letters with a former student of Wüst's, Stommel learned that the heat content calculation indeed showed that heat is transported equatorward. Wüst considered this to be the wrong direction and the results were not published along with the other flux values. The point of this story is to illustrate the way in which our prejudice can lead us to reject significant results. In such cases, there is no hard rule as to how this decision is made and a great deal of subjectivity will always be inherent in this level of data interpretion.

The need to examine all the data to detect errors presents a difficult task because of the large numbers of values and the difficulty of looking at unprocessed data. In this case, it is more important to think of ways in which we can present the data so as to ask and answer the questions regarding consistency of the measurements. A compact over-view of all the data is the best solution. This presentation may be as simple as a scatter diagram of the observations versus some independent variable, or a scatter diagram relating two concurrently measured parameters. While scatter diagrams cannot be used to to resolve visually individual points, they do reveal groupings of points which relate to the physical processes expressed by the data. As an example, consider a temperature–salinity scatter diagram (Figure 3.15) computed using a large number of hydrographic data collected from bottle casts. Here, the groups of dots labelled "*a*", "*b*" refer to different water masses present in the 5° square 35–40°N, 15–20°W where the data were collected. The data labelled "*c*" clearly represent a distinct water mass since the points lie along a line divergent from the rest of the scatter values. If we look at other similar *TS* scatter plots, we recognize that this line is consistent with the *TS* relationship from a corresponding square at this same longitude but south of the equator. Thus, it is likely that the latitude recorded was incorrect and that these data are simply misplaced. We correct this by eliminating the points "c" from our square. However, we can't be sufficiently confident of our assumption to add the points to the other square even though the data coverage there is not very good.

Often it is not possible to develop a simple summary presentation of all the data. In the case of current meter data, a time-series presentation is the most appropriate way of looking at the data. As noted by Pillbury *et al.* (1974), error detection using this technique is very time consuming. They note that this procedure can be used successfully for speed, pressure, salinity, and temperature but not for direction, which varies widely. This is due to the fact that direction is limited to the range 0–360° and shows no extreme values. Because of the wrap-around (2π discontinuity) problem, in which $0° = 360°$ (or $-180° = +180°$), direction records tend to be very "spiky", especially in regions of strong tidal flow. A scatter diagram of speed versus direction can be used to detect systematic errors between the speed and direction sensors and to pinpoint those times when the current speed is below the threshold recording level of the instrument. This would be displayed by the direction readings at speeds below threshold and would be easier to identify on the scatter plot than in the individual time series. The only way around the problem with the direction channel is to transform the recorded time series of speed and direction (U, θ) to orthogonal components of velocity (u, v). In particular, separate plots of the east–west (u) and the north–south (v) velocity components (or alongshore and cross-shore components for data collected near the coast) quickly reveal any erroneous values in the data (Figure 3.16).

Pillsbury *et al.* (1974) report that, for Aanderaa RCM4 and RCM5 current meters, there are several sources of large errors. We will discuss these as typical of the errors inherent in moored current meter data since many of these instruments remain in use. One source of error is the current meter's encoder which might encounter a small electrical resistance. The probability of this occurring is considered small. A more likely error is due to nonuniformity in the 1/4-in magnetic tape which may have variations in the coating or carry residual magnetism. The tape transcriber is also a possible error source since it occasionally drops a bit. An error particular to the speed parameter where the speed is seen to abnormally increase, may be caused by nonuniformities in the speed potentiometer winding. A less frequent error type is that associated with clock and trigger malfunctions. Instances have been observed where a meter has cycled several times in rapid succession or conversely missed one or more cycles. These problems are addressed here under the section on timing errors. Direction errors are due to mechanical failures in the compass itself. In some cases the compass needle failed to contact the resistance ring around the compass while in others direction readings in one range all were recorded in a different range. Many of these compass problems were apparent in the raw data while others were only discovered later by looking at the direction histograms.

Other problems with Aanderaa RCM4/5 current meters have been noted over the years. These can be minimized if the following protocol is observed (assuming that the instrument is operational and calibrated):

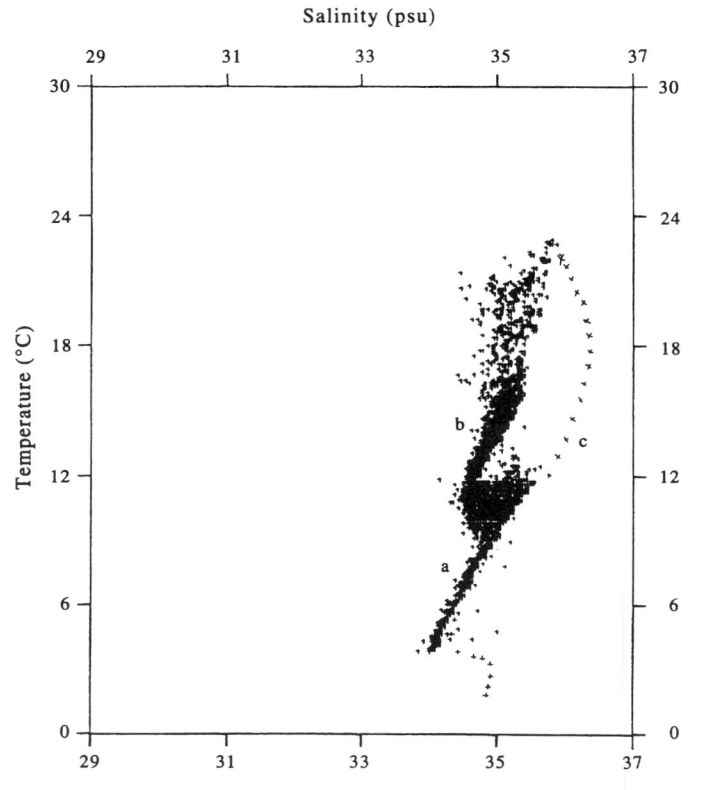

Figure 3.15. TS relationship computed using a large number of hydrographic data collected from bottle casts. Groups labeled "a", "b" refer to different water masses present in the 5° square (35–40°N, 15–20°W) where the data were collected. The data labeled "c" clearly represent a distinct water mass since the points lie along a line divergent from the rest of the scatter values.

(1) Use a new nonmagnetic battery and load test with a 100 ohm resistor to ensure that it meets the manufacturer's specification. Keep in mind that battery amp-hours decrease with decreasing water temperature.
(2) Do not overfill the supply spool with magnetic tape. Leave a 2 mm space so that the tape will not spill off the spool and jam the mechanical mechanism when the instrument is tilted or laid on its side.
(3) Check the tape take-up spool clearance between pinch-rollers spring, circlip, and frame. Spin spool by hand. Check for space between the feed spool and pressure sensor (if installed). Wrap 20 turns of leader on the take-up spool and check the clutch tension.
(4) Check that both spool nuts are in place and do not over-tighten. Do not over-tighten the nylon rotor pivot screw.
(5) Ensure that no ferrous metal screws are used near the compass. Replace these with stainless steel or brass. Also, do not use a ferrous bar to balance the direction vane—it may be close enough to cause the compass to "stick" and ruin the directional data.

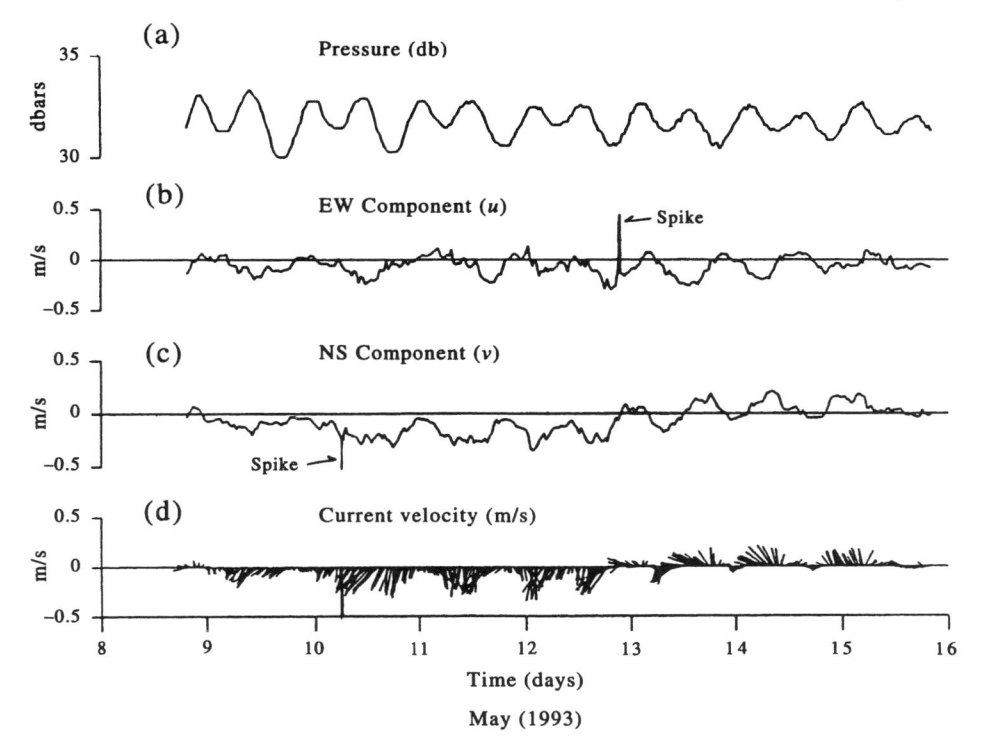

Figure 3.16. A plot of hourly data obtained from an Aanderra RCM4 current meter moored at 30 m depth data in 250 m of water near the entrance to Juan de Fuca Strait (48° 3.3'N, 125° 18.8'W) during the period 8–16 May 1993. (a) Ambient pressure ($\approx$ instrument depth in meters); (b) East–west (u) component of velocity (m/s); (c) North–south (v) component of velocity (m/s); (d) Velocity stick vector (m/s). Erroneous current velocity values ("spikes") stand out in the (u, v) records. Flow consisted of moderate tidal currents superimposed on a surface estuarine outflow that weakened around May 13.

(6) Inspect the O-ring for cuts or nicks and do not trap loose wiring under the ring seat when closing the case. Leakage of small amounts of water to the bottom of the instrument case can cause electrical malfunctions when instrument tilts.

(7) Do not jam a spinning rotor with tissue paper or other material prior to deployment. It is better to shield rotor from wind while on deck. Too often the instrument is recovered with the material still jammed in place.

(8) It is essential to hand-record accurate times for the first and last data records. Make sure the time zone is recorded. Record the time the instrument enters the water on deployment and leaves the water on recovery. More problems can be linked to poor bookkeeping than any other cause.

(9) Spin the rotor in multiples of 24 times (or some multiple of four) to ensure that sampling interval and rotor counter switch (if applicable) are correct.

Another standard method for isolating large errors is to compute a histogram of the sample values. This amounts to completing step 1 in a goodness of fit calculation since a histogram is nothing more than a diagram showing the frequencies of occurrence of sample values. While this is a very straightforward procedure some care must be taken in selecting the parameter intervals, or bins, over which the sample frequencies are calculated. If the bins are too large, the histogram will not resolve the character of the sample PDF and the effects of large error values will be suppressed by being grouped with more commonly occurring values. On the other hand, if the bins are too narrow, individual values take on more influence and the resulting distribution will not appear smooth. This makes it difficult to "see" the real shape of the distribution.

The use of a histogram in locating large errors is that it readily identifies the number of widely differing values that occur and shows whether these divergent values fit into the assumed PDF for the assumed variable. In other words, we can not only see how many values ("outliers") differ widely from the mean values, but also determine if the number of large values in the sample is consistent with the expected distribution of large values for the population. Thus, we have an added guideline for deciding whether the sample values should be retained or eliminated for subsequent analysis. Both PDFs and histograms use visual means of detecting large error values. It is possible to use more automated and objective techniques, such as eliminating all values that exceed a specified standard deviation (e.g. $\pm 3\sigma$). However, these approaches have the weakness that they must first consider all data points, including the extreme values, as valid in order to determine decision levels for selecting or rejecting data. Here, we could use an iterative process in which the values outside the accepted range are omitted from each subsequent recalculation of the mean and standard deviation, until the remaining data have near constant statistics with each new iteration. Large errors, which are usually easy to spot using visual editing techniques, should be removed before proceeding to a more objective step involving the detection of less obvious random deviations. An objective technique for identifying outlier values is to compute a function which selects extremes of the population such as the first derivative of the measured variable with respect to an independent parameter. An example would be a time series of temperature measured from a line in a satellite image. After the extreme gradients are identified in the first derivative calculation, there is still the question of how widely the extremes should be allowed to differ from the rest of the population and whether a value should be considered as an error value or as simply as a maximum (or minimum) of the process being observed.

In making such a decision, it is necessary to have an estimate of the variability of the process. As discussed above, the dispersion of the population distribution is best represented by the variance or the standard deviation. If we are dealing with a normal population, we know that the standard deviation specifies the spread of the distribution and that 66% of the population values lie within $\mu \pm \sigma$ while 95% of these values are in the interval $\mu \pm 2\sigma$. Beyond $\mu \pm 3\sigma$ lie only 0.26% of the total frequency of occurrence, leaving 99.74% within this interval. Thus, it is again a matter of probabilities and significance level; and we must choose at what level we will reject deviations from the mean as errors. If we choose to discard all measurements beyond 2σ, we will have retained 95% of the sample population as our new sample population for which we will repeat out estimation of the statistics. This suggests that we will make our statistical estimate twice; first to decide what data to retain, and second to make statistical inferences about the behavior exhibited by the revised sample data. It is customary to use a much coarser sub-sampling interval, or to use broadly smoothed data, to compute the initial sample standard deviations for the purposes of editing the data. For our *TS* curve example (Figure 3.15), we might initially have used a computational interval of 1 or 2°C to compute a standard deviation for the first-stage editing and then have used the newly defined sample population (original sample minus large deviations > 2°C) to recalculate the mean and standard deviation with a resolution of 0.1°C, closer to the measurement accuracy for reversing thermometers. In statistical analysis we should not expect to exceed the inherent accuracy and resolution of our data. Modern computing facilities, and even pocket calculators, make it tempting to work with many decimal places despite the fact that higher place values are not at all representative of the ability of the instrument to make the oceanic measurement.

A form of two-step editing is used in the routine processing of CTD data which is typically sampled at ≈ 25 samples/s per channel (≈ 25 Hz/channel). Since these instruments produce many more data than we are capable of examining, both smoothing and editing procedures are often built into the routine processing programs. The steps involved with processing calibrated CTD data at the Institute of Ocean Sciences are as follows:

(1) Write the data to a file for display on a computer screen using an interactive editing program written for the particular data set.
(2) Examine all data for a given set of parameters by displaying the data simultaneously on a computer monitor; as a consistency check, it is important to know if large errors in one paramater such as temperature, are associated with some real feature in another parameter, such as salinity.
(3) With the cursor, eliminate erroneous values collected near the ocean surface where the probe rises in and out of the water with the roll of the ship.
(4) Using the file in (3), calculate the pressure gradient versus depth for the data and eliminate those data values for which the depth is decreasing with time for a downcast and increasing with time for an upcast (wave action eliminator).
(5) Using the file of (4), produce a hardcopy plot of the entire profile plus an expanded version for the upper ocean (say 0 to 300 m depth).
(6) On the hardcopy, "flag" erroneous values and irregularities in all data channels.
(7) Use the interactive screen display to eliminate "bad" data identified in (5). If gaps between data points are small, linearly interpolate between adjacent values.
(8) Smooth the edited file by averaging values over a specified depth range. Typically, 1-m averaged files are generated for profile data and 1-s averaged files for time-series data.

Because of improved CTD technology in recent years, step (8) is often conducted first. This step is then eliminated or replaced with a larger averaging interval such as 5 m.

Fofonoff *et al.* (1974) used a 1/2-s average (15 scans) to smooth the measured pressure series. From this smoothed set, a 10th decibar pressure series was generated. Even with the smoothing, the pressure was oversampled, with roughly two observations for each pressure interval. The goal of this computation was to produce a uniform pressure series that could be used to generate profiles of *T* and *S* with depth. Processing routines could be added that first sorted out spurious extreme *T* and *S* values, based on a running mean standard deviation, and which ensured that the pressure series was monotonically increasing. This would correct for small variations in the depth of the probe due to ship motion or strong current shear. Also, in making these editing decisions we should always keep in mind the instrument characteristics and not discard data well within the noise level of the measurement system.

When editing newly collected data, we should always consider what is already known from similar, or related measurements in order to detect obvious errors. A typical example is the use of *TS* curves to evaluate the performance of sample bottles in a hydrocast. Since *TS* curves are known to remain relatively stationary for many areas, previously sampled *TS* curves for an area can be used to locate data points that may have been caused by the erroneous performance of a water sampler; these are generally due to inadvertant bottle "trips" in which the sampler likely closed before or after the desired depth was achieved. Prior *TS* curves also have served as a means of interpolating a particular hydrocast or perhaps providing salinities to match measured temperatures. This approach is limited, however, to those areas and those parameters for which a sufficient number of existing observations are available to define the mean state and variability. In many areas, and for many parameters, information is too limited for existing data to be of any real use in evaluating the quality of new measurements. As a matter of curiosity, it would be interesting to determine the numbers of deep hydrocast data that were unknowingly collected at hydrothermal venting sites and discarded because they were "erroneous". Anomalously high temperatures would be difficult to justify if one did not know about hydrothermal circulation and associated buoyant plumes.

In contrast to large accidental errors, which lead to large offsets or systematic biases, random errors are generally small and normally distributed. These errors often are the result of inaccuracies in the instrumentation or data collection procedures and therefore represent the limit of our ability to measure the desired variable. Added to this is our inability to completely resolve the inherent variability in a particular parameter. This too may be a limitation of our instrument or of our sampling scheme. In either case, when we cannot directly measure a scale of oceanic variability that contributes to the alias of our measurement, the variability will form part of the uncertainty in the final calculated value.

The theory of random errors is well established (Scarborough, 1966). The fundamental approach is to treat the errors as random numbers with a normal PDF. Basic to this assumption is that positive and negative errors of the same size occur in about equal number and tend to cancel each other. This suggests that the appropriate way to treat data containing random errors is in terms of mean-square (MS) and root-mean-square (RMS) values. Another fundamental assumption is that the probability of an error occurring depends inversely on its magnitude; thus, small errors are more frequent than large ones. Following the first of these two assumptions, the PDF of the random errors might be written as

$$p(\varepsilon_x) = f(\varepsilon_x^2) \qquad (3.16.1)$$

where p is the PDF of the errors ε_x. The second characteristic requires that the probability decreases with increasing ε_x so we can write for any real constant, k

$$p(\varepsilon_x) = C \exp(-k^2 \varepsilon_x^2) \qquad (3.16.2)$$

Using the fact that the integral under the curve of any PDF is unity, we solve for C and get

$$p(\varepsilon_x) = \frac{k}{\sqrt{\pi}} \exp(-k^2 \varepsilon_x^2) \qquad (3.16.3)$$

This expression is known as the probability equation or the error equation. A graph of the function gives the normal or Gaussian probability curve. The term k is a constant called the *index of precision* and sets both the amplitude and the width of the normal curve. As k increases, the normal curve becomes narrower and the errors get smaller, making the measurement more precise. (This description applies only for small random errors and not to systematic errors.)

3.16.2 Propagation of error

Suppose we have a quantity, F, which is calculated from a combination of a number (n) of independently observed variables. For example, F might be oceanic heat transport computed from independent velocity and temperature profiles, x. We can estimate the combined random error of F as the sum of squared errors of the individual variables provided that the errors are independent of the variables and that they are all normally distributed. As a simple example, let F be a linear combination of our measurement variables, x

$$F = a_1 x_1 + a_2 x_2 + \ldots + a_N x_N \qquad (3.16.4)$$

where $a_1, \ldots, a_N$ are constants. The inverse of the squared error or *index of precision* (H) of F can be written

$$\frac{1}{H^2} = \frac{a_1^2}{h_1^2} + \frac{a_2^2}{h_2^2} + \ldots + \frac{a_N^2}{h_N^2} = \sum_{i=1}^{N} \frac{a_i^2}{h_i^2} \qquad (3.16.5)$$

where h_i is the error for the ith measurement, x_i.

A more generalized formula for error calculations for arbitary F for which the contibuting variables are uncorrelated is

$$\begin{aligned}\frac{1}{H^2} &= \frac{(\partial F/\partial x_1)^2}{h_1^2} + \frac{(\partial F/\partial x_2)^2}{h_2^2} + \ldots + \frac{(\partial F/\partial x_N)^2}{h_N^2} \\ &= \sum_{i=1}^{N} \frac{(\partial F/\partial x_i)^2}{h_i^2}\end{aligned} \qquad (3.16.6)$$

where partial derivatives $\partial F/\partial x_i$ are obtained from Taylor expansions of the function F in terms of the independent variables x_i. To convert this expression to one in terms of relative errors, we use the fact that

$$\frac{1}{h^2} = \frac{r_e^2}{\rho^2} \qquad (3.16.7)$$

where r_e is the corresponding relative error and $\rho = 0.4769$ is a constant obtained from the error equation (3.16.3). Using this definition we can write our final error as

$$R_e = \left[(\partial F/\partial x_1)^2 r_1^2 + (\partial F/\partial x_2)^2 r_2^2 + \ldots + (\partial F/\partial x_N)^2 r_N^2\right]^{1/2} \qquad (3.16.8)$$

In this form, R_e is really only the RMS error that describes the equivalent combined error in the equation of interest. This Taylor expansion of the contributing error terms is known as the *propagation of errors formula*. It is limited to small errors and uncorrelated independent variables. Since these principles apply only to small random errors, it is necessary to use some data editing procedure to remove any large errors or biases in the measurements before using this formula. By using a mean-square formulation, we take advantage of the fact that small random errors can be expected to often cancel each other resulting in a far smaller mean-square error than would result if the measurement errors were simply added regardless of sign to yield a maximum "worst case error". The primary application of equation (3.16.8) is in determining the overall error in a quantity derived from a number of component variables all with measurement errors. This is a situation common to many oceanographic problems.

A more complicated propagation of error formula is needed if there is a nonzero correlation between the independent variables, x. In this case, we must also retain the covariance terms in any Taylor expansion of the small error terms. For example, the density ρ is a function of both temperature T and salinity S so that the errors (variances) in density σ_ρ^2 can be related to the measurement errors in temperature σ_T^2 and salinity σ_S^2 by

$$\sigma_\rho^2 = (\partial\rho/\partial T)^2 \sigma_T^2 + (\partial\rho/\partial S)^2 \sigma_S^2 + 2[(\partial\rho/\partial T) \cdot (\partial\rho/\partial S)]C(T,S) \qquad (3.16.9)$$

where $C(T, S)$ is the covariance between temperature and salinity fluctuations. Only when $C(T, S) = 0$ do we get the result (3.16.8). An example of a detailed error calculation is the measurement of flow through trawl nets towed at various angles through the water column is given in Burd and Thomson (1993).

3.16.3 Dealing with numbers: the statistics of roundoff

Since we must represent all measurements in discrete digital form, we are forced to deal with the consequences of numerical roundoff, or truncation. The problem results from the limitations of digital computing machines. For example, the irrational fraction 1/3 is represented in the computer as the decimal equivalent 0.3333 ... 3 with an obvious roundoff effect. This may not seem to be a problem for most applications since most computers carry a minimum of eight decimal places at single precision. The large number of arithmetic operations carried out in a problem lasting for only a few seconds of computer processing time can, however, lead to large errors in due to roundoff and truncation errors. The case of greatest concern is when two nearly identical numbers are subtracted, requiring proper representation to the smallest possible digit. Such differences can easily occur unknowingly in a complicated computational problem. Rather than discuss procedures for estimating this roundoff error, we will discuss the nature of the problem and emphasize the need to avoid roundoff.

General floating-point values (decimal numbers) in a computer follow closely the so-called "scientific notation" and are represented as a mantissa (to the right of the decimal point) and an exponent (the associated power of 10). For example, in a three-digit system, the number 64.282 would be represented as 0.643×10^2 where the roundoff is accomplished by adding five in the thousands' decimal place and then truncating after the third digit. This process of rounding off results in a slight bias because it always rounds up when there is a 5 in the least significant digit. A way to overcome this bias is to use the last digit retained to determine whether to round up or down when the next digit is exactly 5. This rule, which leads to the least possible error, is to round-up if the next to the last digit retained is odd and to round down when it is even. This procedure can be summarized as follows. When rounding a number to k decimals:

(1) if the $k + 1$ decimal is 0, 2, 4, 6, 8 then the k decimal is unchanged;
(2) if the $k + 1$ decimal is 1, 3, 5, 7, 9 then the k decimal is increased by 1.

This system of rounding-off will result in errors that are generally less than 0.5×10^{-k} and maximum roundoff errors of 0.6×10^{-k}. In most applications, the effect of this roundoff bias is too small to justify the added numerical manipulation required to implement this even–odd roundoff scheme.

In computing systems, floating-point numbers are handled in a binary representation having 24 bits (wordlength is 32 bits but eight bits are used for the exponent) which results in seven significant decimal digits. Called *single precision*, this level of accuracy is adequate for many computations. For those problems with repeated calculations, and the subsequent high probability of differencing two nearly identical numbers, a *double-precision* representation is used which has 56 binary bits leading to 16 significant decimal digits. Roundoff, in the case of double precision, results in very small biases which can be completely ignored for most applications. Another approach to the problem of roundoff errors is to consider them to be random variables. In this way, statistical methods can be applied to better understand the effects of roundoff errors. Consider the roundoff of a single number x; for this number, all numbers occurring in the interval $x_o - 1/2 < x < x_o + 1/2$ (measured in units of the last digit) become that number. Thus, the roundoff has a uniform probability distribution in the last digit. We can write the corresponding probability density function $f(x)$ for x as

$$f(x) = \begin{cases} 1 & (x_o - 1/2, x_o + \tfrac{1}{2}) \\ 0, & \text{elsewhere} \end{cases} \quad (3.16.10)$$

and note that

$$\int_{-\infty}^{\infty} f(x)\,dx = 1 \quad (3.16.11)$$

The most common measures of a PDF are its first two moments, the mean and variance. The mean of $f(x)$ in equation (3.16.10) is x_o and the variance is

$$V[f(x)] = \sigma^2 = \int_{x_o - 1/2}^{x_o + 1/2} [x - x_o] f(x)\,dx = \int_{-1/2}^{+1/2} x'^2\,dx' = \frac{1}{12} \quad (3.16.12)$$

Experimental tests have verified the uniform distribution of roundoff in computer systems. In fact, computers generate random numbers by using the overflow value of the mantissa.

We can represent roundoff as an additive random error (ε) superimposed on the true variable (x). In this case, we can write the computer representation of our variable (which we assume is free from measurement and sampling errors) as $x + \varepsilon$. For a floating-point number system, it is better to use

$$x(1 + \varepsilon); |\varepsilon| < \tfrac{1}{2}(10^{-2}) \quad (3.16.13)$$

for the variable with roundoff error ε. This formulation has the effect of focusing attention on the consequences of roundoff for every application in which it appears. For example, the product

$$x_1(1 + \varepsilon_1)x_2(1 + \varepsilon_2) = x_1 x_2(1 + \varepsilon_1 + \varepsilon_2 + \varepsilon_1\varepsilon_2) \quad (3.16.14)$$

demonstrates how roundoff propagates during multiplication. Generally, the product

$\varepsilon_1\varepsilon_2$ is sufficiently small to be ignored. However, in the above multiplication we must include the roundoff for this operation, whereby (3.16.14) becomes

$$x_3(1+\varepsilon_3) = x_1 x_2(1+\varepsilon_1+\varepsilon_2+\varepsilon)$$
$$|\varepsilon| < \tfrac{1}{2}(10^{-2}); \quad \varepsilon_3 = \varepsilon_1+\varepsilon_2+\varepsilon \quad (3.16.15)$$

Similar error propagation results are found for other arithmetical operations.

We can extend this to a generalized product

$$y_1(1+\varepsilon_1)y_2(1+\varepsilon_2)\ldots y_N(1+\varepsilon_N) \quad (3.16.16)$$

which becomes

$$y_1 y_2 \ldots y_N[1+(\varepsilon_1+\varepsilon_2+\ldots+\varepsilon_N)] \quad (3.16.17)$$

By the central limit theorem, the sum of n independent random numbers (the roundoff errors) approaches a normal distribution. The effect for the other operations is much the same; therefore, while individual roundoffs are from a uniform distribution, the result of many arithmetic roundoff operations tends toward a normal distribution. This also can be demonstrated experimentally.

As stated earlier, we will generally ignore roundoff as a source of error in the processing and analysis of oceanographic data. The above discussion has been presented here to make the reader aware of potential problems and provide some familiarity with the problems of using computing systems. In most data applications, the effects of roundoff error are small enough to be ignored. Only in the case of recursive calculations, where each computation depends on the previous one, do we anticipate large roundoff errors. This is usually a problem for numerical modelers who must deal with the repeated manipulation of computer-generated "data". In cases where roundoff errors are of some consequence, statistical methods can be used in which the errors can be treated as variables from a normal population.

3.16.4 Gauss–Markov theorem

The term *Gauss–Markov process* is often used to model certain kinds of random variability in oceanography. To understand the assumptions behind this process, consider the standard linear regression model, $y = \alpha + \beta x + \varepsilon$, developed in the previous sections. As before, α, β are regression coefficients, x is a deterministic variable and ε a random variable. According to the Gauss–Markov theorem, the estimators α, β found from least squares analysis are the *best linear unbiased estimators* for the model for the following conditions on ε:

(1) The random variable ε is independent of the independent variable, x;
(2) ε has a mean of zero; that is $E[\varepsilon] = 0$;
(3) Errors ε_j and ε_k associated with any two points in the population are independent of one another; the covariance between any two errors is zero; $C[\varepsilon_j, \varepsilon_k] = 0, j \neq k$;
(4) ε has a finite variance $\sigma_\varepsilon^2 \neq 0$.

The estimators are *unbiased* since their expected value equals the population values (given 1 and 2) and they are *best* in that they are efficient (if 3 and 4 hold true), the variance of the least-squares estimators being smaller than any other linear unbiased estimator. A further assumption that is often made is that the errors, ε, are normally distributed. In this case, the estimators of α, β, and μ using the least-squares requirements are identical to the estimators resulting from the use of maximum-likelihood estimation. This assumption, combined with the four previous assumptions, provide the rationale for the least-square procedure.

3.17 INTERPOLATION: FILLING THE DATA GAPS

Most analysis procedures used in the physical sciences are designed for comparatively long and densely sampled series with equally spaced measurements in time or space. The wealth of information on time-series analysis primarily applies to regularly spaced and abundant observations. There are two main reasons for this: (1) the mathematical necessity for long, equally-spaced data for the derivation of statistically reliable estimates from modern analytical techniques; and (2) the fact that most modern measurement systems both collect and store data in digital format. Spectral estimates, for example, improve with increased duration of the data series in the sense that one is able to cover an increasing range of the dominant frequency constitutents that make up the record. Digital sampling systems are considerably more economical than analog recording systems in that they cut down on storage space, power consumption and postprocessing effort.

3.17.1 Equally and unequally spaced data

Electronic systems now provide data at regularly spaced sampling increments. Unfortunately, such systems usually operate autonomously and any type of equipment failure generally leads to either data *gaps* or a premature termination of the record. The failure of electronic data logging systems is but one source of gappy records in physical oceanography. Because of their very nature, shipborne measurements are a source of gappy records. Oceanographic research vessels are expensive platforms to operate and must be used in an optimal fashion. As a consequence, it is often impossible to collect observations in time or space of sufficient regularity and spacing to resolve the phenomenon of interest. Efforts are usually made to space measurements as evenly as possible but, for a variety of reasons, station spacings are often considerably greater than desired. Weather conditions, as well as ship and equipment problems, almost invariably lead to unwanted gaps in the data set. Sometimes equipment failures are not detected until the data are examined in the laboratory. In addition, editing out errors produces unwanted gaps in the data record.

The gap problem is even more severe when one is analyzing historical data or data collected from "platforms of opportunity." Historical data are a collection of many different sampling programs all of which had different goals and therefore very different sampling requirements. By its very nature, such collections of data will necessarily be irregularly spaced and variable in terms of accuracy and reliability. Further editing, dictated by the goals of the historical data analysis project, will add new gaps to the set of existing data series.

Monitoring stations, ships of opportunity, and satellite measurements frequently produce data series that are unevenly spaced. The geographic distribution of monitoring stations (e.g. Pacific island sea-level stations) is far from uniform in terms of the spacing between stations. Thus, while the data series collected at each station, may themselves consist of evenly and densely spaced measurements in time, the space intervals between stations will be highly irregular. Open ocean buoys and current meter moorings also fit this classification of densely and evenly spaced temporal observations at widely and often irregularly spaced locations. Here again, any failure in the recording system, whether minor or catastrophic, will lead to gaps in the time-series record. Often these gaps are quite large since unplanned recovery efforts are required to correct the problem. Such a correction effort assumes the telemetering of data which is at present not widely done. Failures of on-board recording systems must wait until the scheduled servicing of the instrument which may then result in relatively large data gaps.

At the other end of the sampling spectrum satellite observing systems provide dense and evenly spaced measurements that are often very irregular in time. A familiar source of temporal gaps, in infrared image series, is cloud cover. Both occasional and persistent cloud cover can interrupt a sequence of images collected to study changes of sea surface temperature. The effects of cloud cover apply also to satellite remote sensing in the optical bands. In addition to the cloud-cover problem, there are often problems with the on-board satellite sensing systems or associated with the ground receiving station that lead to gaps in time series of image data. Microwave sensing of the surface is not as sensitive to cloud attenuation but it is subject to sensor and ground-recording failure problems.

Platforms of opportunity (usually merchant ships) produce uniquely irregular sets of measurements. Most merchant ships repeat the same course with minor adjustments for local weather conditions and season. A seasonal shift in course is generally seen at higher latitudes to take advantage of great circle routes during times of better weather. A return to lower latitudes is seen in winter data as the ships avoid problems with strong storms. Added to the seasonal track changes is the nature of the daily sampling procedure. Usually the ship takes measurements at some specified time interval which, due to variations in ship track, ship speed and weather conditions, may be at very different positions from sailing to sailing. Thus, the merchant ship data will be irregular in both space and time. Systems that operate continuously from ships of opportunity (e.g. injection SST) overcome this problem. These continuous measurements, however, are still subject to variations in ship track.

The net result of all these measurement problems is that oceanographers are often faced with short records of unequally spaced data. Even if the records are long they are often gappy in time or space. It is, therefore, necessary to interpolate these data to produce series of evenly spaced measurements. While some analysis procedures, such as least-squares harmonic analysis, apply directly to uneven or gappy data, it is more often the case that irregularly spaced data are interpolated to yield evenly spaced, regular data. These interpolated records can then be analyzed with familiar methods of time-series analysis.

Interpolation also may be required with evenly spaced data if the subject dynamics apply to smaller space/time scales than are resolved by the measurements. Thus, the data points that are interpolated produce another set of regularly spaced points with a finer resolution. Many interpolation procedures have been developed that only apply to evenly spaced data.

3.17.2 Interpolation methods

Interpolation techniques are needed for both irregularly spaced and evenly spaced data series. Before deciding which interpolation method is most effective, we need to consider the particular application. A series of appropriate questions regarding the selection of the best interpolation procedures are:

(1) What samples (original data series, derivatives, etc.) should we use?
(2) What class of interpolation function (linear, higher-order polynomial, cubic-spline, etc.) best satisifies the dynamical restrictions of the analysis?
(3) What mathematical criteria (exact data-point matching, least-squares fit, continuity of slopes, etc.) do we use to derive the interpolated values?
(4) Where do we apply these criteria?

Answers to these questions serve as guides to the selection of a unique interpolation procedure.

3.17.2.1 Linear interpolation

The type of interpolation scheme to be employed depends on how many data points we want our interpolation curve (polynomial) to pass through. Increasing the number of points we want our curve to fit, increases the order of the polynomial we need to do the fitting. The most straightforward and widely used interpolation procedure is that of *linear interpolation*. This consists of fitting a straight line between two data points and choosing interpolated values at the appropriate positions along that line. For a data series $y(x)$, this linear procedure can be written as

$$y(x) = y(a) + \frac{x-a}{b-a}[y(b) - y(a)]$$
$$= \frac{(b-x)y(a) + (x-a)y(b)}{b-a} \qquad (3.17.1)$$

where $x_{\text{start}} = a$ and $x_{\text{end}} = b$ are the times (positions) of the data collection at the start and end of the sampling increment being interploated, and x represents the corresponding time (position) of the desired interpolated value within the interval $[a, b]$. This is the customary procedure for interpolating between values in most tables. The same formula can be applied to *extrapolation* (extending the data beyond the domain of the observations) where the point x lies beyond the interval $[a, b]$. Equation (3.17.1) is a special case of the Lagrange polynomial interpolation formula discussed in the next section.

3.17.2.2 Polynomial interpolation

If we wish to interpolate between more than two points simultaneously, we need to use higher-order polynomials than the first-order polynomal (straight line) used in the previous section. For example, through three points we can find an unique polynomial of degree 2 (a quadratic); through four points, an unique polynomial of degree 3 (a cubic), and so on. The two methods described below are computationally robust in the sense that they yield reasonable results at most points. Polynomial interpolation techniques such as Vandermonde's method and Newton's method are awkward to program and suffer from problems with roundoff error.

3.17.2.2.1 Lagrange's method

The Lagrange polynomial interpolation formula is a method for finding an interpolating polynomial $y(x)$ of degree N which passes through all of the available data points $(x_i, y_i); i = 1, 2, \ldots, N + 1$. The general form for this polynomial, of which linear interploation is a special case, is given as

$$y(x) = a_o + a_1 x + a_2 x^2 + \ldots + a_N x^N = \sum_{k=0}^{N} a_k x^k$$

$$= \sum_{i=1}^{N+1} \left[y_i \left(\prod_{\substack{k=1 \\ k \neq i}}^{N+1} \frac{x - x_k}{x_i - x_k} \right) \right] \quad (3.17.2)$$

where Π is the product function. Note that in the product function, the ith term—corresponding to the particular data point, x_i, in the denominator—is not included when calculating the product for the term involving x_i. Even though k ranges from 1 to $N + 1$, Π uses only N terms and the final polynomial is of order N, as required.

The goal of the Lagrange interpolation method is to find an Nth degree polynomial which is constrained to pass through the original $N + 1$ data points and which yields a "reasonable" interpolated value for any position x located anywhere between the original data points. To see that the polynomial passes through the original data points, note that the ith product function, Π_i, defined for the data point x_i in the denominator is constructed in such a way that $\Pi_i(x_j; x_i) = \delta_{ij}$ whenever $x = x_j$ is one of the data values (δ_{ij} is the Kronecker delta function). This means that $\Pi_i(x_j; x_i) = 0$ for all x_j except for the specific value $x = x_i$ found in the original data series which matches the term in the denominator. In that case, $\Pi_i(x_j; x_i) = 1$ and $y_i \Pi_i(x_j; x_i) = y_j$

The general polynomial we seek is constructed as a sum of the product functions in equation (3.17.2) which can be expanded to give

$$y(x) = \sum_{i=1}^{N+1} y_i [Q_i(x)/Q_i(x_i)] \quad (3.17.3)$$

in which

$$Q_i(x) = (x - x_1)(x - x_2) \ldots (x - x_{i-1})(x - x_{i+1}) \ldots (x - x_{N+1}) \quad (3.17.4)$$

is the product of all the factors except the ith one. For any x, (3.17.3) can be expanded to give the interpolating polynomial

$$y(x) = y_1 \frac{(x - x_2)(x - x_3) \ldots (x - x_{N+1})}{(x_1 - x_2)(x_1 - x_3) \ldots (x_1 - x_{N+1})} + y_2 \frac{(x - x_1)(x - x_3) \ldots (x - x_{N+1})}{(x_2 - x_1)(x_2 - x_3) \ldots (x_2 - x_{N+1})}$$
$$+ \ldots + y_N \frac{(x - x_1)(x - x_2) \ldots (x - x_N)}{(x_{N+1} - x_1)(x_{N+1} - x_3) \ldots (x_{N+1} - x_N)}$$

$$(3.17.5)$$

Note that, for the original data points, $x = x_i$, the polynomial yields the correct output value $y(x_i) = y_i$, as required.

In the Lagrange interpolation method, the calculation is based on all the known data values. If the user wants to add new data to the series, the whole calculation must be repeated from the start. Although the above formula can be applied directly, programing improvements exist that should be taken into account (Press et al., 1992). Use of Neville's algorithm for contructing the interpolating polynomial is more efficient and allows for an estimate of the errors resulting from the curve fit.

As an example of this interpolation method, consider four points (x_i, y_i), $i = 1, \ldots, 4$ given as (0, 2), (1, 2), (2, 0) and (3, 0) through which we wish to fit a (cubic) polynomial. Substituting these values into equation (3.17.5), we obtain

$$y(x) = 2\frac{(x-1)(x-2)(x-3)}{(0-1)(0-2)(0-3)} + 2\frac{(x-0)(x-2)(x-3)}{(1-0)(1-2)(1-3)} + 0 + 0$$

$$= \frac{2}{3}x^3 - 3x^2 + \frac{7}{3}x + 2$$

The resulting third-order curve is plotted in Figure 3.17.

3.17.2.3 Spline interpolation

In recent years, the method that has received the widest general acceptance is the spline interpolation method. Splines, unlike other polynomial interpolations such as the Lagrange polynomial interpolation formula, apply to a series of segments of the data record rather than the entire data series. This leads to the obvious question to ask in selecting the proper interpolation procedure: Do we want a single, high-order polynomial for the interpolation over the entire domain, or would it be better to use a sequence of lower-order polynomials for short segments and sum them over the domain of interest? This integration is inherently a smoothing operation but one must be careful of discontinuities, or sharp corners, where the segments join together. Spline functions are designed to overcome such discontinuities, at least for the lower-order derivatives. It is because discontinuities are allowed in higher-order derivatives that splines are so effective locally. Constraints placed on the interpolated series in one region have only very small effects on regions far removed. As a result, splines are more effective at fitting nonanalytic distributions characteristic of real data. The term "spline" derives from the flexible drafting tool used by naval architects to draw piecewise continuous curves.

Splines have other favorable properties such as good convergence, highly accurate derivative approximation, and good stability in the presence of roundoff errors. Splines represent a middle ground between a purely analytical description and numerical finite difference methods which break the domain into the smallest possible intervals. The piecewise approximation philosophy represented by splines has given rise to finite element numerical methods.

With spline interpolation, we approximate the interpolation function $y(x)$ over the interval $[a, b]$ by dividing the interval into subregions with the requirement that there be continuity of the function at the joints. We can define a spline function, $y(x)$, of degree N with values at the joints

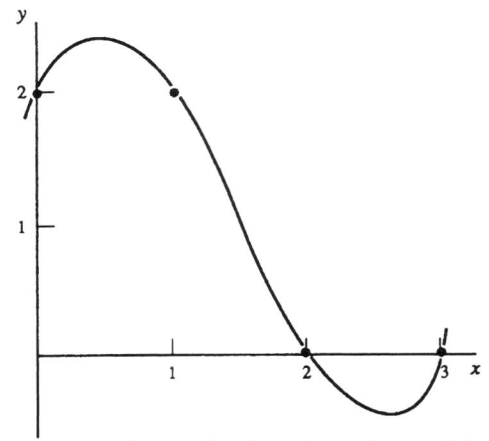

Figure 3.17. Use of Lagrange's method to fit a third-order (cubic) polynomial through the data points (x_i, y_i) given by (0, 2), (1, 2), (2, 0), and (3, 0).

$$a = u_0 \leq u_1 \leq u_2 \ldots \leq u_N = b \tag{3.17.6}$$

and having the properties:

(1) In each interval $u_{i-1} \leq x \leq u_i$ ($i = 1, m$), the function $y(x)$ is a polynomial of degree not greater than N.
(2) At each interior joint, $y(x)$ and its first $N - 1$ derivatives are continuous.

The spline function in widest use is the cubic spline ($N = 3$). To give the reader familiarity with the spline interpolation technique, we will develop the cubic spline equations and work through a simple example. Consider a data series with elements (x_i, y_i), $i = 1, \ldots, N$. Since we are working with a cubic spline interpolation, the first two derivatives $y'(x)$ and $y''(x)$ of the interpolation function, $y(x)$, can be defined for each of the points x_i while the third derivative $y'''(x)$ will be a constant for all x. Here, the prime symbol denotes differentiation with respect to the independent variable x. We write the spline function in the form

$$y(x) = f_i(x); \; x_i \leq x \leq x_{i+1}, \; i = 1, \ldots, N - 1 \tag{3.17.7}$$

and specify the following conditions at the junctions of the segments:

(1) Continuity of the spline function:

$$\begin{aligned} f_i(x_i) &= y(x_i) = y_i, \; i = 1, 2, \ldots, N - 1; \\ f_{i-1}(x_i) &= y(x_i) = y_i, \; i = 2, 3, \ldots, N; \end{aligned} \tag{3.17.8a}$$

(2) continuity of the slope:

$$f'_{i-1}(x_i) = f'_i(x_i), \; i = 1, 2, \ldots, N - 1; \tag{3.17.8b}$$

(3) continuity of second derivative:

$$f''_{i-1}(x_i) = f''_i(x_i), \; i = 1, 2, \ldots, N - 1; \tag{3.17.8c}$$

Statistical Methods and Error Handling 283

Since $y'''(x) = $ constant, $y''(x)$ must be linear, so that

$$f_i''(x_i) = y_i'' \frac{(x_{i+1} - x)}{x_{i+1} - x_i}$$
$$= y_{i+1}'' \frac{(x - x_i)}{x_{i+1} - x_i} \qquad (3.17.9)$$

Integrating twice and selecting integration constants to satisfy the conditions (3.17.8a, b) on $f_i(x_i)$ and $f_{i-1}(x_i)$ gives

$$f_i(x) = y_i \frac{(x_{i+1} - x)}{(x_{i+1} - x_i)} + y_{i+1} \frac{(x - x_i)}{(x_{i+1} - x_i)}$$
$$- \frac{(x_{i+1} - x_i)^2}{6} y_i'' \left\{ \frac{(x_{i+1} - x)}{(x_{i+1} - x_i)} - \left[\frac{(x_{i+1} - x)}{(x_{i+1} - x_i)}\right]^3 \right\} \qquad (3.17.10)$$
$$- \frac{(x_{i+1} - x_i)^2}{6} y_{i+1}'' \left\{ \frac{(x - x_i)}{(x_{i+1} - x_i)} - \left[\frac{(x - x_i)}{(x_{i+1} - x_i)}\right]^3 \right\}$$

which uniquely satisfies the continuity condition for the second derivative but not, in general, for the first derivative (slope). To ensure continuity of the slope at the seams, we expand (3.17.9) by differentiation to get

$$f_i'(x_i) = \frac{(y_{i+1} - y_i)}{x_{i+1} - x_i} - \frac{(x_{i+1} - x_i)}{6}(2y_i'' + y_{i+1}'') \qquad (3.17.11a)$$

$$f_{i-1}'(x_i) = \frac{(y_i - y_{i-1})}{x_{i+1} - x_i} - \frac{(x_{i+1} - x_i)}{6}(y_{i-1}'' + 2y_i'') \qquad (3.17.11b)$$

We then set (3.17.11a) and (3.17.11b) equal in order to satisfy slope continuity (3.17.8b), whereby

$$(x_i - x_{i-1})y_{i-1}'' + 2[(x_{i+1} - x_{i-1})]y_i'' + (x_{i+1} - x_i)y_{i+1}''$$
$$= 6\frac{(y_{i+1} - y_i)}{x_{i+1} - x_i} - \frac{(y_i - y_{i-1})}{x_i - x_{i-1}}, \quad i = 2, ..., N - 1 \qquad (3.17.12)$$

which must be satisfied at $N - 2$ points by the N unknown quantities, y_i''. We require two more conditions on the y_i'' which we get by specifying conditions at the end points x_1 and x_N of the data sequence. After specifying these end values, we have $N - 2$ unknowns which we find by solving the $N - 2$ equations. There are two main ways of specifying the end points: (1) we set one or both of the second derivates, y_1'' and y_N'' at the end points to be zero (this is termed the *natural cubic spline*) so that the interpolating function has zero curvature at one or both boundaries; or (2) we set either y_1'' and y_N'' to values derived from equation (3.17.11) in order that the first derivatives of the interpolating function, y_i', take on specified values at one or both of the termination boundaries.

As a general example, we consider the spline solution for six evenly spaced points with the data interval $h = x_{i+1} - x_i$ and function d_i defined in terms of y_i as

$$d_i = \frac{(y_{i+1} - 2y_i + y_{i-1})}{2h^2} \qquad (3.17.13)$$

284 *Data Analysis Methods in Physical Oceanography*

We can write the equations (3.17.10) for these six equally spaced points in matrix form as

$$\begin{pmatrix} 4 & 1 & 1 & 0 \\ 1 & 4 & 1 & 0 \\ 0 & 1 & 4 & 1 \\ 0 & 0 & 1 & 4 \end{pmatrix} \begin{pmatrix} y_2'' \\ y_3'' \\ y_4'' \\ y_5'' \end{pmatrix} = \begin{pmatrix} 12d_2 - y_1''/h \\ 12d_3 \\ 12d_4 \\ 12d_5 - y_6''/h \end{pmatrix} \quad (3.17.14)$$

If we want to specify y_i' rather than y_i'', we need an equation relating both. If the end conditions are not known, the simplest choice is $y_1'' = 0$ (the *natural spline* noted above). Another, and smoother choice (in the sense of less inflection or curvature at the interpolated point) is $y_1'' = 0.05y_2''$. Although spline interpolation is a global, rather than a local, curve (altering a y_i'' or an end condition affects the overall spline), the dominant diagonal terms in equation (3.17.14) cause the effects to rapidly decrease as the distance from the altered point increases.

We should point out the method of splines offers no advantage over polynomial interpolation when applied to either the approximation of well-behaved mathematical functions or to curve fitting when the experimental data are dense. "Dense" means that the number of data points in a subregion is more than an order of magnitude larger than the number of inflection points in the fitted curve and that there are no abrupt changes in the second derivative. The advantage of splines is their inherent smoothness when dealing with sparse data.

As a numerical example of spline fitting, we consider the six-point fitting of the points represented in equation (3.17.14) for the 11 data points in Table 3.17.1. Using a general polynomial fit yields the curve in Figure 3.18. Here, all but one of the first six points lie on a straight line. Due to this single point, the polynomial curve oscillates with an amplitude that does not decrease. In contrast, the spline amplitude (Figure 3.19) for the same 11 values reduces each cycle by a factor of 3.

Often the first or second derivatives of the interpolated function are important. In Figure 3.18, we see that fitting a polynomial to sparse data can result in large, unrealistic changes in the second derivatives. The spline fit to the same points (Figure 3.19) using the end-point conditions $y_1'' = y_N'' = 0$ demonstrates the smoothness of the spline interpolation. In essence, the spline method sacrifices higher-order continuity to achieve second derivative smoothness.

Spline interpolation is generally accomplished by computer routines that operate on the dataset in question. Computer routines solve for the spline functions by solving the equation

$$\sum_{i=1}^{N} [(g(x_i) - y_i)/\delta y_i]^2 = S \quad (3.17.15)$$

where $g(x_i)$ is composed of cubic parabolas

$$g(x) = a_i + b_i(x - x_i) + c_i(x - x_i)^2 + d_i(x - x_i)^3 \quad (3.17.16)$$

for the interval $x_i \leq x \leq x_i + 1$. The terms δy_i are positive numbers that control the amount of smoothing at each point; the larger δy_i is the more closely the spline fits at each data point. A good choice of δy_i is the standard deviation of the data values.

The S term also controls smoothing, resulting in more smoothing when S increases. As S gets smaller, smoothing decreases and the splines fit the data points more closely.

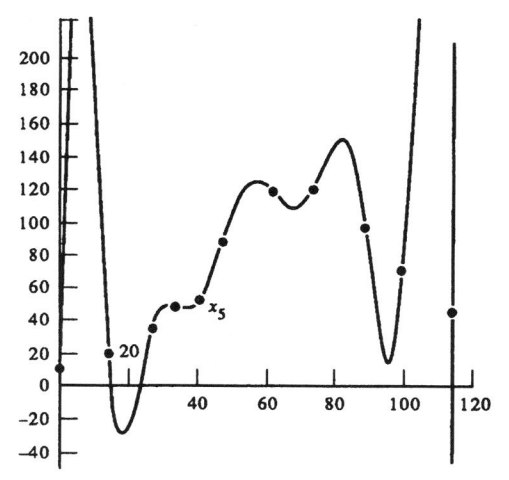

Figure 3.18. A general six-point polynomial fit to the data values in Table 3.17.1. Due to a single point, the polynomial curve oscillates with a amplitude that does not decrease with x.

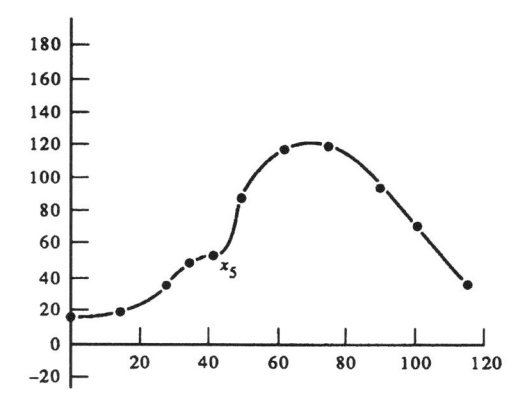

Figure 3.19. Cubic spline fit to the data values in Table 3.17.1. Amplitude of each cycle is reduced by a factor of three compared to Figure 3.18.

When $S = 0$ the data points are fitted exactly by the interpolating spline functions. A recommended value of S is $N/2$, where N is the number of data points. An even smoother interpolation can be achieved using splines under tension. Tension is introduced into the spline procedure to eliminate extraneous inflection points. An iterative procedure is usually used to select the best level for the tension parameter.

3.17.3 Interpolating gappy records: practical examples

Gaps or "holes" occur frequently in geophysical data series. Gaps in a stationary time series are, of course, analogous to gaps in a homogeneous spatial distribution. Small gaps are of little concern and linear interpolation is recommended for filling the gaps. If the gaps are large (of the size of the integral time or space scale), it is generally better to work with the existing short data segments than to "make up" data by pushing interpolation schemes beyond their accepted limitations. For the gray area between these two extremes, one wants to know how large the data loss can be and still

Table 3.17.1. Data pairs (x_i, y_i) used for interpolation schemes in Figures 3.18 and 3.19

i	x_i	y_i
1	0	16
2	14	19
3	27	36
4	33	48
5	41	53
6	48	90
7	62	119
8	74	120
9	89	96
10	99	71
11	114	36

permit reasonable use of standard interpolation techniques and processing methods. The problem of gappy data in oceanography was addressed by Thompson (1971) who suggested that a random sampling of data points might be an optimally efficient approach. Further insight into the problem of missing data can found in Davis and Regier (1977) and Bretherton and McWilliams (1980). In this section, we present two examples of how to deal with gappy data. One is a straightforward analysis by Sturges (1983) who used monthly tide gauge data to investigate what happens to spectral estimates when one punches holes in the data set. The other is a practical guide to the interpolation of satellite-tracked Lagrangian drifter data with its inherently irregular time steps.

3.17.3.1 Interpolating gappy records for time-series analysis

Sturges (1983) used a Monte Carlo technique to poke holes at random in a known time series of monthly mean sea-level. The original record had a "red" spectrum which fell off as f^{-3} at high frequencies and contained a single major spectral peak at a period of 12 months. A total of 120 months of data were used in the analysis. The idea was to reconstruct the gappy series using a cubic spline interpolation method and see how closely the spectrum from the interpolated time series resembled that of the original time series. Data loss was limited to less than 30% of the record length and, for any individual experiment, the holes were all the same length. However, different hole lengths were used in successive runs. The only stipulation was that the length of the data segment before the next gap be at least as long as the gap itself. The program was not allowed to eliminate the first and last data points.

Cross-spectra were computed between the original time series and the interpolated gappy series. For a specified hole size, holes were generated randomly in the data series, the cross-spectra computed and the entire process repeated 1000 times. The magnitudes of the resulting cross-spectra provided estimates of how much power was lost or gained during the interpolation while the corresponding phases was interpreted as the error introduced by the interpolation process (Figure 3.20). Several important conclusions arise from Sturges' analysis:

(1) Gaps have a more adverse effect on weak spectral components (spectral peaks) than on strong ones embedded in the same background spectrum.
(2) The phase can be estimated to roughly 10° at the 90% confidence level for data losses of over 30% for a strong spectral signal; the requirement is that the gaps are

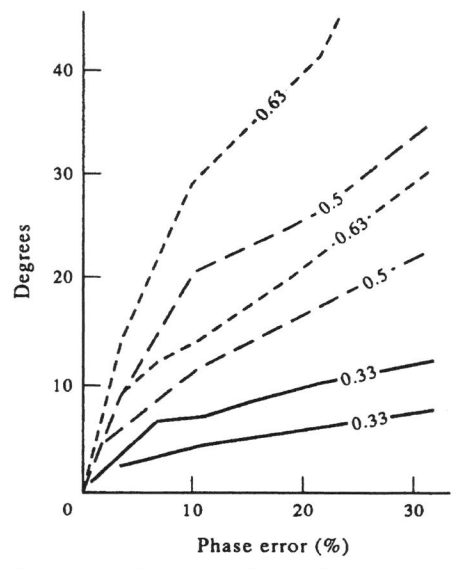

Figure 3.20. Absolute phase errors (°) expressed as a function of percent (%) data lost between the original sea level time series and the series with random holes filled in with a cubic spline fit. On each line, the ratio Δ/T is shown, where Δ is the length of the gap and T is the spectral period of interest; the value 0.5 means that the holes were four units (months) long and the period 8 units long. Results are shown for the 90 and 99% confidence limits (lower and upper lines for each case). (From Sturges, 1983.)

kept to about 1/3 of the period of the signal being examined. If the gaps are 1/2 of the period, the data loss can still be about 20%.
(3) Although correlation functions can be computed for gappy data, it is much more difficult to compute the cross-correlation function for these data.

According to Sturges' analysis, the adverse effects of gaps depends on the length of gaps relative to the length of data set and on the magnitudes of the dominant spectral components in the signal.

3.17.3.2 Interpolating satellite-tracked positional data

The analysis of positional (latitude, longitude) time series collected through the Service Argos satellite-tracking system illustrates some of the problems that may arise with standard interpolation procedures. Because the times that polar orbiting satellites pass over an oceanic region change through the day and because drifters move relative to the orbits of the satellite, the times between satellite fixes are irregular. At mid-latitudes, times between locational fixes can range from less than an hour to as long as 10 h. Typical average times between fixes are around 2–3 h (Thomson et al. 1997). The challenge is to generate regularly spaced time series of latitude (x) and longitude (y) from which one can derive regularly spaced time series of drifter zonal velocity ($u = \Delta x/\Delta t$) and meridional velocity ($v = \Delta y/\Delta t$). This challenge is especially problematic where a "duty cycle" has been programmed into the drifter transmitter to reduce the number (and cost) of transmissions to the passing NOAA satellites. A commonly used duty cycle, consisting of one day continuous transmission followed by two days of no transmission, results in large data gaps that make calculation of mean currents difficult in regions having strong currents in the inertial and tidal frequency bands. The duty cycle of 8 h continuous transmission followed by 16 h of silence is superior for mid-latitude regions with strong inertial or tidal frequency variability.

Because of strong inertial motions in the upper layer of the open ocean and strong tidal motions over continental margins, sampling intervals of 3–4 h, or less, are preferable. A typical time step of 6 h used in many analyses of satellite-tracked drifters is inadequate to resolve inertial motions except in regions equatorward of 30° latitude where the inertial period $T = 1/f_{\text{inertial}}$ exceeds 24 h. (At 50° latitude, $T \approx$ 16.5 h; see *Coriolis frequency*.) To generate time series at a reasonably short time step, say 3 h, we need to interpolate between irregularly spaced data points. To do this, we use a cubic spline interpolation for each of the positional records. After the correct start and end times for the oceanic portion of the record have been determined, the first step in the process is to remove any erroneous points from the "raw" data by calculating speeds over adjacent time steps, t_i; e.g. $u_i = (x_{i+1} - x_i)/(t_{i+1} - t_i)$. One then omits any unrealistic velocity values that exceed some threshold value (say 5 m/s). This "edited" record needs to undergo further editing by averaging successive data positions for which the time step Δt is less than an hour. The reason for this is quite simple: Because positional accuracies Δx and Δy are about 350 m roughly 63% of the time, velocity errors are roughly $\Delta x/\Delta t > 0.1$ m/s when $\Delta t < 1$ h. Such error values are comparable to mean ocean currents and need to be eliminated from the records. Drifters located using GPS transmitters have smaller position errors and better velocity resolution. The time series also need to be examined for drogue-on, drogue-off. If a reliable strain sensor is built into the drogue system, it can be used to determine if and when the drogue fell off. Otherwise, one needs to calculate the speed-squared from the raw data and look for sudden major "jumps" in speed that signal loss of the drogue (Figure 3.21). We recommend this approach for all modern-day drifters since strain gauge sensors appear to be unreliable. At the time this book was being written, drogue loss and not battery or transmitter failure, was the primary cause of drifter "failure" in the open ocean.

Provided there are more than about six accurate satellite fixes per day, the edited positional records can be interpolated to regularly spaced 3-h time series using a cubic spline interpolation algorithm. In general, the spline curve will be well behaved and the fit will resemble the kind of curve one would draw through the data by eye. Inertial and tidal loops in the trajectory will be fairly well resolved. Spurious results will occur where data gaps are too large to properly condition the spline interpolation algorithm. Assuming that the spline interpolation of positions looks reasonable, the next step is to calculate the velocity components from the rate of change of position. It is tempting to equate the coefficient for the linear term in the cubic spline interpolation to the "instantaneous" velocity at any location along the drifter trajectory. That would be a mistake. Although trajectories can look quite smooth, curvatures can be large and resulting velocities unrealistic. In fact, use of the spline coefficients to calculate instantaneous velocity components leads to an increase in the kinetic energy of the motions. The reader can verify this by artificially generating a continuous time series of position consisting of a linear trend and time varying inertial motions. The artifical position record is then decimated to 3-hourly values and a cubic spline interpolation scheme applied. Using instantaneous velocity values at the 3-hourly time steps derived from the interpolation, one finds that the kinetic energy in most frequency bands is increased relative to the original record. The recommended procedure is to calculate

Statistical Methods and Error Handling 289

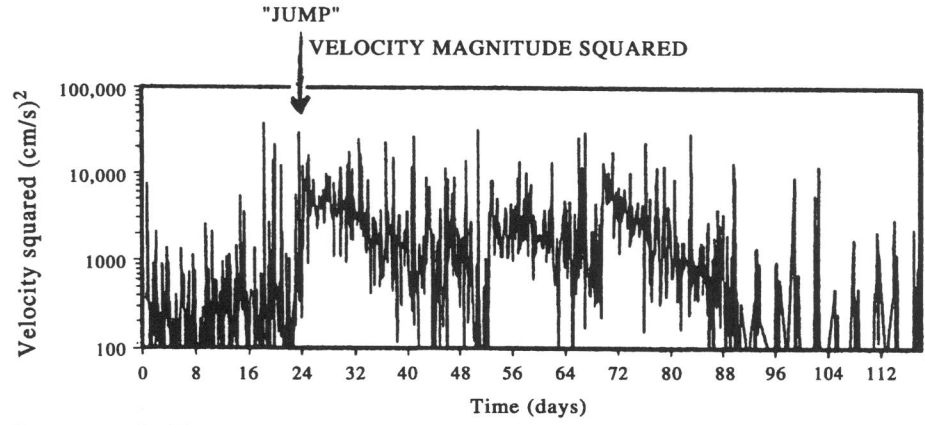

Figure 3.21. *Sudden "jumps" in the speed-squared values from edited satellite-tracked drifter velocity data collected in the North Pacific near 50°N between 133–142°W longitude during the period 4 September to 30 December 1990. The "jump" indicates rapid acceleration of the drifter following probable loss of its drogue.*

the two horizontal velocity components (u, v) from the central differences between three consecutive values of the 3-hourly positional data. From first differences, the velocity components at each point "i" are then: $u_i = (x_{i+1} - x_i)/(t_{i+1} - t_i)$ and $v_i = (y_{i+1} - y_i)/(t_{i+1} - t_i)$ for simple two-point differences or for the recommended centered values, $u_i = (x_{i+1} - x_{i-1})/(t_{i+1} - t_{i-1})$ and $v_i = (y_{i+1} - y_{i-1})/(t_{i+1} - t_{i-1})$. In summary, for those oceanic regions subject to pronounced inertial and tidal frequency motions, we have recommended the use of cubic spline interpolation to generate 2–4-hourly time series for position but simple linear interpolation of positional data to generate the corresponding time series for velocity. The interpolation requires more than 6–8 satellite fixes through the day to be successful.

Trajectories with data gaps that are long relative to the local inertial period require special consideration. For gaps associated with a transmitter duty cycle of 8 h "on" followed by 16 h "off", we can obtain accurate daily mean positional values by least-squares fitting a time-varying continuous function to successive segments of the irregular data and then averaging the resulting function over successive 24-h periods. This filtering processes is as follows (see Bograd et al., 1999):

(1) Use least squares to fit a specified function, $\xi(t)$, to several (N) successive 8-h days of zonal (or meridional) trajectory data. The general model has the form $\xi(t) = a + bt + ct^2 + dt^3 + a_1 \sin(2\pi ft + \phi_1) + a_2 \sin(2\pi \omega_2 t + \phi_2)$ where $a, b, c, d, a_1, \phi_1, a_2$ and ϕ_2 are the unknown coefficients, f is the local Coriolis frequency and ω_2 the semidiurnal frequency (0.081 cph). The phases ϕ_1, ϕ_2 for the two frequencies will vary from segment to segment. We suggest that four to five days ($N = 4$ or 5) of data be used for each segment fit. Shorter segments will have too few data for an accurate least-squares fit; longer segments will result in too much smoothing of the intermittant inertial and tidal motions;

(2) Repeat the least-squares operation for each segment of length N days, shifting forward in time by one day after each set of coefficients is determined. This yields one estimate for the first day $\xi_1 = \xi(t = t_1)$, two estimates for the second day, ξ_2, three estimates for the third day and four estimates for all other days until near the end of

the record when the number of estimates again falls to unity for the last record. Average all the values in each daily segment for each of the multiple curves $\xi_i(t)$ ($i=1$, ..., up to N) to get the average daily latitude $\xi_x(t)$ and longitude $\xi_y(t)$;
(3) The pairs of coefficients a_1, ϕ_1 and a_2, ϕ_2 can be used to give rough reconstructions of the inertial and semidiurnal tidal motions, respectively. Expect the phases to fluctuate considerably from segment to segment due to natural variability in the phases of the motions and from contamination by adjacent frequency bands.

For the duty cycle consisting of one day "on" followed by two days "off", the model is less useful (except at equatorial latitudes) and requires a much longer data segment (say 12 days instead of four) for each least-squares analysis.

3.17.3.3 Interpolation records from nearby stations

Provided that the spatial scales of the processes being examined are large compared to the separation between sampling sites, short gaps in the time series at one location can be filled using an identical type of time series from a nearby location. For example, missing hourly tide heights at one coastal tide gauge station can be filled using hourly tide heights from an adjacent station further along the coast. To do this, we first use coincident data segments to determine the relative amplitudes and phases of the time series at the two locations. A simple cross-correlation analysis can be used to determine the peak time lag between the series while the relative amplitudes can be obtained from the ratio of the standard deviations of the two series. Gaps in one time series (series 1) are then filled by applying the appropriate time lag and amplitude factor to the uninterrupted data series (series 2). A more sophisticated approach would be to first obtain the complex transfer function $H_{12}(\omega) = |H_{12}(\omega)| \exp[i\phi_{12}(\omega)]$ as a function of frequency ω for the two coincident time series. The missing time series values at site 1 could then be filled using the amplitudes $|H_{12}(\omega)|$ and phase differences $\phi_{12}(\omega)$ of the transfer function applied to the uninterrupted data series.

3.18 COVARIANCE AND THE COVARIANCE MATRIX

Covariance, like variance, is a measure of variability. For two variables, the covariance is a measure of the joint variation about a common mean. When extended to a multivariate population, the relevant statistic is the covariance matrix. As we shall see, it is equivalent to what will be introduced later as the "mean product matrix." The covariance and covariance matrix are the fundamental concepts behind the spatial analysis techniques discussed in the next chapter.

3.18.1 Covariance and structure functions

The covariance $C(Y_1, Y_2)$, also written as $\text{cov}[Y_1, Y_2]$, between variables Y_1, Y_2 is

$$C(Y_1, Y_2) = E[(Y_1 - \mu_1)(Y_2 - \mu_2)] \tag{3.18.1}$$

where $\mu_1 = E[Y_1]$ and $\mu_2 = E[Y_2]$. A positive covariance indicates that Y_2 and Y_1 increase and decrease together while a negative covariance has Y_2 decreasing as Y_1

increases, and vice versa. We can expand equation (3.18.1) into a more convenient computational form

$$C(Y_1, Y_2) = E[Y_1 Y_2] - E[Y_1]E[Y_2] \qquad (3.18.2)$$

Note, that if Y_1, Y_2 are independent random variables, then $C[Y_1, Y_2] = 0$.

For a two-dimensional isotropic velocity field, $u_i(y)$, the covariance tensor $C(r)$, also called the *structure function* from earlier studies of turbulence, takes the form

$$\begin{aligned} C_{ij}(r) &= \langle u_i(\mathbf{y}) u_j(\mathbf{y}+\mathbf{r}) \rangle \\ &= \sigma^2 \frac{[f(r) - g(r)] r_i r_j}{r^2 + g(r) \delta_{ij}} \end{aligned} \qquad (3.18.3)$$

where $\langle \cdot \rangle$ denotes an ensemble average, $r \equiv |\mathbf{r}|$, $\mathbf{y} = (y_1, y_2)$ is the position vector, $f(r)$ and $g(r)$ are, respectively, the one-dimensional longitudinal and transverse correlation functions, and $\sigma^2 = \langle u_i(\mathbf{y})^2 \rangle$. The longitudinal and transverse correlation functions are

$$f(r) = \langle u_L(\mathbf{y}) u_L(\mathbf{y}+\mathbf{r}) \rangle \qquad (3.18.4a)$$
$$g(r) = \langle u_P(\mathbf{y}) u_P(\mathbf{y}+\mathbf{r}) \rangle \qquad (3.18.4b)$$

where $u_L(\mathbf{y})$ and $u_P(\mathbf{y})$ are the velocity fluctuations parallel and perpendicular to $\mathbf{r} = (r_1, r_2)$. The velocity fluctuations are normalized so that the correlations equal unity at $r = 0$. If the two-dimensional flow field is horizontally nondivergent, homogenous and isotropic, then $C_{ij}(r) = 0$ and

$$g(r) = \frac{d}{dr}[rf(r)] \qquad (3.18.5)$$

Freeland *et al.* (1975) have used (3.18.5) to test for two-dimensional, nondivergent, homogenous, and isotropic low-frequency velocity structure in SOFAR float data collected in the North Atlantic. Stacey *et al.* (1988) used this relation to test for similar flow structure in the Strait of Georgia. Although close to the error limits in certain cases, the observed structure is generally consistent with horizontal, nondivergent, homogeneous and isotropic flow (Figure 3.22). The dotted lines in Figure 3.22 are the analytical functions

$$f(r) = (1 + br) e^{-br} \qquad (3.18.6a)$$
$$g(r) = (1 + br - b^2 r^2) e^{-br} \qquad (3.18.6b)$$

3.18.2 A computational example

If Y_1, Y_2 have a joint probability density function

$$f(y_1, y_2) = \begin{cases} 2y_1, & 0 \le y_1 \le 1; \ 0 \le y_2 \le 1 \\ 0, & \text{elsewhere} \end{cases} \qquad (3.18.7)$$

what is the covariance of Y_1, Y_2? We first write the expected value of Y_1, Y_2 as

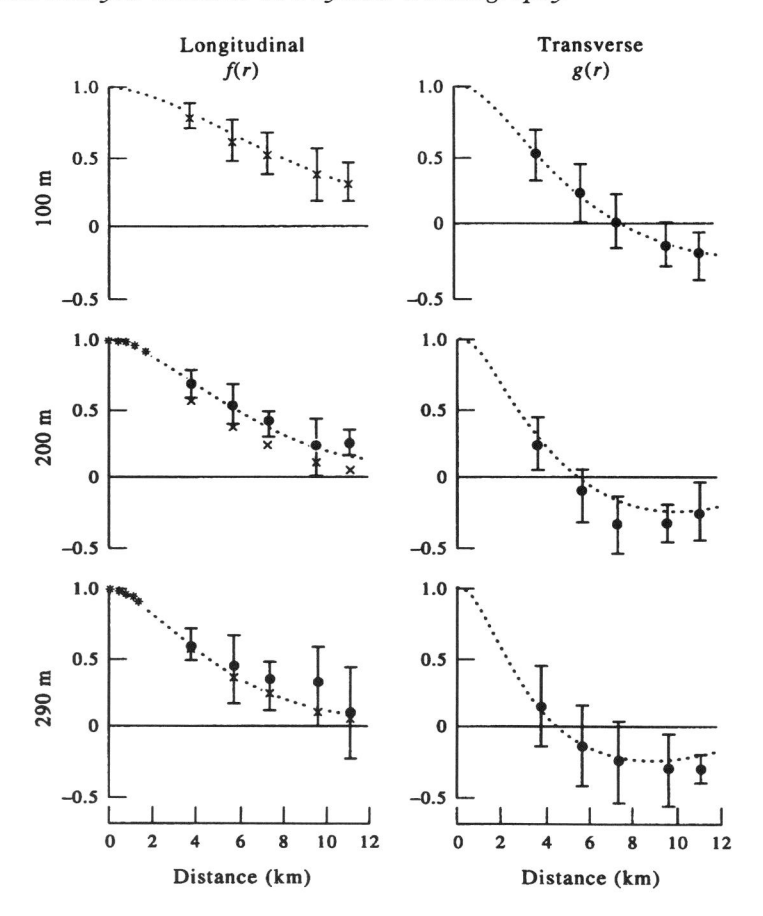

292 *Data Analysis Methods in Physical Oceanography*

Figure 3.22. Longitudinal and transverse correlations at 100, 200, and 280/290 m depths. The dots are measured average values and error bars are the standard deviations. The mean and trend were removed from each time series before calculation of the correlations. The crosses are predicted values of f(r) calculated using (3.18.5) by drawing straight line segments between the average values of g(r) and integrating under the curve. (From Stacey et al., 1988.)

$$E[Y_1 Y_2] = \int_0^1 \int_0^1 y_1 y_2 f(y_1, y_2) \mathrm{d}y_1 \mathrm{d}y_2 = \int_0^1 \int_0^1 y_1 y_2 (2y_1) \mathrm{d}y_1 \mathrm{d}y_2$$

$$= \int_0^1 \frac{1}{3} y_2 (2y_1^3) \Big|_0^1 \mathrm{d}y_2 = \int_0^1 \frac{2}{3} y_2 \mathrm{d}y_2 = \frac{2 y_2^2}{3 \cdot 2} \Big|_0^1 = \frac{1}{3}$$

Recall that, for discrete variables

$$E[g(Y_1, ..., Y_k)] = \sum_{y_k} \cdots \sum_{y_1} g(y_1, ..., y_k) P_1(y_1, ..., y_k)$$

or for continuous variables

$$E[g(Y_1, ..., Y_k)] = \int_{y_k} ... \int_{y_1} g(y_1, ..., y_k) f(y_1, ..., y_k) dy_1 ... dy_k$$

For this example, we find $E[Y_1 Y_2] = 1/3$. Now

$$E[Y_1] = \int_0^1 \int_0^1 y_1(2y_1) dy_1 dy_2 = \int_0^1 \frac{2}{3} y_1^3 \bigg|_0^1 dy_2 = \frac{2}{3} y_2 \bigg|_0^1 = \frac{2}{3}$$

and $E[Y_2] = 1/2$, so that $\text{cov}[Y_1 Y_2] = E[Y_1 Y_2] - \mu_1 \mu_2 = 1/3 - (2/3)(1/2) = 0$. Therefore, Y_1 and Y_2 are independent. Of course, we could have anticipated this result since $f(y_1, y_2)$ in equation (3.18.7) is independent of y_2.

3.18.3 Multivariate distributions

In the case of multivariate distributions, the covariance becomes the *covariance matrix*. If we have n measurements (samples) of N variables (Y), we can describe this as n random variables having a joint N-dimensional probability density function (PDF)

$$f_{1,2,...,N}(Y_1, Y_2, ..., Y_N) \tag{3.18.8}$$

If the random variables, Y, are mutually independent, the joint PDF can be factored in the usual way as

$$f_{1,2,...,N}(Y_1, Y_2, ..., Y_N) = f_1(Y_1) f_2(Y_2) ... f_N(Y_N) \tag{3.18.9}$$

An important multivariate PDF is the multivariate normal PDF

$$f_Y(Y) = \frac{1}{2\pi^{N/2} |\mathbf{W}|^{1/2}} \exp\left[-\frac{1}{2}(Y-\mu)^T \mathbf{W}^{-1}(Y-\mu)\right]$$

where $Y^T = (Y_1, Y_2, ..., Y_N)$, $\mu^T = (\mu_1, \mu_2, Ö, \mu_N)$, are the row vectors and $\mathbf{W}^{-1}$ is the inverse of the covariance matrix $\mathbf{W}$

$$\mathbf{W} = \begin{pmatrix} \sigma_1^2 & \sigma_1 \sigma_2 \rho_{12} & \sigma_1 \sigma_3 \rho_{13} & ... & \sigma_1 \sigma_N \rho_{1N} \\ \sigma_2 \sigma_1 \rho_{12} & \sigma_2^2 & \sigma_2 \sigma_3 \rho_{23} & ... & \sigma_2 \sigma_N \rho_{2N} \\ ... & ... & ... & ... & ... \\ \sigma_N \sigma_1 \rho_{N1} & \sigma_2 \sigma_N \rho_{2N} & ... & ... & \sigma_N^2 \end{pmatrix} \tag{3.18.10}$$

or

$$\mathbf{W} = \begin{pmatrix} V[Y_1] & C[Y_1 Y_2] & ... & ... & C[Y_1 Y_N] \\ C[Y_2 Y_1] & V[Y_2] & ... & ... & C[Y_2 Y_N] \\ ... & ... & ... & ... & ... \\ C[Y_N Y_1] & C[Y_N Y_2] & ... & ... & V[Y_N] \end{pmatrix} \tag{3.18.11}$$

Note that $C[Y_i Y_j] = C[Y_j Y_i]$ and therefore $\mathbf{W}$ is symmetric ($\mathbf{W} = \mathbf{W}^T$). In addition, $\mathbf{W}$ is positive semi-definite; that is, $|\mathbf{W}|$ and all its principal minors are nonnegative. Another way to show this is

$$V[\lambda^T Y] = E[\lambda^T(Y-\mu)(Y-\mu)^T \lambda] = \lambda^T |\mathbf{W}| \qquad (3.18.12)$$

which will always be nonnegative for any λ.

3.19 THE BOOTSTRAP AND JACKKNIFE METHODS

Many data series in the natural sciences are nonreproducible and the researcher is left with only one set of observations with which to work. With only one realization of a series, it is impossible to compare it with a related series to determine if they are drawn from the same, or from different, populations. There are numerous oceanographic examples, including tsunami oscillations recorded by a coastal tide gauge, a single seasonal cycle of monthly mean currents at a mooring location, and a trend in long-term temperature data from a climate monitoring station. Marine biologists face similar limitations when analyzing groups of animal species caught in nets or bottom grab samples. The problem is that empirical observations are prone to error and any interpretation of an event must be devised based on statistical measures of the probability of the event. A fundamental measure for testing the validity of any property of a data set is its variance. Parametric statistical models have been developed which help the investigator decide the degree of faith to be placed in a given statistic. However, data and model are often nonlinear so that it is not usually possible to find an analytical expression for model variance in terms of the data variance.

The *parametric* statistical methods presented in the previous sections were institutionalized long before the time of modern digital computers when use of analytical expressions greatly simplified the laborious hand calculation of statistical properties. During the past few decades, *nonparametric* statistical methods have been developed to take advantage of the increasing computational efficiencies of computers. An advantage of the new methods is that they permit investigations of the statistical properties of a sample which do not conform to a specific analytical model. Equally importantly, they can be applied to small data sets while still providing a reliable estimation of confidence limits on the statistic of interest. "Bootstrapping" and "jackknifing" are two of the more commonly used methods that could not be used effectively until the invention of the digital computer. Both are resampling techniques in which artificial data sets are generated by selection of points from an original set of data. Specifically, we start with a single realization of an "experiment" and from that one set of experimental data we create a multitude of new artificial realizations of the experiment without having to repeat the observations. These realizations are then used to estimate the reliability of the particular statistic of interest, with the underlying assumption that the sample data are representative of the entire population.

In the bootstrapping method, random samples selected during the resampling process are replaced before each new sample is created. As a consequence, any data value can be drawn many times, or not at all. The name bootstrap arises from the expression "to lift oneself up by one's bootstraps". In jackknifing, artificial data sets are created by selectively and systematically removing samples from the original data set. The statistics of interest are recalculated for each resulting truncated data set and the variability among the artificial samples used to describe the variability of these statistics. "Cross-validation" is an older technique. The idea is to split the data into two parts and set one part aside. Curves are fitted to the first part and then tested against values in the second part. Cross-validation consists of seeing how well the fitted curves predict the values in the portion of data set aside. The data can be randomly split in many ways and many times in order to obtain the needed statistical reliability. For additional information on this technique, the reader is referred to Efron and Gong (1983).

3.19.1 Bootstrap method

Introduced by Efron in 1977 (Diaconis and Efron, 1983), bootstrapping provides freedom from two limiting factors that have constrained statistical theory since its beginning: (1) the assumption of normal (Gaussian) data distributions; and (2) the focus on statistical measures whose theoretical properties can be analyzed mathematically. As with other nonparametric methods, bootstrapping is insensitive to assumptions made with respect to the statistical properties of the data and does not need an analytical expression for the connection between model and data statistical properties. Resampling techniques are based on the idea that we can repeat a particular experiment by constructing multiple data sets from the one measured data set. Application of the resampling procedure must be modeled on a testable hypothesis so that the resulting probability can be used to accept or reject the null hypothesis. The methods can be applied just as well to any statistic, simple or complicated. A *bootstrap sample* is a "copy" of the original data that may contain a certain value (datum, x_n) more than once, once, or not at all (i.e. the number of occurrences of x_n lies between 0 and N, where N is the number of independent data points). Introductions into the bootstrapping procedure can be found in Efron and Gong (1983), Diaconis and Efron (1983), and Tichelaar and Ruff (1989). Nemec and Brinkhurst (1988) apply the method to testing the statistical significance of biological species cluster analysis for which there are duplicate or triplicate samples for each location.

Suppose that we have N values of a scalar or vector variable, x_n ($1 \leq n \leq N$), whose statistical properties we wish to investigate in relation to another variable. This could be a univariate variable such as sea-level height $x_n = \eta(t_n)$ at a single location over a period of N time steps, t_n, or the structure of the first mode empirical orthogonal function $\phi_1(x_n)$ as a function of location, x_n. Alternatively, we could be dealing with a bivariate variable (x_{1n}, x_{2n}) such as water temperature versus dissolved oxygen content from a series of vertical profiles. Results apply to any other set of measurements whose statistics we wish to determine. We may want to compare means and standard deviations (variances) of different records to see if they are significantly different. Alternatively, we might want to place confidence limits on the slope of a line derived using a standard least-squares fit to our bivariate data (x_{1n}, x_{2n}), or, determine how much confidence we can have in the coefficients we obtained from the least-squares fit of an annual cycle to a single set of 12 monthly mean current records from a mooring location. Note that if there is a high degree of correlation among the N data values, the N are not statistically independent samples and we are faced with the usual problem of dealing with an effective number of degrees of freedom N^* for the data set.

The procedure is to equate each of our N independent data points with a number produced by a random number generator. We can do this by assigning each of the data values to separate uniform-width bins lying along the line $(-1, +1)$, or $(0, 1)$, depending on the random number generator being used. For N values, there will be N

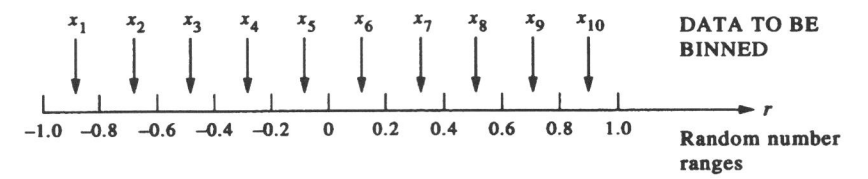

Figure 3.19.1. The assignment (binning) of observed data values x_n ($n = 1, \ldots, 10$) to 10 range values of the random number, r_k ($k = 1, \ldots, 10$). For each bootstrap sample of 10 values, 10 random numbers are selected and located according bin range. The datum values x_n assigned to each range are then used to form the bootstrap sample.

uniform-width bins on the line and each bin will be equated with one of the N data values (Figure 3.19.1). The bin width is $2/N$ if the line -1 to $+1$ is used. A random number generator such as a Monte Carlo scheme is used to randomly select sequences of N bins corresponding to the multiple bootstrap samples. Suppose that the random number generator picks a number, r, from the range $-1 \leq r \leq 1$. If this number falls into the range of bin k, corresponding to the range $[2(k-1)/N] - 1 \leq r_k \leq (2k/N) - 1$, for $k = 1, \ldots, N$, then the data value x_k assigned to bin k is taken to be one of the samples we need to make up our bootstrap data set. In Figure 3.19.1, there are 10 data values and 10 corresponding random number segments of length 0.2, with datum value x_1 assigned to the range -1.0 to -0.8, x_2 assigned to -0.8 to -0.6, and so on. Since bootstrapping works with replacement, it is quite possible to get the same bin several times, or not at all. The first N data values from our resampling constitute the first bootstrap sample. The process is then repeated again and again until hundreds or thousands of bootstrap samples have been generated. Diaconis and Efron (1983) discuss making a billion bootstrap samples. They also take another approach. Instead of generating one bootstrap sample at a time by equating bins along the real line $(-1, 1)$ with N samples, they generate all the needed multiple copies of all the N data values (say one million copies of each of the original data values or data points) and place them all in a rotating "lotto" bin. They then reach in and pull out all the requisite number of N-value bootstrap samples from the shuffled points, being careful to throw each data point back into the bin before selecting the next value. This requires some sort of label for each value in the bin based on a random selection process that can identify a data point that has been selected.

Although bootstrapping has yet to find widespread application in the marine sciences, there are several noteworthy examples in the literature. Enfield and Cid (1990) examined the stationarity of different groupings of El Niño recurrence rates based on the chronology of Quinn *et al.* (1987). For example, group 1 consisted of all strong (S) and very strong (VS) events for the period 1525–1983, while groups 4 and 5 consisted of S/VS events for times of high and low solar activity for this period. Groups 6–10 contained different samples of intensities for the modern period of 1803–1987. Maximum likelihood estimation was used to fit a two-parameter Weibull distribution $f(t)$ to each sample group,

$$f(t) = (\beta t^{\beta-1}/\tau^\beta) \exp\left[-(t/\tau)^\beta\right] \qquad (3.19.1)$$

where β and τ are, respectively, the shape (peakedness) and time scale (RMS return interval) parameters, and t is the random variable for the return interval. For each group, only a single distribution could be fitted. To derive estimates of the mean and

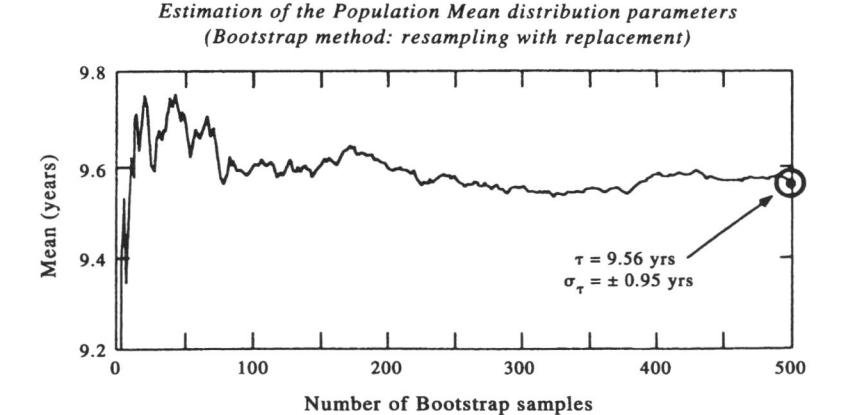

Figure 3.19.2. Estimation of the population mean distribution parameters (mean return time in years) using the bootstrap method for El Niño events taking place during times of low solar activity for the period 1525–1983. τ is the return time and σ_τ its standard deviation. (Enfield and Cid, 1990.)

standard deviations of the parameters for each group, 500 bootstrap samples were generated and the Weibull parameters obtained for each sample. As indicated by Figure 3.19.2, this number of samples provides good convergence to the mean value for the Weibull distribution fit for each group. The distribution of El Niño return events for bootstrap samples for all intensities for the "early modern" period 1803–1891 is shown in Figure 3.19.3 along with its corresponding Weibull distribution. Enfield and Luis use the resampling analysis to show that, for the groups associated with times of low solar activity and those associated with times of high solar activity, there is comparatively little overlap between the bootstrap-derived frequency histograms and mean return time scales, τ (years) (Figure 3.19.4). These results suggest that there is a statistical difference in the return times for the two groups and that return times are nonstationary.

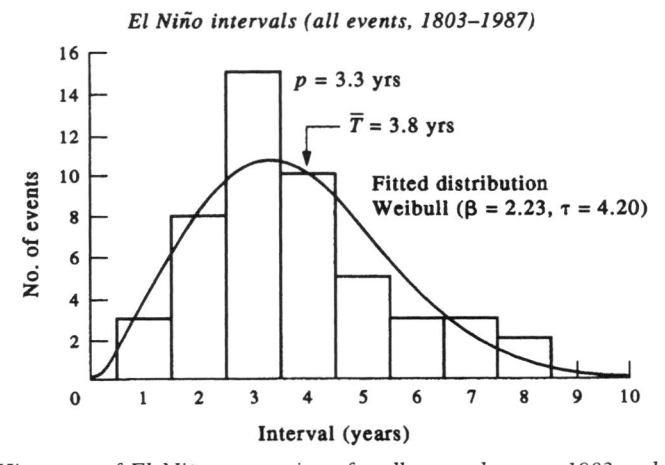

Figure 3.19.3. Histogram of El Niño return times for all events between 1803 and 1987 (group #7) derived using the bootstrapping resampling technique. The solid curve is the Weibull distribution fitted to the histogram. The modal and mean return intervals (3.3 and 3.8 years, respectively) are the derived from the MLE-estimated population parameters. (From Enfield and Cid, 1990.)

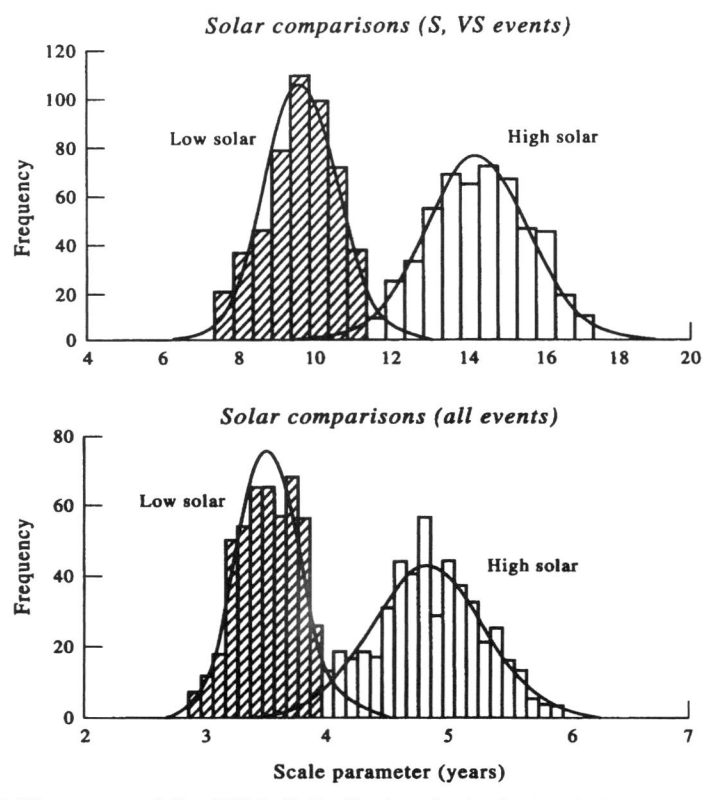

298 *Data Analysis Methods in Physical Oceanography*

Solar comparisons (S, VS events)

Solar comparisons (all events)

Scale parameter (years)

Figure 3.19.4. Histograms and fitted Weibull distributions obtained using the bootstrapping method. Plots show the occurrences of El Niño events for the times of low and high solar flare activity for (a) Strong and very strong El Niño events, only; (b) All El Niño events.

Much of the present evidence for possible global warming is based on Northern Hemispheric annual surface air temperature records over the past 100 years (Jones *et al.*, 1986; Hansen and Lebedeff, 1987; Gruza *et al.*, 1988). Interest in the reliability of the means and trends of these records (labeled H, J, and G) prompted Elsner and Tsonis (1991) to examine differences in means and trends of pairs of these records for the three global mean temperature curves. The data sets have been constructed using different averaging methods and different observational data bases. Data set H contains only observations from land stations whereas data set J uses both land and ship-based observations. Averages for set H are derived using equal-area boxes over the globe whereas data set G is constructed by visual inspection of anomalies from sea-level temperature analyses. The usual assumption is that these time series are representative of the same population, a result that appears to be supported by the statistically significant correlation $r > 0.79$ among the different curves. As pointed out by Elsner and Tsonis, however, the presence of trends in these data means that the linear cross-correlation coefficient may not be a reliable measure of the covariability of the records. Two questions can be addressed using the bootstrapping method: (1) are the three versions of the temperature records significantly different that we can say they are not drawn from the same population? (The null hypothesis is false.); and (2) are the trends in the three records sufficiently alike that they are measuring a true rise in global temperature?

Statistical Methods and Error Handling 299

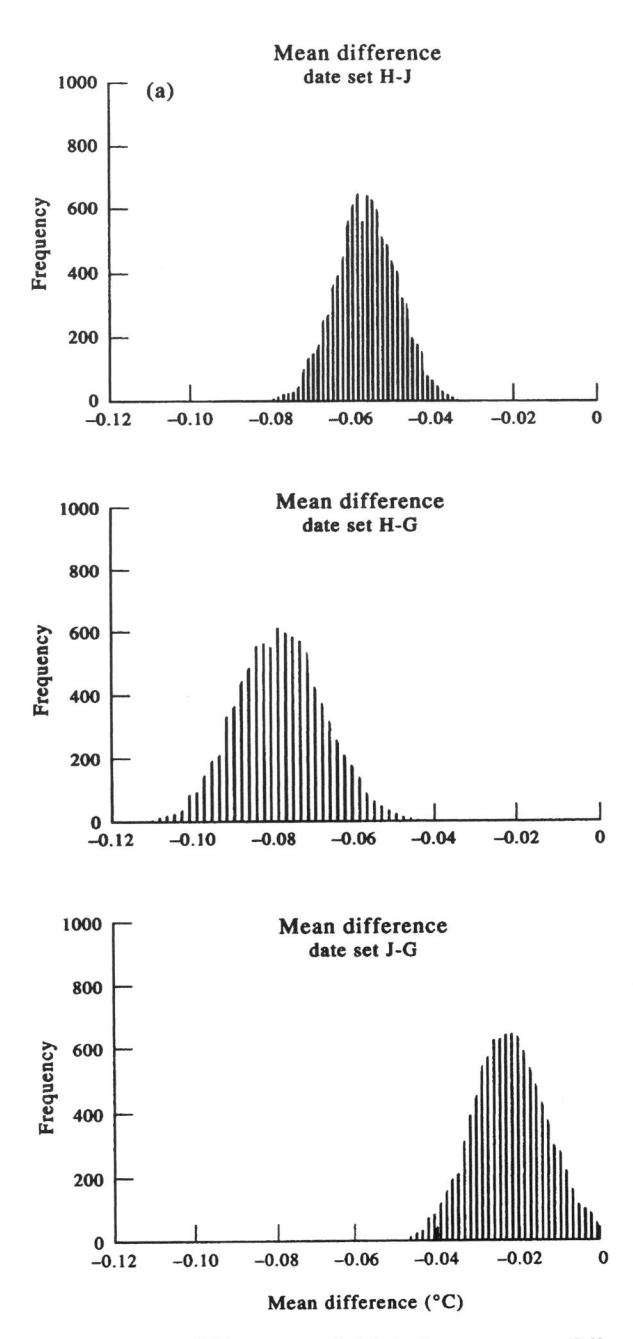

Figure 3.19.5. Bootstrap-generated histograms of global air temperature difference records obtained by subtracting the temperature records of Jones et al. (1986) (J), Hansen and Lebedeff (1987) (H), and Gruza et al. (1988) (G). (a) Frequency distributions of the mean differences plotted for 10^4 bootstrap samples. The x-axis (ordinate) gives the number of times the bootstrap mean fell into a given interval. All three distributions are located to the left of a zero mean difference. (From Elsner and Tsonis, 1991.)

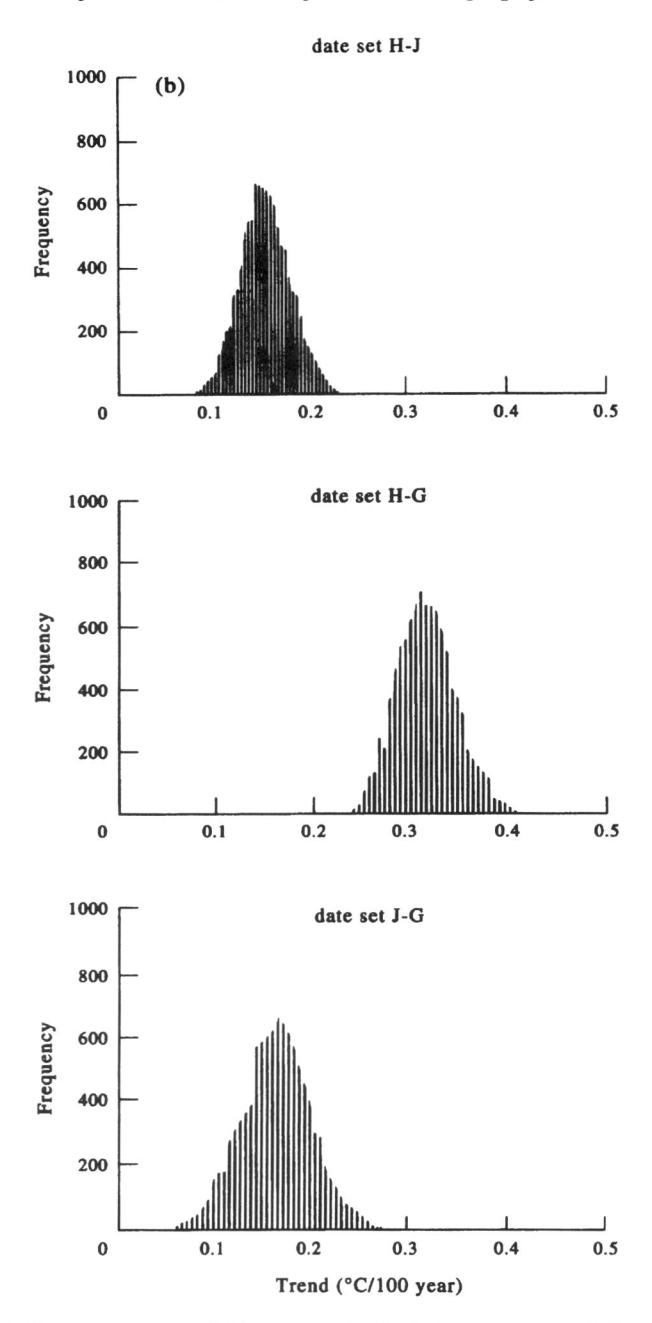

Figure 3.19.5. Bootstrap-generated histograms of global air temperature difference records obtained by subtracting the temperature records of Jones et al. (1986) (J), Hansen and Lebedeff (1987) (H), and Gruza et al. (1988) (G). (b) Same as (a), but for slope (trend) of the temperature difference curves. All three distributions are separated from zero indicating significant differences between long-term surface temperature trends given by each of the three data sets. (From Elsner and Tsonis, 1991.)

Because of the strong linear correlation in the records, the authors work with difference records. A difference record is constructed by subtracting the annual (mean removed) departure record of one data set from the annual departure record of another data set. Although not zero, the cross-correlation for the difference records is considerably less than those for the original departure records, showing that differencing is a form of high-pass filtering that effectively reduces biasing from the trends. The average difference for all 97 years of data used in the analyses (the difference record H–J relative to the years 1951–1980) is $-0.05°C$, indicating that the hemispheric temperatures of Jones *et al.* (1986) are slightly warmer than those of Hansen and Lebedeff (1987). Similar results were obtained for H–G and J–G. To see if these differences are statistically significant, 10,000 bootstrap samples of the difference records were generated. The results (Figure 13.9.5a) suggest that all three hemispheric temperature records exhibit significantly different nonzero means. The overlap in the distributions is quite minimal. The same process was then used to examine the trends in the difference records. For the H–J record, the trend is $+0.15°C$/century so that the trend of Hansen and Lebedeff is greater than that of Jones *et al.* As indicated in Figure 13.9.5(b), the long-term trends were distinct. On the basis of these results, the authors were forced to conclude that at least two of the data sets do not represent the true population (i.e. the truth). More generally, the results bring into question the confidence one can have that the long-term temperature trends obtained from these data are representative of trends over hemispheric or global scales.

Biological oceanographers often have difficult sampling problems that can be addressed by bootstrap methods. For example, the biologist may want to use cluster analyses of animal abundance for different locations to see if species distributions differ statistically from one sampling location (or time) to the next. Cluster analyses of ecological data use dendrograms—linkage rules which group samples according to the relative similarity of total species composition—to determine if the organisms in one group of samples have been drawn from the same or different statistical assemblages of those of another group of samples. Provided there are, at least, replicates for most samples, bootstrapping can be used to derive tests for statistical significance of similarity linkages in cluster analyses (Burd and Thomson, 1994). For further information on this aspect of bootstrapping, the reader is referred to Nemec and Brinkhurst (1988). Finally, in this section, we note that it is possible to vary the bootstrap size by selecting samples smaller than N, the original size of the data set, to compare various estimator distributions obtained from different sample sizes. This allows one to observe the effects of varying sample size on sample estimator distributions and statistical power.

3.19.2 Jackknife method

Several other methods are similar in concept to bootstrapping but differ significantly in detail. The idea, in each case, is to generate artificial data sets and assess the variability of a statistic from its variability over all the sets of artificial data. The methods differ in the way they generate the artificial data. Jackknifing differs from bootstrapping in that data points are not replaced prior to each resampling. This technique was first proposed by Maurice Quenouille in 1949 and developed by John Tukey in the 1950s. The name "jackknife" was used by Tukey to suggest an all-purpose statistical tool.

A jackknife resample is obtained by deleting a fixed number of data points (j) from the original set of N data points. For each resample, a different group of j values is removed so that each resample consists of a distinct collection of data values. In the "delete-j" jackknife sample, there will be $k = N - j$ samples in each new truncated data set. The total number of new artificial data that can be generated is

$$\binom{N}{j}$$

which the reader will recognize as $_N P_j = N!/(N-j)!$, the number of permutations of N objects taken j at a time. Consider the simple delete-1 jackknife. In this case, there are $N-1$ samples per artificial data set and a total of $_N P_j = N$ new data sets that can be created by systematically removing one value at a time. As illustrated by Figure 3.19.6, an original data set of four data values will yield a total of four distinct delete-1 jackknife samples, each of size three (3), which can then be used to examine various statistics of the original data set. The sample average of the data derived by deleting the ith datum, denoted by the subscript (i), is

$$\bar{x}_{(i)} = \frac{N\bar{x} - x_i}{N-1} = \frac{1}{N-1} \sum_{\substack{j \neq i}}^{N} x_j \qquad (3.19.2)$$

where

$$\bar{x} = \frac{1}{N} \sum_{i=1}^{N} x_i$$

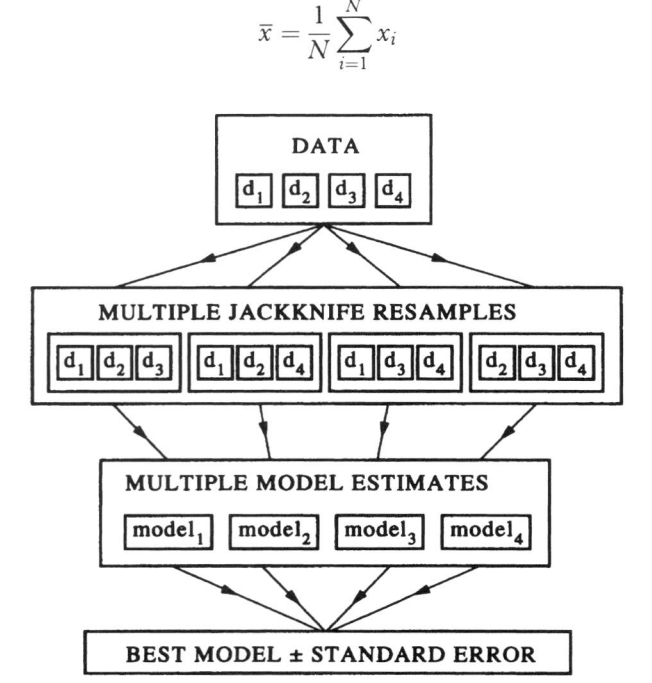

Figure 3.19.6. Schematic representation of the jackknife. The original data vector has four components (samples), labeled d_1 to d_4. The data are resampled by deleting a fixed number of components (here, one) from the original data to form multiple jackknife resamples (in case, four). Each resample defines a model estimate. The multiple model estimates are then combined to a best model and its standard deviation. (From Tichelaar and Ruff, 1989.)

is the mean found using all the original data. The average of the N jackknife averages, $\bar{x}_{(i)}$, is

$$\bar{x}^* = \frac{1}{N}\sum_{i=1}^{N} \bar{x}_{(i)} = \bar{x} \qquad (3.19.3)$$

The last result, namely that the mean of all the jackknife samples is identical to the mean of the original data set, is easily obtained using equation (3.19.2). The estimator for the standard deviation, σ_j, of the delete-1 jackknife is

$$\sigma_j = \sum_{i=1}^{N}\left[(\bar{x}_{(i)} - \bar{x}^*)^2\right]^{1/2} \qquad (3.19.4a)$$

$$= \frac{1}{N-1}\sum_{i=1}^{N}\left[(x_i - \bar{x})^2\right]^{1/2} \qquad (3.19.4b)$$

where (3.19.4b) is the usual expression for the standard deviation of N data values. Our expression differs slightly from that of Efron and Gong (1983) who use a denominator of $1/[(N-1)N]$ instead of $1/(N-1)^2$ in their definition of variance. The advantage of (3.19.4a) is that it can be generalized for finding the standard deviation of any estimator θ that can be derived for the original data. In particular, if θ is a scalar, we simply replace $x_{(i)}$ with $\theta_{(i)}$ and x^* with θ^* where $\theta_{(i)}$ is an estimator for θ obtained for the data set with the ith value removed. Although the jackknife requires fewer calculations than the bootstrap, it is less flexible and at times less dependable (Efron and Gong, 1983). In general, there are N jackknife samples for the delete-1 jackknife as compared with

$$_{2N-1}P_N = \binom{2N-1}{N}$$

bootstrap points.

Our example of jackknifing is from Tichelaar and Ruff (1989) who generated $N = 20$ unequally spaced data values y_i that follow the relation $y_i = cx_i + \varepsilon_i$ ($c = 1.5$, exactly), where ε_i is a noise component drawn from a "white" spectral distribution with a normalized standard deviation of 1.5 and mean of zero. The least squares estimator for the standard deviation of the slope is

$$\hat{\sigma} = \sum_{i=1}^{N}\left[(y_i - \hat{c}x_i)^2\right] \Big/ \left[(N-1)\sum_{i=1}^{N} x_i^2\right] \qquad (3.19.5)$$

where $\hat{c} \sum y_i x_i / \sum x_i^2$. Two jackknife estimators were used: (1) The delete-1 jackknife, for which the artificial sample sizes are $N - 1 = 19$; and (2) The delete-half ($N/2$) jackknife for which the sample sizes are $N - N/2 = 10$. In both cases, the jackknife resamples had equal weighting in the analysis. For the delete-half jackknife, a Monte Carlo determination of 100 subsamples was used since the total samples $_{20}P_{10} = 20!/10!$ is very large. The results are presented in Figure 3.19.7. The last panel gives the corresponding result for the bootstrap estimate of the slope using 100 bootstrap samples. Results showed that the bootstrap standard error of the slope was slightly lower than those for both jackknifing estimates.

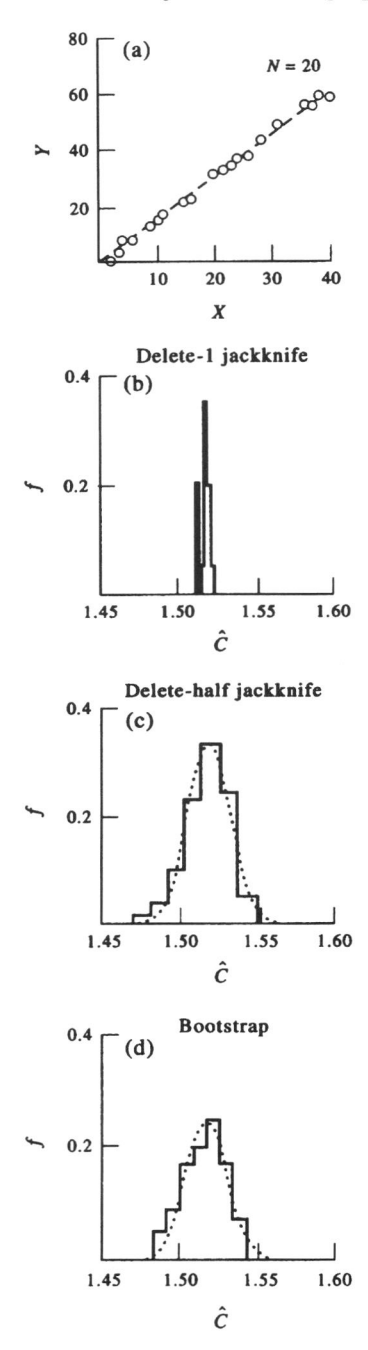

304 *Data Analysis Methods in Physical Oceanography*

Figure 3.19.7. Use of the bootstrapping technique to estimate the reliability of a linear regression line. (a) A least-squares fit through the noisy data, for which the estimated slope $\hat{c} = 1.518 \pm 0.0138$ (± 1 standard error); (b) The normalized frequency of occurrence distribution, f, for the delete-1 jackknife which yields $\hat{c} = 1.518 \pm 0.0136$; (c) As in (b) but for the delete-half jackknife for which $\hat{c} = 1.517 \pm 0.0141$; (d) The corresponding bootstrapping estimate, for which $\hat{c} = 1.517 \pm 0.0132$. Note the scale difference between (b) and (c). The dashed line is the analytical distribution of $\hat{c}$. (From Tichelaar and Ruff, 1989.)

CHAPTER 4

The Spatial Analyses of Data Fields

A fundamental problem in oceanography is how best to represent spatially distributed data (or statistical products computed from these data) in such a way that dynamical processes or their effects can best be visualized. As in most aspects of observational analysis, there has been a dramatic change in the approach to this problem due to the increased abundance of digital data and our ability to process them. Prior to the use of digital computers, data displays were constructed by hand and "contouring" was an acquired skill of the descriptive analyst. Hand contouring is still practiced today although, more frequently, the data points being contoured are averaged values produced by a computer. In other applications, the computer not only performs the averaging but also uses objective statistical techniques to produce both the gridded values and the associated contours.

The purpose of this section is to review data techniques and procedures designed to reduce spatially distributed data to a level that can be visualized easily by the analyst. We will discuss methods that address both spatial fields and time series of spatial fields since these are the primary modes of data distribution encountered by the oceanographer. Our focus is on the more widely used techniques which we present in a practical fashion, stressing the application of the method for interpretive applications.

4.1 TRADITIONAL BLOCK AND BULK AVERAGING

A common method for deriving a gridded set of data is simply to average the available data over an arbitrarily selected rectangular grid. This averaging grid can lie along any chosen surface but is most often constructed in the horizontal or vertical plane. Because the grid is often chosen for convenience, without any consideration to the sampling coverage, it can lead to an unequal distribution of samples per grid "box". For example, because distance in longitude varies as the cosine of the latitude, the practice of gridding data by 5 or 10° squares in latitude and longitude may lead to increasingly greater spatial coverage at low latitudes. Although this can be overcome somewhat by converting to distances using the central latitude of the box (Poulain and Niiler, 1989), it is easy to see that inhomogeneity in the sampling coverage can quickly nullify any of the useful assumptions made earlier about the Gaussian nature of sample populations or, at least, about the set of means computed from these samples. This is less of a problem with satellite-tracked drifter data since satellite ground tracks converge with increasing latitude, allowing the data density in boxes of fixed longitude length to remain nearly constant.

With markedly different data coverage between sample regions, we cannot always fairly compare the values computed in these squares. At best, one must be careful to consider properly the amount of data being included in such averages and be able to evaluate possible effects of the variable data coverage on the mapped results. Each value should be associated with a sample size indicating how many data points, N, went into the computed mean. This will not dictate the spatial or temporal distributions of the sample data field but will at least provide a sample size parameter which can be used to evaluate the mean and standard deviation at each point. While the standard deviation of each grid sample is composed of both spatial and temporal fluctuations (within the time period of the grid sample), it does give an estimate of the inherent variability associated with the computed mean value.

Despite the problems with nonuniform data coverage, it has proven worthwhile to produce maps or cross-sections with simple grid-averaging methods since they frequently represent the best spatial resolution possible with the existing data coverage. The approach is certainly simple and straightforward. Besides, the data coverage often does not justify more complex and computer-intensive data reduction techniques. Specialized block-averaging techniques have been designed to improve the resolution of the corresponding data by taking into account the nature of the overall observed global variability and by trying to maximize the coverage appropriately. For example, averaging areas are frequently selected which have narrow meridional extent and wide zonal extent, taking advantage of the stronger meridional gradients observed in the ocean. Thus, an averaging area covering 2° latitude by 10° longitude may be used to better resolve the meridional gradients which dominate the open ocean (Wyrtki and Meyers, 1975). This same idea may be adapted to more limited regions if the general oceanographic conditions are known. If so, the data can be averaged accordingly, providing improved resolution perpendicular to strong frontal features. A further extension of this type of grid selection would be to base the entire averaging area selection on the data coverage. This is difficult to formalize objectively since it requires the subjective selection of the averaging scheme by an individual. However, it is possible in this way to improve resolution without a substantial increase in sampling (Emery, 1983).

All of these bulk or block-averaging techniques make the assumption that the data being considered in each grid box are statistically homogeneous and isotropic over the region of study. Under these assumptions, area sample size can be based strictly on the amount of data coverage (number of data values) rather than having to know details about processes represented by the data. Statistical homogeneity does not require that all the data were collected by the same instrument having the same sampling characteristics. Thus, our grid-square averaging can include data from many different instruments which generally have the same error limits.

One must be careful when averaging different kinds of measurements, even if they are of the same parameter. It is very tempting, for example, to average mechanical bathythermograph (MBT) temperatures with newer expendable bathythermograph (XBT) temperatures to produce temperature maps at specific depths. Before doing so, it is worth remembering that XBT data are likely to be accurate to 0.1°C, as reported earlier, while MBT data are decidedly less accurate and less reliable. Another marked difference between the two instruments is their relative vertical coverage. While most MBTs stopped at 250 m depth, XBTs are good to 500–1800 m, depending on probe type. Thus, temperature profiles from MBTs can be expected to be different from those collected with XBTs. Any mix of the two will necessarily degrade the average to the quality of the MBT data and bias averages to shallow (< 300 m) depths. In some applications, the level of degraded accuracy will be more than adequate and it is only necessary to state clearly and be aware of the intended application when mixing the data from these instruments. Also, one can expect distinct discontinuities as the data make the transition from a mix of measurements at shallower levels to strictly XBT data at greater depth.

Other important practical concerns in forming block averages have to do with the usual geographic location of oceanographic measurements. Consider the global distribution of all autumn measurements up to 1970 of the most common oceanographic observation, temperature profiles (Figure 4.1.1). It is surprising how frequently these observations lie along meridians of latitude or parallels of longitude. This makes it difficult to assign the data to any particular 5 or 10° square when the border of the square coincides with integer values of latitude or longitude. When the latter occurs, one must decide to which square the borders will be assigned and be consistent in carrying this definition through the calculation of the mean values.

As illustrated by Figure 4.1.1, data coverage can be highly nonuniform. In this example, some areas were frequently sampled while others were seldom (or never) occupied. Such nonuniformity in data coverage is a primary factor in considering the representativeness of simple block averages. It certainly brings into question the assumptions of homogeneity (spatially uniform sampling distribution) and isotropy (uniform sampling regardless of direction) since the sample distribution varies greatly with location and may often have a preferred orientation. The situation becomes even more severe when one examines the quality of the data in the individual casts represented by the dots in Figure 4.1.1. In order to establish a truly consistent data set in terms of the quality of the observations (i.e. the depth of the cast, the number of samples, the availability of oxygen and nutrients, and so on), it is generally necessary to reject many of the available hydrographic casts.

The question of data coverage depends on the kind of scientific questions the data set is being asked to address. For problems not requiring high-quality hydrographic ·stations, a greater number of observations are available, while for more restrictive studies requiring a higher accuracy, far fewer casts would match the qualifications. This is also true for other types of historical data but is less true of newly collected data. However, even now, one must ensure that all observations have a similar level of accuracy and reliability. Variations in equipment performance, such as sensor response or failure,' must be compensated for in order to keep the observations consistent. Also, changes in instrument calibration need to be taken into account over the duration of a sampling program. For example, transmissometer lenses frequently become matted with a biotic film that reduces the amount of light passing between the source and receiver lenses. A nonlinear, time-dependent calibration is needed to correct for this effect.

Despite the potential problems with the block-averaging approach to data presentation, much information can be provided by careful consideration of the data rather than the use of more objective statistical methods to judge data quality. The shift to statistical methods represents a transition from the traditional oceanographic efforts of the early part of the twentieth century when considerable importance was given to every measurement value. In those days, individual scientists were personally responsible for the collection, processing and quality of their data.

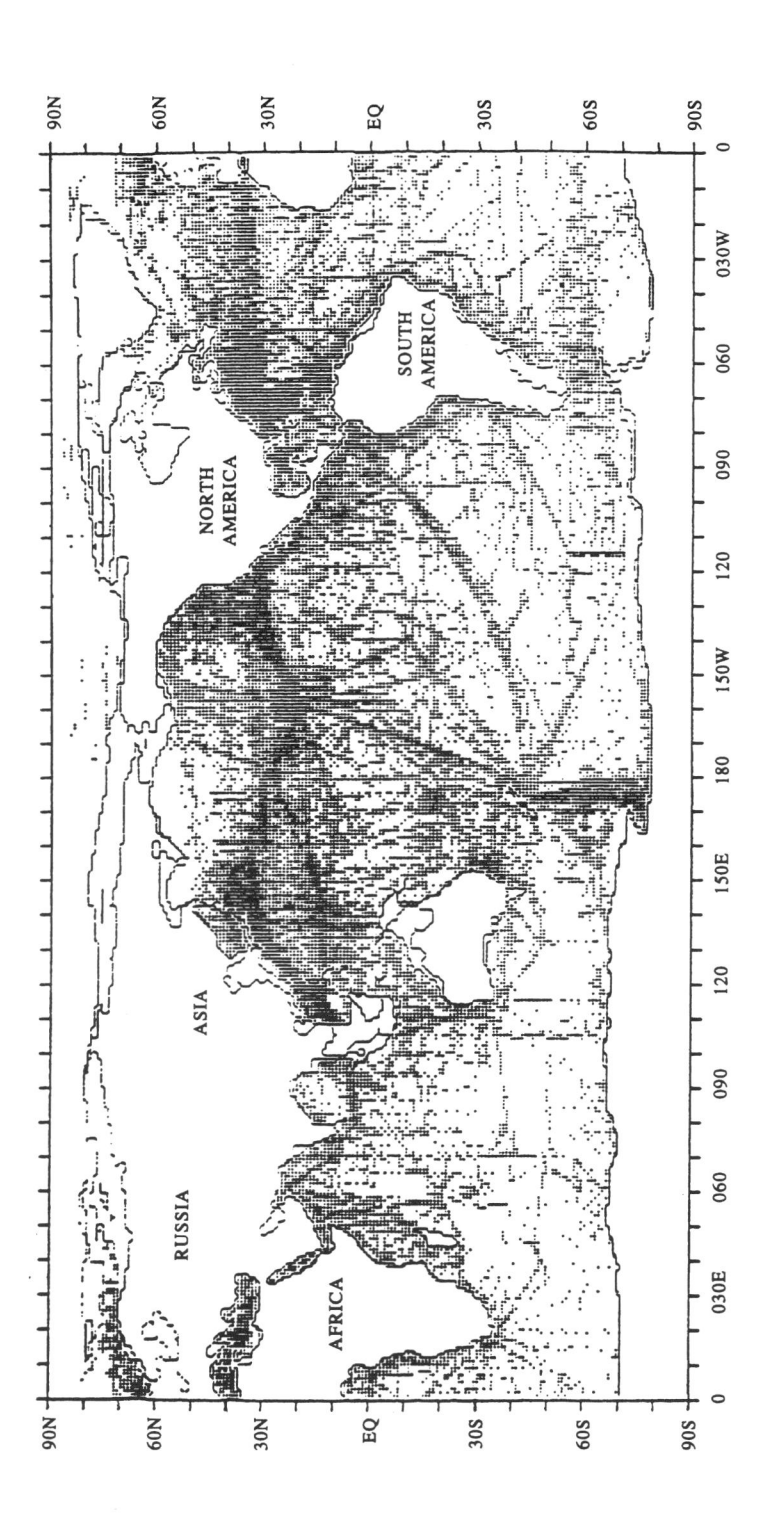

308 *Data Analysis Methods in Physical Oceanography*

Figure 4.1.1. The global distribution of all temperature profiles collected during oceanographic surveys in the fall up to 1970. Sampling is most dense along major shipping routes.

Then, it was a simple task to differentiate between "correct" and "incorrect" samples without having to resort to statistical methods to indicate how well the environment had been observed. In addition, earlier investigations were primarily concerned with defining the mean state of the ocean. Temporal variability was sometimes estimated but was otherwise ignored in order to emphasize the mean spatial field. With today's large volumes of data, it is no longer possible to "hand check" each data value. A good example is provided by satellite-sensed information which generally consists of large groupings of data that are usually treated as individual data values.

In anticipation of our discussion of filtering in Chapter 5, we should point out that block averaging corresponds to the application of a box-car-shaped filter to the data series. This type of filter has several negative characteristics such as a slow filter roll off and large side lobes which distort the information in the original data series.

4.2 OBJECTIVE ANALYSIS

In a general sense, *objective analysis* is an estimation procedure which can be specified mathematically. The form of objective analysis most widely used in physical oceanography is that of least squares optimal interpolation, more appropriately referred to as *Gauss–Markov smoothing*, which is essentially an application of the linear estimation (smoothing) techniques discussed in Chapter 3. Since it is generally used to map spatially nonuniform data to a regularly spaced set of gridded values, Gauss–Markov smoothing might best be called "Gauss–Markov mapping". The basis for the technique is the Gauss–Markov theorem which was first introduced by Gandin (1965) to provide a systematic procedure for the production of gridded maps of meteorological parameters. If the covariance function used in the Gauss–Markov mapping is the covariance of the data field (as opposed to a more *ad hoc* covariance function, as is usually the case), then Gauss–Markov smoothing is optimal in the sense that it minimizes the mean square error of the objective estimates. A similar technique, called Kriging after a South African engineer H. G. Krige, was developed in mining engineering. Oceanographic applications of this method are provided by Bretherton *et al.* (1976), Freeland and Gould (1976), Bretherton and McWilliams (1980), Hiller and Käse (1983), Bennett (1992), and others.

The two fundamental assumptions in optimal interpolation are that the statistics of the subject data field are stationary (unchanging over the sample period of each map) and homogeneous (the same characteristics over the entire data field). A further assumption often made to simplify the analysis is that the statistics of the second moment, or covariance function, are isotropic (the same structure in all directions). Bretherton *et al.* (1976) point out that if these statistical characteristics are known, or can be estimated for some existing data field (such as a climatology based on historical data), they can be used to design optimum measurement arrays to sample the field. Since the optimal estimator is linear and consists of a weighted sum of all the observations within a specified range of each grid point, the objective mapping procedure produces a smoothed version of the original data field that will tend to underestimate the true field. In other words, if an observation point happens to coincide with an optimally interpolated grid point, the observed value and interpolated value will probably not be equal due to the presence of noise in the data. The degree of smoothing is determined by the characteristics of the signal and error covariance functions used in the mapping and increases with increasing spatial scales for a specified covariance function.

The general problem is to compute an estimate $\hat{D}(\mathbf{x}, t)$ of the scalar variable $D(\mathbf{x}, t)$ at a position $\mathbf{x} = (x, y)$ from irregularly spaced and inexact observations $d(\mathbf{x}_n, t)$ at a limited number of data positions $\mathbf{x}_n$ ($n = 1, 2, ..., N$). Implementation of the procedure requires *a priori* knowledge of the variable's covariance function, $C(\mathbf{r})$, and uncorrelated error variance, ε, where $\mathbf{r}$ is the spatial separation between positions. For isotropic processes, $C(\mathbf{r}) \to C(r)$, where $r = |\mathbf{r}|$. Although specification of the covariance matrix should be founded on the observed structure of oceanic variables, selection of the mathematical form of the covariance matrix is hardly an "objective" process even with reliable data (cf. Denman and Freeland, 1985). In addition to the assumptions of stationarity, homogeneity, and isotropy, an important constraint on the chosen covariance matrix is that it must be positive definite (no negative eigenvalues). Bretherton *et al.* (1976) report that objective estimates computed from nonpositive definite matrices are not optimal and the mapping results are poor. In fact, nonpositive definite covariance functions can yield objective estimates with negative expected square error. One way to ensure that the covariance matrix is positive definite is to fit a function which results in a positive definite covariance matrix to the sample covariance matrix calculated from the data (Hiller and Käse, 1983). This results in a continuous mathematical expression to be used in the data weighting procedure. In attempting to specify a covariance function for data collected in continental shelf waters, Denman and Freeland (1985) further required that $\partial^2 C/\partial^2 x$ and $\partial^2 C/\partial^2 y$ be continuous at $r = 0$ (to ensure a continuously differentiable process) and that the variance spectrum, $S(k)$, derived from the transform of $C(\mathbf{r})$ be integrable and nonnegative for all wavenumbers, $\mathbf{k}$ (to ensure a realizable stochastic random process).

Calculation of the covariance matrix requires that the mean and "trend" be removed from the data (the trend is not necessarily linear). In three-dimensional space, this amounts to the removal of a planar or curvilinear surface. For example, the mean density structure in an upwelling domain is a curved surface which is shallow over the outer shelf and deepens seaward. Calculation of the density covariance matrix for such a region first involves removal of the curved mean density surface (Denman and Freeland, 1985). Failure to remove the mean and trend would not alter the fact that our estimates are optimal but it would redistribute variability from unresolved larger scales throughout the wavenumber space occupied by the data. We would then map features that have been influenced by the trend and mean.

As discussed later in the section on time series, there are many ways to estimate the trend. If ample good-quality historical data exist, the trend can be estimated from these data and then subtracted from the data being investigated. If historical data are not available, or the historical coverage is inadequate, then the trend must be computed from the sample data set itself. Numerous methods exist for calculating the trend and all require some type of functional fit to the existing data using a least-squares method. These functions can range from straight lines to complex higher-order polynomials and associated nonlinear functions. We note that, although many candidate oceanographic data fields do not satisfy the conditions of stationarity, homogeneity, and isotropy, their anomaly fields do. In the case of anomaly fields, the trend and mean have already been removed. Gandin (1965) reports that it may be possible to estimate the covariance matrix from existing historical data. This is more often the case in meteorology than in oceanography. In most oceanographic applications, the analyst must estimate the covariance matrix from the data set being studied.

In the following, we present a brief outline of objective mapping procedures. The interested reader is referred to Gandin (1965) and Bretherton et al. (1976) for further details. As noted previously, we consider the problem of constructing a gridded map of the scalar variable $D(\mathbf{x}, t)$ from an irregularly spaced set of scalar measurements $d(\mathbf{x}, t)$ at positions $\mathbf{x}$ and times t. The notation $\mathbf{x}$ refers to a suite of measurement sites, x_n ($n = 1, 2, ...$), each with distinct (x, y) coordinates. We use the term variable to mean directly measured oceanic variables as well as calculated variables such as the density or streamfunction derived from the observations. Thus, the data $d(\mathbf{x}, t)$ may consist of measurements of the particular variable we are trying to map or they may consist of some other variables that are related to D in a linear way. The former case gives

$$d(\mathbf{x}, t) = D(\mathbf{x}, t) + \varepsilon(\mathbf{x}) \tag{4.2.1}$$

where the ε are zero-mean measurement errors which are not correlated with the measurement D. In the latter case

$$d(\mathbf{x}, t) = F[D(\mathbf{x}, t)] + \varepsilon(\mathbf{x}) \tag{4.2.2}$$

in which F is a linear functional which acts on the function D in a linear fashion to give a scalar (Bennett 1992). For example, if $D(\mathbf{x}, t) = \Psi(\mathbf{x}, t)$ is the streamfunction, then the data could be current meter measurements of the zonal velocity field, $u(\mathbf{x}, t) = F[\Psi(\mathbf{x}, t)]$, where

$$d(\mathbf{x}, t) = u(\mathbf{x}, t) + \varepsilon(\mathbf{x}) = -\frac{\partial \Psi(\mathbf{x})}{\partial y} + \varepsilon(\mathbf{x}) \tag{4.2.3}$$

and $\partial \Psi / \partial y$ is the gradient of the streamfunction in the meridional direction.

To generalize the objective mapping problem, we assume that mean values have *not* been removed from the original data prior to the analysis. If we consider the objective mapping for a single "snap shot" in time (thereby dropping the time index, t), we can write linear estimates $\hat{D}(\mathbf{x})$ of $D(\mathbf{x})$ as the summation over a weighted set of the measurements d_i ($i = 1, ..., N$)

$$\hat{D}(\mathbf{x}) = \overline{D}(\mathbf{x}) + \sum_{i=1}^{N} b_i (d_i - \overline{d}) \tag{4.2.4}$$

where the overbar denotes an expected value (mean), $d_i = d(\mathbf{x}) = d(x_i)$, $1 \leq i \leq N$ is shorthand notation for the data values, and the $b_i = b(\mathbf{x}) = b(x_i)$ are, as yet unspecified, weighting coefficients at the data points x_i. The selection of the N data values is made by restricting these values to some finite area about the grid point. The estimates of the parameters b_i in equation (4.2.4) are found in the usual way by minimizing the mean square variance of the error $e(\mathbf{x})^2$ between the measured variable, D, and the linear estimate, $\hat{D}$, at the data location. In particular,

$$\overline{e(\mathbf{x})^2} = \overline{\left[D(\mathbf{x}) - \hat{D}(\mathbf{x})\right]^2} \tag{4.2.5}$$

which on substitution of (4.2.4) yields

$$\overline{e(\mathbf{x})^2} = \overline{[D(\mathbf{x}) - \overline{D}(\mathbf{x})]^2} + \sum_{i=1}^{N}\sum_{j=1}^{N} b_i b_j \overline{(d_i - \overline{d})(d_j - \overline{d})} - 2\sum_{i=1}^{N} b_i \overline{(d_i - \overline{d})(D - \overline{D})} \quad (4.2.6)$$

Note, that if the mean has been removed, we can set $\overline{D}(\mathbf{x}) = \overline{d}(\mathbf{x}) = 0$ in (4.2.6). The mean square difference in equation (4.2.6) is minimized when

$$b_i = \sum_{j=1}^{N} \{[\overline{(d_i - \overline{d})(d_j - \overline{d})}]^{-1} \overline{(d_j - \overline{d})(D - \overline{D})}\} \quad (4.2.7)$$

To calculate the weighting coefficients in (4.2.7), and therefore the grid-value estimates in (4.2.4), we need to compute the covariance matrix by averaging over all possible pairs of data taken at points x_i, x_j; the covariance matrix is

$$\overline{(d_i - \overline{d})(d_j - \overline{d})} = \overline{(d(x_i) - \overline{d})(d(x_j) - \overline{d})} \quad (4.2.8)$$

We do the same for the interpolated value

$$\overline{(d_i - \overline{d})(D_j - \overline{D})} = \overline{(d(x_i) - \overline{d})(d(x_k) - \overline{D})} \quad (4.2.9)$$

where x_k is the location vector for the grid point estimate $\hat{D}(x_k)$.

In general, we need a series of measurements at each location so that we can obtain statistically reliable expected values for the elements of the covariance matrices in (4.2.8) and (4.2.9). The expected values in the above relations could be computed as ensemble averages over spatially distributed sets of measurements. Typically, however, we have only one set of measurements for the specified locations x_i, x_j. As a consequence, we need to assume that, for the region of study, the data statistics are homogeneous, stationary and isotropic. If these conditions are met, the covariance matrix for the data distribution (for example, sea surface temperature) depends only on the distance r between data values, where $r = |x_j - x_i|$. Thus, we have elements i, j, of the covariance matrix given by

$$\begin{aligned}\overline{(d_i - \overline{d})(d_j - \overline{d})} &= C(|x_j - x_i|) + \overline{\varepsilon^2} \\ \overline{(d_i - \overline{d})(D_j - \overline{D})} &= C(|x_j - x_k|) + \overline{\varepsilon^2}\end{aligned} \quad (4.2.10)$$

where $C(|\mathbf{r}|) = \overline{d(\mathbf{x})d(\mathbf{x} + \mathbf{r})}$ is the covariance matrix and the mean square error $\varepsilon(\mathbf{x})^2$ implies that this estimate is not exact and there is some error in the estimation of the correlation function from the data. We note that this is not the same error in (4.2.6) that we minimize to solve for the weights in (4.2.7). The matrix can now be calculated by forming pairs of observed data values separated into bins according to the distance between sample sites, x_i. These are then averaged over the number of pairs that have the same separation distance to yield the product matrix

$$\overline{(d_i - \overline{d})(d_j - \overline{d})}$$

This computation requires us to define some "bin interval" for the separation distances so that we can group the product values together. To ensure that the resulting covariance matrix meets the condition of being positive definite, a smooth function satisfying this requirement can be fitted to the computed raw covariance function. This fitted covariance function is used for

$$\overline{(d_i - \bar{d})(D - \bar{D})}$$

and to calculate

$$\left[\overline{(d_i - \bar{d})(d_i - \bar{d})}\right]^{-1}$$

The weights b_i are then computed from (4.2.7). It is a simple process to then compute the optimal grid value estimates from (4.2.4). Note that, where the data provide no help in the estimate of D (that is, $\varepsilon(\mathbf{x}) \to \infty$), then $b_i = 0$ and the only reasonable estimate is $\hat{D}(\mathbf{x}) = \bar{D}$, the mean value. Similarly, if the data are error free (such that $\varepsilon(\mathbf{x})^2 \to 0$), then $\hat{D}(x_i) = D(x_i)$ for all x_i ($i = 1, \ldots, N$). In other words, the estimated value and the measured data are identical at the measurement sites (within the limits of the noise in the data) and the estimator interpolates between the observations.

The critical step in the objective mapping procedure is the computation of the covariance matrix. We have described a straightforward procedure to estimate the covariance matrix from the sample data. As with the estimate of the mean or overall trend, it is often possible to use an existing set of historical data to compute the covariance matrix. This is frequently the case in meteorological applications where long series of historical data are available. In oceanography, however, the covariance matrix typically must be computed from the sample data. Where historical data are available, it is important to recognize that using these data to estimate the covariance matrix for use with more recently collected data is tantamount to assuming that the statistics have remained stationary since the time that the historical data were collected.

Bretherton et al. (1976) suggest that objective analysis can be used to compute the covariance matrix. In this case, they start with an assumed covariance function, $\hat{F}$, which is then compared with a covariance function computed from data with a fixed distance x_o. The difference between the model $\hat{F}$ and the real F computed from the data is minimized by repeated iteration.

To this point, we have presented objective analysis as it applies to scalar fields. We can also apply optimal (Gauss–Markov) interpolation to vector fields. One approach is to examine each scalar velocity component separately so that for n velocity vectors we have $2n$ velocity components

$$d_r = u_1(\mathbf{x}_r); \quad d_{r+n} = u_2(\mathbf{x}_r) \tag{4.2.11}$$

where u_1 and u_2 are the x, y velocity components at x_r. If the velocity field is nondivergent, we can introduce a scalar streamfunction $\Psi(\mathbf{x})$ such that

$$u_1 = -\frac{\partial \Psi}{\partial y}; \quad u_2 = \frac{\partial \Psi}{\partial x} \tag{4.2.12}$$

and apply scalar methods to Ψ.

Once the optimal interpolation has been executed, there is a need to return to equation (4.2.6) to compute the actual error associated with each optimal interpolation. To this end, we note that we now have the interpolated data from (4.2.4). Thus, we can use $\hat{D}$ computed from (4.2.4) as the value for D in (4.2.6). The product in

314 *Data Analysis Methods in Physical Oceanography*

the last term of (4.2.6) is computed from the covariance in (4.2.9). In this way, it is possible to compute the error associated with each optimally interpolated value. Frequently, this error field is plotted for a specific threshold level, typically 50% of the interpolated values in the mapped field (see following examples). It is important to retain this error estimate as part of the optimal interpolation since it enables us to assess the statistical significance of individual gridded values.

4.2.1 Objective mapping: examples

An example of objective mapping applied to a single oceanographic survey is provided by the results of Hiller and Käse (1983). The data are from a CTD survey grid occupied in the North Atlantic about midway between the Azores and the Canary Islands (Figure 4.2.1). At each CTD station, the geopotential anomaly at 25 db (dBar) relative to the anomaly at 1500 db (written 25/1500 db) was calculated and selected as the variable to be mapped. The two-dimensional correlation function for these data is shown in three-dimensional perspective in Figure 4.2.2(a). A series of different correlation functions were examined and an isotropic, Gaussian function that was positive definite was selected as the best fit (Figure 4.2.2b). Using this covariance function, the authors obtained the objectively mapped 25/1500 db geopotential anomaly shown in Figure 4.2.3(a). Removal of a linear trend gives the objective map shown in Figure 4.2.3(b) and the associated RMS error field shown in Figure 4.2.3(c). Only near the outside boundaries of the data domain does the RMS error increase to around 50% of the geopotential anomaly field (Figure 4.2.3b).

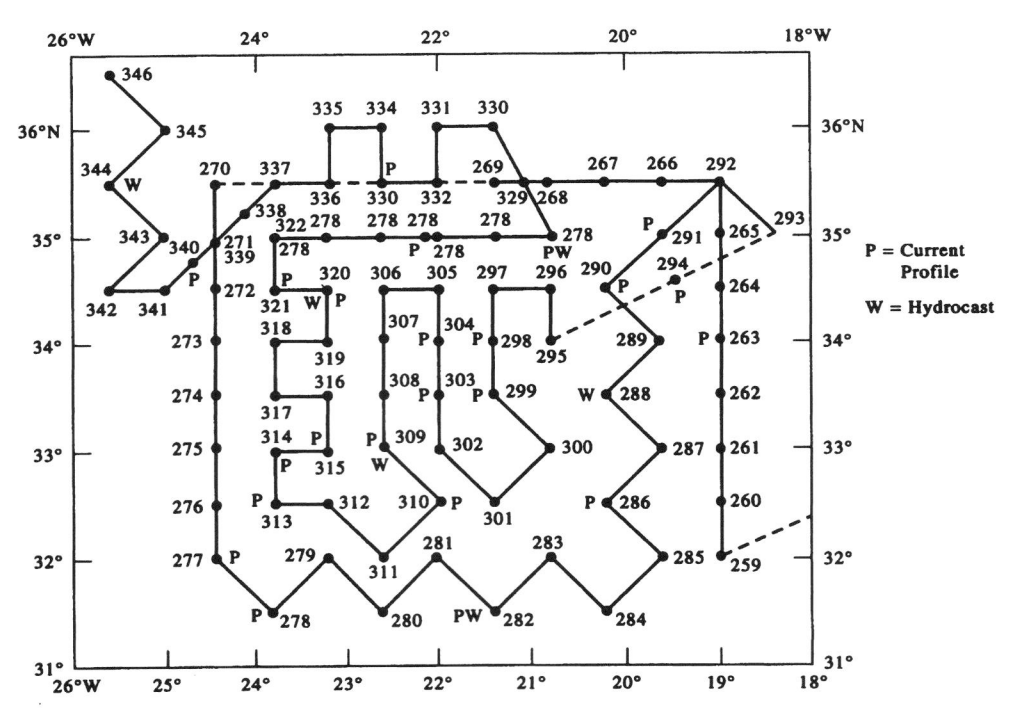

Figure 4.2.1. Locations of CTD stations taken in the North Atlantic between the Azores and the Canary Islands in spring 1982 (experiment POSEIDON 86, Hiller and Käse, 1983). Also shown are locations of current profile (P) and hydrocast (W) stations.

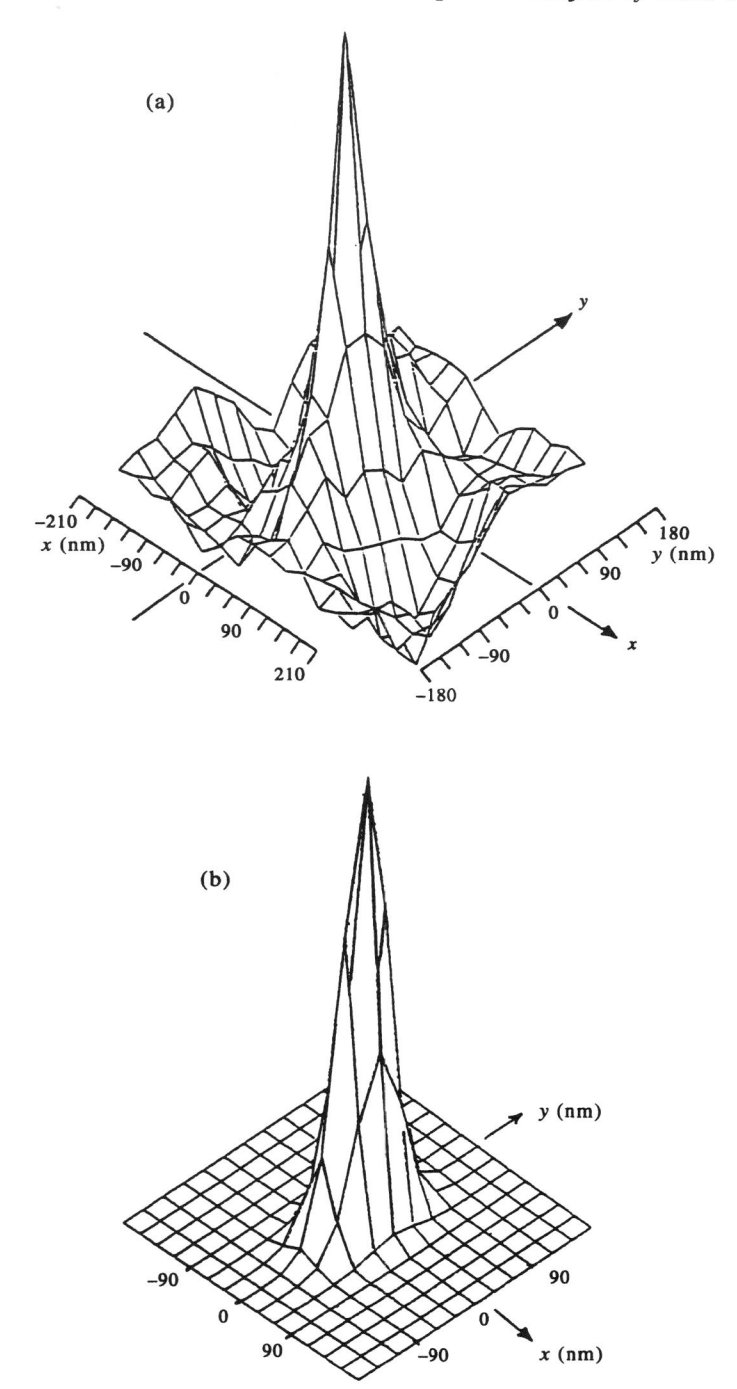

Figure 4.2.2. The two-dimensional correlation function $C(\mathbf{r})$ for the geopotential anomaly field at 25 db referenced to 1500 db (25/1500 dBar) for the data collected at stations shown in Figure 4.2.1 (1 db = 1 dBar = 1 m^2/s^2). Here, $\mathbf{r} = (x, y)$, where x, y are the eastward and northward coordinates, respectively. Distances are in nautical miles. (a) The "raw" values of $C(\mathbf{r})$ based on the observations; (b) A model of the correlation function fitted to (a). (From Hiller and Käse, 1983).

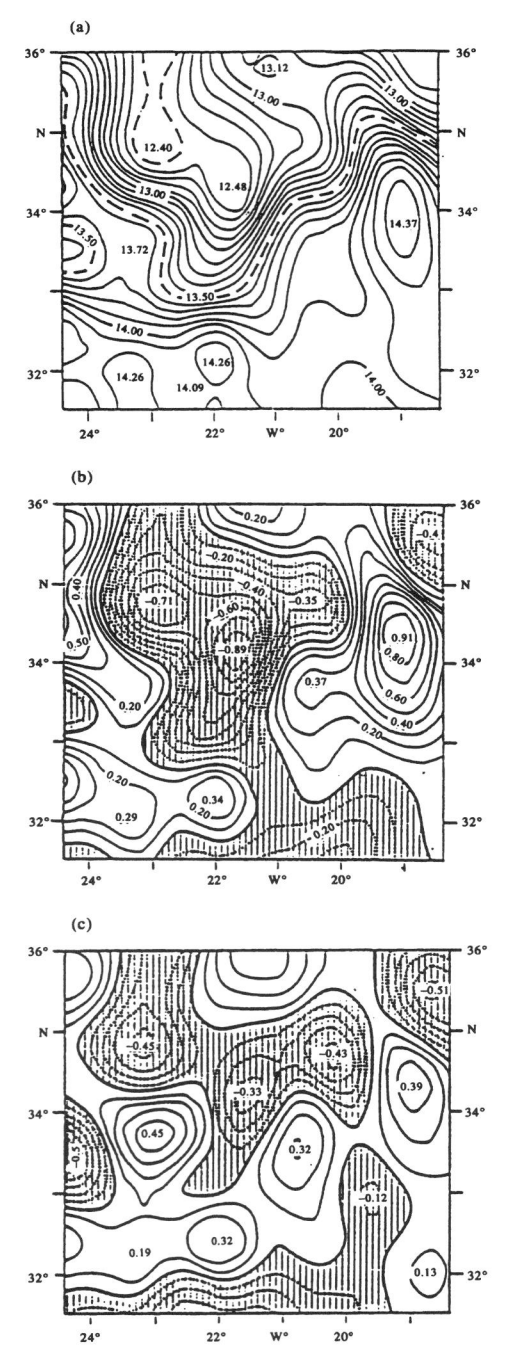

Figure 4.2.3. Objective analysis of the geopotential anomaly field 25/1500 db (m^2/s^2) using the correlation function in equation (4.2.2b). (a) The approximate center of the frontal band in this region of the ocean is marked by the 13.5 db isoline; (b) Same as (a) but after subtraction of the linear spatial trend; (c) Objective analysis of the residual mesoscale perturbation field 25/1500 dBar after removal of the composite mean field. (After Hiller and Käse, 1983.)

The Spatial Analyses of Data Fields 317

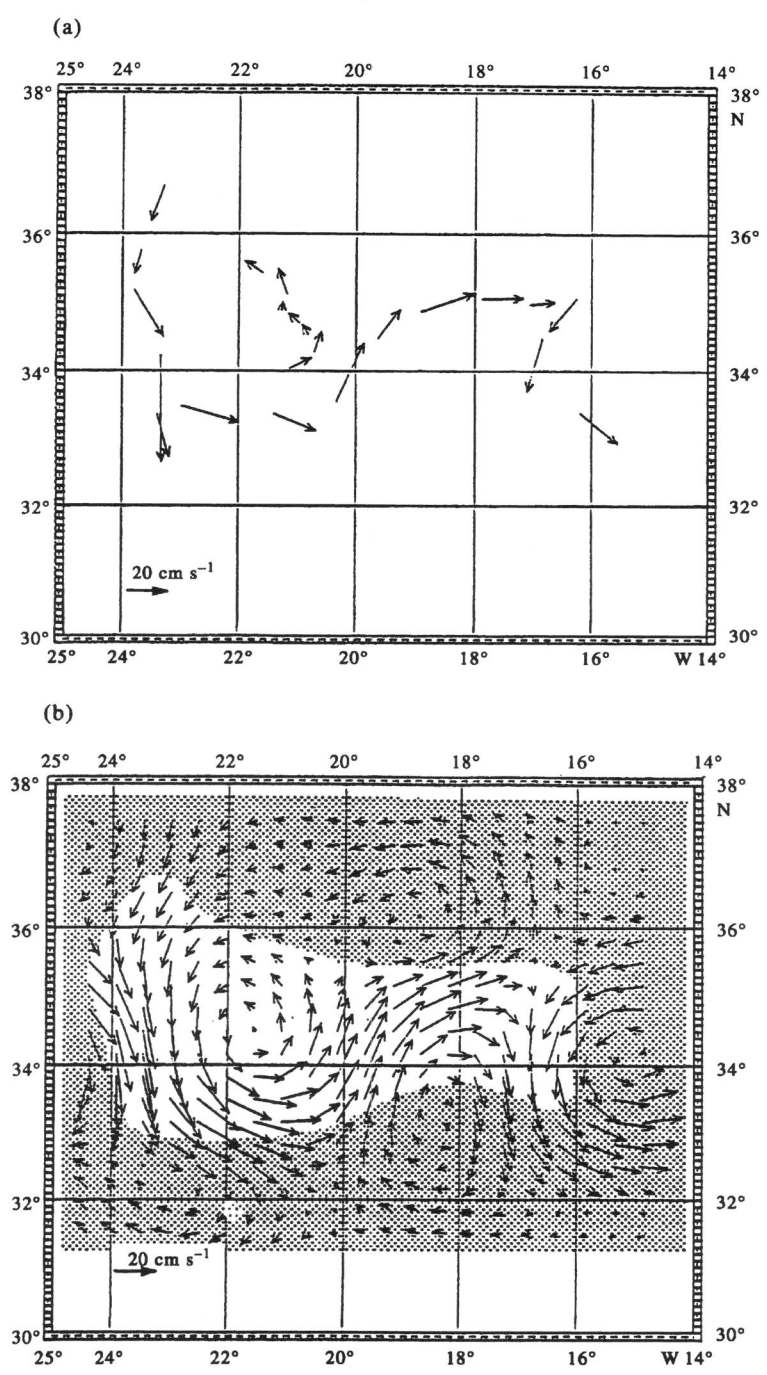

Figure 4.2.4. Analysis of the velocity field for the current profile collected on the grid in Figure 4.2.1. (a) The input velocity field; (b) Objective analysis of the input velocity field with correlation scale $\lambda = 200$ km and assumed noise variance of 30% of the total variance of the field. This approach treats mesoscale variability on scales less than 200 km as noise, which is smoothed out. In the shaded area, the error variance exceeds 50% of the total variance.

As an example of objective mapping applied to a vector field, Hiller and Käse (1983) examined a limited number of satellite-tracked drifter trajectories that coincided with the CTD survey in space and time. Velocity vectors based on daily averages of low-passed finite difference velocities are shown in Figure 4.2.4(a). Rather than compute a covariance function for this relatively small sample, the covariance function from the analysis of the 25/1500 db geopotential anomaly was used. Also, an assumed error level, $\overline{\varepsilon^2}$, was used rather than a computed estimate from the small sample. With the isotropic correlation scale estimated to be 75 km, the objective mapping produces the vector field in Figure 4.2.4(b). The stippled area in this figure corresponds to the region where the error variance exceeds 50% of the total variance. Due to the paucity of data, the area of statistically significant vector mapping is quite limited. Nevertheless, the resulting vectors are consistent with the geopotential height map in Figure 4.2.3(a).

Another example is provided by McWilliams (1976) who used dynamic height relative to 1500 m depth plus deep float velocities at 1500 m to estimate the streamfunction field. The isotropic covariance function for the random fluctuations in streamfunction $\Psi' = \Psi - \overline{\Psi}$ at 1500 m depth was

$$\begin{aligned} C(r) &= \overline{\Psi'(\mathbf{x},z,t)\Psi'(\mathbf{x}+\mathbf{r},z,t)} \\ &= \overline{\Psi'^2}(1-\varepsilon^2)(1-\gamma^2 r^2)\exp(-\tfrac{1}{2}\delta^2 r^2) \end{aligned} \quad (4.2.13)$$

where $\mathbf{r}$ is a horizontal separation vector, $r = |\mathbf{r}|$, ε is an estimate of relative measurement noise ($0 \leq \varepsilon \leq 1$), and γ^{-1}, δ^{-1} are decorrelation length scales found by fitting equation (4.2.13) to prior data. Denman and Freeland (1985) discuss the merits of five different covariance functions fitted to geopotential height data collected over a period of three years off the west coast of Vancouver Island. For other examples, the reader is referred to Bennett (1992).

As a final point, we remark that the requirement of isotropy is easily relaxed by using direction-dependent covariance matrices, $C(r_1, r_2)$ whose spatial structure depends on two orthogonal spatial coordinates, r_1 and r_2 (with $r_2 \geq r_1$). For example, the map of light attenuation coefficient at 20 m depth obtained from transmissometer profiles off the west coast of Vancouver Island (Figure 4.2.5) uses an exponentially decaying, elliptically shaped covariance matrix

$$C(r_1, r_2) = \exp[-a\Delta x^2 - b\Delta y^2 - c\Delta x \Delta y] \quad (4.2.14a)$$

where

$$\begin{aligned} a &= \tfrac{1}{2}\left\{[\cos(\pi\phi/180)/r_1]^2 + [\sin(\pi\phi/180)/r_2]^2\right\} \\ b &= \tfrac{1}{2}\left\{[\sin(\pi\phi/180)/r_1]^2 + [\cos(\pi\phi/180)/r_2]^2\right\} \\ c &= \cos(\pi\phi/180)\sin(\pi\phi/180)\left[r_2^2 - r_1^2\right]/(r_1 r_2)^2 \end{aligned} \quad (4.2.14b)$$

Here, Δx and Δy are, respectively, the eastward and northward distances from the grid point to the data point, and ϕ is the orientation angle (in degrees) of the coastline measured counterclockwise from north. In this case, it is assumed that the alongshore correlation scale, r_2, is twice the across-shore correlation scale, r_1. The idea here is that, like water-depth changes, alongshore variations in coastal water properties such as temperature, salinity, geopotential height, and log-transformed phytoplankton chlorophyll-*a* pigment concentration occur over longer length scales than across-shore variations.

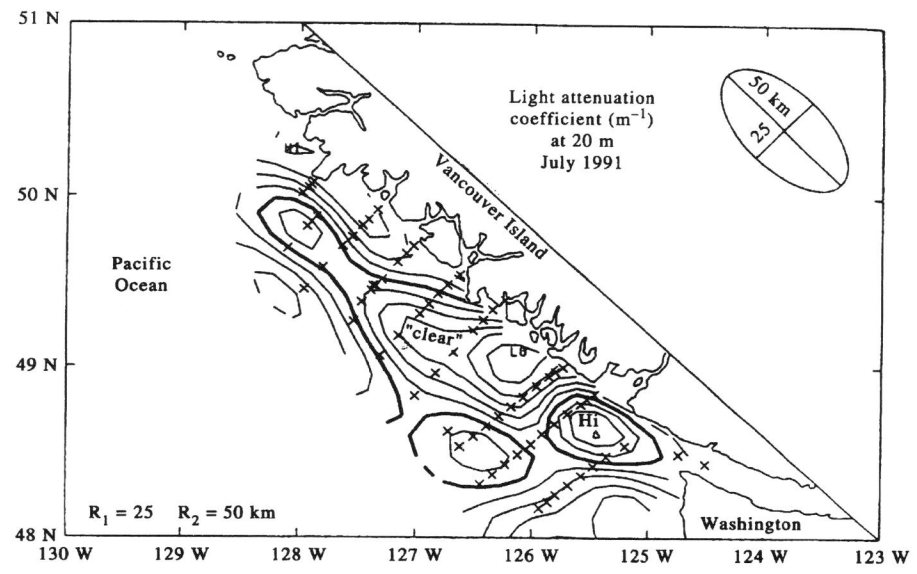

Figure 4.2.5. Objective analysis map of light attenuation coefficient (per meter) at 20 m depth on the west coast of Vancouver Island obtained from transmissometer profiles. The covariance function $C(r_1, r_2)$ given by the ellipse is assumed to decay exponentially with distance with the longshore correlation scale r_2 = 50 km and cross-shore correlation scale r_1 = 25 km.

4.3 EMPIRICAL ORTHOGONAL FUNCTIONS

The previous section dealt with the optimal smoothing of irregularly spaced data onto a gridded map. In other studies of oceanic variability, we may be presented with a large data set from a grid of time-series stations which we wish to compress into a smaller number of independent pieces of information. For example, in studies of climate change, it is necessary to deal with time series of spatial maps, such as surface temperature. A useful obvious choice would involve a linear combination of orthogonal spatial "predictors", or modes, whose net response as a function of time would account for the combined variance in all of the observations. The signals we wish to examine may all consist of the same variable, such as temperature, or they may be a mixture of variables such as temperature and wind velocity, or current and sea level. The data may be in the form of concurrent time-series records from a grid (regular or irregular) of stations $x_i(t), y_i(t)$ on a horizontal plane or time-series records at a selection of depths on an $x_i(t), z_i(t)$ cross-section. Examples of time series from cross-sectional data include those from a single current meter string or from along-channel moorings of thermistor chains.

A useful technique for compressing the variability in this type of time-series data is *principal component analysis* (PCA). In oceanography, the method is commonly known as *empirical orthogonal function* (EOF) analysis. The EOF procedure is one of a larger class of inverse techniques and is equivalent to a data reduction method widely used in the social sciences known as *factor analysis*. The first reference we could find to the application of EOF analysis to geophysical fluid dynamics is a report by Edward Lorenz (1956) in which he develops the technique for statistical weather prediction and coins the term "EOF".

The advantage of EOF analysis is that it provides a compact description of the spatial and temporal variability of data series in terms of orthogonal functions, or statistical "modes." Usually, most of the variance of a spatially distributed series is in the first few orthogonal functions whose patterns may then be linked to possible dynamical mechanisms. It should be emphasized that no direct physical or mathematical relationship necessarily exists between the statistical EOFs and any related dynamical modes. Dynamical modes conform to physical constraints through the governing equations and associated boundary conditions (LeBlond and Mysak, 1979); empirical orthogonal functions are simply a method for partitioning the variance of a spatially distributed group of concurrent time series. They are called "empirical" to reflect the fact that they are defined by the covariance structure of the specific data set being analyzed (as shown below).

In oceanography and meteorology, EOF analysis has found wide application in both the time and frequency domains. Conventional EOF analysis can be used to detect standing oscillations only. To study propagating wave phenomena, we need to use lagged covariance matrix (Weare and Nasstrom, 1982), or complex principal component analysis in the frequency domain (Wallace and Dickinson, 1972; Horel, 1984). Our discussion, in this section, will focus on space/time domain applications. Readers seeking more detailed descriptions of both the procedural aspects and their applications are referred to Lorenz (1956), Davis (1976), and Preisendorfer (1988).

The best analogy to describe the advantages of EOF analysis is the classical vibrating drum problem. Using mathematical concepts presented in most undergraduate texts, we know that we can describe the eigenmodes of drumhead oscillations through a series of two-dimensional orthogonal patterns. These modes are defined by the eigenvectors and eigenfunctions of the drumhead. Generally, the lowest modes have the largest spatial scales and represent the most dominant (most prevalent) modes of variability. Typically, the drumhead has as its largest mode an oscillation in which the whole drumhead moves up and down, with the greatest amplitude in the center and zero motion at the rim where the drum is clamped. The next highest mode has the drumhead separated in the center with one side 180° out of phase with other side (one side is up when the other is down). Higher modes have more complex patterns with additional maxima and minima. Now, suppose we had no mathematical theory, and were required to describe the drumhead oscillations in terms of a set of observations. We would look for the kinds of eigenvalues in our data that we obtain from our mathematical analysis. Instead of the analytical or dynamical solutions that can be derived for the drum, we wish to examine "empirical" solutions based strictly on a measured data set. Since we are ignorant of the actual dynamical analysis, we call the resulting modes of oscillation, empirical orthogonal functions.

EOFs can be used in both the time and frequency domains. For now, we will restrict ourselves to the time domain application and consider a series of N maps at times $t = t_i$ $(1 \leq i \leq N)$, each map consisting of scalar variables $\psi_m(t)$ collected at M locations, $\mathbf{x}_m (1 \leq m \leq M)$. One could think of N weather maps available every 6 h over a total period of $6N$ h, with each map showing the sea surface pressure $\psi_m(t) = P_m(t) (1 \leq m \leq M)$ recorded at M weather buoys located at mooring sites $\mathbf{x}_m = (x_m, y_m)$. Clearly, the subscript m refers to the spatial grid locations in each map.

Alternatively, the N maps might consist of pressure data $P(t)$ from $M - K$ weather buoys plus velocity component records $u(t)$, $v(t)$ from $K/2$ current meter sites. Or, again the time series could be from $M/2$ current meters on a moored string. Any combination of scalars is permitted (remember, this is a statistical analysis not a dynamical analysis). The goal of this procedure is to write the data series $\psi_m(t)$ at any given location $\mathbf{x}_m$ as the sum of M orthogonal spatial functions $\phi_i(\mathbf{x}_m) = \phi_{im}$ such that

$$\psi(\mathbf{x}_m, t) = \psi_m(t) = \sum_{i=1}^{M} [a_i(t)\phi_{im}] \quad (4.3.1)$$

where $a_i(t)$ is the amplitude of the ith orthogonal mode at time $t = t_n$ ($1 \leq n \leq N$). Simply put, equation (4.3.1) says that the time variation of the dependent scalar variable $\psi(\mathbf{x}_m, t)$ at each location $\mathbf{x}_m$ results from the linear combination of M spatial functions, ϕ_i, whose amplitudes are weighted by M time-dependent coefficients, $a_i(t)$, ($1 \leq i \leq M$). The weights $a_i(t)$ tell us how the spatial modes ϕ_{im} vary with time. There are as many (M) basis functions as there are stations for which we have data. Put another way, we need as many modes as we have time-series stations so that we can account for the combined variance in the original time series at each time, t. We can also formulate the problem as M temporal functions whose amplitudes are weighted by M spatially variable coefficients. Whether we partition the data as spatial or temporal orthogonal functions the results are identical.

Since we want the $\phi_i(\mathbf{x}_m)$ to be orthogonal, so that they form a set of basis functions, we require

$$\sum_{m=1}^{M} [\phi_{im}\phi_{jm}] = \delta_{ij} \text{ (orthogonality condition)} \quad (4.3.2)$$

where the summation is over all observation locations and δ_{ij} is the Kronecker delta

$$\delta_{ij} = \begin{cases} 1, & j = i \\ 0, & j \neq i \end{cases} \quad (4.3.3)$$

It is worth remarking that two functions are said to be orthogonal when the sum (or integral) of their product over a certain defined space (or time) is zero. Orthogonality in equation (4.3.2) does not mean $\phi_{im}\phi_{jm} = 0$ for each m. For example, in the case of continuous sines and cosines, $\int \sin\theta \cos\theta \, d\theta = 0$ when the integral is over a complete phase cycle, $0 \leq \theta \leq 2\pi$. By itself, the product $\sin\theta \cdot \cos\theta = 0$ only if the sine or cosine term happens to be zero.

There is a multitude of basis functions, ϕ_i, that can satisfy equations (4.3.1) and (4.3.2). Sine, cosine, and Bessel functions come to mind. The EOFs are determined uniquely among the many possible choices by the constraint that the time amplitudes $a_i(t)$ are uncorrelated over the sample data. This requirement means that the time-averaged covariance of the amplitudes satisfies

$$\overline{a_i(t)a_j(t)} = \lambda_i \delta_{ij} \text{ (uncorrelated time variability)} \quad (4.3.4)$$

in which the overbar denotes the time-averaged value and

$$\lambda_i = \overline{a_i(t)^2} = \frac{1}{N} \sum_{n=1}^{N} [a_i(t_n)^2] \quad (4.3.5)$$

is the variance in each orthogonal mode. If we then form the covariance matrix

322 *Data Analysis Methods in Physical Oceanography*

$\psi_m(t)\psi_k(t)$ for the known data and use (4.3.4), we find

$$\overline{\psi_m(t)\psi_k(t)} = \overline{\sum_{i=1}^{M}\sum_{j=1}^{M}\left[a_i(t)a_j(t)\phi_{im}\phi_{jk}\right]} \quad (4.3.6)$$

$$= \sum_{i=1}^{M}\left[\lambda_i\phi_{im}\phi_{ik}\right]$$

Multiplying both sides of (4.3.6) by ϕ_{ik}, summing over all k and using the orthogonality condition (4.3.2), yields

$$\sum_{k=1}^{M}\overline{\psi_m(t)\psi_k(t)}\phi_{ik} = \lambda_i\phi_{im} \quad (i\text{th mode at the }m\text{th location}; i, m = 1, ..., M) \quad (4.3.7)$$

Equation (4.3.7) is the canonical form for the *eigenvalue problem*. Here, the EOFs, ϕ_{im}, are the ith *eigenvectors* at locations $\mathbf{x}_m$, and the mean-square time amplitudes

$$\lambda_i = \overline{a_i(t)^2}$$

are the corresponding *eigenvalues* of the mean product, $\mathbf{R}$, which has elements

$$R_{mk} = \overline{\psi_m(t)\psi_k(t)}$$

This is equal to the covariance matrix, $\mathbf{C}$, if the mean values of the time series $\psi_m(t)$ have been removed at each site $\mathbf{x}_m$. The total of M empirical orthogonal functions corresponding to the M eigenvalues of (4.3.7) forms a complete basis set of linearly independent (orthogonal) functions such that the EOFs are uncorrelated modes of variability. Assuming that the record means $\psi_m(t)$ have been removed from each of the M time series, equation (4.3.7) can be written more concisely in matrix notation as

$$\mathbf{C}\boldsymbol{\phi} - \lambda\mathbf{I}\boldsymbol{\phi} = 0 \quad (4.3.8)$$

where the covariance matrix, $\mathbf{C}$, consists of M data series of length N with elements

$$C_{mk} = \overline{\psi_m(t)\psi_k(t)}$$

$\mathbf{I}$ is the unity matrix, and $\boldsymbol{\phi}$ are the EOFs. Expanding (4.3.8) yields the eigenvalue problem

$$\begin{pmatrix} \overline{\psi_1(t)\psi_1(t)} & \overline{\psi_1(t)\psi_2(t)} & \cdots & \overline{\psi_1(t)\psi_M(t)} \\ \overline{\psi_2(t)\psi_1(t)} & \overline{\psi_2(t)\psi_2(t)} & \cdots & \overline{\psi_2(t)\psi_M(t)} \\ \cdots & \cdots & \cdots & \cdots \\ \overline{\psi_M(t)\psi_1(t)} & \overline{\psi_M(t)\psi_2(t)} & \cdots & \overline{\psi_M(t)\psi_M(t)} \end{pmatrix} \begin{pmatrix} \phi_1 \\ \phi_2 \\ \cdots \\ \phi_M \end{pmatrix} = \begin{pmatrix} \lambda & 0 & \cdots & 0 \\ 0 & \lambda & \cdots & 0 \\ \cdots & & & \\ 0 & & \cdots & \lambda \end{pmatrix} \begin{pmatrix} \phi_1 \\ \phi_2 \\ \cdots \\ \phi_M \end{pmatrix} \quad (4.3.9\text{a})$$

corresponding to the series of linear system of equations

$$\left[\overline{\psi_1(t)\psi_1(t)} - \lambda\right]\phi_1 + \overline{\psi_1(t)\psi_2(t)}\,\phi_2 + \ldots + \overline{\psi_1(t)\psi_M(t)}\,\phi_M = 0$$

$$\overline{\psi_2(t)\psi_1(t)}\,\phi_1 + \left[\overline{\psi_2(t)\psi_2(t)} - \lambda\right]\phi_2 + \ldots + \overline{\psi_2(t)\psi_M(t)}\,\phi_M = 0 \quad (4.3.9\text{b})$$

$$\cdots$$

$$\overline{\psi_M(t)\psi_1(t)}\,\phi_1 + \overline{\psi_M(t)\psi_2(t)}\,\phi_2 + \ldots + \left[\overline{\psi_M(t)\psi_M(t)} - \lambda\right]\phi_M = 0$$

The eigenvalue problem involves diagonalization of a matrix, which in turn amounts to finding an axis orientation in M-space for which there are no off-diagonal terms in the matrix. When this occurs, the different modes of the system are orthogonal. Since each **C** is a real symmetric matrix, the eigenvalues λ_i are real. Similarly, the eigenvectors (EOFs) of a real symmetric matrix are real. Because $\overline{C(x_m, x_k)}$ is positive, the real eigenvalues are all positive.

If equation (4.3.8) is to have a nontrivial solution, the determinant of the coefficients must vanish; that is

$$\det \begin{vmatrix} C_{11}-\lambda & C_{12} & \cdots & C_{1M} \\ C_{21} & C_{22}-\lambda & \cdots & \cdots \\ \cdots & \cdots & \cdots & \cdots \\ C_{m1} & \cdots & \cdots & C_{MM}-\lambda \end{vmatrix} = 0 \qquad (4.3.10)$$

which yields an Mth order polynomial, $\lambda^M + \alpha \lambda^{M-1} + ...$, whose M eigenvalues satisfy

$$\lambda_1 > \lambda_2 > ... > \lambda_M \qquad (4.3.11)$$

Thus, the "energy" (variance) associated with each statistical mode is ordered according to its corresponding eigenvector. The first mode contains the highest percentage of the total variance, λ_1; of the remaining variance, the greatest percentage is in the second mode, λ_2, and so on. If we add up the total variance in all the time series, we get

$$\sum_{m=1}^{M} \left\{ \frac{1}{N} \sum_{n=1}^{N} [\psi_m(t_n)]^2 \right\} = \sum_{j=1}^{M} \lambda_j$$

Sum of variances in data = sum of variance in eigenvalues (4.3.12)

The total variance in the M time series equals the total variance contained in the M statistical modes. The final piece of the puzzle is to derive the time-dependent *amplitudes* of the ith statistical mode

$$a_i(t) = \sum_{m=1}^{M} \psi_m(t) \phi_{im} \qquad (4.3.13)$$

Equation (4.3.7) provides a computational procedure for finding the EOFs. By computing the mean product matrix, $\overline{\psi_m(t)\psi_k(t)}$ ($m, k = 1, ..., M$) or "scatter matrix" **S** in the terminology of Preisendorfer (1988), the eigenvalues and eigenvectors can be determined using standard computer algorithms. From these, we obtain the variance associated with each mode, λ_j, and its time-dependent variability, $a_i(t)$.

As outlined by Davis (1976), two advantages of a statistical EOF description of the data are: (1) the EOFs provide the most efficient method of compressing the data; and (2) the EOFs may be regarded as uncorrelated (i.e. orthogonal) modes of variability of the data field. The EOFs are the most efficient data representation in the sense that, for a fixed number of functions (trigonometric or other), no other approximate expansion of the data field in terms of $K < M$ functions

$$\hat{\psi}_m(t) = \sum_{m=1}^{K} a_i(t) \hat{\phi}_{im} \qquad (4.3.14)$$

can produce a lower total mean-square error

$$\sum_{m=1}^{K} \overline{\left[\psi_m(t) - \hat{\psi}_m(t)\right]^2} \tag{4.3.15}$$

than would be obtained when the $\hat{\phi}_i$ are the EOFs. A proof of this is given in Davis (1976). Also, as we will discuss later in this section, we could just as easily have written our data $\psi(\mathbf{x}_m, t)$ as a combination of orthogonal temporal modes $\phi_i(t)$ whose amplitudes vary spatially as $a_i(\mathbf{x}_m)$. Since this is a statistical technique, it doesn't matter whether we use time or space to form the basis functions. However, it might be easier to think in terms of spatial orthogonal modes that oscillate with time.

As noted above, EOFs are ordered by decreasing eigenvalue so that, among the EOFs, the first mode, having the largest eigenvalue, accounts for most of the variance of the data. Thus, with the inherent efficiency of this statistical description a very few empirical modes generally can be used to describe the fundamental variability in a very large data set. Often it may prove useful to employ the EOFs as a filter to eliminate unwanted scales of variability. A limited number of the first few EOFs (those with the largest eigenvalues) can be used to reconstruct the data field, thereby eliminating those scales of variability not coherent over the data grid and therefore less energetic in their contribution to the data variance. An EOF analysis can then be made of the filtered data set to provide a new apportionment of the variance for those scales associated with most of the variability in the original data set. In this application, EOF analysis is much like standard Fourier analysis used to filter out scales of unwanted variability. In fact, for homogeneous time series sampled at evenly spaced increments, it can be shown that the EOFs are Fourier trigonometric functions.

The computation of the eigenfunctions $a_i(t)$ in equation (4.3.13) requires the data values $\psi_m(t)$ for all of the time series. Often these time series contain gaps which make it impossible to compute $a_i(t)$ at those times for which the data are missing. One solution to this problem is to fill the gaps in the original data records using one of the procedures discussed in the previous chapter on interpolation. Most consistent with the present approach is to use objective analysis as discussed in the preceding section. While this will provide an interpolation consistent with the covariance of the subject data set, these optimally estimated values of $\psi_m(t)$ often result in large expected errors if the gaps are large or the scales of coherent variability are small.

An alternative method, suggested by Davis (1976), that can lead to a smaller expected error is to estimate the EOF amplitude at time, t, directly from the existing values of $\psi_m(t)$ thus eliminating the need for the interpolation of the original data. Conditions for this procedure are that the available number of sample data pairs is reasonably large (gaps do not dominate) and that the data time series are stationary. Under these conditions, the mean product matrix $\overline{\psi_m(t)\psi_k(t)}$ ($m, k = 1, \ldots, M$) will be approximately the same as it would have been for a data set without gaps. For times when none of the $\psi_m(t)$ values are missing, the coefficients $a_i(t)$ can be computed from equation (4.3.13). For times t when data values are missing, $a_i(t)$ can be estimated from the available values of $\psi_m(t)$

$$\hat{a}_i(t) = b_i(t) \sum_{j=1}^{M'} \psi_j(t)\phi_{ij} \tag{4.3.16}$$

where the summation over j includes only the available data points, $M' \leq M$. From equations (4.3.8), (4.3.14), and (4.3.16), the expected square error of this estimate is

$$\overline{[a_i(t) - \hat{a}_i(t)]^2} = b_i^2(t) \sum_{j=1}^{M'} \left(\lambda_j \gamma_{ji}^2\right) + \lambda_i [1 + b_i(t)(\gamma_{ii}-1)]^2 \qquad (4.3.17)$$

where

$$\gamma_{ji} = \sum_k \phi_j(k)\phi_i(k) \qquad (4.3.18)$$

and the summation over k applies only to those variables with missing data. Taking the derivative of the right-hand side of (4.3.17) with respect to b_i, we find that the expected square error is minimized when

$$b_i(t) = (1 - \gamma_{ii})\lambda_j / \left[(1 - \gamma_{ii})^2 \lambda_j + \sum_j \lambda_j \gamma_{ji}^2\right] \qquad (4.3.19)$$

Applications of this procedure (Davis, 1976, 1978; Chelton and Davis, 1982, Chelton et al., 1982), have shown that the expected errors are surprisingly small even when the number of missing data is relatively large. This is because the dominant EOFs in geophysical systems generally exhibit large spatial scales of variability, leading to a high coherence between grid values. As a consequence, contributions to the spatial pattern from the most dominant EOFs at any particular time, t, can be reliably estimated using a relatively small number of sample grid points.

4.3.1 Principal axes of a single vector time series (scatter plot)

A common technique for improving the EOF analysis for a set of vector time series is to first rotate each data series along its own customized principal axes. In this new coordinate system, most of the variance is associated with a major axis and the remaining variance with a minor axis. The technique also provides a useful application of principal component analysis. The problem consists of finding the principal axes of variance along which the variance in the observed velocity fluctuations $\mathbf{u}'(t) = [u_1'(t), u_2'(t)]$ is maximized for a given location; here u_1' and u_2' are the respective east–west and north–south components of the wind or current velocity obtained by removing the respective means $\overline{u_1}$ and $\overline{u_2}$ from each record; i.e. $u_1' = u_1 - \overline{u_1}$, $u_2' = u_2 - \overline{u_2}$. The amount of data "scatter" is a maximum along the major axis and a minimum along the minor axis (Figure 4.3.1). We also note that principal axes are defined in such a way that the velocity components along the principal axes are uncorrelated.

The eigenvalue problem (4.3.8) for a two-dimensional scatter plot has the form

$$\begin{vmatrix} C_{11} & C_{21} \\ C_{12} & C_{22} \end{vmatrix} \begin{vmatrix} \phi_1 \\ \phi_2 \end{vmatrix} = \begin{vmatrix} \lambda & 0 \\ 0 & \lambda \end{vmatrix} \begin{vmatrix} \phi_1 \\ \phi_2 \end{vmatrix} \qquad (4.3.20)$$

where the C_{ij} are components of the covariance matrix, $\mathbf{C}$, and (ϕ_1, ϕ_2) are the eigenvectors associated with the two possible values of the eigenvalues, λ. To find the principal axes for the scatter plot of u_2' versus u_1', we set the determinant of the covariance matrix equation (4.3.20) to zero

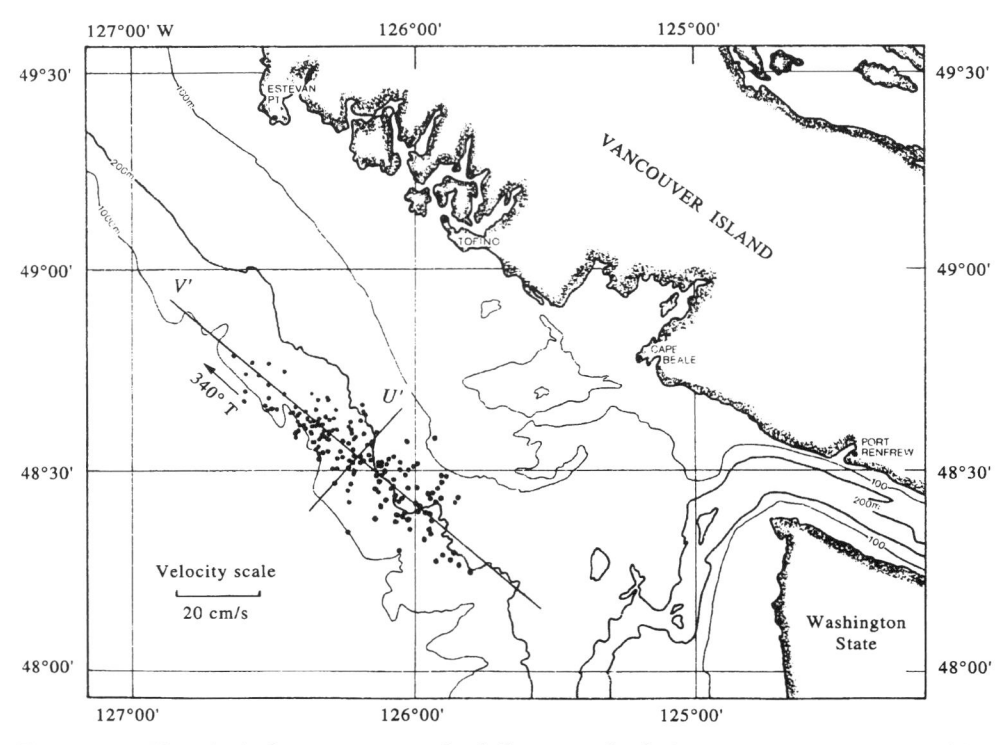

Figure 4.3.1. The principal component axes for daily averaged velocity components u, v measured by a current meter moored at 175 m depth on the west coast of Canada. Here, the north–south component of velocity, v(t), is plotted as a scatter diagram against the east–west component of current velocity, u(t). Data cover the period 21 October 1992 to 25 May 1993. The major axis along 340°T can be used to define the longshore direction, $\mathbf{v}'$.

$$\det |\mathbf{C} - \lambda \mathbf{I}| = \det \begin{vmatrix} C_{11} - \lambda & C_{12} \\ C_{21} & C_{22} - \lambda \end{vmatrix}$$
$$= \det \begin{vmatrix} \overline{u_1'^2} - \lambda & \overline{u_1' u_2'} \\ \overline{u_2' u_1'} & \overline{u_2'^2} - \lambda \end{vmatrix} = 0 \quad (4.3.21a)$$

where (for $i = 1, 2$) the elements of the determinant are given by

$$C_{ii} = \overline{u_i'^2} = \frac{1}{N} \sum_{n=1}^{N} [u_i'(t_n)]^2 \quad (4.3.21b)$$

$$C_{ij} = \overline{u_i' u_j'} = \frac{1}{N} \sum_{n=1}^{N} [u_i' u_j'(t_n)] \quad (4.3.21c)$$

Solution of (4.3.21) yields the quadratic equation

$$\lambda^2 - \left[\overline{u_1'^2} + \overline{u_2'^2}\right] \lambda + \overline{u_1'^2} \, \overline{u_2'^2} - \overline{u_1' u_2'}^2 = 0 \quad (4.3.22)$$

whose two roots $\lambda_1 > \lambda_2$ are the eigenvalues, corresponding to the variances of the velocity fluctuations along the major and minor principal axes. The orientations of the two axes differ by 90° and the principal angles θ_p (those along which the sum of the squares of the normal distances to the data points u'_1, u'_2 are extremum) are found from the transcendental relation

$$\tan(2\theta_p) = \frac{2\overline{u'_1 u'_2}}{\overline{u'^2_1} - \overline{u'^2_2}} \tag{4.3.23a}$$

or

$$\theta_p = \frac{1}{2} \tan^{-1} \left[\frac{2\overline{u'_1 u'_2}}{\overline{u'^2_1} - \overline{u'^2_2}} \right] \tag{4.3.23b}$$

where the principal angle is defined for the range $-\pi/2 \leq \theta_p \leq \pi/2$ (Freeland et al., 1975; Kundu and Allen, 1976; Preisendorfer, 1988). As usual, the multiple $n\pi/2$ ambiguities in the angle that one obtains from the arctangent function must be addressed by considering the quantrants of the numerator and denominator in equation (4.3.23). Preisendorfer (1988; Figure 2.3) outlines the nine different possible cases. Proof of (4.3.23) is given in Section 4.3.5.

The principal variances (λ_1, λ_2) of the data set are found from the determinant relations (4.3.21a) and (4.3.22) as

$$\left. \begin{array}{c} \lambda_1 \\ \lambda_2 \end{array} \right\} = \frac{1}{2} \left\{ \left(\overline{u'^2_1} + \overline{u'^2_2} \right) \pm \left[\left(\overline{u'^2_1} - \overline{u'^2_2} \right)^2 + 4\left(\overline{u'_1 u'_2} \right)^2 \right]^{1/2} \right\} \tag{4.3.24}$$

in which the $+$ sign is used for λ_1 and the $-$ sign for λ_2. In the case of current velocity records, λ_1 gives the variance of the flow along the major axis and λ_2 the variance along the minor axis. The slope, $s_1 = \phi_2/\phi_1$, of the eigenvector associated with the variance λ_1 is found from the matrix relation

$$\begin{vmatrix} \overline{u'^2_1} - \lambda & \overline{u'_1 u'_2} \\ \overline{u'_2 u'_1} & \overline{u'^2_2} - \lambda \end{vmatrix} \begin{vmatrix} \phi_1 \\ \phi_2 \end{vmatrix} = 0 \tag{4.3.25a}$$

Solving (4.3.25a) for $\lambda = \lambda_1$, gives

$$\left[\overline{u'^2_1} - \lambda_1 \right] \phi_1 + \overline{u'_1 u'_2} \phi_2 = 0$$
$$\left[\overline{u'_2 u'_1} \phi_1 \right] + \left[\overline{u'^2_2} - \lambda_1 \right] \phi_2 = 0 \tag{4.3.25b}$$

so that

$$s_1 = \left[\lambda_1 - \overline{u'^2_1} \right] / \overline{u'_1 u'_2} \tag{4.3.25c}$$

with a similar expression for the slope s_2 associated with the variance $\lambda = \lambda_2$. If $\lambda_1 \gg \lambda_2$, then $\lambda_1 \approx \overline{u'^2_1} + \overline{u'^2_2}$, and $s_1 \approx \overline{u'^2_2}/\overline{u'_1 u'_2}$. The usefulness of principal component analysis is that it can be used to find the main orientation of fluid flow at any current meter or anemometer site, or within a "box" containing velocity variances derived from Lagrangian drifter trajectories (Figure 4.3.2). Since the mean and low frequency currents in relatively shallow waters are generally "steered" parallel to the

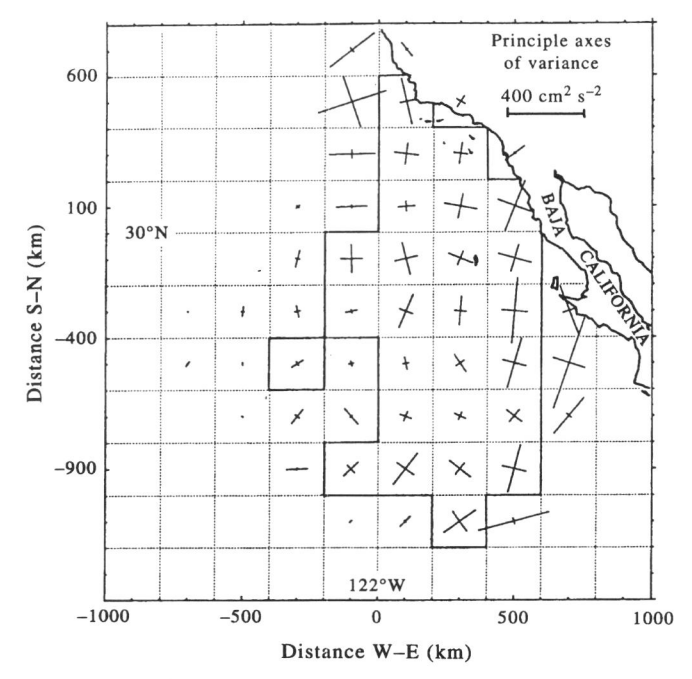

328 *Data Analysis Methods in Physical Oceanography*

Figure 4.3.2. Principal axes of current velocity variance (kinetic energy) obtained from surface satellite-tracked drifter measurements off the coast of southern California during 1985–86. For this analysis, data have been binned into 200 × 200 km² boxes Solid border denotes the region for which there were more than 50 drifter-days and more than two different drifter tracks. (From Poulain and Niiler, 1989).

coastline or local bottom contours, the major principal axis is often used to define the "longshore" direction while the minor axis defines the "cross-shore" direction of the flow. It is this type of information that is vital to estimates of cross-shore flux estimates. In the case of prevailing coastal winds, the major axis usually parallels the mean orientation of the coastline or coastal mountain range that steers the surface winds.

4.3.2 EOF computation using the scatter matrix method

There are two primary methods for computing the EOFs for a grid of time series of observations. These are: (1) The scatter matrix method which uses a "brute force" computational technique to obtain a symmetric covariance matrix **C** which is then decomposed into eigenvalues and eigenvectors using standard computer algorithms (Preisendorfer, 1988); and (2) the computationally efficient singular value decomposition (SVD) method which derives all the components of the EOF analysis (eigenvectors, eigenvalues, *and* time-varying amplitudes) without computation of the covariance matrix (Kelly, 1988). The EOFs determined by the two methods are identical. The differences are mainly the greater degree of sophistication, computational speed, and computational stability of the SVD approach.

Details of the covariance matrix approach can be found in Preisendorfer (1988). This recipe, which is only one of several possible procedures that can be applied, involves the preparation of the data and the solution of equation (4.3.8) as follows:

(1) Ensure that the start and end times for all M time series of length N are identical. Typically, $N > M$.
(2) Remove the record mean and linear trend from each time-series record $\psi_m(t)$, $1 \leq m \leq M$, such that the fluctuations of $\psi_m(t)$ are given by $\psi'_m(t) = \psi_m(t) - [\overline{\psi_m(t)} + b_m(t - \bar{t})]$ where b_m is the slope of the least-squares regression line for each location. Other types of trend can also be removed.
(3) Normalize each de-meaned, de-trended time series by dividing each data series by its standard deviation $s = [1/(N-1)\sum(\psi'_{m'})^2]^{1/2}$ where the summation is over all time, t ($t_n : 1 \leq n \leq N$). This ensures that the variance from no one station dominates the analysis (all stations get an equal chance to contribute). The M normalized time-series fluctuations, ψ'_m, are the data series we use for the EOF analysis. The total variance for each of the M eigenvalues $= 1$; thus, the total variance for all modes, $\sum \lambda_i = M$.
(4) Rotate any vector time series to its principal axes. Although this operation is not imperative, it helps maximize the signal-to-noise ratio for the preferred direction. For future reference, keep track of the means, trends and standard deviations derived from the M time series records.
(5) Construct the $M \times N$ data matrix, $\mathbf{D}$, using the M rows (locations $\mathbf{x}_m$) and N columns (times t_n) of the normalized data series

Time $\rightarrow$

$$D = \begin{pmatrix} \psi'_1(t_1) & \psi'_1(t_2) & \ldots & \psi'_1(t_N) \\ \psi'_2(t_1) & \psi'_2(t_2) & \ldots & \psi'_2(t_N) \\ \ldots & \ldots & \ldots & \ldots \\ \psi'_M(t_1) & \psi'_M(t_2) & \ldots & \psi'_M(t_N) \end{pmatrix} \text{Location} \downarrow \quad (4.3.26)$$

and from this derive the symmetric covariance matrix, $\mathbf{C}$, by multiplying $\mathbf{D}$ by its transpose $\mathbf{D}^T$

$$\mathbf{C} = \frac{1}{N\phi\phi - 1} \mathbf{D}\mathbf{D}^T \quad (4.3.27)$$

where $\mathbf{S} = (N-1)\mathbf{C}$ is the scatter matrix defined by Preisendorfer (1988), and

$$\mathbf{C} = \begin{pmatrix} C_{11} & C_{12} & \ldots & C_{1M} \\ C_{21} & C_{22} & \ldots & C_{2M} \\ \ldots & \ldots & \ldots & \ldots \\ C_{M1} & \ldots & \ldots & C_{MM} \end{pmatrix} \quad (4.3.28)$$

The elements of the real symmetric matrix $\mathbf{C}$ are

$$C_{ij} = C_{ji} = \frac{1}{N-1} \sum_{n=1}^{N} \left[\psi'_i(t_n) \psi'_j(t_n) \right] \quad (4.3.29)$$

The eigenvalue problem then becomes

$$\mathbf{C}\boldsymbol{\phi} = \lambda \boldsymbol{\phi} \quad (4.3.30)$$

where λ are the eigenvalues and $\boldsymbol{\phi}$ the eigenvectors.

330 *Data Analysis Methods in Physical Oceanography*

At this point, we remark that we have formulated the EOF decomposition in terms of an $M \times M$ "spatial" covariance matrix whose time-averaged elements are given by the product $(N - 1)^{-1} \mathbf{DD}^T$ (4.3.27). We could just as easily have formed an $N \times N$ "temporal" covariance matrix whose spatially averaged elements are given by the product $(M - 1)^{-1} \mathbf{D}^T \mathbf{D}$. The mean values we remove in preparing the two data sets are slightly different since preparation of $\mathbf{D}$ involves time averages while preparation of $\mathbf{D}^T$ involves spatial averages. However, in principle, the two problems are identical, and the percentage of the total time-series variance in each mode depends on whether one computes the spatial EOFs or temporal EOFs. As we further point out in the following section, another difference between the two problems is how the singular values are grouped and which is identified with the spatial function and which with the temporal function (Kelly, 1988). The designation of one set of orthogonal vectors as EOFs and the other as amplitudes is quite arbitrary.

Once the matrix $\mathbf{C}$ has been calculated from the data, the problem can be solved using "canned" programs from one of the standard statistical or mathematical computer libraries for the eigenvalues and eigenvectors of a real symmetric matrix. In deriving the values listed in Tables 4.3.1–4.3.6, we have used the double-precision program DEVLSF of the International Math and Science Library (IMSL). The program outputs the eigenvalues λ in increasing order. To obtain λ in decreasing order of importance, we have had to invert the eigenvalue output. For each eigenvector or mode, the program normalizes all values to the maximum value for that mode. The amplitude of the maximum value is unity (= 1). Since there are M eigenvalues, the data normalization process gives a total EOF variance of $M(\sum \lambda_i = M)$. The canned programs also allow for calculation of a "performance index" (PI) which measures the error of the eigenvalue problem (4.3.30) relative to the various components of the problem and the machine precision. The performance of the eigenvalue routine is considered "excellent" if PI < 1, "good" if $1 \leq \text{PI} \leq 100$, and "poor" if PI > 100. As a final analysis, we can conduct an *orthogonality check* on the EOFs by using the relation (4.3.2). Here we look for significant departures from zero in the products of different modes; if any of the products

$$\sum_{m=1}^{M} [\phi_{im} \phi_{jm}]$$

Table 4.3.1. Data matrix $\mathbf{D}^T$. Components of velocity (cm/s) at three different sites at 1700 m depth in the northeast Pacific. Records start 29 September 1985 and are located near 48°N, 129°W. For each of the three stations we list the east–west (u) and north–south component (v). The means and trends have not yet been removed

Time (days)	Site 1 (u_1)	Site 1 (v_1)	Site 2 (u_2)	Site 2 (v_2)	Site 3 (u_3)	Site 3 (v_3)
1	−0.3	0.0	0.4	−0.4	−0.8	−1.4
2	−0.1	0.3	0.4	−0.3	−1.1	0.0
3	−0.1	−0.4	0.0	−0.5	0.0	−2.5
4	0.2	0.6	0.0	−0.6	−0.7	0.4
5	0.3	−0.1	−0.6	−0.3	0.0	−0.3
6	0.5	0.0	0.9	−0.6	0.6	0.3
7	0.2	0.2	−0.1	−0.7	1.2	−2.8
8	−0.5	−0.9	0.0	−0.6	0.0	−1.8

The Spatial Analyses of Data Fields

Table 4.3.2. *Means, standard deviations and trends for each of the time-series components for each of the three current meter sites listed in Table 4.3.1. Means and trends have been removed from the time series prior to calculation of the standard deviations*

Component	Mean (cm/s)	Standard deviation (cm/s)	Trend (cm/s/day)
u_1 (east–west)	0.025	0.328	0.024
v_1 (north–south)	−0.037	0.418	−0.075
u_2 (east–west)	0.125	0.433	−0.038
v_2 (north–south)	−0.500	0.114	−0.040
u_3 (east–west)	−0.100	0.503	0.233
v_3 (north–south)	−1.012	1.250	−0.108

Table 4.3.3. *Principal axes for the current velocity at each site in Table 4.3.1. The angle θ is measured counterclockwise from east. Axes (half) lengths are in cm/s*

Station ID	Angle θ (°)	Major axis	Minor axis
Site 1	54.7	0.461	0.185
Site 2	−6.2	0.408	0.098
Site 3	−77.7	1.193	0.406

Table 4.3.4. *Eigenvalues and percentage of variance in each statistical mode derived from the data in Table 4.3.1*

Eigenvalue No.	Eigenvalue	Percentage
1	2.2218	37.0
2	1.7495	29.2
3	1.1787	19.6
4	0.6953	11.6
5	0.1498	2.5
6	0.0048	0.1
Total	6.0000	100.0

Table 4.3.5. *Eigenvectors (EOFs) ϕ_i for the data matrix in Table 4.3.1. Modes are normalized to the maximum value for each mode*

Station ID	Mode 1	Mode 2	Mode 3	Mode 4	Mode 5	Mode 6
Site 1 u_1	1.000	−0.032	−0.430	0.479	−0.599	−0.969
Site 1 v_1	0.958	−0.078	−0.162	−0.966	1.000	0.085
Site 2 u_2	0.405	0.230	1.000	0.910	0.517	−0.295
Site 2 v_2	−0.329	−0.898	−0.525	1.000	0.784	−0.111
Site 3 u_3	0.349	1.000	−0.474	0.812	0.124	0.907
Site 3 v_3	0.654	−0.964	0.263	0.190	−0.539	1.000

are significantly different from zero for $i \neq j$, then the EOFs are not orthogonal and there are errors in the computation. A computational example is given in Section 4.3.4.

Table 4.3.6. Time series of the amplitudes, $a_i(t)$, for each of the statistical modes

Time	Mode 1	Mode 2	Mode 3	Mode 4	Mode 5	Mode 6
Day 1	0.798	−0.773	0.488	0.089	0.091	0.124
Day 2	−0.076	1.258	0.402	0.126	0.595	−0.089
Day 3	1.153	−1.582	−0.458	0.275	−0.492	−0.094
Day 4	−1.531	0.759	0.363	−1.585	−0.382	0.000
Day 5	0.097	1.647	−2.099	0.509	−0.128	0.039
Day 6	−2.169	−0.142	1.084	1.296	−0.171	0.008
Day 7	−0.721	−1.921	−0.866	−0.534	0.503	0.004
Day 8	2.450	0.754	1.085	−0.176	−0.017	0.008

4.3.3 EOF computation using singular value decomposition

The above method of computing EOFs requires use of covariance matrix, **C**. This becomes computationally impractical for large, regularly spaced data fields such as a sequence of infrared satellite images (Kelly, 1988). In this case, for a data matrix **D** over N time periods (N satellite images, for example), the covariance or mean product matrix is given by (4.3.27)

$$\mathbf{C} = \frac{1}{N-1} \mathbf{D}\mathbf{D}^T \qquad (4.3.31)$$

where $\mathbf{D}^T$ is the transpose of the data matrix **D**. If we assume that all of the spatial data fields (i.e. satellite images) are independent samples, then the mean product matrix is the covariance matrix and the EOFs are again found by solving the eigenvalue problem

$$\mathbf{C}\boldsymbol{\phi} = \boldsymbol{\phi}\boldsymbol{\Lambda} \qquad (4.3.32)$$

where $\boldsymbol{\phi}$ is the square matrix whose columns are eigenvectors and $\boldsymbol{\Lambda}$ is the diagonal matrix of eigenvalues. For satellite images, there may be $M = 5000$ spatial points sampled $N = 50$ times, making the covariance matrix a 5000×5000 matrix. Solving the eigenvalue problem for $\boldsymbol{\phi}$ would take $\max\{O(M^3), O(MN^2)\}$ operations. As pointed out by Kelly (1988), the operation count for the SVD method is $O(MN^2)$ which represents a considerable savings in computations over the traditional EOF approach if M is large. This is primarily true for those cases where M, the number of locations in the spatial data matrix, **D**, are far greater than the number of temporal samples (i.e. images).

There are two computational reasons for using the singular value decomposition method instead of the covariance matrix approach (Kelly, 1988): (1) The SVD formulation provides a one-step method for computing the various components of the eigenvalue problem; and (2) it is not necessary to compute or store a covariance matrix or other intermediate quantities. This greatly simplifies the computational requirements and provides for the use of canned analysis programs for the EOFs. Our analysis is based on the double-precision program DLSVRR in the IMSL. The SVD method is based on the concept in linear algebra (Press et al., 1992) that any $M \times N$ matrix, **D**, whose number of rows M is greater than or equal to its number of columns, N, can be written as the product of three matrices: an $M \times N$ column-orthogonal matrix, **U**, an $N \times N$ diagonal matrix, **S**, with positive or zero elements, and the transpose ($\mathbf{V}^T$) of an $N \times N$ orthogonal matrix, $\mathbf{V}$. In matrix notation, the SVD becomes:

$$\mathbf{D} = \mathbf{U} \begin{pmatrix} s_1 & & & \\ & s_2 & & \\ & & \ldots & \\ & & & s_N \end{pmatrix} \mathbf{V}^T \qquad (4.3.33)$$

For oceanographic applications, the data matrix, $\mathbf{D}$, consists of M rows (spatial points) and N columns (temporal samples). The scalars $s_1 \geq s_2 \geq \ldots \geq s_N \geq 0$ of the matrix $\mathbf{S}$, called the *singular values* of $\mathbf{D}$, appear in descending order of magnitude in the first N positions of the matrix. The columns of the matrix $\mathbf{V}$ are called the left singular vectors of $\mathbf{D}$ and the columns of the matrix $\mathbf{U}$ are called the right singular vectors of $\mathbf{D}$. The matrix $\mathbf{S}$ has a diagonal upper $N \times N$ part, $\mathbf{S}'$, and a lower part of all zeros in the case when $M > N$. We can express these aspects of $\mathbf{D}$ in matrix notation by rewriting equation (4.3.33) in the form

$$\mathbf{D} = [\mathbf{U}|\mathbf{0}] \left| \begin{matrix} \mathbf{S}' \\ \mathbf{0} \end{matrix} \right| \mathbf{V}^T \qquad (4.3.34)$$

where $[\mathbf{U}|\mathbf{0}]$ denotes a left singular matrix and $\mathbf{S}'$ denotes the nonzero part of $\mathbf{S}$ which has zeros in the lower part of the matrix (Kelly, 1988).

The matrix $\mathbf{U}$ is orthogonal, and the matrix $\mathbf{V}$ has only N significant columns which are mutually orthogonal such that,

$$\begin{aligned} \mathbf{V}^T \mathbf{V} &= \mathbf{I} \\ \mathbf{U}^T \mathbf{U} &= \mathbf{I} \end{aligned} \qquad (4.3.35)$$

Returning to equation (4.3.33), we can compute the eigenvectors, eigenvalues and eigenfunctions of the principal component analysis in one single step. To do this, we prepare the data as before following steps 1–5 in Section 4.3.2. We then use commercially available programs such as the double-precision program DLSVRR in the IMSL. The elements of matrix $\mathbf{U}$ are the eigenvectors while those of matrix $\mathbf{S}$ are related to the eigenvalues $s_1 \geq s_2 \geq \ldots \geq s_N \geq 0$. To obtain the time-dependent amplitudes (eigenfunctions), we require a matrix $\mathbf{A}$ such that

$$\mathbf{D} = \mathbf{U} \mathbf{A}^T \qquad (4.3.36)$$

which, by comparison with equation (4.3.33), requires

$$\mathbf{A} = \mathbf{V} \mathbf{S} \qquad (4.3.37)$$

Hence, the amplitudes are simply the eigenvectors of the transposed problem multiplied by the singular values, $\mathbf{S}$. Solutions of (4.3.33) are identical (within round-off errors) to those obtained using the covariance matrix of the data, $\mathbf{C}$. We again remark that the only difference between the matrices $\mathbf{U}$ and $\mathbf{V}$ is how the singular values are grouped and which is identified with the spatial function and which with the temporal function. The designation of $\mathbf{U}$ as EOFs and $\mathbf{V}$ as amplitudes is quite arbitrary.

The decomposition of the data matrix $\mathbf{D}$ through singular value decomposition is possible since we can write it as a linear combination of functions $F_i(x)$, $i = 1, M$ so

that
$$\mathbf{D} = \mathbf{F}\boldsymbol{\alpha} \quad (4.3.38a)$$

or

$$\begin{pmatrix} D(x_1, t_j) \\ D(x_2, t_j) \\ \cdots \\ \cdots \\ D(x_N, t_j) \end{pmatrix} = \begin{pmatrix} F_1(x_1) \ldots F_N(x_1) \\ F_1(x_2) \ldots F_N(x_2) \\ \cdots \\ \cdots \\ F_1(x_N) \ldots F_N(x_N) \end{pmatrix} \begin{pmatrix} \alpha_1(t_j) \\ \alpha_2(t_j) \\ \cdots \\ \cdots \\ \alpha_N(t_j) \end{pmatrix} \quad (4.3.38b)$$

where the α_i are functions of time only. The functions F are chosen to satisfy the orthogonality relationship

$$\mathbf{F}\mathbf{F}^T = \mathbf{I} \quad (4.3.39)$$

so that the data matrix $\mathbf{D}$ is divided into orthogonal modes

$$\mathbf{D}\mathbf{D}^T = \mathbf{F}\mathbf{a}\mathbf{a}^T\mathbf{F}^T = \mathbf{F}\mathbf{L}\mathbf{F}^T \quad (4.3.40)$$

where $\mathbf{L} = \mathbf{a}\mathbf{a}^T$ is a diagonal matrix. The separation of the modes arises from the diagonality of the $\mathbf{L}$ matrix, which occurs because $\mathbf{D}\mathbf{D}^T$ is a real and symmetric matrix and $\mathbf{F}$ a unitary matrix. To reduce sampling noise in the data matrix $\mathbf{D}$, one would like to describe it with fewer than M functions. If $\mathbf{D}$ is approximated by $\tilde{\mathbf{D}}$, which uses only K functions ($K < M$), then the K functions which best describe the $\mathbf{D}$ matrix in the sense that

$$(\tilde{\mathbf{D}} - \mathbf{D})^T(\tilde{\mathbf{D}} - \mathbf{D})$$

is a minimum are the empirical orthogonal functions which correspond to the largest valued elements of the traditional EOFs found earlier.

4.3.4 An example: deep currents near a mid-ocean ridge

As an example of the different concepts presented in this section, we again consider the eight days of daily averaged currents ($N = 8$) at three deep current meter sites in the northeast Pacific near the Juan de Fuca Ridge (Table 4.3.1). Since each site has two components of velocity, $M = 6$. The data all start on the same day and have the same number of records. Following the five steps outlined in Section 4.3.2, we first removed the average value from each time series. We then calculated the standard deviation for each time series and used this to normalize the time series so that each normalized series has a variance of unity. For convenience, we write the transpose of the data matrix, $\mathbf{D}^T$, where columns are the pairs of components of velocity (u, v) and rows are the time in days.

Time-series plots of the first three eigenmodes are presented in Figure 4.3.3. The performance index (PI) for the scatter matrix method was 0.026, which suggests that the matrix inversion in the eigenvalue solutions was well defined. A check on the orthogonality of the eigenvectors suggests that the singular value decomposition gave vectors which were slightly more orthogonal than the scatter matrix approach. For each combination (i, j) of the orthogonality condition (4.3.2), the products $\sum_{i,j}[\phi_{im}\phi_{jm}]$ were typically of order 10^{-7} for the SVD method and 10^{-6} for the

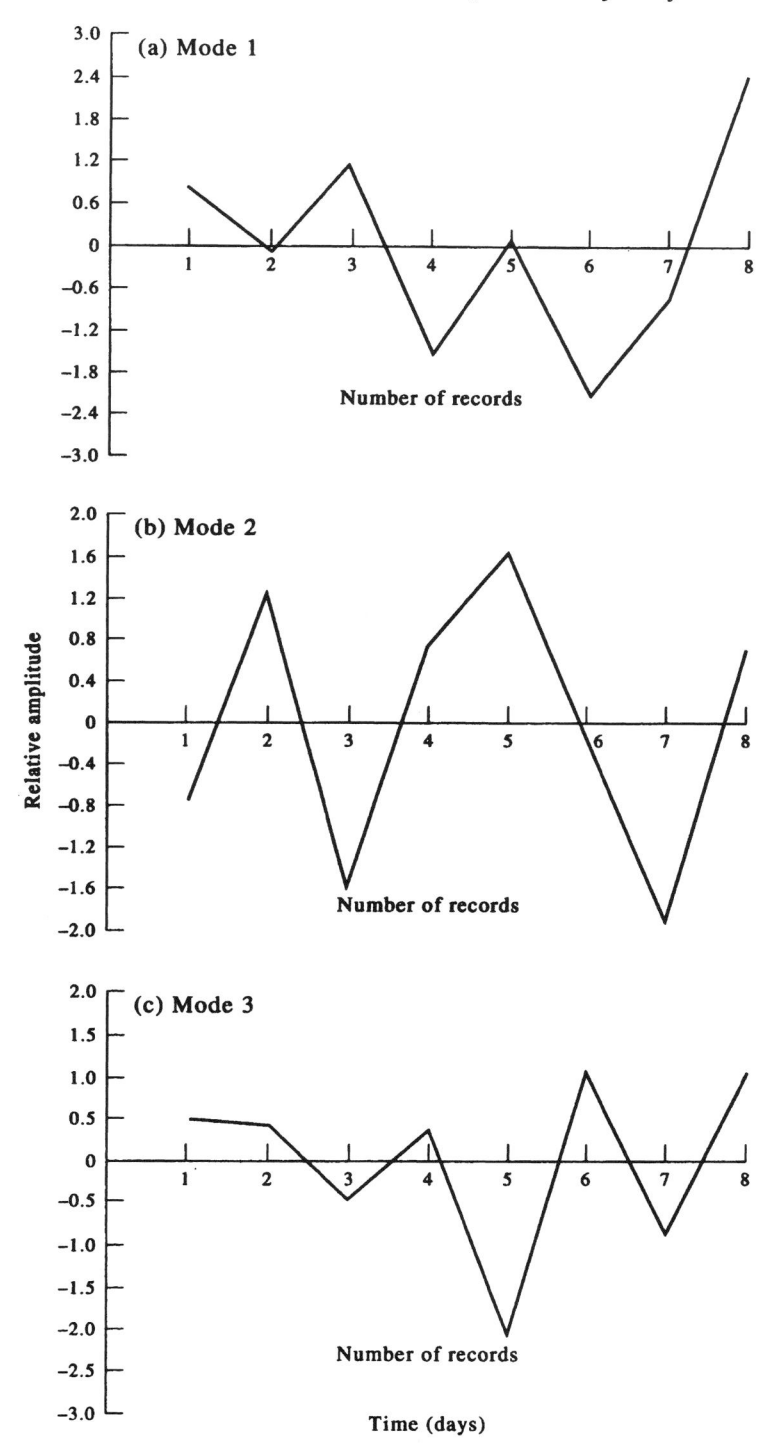

Figure 4.3.3. Eight-day time series for the first three EOFs for current meter data collected simultaneously at three sites at 1700 m depth in the northeast Pacific in the vicinity of Juan de Fuca Ridge, 1985. Modes 1, 2, 3 account for 37.0, 29.2, and 19.6 % of the variance, respectively.

scatter matrix method. Similar results apply to the orthogonality of the eigenmodes given by equation (4.3.4).

Before closing this section, we remark that we also could have performed the above analysis using complex EOFs of the form

$$\psi_m(t) = u_m(t) + iv_m(t)$$

(where $i = \sqrt{-1}$) in which case $M = 3$. This formulation not only allows the EOF vectors to change amplitude with time, as in our previous decomposition using $2M$ real EOFs, but also to rotate in time.

4.3.5 Interpretation of EOFs

In interpreting the meaning of EOFs, we need to keep in mind that, while EOFs offer the most efficient statistical compression of the data field, empirical modes do not necessarily correspond to true dynamical modes or modes of physical behavior. Often, a single physical process may be spread over more than one EOF. In other cases, more than one physical process may be contributing to the variance contained in a single EOF. The statistical construct derived from this procedure must be considered in light of accepted physical mechanisms rather than as physical modes themselves. It often is likely that the strong variability associated with the dominant modes is attributable to several identifiable physical mechanisms. Another possible clue to the physical mechanisms associated with the EOF patterns can be found in the time-series coefficients $a_i(t)$. Something may be known about the temporal variability of a process that might resemble the time series of the EOF coefficients, which would then suggest a causal relationship not readily apparent in the spatial structure of the EOF.

One way to interpret EOFs is to imagine that we have displayed the data as a scatter diagram in an effort to discover if there is any inherent correlation among the values. For example, consider two parameters such as sea surface temperature (SST) and sea-level pressure (SLP) measured at a number of points over the North Pacific. This is the problem studied by Davis (1976) where he analyzed sets of monthly SST and SLP over a period of 30 years for a grid in the North Pacific. If we plot $x = $ SST against $y = $ SLP in a scatter diagram, any correlation between the two would appear as an elliptical cluster of points. A more common example is that of Figure 4.3.1 where we plotted the north–south (y) component of daily mean current against the corresponding east–west (x) component for a continental shelf region. Here, the mean flow tends to parallel the coastline, so that the scatter plot again has an elliptical distribution. To take advantage of this correlation, we want to redefine our coordinate system by rotating x and y through the angle θ to the principal axes representation x', y' discussed in Section 4.3.2. This transformation is given by

$$\begin{aligned} x' &= x\cos\theta + y\sin\theta \\ y' &= -x\sin\theta + y\cos\theta \end{aligned} \quad (4.3.41)$$

What we have done in this rotation is to formulate a new set of axes that explains most of the variance, subject to the assumption that the variance does not change with time. Since the axes are orthogonal, the total variance will not change with rotation. Let $V = \overline{x'^2} = N^{-1}\sum x'^2$ be the particular variance we want to maximize (as usual, the summation is over all time). Note that we have focused on x' whereas the total variance is actually determined by r^2, where r is the distance of each point from the origin. However, we can expand $r^2 = x^2 + y^2$ and associate the variance with a given coordinate. In other words, if we maximize the variance associated with x', we will minimize the variance associated with y'. Using our summation convention, we can write

$$V = \overline{x'^2} = \overline{x^2} \cos^2 \theta + 2\overline{xy} \sin \theta \cos \theta + \overline{y^2} \sin^2 \theta \qquad (4.3.42)$$

and

$$\frac{\partial V}{\partial \theta} = 2\left(\overline{y^2} - \overline{x^2}\right) \sin \theta \cos \theta + 2\overline{xy} \cos 2\theta \qquad (4.3.43)$$

We maximize (4.3.43) by setting $\partial V/\partial \theta = 0$, giving (4.3.24), which we previously quoted without proof

$$\tan(2\theta_p) = \frac{2\overline{xy}}{\overline{x^2} - \overline{y^2}} \qquad (4.3.44)$$

From (4.3.44), we see that if

$$\overline{xy} \ll \max\left(\overline{x^2}, \overline{y^2}\right)$$

then $\tan(2\theta_p) \to 0$ and $\theta_p = 0$, or $\pm 90°$, and we are left with the original axes. If $\overline{x^2} = \overline{y^2}$ and $\overline{xy} \neq 0$, then $\tan(2\theta_p) \to \pm\infty$ and the new axes are rotated $\pm 45°$ from the original axes.

We now find the expression for V. Since $\sec^2(2\theta) = 1 + \tan^2(2\theta)$

$$\cos 2\theta = \left(\overline{x^2} - \overline{y^2}\right)/\pm D$$
$$\sin 2\theta = \left[1 - \cos^2(2\theta)\right]^{1/2} = 2\overline{xy}/\pm D \qquad (4.3.45)$$

where

$$D = \left[\left(\overline{x^2} - \overline{y^2}\right)^2 + 4\overline{xy}^2\right]^{1/2} \qquad (4.3.46)$$

Then, using the identities

$$\cos^2 \theta = \tfrac{1}{2}(1 + \cos 2\theta), \quad \sin^2 \theta = \tfrac{1}{2}(1 - \cos 2\theta) \qquad (4.3.47)$$

we can write the variance as

$$V = \overline{x^2}\frac{(1 + \cos 2\theta_p)}{2} + \overline{y^2}\frac{(1 - \cos 2\theta_p)}{2} + \overline{xy} \sin 2\theta_p$$
$$= \frac{1}{2}\left\{\left(\overline{x^2} + \overline{y^2}\right) \pm \left[\left(\overline{x^2} - \overline{y^2}\right)^2 + 4\overline{xy}^2\right]^{1/2}\right\} \qquad (4.3.48)$$

The two roots of this equation correspond to a maximum and a minimum of V. For a new axis for which $\overline{x'^2}$ is a maximum, we will find $\overline{y'^2}$ a minimum. This follows automatically from the fact that the total variance is conserved. However, we can confirm this mathematically by computing $\partial^2 V/\partial \theta^2 = 0$. From equation (4.3.43) we

338 *Data Analysis Methods in Physical Oceanography*

find

$$\partial^2 V/\partial \theta^2 = 2\left(\overline{y^2} - \overline{x^2}\right)\cos(2\theta_p) - 4\overline{xy}\sin(2\theta_p)$$
$$= -2\left[\left(\overline{x^2} - \overline{y^2}\right) + 4\overline{xy}\right]/\pm D = \pm 2D \quad (4.3.49)$$

The positive sign in equation (4.3.49) corresponds to a maximum (since (4.3.48) is negative); the negative sign corresponds to a minimum. It so happens that the variance solutions given by (4.3.49) are also the eigenvalues of the covariance matrix. Thus, we can return to our previous methods where we used the covariance matrix to compute the EOFs.

A published example of EOF analysis is presented by Davis (1976) who examined monthly maps of SST and SLP for the years 1947–74. The SLP data were originally obtained from the Long-Range Prediction Group of the U.S. National Meteorological Center (NMC) as one-month averages on a 5° diamond-shaped grid (i.e. 20°N–140°W, 20°N–150°W, ... , 25°N–145°W, 25°N–155°W, etc.). The data were transferred to a regular 5°-square grid using linear interpolation from the four nearest diamond grid points to fill in the square grid. The SST data were obtained from the U.S. National Marine Fisheries Service in the form of monthly averages over 2° squares. Because this grid spacing is not a submultiple of 5°, and because sometimes data were missing, the following data analysis scheme was employed. The 2° data were subjectively analyzed to produce maps contoured with a 1°F contour interval. During this stage, missing values were filled in where feasible. The corrected values were then linearly interpolated onto a 1° grid and 25 values were averaged to formulate area averages on the chosen 5° grid coincident with the SLP data. The ship data originated as ship injection temperatures and are subject to all of the problems discussed earlier in the section on SST.

Before carrying out the EOF analysis, the SST and SLP data sets were further averaged onto a grid with a 5° latitude spacing and a 10° longitude spacing (Figure 4.3.4). In those cases where some SST values were missing, the available observations were used to compute the grid average. Even then there were some 5° × 10° regions with missing data in the SST fields. Both fields were then converted to anomalies using the mean of the 28-year data set as the reference field. Thus, each of the individual monthly maps were transformed into anomaly maps, corresponding to the deviation of local values from the long-term mean.

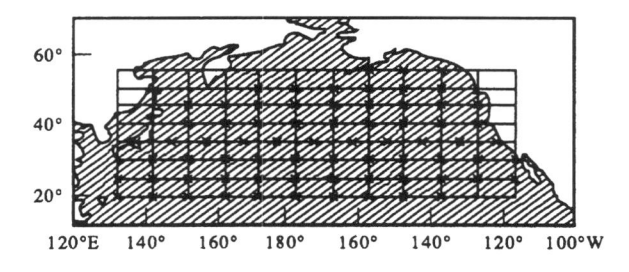

Figure 4.3.4. The grid of sea surface temperature (SST) and sea-level pressure (SLP). The 10° longitude by 5° latitude SLP averages are centered at grid intersections and SST averages are centered at crosses. (From Davis, 1976.)

The Spatial Analyses of Data Fields 339

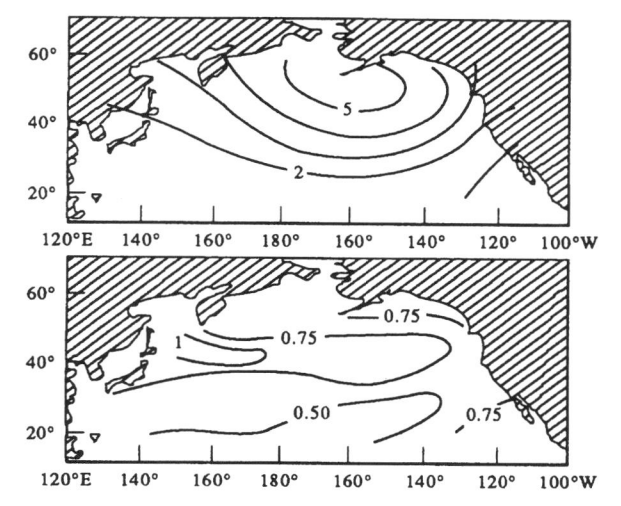

Figure 4.3.5. Standard deviation of: (a) Sea level pressure anomaly (mb); and (b) Sea surface temperature anomaly (°C) for the North Pacific. The anomalies are departures from monthly normal values. Variances are averaged over all months of the 28-year record (1947–1974). (From Davis, 1976.)

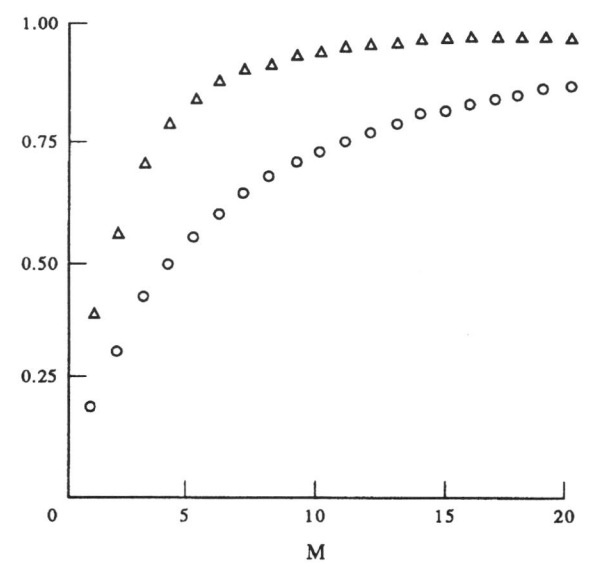

Figure 4.3.6. The fraction of total sea surface temperature (circles, ○) and sea-level pressure (triangles, △) anomaly variance accounted for by the first M empirical orthogonal functions. (From Davis, 1976.)

The standard deviations of both the SLP and SST anomaly fields are shown in Figure 4.3.5. It is interesting to note some of the basic differences between the variability of these two fields. The SLP field has its primary variability in the central northern part of the field just off the tip of the Aleutian Islands. Here, the Aleutian Low dominates the pressure field in winter and becomes the source of the main variability in the SLP data. In contrast, the SST field has near-uniform variance levels except in the Kuroshio Extension region off of northeast Japan where a maximum associated with advection from the Kuroshio is clearly evident.

To compute the EOFs from the anomaly fields, Davis (1976) used the covariance (scatter) matrix method presented in Section 4.3.2. The fraction of total variance accounted for by the EOFs for both the SST and SLP data is presented in Figure 4.3.6 as a function of the number of EOFs. The steep slope of the SLP curve means that fewer SLP EOFs are needed to express the variance. The SST EOF level is consistently below that for the SLP EOF series. As a consequence, Davis presented only the first six SLP EOFs (labeled P_1–P_6 in Figure 4.3.7) but presented the first eight SST EOFs (labeled T_1–T_8 in Figure 4.3.8). The SLP EOFs exhibited fairly simple, large-scale patterns with P_1 having the same basic shape as the SLP standard deviation (Figure 4.3.5). The structural sequence for the first three SLP EOFs was: For P_1, a single maximum; for P_2, two meridionally separated maxima; and for P_3, two zonally separated maxima. Higher modes appear to be combinations of these first three with an increasing number of smaller maxima.

The SST maps obtained by Davis were considerably more complicated than the SLP maps, with large-scale patterns dominating only the first three modes of the temperature field. As with the SLP modes, the sequence seems to be from a central maximum (T_1), to meridionally separated maxima (T_2), and then to zonally separated maxima (T_3). The higher-order EOFs have a number of smaller maxima with no simple structures. The overall scales are much shorter than those for the SLP EOFs. This turns out to be true for the time scales of the EOFs, with the SLP time scales being much shorter than those computed for the SST EOFs.

The goal of the EOF analysis by Davis (1976) was to determine if there is some direct statistical connection between the SLP and SST anomaly fields. By using the EOF procedure he was able to present the primary modes of variability for both fields in the most compact form possible. This is the real advantage of the EOF procedure. In terms of the two anomaly fields, Davis found that there were connections between the variables. First, he found that SST anomalies could be predicted from earlier SST anomaly fields. This is a consequence of the persistence of individual SST patterns as well as the fact that some patterns appear to evolve from earlier patterns through advective processes. Davis also concluded that it was possible to specify the SLP anomaly on the basis of the coincident SST anomaly field. Finally, it was not possible to statistically predict the SST field from the simultaneous SLP field. These conclusions would have been difficult to arrive at without using the EOF procedure.

4.3.6 Variations on conventional EOF analysis

Conventional principal component (EOF) analysis is limited by a number of factors including the dependence of the solution on the domain of analysis, the requirement for orthogonal spatial modes, and the lumping together of variability over all frequency bands. In addition, the method can detect standing waves but not progressive waves. Over the years, several authors have developed what might be called "variations" on the standard EOF theme. For the most part, the methods differ in the types of variances they insert into the algorithms used to determine the empirical orthogonal functions (principal components). Given that EOF analysis is a strictly statistical method, it is irrelevant how the variance is derived, provided that the type of variance used in the analysis is the same for all spatial locations. All that is required is that the matrix **D**, derived from statistical averages (such as the covariance, correlation and cross-covariance functions) of the gridded time series is a Hermitian matrix.

The Spatial Analyses of Data Fields 341

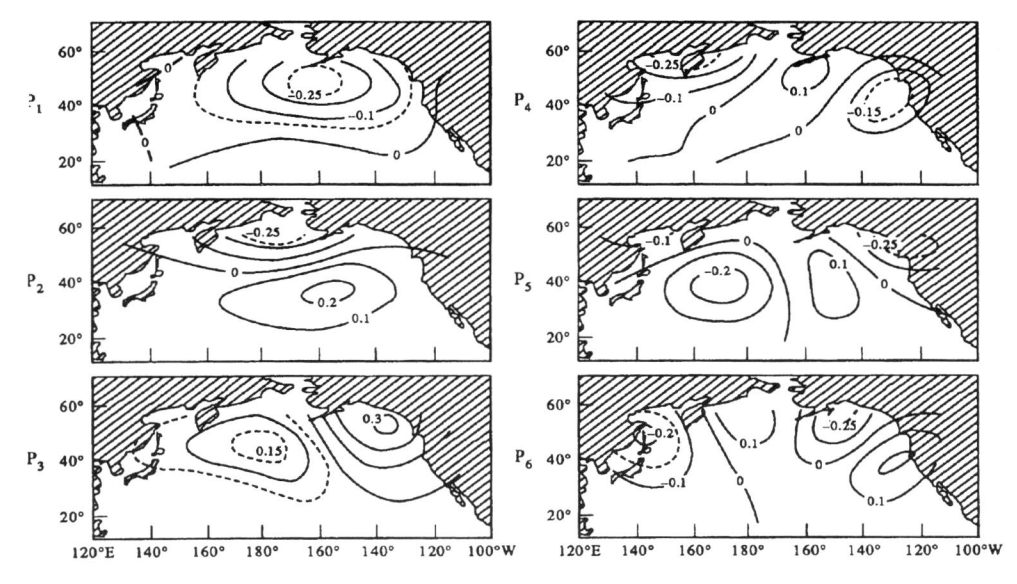

Figure 4.3.7. The six principal empirical orthogonal functions P_1–P_6 describing the sea level pressure anomalies. (From Davis, 1976.)

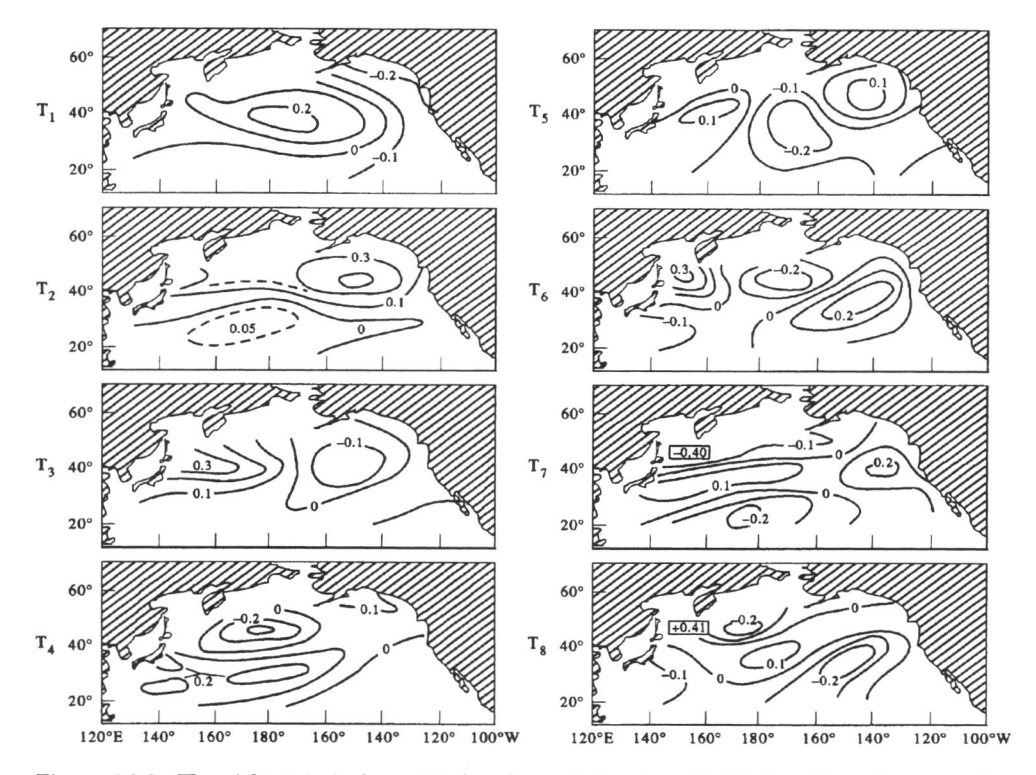

Figure 4.3.8. The eight principal empirical orthogonal functions T_1–T_8 describing the sea surface temperature anomalies. (From Davis, 1976.)

Departure from standard EOF analysis can have numerous forms. For example, one may choose to work in the frequency domain instead of the time domain by using spectral analysis to calculate the spectral "energy" density for specific frequency bands. In this case, the matrix **D** is complex, consisting of the cross-spectra between the gridded time series over a specific frequency band. The spectral densities represent the data variances which are used to determine the empirical orthogonal functions. Thus, the method is equally at home with variances obtained in the time or frequency domains. Regardless of variance-type, principal component methods are simply techniques for compressing the variability of the data set into the fewest possible number of modes.

Returning to the time domain, suppose that we are examining the statistical structure of longshore wind and current fluctuations over the continental shelf and that we have reason to believe that current response to wind forcing is delayed by one or more time steps in the combined data series. A delay of half a pendulum day ($\approx$12 h at mid-latitudes) is not unreasonable. From a causal point of view, the best way to examine the EOF modes for the combined wind and current data is to first create new time series in which the wind records are lagged (shifted forward in time) relative to the current records. Suppose we want a delay of one time step. Then, longshore wind velocity values $V_k(t_j)$ at site k at times t_j ($j = 2, 3, ...$) get replaced with the earlier records at times t_{j-1}. That is, $V_k(t_j) \to V_k(t_{j-1}) = V_k^*(t_j)$, while the current record remains unchanged, $v_k(t_j) = v_k^*(t_j)$. The asterisk (*) denotes the new time series. Optimal empirical modes are those for which the wind and current records are properly "tuned" with the correct time lags. For large spatial regions with variable wind response times, this can get a little tricky so caution is advised.

A departure from conventional EOF analysis was presented by Kundu and Allen (1976) who combined the zonal (u) and meridional (v) time series of currents into complex time series $w = u + iv$, where each scalar series is defined for times t_j and locations x_k. The method was applied to current data collected during the Coastal Upwelling Experiment (CUE-II) off the Oregon coast in the summer of 1973. The complex covariance matrix obtained from these time series were then decomposed into complex eigenvectors by solving a standard complex eigenvalue problem. Unlike the scalar approach to the problem, this complex EOF technique can be used to describe rotary current variability within selected frequency bands. A further variation on conventional EOF analysis, which is related to complex EOF analysis, was provided by Denbo and Allen (1984). Using a technique we describe in Chapter 5, the current fluctuations in each of the time series (u, v) records collected during CUE-II were decomposed into clockwise (S^+) and counterclockwise (S^-) rotary spectra. The spectra (or variance per unit frequency range) for the dominant spectral components, which is typically $S-$ in the ocean, were then decomposed into empirical orthogonal functions by solving the standard complex eigenvalue problem. This *rotary empirical orthogonal function analysis* is best suited to flows with strong rotary signals such as continental shelf waves and near-inertial motions, but is not well suited to highly rectilinear flows such as those in tidal channels for which S^+ and S^- are of comparable amplitude (see Hsieh, 1986; Denbo and Allen, 1986).

The first use of *complex empirical orthogonal functions* in the frequency domain was described by Wallace and Dickinson (1972) and subsequently used by Wallace (1972) to study long-wave propagation in the tropical atmosphere. Early oceanographic applications are provided by Hogg (1977) for long waves trapped along a continental rise and by Wang and Mooers (1977) for long, coastal trapped waves along a continental margin. In this approach, complex eigenvectors are computed from the cross-spectral matrices for specified frequency bands. This is the most general technique for studying propagating wave phenomena. As noted by Horel (1984), however, EOF analysis in the frequency domain can be cumbersome if applied to time series in which the power of a principal component is spread over a wide range of frequencies as a result of nonstationarity in the data. Horel presents a version of complex EOF analysis in the time domain in which complex time series of a scalar variable are formed from the original time series and their Hilbert transforms. The complex eigenvectors are then determined from the cross-correlation or cross-covariance matrices derived from the complex time series. The Hilbert transform $u_m^H(t)$ of the original time series $u_m(t)$ represents a filtering operation in which the amplitude of each spectral component remains unchanged but the phase of each component is shifted by $\pi/2$. Expanding the scalar time series

$$u_m(t) = \sum_\omega [a_m(\omega) \cos(\omega t) + b_m(\omega) \sin(\omega t)] \qquad (4.3.50)$$

as a Fourier series over all frequencies, ω, the Hilbert transform $u_m^H(t)$ is

$$u_m^H(t) = \sum_\omega [b_m(\omega) \cos(\omega t) - a_m(\omega) \sin(\omega t)] \qquad (4.3.51)$$

In practice, the Hilbert transform can be derived directly from the coefficients of the Fourier transform of $u_m(t)$, although with the usual problems caused by aliasing and truncations effects. The complex covariance matrix $r_{mk} = \overline{U_m(t)U_k(t)^*}$ obtained for the series $U_m(t) = u_m(t) + iv_m(t)$ and its complex conjugate, $U_k(t)^*$, are shown to be useful for identifying traveling and standing wave modes; here, (u, v) are the zonal and meridional components of velocity. In the extreme case where the data set is dominated by a single frequency, the frequency domain EOF technique and complex time domain EOF technique are identical. According to Merrifield and Guza (1990), the Hilbert transform complex EOF only makes sense if the frequency distribution in the original time series $u(t)$ is narrow band.

In summary, conventional EOF analysis in the time domain works best when the variance is dominated by standing waves and spread over a wide range of frequencies and wavenumbers. Frequency domain EOF analysis should be used when the dominant variability within the data set is concentrated into narrow frequency bands. Rotary spectral EOF analysis is best used for data sets in which the variance is in narrow frequency bands and dominated by either the clockwise or counterclockwise rotating component of velocity. Complex time domain principal component analysis allows for the detection of propagating wave features (if the process has a narrow frequency band) and the identification of these motions in terms of their spatial and temporal behavior. However, regardless of which method is applied, the best test of a method's validity is whether the results make sense physically and whether the variability is readily visible in the raw time series.

4.4 NORMAL MODE ANALYSIS

In the previous sections, we were concerned with the partition of data variance into an ordered set of spatial and temporal statistical modes. The eigenvalue problem associated with these EOF modes was solved without any consideration given to the underlying physics of the oceanic system. In contrast, normal mode decomposition takes into account the physics and associated boundary conditions of the fluid motion. A common approach is to separate the vertical and horizontal components of the motion and to isolate the forced component of the response from the freely propagating response. As illustrations of these techniques, we consider two basic types of normal mode, eigenvalue problem:

(1) The calculation of vertical normal modes (eigenfunctions), $\psi_k(z)$, for a stratified, hydrostatic fluid with specified top and bottom boundary conditions; and
(2) the derivation of the cross-shore orthogonal modes (eigenfunctions), $\phi_k(x, z)$, for coastal-trapped waves over a variable depth, stratified ocean with or without a coastal boundary.

The first problem can be solved without including the earth's rotation, f, while the second problem requires specification of f. Both eigenvalue problems yield solutions only for certain eigenvalues, λ_k, of the parameter, λ.

4.4.1 Vertical normal modes

A common oceanographic problem is to find the amplitudes (a_k) and phases (θ_k) of a set of K orthogonal basis functions, or modes, by fitting them to a profile of M ($> K$) observed values of amplitude and phase. For instance, one might have observations from $M = 5$ depths and want to find the modal parameters (a_k, θ_k) for the first three theoretical modes, $k = 1, 2, 3$, derived from an analysis of the equations of motion. Once the set of theoretical modes are derived, they can be fitted using a least-squares technique to observations of the along-channel current amplitude and phase. This yields the required estimates, (a_k, θ_k), for $k = 1, 2, 3$.

To obtain the vertical normal modes for a nonrotating fluid ($f = 0$), we assume that the pressure, p, density, ρ, and horizontal and vertical components of velocity (u, v) and w, respectively, can be separated into vertical and horizontal components. This separation of variables has the form

$$[u(\mathbf{x}, t), v(\mathbf{x}, t), p(\mathbf{x}, t)/\rho_o] = \sum_{k=0}^{\infty} p_k(x, y, t)\psi_k(z) \quad (4.4.1a)$$

$$w = \sum_{k=0}^{\infty} w_k \int_{-H}^{z} \psi_k(z)\, dz \quad (4.4.1b)$$

$$\rho = \sum_{k=0}^{\infty} \rho_k \frac{d\psi_k(z)}{dz} \quad (4.4.1c)$$

where $k = 0, 1, 2, \ldots$ is the vertical mode number and the variables without subscripts are functions of $(\mathbf{x}, t) = (x, y, t)$. Substituting these expressions into the usual equations of motion (see LeBlond and Mysak, 1979; Kundu, 1990), we obtain the *Sturm–Liouville equation*

$$\frac{d}{dz}\left(\frac{1}{N^2}\frac{d\psi_k}{dz}\right) + \frac{1}{c_k^2}\psi_k = 0 \qquad (4.4.2)$$

where $N(z) = [-(g/\rho)\,d\rho/dz]^{1/2}$ is the Brunt–Väisälä frequency, c_k^2 is the separation constant and $1/c_k^2$ the eigenvalues, λ_k. For a rotating fluid ($f \neq 0$), we assume $N(z)$ is uniform with depth and replace N^2/c_k^2 in equation (4.4.2) as follows:

$$N^2/c_k^2 \rightarrow (N^2 - \omega^2)/gh_k, \quad k = 1, 2, \ldots \qquad (4.4.3a)$$

where h_k is an "equivalent depth", ω is the wave frequency

$$gh_k = (\omega^2 - f^2)/(l^2 + q^2) = c_k^2 - f^2/l^2 \qquad (4.4.3b)$$

and (l, q) are the wavenumbers in the horizontal (x, y) directions. Wave-like solutions are possible provided that $f^2 < \omega^2 < N^2$. For a rectangular channel of width L, the cross-channel wavenumber $q \rightarrow q_m = m\pi/L$ and solutions must be considered for both $k, m = 1, 2, \ldots$ (Thomson and Huggett, 1980). For both the rotating and nonrotating case, solutions to the eigenvalue problem (4.4.2) are subject to specified boundary conditions at the seafloor ($z = -H$) and the upper free surface ($z = 0$) of the fluid. These end-point boundary conditions are:

$$\frac{d\psi_k}{dz} = 0 \text{ (i.e. } w = 0\text{) at } z = -H \qquad (4.4.4a)$$

$$\frac{d\psi_k}{dz} + \frac{N^2}{g}\psi_k = 0 \text{ (i.e. } \frac{\partial p}{\partial t} = \rho g w\text{) at } z = 0 \qquad (4.4.4b)$$

Modal analysis of the type described by (4.4.2)–(4.4.4) is valid only for an inviscid hydrostatic fluid in which oscillations occur at frequencies much lower than the local buoyancy frequency, N, and for which the vertical length scale is much smaller than the horizontal length scale. In addition, the ocean must be of uniform depth and have no mean current shear. (For sloping bottoms, the horizontal cross-slope velocity component, u, is linked to the vertical boundary, w, through the bottom boundary condition $u = -w\,dH/dx$ and separation of variables is not possible.) The method can be applied to an ocean with zero rotation or with rotation that changes linearly with latitude, y. Solutions to (4.4.2) are obtained for specified values of $N(z)$ subject to the surface and bottom boundary conditions. Although the individual orthogonal modes propagate horizontally, the sum of a group of modes can propagate vertically if some of the modes are out of phase.

Analytical solutions: Simple analytical solutions to the Sturm–Liouville equation are obtained with and without rotation when N = constant (density gradient constant with depth). Assuming the rigid lid condition (i.e. no surface gravity waves so that $w = 0$ at $z = 0$), the vertical shapes of the orthogonal eigenfunctions $\psi_k(z)$ in (4.4.2) are given by

$$\psi_k(z) = \cos(k\pi z/H), \quad k = 0, 1, 2, \ldots \qquad (4.4.5)$$

where $k = 0$ is the depth-independent barotropic mode, and $k = 1, 2, \ldots$ are the depth-dependent baroclinic modes. The kth mode has k zero crossings over the depth range

$-H \leq z \leq 0$ and satisfies the boundary conditions $w = 0$ (cf. 4.4.1b). Phase speeds (eigenvalues) of the modes are given by

$$c_o = (gH)^{1/2}, \quad k = 0 \text{ (barotropic mode)} \tag{4.4.6a}$$
$$c_k = NH/k\pi, \quad k = 1, 2, \ldots \text{ (baroclinic modes)} \tag{4.4.6b}$$

In general, $N(z)$ is nonuniform with depth and, for a given k, the solutions will have the form

$$c_k = (gh_k)^{1/2} \tag{4.4.7}$$

where the "equivalent depth" h_k is used in analogy with H in (4.4.6a). For an ocean of depth $H \approx 2500$ m and buoyancy frequency $N \approx 2 \times 10^{-3}$/s, the eigenvalue for the first baroclinic mode has a phase speed $c_1 \approx 1.6$ m/s and the equivalent depth $h_k = c_1^2/g \approx 0.26$ m. For the 400-m deep tidal channel, we find $N \approx 5 \times 10^{-3}$ m/s, $c_1 \approx 0.8$ m/s and $h_k \approx 0.06$ m.

General solutions: To solve the general eigenvalue problem (4.4.2)–(4.4.4) for variable buoyancy frequency, $N(z)$, we resort to numerical integration techniques for ordinary differential equations with two-point boundary conditions. That is, given the start and end values of the function $\psi_k(z)$, and variable coefficient $N(z)$ we seek values at all points within the domain ($-H \leq z \leq 0$). Fortunately, there exist numerous packaged programs for finding the eigenvectors and eigenvalues of the Sturm-Liouville equation for specified boundary conditions. The NAG routine D02KEF (Nag Library Routines, 1986) finds the eigenvalues and eigenfunctions (and their derivatives) of a regular singular second-order Sturm–Liouville system of the form

$$\frac{d}{dz}\left[F(z)\frac{d\psi_k}{dz}\right] + G(z; \lambda)\psi_k = 0 \tag{4.4.8}$$

together with boundary conditions

$$z_{a2}\psi_k(z_a) = z_{a1}F(z_a)\,d\psi_k(z_a)/dz \tag{4.4.9a}$$
$$z_{b2}\psi_k(z_b) = z_{b1}F(z_b)\,d\psi_k(z_b)/dz \tag{4.4.9b}$$

for real-valued functional coefficients F and G on a finite or infinite range, $z_a < z < z_b$. Provision is made for discontinuities in F and G and their derivatives. The following conditions hold on the function coefficients:

(1) The function $F(z)$, which equals $1/N^2(z)$ in the case of (4.4.2), must be nonzero and of one sign throughout the closed interval $z_a < z < z_b$. This is certainly true in a stable oceanic environment where $N^2 > 0$; for $N^2 < 0$, the fluid is gravitationally unstable and vertical modes are not possible;
(2) $\partial G/\partial \lambda$ must be of constant sign and nonzero throughout the interval $z_a < z < z_b$ and for all relevant values λ, and must not be identically zero as z varies for any relevant value of λ.

Numerical solutions to the Sturm–Liouville equation are obtained through a Pruefer transformation of the differential equations and a shooting method. (The shooting method and relaxation methods for the solution of two-point boundary value problems are described in *Numerical Methods* (Press et al., 1992)). The computed eigenvalues are correct to a certain error tolerance specified by the user. Eigenfunctions $\psi_k(z)$ for the problem have increasing numbers of inflection points and zero crossings within the domain $z_a < z < z_b$ as the eigenvalue increases. When the final estimate of λ_k is found by the shooting method, the routine D02KEF integrates the differential equation once more using that value of λ_k and with initial conditions chosen such that the integral

$$I_k = \int_{z_a}^{z_b} [\psi_k(z)]^2 \partial G/\partial \lambda(z; \lambda) \, dz \qquad (4.4.10)$$

is roughly unity. When $G(z; \lambda)$ is of the form $\lambda w(z) + \psi(z)$, which is the most common case, I_k represents the square of the norm of ψ_k induced by the inner product

$$\overline{\psi_k(z)\psi_m(z)} = \int_{z_a}^{z_b} \psi_k(z)\psi_m(z)w(z) \, dz \qquad (4.4.11)$$

with respect to which the eigenfunctions are mutually orthogonal if $k \neq m$. This normalization of ψ for $k = m$ is only approximate but typically differs from unity by only a few percent.

If one is working with observed density (σ_t) profiles for the region of interest, a useful approach is to solve the Sturm–Liouville equation using an analytical expression for $N(z)$ by fitting a curve of the type $\sigma_t(z) = [\rho(z) - 1]10^3 = \sigma_o \exp[a/(z+b)]$ or other exponential form, to the data. The eigen (modal) analysis is fairly insensitive to small changes in density so that, even though changes in $N(z)$ are large in the upper oceanic layer, we usually can get away with a simple analytical curve fit. Alternatively, we can specify the actual density on a numerical grid for which modes are to be calculated. Once $N(z)$ is available, we can use numerical methods to solve (4.4.2) subject to the boundary conditions (4.4.4), allowing for specified error bounds or degree of convergence on the final boundary estimate. Based on the analytical solutions (4.4.5), we can expect solutions ψ_k to resemble cosine functions whose vertical structure has been distorted by the nonuniform distribution of density along the vertical profile. There is a direct analogy here with the modes of oscillation of a taut string clamped at either end and having a nonuniform mass distribution along its length.

The normal modes are normalized relative to their maximum value and then fitted to the data in a least-squares sense (Table 4.4.1). If there are M current meters on a mooring string, the maximum possible number of normal baroclinic modes is $M - 1$. By comparing the normal modes with the data, we can derive the absolute values of the barotropic mode and a maximum of $M - 1$ baroclinic modes. Solutions to the least-squares fitting are described in (Press et al., 1992).

4.4.2 An example: normal modes of semidiurnal frequency

Suppose that the along-axis semidiurnal currents, u, in a tidal channel have the form $u_m = a_m \cos(\omega_t + \theta_m)$, where a_m, θ_m ($m = 1, \ldots, M$) are the observed current amplitude and phase, respectively. In terms of tidal current ellipses, we can think of u as the major axis of the current ellipse for each current meter on the mooring line. The oscillations have frequency $\omega = \omega_{M_2}$ corresponding to M_2 semidiurnal tidal currents and the phase θ is referenced to some time zone or meridian of longitude so that we can intercompare values for different current meters and the surface tides. The values a_m, θ_m for the different current meter records can be determined using harmonic analysis techniques (Foreman, 1976) provided the measured data are at hourly intervals over a period of seven days or longer so that the M_2 and K_1 constituents are separable. We next rewrite the above expression for u in the usual way as $u_m = A_m \cos(\omega_t) + B_m \sin(\omega_t)$, where $\tan \theta_m = A_m/B_m$ and $a_m^2 = (A_m^2 + B_m^2)$. This allows us to examine the sine and cosine components separately. The observed magnitudes A_m and B_m at each current meter depth z_m, $m = 1, \ldots, M$ are then used to compute the amplitudes and phases of the basis functions $\psi_k(z_m)$, for a maximum of K different modes ($K < M$). At best, we can obtain the amplitudes and phases of the barotropic mode ($k = 0$) and up to $M - 1$ baroclinic modes.

Details of the modal analysis at semidiurnal frequency using current meter data from a tidal channel are presented by Thomson and Huggett (1980). The first step is to obtain an exponential functional fit (Figure 4.4.1a) to the observed mean density structure, $N(z)$. This structure is then used with the local water depth H (assuming a flat bottom), the Coriolis parameter, f, and the wave frequency, ω, to calculate the theoretical dynamic modes (Figure 4.4.1b). A finite sum of these theoretical modes $\sum \psi_k(z)$ is then least-squares fitted to the observed cosine component $A_m(z)$ to obtain estimates of the contributions A_k from each mode, k. This operation is repeated for the sine component B_k. (Recall that the maximum total of barotropic plus baroclinic modes allowed in the summation is fewer than the number of current meter records per mooring string and that the vertical structure of each mode is found through the products $(A_k, B_k)\psi_k(z)$ where the coefficients are constant.) Using the relationships $\tan \theta_k = A_k/B_k$ and $a_k^2 = (A_k^2 + B_k^2)$, we get the amplitudes and phases of the various modes. In their analysis, Thomson and Huggett (1980) typically had only three reliable current meter records per mooring string. Normally, this would be enough to obtain the first two baroclinic modes. However, the bottom current meter in most instances was within a few meters of the bottom and therefore strongly affected by benthic boundary layer effects. To include a mode-2 solution in the estimates, the observed phase and amplitude of the bottom current meter record had to be adjusted for frictional effects via the added term $\exp(-z')\cos(\omega t + \theta - z')$, where

Table 4.4.1. *Modal amplitudes (cm/s) and phases (degrees relative to 120°W longitude) for Johnstone Strait M_2 tidal currents computed from nine-day current meter records. Column 2 gives the number of current meters (M) on the string. The first column for the barotropic mode (a_0, θ_0) and each of the two baroclinic modes (a_k, θ_k), $k = 1, 2$, gives the amplitude and phase (a, θ) before and after the bottom current meter is included in the analysis. The bottom current is included after its amplitude and phase are corrected for bottom boundary layer friction. The vertical eddy viscosity K_v is that value which gives the minimum ratio between the first and second baroclinic modes when the frictionally corrected bottom current meter is included. NC means "no change", implying perfect modal fit for all depths with and without the bottom current meter record. At CM04, no near-surface current meter was deployed and the records were only five days long and therefore suspect*

Site	M	K_v (cm²/s)	Before (a_0,θ_0)	After (a_0, θ_0)	Before (a_1, θ_1)	After (a_1, θ_1)	Before (a_2, θ_2)	After (a_2, θ_2)
CM13	3	15	42, 55°	42, 55°	12, 172°	25, 171°	–	19, −10°
CM14	3	8	35, 51°	35, 51°	11, 169°	15, 171°	–	8, −4°
CM15	3	13	32, 35°	32, 36°	18, 175°	12, 166°	–	7, −31°
CM02	5	0	36, 42°	NC	21, 220°	NC	9, 13°	NC
CM04	4	7	29, 45°	50, 24°	13, 215°	79, 174°	2, −34°	70, 0°

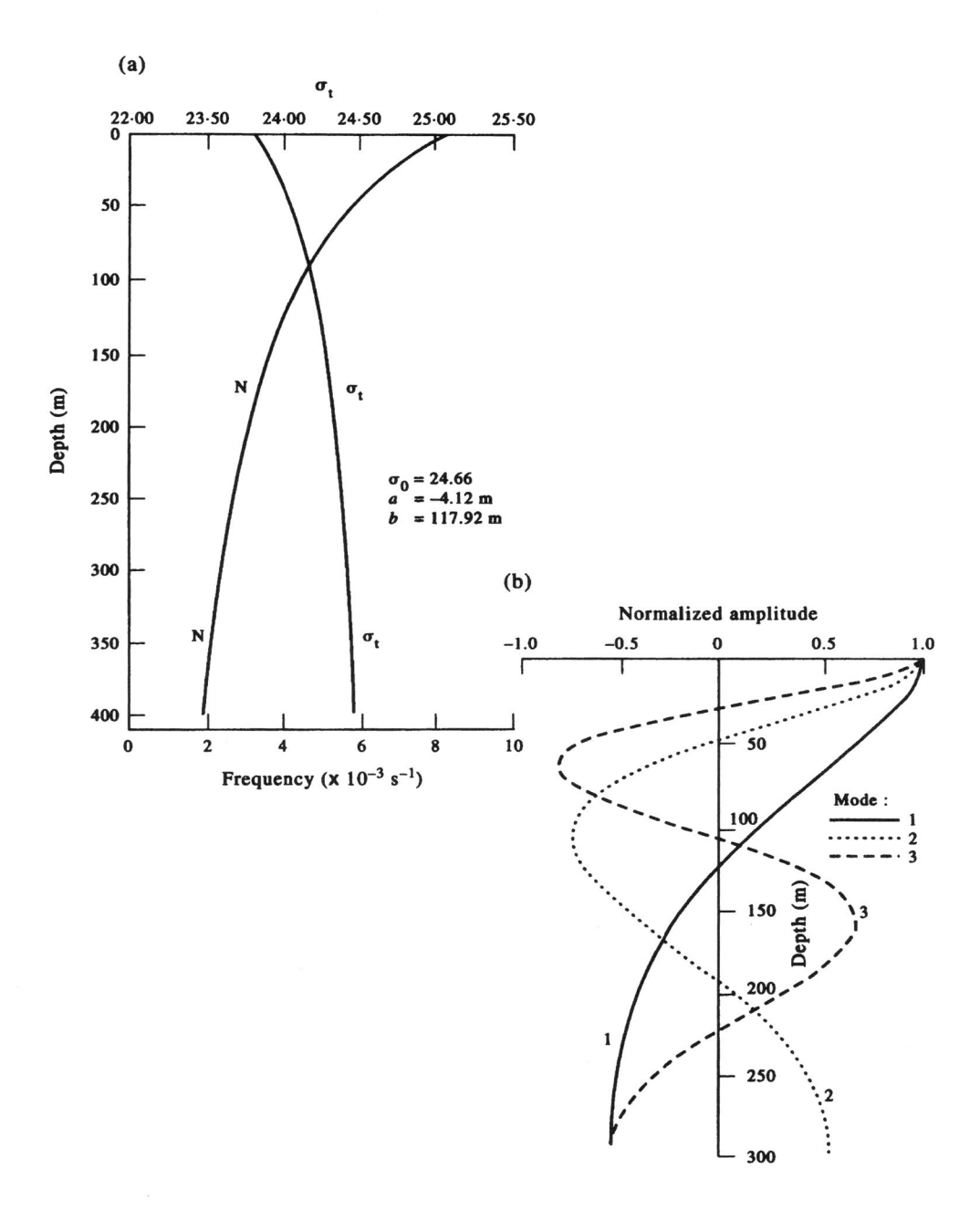

Figure 4.4.1. Baroclinic modes for semidiurnal frequency (ω_{M_2}) in a uniformly rotating, uniform depth channel. (a) The mean density structure (σ_t) and corresponding buoyancy frequency $N(z)$ used to calculate the eigenvalues; (b) Eigenvectors for the first three baroclinic modes. The barotropic mode (not plotted) has a magnitude of unity at all depths. Phase speeds for the modes fitted to the current meter data are $c_1 \approx 34$ cm/s; $c_2 \approx 20$ cm/s. (From Thomson and Huggett, 1980.)

$z' = (z+H)/\delta$, and $\delta \approx (2K_v/\omega)^{1/2}$ is the boundary layer thickness for eddy viscosity K_v. Since K_v is not known *a priori*, the final solution required finding that value of K_v which minimized the ratio formed by the first mode calculated with and without the bottom current meter included in the analysis (Table 4.4.1). In the case where five current meters were available, Thomson and Huggett found that there was no difference in the value of the second mode estimate with and without inclusion of the bottom current meter record in the analysis, suggesting that the three-mode decomposition was representative of the actual current variability with depth.

4.4.3 Coastal-trapped waves (CTWs)

Stratified or nonstratified oceanic regions characterized by abrupt bottom topography adjacent to deeper regions of uniform depth support the propagation of trapped ocean waves with frequencies, ω, which are lower than the local inertial frequency, f. Trapped sub-inertial motions ($\omega < f$) typically are found along continental margins where the coastal boundary is bordered by a marked change in water depth consisting of a shallow (< 200 m) continental shelf, a steep continental slope, and a deep (> 2000 m) weakly sloping continental rise. The longshore wavelengths vary from tens to thousands of kilometers while the cross-shore trapping scale is determined by the density structure and length scale of the topography. For baroclinic waves, the *internal deformation radius* $r = NH/f$ provides an estimate of the cross-shelf trapping scale while the *stratification parameter* $S = (N^2_{max} H^2_{max})/f^2 L^2$ measures the importance of stratification for a shelf-slope region of width L. For a mid-latitude ocean of depth $H \approx 2500\,m$ and buoyancy frequency $N \approx 2 \times 10^{-3}/s$, we find $r \approx 50\,km$. For wide shelves ($L > 100$ km), the motions are confined mainly to the continental slope, while for narrower shelf regions the motions extend to the coast where they "lean" up against the coastal boundary. For $S \gg 1$ the CTWs are strongly baroclinic, while for $S \ll 1$, they are mainly barotropic (Chapman, 1983). The case $S \approx 1$ corresponds to barotropic shelf waves modified by stratification.

In addition to continental shelf regions, coastal-trapped waves can occur along mid-ocean ridges and in oceanic trenches (where they are known as *trench waves*), as well as around isolated seamounts. Phase propagation, in all cases, is with the coastal boundary to the right of the direction of propagation in the Northern Hemisphere and to the left of the direction of propagation in the Southern Hemisphere. For strongly baroclinic waves, energy propagation is always in the direction of phase propagation; for barotropic motions, short waves can propagate energy in the opposite direction to phase propagation.

The general coastal-trapped wave solution consists of a Kelvin wave mode ($k = 0$), for which the cross-shore velocity component is identically zero at the coast ($U \equiv 0$ at $x = 0$), together with a hierarchy of higher mode shelf waves ($k = 1, 2, ...$) whose cross-shore velocity structures have increasing numbers of zero crossings (sign changes) normal to coast. The first shelf wave mode will have one zero crossing in elevation ζ over the continental margin, the second mode will have two crossings, and so on. For the current component, U, the first mode shelf wave will have no zero crossing, the second mode will have one crossing, and so on. The condition of no normal flow through the coastal boundary requires $U = 0$ at $x = 0$.

Computer programs that calculate the frequencies and cross-shore modal structure of coastal-trapped waves of specified wavelength are available in reports written by Brink and Chapman (1987) and Wilkin (1987). We confine ourselves to a general outline of the programs for the interested reader. Practical difficulties with the numerical solutions to the equations are provided in these comprehensive reports. The programs of Brink and Chapman use linear wave dynamics in which the water depth, $h(x)$, is assumed to be a function of the cross-shore coordinate, x, alone. Similarly, the buoyancy frequency, $N(z)$, is a function of depth alone. The one profile that can be used in the analysis is best obtained by least-squares fitting a function (such as a polynomial or exponential) to a series of observed profiles. The wave parameters such as velocity, pressure and density are assumed to be sinusoidal in time (t) and longshore direction (y) such that for any particular wave parameter, ξ, we have

$$\xi(x, y, t) = \xi_o(x)\exp[i(\omega t + ly)] \qquad (4.4.12)$$

where ω is the wave frequency and l is the alongshore wavenumber. This gives rise to a two-dimensional eigenvalue problem in (ω, l) of the form

$$L[\xi_o(x; \omega, l)] = 0 \qquad (4.4.13)$$

where L is a linear operator. The problem is solved for arbitrary forcing and a fixed l. In particular, for a given wavenumer, k, the frequency ω is varied until the algorithm finds the free-wave mode resonance. Resonance is defined as the frequency at which the square of the spatially integrated wave variable

$$I_v = \int_0^\infty \xi_o^2 \, dx, \quad \text{or } I_p = \int_0^\infty \int_{-h}^0 p^2 \, dz \, dx \qquad (4.4.14)$$

is at a maximum. The suite of programs tackle the following problems for which the user provides the bottom profile $h(x)$, a mean flow profile (if needed) and a selection of boundary conditions:

(1) The program BTCSW yields the dispersion curves $\omega = \omega(l)$ (the frequency as a function of wavenumber), the cross-shore modal structure for velocity $U(x)$ and/or surface elevation $\zeta(x)$, and wind coupling coefficients for barotropic coastal-trapped waves—including continental shelf waves and trench waves—for arbitrary topography and mean longshore current. Options for the long-wave and rigid-lid approximations are included in the program. The user can specify one of two geometries corresponding to topography with and without a coastal boundary. The outer boundary $x = x_{\max}$ is set as $-2L$, where L is the width of the typographically varying domain in the cross-shore direction. Thus, about half the domain has a flat bottom. The outer boundary condition is specified as $\partial U/\partial x = 0$. To obtain solutions for both ζ and U, the depth at the coast should be given a nonzero value $h(0) \geq 1$ m.
(2) For wave frequencies $\omega \leq 0.9f$, the program BIGLOAD2 yields dispersion curves $\omega = \omega(l)$, the horizontal modal structure, and wind-coupling coefficients for an ocean with continuous, horizontally uniform stratification and arbitrary topography. Density in the model has the form $\rho^*(x, y, z, t) = \rho_o(z) + \rho(x, y, z, t)$, where ρ_o is background density and ρ is the density perturbation. Since $\rho \ll \rho_o$, the Boussinesq approximation is assumed throughout (i.e. density perturbations are ignored except where they multiply gravity, g, the acceleration due to gravity). The program allows for the component of the β-effect normal to the coast and for both the free surface and rigid lid boundary conditions at the

ocean surface. Solutions are obtained using the coordinate transformation $\theta = z/h(x)$ and assuming a linear bottom friction drag. A total of 17 vertical and 25 horizontal grids (rectangles) are generated so that the vertical resolution is much better near shore than in deep water. Problems with singularities are avoided by setting $h(x) \geq 1$ m at the coast, $x = 0$. The program does not work well when the shelf-slope width (or width of a trench at the base of the shelf) is small relative to the internal deformation radius for the first mode in the deep ocean. Spurious features appear in unexpected places and force the user to increase the density of horizontal grids over regions of rapidly varying topography. In addition, a spurious mode occurs in the pressure equation for $\beta = 0$ at the local inertial frequency $\omega = f$, making the overall solution suspect. As noted by the authors, the user will have difficulty finding the barotropic Kelvin wave parameters.

(3) The program CROSS is used to find baroclinic coastal-trapped modes for $\omega \leq f$ for arbitrary stratification and uniform depth.

(4) The program BIGDRV2 is used to obtain the velocity, pressure, and density fluctuations over a continental shelf-slope region of arbitrary depth, stratification, and bottom friction and is driven by a longshore wind stress of the form $\tau(x) = \tau_o \exp\left[i(\omega t + ly)\right]$. Specification of a linear friction coefficient of zero ($r = 0$) results in a divide by zero error. As a result, inviscid solutions should not be attempted. As with (2), solutions are obtained on a 25×17 stretched grid. In practice, it is generally best to start a study of coastally trapped waves using BTCSW since it gives first-order insight into the type of modal structure one can expect. However, if the barotropic dispersion curves do not fit the data (e.g. observations reveal strong diurnal-period shelf waves but the first-mode dispersion curves consistently remain below the diurnal frequency band for realistic topography), then density and mean currents should be introduced using BIGLOAD2 and CROSS.

The Brink and Chapman programs have been used by Crawford and Thomson (1984) to examine free wave propagation along the west coast of Canada and by Church *et al.* (1986) and Freeland *et al.* (1986) to examine wind-forced coastal-trapped waves along the southeast coast of Australia (Figure 4.4.2). In all cases, model results are compared with longshore sea-level records and current meter observations from cross-shore mooring lines. The cross-shore depth profiles $h(x)$ and associated buoyancy frequencies $N^2(z)$ used in the Australian model are presented in Figures 4.4.3(a, b). From these input parameters, the program was used to generate eigenvalues and eigenfunctions for the first three CTW wave modes (Figure 4.4.4) and the theoretical dispersion curves (Figure 4.4.5) relating wave frequency, ω, to longshore wavenumber, l. The slopes of the (ω, l) curves give the phase speeds c_k for the given modes ($k = 1, 2, 3$) listed on the figure.

Wilkin (1987) presents a series of FORTRAN programs for computing the frequencies and cross-shore modal structure of free coastal-trapped waves in a stratified, rotating channel with arbitrary bottom topography. The programs solve the linearized, inviscid, hydrostatic equations of motion using the Boussinesq approximation. The Brunt-Väisälä frequency $N(z)$ is a function of the vertical coordinate only. As with Brink and Chapman (1987), the eigenvalue problem is solved using resonance iteration and finite difference equations. The cross-shore perturbation fields returned by the model include velocity, pressure, and density. The difference with Wilkin's model

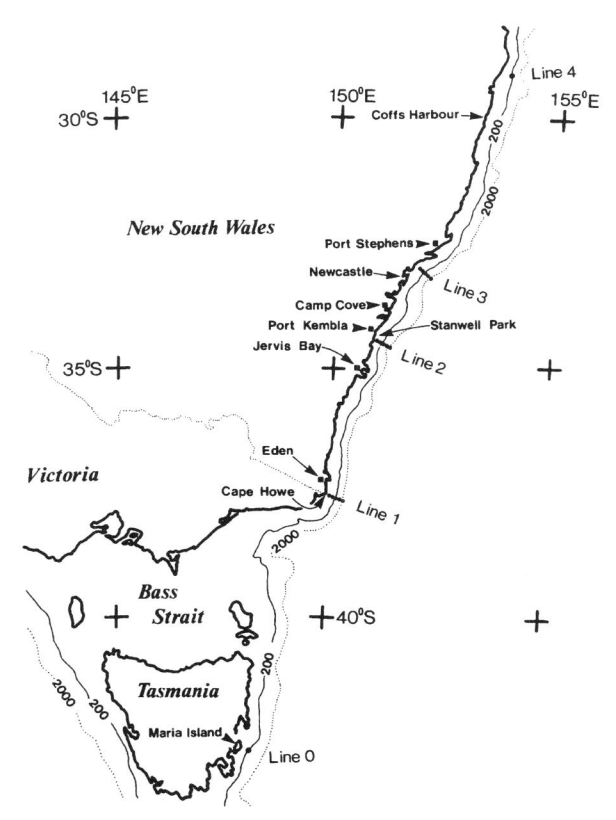

Figure 4.4.2. Southwest coast of Australia showing the locations of the tide gauge stations (■) and current meter lines (0, 1, 2, 3) occupied during the Australian Coastal Experiment (ACE). (From Freeland et al., 1986.)

is that it uses a staggered horizontal (Arakawa "C") grid for which the usual horizontal Cartesian coordinates (x, y) have been mapped to orthogonal curvilinear coordinates (ξ, η). Instead of using finite differencing, the vertical structures of the modes are determined through modified sigma coordinates with expansion of the field variables in terms of Chebyshev polynomials of the first kind. The program has the option of specifying wavenumber, l, and searching for the corresponding free wave frequency, $\omega(l)$, as in Brink and Chapman, or specifying ω and searching for l. For reasons explained by Wilkin, the model is designed to be compatible with the primitive equation ocean circulation model developed by Haidvogel et al. (1988).

In the curvilinear coordinate system, a line element of length ds in the Wilkin model satisfies

$$ds^2 = dx^2 + dy^2 = d\xi^2/dm^2 + d\eta^2/dn^2 \qquad (4.4.15)$$

and the metric coefficients m, n are defined by

$$m = \left[(\partial x/\partial \xi)^2 + (\partial y/\partial \xi)^2\right]^{-1/2} \qquad (4.4.16a)$$

$$n = \left[(\partial x/\partial \eta)^2 + (\partial y/\partial \eta)^2\right]^{-1/2} \qquad (4.4.16b)$$

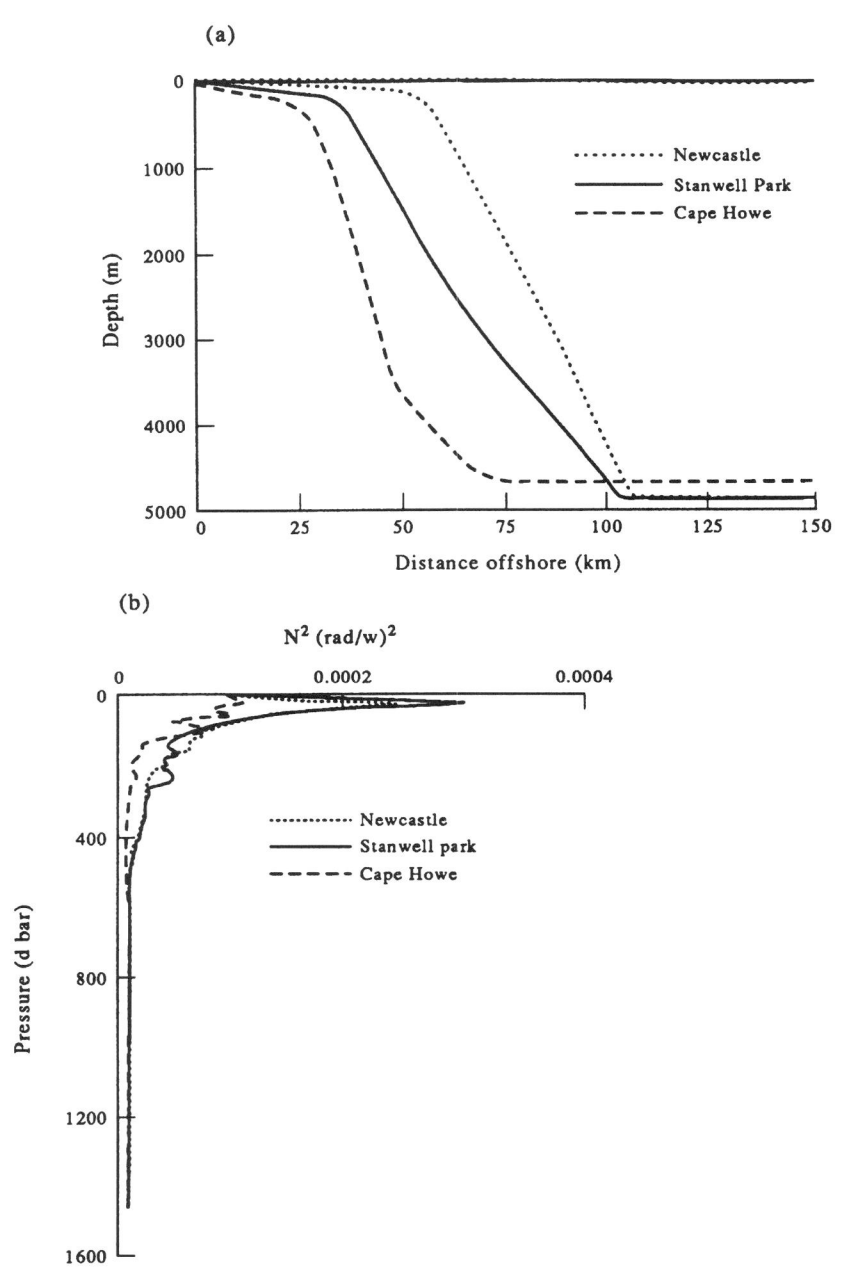

Figure 4.4.3. Parameters used in determining the coastal-trapped wave eigenfunctions at Cape Howe Stanwell Park and Newcastle; (a) The cross-shore depth profiles $h(x)$; (b) The $N(z)^2$ profiles. Below 600 db ($\approx$ 590 m) all curves are similar so that only one is drawn. (From Church et al., 1986.)

The Spatial Analyses of Data Fields 355

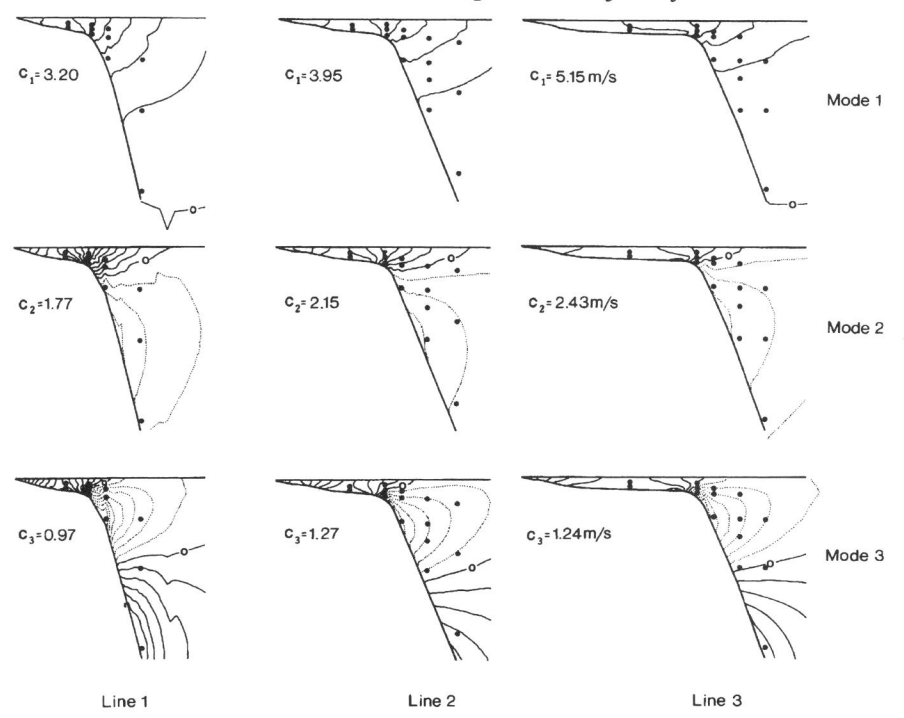

Figure 4.4.4. The eigenfunctions U(x; z) for the first three baroclinic longshore current modes for the three lines in Figures 4.4.2 and parameters in Figure 4.4.3. The contouring is in arbitrary units. Phase speeds c_k (eigenvalues) of each mode for each of the three lines also are shown. (From Church et al., 1986.)

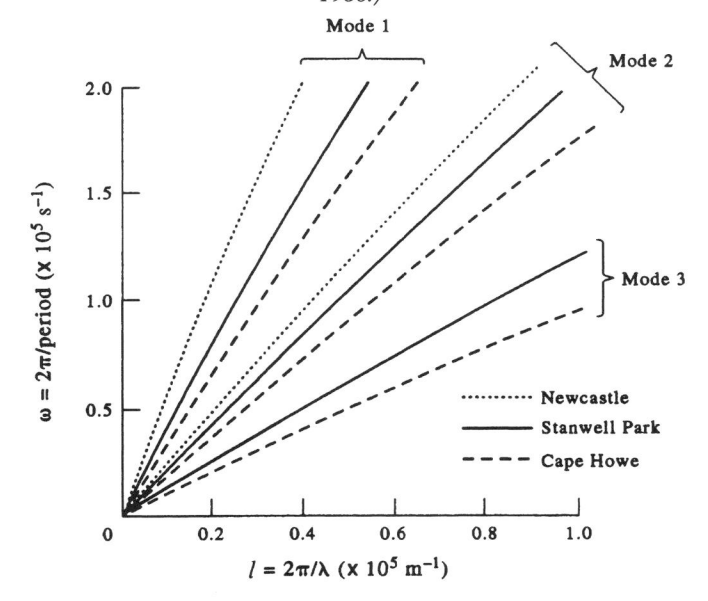

Figure 4.4.5. The theoretical dispersion curves $\omega = \omega(l)$ relating the longshore wavenumber, l, to the wave frequency, ω (here, λ is the wavelength). Curves correspond to the first three baroclinic modes for each mooring location. For mode 3, the dispersion curve at Stanwell Park and Newcastle are almost identical. The slopes of the lines are the theoretical phase speeds, c_k. (From Church et al., 1986.)

356   Data Analysis Methods in Physical Oceanography

The velocity perturbations for time-dependent solutions of the form $\exp(-i\omega t)$ are then

$$u = \frac{1}{f^2 - \omega^2}\left(i\omega m \frac{\partial \phi}{\partial \xi} - fn \frac{\partial \phi}{\partial \eta}\right) \quad (4.4.17a)$$

$$v = \frac{1}{f^2 - \omega^2}\left(i\omega n \frac{\partial \phi}{\partial \eta} - fm \frac{\partial \phi}{\partial \xi}\right) \quad (4.4.17b)$$

$$w = \frac{i\omega}{N^2}\frac{\partial \phi}{\partial z} \quad (4.4.17c)$$

where (u, v, w) are the usual velocity components and $\phi = p/\rho_o$ is the perturbation pressure. Solutions are then sought for the resulting pressure equation

$$mn\frac{\partial}{\partial \eta}\left(\frac{n}{m}\frac{\partial \phi}{\partial \eta}\right) + (f^2 - \omega^2)\frac{\partial}{\partial z}\left(\frac{1}{N^2}\frac{\partial \phi}{\partial z}\right) + mn\frac{\partial}{\partial \xi}\left(\frac{m}{n}\frac{\partial \phi}{\partial \xi}\right) = 0 \quad (4.4.18)$$

For a straight coastline, $m\partial/\partial \xi = \partial/\partial x$ and we arrive at the usual solutions for longshore (x-direction) propagation of progressive waves of the form $F(y)\exp[i(lx - \omega t)]$.

The Wilkin model is less general than the Brink and Chapman model in that application of the rigid-lid approximation does not allow for the barotropic (longwave) Kelvin wave solution and a "slippery" solid wall is placed at the offshore boundary. The new vertical coordinate variable, σ, is defined by

$$\sigma = 1 + 2z/h(\eta) \quad (4.4.19)$$

so that the ocean surface is located at $\sigma = 1$ and the (now flattened) seafloor at $\sigma = -1$. Application of this model to the west coast of New Zealand (South Island) is presented by Cahill *et al.* (1991). Modes 1 and 2 of the longshore current for the northern portion of this region based on Wilkin's program CTWEIG are reproduced in Figure 4.4.6. Similar results for the southern region are presented in Figure 4.4.7. Notice that the coastal-trapped waves are nearly barotropic over the shallow shelf immediately seaward of the coast in both sections but are more baroclinic in the offshore region off the southwest coast.

4.5 INVERSE METHODS

4.5.1 General inverse theory

General inverse methods have become a sophisticated analysis tool in the earth sciences. For example, in the field of geophysics, a goal of this technique is to infer the internal structure of the earth from the measurement of seismic waves. The essence of the geophysical *inverse problem* is to find an earth structure model which could have generated the observed acoustic travel-time data. This is in contrast to the *forward problem* which uses a known input and an understood physical system to predict the output. In the inverse problem, the input and output are known and the result is the *model* required to translate one set of data into the other.

In oceanography, inverse methods are used for a variety of applications, including the inference of absolute ocean currents using known tracer distributions and

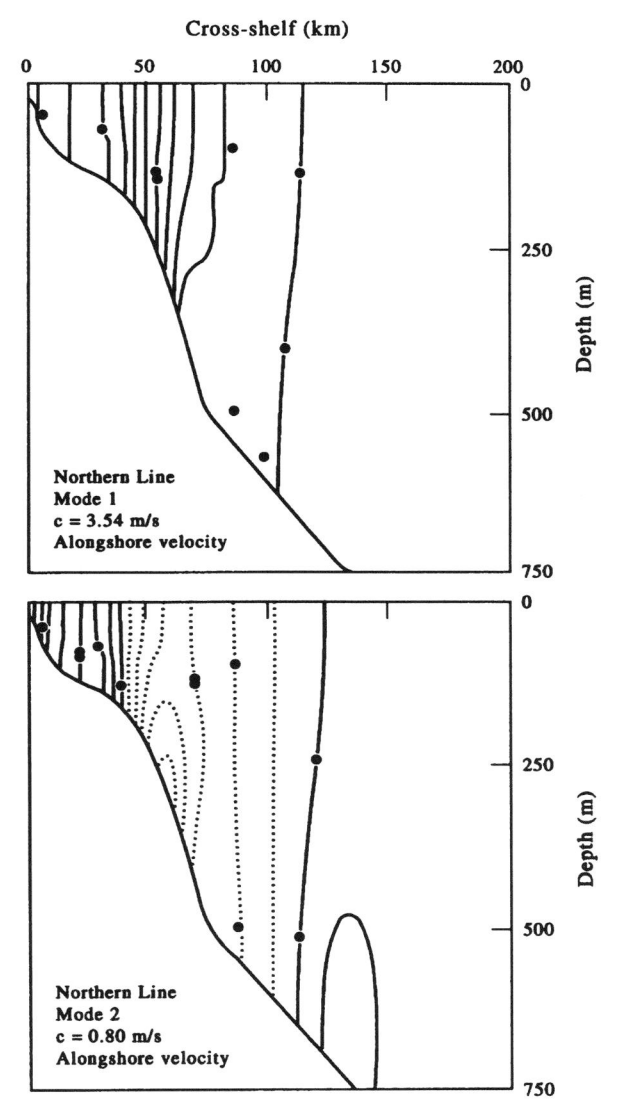

Figure 4.4.6. The longshore velocity structure of coastal-trapped waves for the northwestern shelf-slope region of South Island, New Zealand. (a) Mode 1; (b) Mode 2. Contour lines when multiplied by 10^{-7} correspond to the longshore velocities in m/s for unit energy flux in watts. Negative values are dashed. Current meter locations are given by the dots. (From Cahill et al., 1991.)

geostrophic flow dynamics (Wunsch, 1978, 1988). Another application uses underwater acoustic travel times to determine the average temperature of the global ocean for long-term climate studies (Worchester *et al.*, 1988). A study by Mackas *et al.* (1987) used inverse techniques to determine the origins and mixing of water masses off the coast of British Columbia. In these oceanographic applications, the "solutions" are what we previously called the "models" in the geophysical problem. The kernel functions are formulated from the physics of the problem in question and the result is found by matching the "solution" to the input data. A cursory look at the problem is

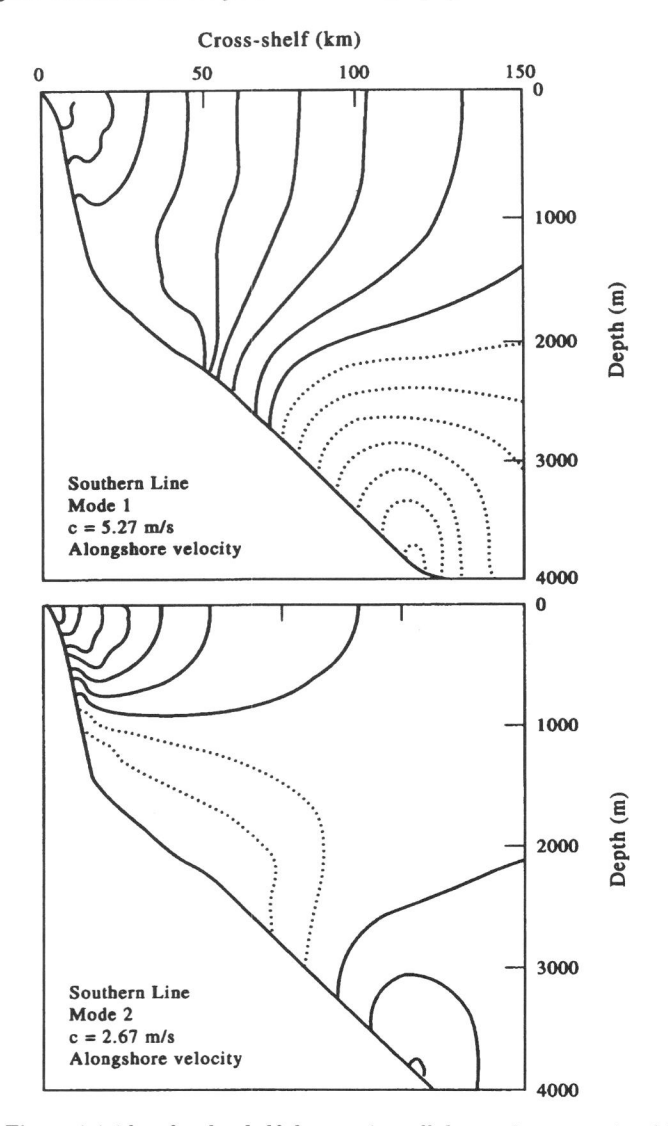

Figure 4.4.7. As for Figure 4.4.6 but for the shelf-slope region off the southwestern tip of South Island. Note the change in depth and offshore distance scale in the two figures. This line is roughly 500 km to the south of the line in Figure 4.4.6.

provided in this section. The interested reader is referred to Bennett (1992) for detailed insight into the theory and application of inverse methods in oceanography.

In general, the inverse problem takes the form

$$e(t) = \int_a^b C(t, \xi) m(\xi)\, d\xi \qquad (4.5.1)$$

where $e(t)$ are the input data, $m(\xi)$ is the model and $C(t, \xi)$ is the kernel function for the variable ξ. The kernel functions are determined from the relevant physical equations for the problem and are assumed to be known (Oldenburg, 1984). It is the judicious selection of these kernel functions that makes the inverse problem a complex exercise requiring physical insight from the oceanographer. In order to extract information about the model, $m(\xi)$, we will restrict our consideration of inverse theory to linear inverse methods applied to a set of observations. This is referred to as "finite dimensional inverse theory" by Bennett (1992). In his discussion of this form of inverse theory, Bennett suggests that it applies to:

(1) An incomplete ocean model, based on physical laws but possessing multiple solutions.
(2) Measurements of quantities not included in the original model but related to the model by additional physical laws.
(3) Inequality constraints on the model fields or the data.
(4) Prior estimates of errors in the physical laws and the data.
(5) Analysis of the level of information in the system of physical laws, measurements, and inequalities.

Equation (4.5.1) is a *Fredholm equation* of the first kind. Inverse theory is centered around solving this equation in such a way as to extract information about the model, $m(\xi)$, when information is available for the data, $e(t)$. It is important to realize that the inverse problem cannot be solved unless the physics and the geometry of the problem are known (i.e. equation (4.5.1) has been set up). It is, therefore, impossible to consider a solution to the inverse problem unless the forward problem can be solved. The physics of the forward problem may be ill-posed, in which case not all of the solutions will match or, if they do, it is a coincidence and not a solution to (4.5.1). Thus, the basic questions to ask regarding a solution of the inverse problem are: (1) Does a solution exist? In other words, is there an $m(\xi)$ which produces $e(t)$?; (2) How does one construct a solution?; (3) Is the solution unique?; and (4) How is the nonuniqueness appraised?

The answers to the above questions will depend on the data, $e(t)$. In theory, there exist three types of data:

(1) An infinite amount of accurate data;
(2) a finite amount of accurate data;
(3) a finite amount of inaccurate data.

In reality, only option (3) occurs as we are forced to work with observations which contain a variety of measurement and sampling errors. While perfect data are limited to the realm of the mathematical, it is often instructive to consider analytic "inverses". For example, the analytical inverse to

$$x(f) = \int_{-\infty}^{\infty} x(t)e^{-i2\pi ft} \, dt \qquad (4.5.2)$$

is

$$x(t) = \frac{1}{2\pi} \int_{-\infty}^{\infty} x(f)e^{i2\pi ft} \, dt \qquad (4.5.3)$$

Similarly, the inverse of

$$\phi(x) = 2/\lambda \int_x^a \left[r\varepsilon(r)/(r^2 - x^2)^{1/2}\right] dr \qquad (4.5.4)$$

is

$$\varepsilon(r) = -\lambda/\pi \int_r^a \left[(d\phi/dx)/(x^2 - r^2)^{1/2}\right] dx \qquad (4.5.5)$$

In the second case, we require knowledge of $d\phi/dx$ to find $\varepsilon(r)$, which is easy to do for ideal continuous data (Figure 4.5.1a), or even for a finite sample of accurate data (Figure 4.5.1b). If, however, we have a finite sample of inaccurate data (Figure 4.5.1c), we have difficulty estimating $d\phi/dx$.

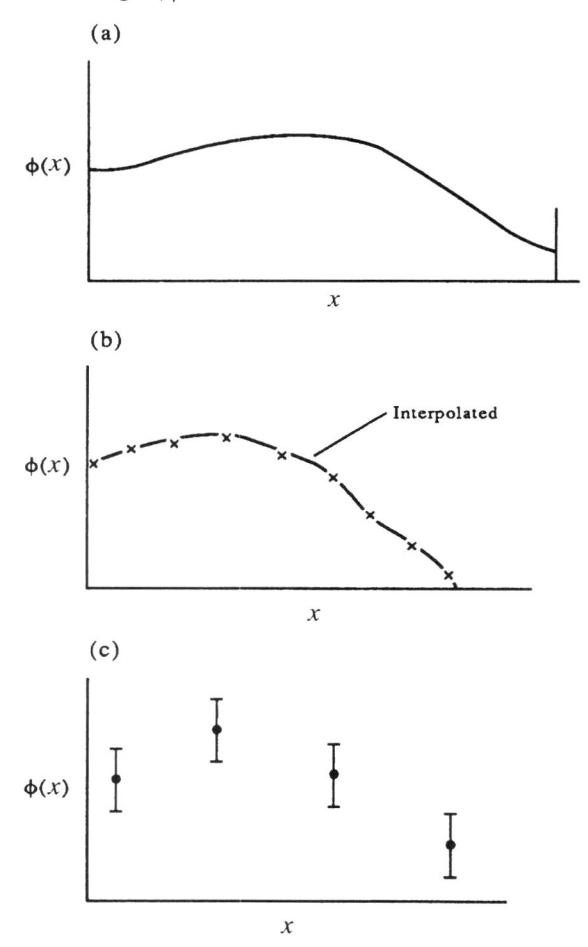

Figure 4.5.1. Three examples of the function $\phi(x)$ required for the inverse solution, $\varepsilon(r)$, of equation (4.5.5). Analytical (a) and digital (b) versions of $\phi(x)$ for which inversion is readily possible. (c) A typical "observed" version of $\phi(x)$, consisting of four mean values (plus standard deviations) for which inversion is considerably less accurate.

The problem of dealing with a limited sample of inaccurate measurements is the most common obstacle to the application of inverse methods. Usually, these inaccuracies can be treated as additive noise superimposed on the true data and, therefore, can be handled with statistical techniques. These additive errors have the effect of "blurring" or distorting our picture of the solution (model). Unfortunately, one cannot conclude that if the error noise is small that the model distortions also will be small. The reason for this is that most geophysical kernel functions act to smooth the model, thus changing the length scale of the response for both the forward and inverse problems. In other words, the solution obtained with inaccurate data using the inverse procedure may be very different from the model which actually generated the data. In addition, particular solutions to the model are not unique and a wide variety of solutions is equally possible.

In most oceanographic applications of inverse methods, we are primarily interested in finding a model which reproduces the observations. Here, the fundamental problem is the nonuniqueness of any inverse solution which is one of infinitely many functions that can reproduce a finite number of observations. This nonuniqueness becomes more severe when the data are inaccurate, as they must be in any practical oceanographic application. The key to the application of inverse methods in oceanography is to select the "correct" (by which we mean the most probable or the most reasonable) inverse model-solution.

Inverse construction in oceanography may take the form of parametric modeling. In this case, we write our model as $m = f(a_1, a_2, ... , a_N)$ and a numerical scheme is sought to find appropriate values of the parameters, a_i ($i = 1, ... , N$). Parameterization is justified when the physical system actually has this form and depends on a number of input parameters. The model is solved by collecting more than N data points and finding the parameters through a least-squares minimization of

$$\phi = \sum_{i=1}^{N} (e_i - e_i')^2 \qquad (4.5.6)$$

where

$$e_i' = f(a_1, a_2, ... , a_N; \varepsilon_i) \qquad (4.5.7)$$

In (4.5.7) ε_i is the ith kernel function.

4.5.2 Inverse theory and absolute currents

As reviewed by Bennett (1992), an important application of inverse theory to ocean processes was the computation of absolute currents for large-scale ocean circulation. In the 1970s, two different approaches to this problem were proposed. The first by Stommel and Schott (1977) was called the "beta spiral" technique, which demonstrated that the vertical structure of large-scale, open-ocean velocity fields could be explained using simple equations expressing geostrophy and continuity (conservation of mass). The second method, introduced by Wunsch (1977), showed that reference velocities could be estimated simultaneously around a closed path in the ocean. The resultant absolute velocities were consistent with geostrophy and the conservation of heat and salt at various levels. As a guide to oceanographic applications of inverse techniques, we provide succinct reviews of both applications.

4.5.2.1 The beta spiral method

Good reviews of the Stommel and Schott (1977) beta spiral method are provided by Olbers et al. (1985) and Bennett (1992). The basic equations for this application are the usual linearized beta (β)-plane equations for horizontal geostrophic flow (u, v) in a Boussinesq fluid

$$-\rho_o f v = -\partial p/\partial x \qquad (4.5.8a)$$
$$\rho_o f u = -\partial p/\partial y \qquad (4.5.8b)$$

the hydrostatic equation

$$0 = -\partial p/\partial z - \rho g \qquad (4.5.9)$$

which relate pressure perturbations, $p(\mathbf{x}, t)$, to density fluctuations, $\rho(z, t)$, and the conservation of mass (or continuity) relation

$$\nabla \cdot \mathbf{u} + \partial w/\partial z = 0 \qquad (4.5.10)$$

In these equations, f is the Coriolis parameter, u, v, and w are, respectively, the eastward (x), northward (y) and upward (z) components of current velocity, and $\rho = \rho(x, y, z)$ is the density perturbation about the mean density $\rho_o = \rho_o(z)$. Following Bennett (1992), we will reserve vector notation for horizontal fields and operators ($\mathbf{x} = (x, y)$, $\mathbf{u} = (u, v)$, etc.).

Using the above equations, we can derive the well-known "thermal wind" relation, whose vertically integrated velocity components are

$$u(\mathbf{x}, z) = u_o(\mathbf{x}) + (g/f\rho_o) \int_{z_o}^{z} \rho_y(x, \zeta) \, d\zeta \qquad (4.5.11a)$$

$$v(\mathbf{x}, z) = v_o(\mathbf{x}) - (g/f\rho_o) \int_{z_o}^{z} \rho_x(x, \zeta) \, d\zeta \qquad (4.5.11b)$$

where subscripts x, y refer to partial differentiation and $u_o(\mathbf{x})$, $v_o(\mathbf{x})$ are the velocity components at some reference depth. Equations (4.5.8–4.5.10) also give rise to the well-known Sverdrup interior vorticity balance

$$w_z = \beta v/f \qquad (4.5.12)$$

where β is the northward (y) gradient of the Coriolis parameter, and $f = f(y) = f_o + \beta y$ in the beta-plane approximation.

These equations cannot be used alone to determine the full absolute velocity field ($\mathbf{u}$, w), even if the density field ρ were known. However, to resolve this indeterminacy, all we need is the velocity field at a particular depth where $\mathbf{u} = \mathbf{u}(\mathbf{x}, z_o)$ and $w = w(\mathbf{x}, z_o)$. Stommel and Schott (1977) demonstrated that these unknown reference values may be estimated by assuming the availability of measurements of some conservative tracer ϕ which satisfy the steady-state conservation law

$$\mathbf{u} \cdot \nabla \phi + w \phi_z = 0 \qquad (4.5.13)$$

This tracer might be salinity (S) or potential temperature (θ), or some function of both

The Spatial Analyses of Data Fields

S and θ. Combining the vertical derivative of equation (4.5.13) with equations (4.5.11) and (4.5.12), yields

$$\left(\mathbf{u} \cdot \nabla + w \frac{\partial}{\partial z}\right)(f\phi_z) = (g/\rho_o)\mathcal{J} \tag{4.5.14}$$

where $\mathcal{J}$ is the Jacobian $\mathcal{J}(\rho, \phi) = \rho_x\phi_y - \rho_y\phi_x$. In equation (4.5.14), $f\phi_z$ represents the potential vorticity which would be conserved if density ρ were itself conserved. The tracer equation can be used again to eliminate the vertical velocity w

$$\mathbf{u} \cdot \mathbf{a} = (g/\rho_o)\mathcal{J}(\rho, \phi) \tag{4.5.15}$$

where the vector $\mathbf{a}$ is given by

$$\mathbf{a}(\mathbf{x}, z) = \nabla(f\phi_z) - \frac{\nabla\phi}{\phi_z}f\phi_{zz} \tag{4.5.16}$$

Using the integrated thermal wind equations (4.5.11) yields

$$\mathbf{u}_o \cdot \mathbf{a} = c \tag{4.5.17}$$

where $\mathbf{u}_o$ is the horizontal velocity at depth z_o and c is given by

$$c(\mathbf{x}, z) = -\mathbf{u}' \cdot \mathbf{a} + (g/\rho_o)\mathcal{J}(\rho, \phi) \tag{4.5.18}$$

In equation (4.5.18), the $\mathbf{u}'$ is that part of the horizontal velocity in the thermal wind relation that depends on the density field.

Since $\mathbf{a}$ and c depend on $g, \rho, f, \nabla\rho, \nabla f, \phi_z$ and ϕ_{zz}, they can be determined using closely spaced hydrographic stations through measurements of $T(z)$ and $S(z)$. Thus, from (4.5.17), we can calculate $\mathbf{u}_o$ using the hydrographic data. Equation (4.5.17) holds at all levels so that two different levels can be used to specify u_o and v_o. We can then calculate the vertical velocity w from (4.5.14). The full velocity solution should be independent of the levels chosen for these computations. In reality, (4.5.17) is not an exact relation as it was derived from approximate dynamical laws and computed from data that contain measurement and sampling errors. As a consequence, our estimate of $\mathbf{u}_o$ from (4.5.17) should be done as a best fit to the data from the two levels chosen.

Suppose that N levels are chosen from the hydrographic data ($N \geq 2$). Let $c_n = x(\mathbf{x}, z_n)$ and $\mathbf{a}_n = \mathbf{a}(\mathbf{x}, z_n)$ for $1 \leq n \leq N$. The simple least-squares best fit minimizes

$$R^2 = \sum_{n=1}^{N} R_n^2 = \sum_{n=1}^{N} (c_n - \mathbf{u}_o \cdot \mathbf{a}_n)^2 \tag{4.5.19}$$

where R_n is the residual at level n and R is the root-mean-square (RMS) total error. R^2 is a minimum if $\mathbf{u}_o$ satisfies a simple linear system

$$\mathbf{M}\mathbf{u}_o = \mathbf{d} \tag{4.5.20}$$

where the 2×2 systematic, nonnegative matrix $\mathbf{M}$ depends on the components of $\mathbf{a}_n$, while $\mathbf{d}$ depends on $\mathbf{a}_n$ and c. If $\mathbf{a}$ or c varies with depth, equation (4.5.15) implies that the total velocity vector $\mathbf{u}$ must also depend on depth. For the β-spiral problem, we find that the large-scale ocean currents constitute a spiral with depth at each station. The β-spiral in Figure 4.5.2 is from the study by Stommel and Schott (1977) who used

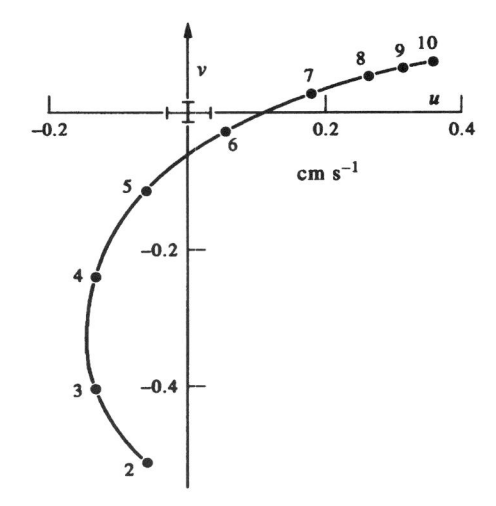

Figure 4.5.2. The β-spiral in horizontal velocity $\mathbf{u} = (u(z), v(z))$ *at 28°N, 36°W, with depths in hundreds of meters. Error bars for the two components of velocity are given at the origin. (After Stommel and Schott, 1977.)*

hydrographic data from the North Atlantic to estimate $\mathbf{u}_o$ for a reference level of $z_o = 1000$ m depth. In this application they found, $u_o = 0.0034 \pm 0.00030$ m/s and $v_o = 0.0060 \pm 0.00013$ m/s at 28°N, 36°W.

The β-spiral problem includes two of the basic concepts common to inverse methods. First, we deal with an incomplete set of physical laws (4.5.8–4.5.10), or their rearrangement, as in the case of the thermal wind equations (4.5.11a, b) which includes the unknown reference velocity. Second, we often resort to the indirect measurement of an additional quantity which, in the case of the present example, is a conservative tracer. This application could have benefited from the inclusion of prior estimates of the errors in the dynamical equations and in the hydrographic data.

4.5.2.2 Wunsch's method

In a parallel development to the β-spiral technique, Wunsch (1977) used inverse methods to estimate reference velocities simultaneously around a closed path in the ocean (Bennett, 1992). As discussed by Davis (1978), both Wunsch's method and the β-spiral method are closely related. Both approaches assume the vertically integrated thermal wind equations (4.5.11) and both provide estimates for the reference velocity $\mathbf{u}_o$. In Wunsch's method, the thermal wind velocity, u', is assumed to be zero at the reference level z_o, which in general may be a function of position [$z_o = z_o(x)$]. Wunsch chose the reference level to be the ocean bottom at $z_o(\mathbf{x}) = H(\mathbf{x})$, with $\mathbf{u}_o(\mathbf{x})$ defined to be the bottom velocity. He then divided the water column into a number of layers defined by temperature ranges. This is consistent with the general water mass structure of the North Atlantic as defined by Worthington (1976). These layers need not be uniform in depth at each hydrographic station. Together with the coastline of the U.S., the hydrographic stations formed a closed path in the western North Atlantic (Figure 4.5.3).

The Spatial Analyses of Data Fields 365

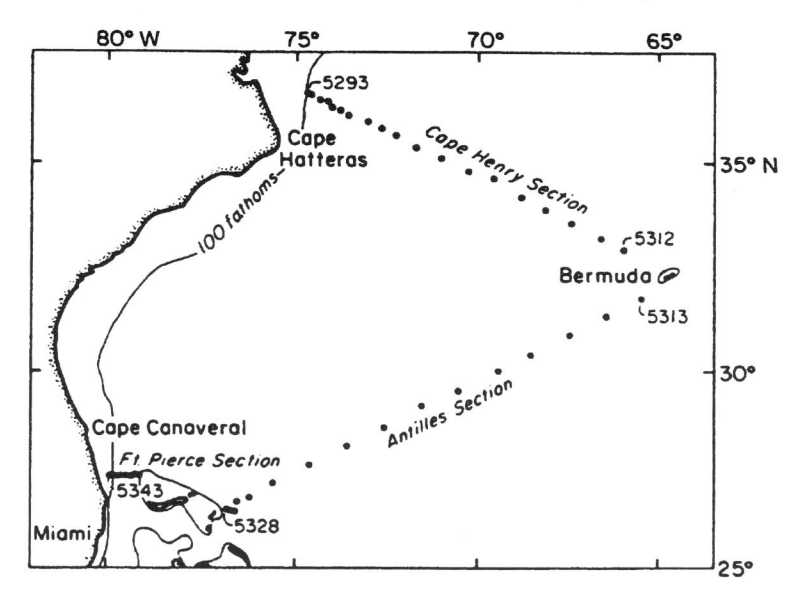

Figure 4.5.3. The locations of hydrographic stations in the North Atlantic used by Wunsch to obtain absolute current estimates using inverse theory. (After Wunsch, 1977.)

We now let v denote the outward component of velocity across the closed triangle formed by the lines of hydrographic stations in Figure 4.5.3. That is, $v = \mathbf{u} \cdot \mathbf{n}$ where $\mathbf{n}$ is the outward unit normal to the sections. We can further let $v' = \mathbf{u}' \cdot \mathbf{n}$ be the outward thermal wind velocity and $b = \mathbf{u}_o \cdot \mathbf{n}$ be the outward horizontal velocity at the seafloor. Let $v'_n(z)$ and b_n denote the thermal wind velocity estimate and unknown bottom velocity midway between the nth station pair, where $1 \leq n \leq N$, and let v'_{mn} denote the average value of v'_n in the mth layer of the water column, where $1 \leq m \leq M$. Wunsch chose the Mth layer to be the total water column, thus the Mth tracer is the total mass of the water column. The assumption of tracer conservation within each layer can be written as

$$\sum_{n=1}^{N} (v'_{mn} + b_n) \Delta z_{mn} \Delta x_n = 0, \quad 1 \leq m \leq M \tag{4.5.21}$$

where Δz_{mn} is the thickness of the mth layer at the nth station pair, and Δx_{mn} is the separation distance between the nth station pair. This system of M equations for N unknowns b_n, $1 \leq n \leq N$, may be written in matrix notation as

$$\mathbf{A}\mathbf{b} = \mathbf{c} \tag{4.5.22}$$

where $\mathbf{A}$ is an $M \times N$ matrix and $\mathbf{c}$ is a column vector of length M with elements

$$A_{mn} = \Delta z_{mn} \Delta x_n \tag{4.5.23a}$$

$$c_m = -\sum_{n=1}^{N} \overline{v'_{mn}} A_{mn} \tag{4.5.23b}$$

Wunsch used $M = 5$ layers as defined by the ranges 12–17°C, 4–7°C, 2.5–4°C, and the entire water column (total mass). The hydrographic data were from $N = 43$ station pairs. For this problem, the matrix equation (4.5.23) represents five equations for 43 unknown velocities, so that the system is underdetermined and has many different solutions.

As reported by Bennett (1992), Wunsch (1977) somewhat arbitrarily selected the vector **b** with the shortest length. This was found by minimizing

$$t_1 = \mathbf{b}^T\mathbf{b} + 2\mathbf{I}^T(\mathbf{Ab} - \mathbf{c}) \tag{4.5.24}$$

where the superscript T denotes the transpose of the matrix and **I** is an unknown Lagrange multiplier consisting of a column vector of length M. It can be shown that t_1 is a minimum when

$$\mathbf{b} + \mathbf{A}^T\mathbf{I} = \mathbf{0} \tag{4.5.25}$$

which gives the minimum solution

$$\mathbf{b} = \mathbf{A}^T(\mathbf{AA}^T)^{-1}\mathbf{c} \tag{4.5.26}$$

which satisfies (4.5.22). The symmetric matrix $\mathbf{AA}^T$ has dimensions $M \times M$ and is nonnegative (Bennett, 1992). However, $\mathbf{AA}^T$ may be singular. These singularities may be overcome by allowing errors in the hydrographic data and conservation laws; that is, by not seeking exact solutions of (4.5.22). We can instead write (4.5.22) in a quadratic form adding weights to each term. It can be shown that for positive weights, we are able to define an exact solution of the problem. This transfers the problem to the selection of these weights.

This cursory presentation of Wunsch's method for computing reference velocities demonstrates, once again, some of the basic elements of inverse methods: A system of incomplete physical laws and inexact measurements of related fields. It is necessary to admit errors into the equations and data values in order to stabilize the solution and to derive a unique solution. In his review, Davis (1978) concluded that both the underdetermined problem of Wunsch's method and the overdetermined problem of the β-spiral method are consequences of tacit assumptions made about noise levels and fundamental scales of motion. Davis suggested that a more orderly approach would be based on Gauss–Markov smoothing (Bennett, 1992) which should be an improvement, assuming explicit and quantitative estimates of the noise and its structure.

4.5.3 The IWEX internal wave problem

Another oceanographic example of the inverse method is found in Olbers *et al.* (1976) and Willebrand *et al.* (1977). Here, inverse theory is used to determine the three-dimensional internal wave spectrum from an array of moored current meters (Figure 4.5.4). In this example, the Fredholm equation (4.5.1) is written in matrix form and becomes

$$y_i = A_{ij}x_j; \quad 1 \leq i \leq N; \, 1 \leq j \leq K \tag{4.5.27}$$

where y_i are N observed velocity cross-spectra (the data), A_{ij} are the kernel functions (for matrix **A**) representing the physical relations from internal wave theory and x_j are the K internal wave parameters to be determined by the inverse method. The inverse

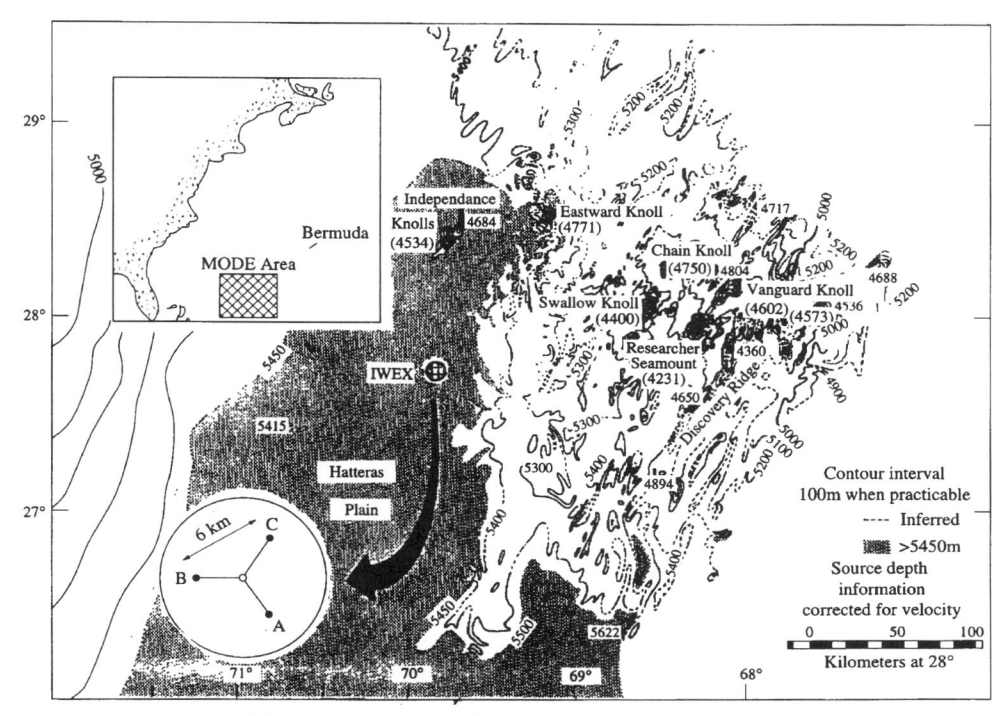

Figure 4.5.4. Location of the IWEX study area showing the positions of the three current meter moorings on the Hatteras Plain in the western North Atlantic. (From Briscoe, 1975.)

problem is to find the K parameters of the theoretical internal wave energy density cross-spectra using the N observed cross-spectra from the current meter array. We achieve this by using the least-squares method to minimize

$$\varepsilon^2(a) = [\hat{y} - y(a)]W[\hat{y} - y(a)]^* \tag{4.5.28}$$

where a represents a set of trial values used to find the minimum and the asterisk (*) denotes the complex conjugate. In equation (4.5.28), W is a weighting matrix used to scale the problem and to produce statistical independence (Jackson, 1972).

It is common to expand the kernel function matrix **A** into eigenvectors (Jackson, 1972). Thus, we write

$$\mathbf{A}V_j = \lambda_j u_j, \quad \mathbf{A}^T u_j = \lambda_i V_i \tag{4.5.29}$$

Following the singular value decomposition we conducted in the EOF analysis (Section 4.3.2), we can factor the matrix **A** as

$$\mathbf{A} = \mathbf{UBV}^T \tag{4.5.30}$$

where **U** is an $N \times P$ matrix whose columns are the eigenvectors u_i, $i = 1, \ldots, P$; **V** is the $M \times P$ matrix whose columns are the eigenvectors v_i, $i = 1, \ldots, P$, and **B** is the diagonal matrix of eigenvalues. After **U** and **V** are formed from the eigenvectors corresponding to the P nonzero eigenvalues of **A**, there remain $(N - P)$ eigenvectors U_j and $(K - P)$ eigenvectors V_j which correspond to zero eigenvalues. If we assemble these into columns of matrices, we have U_o (an $N \times (N - P)$ matrix) and V_o (a K

$\times (K - P)$ matrix). This is called *annihilator space* and reveals that our model is composed of both real model space (which corresponds to the data) and annihilator space which is linked to zeros in the data field. When we perform an inverse calculation, we usually recover a solution which lies in real model space. We must remember, however, that any function in space a can be added to the solution and still produce a solution that fits the data. With the kernel functions transformed into an orthogonal framework (expanded into eigenvectors) we construct the "smallest" or minimum energy model-solution.

When $P = N$, there is a solution to (4.5.28) and $P = M$ guarantees that a solution, if it exists, is unique. For $P < N$, the system is said to be overconstrained, while if $P < M$, the system is both overconstrained and underdetermined. In the latter case, an exact solution may not exist but there will be an infinite number of solutions satisfying the least squares criterion. This is the case for the present internal wave example, which is both overconstrained and underdetermined.

Returning to our internal wave problem, we find W in equation (4.5.28) using the least-squares method which produces the maximum likelihood estimator for a Gaussian distribution. This estimator is defined to be the inverse of the data covariance matrix. From the current meter array, 60 time series were divided into 25 overlapping segments. For each segment, cross-spectral estimates were computed for each of 600 equidistant frequencies. Averaging over segments and frequency bands to increase statistical significance, resulted in 3660 cross-spectra. The resultant 3660×3660 covariance matrix is difficult to invert. The diagonal of the weight matrix was selected to be

$$W = \mathrm{diag}[1/\mathrm{var}(y_i)] \qquad (4.5.31)$$

which reproduces the main features of the maximum likelihood weight matrix (Olbers et al., 1976). We note that, again for this problem, there are many more data points than parameters so that the system is overconstrained.

The least-squares solution procedure for this internal wave example is as follows:

(a) first find a parameter estimate $\hat{a}$ (the best guess);
(b) linearize at the value $a = \hat{a}$, such that

$$\hat{y}(a) = \hat{y}(\hat{a}) + \mathbf{D}(a - \hat{a}) + \ldots \qquad (4.5.32)$$

where

$$\mathbf{D} = \{\delta\hat{y}_i/\delta a_j\}|_{a=\hat{a}} \qquad (4.5.33)$$

(c) improve the parameter estimate by using

$$a - \hat{a} = \mathbf{H}[\hat{y}(a) - \hat{y}(\hat{a})] \qquad (4.5.34)$$

where the $N \times K$ matrix $\mathbf{H}$ is the generalized inverse of $\mathbf{D}$ derived from the linear terms of (4.5.32). If the matrix $\mathbf{D} < \mathbf{TWD}$ is nonsingular and well conditioned then

$$\mathbf{H} = (\mathbf{D}^T\mathbf{W}\mathbf{D})^{-1}\mathbf{D}^T\mathbf{W} \qquad (4.5.35)$$

and equation (4.5.11) becomes the least-squares solution of (4.5.29). Since $\mathbf{D}^T\mathbf{W}\mathbf{D}$ is an $K \times K$ matrix, it can be easily inverted using standard diagonalization routines.

The Spatial Analyses of Data Fields 369

Having now arrived at a solution, $\mathbf{A} = \{A_{ij}\}$ of our problem in (4.5.27), we are left with two additional questions: (1) How well are the data reproduced by our solution? and (2) How accurately do we know our parameters $a_{\min}$? Since our data are subject to random errors, we can treat y as a statistical quantity and test the hypothesis that y and the model estimate $\hat{y}(a_{\min})$ are the same with a 95% probability (inverse estimate must be within the 95% confidence interval of our data point). Using the central limit theorem for our segment and frequency-averaged spectral values, we can approximate the 95% confidence interval on y as

$$\varepsilon^2_{95\%} = \overline{\delta y W \delta y}\left[1 + O(L^{-1})\right] = L \qquad (4.5.36)$$

where $\delta y = y - \bar{y}$, and $O(\cdot)$ indicates the order of magnitude. Now if

$$\varepsilon^2(a_{\min}) \leq \varepsilon^2_{95\%} \qquad (4.5.37)$$

the model is a statistically consistent representation of the data. The consistency of the IWEX model is provided by the results in Figure 4.5.5, where we have plotted the measured, $\varepsilon^2(a)$, and expected, ε^2, values of the parameter ε^2. In this case, all values have been normalized so that magnitudes provide some indication of the percentage to which the observed and estimated (modeled) values of the data, y, coincide. For the most part, the measured values of ε^2 are scattered about the expected values of this parameter. Except at the M_2 tidal frequency and for frequencies greater than 1 cph, the hybrid IWEX model gives a consistent description of the IWEX data set to the 95% level.

Our second question regarding the accuracy of the parameter solution $a_{\min}$, can be answered by calculating the covariance matrix of the parameters. Using equation (4.5.30), we obtain the $K \times K$ covariance matrix of the parameters,

$$\overline{\delta a \delta a} = \mathbf{H}\overline{\delta y \delta y}\mathbf{H}^T \qquad (4.5.38)$$

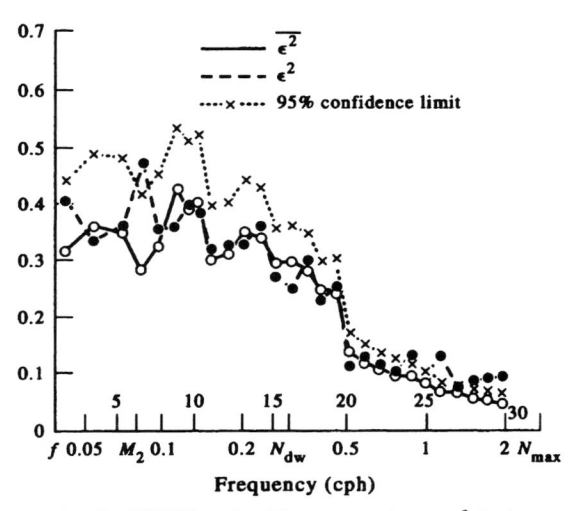

Figure 4.5.5. Consistency for the IWEX study. The error estimate ε^2 is the squared difference between the observed data and the modeled data obtained by inverse methods. Except for motions in the M_2 tidal band and at frequencies greater than about 1 cph, the results are within the 95% confidence level. N_{max} and N_{dw} are the maximum Nyquist frequency and the Nyquist frequency for the deep water, respectively (Briscoe, 1975).

from the data covariance matrix $\overline{\delta y \delta y}$. As usual, there is a reciprocal relation between the variance and the resolution of the parameters. Statistically uncorrelated parameters can be found by diagonalizing the matrix in (4.5.38).

4.4.4 Summary of inverse methods

In this section we have presented the basic concepts of the general inverse problem and have set up the solution system for two different applications in physical oceanography. Our treatment is by no means comprehensive and is intended to serve only as a guide to understanding the process of forming linear inverse solutions to fit observed oceanographic data.

The first example we treated is the computation of absolute geostrophic velocity by specifying an unknown reference velocity. Both the β-spiral (Stommel and Schott, 1977) and Wunsch's method are discussed. The dynamics are restricted to geostrophy and the conservation of mass. The second example was the specification of parameters in theoretical internal wave cross-spectra to reproduce the velocity cross-spectra of an array of moored current meters. The statistical nature of both the data and the model are considered and the accuracy of the results are expressed in probabilistic terms. Readers interested in further discussion of these and other related applications of inverse methods are referred to Bennett (1992). This book contains a complete review of inverse methods along with discussion of most of the popular applications of inverse techniques in physical oceanography. We also direct the interested reader to a recent paper by Egbert et al. (1994) in which a generalized inverse method is used to determine the four principal tidal constituents (M_2, S_2, K_1, O_1) for open ocean tides. The tides are constrained (in a least squares sense) by the hydrodynamic equations and by observational data. In the first example, solutions are obtained using inversion of the harmonic constants from a set of 80 open ocean tide gauges. The second example uses cross-over data from TOPEX/POSEIDON satellite altimetry. According to the authors, "The inverse solution yields tidal fields which are simultaneously smoother, and in better agreement with altimetric and ground truth data, than previously proposed tidal models."

CHAPTER 5

Time-series Analysis Methods

The advent of high-density storage devices and long-term mooring capability has enabled oceanographers to collect long time series of oceanic and meteorological data. Similarly, the use of rapid-response sensors on moving platforms has made it possible to generate snapshots of spatial variability over extensive distances. Time-series data are collected from moored instrument arrays or by repeated measurements at the same location using ships, satellites, or other instrumented packages. Quasi-synoptic spatial data are obtained from ships, manned-submersibles, remotely operated vehicles (ROVs), autonomous underwater vehicles (AUVs), satellites, and satellite-tracked drifters.

As discussed in Chapters 3 and 4, the first stage of data analysis following data verification and editing usually involves estimates of arithmetic means, variances, correlation coefficients, and other sample-derived statistical quantities. These quantities tell us how well our sensors are performing and help characterize the observed oceanographic variability. However, general statistical quantities provide little insight into the different types of signals that are blended together to make the recorded data. The purpose of this chapter is to present methodologies that examine data series in terms of their frequency content. With the availability of modern high-speed computers, frequency-domain analysis has become much more central to our ability to decipher the cause and effect of oceanic change. The introduction of fast Fourier transform (FFT) techniques in the 1960s further aided the application of frequency-domain analysis methods in oceanography.

5.1 BASIC CONCEPTS

For historical reasons, the analysis of sequential data is known as *time series analysis*. As a form of data manipulation, it has been richly developed for a wide assortment of applications. While we present some of the latest techniques, the emphasis of this chapter will be on those "tried and proven" methods most widely accepted by the general oceanographic community. Even these established methods are commonly misunderstood and incorrectly applied. Where appropriate, references to other texts will be given for those interested in a more thorough description of analysis techniques. As with previous texts, the term "time series" will be applied to both temporal and spatial data series; methods which apply in the time domain also apply in the space domain. Similarly, the terms *frequency domain* and *wavenumber domain* (the formal transforms of the time and spatial series, respectively) are used interchangeably.

A basic purpose of time series analysis methods is to define the variability of a data series in terms of dominant periodic functions. We also want to know the "shape" of the spectra. Of all oceanic phenomena, the barotropic astronomically forced tides most closely exhibit deterministic and stationary periodic behavior, making them the most readily predictable motions in the sea. In coastal waters, tidal observations over a period as short as one month can be used to predict local tidal elevations with a high degree of accuracy. Where accurate specification of the boundary conditions is possible, a reasonably good hydrodynamic numerical model that has been calibrated against observations can reproduce the regional tide heights to an accuracy of a few centimeters. Tidal currents are less easily predicted because of the complexities introduced by stratification, nonlinear interactions, and basin topography. Although baroclinic (internal) tides generated over abrupt topography in a stratified ocean have little impact on surface elevations, they can lead to strong baroclinic currents. These currents are generally stochastic (i.e. nondeterministic) and hence only predictable in a statistical sense.

Surface gravity waves are periodic and quasi-linear oceanic features but are generally stochastic due to inadequate knowledge of the surface wind fields, the air–sea momentum transfer, and oceanic boundary conditions. Refraction induced by wave–current interactions can be important but difficult to determine. Other oceanic phenomena such as coastal-trapped waves and near-inertial oscillations have marked periodic signatures but are intermittent because of the vagaries of the forcing mechanisms and changes in oceanic conditions along the direction of propagation. Other less obvious regular behavior can be found in observed time and space records. For instance, oceanic variability at the low-frequency end of the spectrum is dominated by fluctuations at the annual to decadal periods, consistent with baroclinic Rossby waves and short-term climate change, while that at the ultra-low frequencies is dominated by ice-age climate scale variations associated with Milankovitch-type processes (changes in the caloric summer insolation at the top of the atmosphere arising from changes in the earth's orbital eccentricity, and tilt and precision of its rotation axis).

Common sense should always be a key element in any time-series analysis. Attempts to use analytical techniques to find "hidden" signals in a time series often are not very convincing, especially if the expected signal is buried in the measurement noise. Because noise is always present in real data, it should be clear that, for accurate resolution of periodic behavior, data series should span at least a few repeat cycles of the time scale of interest, even for stationary processes. Thus, a day-long record of hourly values will not fully describe the diurnal cycle in the tide nor will a 12-month series of monthly values fully define the annual cycle of sea surface temperature. For these short records, modern spectral analysis methods can help us pin-point the peak frequencies. As we noted in Chapter 1, a fundamental limitation to resolving time-series fluctuations is given by the "sampling theorem" which states that the highest detectable frequency or wavenumber (the Nyquist frequency or wavenumber) is determined by the interval between the data points. For example, the highest frequency that we can resolve by an hourly time series is one cycle per 2 h, or one cycle per $2\Delta t$, where Δt is the interval of time between points in the series.

For the most part, we fit series of well-known functions to the data in order to transform from the time domain to the frequency domain. As with the coefficients of the sine and cosine functions used in Fourier analysis, we generally assume that the functions have slowly varying amplitudes and phases, where "slowly" means that coefficients change little over the length of the record. Other linear combinations of orthogonal functions with similar limitations on the coefficients can be used to describe the series. However, the trigonometric functions are unique in that uniformly spaced samples covering an integer number of periods of the function form orthogonal sequences. Arbitrary orthogonal functions, with a similar sampling scheme, do not necessarily form orthogonal sequences. Another advantage of using common functions in any analysis is that the behavior of these functions is well understood and can be used to simplify the description of the data series in the frequency or wavenumber domain. In this chapter, we consider time series to consist of periodic and aperiodic components superimposed on a secular (long-term) trend and uncorrelated random noise. Fourier analysis and spectral analysis are among the tools used to characterize oceanic processes. Determination of the Fourier components of a time series can be used to determine a *periodogram* which can then be used to define the spectral density (*spectrum*) of the time series. However, the periodogram is not the only way to get at the spectral energy density. For example, prior to the introduction of the fast Fourier transform (FFT), the common method for calculating spectra was through the Fourier transform of the autocorrelation function. More modern spectral analysis methods involve autoregressive spectral analysis (including use of maximum entropy techniques), wavelet transforms, and fractal analysis.

5.2 STOCHASTIC PROCESSES AND STATIONARITY

A common goal of most time-series analysis is to separate deterministic periodic oscillations in the data from random and aperiodic fluctuations associated with unresolved background noise (unwanted geophysical variability) or with instrument error. It is worth recalling that time-series analyses are typically statistical procedures in which data series are regarded as subsets of a stochastic process. A simple example of a stochastic process is one generated by a linear operation on a purely random variable. For example, the function $x(t_i) = 0.5x(t_{i-1}) + \varepsilon(t_i)$, $i = 1, 2, ...$, for which $x(t_0) = 0$, say, is a linear random process provided that the fluctuations $\varepsilon(t_i)$ are statistically independent. Stochastic processes are classified as either discrete or continuous. A continuous ("analog") process is defined for all time steps while a discrete ("digital") process is defined only at a finite number of points. The data series can be scalar (univariate series) or a series of vectors (multivariate series). While we will deal with discrete data, we assume that the underlying process is continuous.

If we regard each data series as a realization of a stochastic process, each series contains an infinite ensemble of data having the same basic physical properties. Since a particular data series is a sample of a stochastic process, we can apply the same kind of statistical arguments to our data series as we did to individual random variables. Thus, we will be making statistical probability statements about the results of frequency transformations of our data series. This fact is important to remember since there is a great temptation to regard transformed values as inherently independent data points. Since many data collected in time or space are highly correlated because of the presence of low-frequency, nearly deterministic components, such as long-period tides and the seasonal cycle, standard statistical methods do not really apply. Contrary to the requirements of stochastic theory, the values are not statistically independent. "What constitutes the ensemble of a possible time series in any given situation is dictated by good scientific judgment and not by purely statistical matters" (Jenkins and Watts, 1968). A good example of this problem is presented by Chelton (1982) who showed that the high correlation between the integrated transport through Drake Passage in the Southern Ocean and the circumpolar-averaged zonal wind stress "may largely be due to the presence of a strong semi-annual signal in both time series." A strong statistical correlation does not necessarily mean there is a cause and effect relationship between the variables.

As implied by the previous section, the properties of a stochastic process generally are time dependent and the value $y(t)$ at any time, t, depends on the time elapsed since the process started. A simplifying assumption is that the series has reached a steady state or equilibrium in the sense that the statistical properties of the series are independent of absolute time. A minimum requirement for this condition is that the probability density function (PDF) is independent of time. Therefore, a stationary time series has constant mean, μ, and variance, σ^2. Another consequence of this equilibrium state is that the joint PDF depends only on the time difference $t_1 - t_2 = \tau$ and not on absolute times, t_1 and t_2. The term *ergodic* is commonly used in association with stochastic processes for which time averages can be used in place of ensemble averages (see Chapter 3). That is, we can average over "chunks" of a time series to get the mean, standard deviation, and other statistical quantities rather than having to produce repeated realizations of the time series. Any formalism involving ensemble averaging is of little value as the analyst rarely has an ensemble at his disposal and typically must deal with a single realization. We need the ergodic theorem to enable us to use time averages in place of ensemble averages.

5.3 CORRELATION FUNCTIONS

Discrete or continuous random time series, $y(t)$, have a number of fundamental statistical properties that help characterize the variability of the series and make it easily possible to compare one time series against another. However, these statistical measures also contain less information than the original time series and, except in special cases, knowledge of the these properties is insufficient to reconstruct the time series.

(1) *Mean and variance.* If y is a stochastic time series consisting of N values $y(t_i) = y_i$ measured at discrete times t_i $\{t_1, t_2, ... , t_N\}$, the true mean value μ for the series can be estimated by

$$\mu \equiv E[y(t)] = \frac{1}{N}\sum_{i=1}^{N} y_i \qquad (5.3.1)$$

where $E[y(t)]$ is the expected value and $E[|y(t)|] < \infty$ for all t. The estimated mean value is not necessarily constant in time; different segments of a time series can have different mean values if the series is nonstationary. If $E[y^2(t)] < \infty$ for all t, an estimate of the true variance function is given by

$$\sigma^2 \equiv E[\{y(t) - \mu\}^2] = \frac{1}{N}\sum_{i=1}^{N}[y_i - \bar{y}]^2 \quad (5.3.2)$$

The positive square root of the variance is the standard deviation, σ, or root-mean-square (RMS) value. See Chapter 3 for further discussion on the mean and variance.

(2) *Covariance and correlation functions*: These terms are used to describe the covariability of given time series as functions of two different times, $t_1 = t$ and $t_2 = t + \tau$, where τ is the lag time. If the process is *stationary* (unchanging statistically with time) as we normally assume, then absolute time is irrelevant and the covariance functions depend only on τ.

Although the terms "covariance function" and "correlation function" are often used interchangeably in the literature, there is a fundamental difference between them. Specifically, covariance functions are derived from data series following removal of the true mean value, μ, which we typically approximate using the sample mean, $\overline{y(t)}$. Correlation functions use the "raw" data series before removal of the mean. The confusion arises because most analysts automatically remove the mean from any time series with which they are dealing. To further add to the confusion, many oceanographers define correlation as the covariance normalized by the variance.

For a stationary process, the *autocovariance function*, C_{yy}, which is based on lagged correlation of a function with itself, is estimated by

$$C_{yy}(\tau) \equiv E[\{y(t) - \mu\}\{y(t + \tau) - \mu\}]$$
$$= \frac{1}{N-k}\sum_{i=1}^{N-k}[y_i - \bar{y}][y_{i+k} - \bar{y}] \quad (5.3.3)$$

where $\tau = \tau_k = k\Delta t$ ($k = 0, \ldots, M$) is the lag time for k sampling time increments, Δt, and $M \ll N$. The corresponding expression for the *autocorrelation function* R_{yy} is

$$R_{yy}(\tau) \equiv E[y(t)y(t + \tau)]$$
$$= \frac{1}{N-k}\sum_{i=1}^{N-k}(y_i y_{i+k}) \quad (5.3.4)$$

At zero lag ($\tau = 0$)

$$C_{yy}(0) = \sigma^2 = R_{yy}(0) - \mu^2 \quad (5.3.5)$$

where we must be careful to define σ^2 in equation (5.3.2) in terms of the normalization factor $1/N$ rather than $1/(N-1)$ (see Chapter 3). From the above definitions, we find

$$C_{yy}(\tau) = C_{yy}(-\tau); \quad R_{yy}(\tau) = R_{yy}(-\tau) \quad (5.3.6)$$

indicating that the autocovariance and autocorrelation functions are symmetric with respect to the time lag τ.

The autocovariance function can be normalized using the variance (5.3.2) to yield the normalized autocovariance function

$$\rho_{yy}(\tau) = \frac{C_{yy}(\tau)}{\sigma^2} \tag{5.3.7}$$

(Note: some oceanographers call (5.3.7) the autocorrelation function.)
The basic properties of the normalized autocovariance function are:

(a) $\rho_{yy}(\tau) = 1$, for $\tau = 0$;
(b) $\rho_{yy}(\tau) = \rho_{yy}(-\tau)$, for all τ;
(c) $|\rho_{yy}(\tau)| \leq 1$, for all τ;
(d) If the stochastic process is continuous, then $\rho_{yy}(\tau)$, must be a continuous function of τ.

If we now replace one of the $y(t)$ in the above relations with another function $x(t)$, we obtain the *cross-covariance function*

$$C_{xy}(\tau) \equiv E[\{y(t) - \mu_y\}\{x(t+\tau) - \mu_x\}]$$
$$= \frac{1}{N-k} \sum_{i=1}^{N-k} [y_i - \bar{y}][x_{i+k} - \bar{x}] \tag{5.3.8}$$

and the *cross-correlation function*

$$R_{xy}(\tau) \equiv E[y(t)x(t+\tau)]$$
$$= \frac{1}{N-k} \sum_{i=1}^{N-k} y_i x_{i+k} \tag{5.3.9}$$

The normalized cross-covariance function (or *correlation coefficient function*) for a stationary process is

$$\rho_{xy} \equiv \frac{C_{xy}(\tau)}{\sigma_x \sigma_y} \tag{5.3.10}$$

Here, $y(t)$ could be the longshore component of daily mean wind stress and $x(t)$ the daily mean sea level elevation at the coast. Typically, sea level lags the longshore wind stress by one to two days.

One should be careful interpreting covariance and correlation estimates made for large lags. Problems arise if low-frequency components are present in the data since the averaging inherent in these functions becomes based on fewer and fewer samples and loses its statistical reliability as the lag increases. For example, at lag $\tau = 0.1T$ (i.e. 10% of the length of the time series) there are roughly 10 independent cycles of any variability on a time scale, $T_{0.1} = 0.1T$, while at lags of $0.5T$ there are only about two independent estimates of the time scale $T_{0.5}$. In many cases, low-frequency components in geophysical time series make it pointless to push the lag times much beyond 10–20% of the data series. Some authors argue that division by N rather than by $N - k$ reduces the bias at large lags. Although this is certainly true ($N \gg N - k$ at large lags), it doesn't mean that the result has anything to do with reality. In essence, neither of these estimators are optimal. Ideally one should write down the likelihood function of the observed time series, if it exists. Differentiation of this likelihood function would then give a set of equations for the maximum likelihood estimates of the autocovariance function. Unfortunately, the derivatives are in general untraceable and one must work with estimators given above. Results for this section are summarized as follows:

(a) Estimators with divisors $T = N\Delta t$ usually have smaller mean square errors than those based on $T - \tau$; also, those based on $1/T$ are positive definite while those based on $1/(T - \tau)$ may not be.
(b) Some form of correction for low-frequency trends is required. In simple cases, one can simply remove a mean value while in others the trend can be removed. Trend removal must be done carefully so that erroneous data are not introduced into the time series during the subtraction of the trend.
(c) There will be strong correlations between values in the autocorrelation function if the correlation in the original series was moderately strong; the autocorrelation function, which can be regarded as a new time series derived from $y(t)$, will, in general, be more strongly correlated than the original series.
(d) Due to the correlation in (c), the autocorrelation function may fail to dampen according to expectations; this will increase the basic length scale in the function.
(e) Correlation is a relative measure only.

In addition to its direct application to time-series analysis, the autocorrelation function was critical to the development of early spectral analysis techniques. Although modern methods typically calculate spectral density distributions directly from the Fourier transforms of the data series, earlier methods determined spectral estimates from the Fourier transform of the autocorrelation function. An important milestone in time-series analysis was the proof by N. Wiener and A. Khinchin in the 1930s that the correlation functions are related to the spectral density functions function through Fourier transform relationships. According to the Wiener–Khinchin relations, the autospectrum of a time series is the Fourier transform of its autocorrelation function.

(3) *Analytical correlation/covariance functions*: The autocorrelation function of a zero-mean random process $\varepsilon(t)$ ("white noise") can be written as

$$R_{\varepsilon\varepsilon}(\tau) = \sigma_\varepsilon^2 \rho_{\varepsilon\varepsilon}(\tau) = \sigma_\varepsilon^2 \delta(\tau) \tag{5.3.11}$$

where $\delta(\tau)$ is the Dirac delta function. In this example, σ_ε^2 is the variance of the data series. Another useful function is the cross-correlation between the time-lagged stationary signal $y(t) = \alpha x(t - \tau) + \varepsilon$ and the original signal $x(t)$. For constant α

$$R_{xy}(\tau) = \alpha R_{xx}(\tau - \tau_o) + \sigma_\varepsilon^2 \tag{5.3.12a}$$

which, for low noise, has a peak value

$$R_{xy}(\tau_o) = \alpha R_{xx}(0) = \alpha \sigma_x^2 \tag{5.3.12b}$$

Functions of the type (5.3.12) have direct use in ocean acoustics where the time lag, τ_o, at the peak of the zero-mean autocorrelation function can be related to the phase speed c and distance of travel d of the transmitted signal $x(t)$ through the relation $\tau_o = d/c$. It is through calculations of this type that modern acoustic Doppler current meters (ADCMs) and scintillation flow meters determine oceanic currents. In the case of ADCMs, knowing τ_o and d gives the speed c and hence the change of the acoustic signal by the currents during the two-way travel time of the signal. Scintillation meters measure the delay τ_o for acoustic signals sent between a transmitter–receiver pair along two parallel acoustic paths separated by a distance d. The relation $\tau_o = d/v$ then gives the mean flow speed, v, normal to the direction of the acoustic path. Sending the signals both ways in the transmitter–receiver pairs gets around the problem of knowing the sound speed c in detail.

Although the calculation of autocorrelation and autocovariance functions is fairly straightforward, one must be very careful in interpreting the resulting values. For example, a stochastic process is said to be Gaussian (or normal) if the multivariate probability density function is normal. Then the process is completely described by its mean, variance, and autocovariance function. However, there is a class of nonGaussian processes which have the same normalized autocovariance function, ρ, as a given normal process. Consider the linear system

$$\tau_o \frac{dy}{dt} + y(t) = z(t) \qquad (5.3.13)$$

where $z(t)$ is white-noise input and $y(t)$ is the output. Here, $y(t)$ is called a "first-order autoregressive process" which has the normalized autocorrelation function

$$\rho_{yy}(\tau) = e^{-|\tau|/\tau_o} \qquad (5.3.14)$$

Thus, if the input to the first-order system has a normal distribution then by an extension of the central limit theorem it may be shown that the output is normal and is completely specified by the autocorrelation function.

Another process with an exponential autocorrelation function which differs greatly from the normal process is called the *random telegraph signal* (Figure 5.3.1). Alpha particles from a radioactive source are used to trigger a flip-flop between $+1$ and -1. Assuming the process was started at $t = -\infty$ we can derive the normalized autocorrelation function as

$$\rho_{yy}(\tau) = e^{-2\lambda|\tau|} \qquad (5.3.15)$$

If $\lambda = 1/2\tau_o$ then this is the same as the autocorrelation function of a normal process, which is characteristically different from the flip-flop time series. Again, one must be careful when interpreting autocorrelation functions.

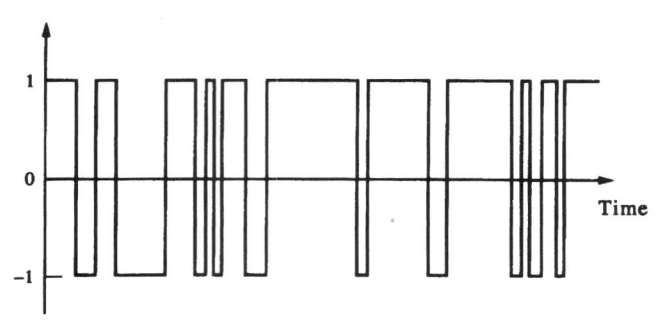

Figure 5.3.1. A realization of a random telegraph signal with digital amplitudes of ± 1 as a function of time.

Table 5.1. *Acoustic backscatter anomaly (decibels) measured in bin #1 from two adjacent transducers on a towed 150 kHz ADCP. The data cover a depth range of 75–230 m at increments of 5 m (32 values). The two vertical profiles are separated horizontally by a distance of 3.5 m. The means have not been removed from the data*

Beam	75 m	80	85	90	95	100	105	110	115	120
1	11.56	0.67	−8.33	−9.82	−13.91	−18.00	3.67	−2.00	−12.29	−13.71
2	14.67	3.00	−5.67	−9.64	−12.82	−16.00	−8.50	−11.00	−15.29	−16.71

125 m	130	135	140	145	150	155	160	165	170	175
−11.33	−8.00	24.14	38.13	40.00	35.00	29.63	24.00	26.50	28.75	30.63
−10.33	−2.00	23.71	36.63	41.00	33.14	24.38	15.00	20.63	26.25	31.88

180 m	185	190	195	200	205	210	215	220	225	230
30.50	31.00	36.00	31.63	21.00	12.25	3.00	−7.00	−4.43	−0.50	0.75
31.00	29.13	29.75	24.75	16.00	7.25	3.25	6.38	11.57	12.25	5.38

(4) *Observed covariance functions*: To see what autocorrelation functions look like in practice, consider the data in Table 5.1. Here, we have tabulated the calibrated acoustic backscatter anomaly measured at 5-m depth increments in the upper ocean using a towed 150 kHz acoustic Doppler current profiler (ADCP). These "time series" data are from the first bin of adjacent beams 1 and 2 of a four-beam ADCP, and represent the backscatter intensity from zooplankton located at a distance of 5 m from the instrument. Since the transducers are tilted at an angle of 30° to the vertical, the two profiles are separated horizontally by only 3.5 m and the autocorrelations should be nearly identical at all lags. In this case, we use the normalized covariance (5.3.7) derived from equation (5.3.3) in which the sum is divided by the number of lag values, $N - k$, for lag $\tau = k\Delta t$.

As indicated by the autocorrelation functions in Figure 5.3.2, the functions are similar at small lags where statistical reliability is large but diverge significantly at higher lags with the decrease in the number of independent covariance estimates.

(5) *Integral time scales*: The integral time scale, T^*, is defined as the sum of the normalized autocorrelation function (5.3.7) over the length $L = N\Delta\tau$ of the time series for N lag steps, $\Delta\tau$. Specifically, the estimate

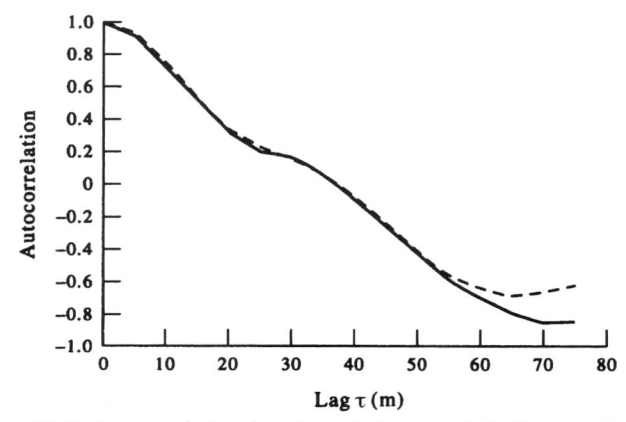

Figure 5.3.2. *Autocorrelation functions of the acoustic backscatter data in Table 5.1.*

$$T^* = \frac{\Delta\tau}{2} \sum_{i=0}^{N'} [\rho(\tau_i) + \rho(\tau_{i+1})]$$
$$= \frac{\Delta\tau}{2\sigma^2} \sum_{i=0}^{N'} [C(\tau_i) + C(\tau_{i+1})]$$
(5.3.16)

for $N' \leq N - 1$ gives a measure of the dominant correlation time scale within a data series—for times longer than T^*, the data become decorrelated. There are roughly $\Delta\tau N/T^*$ actual degrees of freedom within the time series. In reality, the summation typically is limited to $N' \ll N$ since low frequency components within the time series prevent the summation from converging to a constant value over the finite length of the record. In general, one should continue the summation until it reaches a near-constant value which we take as the value for T^*. If no plateau is reached within a reasonable number of lags, no integral time scale exists. In that case, the integral time scale can be approximated by integrating only to the first zero crossing of the autocorrelation function (cf. Poulain and Niiler, 1989).

(6) *Correlation analysis versus linear regression*: Geophysical data are typically obtained from random temporal sequences or spatial fields that cannot be regarded as mutually independent. Because the data series depend on time and/or spatial coordinates, the use of linear regression to study relationships between data series may lead to incomplete or erroneous conclusions. As an example, consider two time series: A white-noise series, consisting of identically distributed and mutually independent random variables, and the same series but with a time shift. As the values of the time series are statistically independent, the cross-correlation coefficient will be zero at zero lag, even though the time series are strictly linearly related. Regression analysis would show no relationship between the two series. However, cross-correlation analysis would reveal the linear relationship (a coefficient of unity) for a lag equal to the time shift. Correlation analysis is often a better way to study relations among time series than traditional regression analysis.

5.4 FOURIER ANALYSIS

For many applications, we can view time series as linear combinations of periodic or quasi-periodic components that are superimposed on a long-term trend and random high-frequency noise. The periodic components are assumed to have fixed, or slowly varying amplitudes and phases over the length of the record. The trends might include a slow drift in the sensor characteristics or a long-term component of variability that cannot be resolved by the data series. "Noise" includes random contributions from the instrument sensors and electronics, as well as frequency components that are outside the immediate range of interest (e.g. small-scale turbulence). A goal of time-series analysis in the frequency domain is to reliably separate periodic oscillations from the random and aperiodic fluctuations. Fourier analysis is one of the most commonly used methods for identifying periodic components in near-stationary time-series oceanographic data. (If the time series are strongly nonstationary, more localized transforms such as the Hilbert and Wavelet transforms should be used.)

The fundamentals of Fourier analysis were formalized in 1807 by the French mathematician Joseph Fourier (1768–1830) during his service as an administrator under Napoleon. Fourier developed his technique to solve the problem of heat conduction in a solid with specific application to heat dissipation in blocks of metal being turned into cannons. Fourier's basic premise was that any finite length, infinitely repeated time series, $y(t)$, defined over the principal interval $[0, T]$ can be reproduced using a linear summation of cosines and sines, or *Fourier series*, of the form

$$y(t) = \overline{y(t)} + \sum_p [A_p \cos(\omega_p t) + B_p \sin(\omega_p t)] \quad (5.4.1)$$

in which $\bar{y}$ is the mean value of the record, A_p, B_p are constants (the Fourier coefficients), and the specified angular frequencies, ω_p, are integer ($p = 1, 2, ...$) multiples of the fundamental frequency, $\omega_1 = 2\pi f_1 = 2\pi/T$, where T is the total length of the time series. Provided enough of these Fourier components are used, each value of the series can be accurately reconstructed over the principal interval. By the same token, the relative contribution a given component makes to the total variance of the time series is a measure of the importance of that particular frequency component in the observed signal. This concept is central to spectral analysis techniques. Specifically, the collection of Fourier coefficients having amplitudes A_p, B_p form a *periodogram* which then defines the contribution that each oscillatory component ω_p makes to the total "energy" of the observed oceanic signal. Thus, we can use the Fourier components to estimate the power spectrum (energy per unit frequency bandwidth) of a time series. Since both A_p, B_p must be specified, there are two degrees of freedom per spectral estimate derived from the "raw" or unsmoothed periodogram.

5.4.1 Mathematical formulation

Let $y(t)$ denote a continuous, finite-amplitude time series of finite duration. Examples include hourly sea-level records from a coastal tide gauge station or temperature records from a moored thermistor chain. If y is periodic, there is a period T such that $y(t) = y(t + T)$ for all t. In the language of Fourier analysis, the periodic functions are sines and cosines, which have the important properties that:

(1) A finite number of Fourier coefficients achieves the minimum mean square error between the original data and a functional fit to the data series;
(2) the functions are orthogonal so that coefficients for a given frequency can be determined independently.

Suppose that the time series is specified only at discrete times by subsampling the continuous series $y(t)$ at a sample spacing of Δt (Figure 5.4.1). Since the series has a duration T, there are a total of $N = T/\Delta t$ sample intervals and $N + 1$ sample points located at times $y(t_n) = y(n\Delta t) \equiv y_n$ ($n = 0, 1, ..., N$). Using Fourier analysis, it is possible to reproduce the original signal as a sum of sine or cosine waves of different amplitudes and phases. In Figure 5.4.1, we show a time series $y(n\Delta t)$ of 41 data points followed by plots of the first, second, and sixth harmonics that were summed to create the time series. The frequencies of these harmonics are $f = 1/T$, $2/T$, and $6/T$, respectively, and each harmonic has the form $y_k(n\Delta t) = C_k \cos[(2\pi k n/N + \phi_k]$ where (C_k, ϕ_k) are the amplitudes and phases of the harmonics for $k = 1, 2, 6$. Here, $T = 40\Delta t$ and we have arbitrarily chosen $(C_1, \phi_1) = (2, \pi/4)$, $(C_2, \phi_2) = (0.75, \pi/2)$,

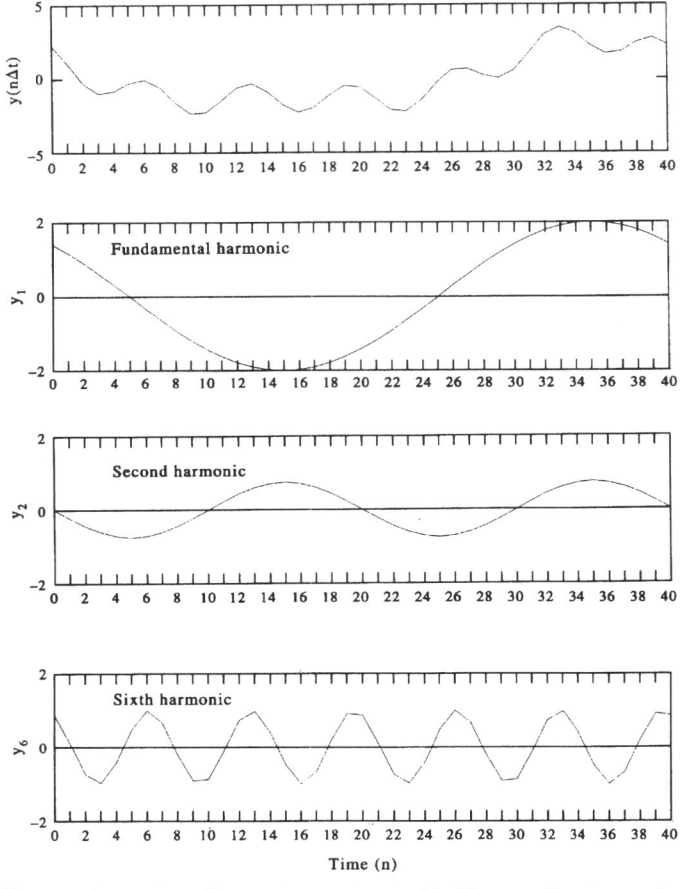

382 *Data Analysis Methods in Physical Oceanography*

Figure 5.4.1. Discrete subsampling of a continuous signal y(t). The sampling interval is $\Delta t = 1$ time unit and the fundamental frequency is $f_1 = 1/T$ where $T = N\Delta t$ is the total record length and $N = 40$. The signal y(t) is the sum of the first, second, and sixth harmonics which have the form
$$y_k(n\Delta t) = C_k \cos[(2\pi k n/N) + \phi_k]; \quad k = 1, 2, 6; \quad n = 0, 1, ..., 40.$$

and $(C_6, \phi_6) = (1.0, \pi/6)$. The $N/2$ harmonic, which is the highest frequency component that can be resolved by this sampling, has a frequency $f_N = (N/2)/N\Delta t = 1/2\Delta t$ cycles per unit time and a period of $2\Delta t$. Called the *sampling* or *Nyquist* frequency, this represents the highest frequency resolved by the sample series in question. (In Chapter 5, we have used the subscript N to denote the Nyquist frequency and there is no confusion between this property and the integer N, as in $n = 1, 2, ..., N$, or the buoyancy frequency $N(z)$.) (In this chapter, we have used the subscript N to denote the Nyquist frequency f_N, and there should be no confusion between this property and the integer N, as in $n = 1, 2, ..., N$, or the buoyancy frequency, $N(z)$.)

The fundamental frequency, $f_1 = 1/T$, is used to construct $y(t)$ through the infinite Fourier series

$$y(t) = \tfrac{1}{2}A_0 + \sum_{p=1}^{\infty} [A_p \cos(\omega_p t) + B_p \sin(\omega_p t)] \tag{5.4.2}$$

in which
$$\omega_p = 2\pi f_p = 2\pi p f_1 = 2\pi p/T; \quad p = 1, 2, \ldots \quad (5.4.3)$$

is the frequency of the pth constituent in radians per unit time (f_p is the corresponding frequency in cycles per unit time) and $A_0/2$ is the mean, or "DC" offset, of the time series. The factor of 1/2 multiplying A_0 is for mathematical convenience. Note that the mean value is synonymous with the zero-frequency component obtained in the limit $\omega \to 0$. Also, the length of the data record, T, defines both the lowest frequency, f_1, resolvable by the data series and the maximum frequency resolution, $\Delta f = 1/T$, one can obtain from discretely sampled data.

To obtain the coefficients A_p, we simply multiply equation (5.4.2) by $\cos(\omega_p t)$ then integrate over all possible frequencies. The coefficients B_p, are obtained in the same way by multiplying by $\sin(\omega_p t)$. Using the orthogonality condition for the product of trigonometric functions (which requires that the trigonometric arguments cover an exact integer number of 2π cycles over the interval $(0, T)$), we find

$$A_p = \frac{2}{T}\int_0^T y(t)\cos(\omega_p t)\,dt, \quad p = 0, 1, 2, \ldots \quad (5.4.4a)$$

$$B_p = \frac{2}{T}\int_0^T y(t)\sin(\omega_p t)\,dt, \quad p = 1, 2, \ldots \quad (5.4.4b)$$

where the integral for $p = 0$ in (5.4.4a) yields $A_0 = 2\bar{y}$, twice the mean value of $y(t)$ for the entire record. Since each pair of coefficients (A_p, B_p) is associated with a frequency ω_p (or f_p), the amplitudes of the coefficients provide a measure of the relative importance of each frequency component to the overall signal variability. For example, if $(A_6^2 + B_6^2)^{1/2} \gg (A_2^2 + B_2^2)^{1/2}$ we expect there is much more "spectral energy" at frequency ω_6 than at frequency ω_2. Here, spectral energy refers to the amplitudes squared of the Fourier coefficients which represent the variance, and therefore the energy, for that portion of the time series.

We can also express our Fourier series as amplitude and phase functions in the compact Fourier series form

$$y(t) = \tfrac{1}{2}C_0 + \sum_{p=1}^{\infty} C_p \cos(\omega_p t - \theta_p) \quad (5.4.5)$$

in which the amplitude of the pth component is

$$C_p = (A_p^2 + B_p^2)^{1/2}, \quad p = 0, 1, 2, \ldots \quad (5.4.6)$$

where $C_0 = A_0$ ($B_0 = 0$) is twice the mean value and

$$\theta_p = \tan^{-1}[B_p/A_p], \quad p = 1, 2, \ldots \quad (5.4.7)$$

is the phase angle of the constituent at time $t = 0$. The phase angle gives the relative "lag" of the component in radians (or degrees) measured counterclockwise from the real axis ($B_p = 0$, $A_p > 0$). The corresponding time lag for the pth component is then $t_p = \theta_p/2\pi f_p$ in which θ_p is measured in radians.

The discrimination of signal amplitude as a function of frequency given by equations (5.4.2) and (5.4.5) provides us with the beginnings of spectral analysis. Notice that neither of these expressions allows for a trend in the data. If any trend is not first removed from the record, the analysis will erroneously blend the variance from the trend into the lower frequency components of the Fourier expansion. Moreover, we now see the need for the factor of 1/2 in the leading terms of (5.4.2) and (5.4.5). Without it, the $p = 0$ components would equal twice the mean component, $\bar{y} = \frac{1}{2}A_0 = \frac{1}{2}C_0$.

Up to now we have assumed that $y(t)$ is a scalar quantity. We can also expand the time series of a vector property, $\mathbf{u}(t)$. Included in this category are time series of current velocity from moored current meter arrays and wind velocity from moored weather buoys. Expressing vector time series in complex notation, we can write

$$\mathbf{u}(t) = u(t) + iv(t) \tag{5.4.8}$$

where, for example, u and v might be the north–south and east–west components of current velocity in Cartesian coordinates. An individual vector can be expressed as

$$\mathbf{u}(t) = \overline{\mathbf{u}(t)} + \sum_{p=1}^{\infty} [A_p \cos(\omega_p t + \alpha_p) + iB_p \sin(\omega_p t + \beta_p)] \tag{5.4.9}$$

Here, $\overline{\mathbf{u}(t)}$ is the mean (time averaged) vector, $\bar{\mathbf{u}} = \bar{u} + i\bar{v}$, and (α_p, β_p) are phase lags or relative phase differences for the separate velocity components.

Vector quantities also can be defined through expressions of the form

$$\mathbf{u}(t) = \bar{\mathbf{u}} + \sum_{p=1}^{\infty} \{\exp[i(\varepsilon_p^+ + \varepsilon_p^-)/2][(A_p^+ + A_p^-)\cos[\omega_p t + (\varepsilon_p^+ - \varepsilon_p^-)/2] \\ + i(A_p^+ - A_p^-)\sin[\omega_p t + (\varepsilon_p^+ - \varepsilon_p^-)/2]\} \tag{5.4.10}$$

in which A_p^+ and A_p^- are, respectively, the lengths of the counterclockwise (+) and clockwise (−) rotary components of the velocity vector, and ε_p^+ and ε_p^- are the angles that these vectors make with the real axis at $t = 0$. The resultant time series is an ellipse with major axis of length $L_M = A_p^+ + A_p^-$ and minor axis of length $L_m = |A_p^+ - A_p^-|$. The major axis is oriented at angle $\theta_p = \frac{1}{2}(\varepsilon_p^+ + \varepsilon_p^-)$ from the u-axis and the current rotates counterclockwise when $A_p^+ > A_p^-$ and clockwise when $A_p^+ < A_p^-$. The velocity vector is aligned with the major axis direction θ_p when $\omega_p t = -\frac{1}{2}(\varepsilon_p^+ - \varepsilon_p^-)$. Motions are said to be *linearly polarized* (rectilinear) if the two oppositely rotating components are of the same magnitude and *circularly polarized* if one of the two components is zero. In the northern (southern) hemisphere, motions are predominantly clockwise (counterclockwise) rotary. Further details on rotary decomposition are presented in Sections 5.6 and 5.8.

5.4.2 Discrete time series

Most oceanographic time or space series, whether they were collected in analog or digital form, are eventually converted to digital data which may then be expressed as series expansions of the form (5.4.2) or (5.4.5). These expansions are then used to compute the Fourier transform (or periodogram) of the data series. The basis for this transform is Parseval's theorem which states that the mean square (or average) energy of a time series $y(t)$ can be separated into contributions from individual harmonic components to make up the time series. For example, if $\bar{y}$ is the sample mean value of the time series, y_p is the contribution from the nth data value and N is the total number of data values in the time series, then the mean square value of the series about its mean (i.e. the variance of the time series)

$$\sigma^2 = \frac{1}{N-1} \sum_{n=1}^{N} [y_n - \bar{y}]^2 \qquad (5.4.11)$$

provides a measure of the total energy in the time series. The variance (5.4.11) also can be obtained by summing the contributions from the individual Fourier harmonics. This kind of decomposition of discrete time series into specific harmonics leads to the concept of a Fourier line spectrum (Figure 5.4.2).

To determine the energy distribution within a time series, $y(t)$, we need to find its Fourier transform. That is, we need to determine the coefficients A_p, B_p in the Fourier series (5.4.2) or, equivalently, the amplitudes and phase lags, C_p, θ_p in the Fourier series (5.4.5). Suppose that we have first removed any trend from the data record. For any time t_n, the Fourier series for a finite length, de-trended digital record having N (even) values at times $t_n = t_1, t_2, \ldots, t_N$, is

$$y(t_n) = \tfrac{1}{2}A_0 + \sum_{p=1}^{N/2} [A_p \cos(\omega_p t_n) + B_p \sin(\omega_p t_n)] \qquad (5.4.12)$$

where the angular frequency $\omega_p = 2\pi f_p = 2\pi p/T$. Using $t_n = n \cdot \Delta t$ together with (5.4.6) and (5.4.7), the final form for the discrete, finite Fourier series becomes

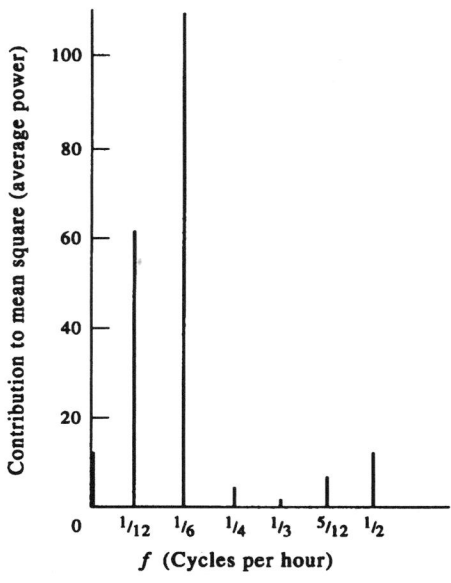

Figure 5.4.2. An example of a Fourier line spectrum with power at discrete frequencies, f, for a 24-h duration record with 1-h sampling increment.

$$y(t_n) = \tfrac{1}{2}A_0 + \sum_{p=1}^{N/2}[A_p \cos(2\pi pn/N) + B_p \sin(2\pi pn/N)]$$

$$= \tfrac{1}{2}C_0 + \sum_{p=1}^{N/2} C_p \cos[(2\pi pn/N) - \theta_p] \qquad (5.4.13)$$

where the leading terms, $\tfrac{1}{2}A_0$ and $\tfrac{1}{2}C_0$, are the mean values of the record. The coefficients are again determined using the orthogonality condition for the trigonometric functions. In fact, the main difference between the discrete case and the continuous case formulated in the last section (aside from the fact we can no longer have an infinite number of Fourier components) is that coefficients are now defined through the summations rather than through integrals

$$A_p = \frac{2}{N}\sum_{n=1}^{N} y_n \cos(2\pi pn/N), \quad p = 0, 1, 2, ..., N/2$$

$$A_0 = \frac{2}{N}\sum_{n=1}^{N} y_n, \quad B_0 = 0$$

$$A_{N/2} = \frac{1}{N}\sum_{n=1}^{N} y_n \cos(n\pi), \quad B_{N/2} = 0$$

$$B_p = \frac{2}{N}\sum_{n=1}^{N} y_n \sin(2\pi pn/N), \quad p = 1, 2, ..., (N/2) - 1 \qquad (5.4.14)$$

Notice that the summations in equations (5.4.14) consist of multiplying the data record by sine and cosine functions which "pick out" from the record those frequency components specific to their trigonometric arguments. Remember, the orthogonality condition requires that the arguments in the trigonometric functions be integer multiples of the total record length, $T = N\Delta t$, as they are in equation (5.4.14). If they are not, the sines and cosines do not form an orthonormal set of basis functions for the Fourier expansion and the original signal cannot be correctly replicated.

The arguments $2\pi pn/N$ in the above equations are based on a hierarchy of equally spaced frequencies $\omega_p = 2\pi p/(N\Delta t)$ and time increment "n". The summation goes to $N/2$ which is the limit of coefficients we can determine; for $p > N/2$ the trigonometric functions simply begin to cause repetition of coefficients already obtained for the interval $p \leq N/2$. Furthermore, it should be obvious that because there are as many coefficients as data points and because the trigonometric functions form an orthogonal basis set, the summation over the $2(N/2) = N$ discrete coefficients provides an exact replication of the time series, $y(t)$. Small differences between the original data and the Fourier series representation arise because of roundoff errors accumulated during the arithmetic calculations (see Chapter 3).

The steps in computing the Fourier coefficients are as follows. Step 1: calculate the arguments $\Phi_{pn} = 2\pi pn/N$ for each integer p and n. Step 2: for each $n = 1, 2, ..., N$, evaluate the corresponding values of $\cos\Phi_{pn}$ and $\sin\Phi_{pn}$, and collect sums of $y_n \cdot \cos\Phi_{pn}$ and $y_n \cdot \sin\Phi_{pn}$. Step 3: Increment p and repeat steps 1 and 2. The procedure requires roughly N^2 real multiply-add operations. For any real data sequence, roundoff errors plus errors associated with truncation of the total allowable number of desired Fourier components (maximum $f_p < f_{N/2}$) will give rise to a less than perfect fit to the data. The residual $\Delta y(t) = y(t) - y_{FS}(t)$ between the observations $y(t)$ and the calculated Fourier series $y_{FS}(t)$ will diminish with increased computational precision and increased numbers of allowable terms used in the series expansion. When computing the phases $\theta_p = \tan^{-1}[B_p/A_p]$ in the formulation (5.4.13), one must take care to examine in which quadrants A_p and B_p are situated. For example, $\tan^{-1}(0.2/0.7)$ differs from $\tan^{-1}(-0.2/-0.7)$ by 180°. The familiar ATAN2 function in FORTRAN is especially designed to take care of this problem.

5.4.3 A computational example

The best way to demonstrate the computational procedure for Fourier analysis is with an example. Consider the two-year segment of monthly mean sea surface temperatures measured at the Amphitrite light station off the southwest coast of Vancouver Island (Table 5.2). Each monthly value is calculated from the average of daily surface thermometer observations collected around noon local time and tabulated to the nearest 0.1°C. These data are known to contain a strong seasonal cycle of warming and cooling which is modified by local effects of runoff, tidal stirring and wind mixing.

The data in Table 5.2 are in the form $y(t_n)$, where $n = 1, 2, \ldots, N$ ($N = 24$). To calculate the coefficients A_p and B_p for these data, we use the summations (5.4.14) for each successive integer p, up to $p = N/2$. These coefficients are then used in (5.4.6) to calculate the magnitude $C_p = (A_p^2 + B_p^2)^{1/2}$ for each frequency component, $f_p = p/T$. Since C_p^2 is proportional to the variance at the specified frequency, the C_p enable us to rate the order of importance of each frequency component in the data series.

The mean value $\overline{y(t)} = \frac{1}{2}A_0$ and the 12 pairs of Fourier coefficients obtainable from the temperature record are listed in Table 5.3 together with the magnitude C_p. Values have been rounded to the nearest 0.01°C. The Nyquist frequency, f_N, is 0.50 cycles per month (cpmo, $p = 12$) and the fundamental frequency, f_1, is 0.042 cpmo ($p = 1$). As we would anticipate from a visual inspection of the time series, the record is dominated by the annual cycle (period = 12 months) followed by weaker contributions from the bi-annual cycle (24 months) and semi-annual cycle (six months). For periods shorter than six months, the coefficients C_p have similar amplitudes and likely represent the roundoff errors and background "noise" in the data series. This suggests that we can reconstruct the original time series to a high degree of accuracy using only the mean value ($p = 0$) and the first three Fourier coefficients ($p = 1, 2, 3$).

Figure 5.4.3 is a plot of the original sea surface temperature (SST) time series and the reconstructed Fourier fit to this series using only the first three Fourier components from Table 5.3. Comparison of these two time series, shows that the reconstructed series does not adequately reproduce the skewed crest of the first year nor the high-frequency "ripples" in the second year of the data record. There also is a

Table 5.2. Monthly mean sea surface temperatures SST (°C) at Amphitrite Point (48°55.16'N, 125°32.17'W) on the west coast of Canada for January 1982 through December 1983

Year 1982												
n	1	2	3	4	5	6	7	8	9	10	11	12
SST	7.6	7.4	8.2	9.2	10.2	11.5	12.4	13.4	13.7	11.8	10.1	9.0
Year 1983												
n	13	14	15	16	17	18	19	20	21	22	23	24
SST	8.9	9.5	10.6	11.4	12.9	12.7	13.9	14.2	13.5	11.4	10.9	8.1

388 *Data Analysis Methods in Physical Oceanography*

Table 5.3. Fourier coefficients and frequencies for the Amphitrite Point monthly mean temperature data. Frequency is in cycles per month (cpmo). $A_0/2$ is the mean temperature and θ_p is the phase lag for the pth component taken counterclockwise from the positive A_p axis

p	Freq. (cpmo)	Period (month)	Coeff. A_p (°C)	Coeff. B_p (°C)	Coeff. C_p (°C)	Phase θ_p (degrees)
0	0	—	21.89	0	21.89	0
1	0.042	24	−0.55	−0.90	1.05	−121.4
2	0.083	12	−1.77	−1.99	2.67	−131.7
3	0.125	8	0.22	−0.04	0.23	−10.3
4	0.167	6	−0.44	−0.06	0.45	−172.2
5	0.208	4.8	0.09	−0.07	0.11	−37.9
6	0.250	4	0.08	−0.04	0.09	−26.6
7	0.292	3.4	0.01	−0.16	0.16	−58.0
8	0.333	3	−0.03	−0.16	0.16	−100.6
9	0.375	2.7	−0.14	0.05	0.15	160.3
10	0.417	2.4	−0.09	−0.07	0.11	−142.1
11	0.458	2.2	−0.08	−0.12	0.14	−123.7
12	0.500	2	−0.15	0	0.15	0

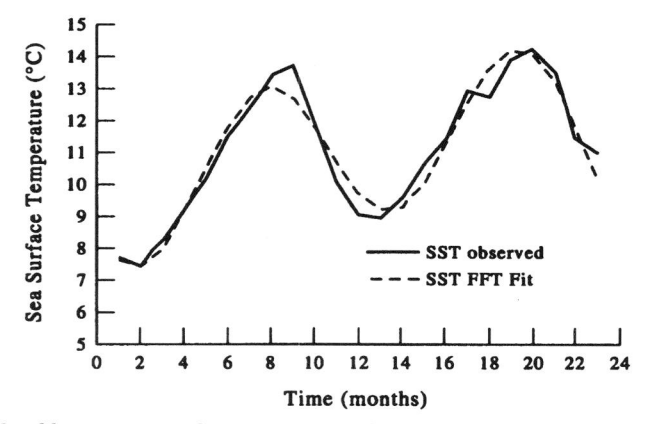

Figure 5.4.3. Monthly mean sea surface temperature (SST) record for Amphitrite Point on the west coast of Vancouver Island (see Table 5.1). The bold line is the original 24-month series; the dashed line is the SST time series generated using the first three Fourier components, f_p, $p = 0, 1, 2$, corresponding to the mean, 24-month, and 12-month cycles (Fourier components appear in Table 5.2).

slight mismatch in the maxima and minima between the series. Differences between the two curves are typically around a few tenths of a degree. In contrast, if we use all 12 components in Table 5.3, corresponding to 24 degrees of freedom, we get an exact replica of the original time series to within machine accuracy.

5.4.4 Fourier analysis for specified frequencies

Analysis of time series for specific frequencies is a special case of Fourier analysis that involves adjustment of the record length to match the periods of the desired Fourier components. As we illustrate in the following sections, analysis for specific frequency components is best conducted using least-squares fitting methods rather than Fourier analysis. Least-squares analysis requires that there be many fewer constituents than data values, which is usually the case for tidal analysis at the well-defined frequencies of the tide-generating potential. Problems arise if there are too few data values. For example, suppose that we have a few days of hourly water level measurements and we want to use Fourier analysis to determine the amplitudes and phases of the daily tidal constituents, f_k. To do this, we need to satisfy the orthogonality condition for the trigonometric basis functions for which terms like $\int \cos(2\pi f_j t) \cos(2\pi f_k t) \, dt$ are zero except where $f_j = f_k$ (the integral is over the entire length of the record, T). The approach is only acceptable when the length of the data set is an integer multiple of all the harmonic frequencies we are seeking. That is, the specified tidal frequencies f_k must be integer multiples of the fundamental frequency, $f_1 = 1/T$, such that $f_k \cdot T = 1, 2, \ldots$, If this holds, we can use Fourier analysis to find the constituent amplitudes and phases at the specified frequencies. In fact, this integer constraint on $f_k \cdot T$ is a principal reason why oceanographers prefer to use record lengths of 14, 29, 180, or 355 days when performing analyses of tides. Since the periods of most of the major tidal constituents (K_1, M_2, etc.) are integer multiples of the fundamental tidal periods (one lunar day, one lunar month $\approx$ 29 days, one year, 8.8 years, 18.6 years, etc.) of the above record lengths, the analysis is aided by the orthogonality of the trigonometric functions.

A note for those unfamiliar with tidal analysis terminology: Letters of tidal harmonics identify the different types ("species") of tide in each frequency band. Harmonic components of the tide-producing force that undergo one cycle per lunar day (≈ 25 h) have a subscript 1 (e.g. K_1), those with two cycles per lunar day have subscript 2 (e.g. M_2), and so on. Constituents having one cycle per day are called diurnal constituents, those with two cycles per day, semidiurnal constituents. The main daily tidal component, the K_1 constituent, has a frequency of 0.0418 cph (corresponding to an angular speed of 15.041° per mean solar hour) and is associated with the cyclic changes in the luni-solar declination. The main semidiurnal tidal constituent, the M_2 constituent, has a frequency of 0.0805 cph (corresponding to an angular speed of 28.984° per mean solar hour) and is associated with cyclic changes in the lunar position relative to the earth. Other major daily constituents are the O_1, P_1, S_2, N_2, and K_2 constituents. In terms of the tidal potential, the hierarchy of tidal constituents is M_2, K_1, S_2, O_1, P_1, N_2, K_2, ... Other important tidal harmonics are the lunar fortnightly constituent, M_f, the lunar monthly constituent, M_m, and the solar annual constituent, S_a. For further details the reader is referred to Thomson (1981) and Foreman (1977).

Returning to our discussion concerning Fourier analysis at specified frequencies, consider the 32-h tide gauge record for Tofino, British Columbia presented in Figure 5.4.4. As we show in Section 5.5, least-squares analysis can be used to reproduce this short record quite accurately using only the K_1 tidal constituent and the M_2 constituent. These are the dominant tidal constituents in all regions of the ocean except near amphidromic points. Because the record is 32 h long, the diurnal and semidiurnal frequencies are not integer multiples of the fundamental frequency $f_1 = 1/T = 0.031$ cph and are not among the sequence of 16 possible frequencies generated from the Fourier analysis. In order to have frequency components centered more exactly at the K_1 and M_2 frequencies, we would need to shorten the record to 24 h or pad the existing record to 48 h using zeros. In either case, the $f_k \cdot T$ for the tides would then be close to integers and a standard Fourier analysis would give an accurate fit to the observed time series. If we stick with the 32-h series, we find that the tidal energy in the diurnal and semidiurnal bands is partitioned among the first three Fourier components at frequencies $f_1 = 0.031$, $f_2 = 0.062$, and $f_3 = 0.093$ cph. These

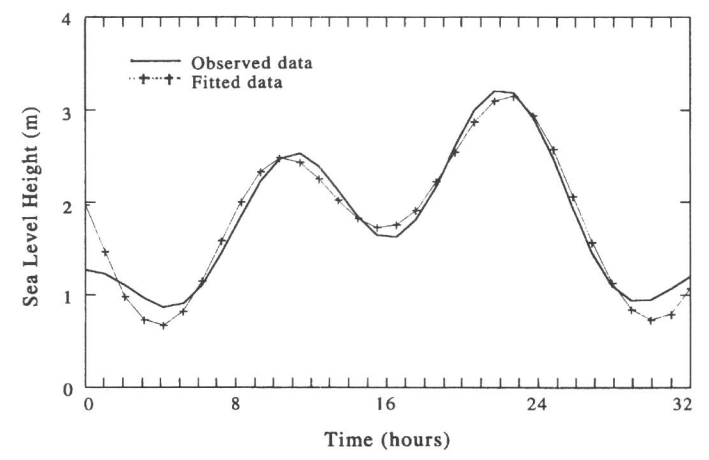

Figure 5.4.4. Hourly sea-level height (SLH) recorded at Tofino on the west coast of Vancouver Island (see Table 5.7). The bold line is the original 32-h series; the dotted line is the SLH series generated using the mean (p = 0) plus the next three Fourier components, f_p, p = 1, 2, 3 having nontidal periods T_p of 32, 16, and 8 h, respectively.

frequencies are only vaguely close to those of the diurnal and semidiurnal constituents but do span the energy-containing frequency bands. As a result, the time series generated from the record mean combined with the first three Fourier components (p = 1, 2, 3) closely approximates the time series obtained using the true tidal frequencies (see Figure 5.5.2).

5.4.5 The fast Fourier transform

One of the main problems with both the autocovariance and the direct Fourier methods of spectral estimation is low computational speed. The Fourier method requires the expansion into series of sine and cosine terms—a time-consuming procedure. The fast Fourier transform (FFT) is a way to speed up this computation while retaining the accuracy of the direct Fourier method. This makes the Fourier method computationally more attractive than the autocovariance approach.

To illustrate the improved efficiency of the FFT method, consider a series of N values for which $N = 2^p$ (p is a positive integer). The discrete Fourier transform of this series would require N^2 operations whereas the FFT method requires only $8N\log_2 N$ operations. The savings in computer time can be substantial. For example, if $N = 8192$, $N^2 = 67,108,864$ while $8N\log_2 N = 851,968$. Computers are much faster now than when the FFT method was introduced but the relative savings in computational efficiency remains the same. Bendat and Piersol (1986) define the speed ratio between the FFT and discrete Fourier method as $N/4p$. This becomes increasingly more important as the number of terms increases since the direct method computational time is $O(N^2)$ while for the FFT method it is $O(N)$. If one is seeking a smoothed power spectrum, it is often more efficient to compute the spectrum using the FFT technique and then smooth in spectral space by averaging over adjoining frequency bands rather than smoothing with an autocovariance lag window in the time domain.

To understand the FFT algorithm, we follow the derivation of Danielson and Lanczos (1942) who first helped pioneer the method. Consider a time series of x_t, where $t = 1, 2, \ldots, N$. We want to find the Fourier transform $X_m = X(m/N\Delta t)$, where $m = 0, 1, \ldots, N-1$. To do this, we first partition x_t into two half-series y_t and z_t, where $y_t = x_{2t-1}, z_t = x_{2t}, t = 1, 2, \ldots, N/2$. The series y_t contains values at the odd number times $(x_1, x_3, \ldots)$ while the function z_t contains values at the even number times $(x_2, x_4, \ldots)$. Both functions have $N/2$ values and their Fourier transforms are

$$Y_m^{(N/2)} = \frac{2}{N} \sum_{t=1}^{N/2} y_t \exp\left[\frac{(-i4\pi tm)}{N}\right] \quad (5.4.15a)$$

$$Z_m^{(N/2)} = \frac{2}{N} \sum_{t=1}^{N/2} z_t \exp\left[\frac{(-i4\pi tm)}{N}\right] \quad (5.4.15b)$$

where the superscript is used to denote the number of terms used in the expansion. But $X_m^{(N)}$, $Y_m^{(N/2)}$, and $Z_m^{(N/2)}$ are related since

$$\begin{aligned} X_m^{(N)} &= \frac{2}{N} \sum_{t=1}^{N/2} x_t \exp\left[\frac{-i4\pi tm}{N}\right] \\ &= \frac{1}{N} \sum_{t=1}^{N/2} \left\{ y_t \exp\left[\frac{-i4\pi tm}{N}(2t-1)\right] + z_t \exp\left[\frac{-i4\pi tm}{N}(2t)\right] \right\} \quad (5.4.16) \\ &= \tfrac{1}{2}\exp\left[\frac{(i2\pi m)}{N}\right] Y_m^{(N/2)} + \tfrac{1}{2} Z_m^{(N/2)}, \quad 0 \le m \le (N/2) - 1 \end{aligned}$$

Also

$$\begin{aligned} Y_{m+N/2}^{(N/2)} &= Y_m^{(N/2)}; \quad 0 \le m \le N/2 - 1 \\ Z_{m+N/2}^{(N/2)} &= Z_m^{(N/2)}; \quad 0 \le m \le N/2 - 1 \end{aligned} \quad (5.4.17)$$

so that

$$\begin{aligned} X_{m+N/2}^{(N)} &= \tfrac{1}{2} \exp\left[i\left(\frac{2\pi}{N}\right)\left(m + \frac{N}{2}\right)\right] Y_m^{(N/2)} + \tfrac{1}{2} Z_m^{(N/2)} \\ &= -\tfrac{1}{2} \exp\left(\frac{i2\pi m}{N}\right) Y_m^{(N/2)} + \tfrac{1}{2} Z_m^{(N/2)}, \quad 0 \le m \le (N/2)-1 \end{aligned} \quad (5.4.18)$$

thus

$$X_m^{(N)} = \tfrac{1}{2} \exp\left[i\frac{(2\pi m)}{N}\right] Y_m^{(N/2)} + \tfrac{1}{2} Z_m^{(N/2)}, \quad 0 \le m \le N/2 - 1 \quad (5.4.19)$$

and

$$X_{m-N/2}^{(N)} = -\tfrac{1}{2} \exp\left[i\frac{(2\pi m)}{N}\right] Y_m^{(N/2)} + \tfrac{1}{2} Z_m^{(N/2)}, \quad 0 \le m \le N/2 - 1 \quad (5.4.20)$$

Thus, the Fourier transform for the series x_t is found from the Fourier series of the half series y_t, z_t. Since $N/2$ is even, this can be repeated. If the length of the data is not a power of 2, it should be padded with zeros up to the next power of two. For a series of length $N = 2p$ (p a positive integer), the procedure is followed until partitions consist of only one term whose Fourier transform equals itself, or the procedure is followed until N becomes a prime number, i.e. $N = 3$. The Fourier transform is then found directly for the remaining short series.

5.5 HARMONIC ANALYSIS

Standard Fourier analysis involves the computation of Fourier amplitudes at equally spaced frequency intervals determined as integer multiples of the fundamental frequency, f_1. That is, for frequencies f_1, $2f_1$, $3f_1$, ... , f_N (f_N = Nyquist frequency). However, as we have shown in the previous section, standard Fourier analysis is not much use when it comes to the analysis of data series in terms of predetermined frequencies. In the case of tidal motions, for example, it would be silly to use any frequencies except those of the astronomical tidal forces. Equally important, we want to determine the amplitudes and phases of as many frequency components as possible by using as short a time series as possible. Since there are typically many more data values than there are prescribed frequencies, we have to deal with an overdetermined problem. This leads to a form of signal demodulation known as *harmonic analysis* in which the user specifies the frequencies to be examined and applies least-squares techniques to solve for the constituents. Harmonic analysis was originally designed for the analysis of tidal variability but applies equally to analysis at the annual and semi-annual periods or any other well-defined cyclic oscillation. The familiar hierarchy of "harmonic" tidal constituents is dominated by diurnal and semidiurnal motions, followed by fortnightly, monthly, semi-annual, and annual variability. In this section, we present a general discussion of harmonic analysis. The important subject of harmonic analysis of tides and tidal currents is treated separately in Section 5.5.3.

The harmonic analysis approach yields the required amplitudes and phase lags of the harmonic tidal coefficients or any other constituents we may wish to specify. Once these coefficients have been determined, we can use them to reconstruct the original time series. In the case of tidal motions, subtraction of the reconstructed tidal signal from the original record yields a time series of the nontidal or *residual* component of the time series. In many cases, it is the residual or "de-tided" signal that is of primary interest. If we break the original time series into adjoining or overlapping segments, we can apply harmonic analysis to the segments to obtain a sequence of estimates for the amplitudes and phase lags of the various frequencies of interest. This leads to the notion of signal *demodulation*.

5.5.1 A least-squares method

Suppose we wish to determine the harmonic constituents A_q and B_q for M specified frequencies which, in general, will differ from the Fourier frequencies defined by (5.4.3). In this case, $q = 0, 1, ... , M$ and $B_0 = 0$ so that there are a total of $2M + 1$ harmonic coefficients. Assume that there are many more observations, N, than specified coefficients (i.e. that $2M + 1 \ll N$). The problem of fitting M harmonic curves to the digital time series is then overdetermined and must be solved using an optimization technique. Specifically, we estimate the amplitudes and phases of the various components by minimizing the squared difference (i.e. the least squares) between the original data series and our fit to that series. The coefficients for each of the M resolvable constituents are found through solution of a $(M + 1) \times (M + 1)$ matrix equation.

For M possible harmonic constituents, the time series $x(t_n)$, $n = 1, \ldots, N$ can be expanded as

$$x(t_n) = \bar{x} + \sum_{q=1}^{M} C_q \cos(2\pi f_q t_n - \phi_q) + x_r(t_n) \tag{5.5.1}$$

in which $\overline{x(t)}$ is the mean value of the record, x_r is the residual portion of the time series (which may contain other kinds of harmonic constituents), $t_n = n\Delta t$, and where C_q, f_q and ϕ_q are respectively the constant amplitude, frequency and phase of the qth constituent that we have specified. In the present configuration, we assume that the specified frequencies have the form $f_q = q/N\Delta t$ so that the argument $2\pi f_q t_n = 2\pi q n/N$. Reformulation of equation (5.5.1) as

$$x(t_n) = \bar{x} + \sum_{q=1}^{M} [A_q \cos(2\pi f_q t_n) + B_q \sin(2\pi f_q t_n)] + x_r(t_n) \tag{5.5.2}$$

yields a representation in terms of the unknown coefficients A_q, B_q where

$$\begin{aligned} C_q &= (A_q^2 + B_q^2)^{1/2}, \quad \text{(frequency component amplitude)} \\ \phi_q &= \tan^{-1}(B_q/A_q), \quad \text{(frequency component phase lag)} \end{aligned} \tag{5.5.3}$$

for $q = 0, \ldots, M$. To reduce roundoff errors (Section 3.17.3), the mean value, $\bar{x}$, should be subtracted from the record prior to the computation of the Fourier coefficients.

The objective of the least-squares analysis is to minimize the variance, e^2, of the residual time series $x_r(t_n)$ in equation (5.5.2), where

$$e^2 = \sum_{n=1}^{N} x_r^2(t_n) = \sum_{n=1}^{N} \left\{ x(t_n) - \left[\bar{x} + \sum_{q=1}^{M} M(t_n) \right] \right\}^2 \tag{5.5.4}$$

and where for convenience we define $\sum M$ as

$$\begin{aligned} \sum_{q=1}^{M} M(t_n) &= \sum_{q=1}^{M} [A_q \cos(2\pi f_q t_n) + B_q \sin(2\pi f_q t_n)] \\ &= \sum_{q=1}^{M} [A_q \cos(2\pi q n/N) + B_q \sin(2\pi q n/N)] \end{aligned} \tag{5.5.5}$$

Taking the partial derivatives of (5.5.4) with respect to the unknown coefficients A_q and B_q, and setting the results to zero, yields $2M + 1$ simultaneous equations for the $M + 1$ constituents

$$\begin{aligned} \frac{\partial e^2}{\partial A_q} &= 0 = 2\sum_{n=1}^{N} \left\{ \left[x_n - \left(\bar{x} + \sum M \right) \right] [-\cos(2\pi q n/N)] \right\}, \quad k = 0, \ldots, M \\ \frac{\partial e^2}{\partial B_q} &= 0 = 2\sum_{n=1}^{N} \left\{ \left[x_n - \left(\bar{x} + \sum M \right) \right] [-\sin(2\pi q n/N)] \right\}, \quad k = 1, \ldots, M \end{aligned} \tag{5.5.6}$$

Derivation of the coefficients in (5.5.6) requires solution of a matrix equation of the form $\mathbf{Dz} = \mathbf{y}$ in which $\mathbf{D}$ is an $(M + 1) \times (M + 1)$ matrix involving sine and cosine summation terms, $\mathbf{y}$ is a vector (column matrix) incorporating summations over the data series and $\mathbf{z}$ is a column matrix containing the required coefficients A_q and B_q. Gaps in the data are still permitted at this stage since the observation times, t_n, used in the least-squares method are not required to be evenly spaced.

Details on the matrix inversion and related problems can be found in Foreman (1977). To simplify the summations (5.5.6), trigonometric identities are often used. This requires that the data be evenly spaced and that the matrix terms be calculated over segments of the time series with no gaps. The resultant matrix $\mathbf{D}$ is symmetric so that only the upper triangle consisting of $2M + 3M + 1$ elements needs to be stored during the computations. We then seek solutions $\mathbf{z}$ through the matrix equation

$$\mathbf{z} = \mathbf{D}^{-1}\mathbf{y} \tag{5.5.7}$$

where $\mathbf{D}^{-1}$ is the inverse of the matrix

$$\mathbf{D} = \begin{pmatrix} N & c_1 & c_2 & \ldots & c_M & s_1 & s_2 & \ldots & s_M \\ c_1 & cc_{11} & cc_{12} & \ldots & cc_{11M} & cs_{11} & cs_{12} & \ldots & cs_{1M} \\ c_2 & cc_{21} & cc_{22} & \ldots & cc_{2M} & cs_{21} & cs_{22} & \ldots & cs_{2M} \\ \ldots & \ldots & \ldots & \ldots & \ldots & \ldots & \ldots & \ldots & \ldots \\ \ldots & \ldots & \ldots & \ldots & \ldots & \ldots & \ldots & \ldots & \ldots \\ c_M & cc_{M1} & cc_{M2} & \ldots & cc_{MM} & cs_{M1} & cs_{M2} & \ldots & cs_{MM} \\ \ldots & \ldots & \ldots & \ldots & \ldots & \ldots & \ldots & \ldots & \ldots \\ s_1 & sc_{11} & sc_{12} & \ldots & sc_{1M} & ss_{11} & ss_{12} & \ldots & ss_{1M} \\ s_2 & sc_{21} & sc_{22} & \ldots & sc_{2M} & ss_{21} & ss_{22} & \ldots & ss_{2M} \\ \ldots & \ldots & \ldots & \ldots & \ldots & \ldots & \ldots & \ldots & \ldots \\ s_M & sc_{M1} & sc_{M2} & \ldots & sc_{MM} & ss_{M1} & ss_{M2} & \ldots & ss_{MM} \end{pmatrix} \tag{5.5.8}$$

and $\mathbf{y}$ and $\mathbf{z}$ are column vectors

$$\mathbf{y} = \begin{pmatrix} yc_o \\ yc_1 \\ yc_2 \\ \ldots \\ \ldots \\ yc_M \\ ys_1 \\ \ldots \\ ys_M \end{pmatrix} \quad \text{and} \quad \mathbf{z} = \begin{pmatrix} A_0 \\ A_1 \\ A_2 \\ \ldots \\ \ldots \\ A_M \\ B_1 \\ \ldots \\ B_M \end{pmatrix} \tag{5.5.9}$$

The elements of $\mathbf{z}$ yield the required coefficients A_q, B_q for each specified harmonic constituent. To find these solutions, we substitute the elements of $\mathbf{D}$ for times $t_n = n\Delta t$ and, using $\alpha_k = f_k T$, $\alpha_j = f_j T$, where f_k and f_j are frequency units of Δt^{-1} and $T = N\Delta t$ is the record length.

$$c_k = \sum_{n=1}^{N} \cos(2\pi\alpha_k n/N), \quad s_k = \sum_{n=1}^{N} \sin(2\pi\alpha_k n/N)$$

$$cc_{kj} = cc_{jk} = \sum_{n=1}^{N} \left[\cos(2\pi\alpha_k n/N)\cos(2\pi\alpha_j n/N)\right]$$

$$ss_{kj} = ss_{jk} = \sum_{n=1}^{N} \left[\sin(2\pi\alpha_k n/N)\sin(2\pi\alpha_j n/N)\right] \quad (5.5.10)$$

$$cs_{kj} = sc_{jk} = \sum_{n=1}^{N} \left[\cos(2\pi\alpha_k n/N)\sin(2\pi\alpha_j n/N)\right]$$

where $\alpha_k n/N = (\alpha_k/N\Delta t)(n\Delta t)$, and the elements of **y** are given by

$$yc_k = \sum_{n=1}^{N} x_n \cos(2\pi\alpha_k n/N), \quad ys_k = \sum_{n=1}^{N} x_n \sin(2\pi\alpha_k n/N) \quad (5.5.11)$$

5.5.2 A computational example

We can illustrate the power of the least-squares method by again using the monthly mean sea surface temperature record of Table 5.2. Our purpose is to estimate the amplitudes and phases of the dominant annual and semi-annual constituents in the Tofino temperature record and compare the results with those we obtained using Fourier analysis in Section 5.4.3. This is also the approach we would use if we wanted to subtract these particular components from the original data record, as we might want to do prior to consideration of less dominant higher frequency variability or before cross-correlation with another data set. We let $f_1 = 1/12$ month ($= 0.0833$ cpmo) and $f_2 = 1/6$ month ($= 0.1667$ cpmo) represent the frequencies of interest. From (5.5.8) and (5.5.10), we find for $\alpha_1 = f_1 T = \frac{1}{12} \times 24 = 2$, and $\alpha_2 = f_2 T = \frac{1}{6} \times 24 = 4$ that

$$\mathbf{D} = \begin{pmatrix} N & c_1 & c_2 & s_1 & s_2 \\ c_1 & cc_{11} & cc_{12} & cs_{11} & cs_{12} \\ c_2 & cc_{21} & cc_{22} & cs_{21} & cs_{22} \\ s_1 & sc_{11} & sc_{12} & ss_{11} & ss_{12} \\ s_2 & sc_{21} & sc_{22} & ss_{21} & ss_{22} \end{pmatrix} \quad (5.5.12)$$

$$= \begin{pmatrix} 24 & 0 & 0 & 0 & 0 \\ 0 & 12 & 0 & 0 & 0 \\ 0 & 0 & 12 & 0 & 0 \\ 0 & 0 & 0 & 12 & 0 \\ 0 & 0 & 0 & 0 & 12 \end{pmatrix} \quad (5.5.13)$$

396 Data Analysis Methods in Physical Oceanography

and from (5.5.9) and (5.5.11)

$$\mathbf{y} = \begin{pmatrix} yc_0 \\ yc_1 \\ yc_2 \\ ys_1 \\ ys_2 \end{pmatrix} = \begin{pmatrix} 262.70 \\ -21.30 \\ -5.30 \\ -23.87 \\ -0.69 \end{pmatrix} \tag{5.5.14}$$

where the elements of $\mathbf{y}$ have units of °C. The solution $\mathbf{z} = \mathbf{D}^{-1}\mathbf{y}$ is the vector

$$\mathbf{z} = \begin{pmatrix} A_0 \\ A_1 \\ A_2 \\ B_1 \\ B_2 \end{pmatrix} = \begin{pmatrix} 10.95 \\ -1.77 \\ -0.44 \\ -1.99 \\ -0.06 \end{pmatrix} \tag{5.5.15}$$

with units of °C. The results are summarized in Table 5.4. As required, the amplitudes and phases of the annual and semi-annual constituents are identical to those obtained using Fourier analysis (see Table 5.3). A plot of the original temperature record and the least-squares fitted curve using the annual and semi-annual constituents is presented in Figure 5.5.1. The standard deviation for the original record is 2.08°C

Table 5.4. Coefficients for the annual and semi-annual frequencies from a least-squares analysis of the Amphitrite Point monthly mean temperature series (Table 5.2). Frequency units are cycles per month (cpmo). $q = 0$ gives the mean value for the 24-month record. Other coefficients are defined through equation (5.5.3)

q	Frequency (cpmo)	Period (month)	A_q (°C)	B_q (°C)	C_q (°C)
0	–	–	10.95	0.0	10.95
2	0.083	12	-1.77	-1.99	2.67
4	0.167	6	-0.44	-0.06	0.45

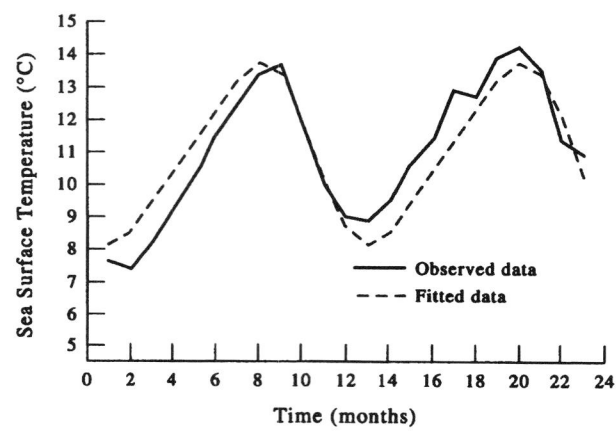

Figure 5.5.1. Monthly mean sea surface temperature (SST) record for Amphitrite Point on the west coast of Vancouver Island (see Table 5.2). The bold line is the original 24-month series. The dashed line is the SST time series obtained from a least-squares fit of the annual (12 month) and semi-annual (six month) cycles to the mean-removed data (see Table 5.3).

while that for the fitted record is 1.91°C. For this short segment of the data record, the two constituents account for 91.7% of the total variance.

5.5.3 Harmonic analysis of tides

Harmonic analysis is most useful for the analysis and prediction of tide heights and tidal currents. The use of this technique for tides appears to have originated with Lord Kelvin (1824–1907) around 1867. Lord Kelvin (Sir William Thomson) is also credited with inventing the first tide-predicting machine, although the first practical use of such a device was not made until several years later. A discussion of tidal harmonic analysis can be found in the *Admiralty Manual of Tides* (Doodson and Warburg, 1941) and Godin (1972). Definitive reports on the least-squares analysis of current and tide-height data were presented by Foreman (1977, 1978).

The least-squares harmonic analysis method has a variety of attractive features. It permits resolution of several hundred tidal constituents of which 45 are typically astronomical in origin and identified with a specific frequency in the tidal potential. The remaining constituents include shallow water constituents associated with bottom frictional effects and nonlinear terms in the equations of motion as well as radiational constituents originating with atmospheric effects. Both scalar and vector time series can be analyzed, with processing of vector series such as current velocity considerably more complex than processing of scalar time series such as sea level and water temperature. If the record is not sufficiently long to permit the direct resolution of neighboring components in the diurnal and semidiurnal frequency bands, the analysis makes provision for the "inference" and subsequent inclusion of these components in the analysis. For example, in the case of the diurnal constituent, P_1, associated with the sun's declination, the phase and amplitude are obtained by lowering the resolution criterion (called the *Rayleigh criterion*) for the separation of frequencies until P_1 is just resolved. The amplitude ratio (amp P_1/amp K_1) and phase difference (phase P_1–phase K_1) relative to the readily resolved diurnal constituent K_1 can then be calculated and used to calculate the P_1 constituent for the original record. Equally importantly, the method allows for gaps in the time series by ignoring those times for which there are no data. Major features of the least-squares optimization procedure for tidal analysis are outlined below.

The aim of least-squares analysis is to estimate the tidal harmonic constituent amplitudes and phases which can then be used for long-term tidal predictions. The commonly used sampling interval for tidal analysis is 1 h, so that even data collected at shorter time intervals are usually averaged to 1 h intervals for standard analysis packages. Records must have a minimum length of 13 h in order that they incorporate at least one cycle of the M_2 tidal frequency (period, 12.42 h). The mean component Z_0 is also included. As the length of the record is increased, additional constituents can be added to the analysis. (As noted in Chapter 1, our ability to resolve adjacent frequencies improves with the length of the time series. Aside from the degree of noise in the data, the main factor limiting the number of derived tidal constituents is the length of the record.) For example, the K_1 constituent (period, 23.93 h) can be adequately determined once the record length exceeds 24 h, although less reliable estimates can be made for shorter record lengths. The criteria for deciding which constituents can be included is discussed in the next section. In essence, inclusion requires that the difference in frequency, Δf, between a given constituent and its so-called *Rayleigh reference constituent* be greater than the fundamental frequency for the record; i.e. $\Delta f \geq f_1 = 1/T$ (see following discussion).

5.5.4 Choice of constituents

The least-squares method can be applied to any combination of tidal frequencies. However, the rational approach is to pick the allowable frequencies on the basis of two factors: (1) their relative contribution to the tide-generating potential; and (2) their resolvability in relation to a neighboring principal tidal constituent. In other words, the constituent should be one that makes a significant contribution to the tide-generating force and the record should be of sufficient duration to permit accurate separation of neighboring frequencies. Consideration should also be given to the required computational time, which increases roughly as the square of the number of constituents used in the analysis. Due to noise limitations, the amplitudes of many constituents are too small to be adequately resolved by most oceanic data sets.

To determine whether a specific constituent should be included in the tidal analysis, the frequency f_m of the constituent is compared to the frequency of the neighboring Rayleigh comparison constituent, f_R. The constituent can be included provided

$$|f_m - f_R|T = |\Delta f|T > R \qquad (5.5.16)$$

where T is the record length and R is typically equal to unity (depending on background noise). In effect, equation (5.5.16) states that f_m should be included if f_R is an included frequency *and* the ratio of the frequency difference Δf to the fundamental frequency $f_1 = 1/T$ is greater than unity. This implies that the fundamental frequency, which corresponds to the best resolution (separation) achievable on the frequency axis, is less than the frequency separation between constituents. Values of $R < 1$ are permitted in the least-squares program to allow for approximate estimates of neighboring tidal frequencies for record lengths T shorter than $1/\Delta f$. Obviously, the longer the record, the more constituents are permitted.

The choice of f_R is determined by the hierarchy of constituents within the tidal band of interest and level of noise in the observations. The hierarchy is in turn based on the contribution a particular constituent makes to the equilibrium tide, with the largest contribution usually coming from the M_2 tidal constituent (Cartwright and Edden, 1973). For the major contributors to the equilibrium tide, the magnitude ratios relative to M_2 in descending order are: $K_1/M_2 = 0.584$, $S_2/M_2 = 0.465$, and $O_1/M_2 = 0.415$. Depending on the level of noise in the observations, the principal semidiurnal constituent M_2 (0.0805 cph) and the record mean Z_0 can be determined for records longer than about 13 h duration, while the principal diurnal component K_1 (0.0418 cph) can be determined for records longer than about 24 h. As a rough guide, separation of the next most significant semidiurnal constituent S_2 (0.0833 cph) from the principal component M_2 requires a record length $T > 1/|f(M_2) - f(S_2)| = 355$ h (14.7 days). Similarly, separation of the next most significant diurnal constituent, O_1 (0.0387 cph), from the principal component, K_1, requires an approximate record length $T > 1/|f(K_1) - f(O_1)| = 328$ h (13.7 days). The frequencies $f(K_1)$ and $f(O_1)$ then become the Rayleigh comparison frequencies for other neighboring tidal constituents in the diurnal band while the frequencies $f(M_2)$ and $f(S_2)$ become the comparison frequencies for neighboring frequencies in the semidiurnal band. Extension of this procedure to longer and longer records eventually encompasses all the significant tidal constituents within the diurnal and semidiurnal bands. The first long-term constituent to be included in the analysis is the lunar–solar fortnightly cycle M_{sf} (0.00282 cph), requiring an approximate record duration $T > 14.8$ days, followed by the lunar monthly constituent M_m (0.00151 cph), duration $T > 31.8$ days, and the lunar fortnightly cycle M_f (0.00305 cph), $T > 182.6$ days. These record length requirements are based on stochastic processes; shorter records can be used for deterministic processes such as tides provided that noise levels are low. Thus, in all cases, shorter record lengths can be used if the data are highly noise free. By the same token, longer records are often needed to resolve the longer period tides because of contamination from atmospheric effects.

A summary of the required record lengths for inclusion of the more important constituents is provided in Tables 5.5–5.7 together with a comparison of the constituents tidal potential magnitude relative to that of the principal component in the frequency band. Where possible, a candidate constituent is compared to the particular neighboring constituent which has already been selected and is nearest in frequency.

5.5.5 A computational example for tides

As a simple example of the least-squares method of harmonic tidal analysis, consider the 32-hourly sea-level heights measured at Tofino, British Columbia during 10–11 September 1986 (Table 5.8). As indicated by Tables 5.5 and 5.6, we can at most resolve the K_1 and M_2 constituents. This problem is similar to that considered in Section 5.5.2 where we used the least-squares technique to fit the annual and semi-annual components to a 24-month record of sea surface temperature. Following the analysis in that section, the various matrices are written in terms of a mean component plus the contributions from the K_1 and M_2 frequencies, $f(K_1) = 0.0418$ cph and $f(M_2) = 0.0805$ cph, respectively. From (5.5.8) and (5.5.9), we find

$$\mathbf{D} = \begin{pmatrix} N & c_1 & c_2 & s & s_2 \\ c_1 & cc_{11} & cc_{12} & cs_{11} & cs_{12} \\ c_2 & cc_{21} & cc_{22} & cs_{21} & cs_{22} \\ s_1 & sc_{11} & sc_{12} & ss_{11} & ss_{12} \\ s_2 & sc_{21} & sc_{22} & ss_{21} & ss_{22} \end{pmatrix} \tag{5.5.17}$$

$$= \begin{pmatrix} 32 & 2.476 & -1.836 & 6.183 & 3.420 \\ 2.476 & 14.809 & 1.450 & 1.136 & 2.117 \\ -1.836 & 1.450 & 16.263 & -2.197 & 0.397 \\ 6.183 & 1.136 & -2.1 & 17.191 & 2.163 \\ 3.420 & 2.117 & 0.397 & 2.163 & 15.737 \end{pmatrix} \tag{5.5.18}$$

and from (5.5.9) and (5.5.11)

$$\mathbf{y} = \begin{pmatrix} yc_0 \\ yc_1 \\ yc_2 \\ ys_1 \\ ys_2 \end{pmatrix} = \begin{pmatrix} 57.640 \\ 6.514 \\ 6.138 \\ -0.199 \\ -3.335 \end{pmatrix} \tag{5.5.19}$$

400 *Data Analysis Methods in Physical Oceanography*

Table 5.5. Record lengths to resolve main tidal constituents in the semidiurnal tidal band assuming a Rayleigh coefficient R = 1. Also listed are the comparison constituents and ratios of tidal potential to that of the principal semidiurnal constituent M_2

Tidal constituent	Frequency (cph)	Comparison constituent	Magnitude ratio	Record length (h)
M_2 (principal lunar)	0.0805	–	1	13
S_2 (principal solar)	0.0833	M_2	0.465	355
N_2 (larger lunar elliptic)	0.0790	M_2	0.192	662
K_2 (luni-solar)	0.0836	S_2	0.029	4383

Table 5.6. Record lengths to resolve main tidal constituents in the diurnal tidal band assuming a Rayleigh coefficient R = 1. Also listed are the comparison constituents and ratios of tidal potential to that of the principal semidiurnal constituent, M_2

Tidal constituent	Frequency (cph)	Comparison constituent	Magnitude ratio	Record length (h)
K_1 (luni-solar)	0.0418	–	0.584	24
O_1 (principal lunar)	0.0387	K_1	0.415	328
P_1 (principal solar)	0.0416	K_1	0.193	4383
Q_1	0.0372	O_1	0.079	662

Table 5.7. Record lengths to resolve main tidal constituents in the long-period tidal band assuming a Rayleigh coefficient R = 1. Also listed are the comparison constituents and ratios of tidal potential to that of the principal semidiurnal constituent, M_2

Tidal constituent	Frequency (cph)	Comparison constituent	Magnitude ratio	Record length (h)
M_{sf} (mixed solar fortnightly)	0.002822	M_f	0.015	355
M_f (lunar fortnightly)	0.003050	–	0.172	4383
M_m (lunar monthly)	0.001512	M_{sm}	0.091	764
M_{sm} (solar monthly)	0.001310	–	0.017	4942
S_{sa} (solar semi-annual)	0.000228	S_a	0.080	4383
S_a (solar annual)	0.000114	–	0.013	8766

Table 5.8. Hourly values of sea-level height (SLH) measured at Tofino, British Columbia (49°09.0′ N, 125°54.0′ W) on the west coast of Canada starting 10 September 1986. Heights are in meters above the local datum

n	1	2	3	4	2	6	7	8	9	10	11
SLH	1.97	1.46	0.98	0.73	0.67	0.82	1.15	1.58	2.00	2.33	2.48

n	12	13	14	15	16	17	18	19	20	21	22
SLH	2.43	2.25	2.02	1.82	1.72	1.75	1.91	2.22	2.54	2.87	3.10

n	23	24	25	26	27	28	29	30	31	32
SLH	3.15	2.94	2.57	2.06	1.56	1.13	0.84	0.73	0.79	1.07

where the elements of **D** and **y** have units of meters. The solution $z = D^{-1}y$ is the vector

$$\mathbf{z} = \begin{pmatrix} A_0 \\ A_1 \\ A_2 \\ B_1 \\ B_2 \end{pmatrix} = \begin{pmatrix} 1.992\,m \\ 0.186\,m \\ 0.523\,m \\ -0.574\,m \\ -0.604\,m \end{pmatrix} \tag{5.5.20}$$

The results are summarized in Table 5.9. A plot of the original sea-level data and the fitted sea-level curve are presented in Figure 5.5.2. The standard deviation for the original record is 0.741 m while that for the fitted record is 0.736 m. For this short segment of the data record, the sum of the two tidal constituents accounts for over 99% of the total variance in the record. As a comparison, we have used the full analysis package without inference to analyze 29 days of the Tofino sea-level record beginning at 2000 on 10 September 1986. The program finds a total of 30 constituents, including the mean, Z_0, with the sum of the tidal constituents accounting for 98% of the original variance in the signal. The record mean for the month is 2.05 m, and the K_1 and M_2 constituents have amplitudes of 0.286 and 0.986 m, respectively. As expected, these are quite different to the values derived on only 32 h of data (Table 5.9). Phases for the two constituents for the 29-day records are 122.0° and 12.5° compared with 107.9° and 130.9° for the same two constituents based on the 32-h records (angles in both cases are measured counterclockwise from the positive x axis).

Table 5.9. Least-squares estimates of the amplitude and phase of the K_1 and M_2 tidal constituents for the 32-h Tofino sea level starting at 2000, 10 September 1986. The mean is $\frac{1}{2}A_0$. The last column, C'_q, gives the constituent amplitudes for a more extensive analysis that used a 29-day (685 h) data segment that had the same start time as the 32-h segment used to derive C_q.

q	Frequency (cph)	Period (h)	A_q (m)	B_q (m)	C_q (m)	C'_q (m)
0	—	—	3.984	0	3.984	4.100
1	0.042	24	0.186	−0.574	0.365	0.286
2	0.081	12	0.523	−0.604	0.638	0.986

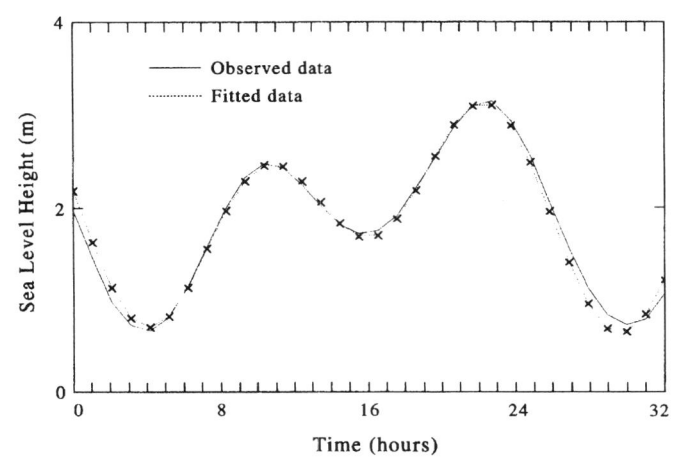

402 *Data Analysis Methods in Physical Oceanography*

Figure 5.5.2. Hourly sea-level height (SLH) recorded at Tofino on the west coast of Vancouver Island (see Table 5.7). The solid line is the original 32-h series; the dotted line is the SLH series obtained from a least-squares fit of the main diurnal (K_1, 0.042 cph) and main semidiurnal (M_2, 0.081 cph) tidal frequencies to the mean-removed data (see Table 5.8).

5.5.6 Complex demodulation

In many applications, we seek to determine how the signal characteristics at a specific frequency, ω, change throughout the duration of a time series. For example, we might ask how the amplitude, phase, and orientation of the semidiurnal tidal current ellipses at different depths at a mooring location change with time. Wave packets associated with passing internal tides would be revealed through rapid changes in ellipse parameters at the M_2 and/or S_2 frequencies. The method for determining the temporal change of a particular frequency component for a velocity or scalar time series is called *complex demodulation*.

A common technique for finding the demodulated signal is to fit the desired parameters to sequential segments of the data series using least-squares algorithms. The analysis requires that there be many more data points than frequency components and each segment must span at least one cycle of the frequency of interest. As with any least-squares analysis, the observations do not have to be at regular time intervals. Inputs to complex demodulation algorithms require specification of the start time of the first segment, the length of each segment, and the time between computation interval start times. Computation intervals may overlap, be end-to-end, or be interspersed with unused data. Following the least-squares analysis described under the section on harmonic analysis, the time increment between each estimate can be as short as one time step, Δt, thereby providing the maximum number of estimates for a given segment length, or as long as the entire record, thereby yielding a single estimate of the signal parameters.

For each segment of current velocity data, the fluctuating component of velocity at frequency ω can be expressed as

$$\mathbf{u}(t) - \overline{\mathbf{u}(t)} = \left[u(t) - \overline{u(t)}\right] + i\left[v(t) - \overline{v(t)}\right]$$
$$= A^+ \exp\left[i(\omega t + \varepsilon^+)\right] + A^- \exp\left[-i(\omega t + \varepsilon^-)\right]$$

(5.5.21)

where $\overline{u(t)}, \overline{v(t)}$ are mean components of the velocity, and (A^+, A^-) are the amplitudes and $(\varepsilon^+, \varepsilon^-)$ the phases of the counterclockwise (+) and clockwise (−) rotating components. Data are at times t_k ($k = 1, \ldots, N$) and solutions are found from the matrix equation

$$\mathbf{z} = \mathbf{D}^{-1}\mathbf{y} \tag{5.5.22}$$

where

$$\mathbf{y} = \begin{pmatrix} u(t_1) \\ u(t_2) \\ \ldots \\ u(t_n) \\ v(t_1) \\ \ldots \\ v(t_n) \end{pmatrix}; \quad \mathbf{z} = \begin{pmatrix} A^+ \cos(\varepsilon^+) \\ A^+ \sin(\varepsilon^+) \\ A^- \cos(\varepsilon^-) \\ A^- \sin(\varepsilon^-) \end{pmatrix} \equiv \begin{pmatrix} ACP \\ ASP \\ ACM \\ ASM \end{pmatrix} \tag{5.5.22a}$$

and

$$\mathbf{D} = \begin{pmatrix} \cos(\omega t_1) & -\sin(\omega t_1) & \cos(\omega t_1) & \sin(\omega t_1) \\ \cos(\omega t_2) & -\sin(\omega t_2) & \cos(\omega t_2) & \sin(\omega t_2) \\ \ldots & \ldots & \ldots & \ldots \\ \cos(\omega t_n) & -\sin(\omega t_n) & \cos(\omega t_n) & \sin(\omega t_n) \\ \sin(\omega t_1) & \cos(\omega t_1) & -\sin(\omega t_1) & \cos(\omega t_1) \\ \ldots & \ldots & \ldots & \ldots \\ \sin(\omega t_n) & \cos(\omega t_n) & -\sin(\omega t_n) & \cos(\omega t_n) \end{pmatrix} \tag{5.5.22b}$$

Once the elements of $\mathbf{z}$ are found from the least-squares solution to the matrix equation (for example, using IMSL routine LLSQAR), we can find the various ellipse parameters from

$$A^+ = (ASP^2 + ACP^2)^{1/2}; \quad A^- = (ASM^2 + ACM^2)^{1/2} \tag{5.5.23a}$$

$$\tan(\varepsilon^+) = \frac{ASP}{ACP}; \quad \tan(\varepsilon^-) = \frac{ASM}{ACM} \tag{5.5.23b}$$

For example, we could obtain the demodulated current amplitude and phase for near-inertial motions observed at a mid-latitude mooring by setting $\omega = 2\Omega \sin\theta$ and obtaining least-squares solutions for a series of adjoining 24-h segments with no overlap (here, Ω is the angular earth rotation rate and θ is latitude). For the least-squares technique to be applicable, data would need to sampled at roughly hourly intervals so that there were more data points per segment than parameters being estimated. Equatorward of $\theta = \pm 30°$ the period of inertial motions exceeds 24 h and the lengths of individual segments must be increased accordingly. Complex demodulation also can be used to examine inertial motions in Lagrangian-type data. In Figure 5.5.3(a), we have plotted the original and demodulated positions of a satellite-tracked drifter launched in the Canadian Arctic in the fall of 1988. The time series covers 60 days and was analyzed using overlapping 24-h subsections with the assumption that displacements occurred at the inertial period of 12.73 h for 70°N latitude. Figure 5.5.3(b) presents a detailed analysis of the trajectory record for the 20 days ending 11 October when the buoy became trapped in growing sea ice. Note the intense inertial currents starting on 30 September, the prevalence of the clockwise

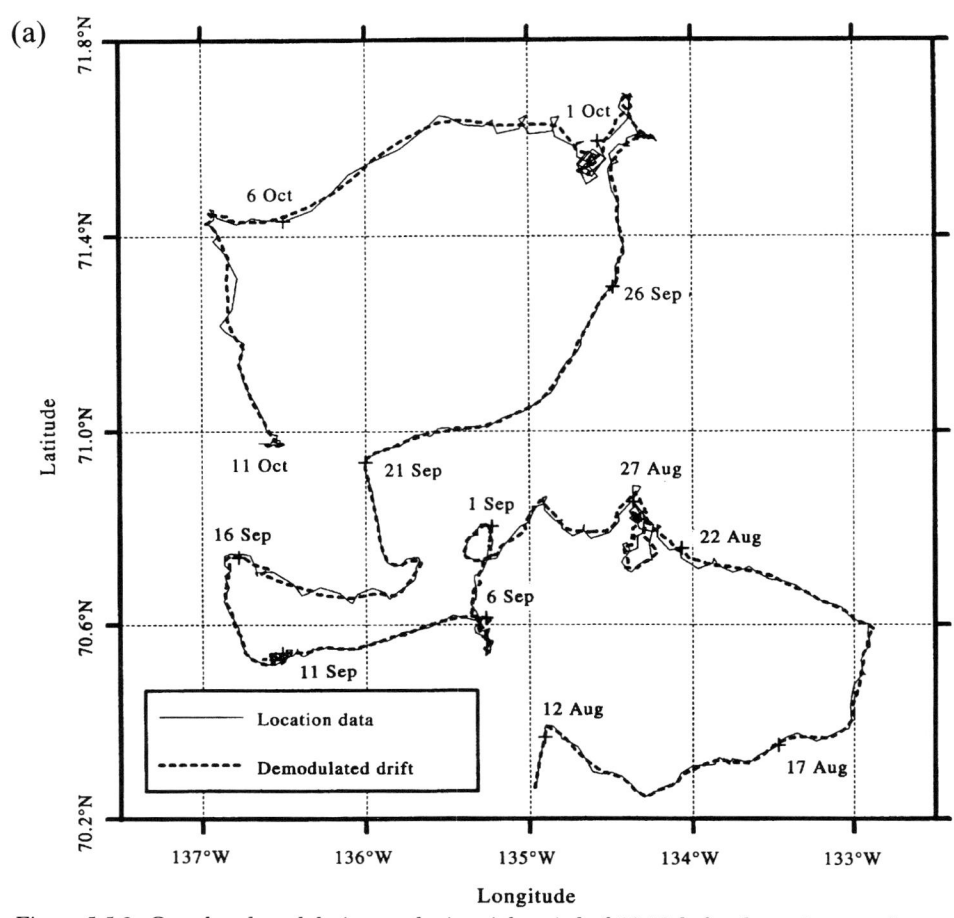

Figure 5.5.3. Complex demodulation at the inertial period of 12.73 h for the trajectory of a satellite-tracked drifter deployed in the Beaufort Sea in August 1988. (a) Original (solid line) and demodulated version (dashed line) of the drifter track. (Courtesy of Humfrey Melling.)

component of rotation, and the roughly −6.4° per day drift in phase of the clockwise component of the current due to the changing latitude of the drifter relative to the reference latitude of 70°N.

5.6 SPECTRAL ANALYSIS

Spectral analysis is used to partition the variance of a time series as a function of frequency. For stochastic time series such as wind waves, contributions from the different frequency components are measured in terms of the *power spectral density* (PSD). For deterministic waveforms such as surface tides, either the PSD or the *energy spectral density* (ESD) can be used. Here, power is defined as energy per unit time. The need for two different spectral definitions lies in the boundedness of the integral of signal variance for increasing record length. In practice, the term *spectrum* is applied to all spectral functions including commonly used terms such as autospectrum and

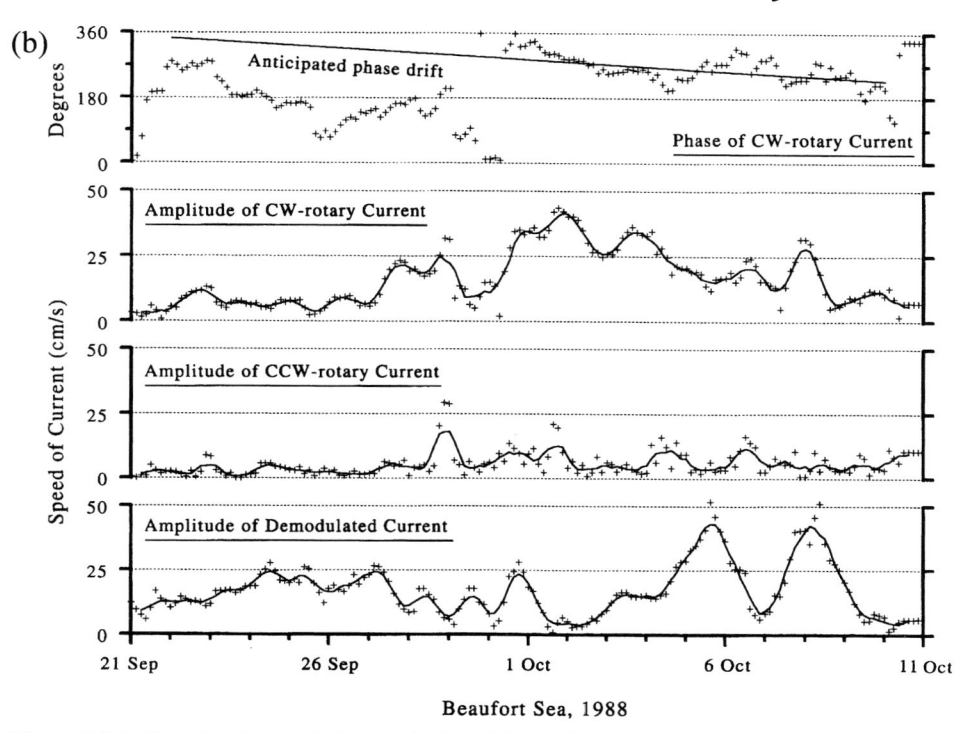

Figure 5.5.3. Complex demodulation at the inertial period of 12.73 h for the trajectory of a satellite-tracked drifter deployed in the Beaufort Sea in August 1988. (b) Parameters of the demodulation over a 20-day period of strong inertial motions. Top panel: phase of the clockwise (CW) rotary component (degrees). Remaining panels: amplitudes of the CW rotary, CCW rotary, and speed of the demodulated current. (Courtesy of Humfrey Melling.)

power spectrum. The term *cross-spectrum* is reserved for the "shared" power between two coincident time series. We also distinguish between *nonparametric* and *parametric* spectral methods. Nonparametric methods, which are based on conventional Fourier transforms, are not data-specific while parametric techniques are data-specific and assign a predetermined model to the time series. In general, we use parametric methods for short time series (few cycles of the oscillations of interest) and non-parametric methods for long time series (many cycles of oscillations of interest).

The word spectrum is a carry-over from optics. The colors red, white, and blue of the electromagnetic spectrum are often used to describe the frequency distribution of oceanographic spectra. A spectrum whose spectral density decreases with increasing frequency is called a "red" spectrum, by analogy to visible light where red corresponds to longer wavelengths (lower frequencies). Similarly, a spectrum whose magnitude increases with frequency is called a "blue" spectrum. A "white" spectrum is one in which the spectral constituents have near-equal amplitude throughout the frequency range. In the ocean, long-period variability (periods greater than several days) tend to have red spectra while instrument noise tends to have white spectra. Blue spectra are confined to certain frequency bands such as the low-frequency portion of wind–wave spectra and within the weather band (2 < period < 10 days) for deep wind-generated currents.

In the days before modern computers it was customary to compute the spectrum of discrete oceanic data from the Fourier transform of the autocorrelation function using a small number of lag intervals, or "lags". First formalized by Blackman and Tukey (1958), the autocorrelation method lacks the wide range of optional improvements to the computations and generalized "tinkering" permitted by more modern techniques. From a historical perspective, the autocorrelation approach has importance for the direct mathematical link it provides through the Wiener–Khinchin relations that link variance functions in the time domain to those in the frequency domain. Today, it is the spectral *periodogram* generated using the fast Fourier transform (FFT) or the Singleton Fourier transform that is most commonly used to estimate oceanic spectra.

Other methods have been developed over the years as a result of fundamental performance limitations with the periodogram method. These limitations are: (1) restricted frequency resolution when distinguishing between two or more signals, with frequency resolution dictated by the available record length independent of the characteristics of the data or its signal-to-noise ratio (SNR); (2) energy "leakage" between the main lobe of a spectral estimate and adjacent side-lobes, with a resulting distortion and smearing of the spectral estimates, suppression of weak signals, and the need to use smoothing windows; (3) an inability to adequately determine the spectral content of short time series; and (4) an inability to adjust to rapid changes in signal amplitude or phase. Other techniques, such as the maximum entropy method (best suited to short time series) and the wavelet transform (best suited to event-like signals), are addressed in this chapter.

Fundamental concepts: Several basic concepts are woven into the fabric of this chapter. First of all, the sample data we collect are subsets of either stochastic or deterministic processes. Deterministic processes are predictable, stochastic ones are not. Secondly, the very act of sampling to generate a time series of finite duration is analogous to viewing an infinitely long time series through a narrow "window" in the shape of a rectangular box-car function (Figure 5.6.1a). The characteristics of this window in the frequency domain can severely distort the frequency content of the original data series from which the sample has been drawn. As illustrated by Figure 5.6.1(b), the sampling process results in spectral energy being "rippled" away from one frequency (the central lobe of the response function) to a wide number range of adjacent frequencies. The large side-lobes of the rectangular window are responsible for the leakage of spectral energy from the central frequency to nearby frequencies.

A third point is that the spectra of random processes are themselves random processes. Therefore, if we are to determine the frequency content of a data series with some degree of statistical reliability (i.e. to be able to put confidence intervals on spectral peaks), we need to precondition the time series and average the raw periodogram estimates. Averaging can be done in the time domain by using specially designed windows or in the frequency domain by averaging together adjacent spectral estimates. Windows (which are discussed in detail in Section 5.6.6) suppress Gibbs' phenomenon associated with finite length data series and enable us to increase the number of *degrees of freedom* used in each spectral estimate. (Here, the term "degrees of freedom" refers to the number of statistically independent variables or values used in a particular estimate.) We can also improve spectral estimates by partitioning a time series into a series of segments and then conducting spectral analysis on the separate pieces. Spectral values in each frequency band for each piece are then averaged as a block to improve statistical reliability. The penalty for doing this is a loss in frequency resolution. The alternative—calculating a single periodogram and then smoothing in the frequency domain—suffers the same loss of frequency resolution for a smoothing that gives the same degrees of freedom.

(a)

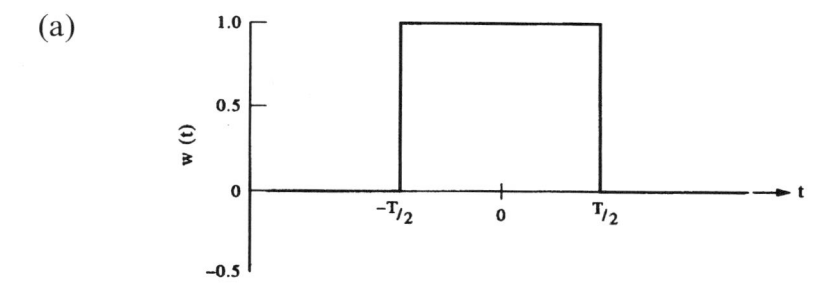

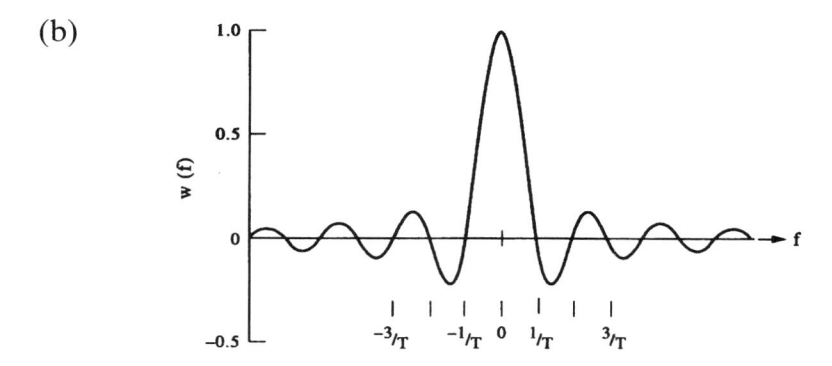

(b)

Figure 5.6.1. The box-car (rectangular) window which creates a sample time series from a "long" time series. (a) The box-car window in the time (t) domain. Here, $w(t) = 1$, $-T/2 \leq t \leq T/2$, and $w = 0$ otherwise. (b) Frequency (f) response of the box-car window in (a). The central lobe straddles each spectral (frequency) component within the time series and has a width $\Delta f = 2/T$. Zeros occur at $f = \pm m/T$, where $m = 1, 2, \ldots$.

Regardless of which averaging approach we choose, the results will be tantamount to viewing the data through another window in the frequency domain. Any smoothing window used to improve the reliability of the spectral estimates will again distort the results and impose structure on the data, such as periodic behavior, when no such structure may exist in the original time series. In addition, conventional methods make the implicit assumption that the unobserved data or correlation lag-values situated outside the measurement interval are zero, which is generally not the case. The smoothing window results in smeared spectral estimates. The more modern parametric methods allow us to make more realistic assumptions about the nature of the process outside the measurement interval, other than to assume it is zero or cyclic. This eliminates the need for window functions. The improvement over conventional FFT spectral estimates can be quite dramatic, especially for short records. However, even then, there remain pitfalls which have tended to detract from the usefulness of these methods to oceanography. Each new method has its own advantages and disadvantages that must be weighed in context of the particular data set and the way it has been collected. For time series with low signal-to-noise ratio (SNR), most of the modern methods are no better than the conventional FFT approach.

Means and trends: Prior to spectral analysis, the record mean and trend are generally removed from any time series (Figure 5.6.2). Unless stated otherwise, we will assume that the time series $y(t)$ we wish to process has the form $y'(t) = y(t) - \overline{y(t)}$ where $\overline{y(t)} = y_o + \alpha t$ is the mean value and αt is the linear trend (y_o and α are constants). If

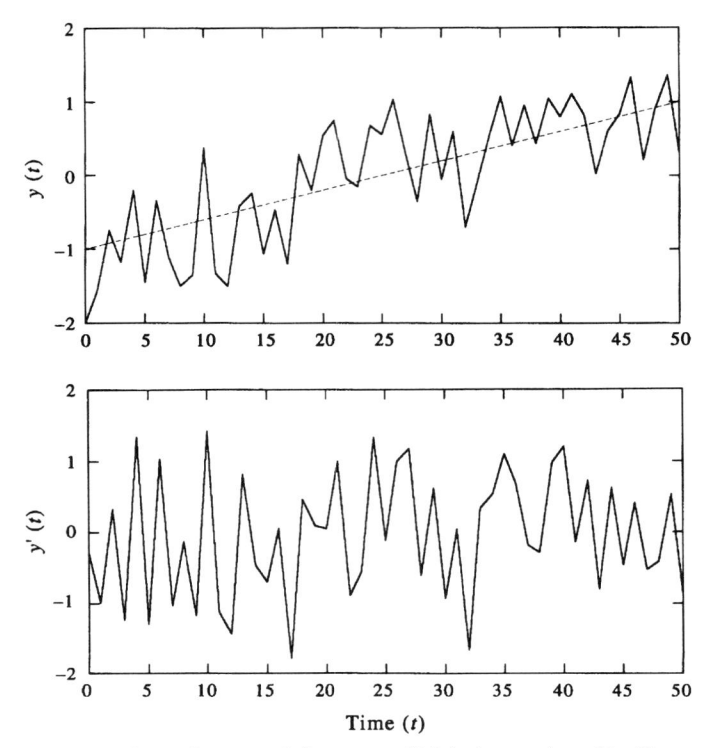

408 *Data Analysis Methods in Physical Oceanography*

Figure 5.6.2. Mean and trend removal for an artificial time series y(t). Here, $y_o = -1.0$, trend $\alpha = 0.025$ and the fluctuating component, y', was obtained using a uniformly distributed random number generator. (a) Original time series, showing the linear trend; (b) Time series with the mean and linear trend removed.

the mean and trend are not removed prior to spectral analysis, they can distort the low-frequency components of the spectrum. Packaged spectral programs often include record mean and linear trend removal as part of the data preconditioning. Nonlinear trends are more difficult to remove, especially since a single function may not be appropriate for the entire data domain. The latter may apply also to linear trends.

The mean value removed from a record is not always the average for the entire record. For example, to examine interannual variability in the monthly time series of sea-level height, $\eta(t_m)$, at Cristobal on the Caribbean end of the Panama Canal, Thomson et al. (1985) first calculated mean-monthly values $\overline{\eta(t_m)}_m$ for each month (e.g. the individual means for January, February, etc.). These mean monthly values, rather than the average value for the entire record, were then subtracted from the original data for the appropriate month to obtain monthly anomalies of sea level, $\eta'(t_m) = \eta(t_m) - \overline{\eta(t_m)}_m$. Trend removal was then applied to the monthly anomalies to obtain the final sea-level anomaly record. As a final comment, we note that certain records, such as those from moored near-surface transmissometers, will contain nonlinear trends that should be removed from the data record prior to spectral analysis. This must be done with care. Unless one has a justified physical model for a particular trend (including a linear trend), removal of the trend may itself add spurious frequency components to the de-trended signal.

5.6.1 Spectra of deterministic and stochastic processes

Time-series data can originate with deterministic or stochastic processes, or a mixture of the two. Turbulence arising from eddy-like motions generated by strong tidal currents in a narrow coastal channel provides an example of mixed deterministic and stochastic processes. To see the difference between the two types of processes in terms of conventional spectral estimation, consider the case of a continuous *deterministic* signal, $y(t)$. If the total signal energy, E, is finite

$$E = \int_{-\infty}^{\infty} |y(t)|^2 \, dt < \infty \tag{5.6.1}$$

then $y(t)$ is absolute-integrable over the entire domain and the Fourier transform $Y(f)$ of $y(t)$ exists. This leads to the standard transform pair

$$Y(f) = \int_{-\infty}^{\infty} y(t) e^{-i2\pi ft} \, dt \tag{5.6.2a}$$

$$y(t) = \int_{-\infty}^{\infty} Y(f) e^{i2\pi ft} \, df = \frac{1}{2\pi} \int_{-\infty}^{\infty} Y(f) e^{i\omega t} \, d\omega \tag{5.6.2b}$$

where $e^{\pm i2\pi ft} = \cos(2\pi ft) \pm i \sin(2\pi ft)$, f is the frequency in cycles per unit time, and $\omega = 2\pi f$ is the angular frequency in radians per unit time. The square of the modulus of the Fourier transform for all frequencies

$$S_E(f) = Y(f) Y^*(f) = |Y(f)|^2 \tag{5.6.3}$$

is then the energy spectral density (ESD), $S_E(f)$, of $y(t)$. (As usual, the asterisk denotes the complex conjugate.) To see equation (5.6.3) that is an energy density, we use Parseval's theorem

$$\int_{-\infty}^{\infty} |y(t)|^2 \, dt = \int_{-\infty}^{\infty} |Y(f)|^2 \, df \tag{5.6.4}$$

which states that the total energy, E, of the signal in the time domain is equal to the total energy, $\int S_E(f) \, df$, of the signal in the frequency domain. Thus, $S_E(f)$, is an energy density (energy per unit frequency) which, when multiplied by df, yields a measure of the total signal energy in the frequency band centered near frequency f. The "power" of a deterministic signal, E/T, is zero in the limit of very long time series ($T \to \infty$).

Now, suppose that $y(t)$ is a stationary *random* process rather than a deterministic waveform. Unlike the case for the finite energy deterministic signal, the total energy in the stochastic process is unbounded (the characteristics of the process remain unchanged over time) and functions of the form (5.6.2) do not exist. In other words, the Fourier transform method introduced earlier fails in the sense that the total energy, as defined by equation (5.6.1), does not decrease as the length of the time series increases without bound. To get around this problem, we must deal with the frequency distribution of the signal *power* (the time average of energy or energy per unit time, E/T) which is a bounded function. The basis for spectral analysis of random processes is the autocorrelation function $R_{yy}(\tau) = E[y(t)y(t+\tau)]$. Using the Wiener–Khinchin relation, the power spectral density, $S(f)$, becomes

$$S(f) = \int_{-\infty}^{\infty} R_{yy}(\tau) e^{-i2\pi f \tau} \, d\tau \qquad (5.6.5a)$$

For an ergodic random process, for which ensemble averages can be replaced by time averages, R_{yy} has the form

$$R_{yy}(\tau) = \lim_{T \to \infty} \frac{1}{T} \int_{-T/2}^{T/2} [y(t) y^*(t+\tau)] \, dt \qquad (5.6.5b)$$

By definition, the energy and power spectral density functions quantify the signal variance per unit frequency. For example, in the case of a stationary random process, integration of $S(f)$ gives the relation

$$s^2 = \int_{f - \Delta f/2}^{f + \Delta f/2} S(f) \, df \qquad (5.6.6)$$

where s^2 is the integrated signal variance in the narrow frequency range $\Delta f = [f - 1/2\Delta f, f + 1/2\Delta f]$. If we assume that the spectrum is nearly uniform over this frequency range, we find

$$S(f) \approx \frac{s^2}{\Delta f} \qquad (5.6.7)$$

which defines the spectrum for a stochastic processes in terms of a power density, or variance per unit frequency. The product $S(f)\Delta f$ is the total signal variance within the frequency band Δf centered at frequency f.

At this point, there are several other basic concepts worth mentioning. First of all, a waveform whose autocorrelation function $R(\tau)$ attenuates slowly with time lag, τ, will have a narrow spectral distribution (Figure 5.6.3a) indicating that there are relatively few frequency components to destructively interfere with one another as τ increases from zero. In the limiting case of only one frequency component, f_a, we find $R(\tau) \approx \cos(2\pi f_a \Delta t)$ and Fourier *line spectra* appear at frequencies $\pm f_a$ (Figure 5.6.3b). Because they consist of near monotone signals, tidal motions are highly autocorrelated and produce sharp spectral lines. In contrast, a rapidly decaying autocorrelation function implies a broad spectral distribution (Figure 5.6.4a) and a large number of frequency components in the original waveform. In the limit $R(\tau) \to \delta(\tau)$ (Figure 5.6.4b), there is an infinite number of equal-amplitude frequency components in the waveform and the spectrum $S(f) \to$ constant (white spectrum).

Figure 5.6.5 provides an example of time-series data generated by the relation $y(k) = A \cdot \cos(2\pi n k/N) + \varepsilon(k)$, where $k = 0, \ldots, N$ is time in units of $\Delta t = 1$, $n/N\Delta t = 0.25$ is the frequency in units of Δt^{-1}, and $\varepsilon(k)$ is a random number between

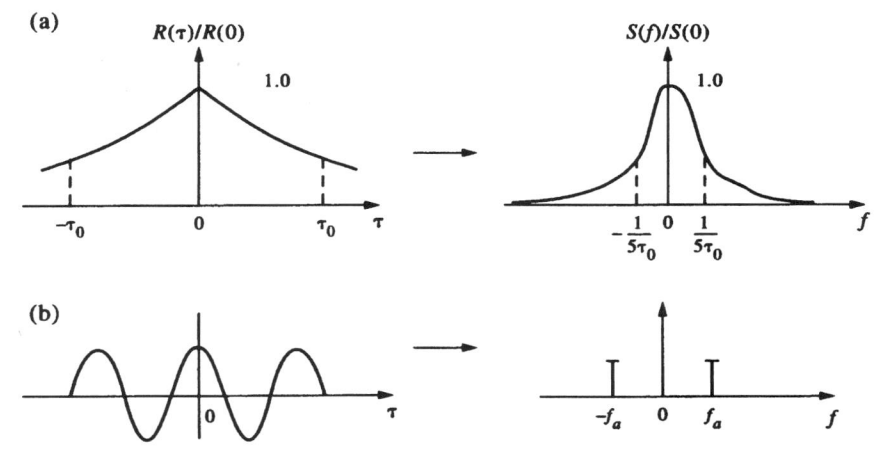

Figure 5.6.3. Examples of slowly decaying autocorrelation functions, $R(\tau)$, as a function of time lag, τ. Functions are normalized by their peak values. (a) The correlation function for a highly correlated signal leads to a relatively narrow power spectra density distribution, $S(f)$; (b) the case for autocorrelation $R(\tau) \approx \cos(2\pi f_a \Delta t)$ for a single frequency component, f_a, and corresponding line spectra at frequencies $\pm f_a$. (From Konyaev, 1990.)

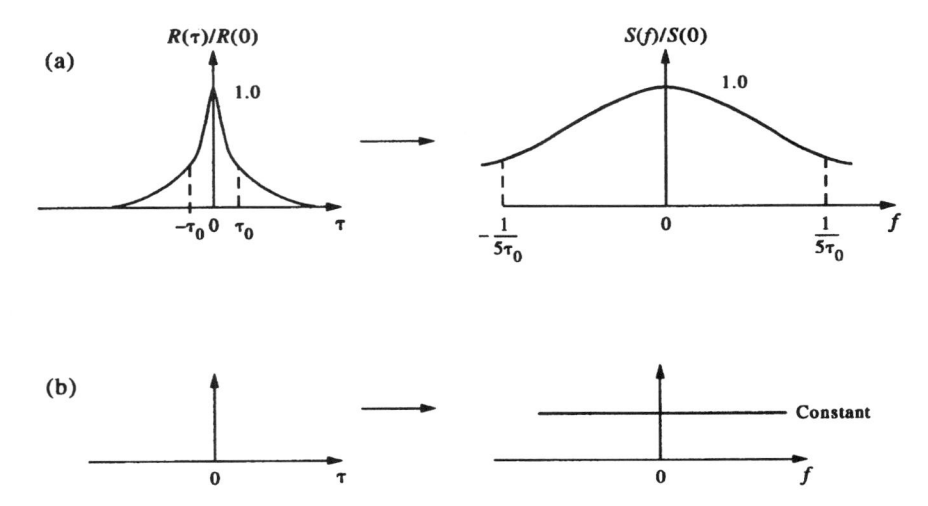

Figure 5.6.4. As for Figure 5.6.3 but for rapidly decaying autocorrelation functions, $R(\tau)$. (a) Correlation function for a weakly correlated signal leading to a broad power spectra density distribution. (b) The limiting case $R(\tau) \approx \delta(\tau)$ and the related spectrum $S(f) =$ constant (a white spectrum). (From Konyaev, 1990.)

−1 and +1. (We will often use this type of generic example rather than a specific example from the oceanographic literature. That way, readers can directly compare their computational results with ours. In the present case, if we set $\Delta t = 1$ day, then the time series $y(k)$ could represent east–west current velocity oscillations of a synoptic (three to 10-day) period associated with wind-forced motions (cf. Cannon and Thomson, 1996). Here, we set $A = 1$ and $\varepsilon(k) \neq 0$ for mostly deterministic data (Figure 5.6.5a) and $A = 0$ for random data (Figure 5.6.5b). In the analysis, the record has been padded with zeros up to time $k = 2N$. For the mostly deterministic case, the noise causes partial decorrelation of the signal with lag, but the spectral peak remains prominent.

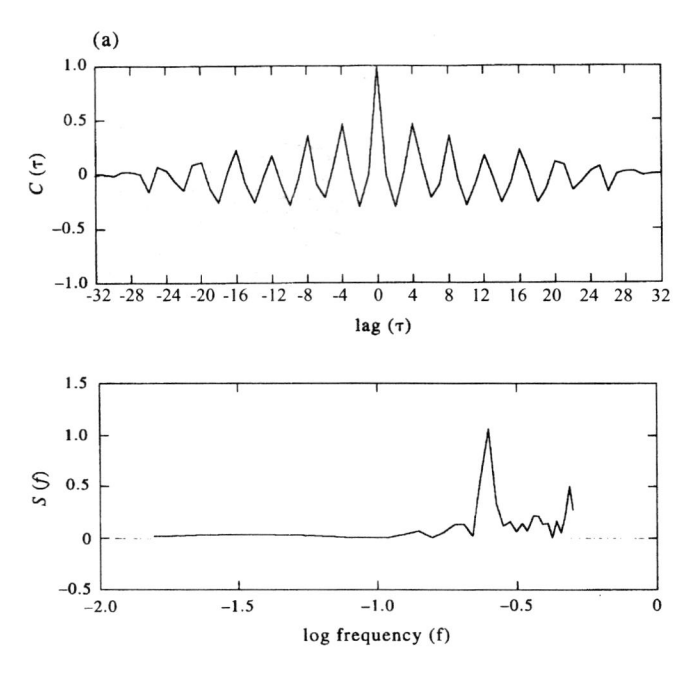

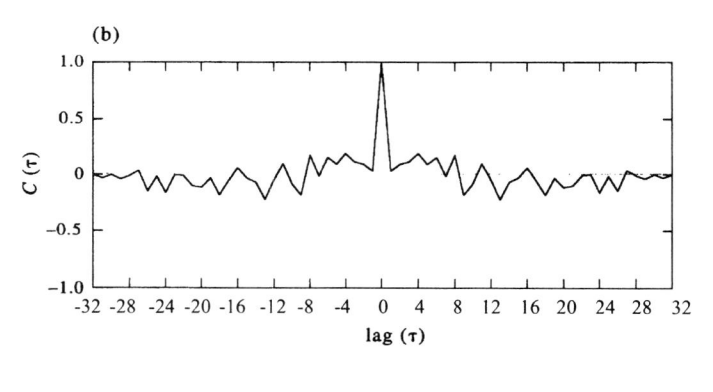

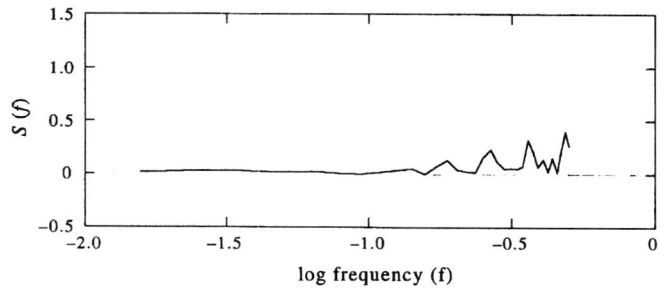

Figure 5.6.5. Autocovariance function $C(\tau)$ and corresponding spectrum $S(f)$ for the time series $y(k) = A\cos(2\pi nk/N) + \varepsilon(k)$; $k = 0, ..., N$, $\Delta t = 1$, $n/N = 0.25$ is the frequency, and $\varepsilon(k)$ is a random number between -1 and $+1$. (a) $C(\tau)$ and $S(f)$ for $A = 1$ and $\varepsilon \neq 0$ (mostly deterministic data); and (b) for $A = 0$ (purely random data). Records have been padded with zeros up to time $k = 2N = 32$.

For the purely random case, the spectrum resembles white noise but with isolated spectral peaks that one might mistake as originating with some physical process. The latter result is a good example of why we need to attach confidence limits to the peaks of spectral estimates (see Section 5.6.8).

5.6.2 Spectra of discrete series

Consider an infinitely long time series $y(t_n) = y_n$ sampled at equally spaced time increments $t_n = n\Delta t$, where Δt is the sampling interval and n is an integer, $-\infty < n < \infty$. From sampling theory, we know that a continuous representation of the discrete times series $y_s(t)$, can be represented as the product of the continuous time series $y(t)$ with an infinite set of delta functions, $\delta(t)$, such that

$$y_s(t) = y(t) \sum_{n=-\infty}^{\infty} \delta(t - n\Delta t)$$
$$= y(t) \frac{\Xi(t/\Delta t)}{\Delta t} \quad (5.6.8a)$$

where Ξ is the "sampling function" and for which the Fourier transform is

$$Y(f) = \int_{-\infty}^{\infty} \left[\sum_{n=-\infty}^{\infty} y(t)\delta(t - n\Delta t)\Delta t \right] e^{-i2\pi ft} \, dt$$
$$= \Delta t \sum_{n=-\infty}^{\infty} y_n e^{-i2\pi ft} \quad (5.6.8b)$$

In effect, the original time series is multiplied by a "picket fence" of delta functions $\Xi(t/\Delta t) \approx \sum_{n=-\infty}^{\infty} \delta(t - n\Delta t)$ which are zero everywhere except for the infinitesimal rectangular region occupied by each delta function (Figures 5.6.6a, b). Comparison of the above expression with equation (5.6.2) shows that retention of the time step Δt ensures conservation of the rectangular area in the two expressions as $\Delta t \to 0$. Provided that the time series $y(t)$ has a limited number of frequencies (i.e. is band-

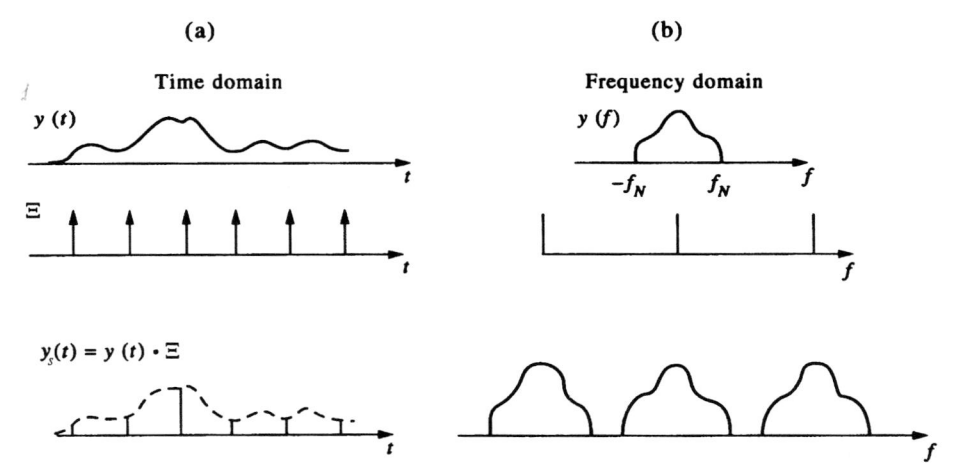

Figure 5.6.6. (a) A "picket fence" of delta functions $\delta(t - n\Delta t)$ used to generate a discrete data series from a continuous time series. (b) The Fourier transform (schematic only) of the different functions.

limited), whereby all frequencies are contained in the Nyquist interval

$$-f_N \leq f_k \leq f_N \tag{5.6.9}$$

in which $f_N = 1/(2\Delta t)$ is the Nyquist frequency, the energy spectral density

$$S_E(f) = |Y(f)|^2 \tag{5.6.10}$$

is identical to that for a continuous function. Conversely, if $Y(f) \neq 0$ for $|f| > f_N$ then the sampled and original times series do not have the same spectrum for $|f| < f_N$. The spectrum (5.6.10) obtained by Fourier analysis of discrete time series is called a *periodogram* spectral estimate, a term first coined by Schuster (1898) in a study of sunspot cycles.

Real oceanographic time-series data are discrete and have finite duration, $T = N\Delta t$. Returning to (5.6.8), this means that the summation is over a limited range $n = 1$ to N, and the spectral amplitude for the sample must be defined in terms of the discrete Fourier transform

$$\begin{aligned} Y_k &= \Delta t \sum_{n=1}^{N} y_n e^{-i2\pi f_k n \Delta t} \\ &= \Delta t \sum_{n=1}^{N} y_n e^{-i2\pi kn/N}; \quad f_k = k/N\Delta t, \ k = 0, \ldots, N \end{aligned} \tag{5.6.11}$$

The frequencies f_k are confined to the Nyquist interval, with positive frequencies, $0 \leq f_k \leq f_N$, corresponding to the range $k = 0, \ldots, N/2$ and negative frequencies, $-f_N \leq f_k \leq 0$, to the range $k = N/2, \ldots, N$. Since $f_{N-k} = f_k$, only the first $N/2$ Fourier transform values are unique. Specifically, $Y_k = Y_{N-k}$ so that we will generally confine our attention to the positive interval only.

The inverse Fourier transform is defined as

$$y_n = \frac{1}{N\Delta t} \sum_{k=0}^{N-1} Y_k e^{i2\pi kn/N}, \quad n = 1, \ldots, N \tag{5.6.12}$$

As indicated by equation (5.6.11), the Fourier transforms, Y_k, are specified for the discretized frequencies f_k, where $f_k = kf_1$ and $f_1 = 1/N\Delta t = 1/T$ characterizes both the fundamental frequency and the bandwidth, Δf, for the time series. The energy spectral density for a discrete, finite-duration time series is then

$$S_E(f_k) = |Y_k|^2, \quad k = 0, \ldots, N-1 \tag{5.6.13}$$

and Parseval's energy conservation theorem (5.6.4) becomes

$$\Delta t \sum_{n=1}^{N} |y_n|^2 = \Delta f \sum_{k=0}^{N-1} |Y_k|^2$$

where we have used $\Delta f = 1/(N\Delta t)$. A plot of $|Y_k|^2$ versus frequency, f_k, gives the discrete form of the periodogram spectral estimate.

Any geophysical data set we collect is subject to discrete sampling and windowing. As noted earlier, a time series of geophysical data, $y(t_n)$, sampled at time steps Δt can be considered the product of an infinitely long time series with a rectangular window which spans the duration ($T = N\Delta t$) of the measured data. The discrete spectrum $S(f_k)$ is the then the *convolution* of the true spectrum, $S(f)$, with the Fourier transform of the rectangular window (Figure 5.6.1b). Since the window allows us to see only a segment of the infinite time series, the spectrum $S(f_k)$ provides a distorted picture of the actual underlying spectrum. This distortion, created during the Fourier transform of the rectangular window, consists of a broadening of the central lobe and leakage of power from the central lobe into the side-lobes. (The "ripples" on either side of the central lobe in Figure 5.6.1(b) are side lobes.) A further problem is that the function Y_k and its Fourier transform now become periodic with period N, although the original infinite time series $y(t)$, of which our sample data are a subset, may have been nonperiodic.

As noted in the previous section, the convergence of $|Y(f)|^2$ to $S(f)$ is smooth for deterministic functions in that the function $|Y'(f)|^2$, obtained by increasing the sample record length from T to T', would be a smoother version of $|Y(f)|^2$. For stochastic signals, the function $|Y'(f)|^2$ obtained from the longer time series (T') is just as erratic as the function for the shorter series. The sample spectra of a stochastic process do not converge in any statistical sense to a limiting value as T tends to infinity. Thus, the sample spectrum is not a consistent estimator in the sense that its PDF does not tend to cluster more closely about the true spectrum as the sample size increases. To show what we mean, consider the spectrum of a process consisting of $N = 400$ random, normally distributed deviates (Gaussian white noise) sampled at 1-s intervals. (True white noise is a mathematical construct and is as physically impossible as the spike of an impulse function.) The highest frequency we can hope to measure with these data is the Nyquist frequency, $f_N = 0.5$ cps. The spectra computed from 50 and then from 100 values of the fully white noise signal are presented in Figure 5.6.7(a). Also shown is the theoretical sample spectrum, corresponding to a uniform amplitude of 1.0. The shorter the sample used for the discrete spectral estimates, the greater the amplitude spikes in the power spectrum. This same tendency also is apparent in Table 5.6.1 which lists the means, variances, and mean square errors computed from various subsamples of the white noise signal. Here, mean square error (MSE) is defined as the variance plus bias of an estimator $\hat{y}(t)$ of the true signal $y(t)$; that is

$$\text{MSE} = E[(\hat{y} - y)^2] = V[\hat{y}] + B^2 \qquad (5.6.14)$$

where $B = E[\hat{y}] - y$ is the bias of the estimator. The mean is lower in both the $N = 50$ and $N = 400$ cases while it is greater in the case where $N = 100$ and is exactly 1.0 for $N = 200$. The variance increases as N increases, as does the MSE. However, if this were a purely random discrete process (discrete white noise), the sample spectral estimator of the variance would be independent of the number of observations.

Now consider the spectrum of a second-order autoregressive process for a sample of $N = 400$ measured at 1-s increments (Figure 5.6.7b). (An autoregressive process of order p is one in which the present value of y depends on a linear combination of the

Table 5.6.1. Behavior of sample spectra of white noise as the record length is increased. (After Jenkins and Watts, 1968)

Record length (N)	50	100	200	400
Mean	0.85	1.07	1.00	0.95
Variance	0.630	0.777	0.886	0.826
Mean square error	0.652	0.782	0.886	0.828

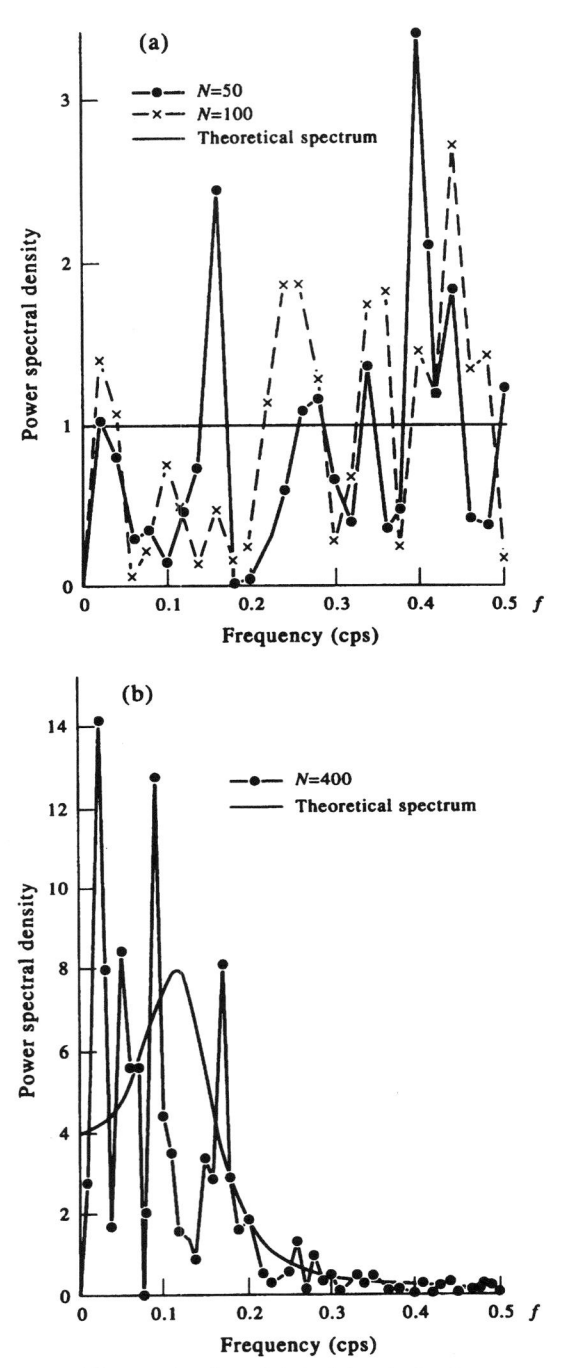

Figure 5.6.7. Power spectra of discrete signals and their theoretical values. Frequency in cycles per second (cps); spectra are in units of amplitude-squared/cps. (a) Power spectrum for the first half (N = 50) and full (N = 100) realization of a discrete normal white-noise process measured at 1-s intervals. (b) Power spectrum for one realization of a second-order autoregressive process of N = 400 values measured at 1-s increments. $f_N = 0.5$ cps is the Nyquist frequency and the maximum bandwidth of the spectral resolution $\Delta f = 1/N\Delta t = 0.0025 /s$. (From Jenkins and Watts, 1968.)

previous p values of y. See Section 5.7.2.) The Nyquist frequency is again 0.5 cps and the maximum bandwidth of the spectral resolution, $\Delta f = 1/N\Delta t$, is equal to 0.0025 cps. At the higher frequencies, the sample spectrum appears to be a good estimator of the theoretical spectrum (the smooth solid line), while for the lower frequencies there are large spikes in the sample spectrum which are not characteristic of the true spectrum. This misleading appearance is largely a consequence of the fact that the theoretical spectrum has most of its energy at the lower frequencies. In reality, the computed raw spectrum (i.e. with no smoothing) can fluctuate by 100% about the mean spectrum. The fluctuations are much smaller at higher frequencies simply because the actual spectral level is correspondingly smaller.

The basic reason why Fourier analysis breaks down when applied to real time series is that it is based on the assumption of fixed (stationary) amplitudes, frequencies, and phases. Stochastic series are instead characterized by random changes in frequency, amplitude, and phase. Thus, our treatment must be a statistical approach that makes it possible to accommodate these types of changes in our computation of the power spectrum.

5.6.3 Conventional spectral methods

The two spectral estimation techniques founded on Fourier transform operations are the indirect autocorrelation approach popularized by Blackman and Tukey in the 1950s and the direct periodogram approach presently favored by the oceanographic community. The fast Fourier transform (FFT) is the most common algorithm for determining the periodogram. The autocorrelation approach is mainly included for completeness. These methods fall into the category of nonparametric techniques which are defined independently of any specific time series. Parametric techniques, described later in this chapter, make assumptions about the variability of the time series and rely on the series for parameter determination.

The following sections first describe the two conventional spectral analysis methods without providing details on how to improve spectral estimates. We wish to first outline the procedures for calculating spectra before describing how to improve the statistical reliability of the spectral estimates. Once this is done, we give a thorough description of windowing, frequency-band averaging, and other spectral improvement techniques.

5.6.3.1 The autocorrelation method

In the Blackman–Tukey method, the autocovariance function, $C_{yy}(\tau)$ (which equals the autocorrelation function, $R_{yy}(\tau)$, if the record mean has been removed), is first computed as a function of lag, τ, and the Fourier transform of $C_{yy}(\tau)$ used to obtain the PSD as a function of frequency. An unbiased estimator for the autocovariance function for a data set consisting of N equally spaced values $\{y_1, y_2, \dots, y_N\}$ is

$$C_{yy}(\tau_m; N-m) = \frac{1}{N-m} \sum_{n=1}^{N-m} y_n y_{n+m} \qquad (5.6.15a)$$

where $m = 0, \dots, M$ is the number of lags ($\tau_m = m\Delta t$) and $M < N$. In place of this estimator, some authors (cf. Kay and Marple, 1981) argue for the use of

$$C_{yy}(\tau_m; N) = \frac{1}{N} \sum_{n=1}^{N-m} y_n y_{n+m} \qquad (5.6.15b)$$

which typically has a lower mean square error than $C_{yy}(\tau_m; N-m)$ for most finite data sets. Because $E[C_{yy}(\tau_m; N)] = [(N-m)/N]C_{yy}(\tau_m; N-m)$, the function $C_{yy}(\tau; N)$ is a biased estimator for the autocovariance function. Despite this, we will often use the relation (5.6.15b) for the autocovariance function since it yields a power spectral density (PSD) that is equivalent to the PSD obtained from the direct application of the FFT, as discussed in the next section. The weighting $(N-m)/N$ acts like a triangular (Bartlett) smoothing window to help reduce spectral leakage. We will use equation (5.6.15a) when we want a "stand-alone" unbiased estimator of the covariance function.

The one-sided power spectral density, G_k, for an autocovariance function with a total of M lags is found from the Fourier transform of the autocovariance function

$$G_k = 2\Delta t \sum_{m=0}^{M} C_{yy}(\tau_m) e^{-i2\pi km/M}, \quad k = 0, ..., (M/2) \qquad (5.6.16a)$$

where $\tau_m = m\Delta t$ and $2\Delta t = 1/f_N$. Since $C_{yy}(\tau_m)$ is an even function, the spectrum of $\{y_n\}$ can be calculated from the cosine transform

$$G_k = 2\Delta t \left[C_{yy}(0) + 2\sum_{m=1}^{M} C_{yy}(\tau_m) \cos \frac{2\pi km}{N} \right], \quad k = 0, ..., (M/2) \qquad (5.6.16b)$$

where $G_k = 2S_k$ is centered at positive frequencies $f_k = k/N\Delta t$ and the Nyquist interval $0 \leq f_k \leq f_N$ is divided into $N/2$ segments (N is even). For the two-sided spectrum, S_k, the first $(N/2) + 1$ frequencies are identical to those for the one-sided spectrum and correspond to positive frequencies in the range $0 \leq f_k \leq f_N$. The last $(N/2) - 1$ spectral values for the two-sided spectral density, defined for $k = (N/2) + 1, (N/2) + 2, ..., N-1$, correspond to spectral density estimates for negative frequencies in the range $-f_N \leq f_k \leq 0$.

The solid line in Figure 5.6.8 shows spectra of monthly mean sea surface temperatures derived from the cosine transform using the Blackman–Tukey autocorrelation method for the version (5.6.15b) of the autocovariance function. The temperature data span the 36-month period from January 1982 to December 1984 for Amphitrite Point (Table 5.6.2). Since, in the next section, we wish to compare these spectra directly with those derived from the data series using a packaged FFT routine (the dashed curve in Figure 5.6.8), the lags were computed for the first 32 (2^5) points only, four fewer points than used in the Blackman–Tukey approach. In this case, extending the lag correlation beyond 10–20% of the data, as recommended earlier, is a necessity if we are to obtain reasonable estimates of the spectra. As expected, results reveal a strong spectral peak centered near, but not at, the annual frequency ($f = 1.0$ cycles per year = 0.083 cpmonth). There are too few data to enable us to accurately resolve the location of the frequency peak. In the present example, all spectral estimates are positive. However, the autocorrelation method can yield erroneous negative spectra for weak frequency components when there are gaps in the data record.

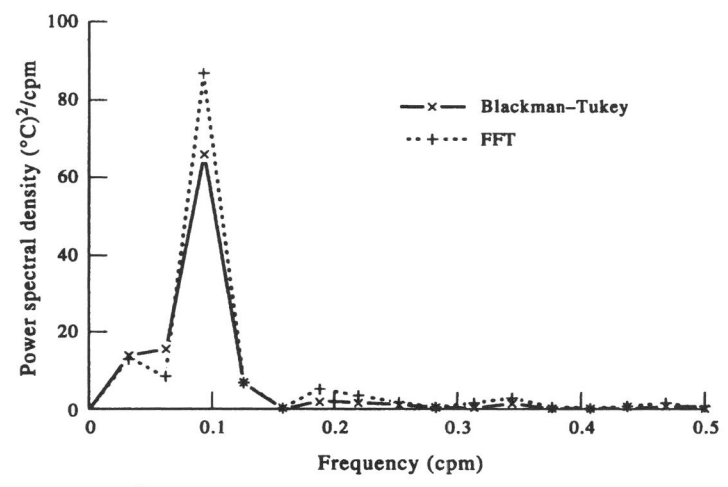

Figure 5.6.8. Spectra $(°C)^2/cpm$ (cpm = cycles per month) versus frequency (per month) for monthly mean sea surface temperatures collected at a coastal station in the northeast Pacific for the period January 1982 to December 1984 (cf. Table 5.6.2). (a) The solid line is the unsmoothed spectrum from the Blackman–Tukey autocorrelation method (the cosine transform of the autocovariance function (5.6.15b)); dashed line is the unsmoothed spectrum from the FFT method based on the first 2^5 (= 32) data values. Spectral peaks span the annual period (f = 0.083 /month).

Table 5.6.2. *Monthly mean sea surface temperatures SST (°C) at Amphitrite Point (48°55.16' N, 125°32.17' W) on the west coast of Canada for January 1982 through December 1984*

Year 1982												
n	1	2	3	4	5	6	7	8	9	10	11	12
SST	7.6	7.4	8.2	9.2	10.2	11.5	12.4	13.4	13.7	11.8	10.1	9.0
Year 1983												
n	13	14	15	16	17	18	19	20	21	22	23	24
SST	8.9	9.5	10.6	11.4	12.9	12.7	13.9	14.2	13.5	11.4	10.9	8.1
Year 1984												
n	25	26	27	28	29	30	31	32	33	34	35	36
SST	7.9	8.4	9.3	9.9	11.0	11.1	12.6	14.0	13.0	11.7	9.8	8.0

We emphasize that the spectra in Figure 5.6.8 have been constructed without any averaging or windowing. This means that each spectral estimate has the minimum possible two degrees of freedom so that the error in each estimate is equal to the value of the estimate itself. Some form of averaging is needed if we are to place confidence limits on our spectra (see Sections 5.6.6 and 5.6.7). The two spectra are slightly different because the record used for the FFT method is shorter than that used for the autocovariance method.

5.6.3.2 The periodogram method

The preferred method for estimating the power spectral density of a discrete sample $\{y_1, y_2, ..., y_N\}$ is the direct or periodogram method. Instead of first calculating the autocorrelation function, the data are transformed directly to obtain the Fourier components $Y(f)$ using (5.6.11). To help avoid end effects (Gibbs' phenomenon) and wrap-around problems, the original time series can be padded with $K \leq N$ zeros after the mean has been removed from the time series. The padding will also increase the frequency resolution of the periodogram (see Section 5.6.9). Although use of $K = N$ zeros is not recommended for computational reasons, it has one advantage: The N-lag covariance function obtained from the inverse Fourier transform of the $2N$-point power spectral density is identical to the N-lag covariance function (5.6.15b). As with the autocorrelation method, improvements in the statistical reliability of the spectral estimates would be provided by "windowing" the time series prior to spectral estimation or by averaging over the raw periodogram estimates over adjacent frequency bands (see Sections 5.6.6 and 5.6.7).

The two-sided power spectral (or autospectral) density for frequency f in the Nyquist interval $-1/(2\Delta t) \leq f \leq 1/(2\Delta t)$ and a padding of K zeros is

$$S_{yy}(f) = \frac{1}{(N+K)\Delta t} \left| \Delta t \sum_{n=0}^{N+K-1} y_n e^{-i2\pi f n \Delta t} \right|^2$$
$$= \frac{1}{(N+K)\Delta t} |Y(f)|^2 \qquad (5.6.17a)$$

while the one-sided power spectral density for the positive frequency interval only, $0 \leq f \leq 1/(2\Delta t)$, is

$$G_{yy}(f) = 2S_{yy}(f) = \frac{2}{(N+K)\Delta t} |Y(f)|^2 \qquad (5.6.17b)$$

Division by Δt transforms the energy spectral density of (5.6.13) into a power spectral density, $S_{yy}(f)$.

Evaluation of (5.6.17a) using the fast Fourier transform defines $Y(f)$ in terms of the discrete Fourier transform estimates, $Y(f_k) = Y_k$, where the f_k form a discrete set of $(N+K)/2$ equally spaced frequencies $f_k = \pm k/[(N+K)\Delta t]$, $k = 0, 1, \ldots, [(N+K)/2] - 1$ in the Nyquist interval, $-1/2\Delta t \leq f_k \leq 1/2\Delta t$. The case $k = 0$ represents the mean component. The two-sided PSD is then

$$S_{yy}(0) = \frac{1}{(N+K)\Delta t} |Y_0|^2, \quad k = 0$$

$$S_{yy}(f_k) = \frac{1}{(N+K)\Delta t} \left[|Y_k|^2 + |Y_{N+K-k}|^2 \right], \quad k = 1, \ldots, \frac{(N+K)}{2} - 1 \qquad (5.6.18a)$$

$$S_{yy}(f_N) = S_{yy}(f_{(N+K)/2-k}) = \frac{1}{(N+K)\Delta t} |Y_{(N+K)/2}|^2, \quad k = \frac{(N+K)}{2}$$

and the one-sided PSD is

$$G_{yy}(0) = \frac{1}{(N+K)\Delta t} |Y_0|^2, \quad k = 0$$

$$G_{yy}(f_k) = \frac{2}{(N+K)\Delta t}|Y_k|^2, \quad k = 1, \ldots, \frac{(N+K)}{2} - 1 \tag{5.6.18b}$$

$$G_{yy}(f_N) = G_{yy}(f_{(N+K)/2-k}) = \frac{1}{(N+K)\Delta t}|Y_{(N+K)/2}|^2, \quad k = \frac{(N+K)}{2}$$

Multiplication of $S_{yy}(f) \equiv S_k$ (or G_k) by the bandwidth of the signal $\Delta f = 1/(N+K)\Delta t$ gives the estimated signal variance, σ_k^2, in the kth frequency band; i.e. $\sigma_k^2 = S'_k = S_k \Delta f$. The summation

$$\sum_{n=0}^{N+K-1} S'_k = \sum_{n=0}^{N+K-1} S_k \Delta f \tag{5.6.19}$$

gives the variance and total power of the signal. The quantity

$$\begin{aligned} S'_k &= \frac{1}{[(N+K)\Delta t]^2}\left[|Y_k|^2 + |Y_{N+K-k}|^2\right] \\ &= \frac{1}{(N+K)^2}\left|\sum_{n=0}^{N+K-1} y_n e^{-i2\pi f n \Delta t}\right|^2 \end{aligned} \tag{5.6.20}$$

is often computed as the periodogram. However, this is not correctly scaled as a power spectral density but represents the "peak" in the spectral plot rather than the "area" under the plot of S_k versus Δf. The representation (5.6.20) is sometimes useful although most oceanographers are more familiar with the power spectral density form of the periodogram. It bears repeating that the use of Fourier transforms assumes a periodic structure to the sampled data when no periodic structure may actually exist in the time series. That is, the FFT of a finite length data record is equivalent to assuming that the record is periodic. We again note that autospectral functions are always real so that $S'_{yy}(f_k) = S'_{yy}(2f_N - f_k)$, and the one-sided autospectral periodogram estimate becomes

$$G'_{yy}(f_k) = 2S'_k = \frac{2}{[(N+K)\Delta t]^2}|Y(f_k)|^2 \tag{5.6.21}$$

Until the 1960s, the direct transform method first used by Schuster (1898) to study "hidden periodicities" in measured sun-spot numbers was seldom used due to difficulties with statistical reliability and extensive computational time. The introduction of the first practical FFT algorithms for spectral analysis (Cooley and Tukey, 1965) greatly reduced the computational time by taking advantage of patterns in discrete Fourier transform functions. Problems with the statistical reliability of the spectral estimates are resolved through appropriate windowing and averaging techniques which we discuss in Sections 5.6.6 and 5.6.7. Figure 5.6.8 compares the unsmoothed periodogram spectral estimate for the monthly mean sea surface temperature data at Amphitrite Point (Table 5.6.2) with the corresponding spectrum obtained from the Blackman–Tukey method. As mentioned earlier, the FFT requires data lengths equal to powers of 2 so that we have shortened the series to $2^5 = 32$

months. As we found with the Blackman–Tukey autocorrelation method, the FFT spectrum of coastal temperatures has a strong peak near the annual period, albeit with a slightly different spectral amplitude.

5.6.3.3 The power spectral density for periodic data

For a strictly periodic digital time series $y(t)$ having an exact integer number of oscillations over the interval $[0, T]$, we can use the Fourier series expansion (5.4.12) and write

$$y(t) = \frac{1}{2}A_0 + \sum_{n=1}^{N}[A_n \cos(\omega_n t) + B_n \sin(\omega_n t)] = \frac{1}{2}C_0 + \sum_{n=1}^{N}[C_n \cos(\omega_n t + \phi_n)] \quad (5.6.22)$$

in which the constants A_n, B_n are given by equation (5.4.14) and where

$$C_n = (A_n^2 + B_n^2)^{1/2}$$
$$\phi_n = \tan^{-1}(B_n/A_n) \quad (5.6.23)$$

are the amplitude and phase of the complex Fourier coefficient for the nth frequency component, $\omega_n = 2\pi f_n$. Since the data record contains periodic components only, a plot of $2|C_n|^2$ against n ($n = 0, \ldots, N-1$) yields a series of distinct "spikes" or line spectra, S_n, with the variance divided equally between negative and positive frequencies

$$S_n = \frac{(\Delta t)^2}{T}[|C_n|^2 + |C_{N-n}|^2]$$
$$= \frac{2\Delta t}{N}|C_n|^2 \quad (5.6.24)$$

where the record mean value C_0 has been subtracted from the record $y(t)$. Here we have assumed that $y(t)$ is a real function. The squared Fourier components $|C_n|^2$ give the contribution of the nth frequency component to the total variance and the various frequency components contribute additively to the total power of the time series. The contribution from each component is assumed to be independent of that from all other components.

5.6.3.4 Variance-preserving spectra

Because the power spectral density, $S_{yy}(f)$, of a time series often ranges over orders of magnitude, spectral distributions are usually plotted as the logarithm of $S_{yy}(f)$ versus frequency or the logarithm of frequency; i.e. $\log[S_{yy}(f)]$ versus f or $\log(f)$. The latter is especially useful where a spectrum has a power law dependence of the form $S_{yy}(f) \sim f^{-p}$. In this case, the slope of the spectrum is given as $p = -\log[S_{yy}(f)]/\log(f)$.

An example of $\log[S_{yy}(f)]$ versus f (a log–linear plot) is presented in Figure 5.6.9(a) where we have used time-series data generated by the relation $y(k) = A \cdot \cos(2\pi nk/N) + \varepsilon(k)$ from Section 5.6.1 (Figure 5.6.5a). The spectral density has units of energy/frequency for the same units used for f. For example, the PSD of a current velocity record are typically in units of (cm/s)²/cph or (cm/s)²/cpd plotted against frequency in cph (cycles per hour) or cpd (cycles per day), respectively. Sometimes, m/s are used in place of cm/s. Since the integration proceeds over frequency bands of width Δf centered at frequency f_c, the area under each small rectangular segment of the spectral curve is equal to a pseudo-variance

$$\sigma_*^2(f_c) = \int_{f_c-\Delta f/2}^{f_c+\Delta f/2} \log[S_{yy}(f)]\, df \qquad (5.6.25)$$

Although log spectra plots have an appealing shape, the integral (5.6.25) is certainly not variance-preserving. To preserve the signal variance, $\sigma^2(f_c)$, under the spectral curve, we need to plot $fS_{yy}(f)$ versus $\log(f)$ (Figure 5.6.9b). Replacing df in (5.6.25) with $d[\log(f)]$, the true *variance-preserving* form of the spectrum becomes

$$\sigma^2(f_c) = \int_{f_c-\Delta f/2}^{f_c+\Delta f/2} fS_{yy}(f)\, d[\log(f)] = \int_{f_c-\Delta f/2}^{f_c+\Delta f/2} S_{yy}(f)\, df \qquad (5.6.26)$$

where we have used the fact that $d[\log(f)] = df/f$. Equation (5.6.26) gives the true signal variance within the band Δf. In particular, if $S_{yy}(f) \approx S_c$ is nearly constant over the frequency increment Δf, then $\sigma^2(f_c) \approx S_c \Delta f$ is the signal variance in band Δf centered at frequency f_c. In this format, there is a clear spectral peak at $f = 0.25$ cycles per unit time that is associated with the term $\cos(2\pi nk/N)$ in the original analytical expression.

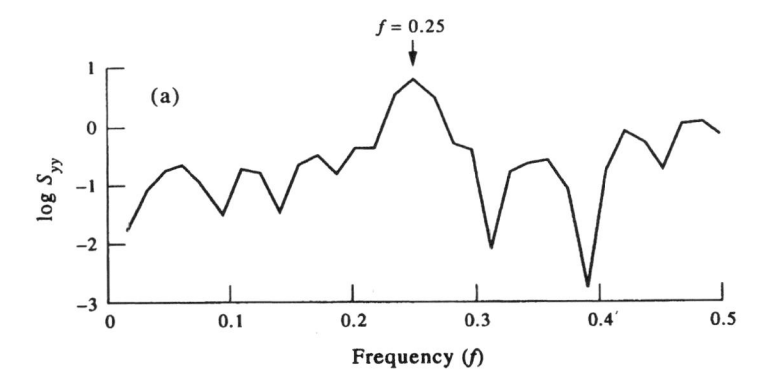

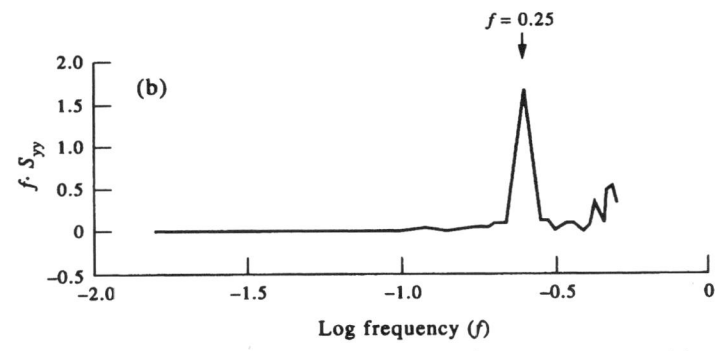

Figure 5.6.9. Two common types of spectral plot derived for the time series $y(k) = A\cos(2\pi nk/N) + \varepsilon(k)$ (see Figure 5.6.5). (a) A plot of log power spectral density, $\log[S_{yy}(f)]$, versus frequency, f; (b) A variance-preserving plot, $f \cdot ([S_{yy}(f)]$ versus $\log(f)$.

5.6.3.5 The chi-squared property of spectral estimators

Throughout this chapter, we have claimed that each spectral estimate for maximum frequency resolution, $1/T$, obtained from Fourier transforms of stochastic time series have two degrees of freedom. We now present a more formal justification for that claim for discrete spectral estimators by showing that each estimate is a stochastic chi-square variable with two degrees of freedom (i.e. there are two independent squares entering the expression for the chi-square variable). Consider any stochastic white noise process $\eta(t)$, for which $E[\eta(t)] = 0$. The Fourier components are

$$A(f) = \sum_{n=-N}^{N-1} \eta(n\Delta t) \cos(2\pi f n \Delta t)$$
$$B(f) = \sum_{n=-N}^{N-1} \eta(n\Delta t) \sin(2\pi f n \Delta t)$$
(5.6.27)

where as usual, $-1/(2\Delta t) \leq f \leq 1/(2\Delta t)$, and it follows that $E[A(f)] = 0 = E[B(f)]$. Thus, at the harmonic frequencies $f_k = k/N\Delta t$, the variance is

$$\begin{aligned} V[A(f_k)] &= E[A^2(f_k)] = \sigma_\eta^2 \sum_{n=-N}^{N-1} \cos^2(2\pi f_k n \Delta t) \\ &= \tfrac{1}{2} N \sigma_\eta^2, \quad k = \pm 1, \pm 2, \ldots, \pm(N-1) \\ &= N \sigma_\eta^2, \quad k = 0, -N \end{aligned}$$
(5.6.28a)

Similarly

$$\begin{aligned} V[B(f_k)] &= \tfrac{1}{2} N \sigma_\eta^2, \quad k = \pm 1, \pm 2, \ldots, \pm(N-1) \\ &= 0, \quad k = 0, -N \end{aligned}$$
(5.6.28b)

When $k \neq j$, the covariance is

$$C[A(f_k), A(f_j)] = \sigma_\eta^2 \sum_{n=-N}^{N-1} \cos(2\pi f_k n \Delta t) \cos(2\pi f_j n \Delta t) = 0$$
(5.6.29a)

and

$$C[A(f_k), B(f_j)] = 0 \text{ (orthogonality condition)}$$
(5.6.29b)

Because $A(f_k)$ and $B(f_k)$ are linear functions of normal random variables, $A(f_k)$ and $B(f_k)$ are also distributed normally. Hence, the random variables

$$\begin{aligned} \frac{A(f_k)^2}{V[A(f_k)]} &= \frac{2A(f_k)^2}{N\sigma_\eta^2} \\ \frac{B(f_k)^2}{V[B(f_k)]} &= \frac{2B(f_k)^2}{N\sigma_\eta^2} \end{aligned}$$
(5.6.30)

are each distributed as χ_1^2, which is a chi-square variable with one degree of freedom.

Since the normal distributions $A(f_k)$ and $B(f_k)$ are independent random variables, the sum of their squares

$$\frac{2}{\sigma_\eta^2}[A(f_k)^2 + B(f_k)^2] = \frac{2}{\Delta t \sigma_\eta^2} S_{yy}(f_k) \qquad (5.6.31)$$

is distributed as χ_2^2, which is chi-square variable with two degrees of freedom. Here, $S_{yy}(f_k)$ is the sample spectrum. Thus

$$\frac{E[2S_{yy}(f_k)]}{\Delta t \sigma_\eta^2} = 2 \qquad (5.6.32)$$

and

$$E[S_{yy}(f_k)] = \sigma_\eta^2 \Delta t \qquad (5.6.33)$$

which is the spectrum. At the harmonic frequencies (set by the record length), the sample spectrum is an unbiased estimator of the white-noise spectrum of $\eta(t)$. Also, at these frequencies, the variance of the estimate is constant and independent of sample size. This explains the failure of the sample estimates of the variance to decrease with increasing sample size. We remark further that, even if $\eta(t)$ is not normally distributed, the random variables $A(f_k)$ and $B(f_k)$ are very nearly normally distributed by the central limit theorem. Hence, the distribution of the $S_{yy}(f)$ will be very nearly distributed as χ_2^2 regardless of the PDF of the $\eta(t)$ process.

5.6.4 Spectra of vector series

To calculate the spectra of vector time series such as current and wind, we first need to resolve the data into orthogonal components. Spectral analysis is then applied to the combined series of components and the results stored as a complex quantity in the computer. Raw data are recorded as speed and direction by rotor-type meters and as orthogonal components by acoustic and electromagnetic meters. The usual procedure is to convert recorded time series to an earth-referenced Cartesian coordinate system consisting of two orthogonal horizontal components and a vertical component. In the open ocean, horizontal velocities typically are resolved into components of eastward (zonal; u) and northward (meridional; v) time series, whereas in the coastal ocean it is preferable to resolve the vector components into cross-shore (u') and longshore (v') components through the rotation

$$\begin{pmatrix} u' \\ v' \end{pmatrix} \begin{pmatrix} \cos\theta & \sin\theta \\ -\sin\theta & \cos\theta \end{pmatrix} \begin{pmatrix} u \\ v \end{pmatrix} \qquad (5.6.34a)$$

or

$$\begin{aligned} u' &= u\cos\theta + v\sin\theta \\ v' &= -u\sin\theta + v\cos\theta \end{aligned} \qquad (5.6.34b)$$

where the angle θ is the orientation of the coastline (or the local bottom contours) measured counterclockwise from the eastward direction (Figure 5.6.10). Alternatively, one can let the current data define θ as the direction of the major axis obtained from principal component analysis; that is, the axis which maximizes the variance in a scatter plot of u versus v (see Figure 4.3.1).

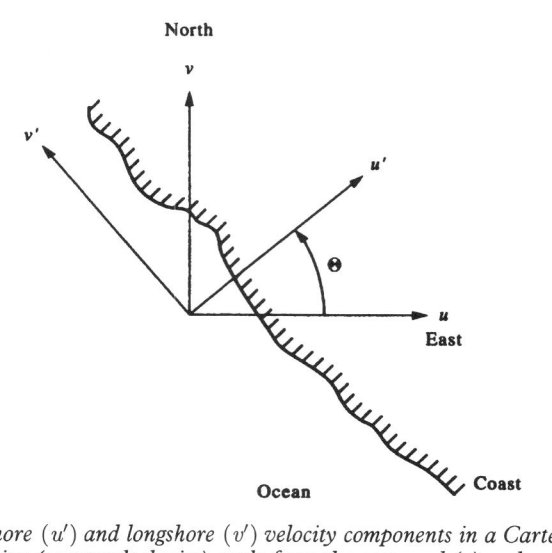

Figure 5.6.10. Cross-shore (u′) and longshore (v′) velocity components in a Cartesian coordinate system rotated through a positive (counterclockwise) angle from the eastward (u) and northward (v) directions.

In coastal regions, the principal axis is usually closely parallel to the coastline. For studies of highly circularly polarized motions, such as inertial waves and tidal currents, resolution into clockwise and counterclockwise rotary components is often more useful. The choice of representation depends on the preference of the investigator and the type of process being investigated. More will be said on this subject in Section 5.6.4.2.

5.6.4.1 Cartesian component rotary spectra

The horizontal velocity vector can be represented in Cartesian coordinates as a complex function $w(t)$ whose real part, $u(t)$, is the projection of the vector on the zonal (or cross-shelf) axis and whose imaginary part, $v(t)$, is the projection of the vector on the meridional (or longshelf) axis (Figure 5.6.11)

$$w(t) = u(t) + \mathrm{i}v(t) \tag{5.6.35}$$

(The use of vector $w(t)$ follows the convention of Gonella (1972), Mooers (1973) and others in their discussion of rotary spectral analysis and is not to be confused with the weights $w(t)$ used in the sections on data windowing or the vertical component of velocity. Gonella (1972) used u_1 and u_2 for the two velocity components.) A complete description of the time variability of a three-dimensional vector at a single point consists of six functions of frequency: Three autospectra for the three velocity components and three cross-spectra. For the two-dimensional vectors considered in this section, there are two autospectra and one cross-spectrum. The discrete Fourier transform, $W(f_k) = U(f_k) + \mathrm{i}V(f_k)$, $(f_k = k/N\Delta t,\ k = 1, \ldots, N;\ k = 0$ is the mean flow) is

$$\begin{aligned} W(f_k) &= \Delta t \sum_{n=0}^{N-1} w(t) \mathrm{e}^{-\mathrm{i}2\pi k n/N} \\ &= \Delta t \sum_{n=0}^{N-1} [u(t) + \mathrm{i}v(t)] \mathrm{e}^{-\mathrm{i}2\pi k n/N} \end{aligned} \tag{5.6.36}$$

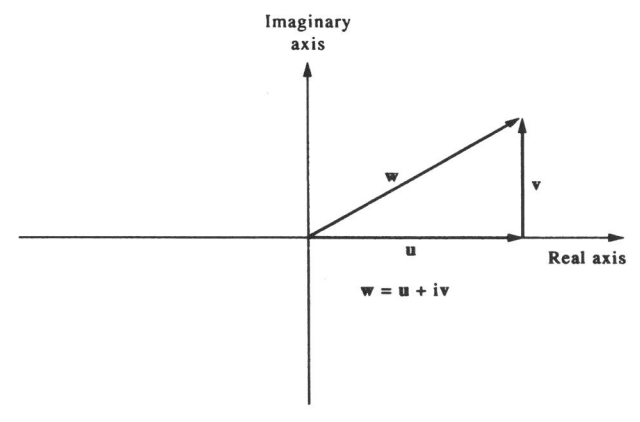

Figure 5.6.11. Horizontal velocity, w, represented as a complex vector $w = u + iv$ with components (u, v) along the real and imaginary axis, respectively.

where $U(f_k)$ and $V(f_k)$ are the Fourier transforms of $u(t)$ and $v(t)$, respectively. If the original record is separated into M blocks of length N', where $N = MN'$ is the total record length if no overlapping of segments is used, the spectral density function is given in terms of the number of segments used to form the block-averaged, one-sided autospectrum ($0 \leq f'_k < \infty$)

$$\begin{aligned} G_{ww}(f'_k) &= \frac{2}{N\Delta t} \sum_{m=1}^{M} |W_m(f'_k)|^2 \\ &= \frac{2}{N\Delta t} \sum_{m=1}^{M} \left\{ [W_{Rm}(f'_k)]^2 + [W_{Im}(f'_k)]^2 \right\} \\ &= \frac{2}{N\Delta t} \sum_{m=1}^{M} \left\{ [U_{Rm}(f'_k) - V_{Im}(f'_k)]^2 + [U_{Im}(f'_k) + V_{Rm}(f'_k)]^2 \right\} \end{aligned} \quad (5.6.37)$$

where $f'_k = k/N'\Delta t$, $k = 0, 1, ..., N'/2$ ($k = 0$ is the mean flow) and for FFT analysis, $N' = 2p$ (positive integer p), and where the subscripts R and I stand for the real and imaginary parts of the given Fourier components.

5.6.4.2 Rotary component spectra

Rotary analysis of currents involves the separation of the velocity vector for a specified frequency, ω, into clockwise and counterclockwise rotating circular components with amplitudes A^-, A^+ and relative phases θ^-, θ^+, respectively. Thus, instead of dealing with two Cartesian components (u, v) we deal with two circular components $(A^-, \theta^-; A^+, \theta^+)$. Several reasons can be given for using this approach: (1) the separation of a velocity vector into oppositely rotating components can reveal important aspects of the wave field at the specified frequencies. The method has proven especially useful for investigating currents over abrupt topography, wind-generated inertial motions, diurnal frequency continental shelf waves, and other forms of narrow-band oscillatory flow; (2) in many cases, one of the rotary components (typically, the clockwise component in the northern hemisphere and counterclockwise

428 *Data Analysis Methods in Physical Oceanography*

component in the southern hemisphere) dominates the currents so that we need only deal with one scalar quantity rather than two. Inertial motions, for example, are almost entirely clockwise rotary in the northern hemisphere so that the counter-clockwise component can be ignored for most applications; (3) many of the rotary properties, such as spectral energy $S^-(\omega)$ and $S^+(\omega)$ and rotary coefficient, $r(\omega)$, are invariant under coordinate rotation so that local steering of the currents by bottom topography or the coastline are not factors in the analysis.

The vector addition of the two oppositely rotating circular vectors (Figure 5.6.12a, b) causes the tip of the combined vector (Figure 5.6.12c) to trace out an ellipse over one complete cycle. The eccentricity, e, of the ellipse is determined by the relative amplitudes of the two components. Motions at frequency ω are circularly polarized if one of the two components is zero; motions are rectilinear (back-and-forth along the same line) if both circularly polarized components have the same magnitude. In rotary spectral format, the current vector $w(t)$ can be written as the Fourier series

$$w(t) = \overline{u(t)} + \sum_{k=1}^{N} U_k \cos(\omega_k t - \phi_k) + i\left[\overline{v(t)} + \sum_{k=1}^{N} V_k \cos(\omega_k t - \theta_k)\right]$$
$$= [\overline{u(t)} + i\overline{v(t)}] + \sum_{k=1}^{N} [U_k \cos(\omega_k t - \phi_k) + iV_k \cos(\omega_k t - \theta_k)]$$
(5.6.38)

in which $\overline{u(t)} + i\overline{v(t)}$ is the mean velocity, $\omega_k = 2\pi f_k = 2\pi k/N\Delta t$ is the angular frequency, $t \,(= n\Delta t)$ is the time and (U_k, V_k) and (ϕ_k, θ_k) are the amplitudes and phases, respectively, of the Fourier constituents for each frequency for the real and imaginary components. Subtracting the mean velocity and expanding the trigonometric functions, we find

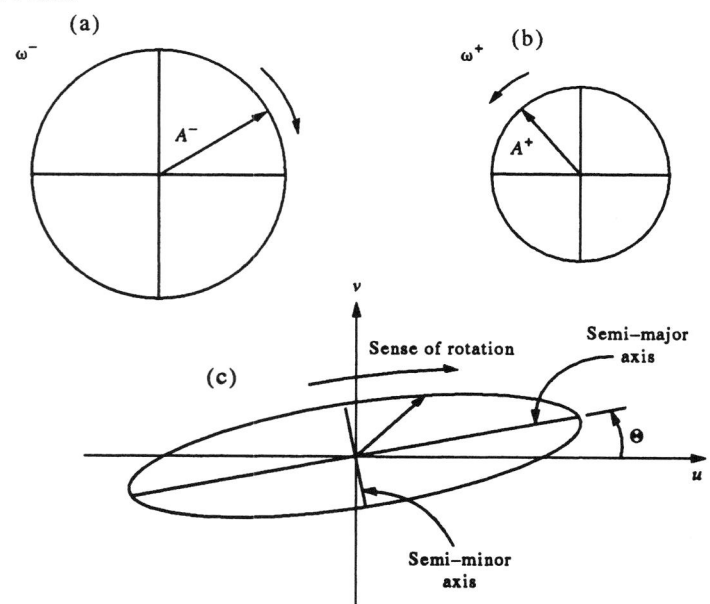

Figure 5.6.12. Current ellipses formed by the vector addition of two oppositely rotating vectors. (a) Clockwise component (ω^-) and (b) counterclockwise component (ω^+) with amplitudes A^- and A^+, respectively. (c) General case of elliptical motion with major axis tilted at an angle θ counterclockwise from east. ε^- and ε^+ (not shown) are the angles of the two circular components at time $t = 0$.

$$w'(t) = w(t) - \left[\overline{u(t)} + i\overline{v(t)}\right]$$
$$= \sum_{k=1}^{N} \{U_{1k} \cos(\omega_k t) + U_{2k} \sin(\omega_k t) + i[V_{1k} \cos(\omega_k t) + V_{2k} \sin(\omega_k t)]\} \quad (5.6.39)$$

in which we have defined the even (U_{1k}, V_{1k}) and odd (U_{2k}, V_{2k}) functions as
$$U_{1k} = U_k \cos \phi_k, \quad U_{2k} = U_k \sin \phi_k \quad (5.6.40a)$$
$$V_{1k} = V_k \cos \theta_k, \quad V_{2k} = V_k \sin \theta_k \quad (5.6.40b)$$

Dropping the prime notation for $w'(t)$ and following some reorganization, we can write the kth frequency component of the series as the sum of counterclockwise (+) and clockwise (−) components

$$\begin{aligned} w_k(t) &= w_k^+(t) + w_k^-(t) \\ &= A_k^+ \exp(i\varepsilon_k^+) \exp(i\omega_k t) + A_k^- \exp(i\varepsilon_k^-) \exp(-i\omega_k t) \\ &= \exp\left[\frac{i(\varepsilon_k^+ + \varepsilon_k^-)}{2}\right] \left\{[A_k^+ + A_k^-] \cos\left[\frac{\varepsilon_k^+ - \varepsilon_k^-}{2} + \omega_k t\right] \right. \\ &\quad \left. + i[A_k^+ - A_k^-] \sin\left[\frac{\varepsilon_k^+ - \varepsilon_k^-}{2} + \omega_k t\right]\right\} \end{aligned} \quad (5.6.41)$$

where the counterclockwise and clockwise rotary component amplitudes are given by
$$A_k^+ = \frac{1}{2}\left\{[(U_{1k} + V_{2k})]^2 + [(U_{2k} - V_{1k})]^2\right\}^{1/2} \quad (5.6.42a)$$
$$A_k^- = \frac{1}{2}\left\{[(U_{1k} - V_{2k})]^2 + [(U_{2k} + V_{1k})]^2\right\}^{1/2} \quad (5.6.42b)$$

and the corresponding phase angles for time $t = 0$, by
$$\varepsilon_k^+ = \tan^{-1}[(V_{1k} - U_{2k})/(U_{1k} + V_{2k})] \quad (5.6.43a)$$
$$\varepsilon_k^- = \tan^{-1}[(U_{2k} + V_{1k})/(U_{1k} - V_{2k})] \quad (5.6.43b)$$

Each of the constituents contributing to equation (5.6.39) have the form of an ellipse with major semi-axis of length $L_M = (A_k^+ + A_k^-)$ and minor semi-axis of length $L_m = |A_k^+ - A_k^-|$ (Figure 5.6.12c). The ellipse is tilted at an angle of $\theta = \frac{1}{2}(\varepsilon_k^+ + \varepsilon_k^-)$ from the u-axis and the vector is along the major axis of the ellipse at time $t = (\varepsilon_k^- - \varepsilon_k^+)/(4\pi f_k)$. The one-sided spectra $(G_k^+, G_k^-) = (S_k^+, S_k^-)$ for the two oppositely rotating components for frequencies $f_k = \omega_k/2\pi$ are

$$S(f_k^+) = S_k^+ = \frac{(A_k^+)^2}{N\Delta t}, \quad f_k = 0, \ldots, 1/(2\Delta t) \quad (5.6.44a)$$

$$S(f_k^-) = S_k^- = \frac{(A_k^-)^2}{N\Delta t}, \quad f_k = -1/(2\Delta t), \ldots, 0 \quad (5.6.44b)$$

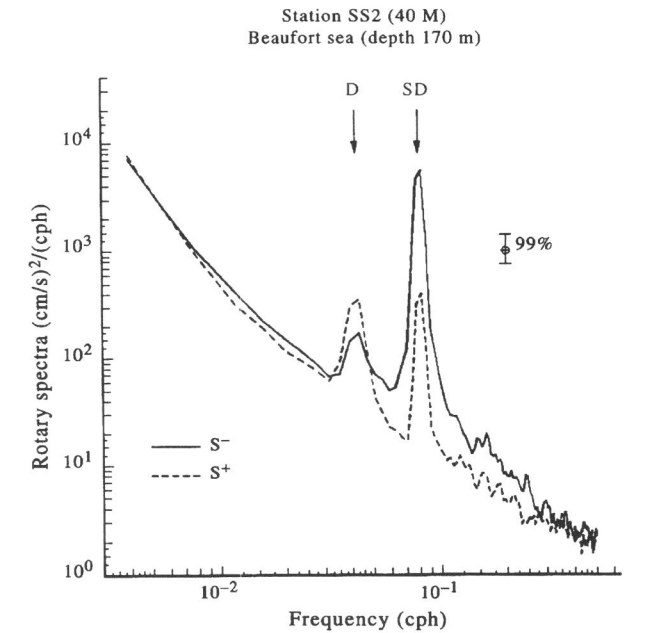

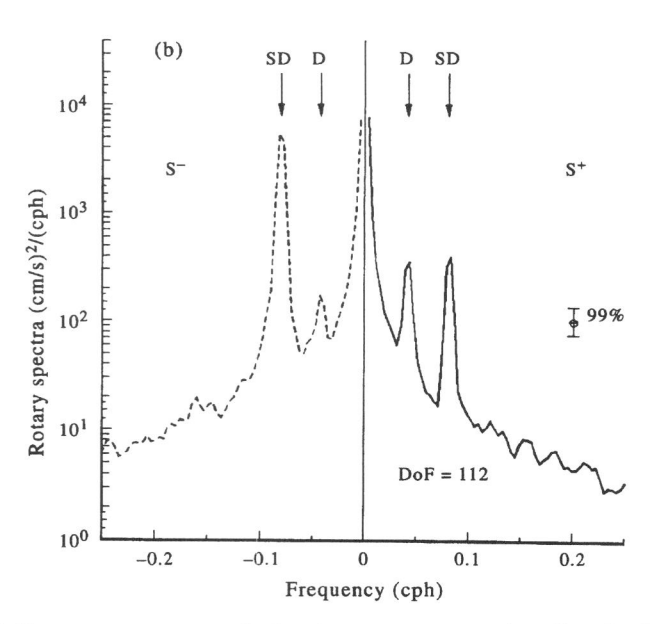

Figure 5.6.13. Rotary current spectra for hourly currents measured at 40-m depth in the Beaufort Sea, Arctic Ocean (water depth = 170 m). Peaks are at the diurnal (D) and semidiurnal (SD) tidal frequencies. Frequency resolution is 0.0005 cph and there are 112 degrees of freedom per spectral band. Vertical bar gives the 99% level of confidence. (a) One-sided rotary spectra, $S^-(f)$ and $S^+(f)$, versus f for positive frequency, f; (b) two-sided rotary spectra, $S(f_k^+)=S^+$ and $S(f_k^-)=S^-$ versus $\log f$ for positive and negative frequencies, $f_k^\pm$. (Courtesy E. Carmack, A. Rabinovich, and E. Kolikov.)

Plots of rotary spectra are generally presented in two ways. In Figure 5.6.13(a), both S^- and S^+ are plotted as functions of frequency magnitude, $|f| \geq 0$, with solid and dashed lines used for the clockwise and counterclockwise spectra, respectively. In Figure 5.6.13(b), we use the fact that clockwise spectra are defined for negative frequencies and counterclockwise spectra for positive frequencies. The spectra $S(f_k^+)$ and $S(f_k^-)$ used in Figure 5.6.13(a) are then plotted on opposite sides of zero frequency. In these spectra, peak energy occurs at the diurnal and semidiurnal periods. The predominantly clockwise rotary motions at semidiurnal periods suggest a combination of tidal and near-inertial motions (at this latitude the inertial period is close to the semidiurnal tidal period).

Another useful property is the rotary coefficient

$$r(\omega) = \frac{S_k^+ - S_k^-}{S_k^+ + S_k^-} \qquad (5.6.45)$$

which ranges from $r = -1$ for clockwise motion, to $r = 0$ for unidirectional flow, to $r = +1$ for counterclockwise motion. The rotary nature of the flow can change considerably with position, depth and time. As indicated by Figure 5.6.14, the observed diurnal tidal currents over Endeavour Ridge in the northeast Pacific change from moderately positive to strongly negative rotation with depth. In contrast, the semidiurnal currents change from strongly negative near the surface to strongly rectilinear at depth. (Data, in this case, are from a string of current meters moored for a period of nine months in the northeast Pacific.) We remark that the definition (5.6.45) differs in sign from that of Gonella (1972) who used $S_k^- - S_k^+$ rather than $S_k^+ - S_k^-$ in the numerator. Because many types of oceanic flow are predominantly clockwise rotary in the northern hemisphere, Gonella's definition has the advantage that clockwise rotating currents have positive rotary coefficients. However, we find Gonella's definition confusing since clockwise motions, which are linked to negative frequencies, then have positive rotary coefficients.

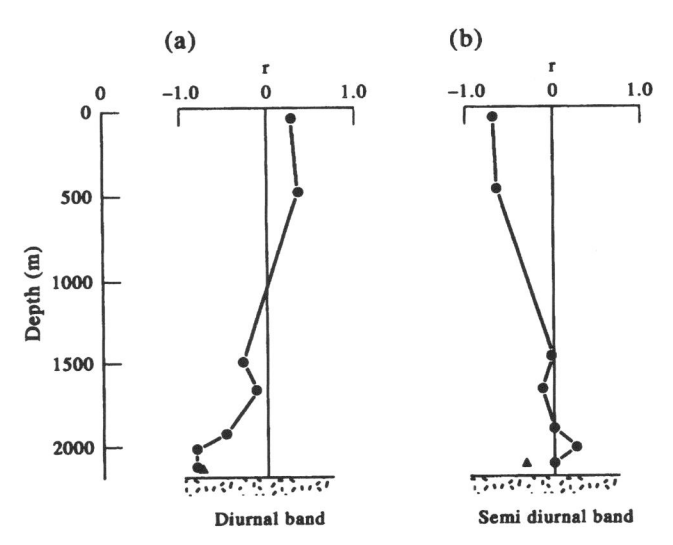

Figure 5.6.14. Rotary coefficient, $r(\omega)$, as a function of depth for current oscillations in (a) the diurnal frequency band ($\omega/2\pi \approx 0.04$ cph) and (b) the semidiurnal band ($\omega/2\pi \approx 0.08$ cph). (From Allen and Thomson, 1993.)

5.6.4.3 Rotary spectra (via Cartesian components)

Gonella (1972) and Mooers (1973) present the rotary spectra in terms of their Cartesian counterparts and provide a number of rotational invariants for analyzing current and wind vectors at specified frequencies. Specifically, the one-side autospectra for the counterclockwise (CCW) and clockwise (CW) rotary components of the vector $w(t) = u(t) + iv(t)$ are, in terms of their Cartesian components

$$G(f_k^+) = \tfrac{1}{2}[G_{uu}(f_k) + G_{vv}(f_k) + Q_{uv}(f_k)], \quad f_k \geq 0 \text{ (CCW component)} \quad (5.6.46a)$$

$$G(f_k^-) = \tfrac{1}{2}[G_{uu}(f_k) + G_{vv}(f_k) - Q_{uv}(f_k)], \quad f_k \leq 0 \text{ (CW component)} \quad (5.6.46b)$$

where $G_{uu}(f_k)$ and $G_{vv}(f_k)$ are the one-sided autospectra of the u and v Cartesian components of velocity and $Q_{uv}(f_k)$ is the quadrature spectrum between the two components, where

$$Q_{uv}(f_k) = -Q_{uv}(-f_k) = (U_{1k}V_{2k} - V_{1k}U_{2k}) \quad (5.6.47)$$

As defined in Section 5.8, the spectrum can be written in terms of co-spectrum (real part) and quadrature spectrum (imaginary part)

$$G_{uv}(f_k) = C_{uv}(f_k) - iQ_{uv}(f_k) \quad (5.6.48)$$

5.6.5 Effect of sampling on spectral estimates

Spectral estimates derived by conventional techniques are limited by two fundamental problems: (1) the finite length, T, of the time series; and (2) the discretization using the sampling interval, Δt. The first problem is inherent to all real datasets while the second is associated with finite instrument response times and/or the need to digitize the time series for purposes of analysis.

Irrespective of the method used to calculate the power spectrum of a waveform, the record duration $T = N\Delta t$ and sampling increment Δt impose severe limitations on the information that can be extracted. Ideally, we would like to sample rapidly enough (small Δt) that no significant frequency component goes unresolved. This also eliminates aliasing problems in which unresolved spectral energy at frequencies higher than the Nyquist frequency is folded back into lower frequencies. At the same time we wish to record for a sufficiently long period (large N) that we capture many cycles of the lowest frequency of interest. Long-term sampling also enables us to better resolve frequencies that are close together and to improve the statistics (confidence intervals) for spectral estimates. In reality, most data series are a compromise based on the frequencies of interest, the response limitations of the sensor, and cost. The choices of the sampling rate and the record duration are tailored to best meet the task at hand.

5.6.5.1 Effect of finite record length

As noted earlier, we can think of a data sample $\{y(t)\}$ of duration $T = N\Delta t$ as the output from an infinite physical process $\{y'(t)\}$ viewed through a finite length window (Figure 5.6.1). The window has the shape of a "box-car" function $w(t_n) = w_n =$

$w(n\Delta t)$ which has unit amplitude and zero phase lag over the duration of the data sequence but is zero elsewhere. That is $y(t_n) = w(t_n)y'(t_n)$ where

$$w_n = 1.0, \quad n = 0, ..., N-1$$
$$w_n = 0, \quad \text{for } n \geq N, n < 0 \tag{5.6.49}$$

Since it is truncated, the dataset has endpoint discontinuities which lead to Gibbs' phenomena "ringing" and the ripple effects in the frequency domain. The discrete Fourier transform $Y(f)$ of the truncated series $y_n = y(n\Delta t)$ is

$$Y(f) = \sum_{n=-\infty}^{\infty} w_n y'_n e^{-i2\pi fn\Delta t} \tag{5.6.50}$$

In frequency space, $Y(f)$ is the convolution (written as $*$) of the Fourier transform of the infinite data set, $Y'(f)$, with the Fourier transform $W(f)$ of the function $w(t)$. That is

$$Y(f) = \int_{-\infty}^{\infty} Y'(f')W'(f-f')\,df$$
$$= Y'(f) * W(f) \tag{5.6.51}$$

where for a box-car function

$$W(f) = T\exp(i\pi fT)\frac{\sin(\pi fN\Delta t)}{(\pi fN\Delta t)}$$
$$= T\exp(i\pi fT)\,\text{sinc}(\pi fN\Delta t) \tag{5.6.52}$$

and $\text{sinc}(x) \equiv \sin(x)/x$. It is the large side-lobes or ripples of the sinc function (Figure 5.6.15) which are responsible for the leakage of spectral power from the main frequency components into neighboring frequency bands (and vice versa). In particular, $Y(f)$ for a specific frequency $f = f_o$ is spread to other frequencies, f, according to the phase and amplitude weighting of the window function. Leakage has the effect of both reducing the spectral power in the central frequency component and contaminating it with spectral energy from adjacent frequency bands. Those familiar with the various mathematical forms for the Dirac delta function, $\delta(f)$, will recognize the formulation

$$\delta(f) = \lim_{f \to 0}\left[\frac{\sin(\pi f\Delta t)}{\pi f\Delta t}\right]$$

Thus, as the frequency resolution increases (i.e. $f \to 0$), $Y(f) \to Y'(f)$.

In addition to distorting the spectrum, the box-car window limits the frequency resolution of the periodogram, independently of the data. The convolution $Y'(f) * W(f)$ means that the narrowest spectral response of the resultant transform is confined to the main-lobe width of the window transform. For a given window, the main-lobe width (the width between the -3 dB levels of the main lobe) determines the frequency resolution, Δf, of a particular window. For most windows, including the box-car window, this resolution is roughly the inverse of the observation time; $\Delta f \approx 1/T = 1/N\Delta t$.

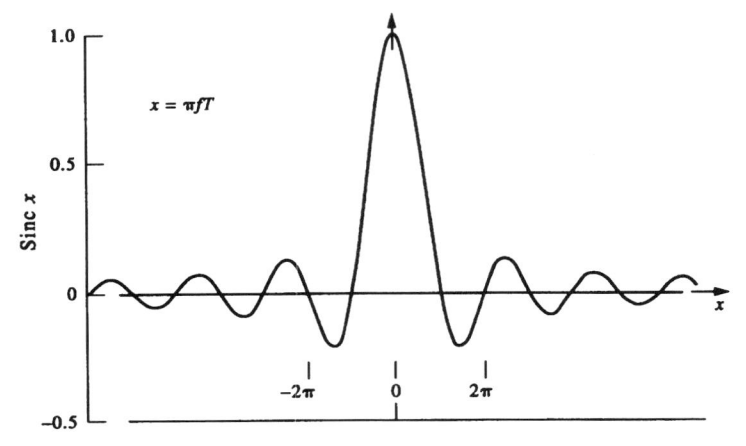

434 *Data Analysis Methods in Physical Oceanography*

Figure 5.6.15. The function sinc (x) = sin (x)/x showing the large side-lobes which are responsible for leakage of spectral power from a given frequency to adjacent frequencies.

5.6.5.2 Aliasing

Poor discretization of time-series data due to limitations in the response time of the sensor, limitations in the recording and data storage rates, or through post-processing methods may cause *aliasing* of certain frequency components in the original waveform (Figure 5.6.16a). An aliased frequency is one that masquerades as another frequency. In Figure 5.6.16(b), for example, the considerable tidal energy at diurnal and semidiurnal periods (1 and 2 cpd) is folded back to lower frequencies of 0.07 and 0.10 cpd that are nowhere near the original frequencies. For a specific sampling interval, it becomes impossible to tell with certainty which frequency out of a large number of possible aliases is actually contributing to the signal variability. This leads to differences in the spectra between the continuous and discrete time series. Since we use the spectra of the discrete series to estimate the spectrum of the continuous series, the sampling interval must be properly selected to minimize the effect of the aliasing. If we know from previous analysis that there is little likelihood of significant energy at the disguised frequencies, then aliasing is not a problem. Otherwise, a degree of smoothing may be required to ensure that higher frequencies do not contaminate the lower frequencies. This smoothing must be performed prior to sampling or digitizing since aliased contributions cannot be recognized once they are present in the discrete data series.

The aliasing problem can be illustrated in a number of ways. To begin with, we note that for discrete data at equally spaced intervals Δt, we can measure only those frequency components lying within the principal frequency range,

$$-\omega_N \leq \omega \leq -\omega_o, \quad \omega_o \leq \omega \leq \omega_N, \quad \omega_N \geq 0 \quad (5.6.53a)$$

$$-f_N \leq f \leq -f_o, \quad f_o \leq f \leq f_N, \quad f_N \geq 0 \quad (5.6.53b)$$

in which $\omega_N = \pi/\Delta t$ and $f_N = 1/(2\Delta t)$ are the usual Nyquist frequencies in radians and cycles per unit time, respectively, and $\omega_o = 2\pi/T$ and $f_o = 1/T$ are corresponding fundamental frequencies for a time series of duration T. The Nyquist frequency is the highest frequency that can be extracted from a time series having a sampling rate of $1/\Delta t$. Clearly, if the original time series has spectral power at frequencies for which

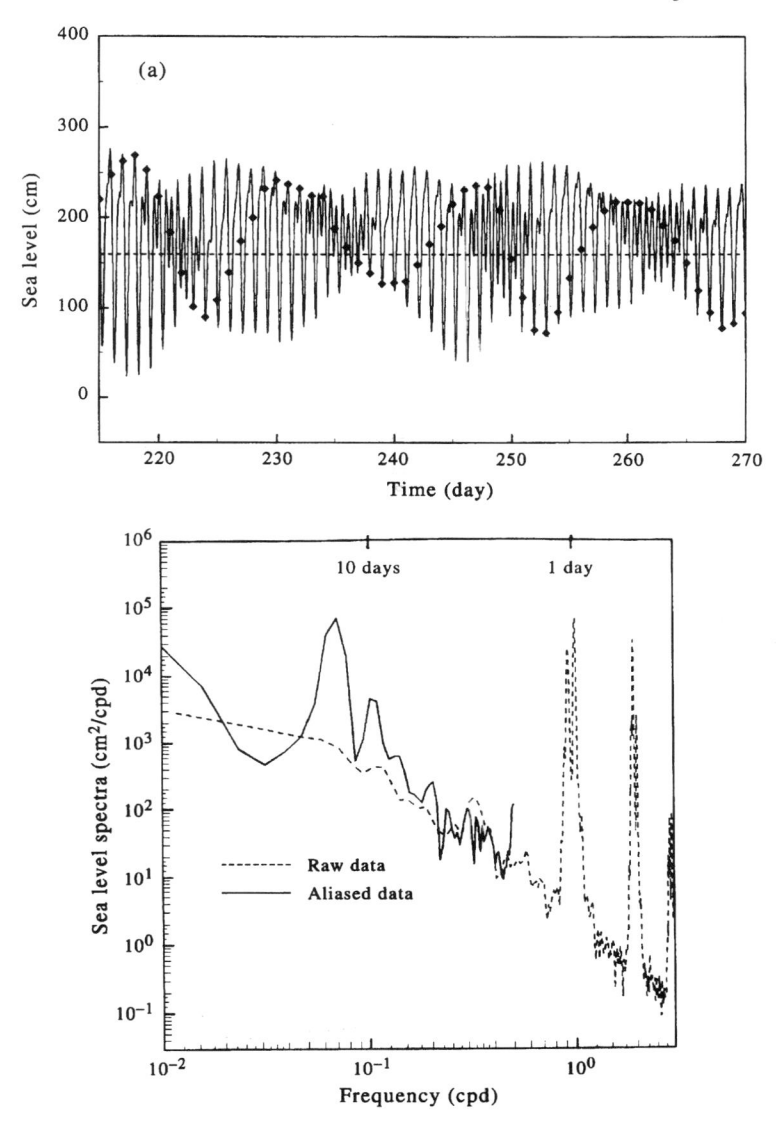

Figure 5.6.16. The origin of aliasing. (a) The solid line is the tide height recorded at Victoria, British Columbia over a 60-day period from 29 July to 27 September 1975 (time in Julian days). The diamonds are the sea-level values one would obtain by only sampling once per day. (b) The power spectrum obtained from the two data series in (a). In this case, the high frequency energy (dashed curve) gets folded back into the spectrum at lower (aliased) frequencies (solid curve).

$|f| \geq f_N$, these spectral contributions are unresolved and will contaminate power associated with frequencies within the principal range (Figure 5.6.17). The unresolved variance becomes lumped together with other frequency components. Familiar examples of aliasing are the slow reverse rotation of stage-coach wheels in classic western movies due to the under-sampling by the frame-rate of the movie camera. Even in modern film, distinguishable features on moving automobile tires often can be seen to rotate rapidly backwards, slow to a stop, then turn forward at the correct rotation speed as the vehicle gradually comes to a stop. Automobile commercials avoid this problem by equipping the wheels with featureless hubcaps.

If $\omega, f \geq 0$ are frequencies inside the principal intervals (5.6.53), the frequencies outside the interval which form aliases with these frequencies are

$$2\omega_N \pm \omega, \quad 4\omega_N \pm \omega, \quad ..., \quad 2p\omega_N \pm \omega \quad (5.6.54a)$$

$$2f_N \pm f, \quad 4f_N \pm f, \quad ..., \quad 2pf_N \pm f \quad (5.6.54b)$$

where p is a positive integer. These results lead to the alternate term *folding* frequency for the Nyquist frequency since spectral power outside the principal range is folded back, accordion-style, into the principal interval. As illustrated by Figure 5.6.17, folding the power spectrum about f_N produces aliasing of frequencies $2f_N - f$ with frequencies f; folding the spectrum at $2f_N$ produces aliasing of frequencies $2f_N + f$ with frequencies $2f_N - f$ which are then folded back about f_N into frequency f, and so forth. For example, if $f_N = 5$ rad/h, the observations at 2 rad/h are aliased with spectral contributions having frequencies of 8 and 12 rad/h, 18 and 22 rad/h, and so on.

We can verify that oscillations of frequency $2p\omega_N \pm \omega$ (or $2pf_N \pm f$) are indistinguishable from frequency ω (or f) by considering the data series $x_\omega(t)$ created by the single frequency component $x_\omega(t) = \cos(\omega t)$. Using the transformation $\omega \to (2p\omega_N \pm \omega)$, together with $t_n = n\Delta t$ and $\omega_N = \pi/\Delta t$, yields

$$\begin{aligned} x_\omega(t_n) &= \cos\left[(2p\omega_N \pm \omega)t_n\right] = \text{Re}\left\{\exp\left[i(2p\omega_N \pm \omega)t_n\right]\right\} \\ &= \text{Re}\left\{\exp\left[i2p\omega_N t_n\right]\exp\left[\pm i\omega t_n\right]\right\} \quad (5.6.55) \\ &= (+1)^{pn}\text{Re}\left[\exp(\pm i\omega t_n)\right] = \cos(\omega t_n) = x_\omega(t_n) \end{aligned}$$

In other words, the spectrum of $x(t)$ at frequency ω will be a superposition of spectral contributions from frequencies $\omega, 2p\omega_N \pm \omega, 4p\omega_N \pm \omega$, and so forth. More specifically, it can be shown that the aliased spectrum $S_a(\omega)$ for discrete data is given by

$$S_a(\omega) = \sum_{n=-\infty}^{\infty} S(\omega + 2n\omega_N) \quad (5.6.56a)$$

$$= S(\omega) + \sum_{n=1}^{\infty} \left[S(2n\omega_N - \omega) + S(2n\omega_N + \omega)\right] \quad (5.6.56b)$$

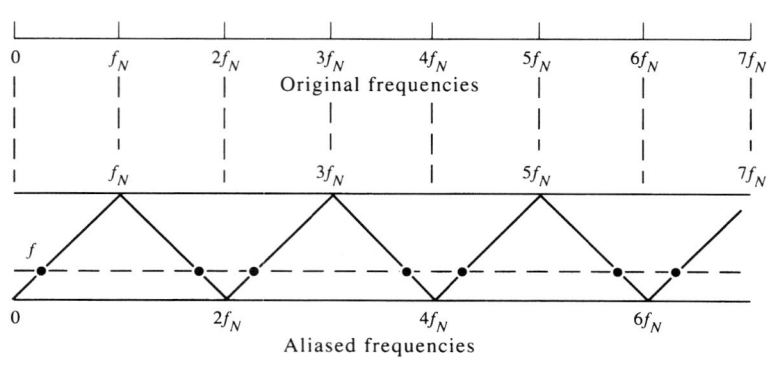

Figure 5.6.17. The spectral energies of all frequencies $f = \omega/2\pi$ at the nodes (•) located along the dotted line are folded back, accordion style, into the spectral estimate for the spectrum $S(f)$ for the primary range $0 \leq f \leq f_N$ ($0 \leq \omega \leq \omega_N$). (Adapted from Bendat and Piersol, 1986.)

The true spectrum, S, gives the distorted spectrum, S_a, caused by the summation of overlapping copies of measured spectra in the principal interval. Only if the original record is devoid of spectral power at frequencies outside the principal frequency range will the spectrum of the observed record equal that of the actual oceanic variability. To avoid aliasing problems, one has no choice but to sample the data as frequently as justifiably possible (i.e. up to frequencies beyond which energy levels become small) or to filter the sampled data before they are recorded (as in the case of a stilling well used to eliminate gravity waves from a tidal record). A further example of spectral contamination by aliased frequencies is illustrated in Figure 5.6.18(a, b). In Figure 5.6.18(b), we have assumed that the wave recorder was inadvertently programmed to record at 0.15 Hz, corresponding to a limiting wave period of 6.67 s. The energy from the shorter period waves were not measured but contaminate the energy of the longer period waves when folded back about the Nyquist frequency.

5.6.5.3 Nyquist frequency sampling

Sampling time series that have significant variability at the Nyquist frequency affords its own set of problems. Suppose we wish to represent $y(t)$ through the usual Fourier relation

$$y(t) = \int_{-\omega_N}^{\omega_N} Y(\omega) e^{i\omega t} \, d\omega \tag{5.6.57}$$

where we have assumed that $Y(\omega) = 0$ for $|\omega| > \omega_N$. In this case, there is no aliasing problem since there is no power at frequencies greater than ω_N. The function $y(t)$ can be constructed from frequency components strictly in the interval $(-\omega_N, \omega_N)$. In discrete form for infinite length data

$$y(t) = \frac{1}{2\omega_N} \sum_{n=-\infty}^{\infty} \left[y_n \int_{-\omega_N}^{\omega_N} e^{i\omega(t-n\Delta t)} \, d\omega \right] \tag{5.6.58a}$$

where the integral has the form of a sinc function such that

$$y(t) = \sum_{n=-\infty}^{\infty} y_n \frac{\sin[\omega_N(t - n\Delta t)]}{t - n\Delta t} \tag{5.6.58b}$$

Given the data $\{y_n\}$, we can construct $y(t)$. However, suppose that $y(t)$ fluctuates with the Nyquist frequency ω_N such that

$$y(t) = y_o \cos(\omega_N t + \theta) \tag{5.6.59}$$

where, for the sake of generality, the phase angle is arbitrary, $0 \leq \theta \leq 2\pi$. Then, using $\sin(n\pi) = 0$ for all n (an integer)

$$\begin{aligned} y_n &= y(n\Delta t) = y_o \cos(n\pi + \theta) = y_o[\cos(n\pi)\cos\theta] \\ &= y_o(-1)^n \cos\theta \end{aligned} \tag{5.6.60}$$

This leads to a component with amplitude $y_n = y_o(-1)^n \cos\theta$ which fluctuates in sign because of the term $(-1)^n$, $-\infty \leq n \leq \infty$. If θ is unknown, the function $y(t)$ cannot be

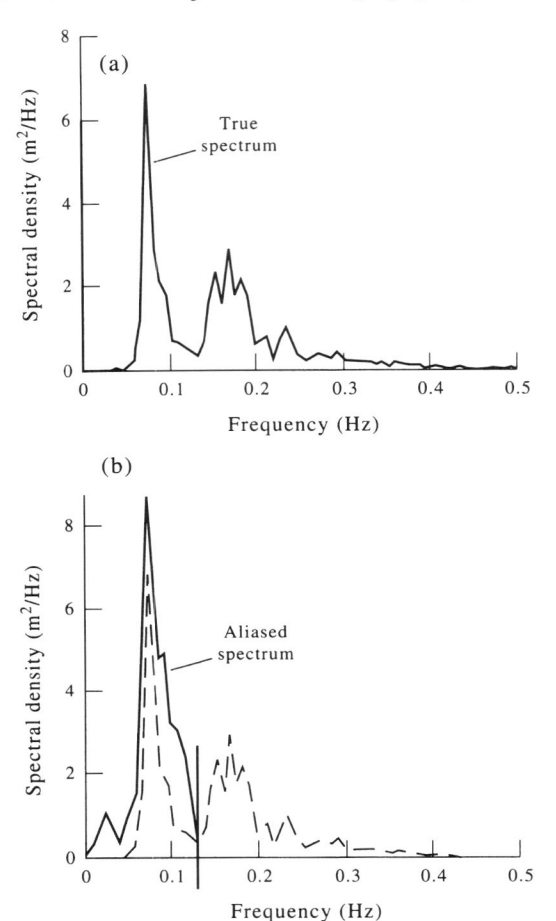

438 Data Analysis Methods in Physical Oceanography

Figure 5.6.18. An aliased autospectrum. (a) The true spectrum, S(f) (m²/cps), of wind-generated waves as a function of frequency (Hz = cycles per second); (b) Aliased spectrum, $S_a(f)$, that would arise from folding about a hypothetical Nyquist frequency $f_N = 0.13$ Hz.

constructed. If $\theta = k\pi/2$, so that $\cos(\omega_N t + \theta) = \sin(\omega_N t)$, the observer will find no signal at all. In general, $0 \leq |\cos \theta| < 1$ and the magnitude will always be less than y_o, resulting in biased data.

According to the above analysis, we should sample slightly more frequently than Δt if we are to fully resolve oscillations at the maximum frequency of interest (assumed to be the Nyquist frequency). A sampling rate of 2.5 samples per cycle of the frequency of interest appears to be acceptable whereby $\Delta t = 1/(2.5 f_N) = (2/5)(1/f_N) = (4/5)\pi/\omega_N$.

5.6.5.4 Frequency resolution

The need to resolve spectral estimates in neighboring frequency bands is an important requirement of time series analysis. Without sufficient resolution, it is not possible to determine whether a given spectral peak is associated with a single frequency, or is a smeared response containing a number of separate spectral peaks. A good example of this for tides is presented by Munk and Cartwright (1966) who show that for long records the main constituents in the diurnal and semidiurnal frequency bands can be resolved into a multitude of other tidal frequencies. How well the peaks can be resolved depends on the frequency differences, Δf, between the peaks and the length, T, of the data set used in the analysis. For an unsmoothed periodogram, the frequency resolution in hertz is roughly the reciprocal of the time duration in seconds of the data.

The distinction between well-resolved and poorly resolved spectral estimates is somewhat subjective and depends on how we wish to define "resolution". As with diffraction patterns in classical optics, we can follow the "Rayleigh criterion" for the separation of spectral peaks (Jenkins and White, 1957). Recall that the diffraction pattern for a given frequency, f, of light varies as $\text{sinc}(\phi) = \sin[(\phi - \phi_f)]/(\phi - \phi_f)$, where ϕ is the angle of the incident light beam to the grating. This also is the functional form for the spectral peak of a truncated time series (see *windowing* in the next section). Two spectral lines are said to be "well resolved" if the separation between peaks exceeds the difference in frequency between the center frequency to the maximum at the first side-lobe and "just resolved" if the spectral peak of one pattern coincides with the first zero of the second pattern (Figure 5.6.19a–c). Here, the separation in frequency is equal to the difference in frequency between the peak of one spectrum and the first zero of the function $\sin(\phi)/\phi$ of the second (where $\phi = \omega T/2$). The spectral peaks are "not resolved" if this separation is less than the separation between the center frequency and the first zero of the $\sin(\phi)/\phi$ functions (Figure 5.6.19d).

Consider an oceanic record consisting of two sinusoidal components, both having amplitude y_o and constant phase lags such that

$$y(t) = y_o[\cos(\omega_1 t + \theta_1) + \cos(\omega_2 t + \theta_2)], \quad -T/2 \le t \le T/2 \tag{5.6.61}$$

where as usual $\omega = 2\pi f$. The one-sided, unsmoothed power spectral density, $S(\omega)$, for these data are then found from the Fourier transform

$$S(\omega) = \tfrac{1}{2} T y_o^2 \left\{ \frac{\sin\left[\tfrac{1}{2}T(\omega - \omega_1)\right]}{\left[\tfrac{1}{2}T(\omega - \omega_1)\right]} + \frac{\sin\left[\tfrac{1}{2}T(\omega - \omega_2)\right]}{\left[\tfrac{1}{2}T(\omega - \omega_2)\right]} \right\}$$

The power spectrum consists of two terms of the form $\sin(\phi)/\phi$ centered at frequencies ω_1 and ω_2. Using the Rayleigh criterion, we can just resolve the two peaks (i.e. determine if there is one or two sinusoids contributing to the spectrum) provided that the frequency separation $\Delta \omega = |\omega_1 - \omega_2|$ ($\Delta f = |f_1 - f_2|$) is equal to the frequency difference for the peak of one frequency and the first zero of $\sin(\phi)/\phi$ for the other frequency. Since zeros of $\sin(\phi)/\phi$ occur at frequencies f equal to $\pm 1/T$, $\pm 2/T$, ..., $\pm p/T$, the frequencies are just resolved when

$$\Delta \omega = \frac{2\pi}{T}; \quad \Delta f = \frac{1}{T} \tag{5.6.63a}$$

and well-resolved for

$$\Delta \omega > \frac{3\pi}{T}; \quad \Delta f > \frac{3}{2T} \tag{5.6.63b}$$

In summary, resolution of two frequencies f_k and $f_{k+1} (= f_k \pm \Delta f)$ using an unsmoothed periodogram or equivalently a rectangular window, requires a record of length T, where $\Delta f = 1/T$ frequency units. Note also that $1/T$ is equal to the fundamental frequency, f_1, which is the lowest frequency that we can calculate for the record. For

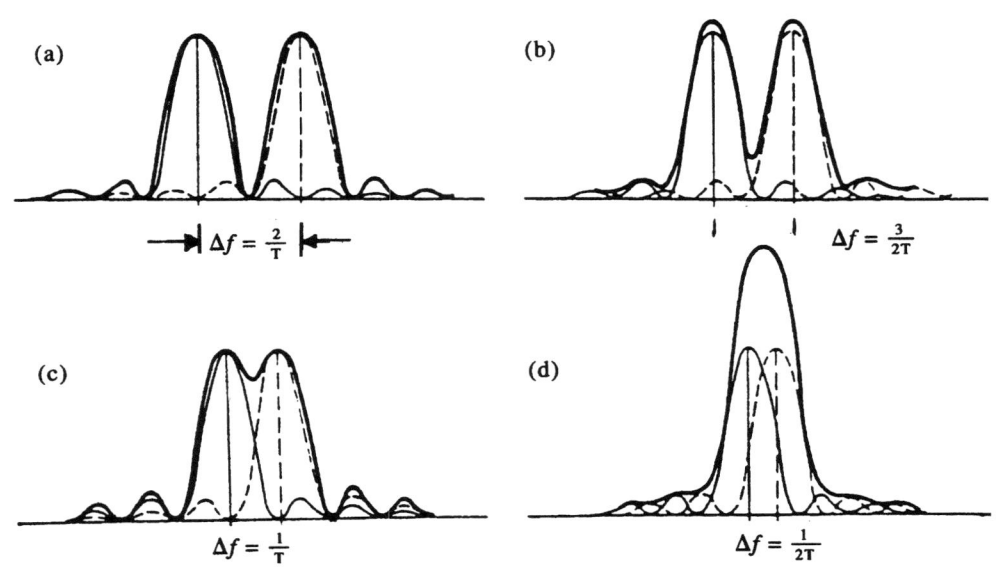

440 *Data Analysis Methods in Physical Oceanography*

Figure 5.6.19. Resolution of spectral lines. (a, b) Well resolved; (c) just resolved; and (d) not resolved. (From Jenkins and White, 1957.)

some nonrectangular windows, the length of the data set must be increased to about $2T = 2/\Delta f$ to achieve the same frequency separation.

In a related study, Munk and Hasselman (1964) discuss the "super-resolution" of tidal frequency variability. The fact that time series of tidal heights vary at precise frequencies and have relatively large signal-to-noise ratios suggests that the traditional requirement (that a minimum record length T is required to separate tidal constituents separated by frequency difference $\Delta f = 1/T$) is "grossly incomplete". The modified resolvable frequency difference is

$$\Delta f = \frac{1}{rT}; \quad \Delta\omega = \frac{2\pi}{rT} \tag{5.6.64}$$

in which $r \equiv$ (signal level/noise level)$^{1/2}$. On this basis, the Rayleigh criterion must be considered a conservative measure of the resolution requirement for deterministic processes.

5.6.6 Smoothing spectral estimates (windowing)

The need for statistical reliability of spectral estimates brings us to the topic of spectral averaging or smoothing. As we have seen, discrete Fourier transforms provide an elegant method for decomposing a data sequence into a set of discrete spectral estimates. For a data sequence of N values, the periodogram estimate of the spectrum can have a maximum of $N/2$ Fourier components. If we use all $N/2$ components to generate the periodogram, there are only two degrees of freedom per spectral estimate, corresponding to the coefficients A_n, B_n of the sine and cosine functions for each Fourier component (see Section 5.6.3.5). Based on the assumption that data are drawn from a normally distributed random sample, we can define the confidence limits for the spectrum in terms of a chi-squared distribution, χ_n^2, where for n degrees of freedom

$$E[\chi_n^2] = \mu^2 = n, \quad E[(\chi_n^2 - \mu^2)] = \sigma^2 = 2n \qquad (5.6.65)$$

Substituting $n = 2$ into these expressions, we find that the standard deviation, σ, is equal to the mean, μ, of the estimate, indicating that results based on two degrees of freedom are not statistically reliable. It is for this reason that some sort of ensemble averaging or smoothing of spectral estimates is required. The smoothing can be applied directly to the time series through convolution with a sliding averaging function or by averaging adjacent spectral estimates. A one-shot smoothing applied to the entire record increases only slightly the number of degrees of freedom per spectral estimate. In most practical applications, the full time series is broken into a series of short overlapping segments and smoothing applied to each of the overlapping segments. We then ensemble average the smoothed spectra from each segment to increase the number of degrees of freedom per spectral estimate. The more smoothing we do, the narrower the confidence limits and the greater the reliability of any observed spectral peaks. The trade-off is a loss of spectral resolution and longer processing time.

A window is a smoothing function applied to finite observations or their Fourier transforms to minimize "leakage" in the spectral domain. Convolution in the time domain and multiplication in the frequency domain are adjoint Fourier functions (see Appendix G regarding convolution). A practical window is one which allows little of the energy in the main spectral lobe to leak into the side-lobes where it can obscure and distort other spectral estimates that are present. In fact, weak signal spectral responses can be masked by higher side-lobes from stronger spectral responses. Skillful selection of tapered data windows can reduce the side-lobe leakage, although always at the expense of reduced resolution. Thus, we want a window that minimizes the side-lobes and maximizes (concentrates) the energy near the frequency of interest in the main lobe. These two performance limitations are rather troublesome when analyzing short data records. Short data occur in practice because many measured processes are of short duration or have slowly time-varying spectra that may be considered constant over only short record segments. The window is applied to data to reduce the order of the discontinuity of the boundary of the periodic extension since few harmonics will fit exactly into the length of the time series.

Signals with frequencies other than those of the basis set are not periodic in the observation window. The periodic extension of a signal, not commensurate with its natural period, exhibits discontinuities at the boundaries of the observational period. Such discontinuities are responsible for spectral contributions or leakage over the entire basis set. In the time domain, the windows are applied to the data as a multiplicative weighting (convolution) to reduce the order of the discontinuities at the boundary of the periodic extensions. The windowed data are brought to zero smoothly at the boundaries so that the periodic extensions of the data are continuous in many orders of the derivatives. The value of $Y(f)$ at a particular frequency f, say f_o, is the sum of all the spectral contributions at each f weighted by the window centered at f_o and measured at f

$$Y(f) = Y'(f) * W(f) \qquad (5.6.66)$$

There exist a multitude of data windows or tapers with different shapes and characteristics ranging from the rectangular (box-car) window discussed in the previous section, to the classic Hanning and Hamming windows, to more sophisticated windows such as the Dolph–Chebyshev window. The type of window used for a given application depends on the required degree of side-lobe suppression, the allowable widening of the central lobe, and the amount of computing one is willing to endure. We will briefly discuss several of the conventional windows plus the Kaiser–Bessel window recommended by Harris (1978).

5.6.6.1 Desired window qualities

Windows affect the attributes of a given spectral analysis method, including its ability to detect and resolve periodic waveforms, its dynamic range, confidence intervals, and ease of implementation. Spectral estimates are affected not only by the broadband noise spectrum of the data but also by narrow-band signals that fall within the bandwidth of the window. Leakage of spectral power from a narrow-band spectral component, f_o, to another frequency component, f_a, produces a bias in the amplitude and position of a spectral estimate. This bias is especially disruptive for the detection of weak signals in the presence of nearby strong signals. To reduce the bias, we need a "good" window. Although there are no universal standards for a good window, we would like it to possess the following characteristics in Fourier transform space:

(1) The central main lobe of the window (which is centered on the frequency of interest) should be as narrow as possible to improve the frequency resolution of adjacent spectral peaks in the dataset, and the first side-lobes should be greatly attenuated relative to the main lobe to avoid contamination from other frequency components. Here, the narrowness of the central lobe is measured by the positions of the −3 dB (half power points) on either side of the lobe. Retention of a narrow central lobe, while suppressing the side-lobes, is not as easy as it sounds since suppression of the side-lobes invariably leads to a broadening of the central lobe;
(2) The window should suppress the amplitudes of side-lobes at frequencies far removed from the central lobe. That is, the side-lobes should have a rapid asymptotic fall-off rate with frequency so that they leak relatively little energy into the spectral estimate at the central lobe (i.e. into the frequency of interest);
(3) The coefficients of the window should be easy to generate for multiplication in the time domain and convolution in the Fourier transform domain.

A good performance indicator (PI) for the time domain window $w(t)$ can be defined as the difference between the equivalent noise bandwidth, ENBW, and the bandwidth, BW, located between the −3 dB levels of the central lobe (Harris, 1978)

$$\text{PI} = \frac{\text{ENBW} - \text{BW}}{\text{BW}} = \frac{\frac{1}{\text{BW}}\sum_n w^2(n\Delta t)}{\left[\sum_n w(n\Delta t)\right]^2} - 1 \qquad (5.6.67)$$

where we have normalized by the bandwidth. The windows that perform well have values for this ratio (×100%) of between 4.0 and 5.5%. A summary of the figures of merit for several well-known windows is presented in Table 5.6.3. PI values are obtained using columns 4 and 5. The choice of window can be daunting; Harris lists more than 44 windows for smoothing spectral estimates.

Time-series Analysis Methods 443

Table 5.6.3. Windows and figures of merit. The last column gives the correlation between adjacent data segments for the specified percentage segment overlap. For completeness, we include the Tukey and Parzen windows. (From Harris, 1978)

Window	Highest side-lobe level (dB)	Side-lobe attenuation (dB/octave)	Equiv. noise BW (BINS)	3.0 dB BW (BINS)	Overlap corr. 75% 50%
Rectangle	−13	−6	1.00	0.89	0.750 0.500
Triangle	−27	−12	1.33	1.28	0.719 0.250
Hanning	−32	−18	1.50	1.44	0.659 0.167
Hamming	−43	−6	1.36	1.30	0.707 0.235
Parzen	−21	−12	1.20	1.16	0.765 0.344
Tukey $\alpha = 0.5$	−15	−18	1.22	1.15	0.727 0.364
Kaiser $\alpha = 2.0$	−46	−6	1.50	1.43	0.657 0.169
Bessel					
$\alpha = 2.5$	−57	−6	1.65	1.57	0.595 0.112
$\alpha = 3.0$	−69	−6	1.80	1.71	0.539 0.074
$\alpha = 3.5$	−82	−6	1.93	1.83	0.488 0.048

5.6.6.2 Rectangular (box-car) and triangular windows

As discussed in Section (5.6.4), a rectangular window has an amplitude of unity throughout the observation interval of duration $T = N\Delta t$, with the weighting given by

$$w(n\Delta t) = 1, \quad n = 0, 1, \ldots, N-1 \ (\text{or} -N/2 \leq n \leq N/2)$$
$$= 0, \quad \text{elsewhere} \tag{5.6.68}$$

(Figure 5.6.20a). Using the relation $\omega T = N\theta$, where $\theta = \omega \Delta t$ and $T = N\Delta t$, the spectral window from the discrete Fourier transform (DFT) is

$$W(\theta) = T e^{-i(N-1)\theta/2} \frac{\sin(N\theta/2)}{N\theta/2} \tag{5.6.69a}$$

$$|W(\theta)|^2 = T^2 \left[\frac{\sin(N\theta/2)}{N\theta/2}\right]^2 \tag{5.6.69b}$$

(Figure 5.6.20b) where the exponential term in equation (5.6.69a) gives the phase shift of the window as a function of the frequency $\omega = \theta/\Delta t$. The function W, the Dirichlet kernal, has strong side-lobes, with the first side-lobe down only 13 dB from the main lobe. The remaining side-lobes fall off weakly at 6 dB per octave, which is the functional rate for a discontinuity (an "octave" corresponds to a factor of two in change frequency). Zeros of $W(\theta)$ occur at integer multiples of the frequency resolution, $f_1 = 1/T$, for which $N\theta/2 = \omega T/2 = \pm p\pi$. That is, where $f = \pm p/T (\pm 1/T, \pm 2/T, \ldots)$.

The triangular (Bartlett) window

$$w(n\Delta t) = \begin{cases} \dfrac{n}{(N/2)}, & n = 0, 1, \ldots, N/2 \\ \dfrac{N-n}{(N/2)}, & n = N/2, \ldots, N-1 \end{cases} \tag{5.6.70a}$$

$$= \frac{N/2 - |n|}{(N/2)}, \quad 0 \leq |n| \leq N/2 \tag{5.6.70b}$$

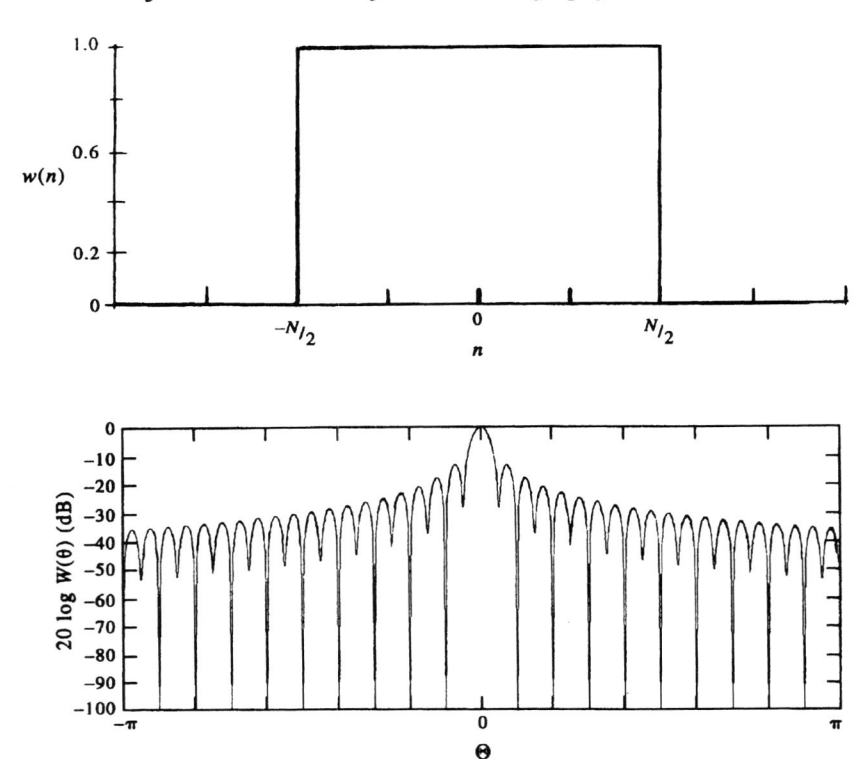

444 *Data Analysis Methods in Physical Oceanography*

Figure 5.6.20. A box-car window for N = 41 weights. (a) Weights, w(n) = 1.0 in the time domain ($-20 \leq n \leq 20$). (b) Fourier transform of the weights, $|W(\theta)|$, plotted as $20 \log |W(\theta)|$ where $\theta = \omega \Delta t / N = 40\pi/N$ is the frequency span of the window.

(Figure 5.6.21a) has the DFT

$$W(\theta) = \frac{2T}{N} e^{-i(N-1)\theta/2} \left[\frac{\sin(N\theta/2)}{N\theta/2} \right]^2 \quad (5.6.71a)$$

$$|W(\theta)|^2 = \frac{4T^2}{N^2} \left[\frac{\sin(N\theta/2)}{N\theta/2} \right]^4 \quad (5.6.71b)$$

(Figure 5.6.21b) which we recognize as the square of the sinc function for the rectangular window. The main lobe between zero crossings is twice that of the rectangular window but the level of the first side-lobe is down by 26 dB, twice that of the rectangular window. Despite the improvement over the box-car window, the side-lobes of the triangular window are still extensive and use of this window is not recommended if other windows are available.

The Parzen window

$$w(n\Delta t) = 1.0 - |n/(N/2)|^2, \quad 0 \leq |n| \leq N/2 \quad (5.6.72)$$

is the squared counterpart to the Bartlett window. This is the simplest of the continuous polynomial windows and has first side-lobes down by -22 dB and falls off as $1/\omega^2$.

5.6.6.3 Hanning and Hamming windows (50% overlap)

The Hann window, or *Hanning window* as it is most commonly known, is named after the Austrian meteorologist Julius von Hann and is part of a family of trigonometric windows having the generic form $\cos^\alpha(n)$, where the exponent, α, is typically an integer from 1 through 4. The case $\alpha = 1$ leads to the *Tukey* (or *cosine-tapered*) *window* (Harris, 1978). As α becomes larger, the window becomes smoother, the side-lobes fall off faster and the main lobe widens. The Hanning window ($\alpha = 2$), also known as the *raised cosine* and *sine-squared* window, is defined in the time domain as

$$w(n\Delta t) = \sin^2(\pi n/N) = \tfrac{1}{2}[1.0 - \cos(2\pi n/N)], \quad n = 0, 1, \dots, N-1 \quad (5.6.73a)$$

$$= \sin^2[\pi(n + N/2)/N]$$
$$= \tfrac{1}{2}[1.0 - \cos[2\pi(n + N/2)/N]], \quad n = -N/2, \dots, N/2 \quad (5.6.73b)$$

(Figure 5.6.22a) which is a continuous function with a continuous first derivative. The DFT of this weighting function is

$$W(\theta) = \tfrac{1}{2}D(\theta) + \tfrac{1}{4}[D(\theta - \theta_1) + D(\theta + \theta_1)] \quad (5.6.74)$$

(Figure 5.6.22b) where $\theta_1 = 2\pi/N$ and

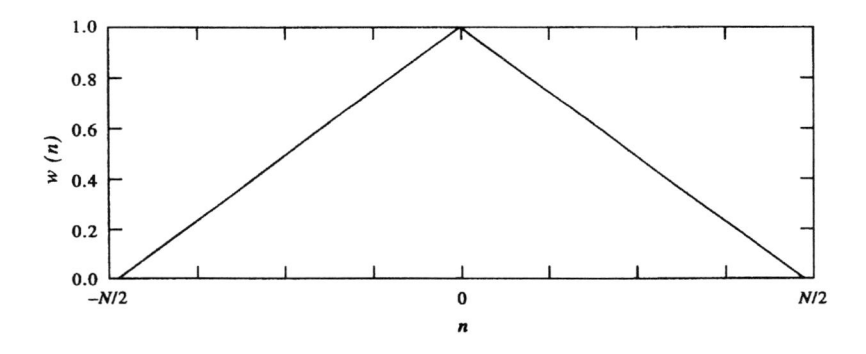

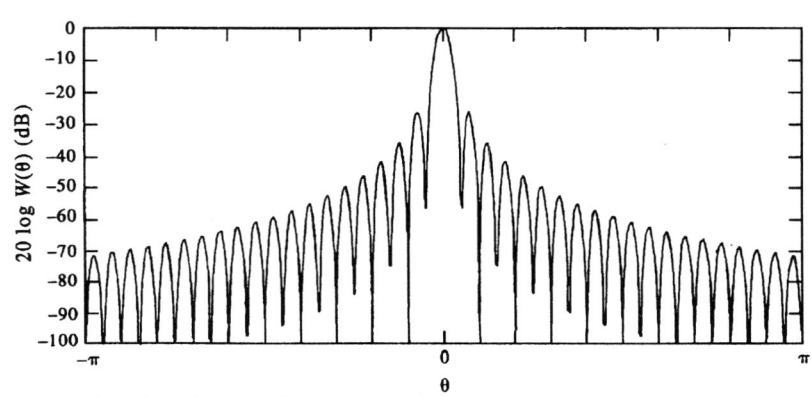

Figure 5.6.21. The triangular (Bartlett) window for $N = 51$ weights. (a) Weights, $w(n)$ in the time domain $(-20 \leq n \leq 20)$. (b) Fourier transform of the weights, $|W(\theta)|$, plotted as $20 \log |W(\theta)|$ (cf. Figure 5.6.20).

$$D(\theta) = Te^{i\theta/2} \frac{\sin(N\theta/2)}{N\theta/2} \qquad (5.6.75)$$

is the standard function (Dirichlet kernal) obtained for the rectangular and triangular windows. Thus, the window consists of the summation of three sinc functions (Figure 5.6.22c), one centered at the origin, $\theta = 0$, and two other translated Dirichlet kernals having half the amplitude of the main kernal and offset by $\theta = \pm 2\pi/N$ from the central lobe. There are several important features of the window response $W(\theta)$. First of all, the functions D are discrete and defined only at points which are multiples of $2\pi/N$, which also correspond to the zero crossings of the central function, $D(\theta)$. Secondly, for all of zero crossings except those at $\theta_{\pm 1} = \pm 2\pi/N$, the translated functions also have zero crossings at multiples of $2\pi/N$. As a result, only values at $-2\pi/N$, 0, and $+2\pi/N$ contribute to the window response. It is the widening of the main lobes of the translated functions that causes them to be nonzero at the first zero-crossings of the central function. Lastly, because the translated functions are out of phase with the central function, they tend to cancel the side-lobe structure. The first side-lobe is down by 32 dB from the main lobe. The remaining side-lobes diminish as $1/\omega^3$ or at about -18 dB per octave.

An attractive aspect of the Hanning window is that smoothing in the frequency domain can be accomplished using only three convolution terms corresponding to $\theta_o, \theta_{\pm 1}$. The Hanning-windowed Fourier transform Y_H for the spectral frequency, f_k, is then obtained from the raw spectra Y for the frequencies f_k and the two adjoining frequencies f_{k-1}, f_{k+1}; that is

$$Y_H(f_k) = \tfrac{1}{2}\{Y(f_k) - \tfrac{1}{2}[Y(f_{k-1}) + Y(f_{k+1})]\} \qquad (5.6.76)$$

The transform $Y(f_k)$ has been rectangular-windowed by the act of collecting the data but is "raw" in the sense that no additional smoothing has been applied. Other processing advantages of the Hanning window are discussed by Harris (1978). Since the squares of the weighting terms $(1/2)^2 + (1/4)^2 + (1/4)^2 = 3/8$, the total energy will be reduced following the application of the Hanning window. To compensate, the amplitudes of the Fourier transforms $Y_H(f)$ should be multiplied by $\sqrt{(8/3)}$ prior to computation of the spectra. Specifically

$$Y_H(f_k) = \Delta t (8/3)^{1/2} \sum_{n=0}^{N-1} y_n [1 - \cos(2\pi n/N)] e^{-i2\pi kn/N} \qquad (5.6.77)$$

where $f_k = k/(N\Delta t)$.

The *Hamming window* is a variation on the Hanning window designed to cancel the first side-lobes. To accomplish this, the relative sizes of the three Dirichlet kernels are adjusted through a parameter, γ where

$$w(n\Delta t) = \gamma + (1-\gamma)\cos(2\pi n/N)], \quad n = -N/2, ..., N/2 \qquad (5.6.78a)$$

$$W(\theta) = \gamma D(\theta) + \tfrac{1}{2}(1-\gamma)[D(\theta - 2\pi/N) + D(\theta + 2\pi/N)] \qquad (5.6.78b)$$

Perfect cancellation of the first side-lobes (located at $\theta_1 = 2.5\pi/N$) occurs when $\gamma = 25/46 \approx 0.543478$. Taking $\gamma = 0.54$ leads to near-perfect cancellation at

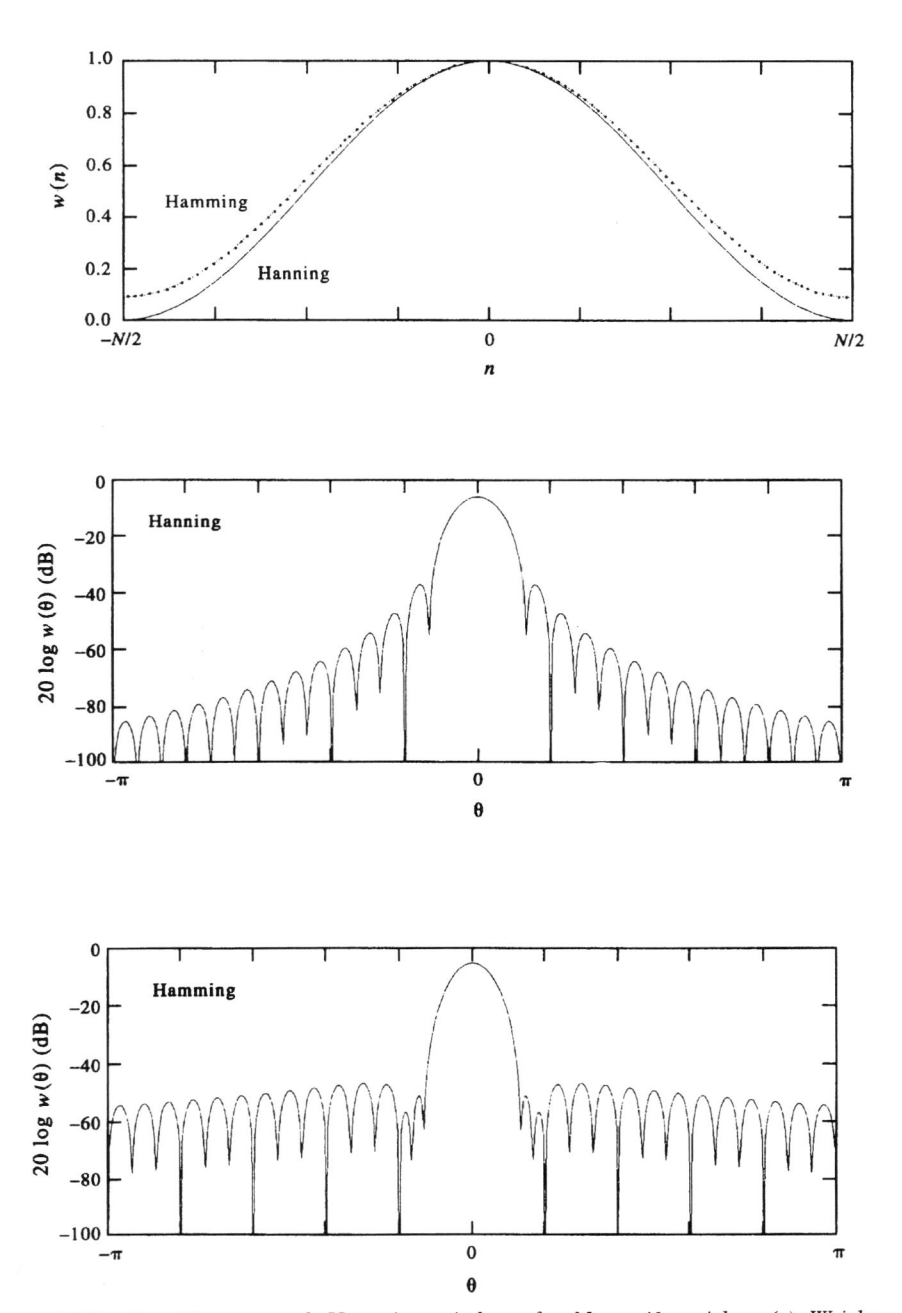

Figure 5.6.22. The Hanning and Hamming windows for N = 41 weights. (a) Weights, w(n), (−20 ≤ n ≤ 20). (b) Fourier transform of the weights, |W(θ)|, plotted as 20 log |W(θ)| (cf. Figure 5.6.20). The response functions have not been re-scaled.

448 Data Analysis Methods in Physical Oceanography

$\theta_1 = 2.6\pi/N$ and a marked improvement in side-lobe level. The Hamming window is defined as

$$w(n\Delta t) = 0.54 + 0.46\cos(2\pi n/N), \quad n = -N/2, ..., N/2 \quad (5.6.79)$$

and has a spectral distribution similar to that of the Hanning window with more "efficient" side-lobe attenuation. The highest side-lobe levels of the Hanning window occur at the first side-lobes and are down by 32 dB from the main lobe. For the Hamming window, the first side-lobe is highly attenuated and the highest side-lobe level (the third side-lobe) is down by 43 dB. To compensate for the filter, the amplitudes of the Fourier transforms $Y_{Ham}(f)$ should be multiplied by $\sqrt{(5/2)}$ prior to computation of the spectra. On a similar note, if you are going to use any of the windows in this section to calculate running mean time series, make sure each estimated value is divided by the sum of the weights used, $\sum_N w_n$.

5.6.6.4 Kaiser–Bessel window (75% overlap)

Harris (1978) identifies the Kaiser–Bessel window as the "top performer" among the many different types of windows he considered. Among other factors, the coefficients of the window are easy to generate and it has a high equivalent noise bandwidth, one of the criteria used to separate good and bad windows. The trade-off is increased main-lobe width for reduced side-lobe levels. In the time domain the filter is defined in terms of the zeroth-order modified Bessel functions of the first kind.

$$w(n\Delta t) = \frac{I_o(\pi\alpha\Omega)}{I_o(\pi\alpha)}, \quad 0 \leq |n| \leq N/2 \quad (5.6.80)$$

where the argument $\Omega = [1.0 - (2n/N)^2]^{1/2}$ and

$$I_o(x) = \sum_{k=0}^{\infty} \left[\frac{(x/2)^k}{k!}\right]^2 \quad (5.6.81)$$

The parameter $\pi\alpha$ is half of the time-bandwidth product, with α typically having values 2.0, 2.5, 3.0, and 3.5. The transform is approximated by

$$W(\theta) \approx [N/I_o(\pi\alpha)] \frac{\sinh\{[\pi^2\alpha^2 - (N\theta/2)^2]^{1/2}\}}{\{[\pi^2\alpha^2 - (N\theta/2)^2]^{1/2}\}} \quad (5.6.82)$$

Plots of the weighting function w and the DFT for W are presented in Figure 5.6.23 for two values of the parameter $\alpha(= 2.0, 3.0)$. The modified Bessel function I_o is defined as follows.

For $|x| \leq 3.75$

$$I_o(x) = \{[\{[(4.5813 \times 10^{-3}Z + 3.60768 \times 10^{-2})Z + 2.659732 \times 10^{-1}]Z + 1.2067492\}Z + 3.0899424]Z + 3.5156229\}Z + 1.0 \quad (5.6.83a)$$

where for real x

$$Z = (x/3.75)^2 \quad (5.6.83b)$$

For $|x| > 3.75$

$$I_o(x) = \exp(|x|)/|x|^{1/2}\{[(\{[\{[(3.92377 \times 10^{-3}Z - 1.647633 \times 10^{-2})Z$$
$$+ 2.635537 \times 10^{-2}]Z - 2.057706 \times 10^{-2}\}Z + 9.16281 \times 10^{-3}]Z$$
$$- 1.57565 \times 10^{-3}\}Z + 2.25319 \times 10^{-3})Z + 1.328592 \times 10^{-2}]Z$$
$$+ 3.9894228 \times 10^{-1}\}$$

(5.6.83c)

where

$$Z = 3.75/|x| \qquad (5.6.83d)$$

The usefulness of the Kaiser–Bessel window is nicely illustrated by Figure 5.6.24. Here, we compare the average spectra (in cm^2/cpd) obtained from a year-long record of hourly coastal sea level following application of a rectangular window (the worst possible window) and a Kaiser–Bessel window (the best possible window) to a series of overlapping data segments. In each case, the window length is 42.7 days and there are $K = 32$ degrees of freedom per spectral estimate, corresponding to roughly 16 separate spectral estimates for 50% window overlaps. Both windows preserve the strong spectra peaks within the tidal frequency bands centered at 1, 2, and 3 cpd. However, unlike the rectangular window, the Kaiser–Bessel window results in little energy leakage from the tidal bands to adjacent frequency bands. The high spectral levels at periods shorter than about two days ($f > 0.5$ cpd) in the nontidal portion of the rectangularly windowed spectra is an artifact of the window. The slightly better ability of the rectangular window to resolve frequency components within the various tidal bands is outweighed by the high contamination of the spectrum at nontidal frequencies.

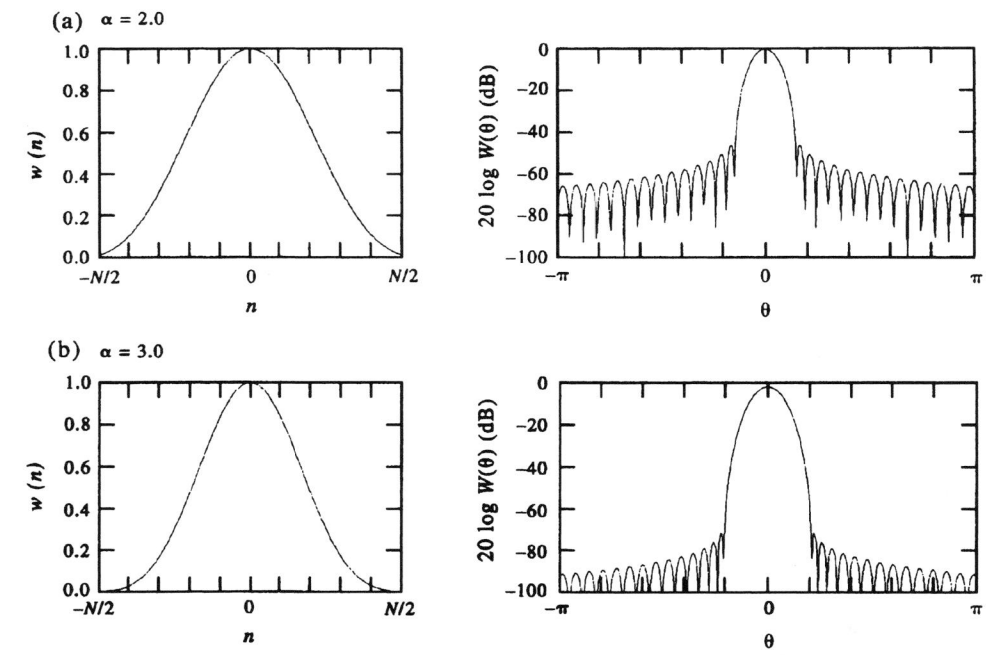

Figure 5.6.23. *The Kaiser–Bessel window for $N = 51$ weights and $\alpha = 2.0$ and 3.0. (a) Weights, $w(n)$, ($-20 \leq n \leq 20$). (b) Fourier transform of the weights, $|W(\theta)|$, plotted as $20 \log |W(\theta)|$ (cf. Figure 5.6.20). (From Harris, 1978.)*

450 *Data Analysis Methods in Physical Oceanography*

5.6.7 Smoothing spectra in the frequency domain

As we noted earlier, each spectral estimator for a random process is a chi-squared function with only two degrees of freedom (DOF). Because of this minimal number of degrees of freedom, some sort of smoothing or filtering is needed to increase the statistical significance of a given spectral estimate. The windowing approach described in the previous section, in which we partitioned the time series into a series of shorter overlapping segments, is one of a number of computational methods used to smooth (average) spectral estimates.

5.6.7.1 Band averaging

For a time series consisting of N data points, one of the simplest forms of smoothing is to use the discrete Fourier transform or fast Fourier transform to calculate individual spectral estimates for the maximum number of frequency bands ($N/2$) and then average together adjacent spectral estimates. The resultant spectral estimate is assigned to the mid-point of the average. Thus, we could average bands 1, 2, and 3, to form a single spectral estimate centered at band 2, then bands 4, 5, and 6 to form an estimate centered at band 5, and so on. It is often useful in this type of *frequency band averaging* to use an odd-numbered smoother so that the center point is easily defined. In particular, if we were to average groups of three adjacent (and different) bands to form each estimate, the number of degrees of freedom per estimate would increase

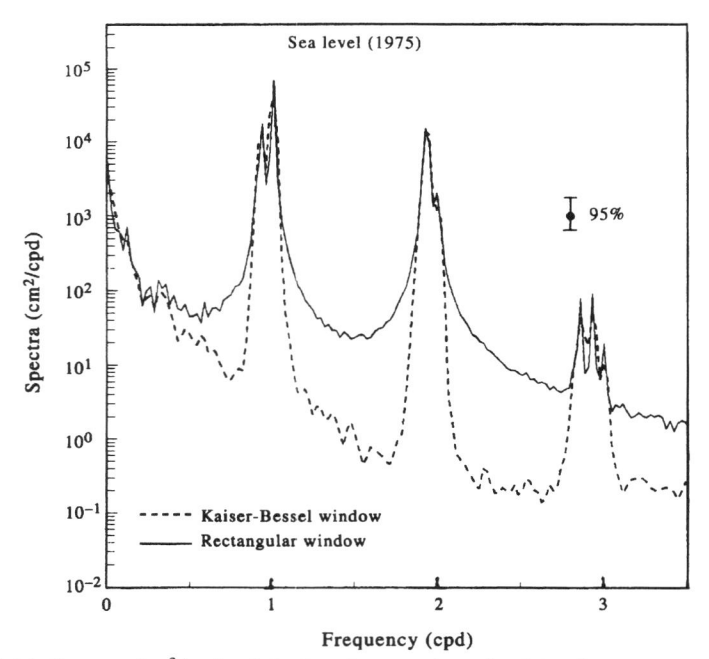

Figure 5.6.24. Spectra (cm^2/cpd) of the hourly coastal sea-level height recorded at Victoria, British Columbia during 1975 following windowing (number of hourly samples, N = 8750). Linear frequency. Solid line: Rectangular window. Dashed line: Kaiser–Bessel window with $\alpha = 3$. Both windows have a length of 1024 h (= 42.67 days) and there are DOF = 32, using a total of 16 50% overlapping data segments. The tidal peak centered at 3 cpd results from nonlinear interactions within the semidiurnal frequency band. Vertical line is the 95% level of confidence. (Courtesy, A. Rabinovich.)

from 2 to 6. In the case of the Blackman–Tukey procedure, an alternative method is to use bigger lag steps in the computation of the autocovariance function before its transform is taken. This is functionally equivalent to smoothing by averaging together the individual spectral estimates.

5.6.7.2 Block averaging

As noted earlier, a common smoothing technique is to segment the time series (of length N) into a series of short, equal-length segments of length N_s (where $N = KN_s$, and K is a positive integer). Spectra are then computed for each of the K segments and the spectral values for each frequency band then *block averaged* to form the final spectral estimates for each frequency band. If there is no overlap between segments, the resulting degrees of freedom for the composite spectrum will be $2K$. This assumes that the individual sample spectra have not been windowed and that each spectral estimate is a chi-squared variable with 2 degrees of freedom. Since the frequency resolution of a time series is inversely proportional to its length, the major difficulty with this approach is that the shorter time series have fewer spectral values than the original record over the same Nyquist frequency range. In other words, the maximum resolvable frequency $1/2\Delta t$ remains the same since Δt is unchanged, but the frequency spacing between adjacent spectral estimates is increased for the short segments because of the reduced record lengths.

However, by not overlapping adjacent segments, we could be overly conservative in our estimate of the number of degrees of freedom. For that reason, most analysts overlap adjacent segments by 30–50% so that more uniform weighting is given to individual data points. The need for overlapping segments is necessary when a window is applied to each individual segment prior to calculation of the spectra. The effect of the window is to reduce the effective length of each segment in the time domain so that, for some sharply defined windows such as the Kaiser–Bessel window, even adjoining segments with 50% overlap can be considered independent time series for spectral analysis. As in Figure 5.6.24, The degrees of freedom of the periodograms averaged together is $4K$, rather than $2K$ for the nonoverlapping segments. Consideration must be given to the correlation among individual estimates (the greater the overlap the higher the correlation). Nuttall and Carter (1980) report that 92% of the maximum number of equivalent degrees of freedom can be achieved for a Hanning window which uses 50% overlap. Clearly, we must sacrifice something to gain improved statistical reliability. That "something" is a loss of frequency resolution due to the broad central lobe that accompanies windows with negligible side-lobes.

As an example, consider the spectrum of a 1-min sampled time series $y(t) = A\cos(2\pi f t) + \varepsilon(t)$ of length 512 min composed of Gaussian white noise $\varepsilon(t) (|\varepsilon| \leq 1)$ and a single cosine component of amplitude, A, and frequency $f = 0.23$ cpmin (period $T = 1/f = 4.3$ min). The magnitude of the deterministic component, A, is five times the standard deviation of the white noise signal and $V[\varepsilon] = (1/\sqrt{2})$ cm^2. The raw periodogram (Figure 5.6.25a) reveals a large narrow peak at the frequency (0.23 cpmin) of the single cosine term plus a large number of smaller peaks associated with the white noise oscillations. In this case, there has been no spectral smoothing and the resultant spectral estimates are chi-squared functions with 2 degrees of freedom. The variances of the spectral peaks are as large as the peaks themselves. If we average together three adjacent spectral components (Figure 5.6.25b), we obtain a much smoother spectrum, $S(f)$. Here, $S_i = S(f_i)$ is defined by $S_i =$

$1/3[S(f_{i-1}) + S(f_i) + S(f_{i+1})]$, $S_{i+3} = 1/3[S(f_{i+2}) + S(f_{i+3}) + S(f_{i+4})]$, and so on. Each of the new spectral estimates now has six degrees of freedom instead of only two. The bottom two panels in this figure show what happens if we increase the number of frequency bands averaged together to seven (Figure 5.6.25c) and then to 15 (Figure 5.6.25d). Note that, with increasing degrees of freedom (DOF), our confidence in the existence of a spectral peak increases but delineation of the peak frequency decreases. With increasing DOF, there is increased smoothing of all spectral peaks (see also Figure 5.6.24). The same effect can be achieved by operating on the autocovariance function rather than on the Fourier spectral estimates. In particular, a spectrum similar to Figure 5.6.25(a) is obtained using the autocovariance transform method on the time series $y(t)$ for a time lag of 1 min (the sampling interval). If we apply a lag of 3 min in computing the autocovariance transform, we obtain a spectrum similar to Figure 5.6.25(b), and so on. Any differences between the two methods will be due to computational uncertainties.

To determine the number of degrees of freedom for any block averaging, we define the normalized standard error $\varepsilon(\tilde{G})$ of the one-sided spectrum, $\tilde{G}_{yy}(f)$, of the time series $y(t)$ of length $T = N\Delta t$, as

$$\varepsilon[\tilde{G}_{yy}(f)] = \frac{V[\tilde{G}_{yy}(f)]^{1/2}}{G_{yy}(f)} \tag{5.6.84}$$

where $V[\tilde{G}]$ is the variance of $\tilde{G}$, the tilde ($\sim$) denotes the raw estimate of the time series, and

$$\tilde{G}_{yy}(f)/G_{yy}(f) = \chi_2^2/2 \tag{5.6.85}$$

is a chi-square variable with $n = 2$ degrees of freedom. For the narrowest possible resolution $\Delta f = 1/T$, we have

$$\varepsilon[\tilde{G}_{yy}(f)] = \frac{(2n)^{1/2}}{n} = (2/n)^{1/2} \tag{5.6.86}$$

For maximum resolution, $n = 2$ and so $\varepsilon(\tilde{G}) = 1$, giving the not-so-useful result that the standard deviation of the estimate is as large as the estimate itself. If, on the other hand, we average the spectral estimates for each frequency for the maximum resolution spectra using a total of N_s separate and independent record segments of length T_s (where $T = N_s \cdot T_s$) we find

$$\tilde{G}_{yy}(f) = \frac{2}{N_s T_s} \sum_{i=1}^{N_s} |Y_i(f_i, T_s)|^2 \tag{5.6.87}$$

so that

$$\varepsilon[\tilde{G}_{yy}(f)] = (2n/2N_s)^{1/2} = (1/N_s)^{1/2} \tag{5.6.88}$$

The resolution (effective) bandwidth is $b_e = N_s/T = 1/T_s$. Since the first estimate gives two degrees of freedom per spectral band, this gives $2N_s$ degrees of freedom per frequency band.

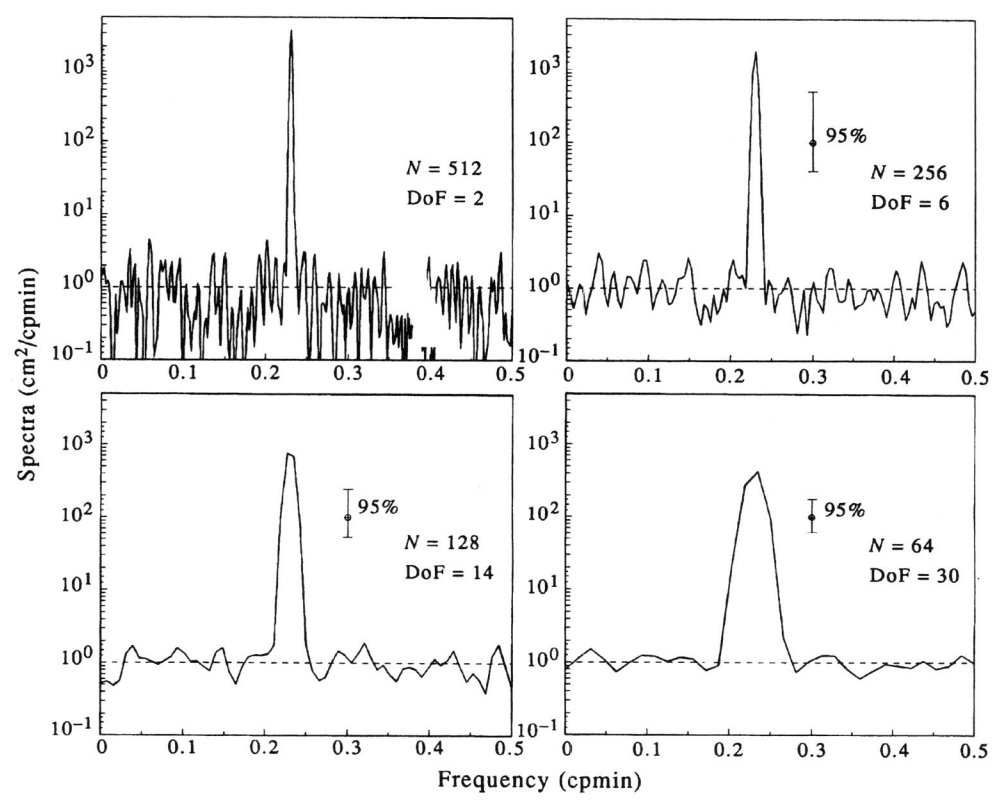

Figure 5.6.25. Periodogram power spectral estimates for a time series composed of Gaussian white noise and a single cosine constituent with a frequency of 0.23 cpmin and amplitude five times that of the white noise component. N = number of spectral bands and vertical lines are the 95% confidence intervals. (a) Raw (unsmoothed) periodogram, with DOF = 2; (b) smoothed periodogram, by averaging three adjacent spectral estimates such that DOF = 6; (c) as with (b) but for seven frequency bands, and DOF = 14; as with (b) but for 15 frequency bands, DOF = 30.

5.6.8 Confidence intervals on spectra

We can generalize equation (5.6.85) by noting that the ratio of the estimated spectrum and the expected values of the spectrum

$$\frac{\nu \tilde{G}_{yy}(f)}{G_{yy}(f)} = \chi_\nu^2 \tag{5.6.89}$$

is distributed as a chi-square variable with ν degrees of freedom. It then follows that

$$P\left[\chi_{\alpha/2,\nu}^2 < \frac{\nu \tilde{G}_{yy}(f)}{G_{yy}(f)} < \chi_{1-\alpha/2,\nu}^2\right] = 1 - \alpha \tag{5.6.90}$$

where

$$P[\chi_\nu^2 \leq \chi_{\alpha/2,\nu}^2] = \alpha/2 \tag{5.6.91}$$

454 *Data Analysis Methods in Physical Oceanography*

Thus, the true spectrum, $G_{yy}(f)$, is expected to fall into the interval

$$\frac{\nu \tilde{G}_{yy}(f)}{\chi^2_{1-\alpha/2,\nu}} < G_{yy}(f) < \frac{\nu \tilde{G}_{yy}(f)}{\chi^2_{\alpha/2,\nu}} \qquad (5.6.92)$$

with $(1-\alpha)100\%$ confidence. In this form, the confidence limit applies only to the frequency f and not to other spectral estimates. We further point out that the degrees of freedom, ν, in the above expressions are different for windowed and nonwindowed time series. For windowed time series, we need to use the "equivalent" degrees of freedom, as presented in Table 5.6.4 for some of the more commonly used windows.

Another way to view these arguments is to equate $\tilde{G}_{yy}(f)$ with the measured standard deviation, $s^2(f)$, of the spectrum and $G_{yy}(f)$ with the true variance, $\sigma^2(f)$. Then

$$\frac{(\nu-1)s^2(f)}{\chi^2_{1-\alpha/2,\nu}} < \sigma^2(f) < \frac{(\nu-1)s^2(f)}{\chi^2_{\alpha/2,\nu}} \qquad (5.6.93)$$

If spectral peaks fall outside the range (5.6.92) then to the $(1-\alpha)100\%$ confidence level they cannot have occurred by chance. The confidence levels are found by looking up the values for $\chi^2_{1-\alpha/2,\nu}$ and $\chi^2_{\alpha/2,\nu}$ in a chi-square table, then calculating the intervals based on the observed standard deviation, s. (Confidence limits on spectral coherency functions are given in Section 5.8.6.1.)

5.6.8.1 Confidence intervals on a logarithmic scale

The confidence intervals derived above apply only to individual frequencies, f. This results from the fact that the confidence interval is determined by the value $G_{yy}(f)$ of the spectral estimate and will be different for each spectral estimate. It would be convenient if we could have a single confidence interval that applies to all of the spectral values at all frequencies. To obtain such a confidence interval, we transform the spectrum using the $\log_{10}$ function. Transforming the above confidence limits we have

$$\log[\tilde{G}_{yy}(f)] + \log[\nu/\chi^2_{1-\alpha/2,\nu}] \leq \log[G_{yy}(f)] \leq \log[\tilde{G}_{yy}(f)] + \log[\nu/\chi^2_{\alpha/2,\nu}] \qquad (5.6.94)$$

or

$$\log[\nu/\chi^2_{1-\alpha/2,\nu}] \leq \log[G_{yy}(f)] - \log[\tilde{G}_{yy}(f)] \leq \log[\nu/\chi^2_{\alpha/2,\nu}] \qquad (5.6.95)$$

When the estimated spectrum is plotted on a log scale, a single vertical confidence

Table 5.6.4. Equivalent degrees of freedom for spectra calculated using different windows. N is the number of data points in the time series and M is the half-width of the window in the time domain. (From Priestley, 1981). $N \neq M$ for the truncated periodogram

Type of window	Equivalent degrees of freedom
Truncated periodogram	N/M
Bartlett window	$3N/M$
Daniell window	$2N/M$
Parzen window	$3.708614(N/M)$
Hanning window	$(8/3)(N/M)$
Hamming window	$2.5164\,(N/M)$

interval is determined for all frequencies by the upper and lower bounds in the above expression (Figure 5.6.26a). The spectral estimate $G_{yy}(f)$ itself is no longer a part of the confidence interval. This aspect, together with the fact that most spectral amplitudes span many orders of magnitude, is a principal reason for presenting spectra as log values. If larger numbers of spectral estimates are averaged together at higher frequencies (i.e. ν is increased), the confidence interval narrows with increasing frequency (Figure 5.6.26b). Note that the length of the confidence interval is longer above the central point than below.

5.6.8.2 Fidelity and stability

The general objective of all spectral analysis is to estimate the function $G_{yy}(f)$ as accurately as possible. This involves two basic requirements:

(1) That the mean smoothed spectrum, $\tilde{G}_{yy}(f)$, be as close as possible to the actual spectrum $G_{yy}(f)$. That is, the bias

$$B(f) = G_{yy}(f) - \tilde{G}_{yy}(f) \qquad (5.6.96)$$

should be small. If this is true for all frequencies, then $\tilde{G}_{yy}(f)$ is said to reproduce $G_{yy}(f)$ with high *fidelity*.

(2) For a time series of length T that has been segmented into M pieces for spectral estimation, the variance of the smoothed spectral estimator for bandwidth b_1 is

$$V[\tilde{G}_{yy}(f)] \approx \frac{(M/b_1)}{T}[G_{yy}(f)]^2 \qquad (5.6.97)$$

and should be small. If this is true, the spectral estimator is said to have high *stability*.

5.6.9 Zero-padding and prewhitening

For logistical reasons, many of the time series that oceanographers collect are too short for accurate definition of certain spectral peaks. The frequency resolution $\Delta f = 1/T$ for a record of length T may not be sufficient to resolve closely spaced spectral components. Also, discrete points in the computed spectrum may be too widely spaced to adequately delineate the actual frequency of the spectral peaks. Unfortunately, the first problem—that of trying to distinguish waveforms with nearly the same frequency—can only be solved by collecting a longer time series; i.e. by increasing T to sharpen up the frequency resolution f of the periodogram. However, the second problem—that of locating the frequency of a spectral peak more precisely—can be addressed by padding (extending) the time series with zeros prior to Fourier transforming. Transforming the data with zeros serves to refine the frequency scale through interpolation between power spectral density estimates within the Nyquist interval $-f_N \leq f \leq f_N$. That is, additional frequency components are added between those that would be obtained with a nonzero-padded transform. Adding zeros helps fill in the shape of the spectrum but in no case is there an improvement in the fundamental frequency resolution. *Zero-padding* is useful for: (1) smoothing the appearance of the periodogram estimates via interpolation; (2) resolving potential ambiguities where the frequency difference between line spectra is greater than the fundamental frequency resolution; (3) helping define the exact frequency of spectral peaks by reducing the "quantization" accuracy error; and (4)

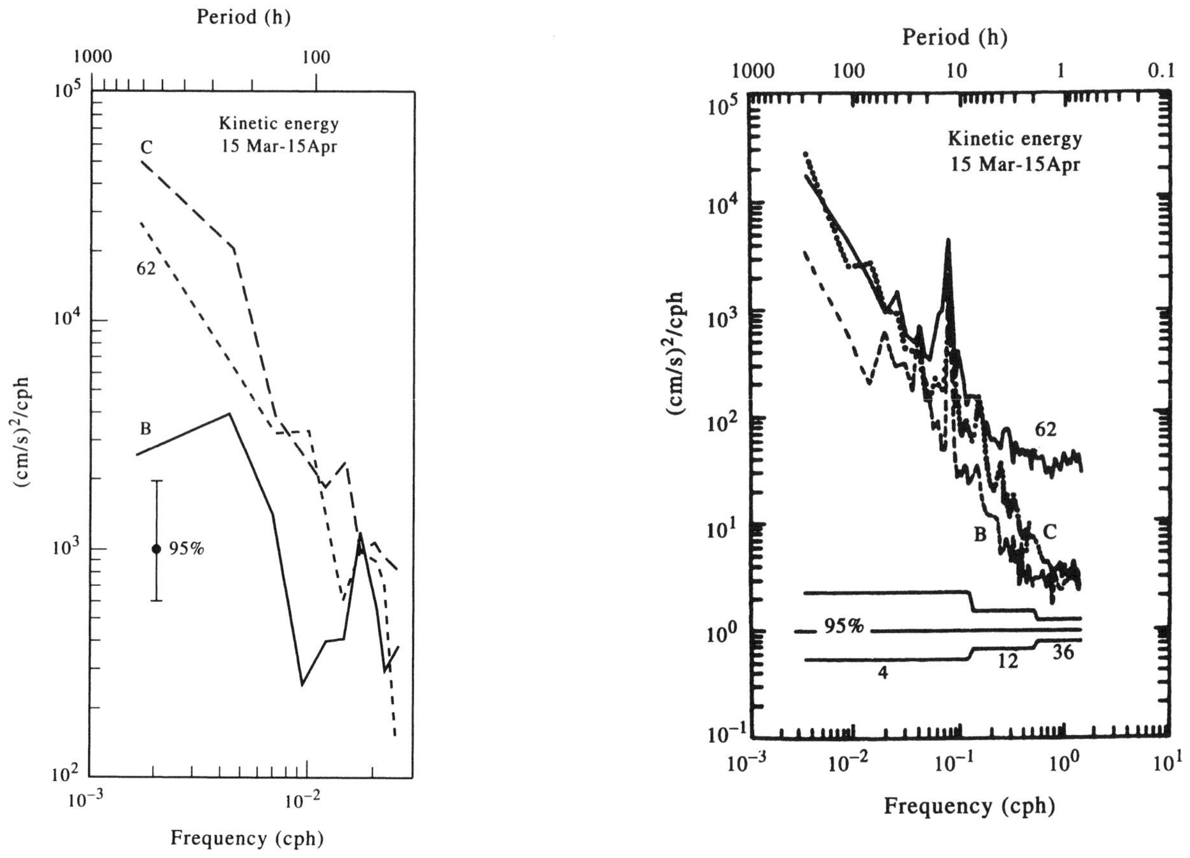

456   *Data Analysis Methods in Physical Oceanography*

Figure 5.6.26. Confidence intervals for current velocity spectra at 50-m depth for three locations (B,C, and 62) on the northeast Gulf of Alaska shelf (59.5°N, 142.2°W), 15 March to 15 April 1976. (a) 95% interval for the low-pass filtered currents. The single vertical bar applies to all frequencies; (b) 95% interval for unfiltered records. Confidence interval narrows at higher frequencies with the increased number of degrees of freedom (from 4 to 36) used in selected frequency ranges. (Adapted from Muench and Schumacher, 1979.)

extend the number of samples to an integer power of 2 for FFT analysis. An example of how zero-padding improves the spectral resolution of a simple digitized data set is provided in Figure 5.6.27. We again emphasize that increased zero-padding helps locate the frequency of discernible spectral peaks, in this case the peaks of the sin x/x function, but cannot help distinguish closely spaced frequency components that were unresolved by the original time series prior to padding.

Prewhitening is a filtering or smoothing technique used to improve the statistical reliability of spectral estimates by reducing the leakage from the most intense spectral components and low-frequency components of the time series that are poorly resolved. To reduce the biasing of these components, the data are smoothed by a window whose spectrum is inversely proportional to the unknown spectrum being considered. Within certain frequency bands, the spectrum becomes more uniformly distributed and approaches that of white noise. Information on the form of the window necessary to construct the white spectrum must be available prior to the application of the smoothing. In effect, the time series $y(n\Delta t)$ is filtered with the weighting function $h(n\Delta t)$ such that the output is

$$y'(n\Delta t) = h(n\Delta t)\cdot(n\Delta t) \tag{5.6.98}$$

has a nearly white spectrum. Once the spectrum $S'_y(\omega)$ is determined, the desired spectrum is derived directly as

$$S_y(\omega) = \frac{S'_y(\omega)}{|H(\omega)|^2} \tag{5.6.99}$$

The best aspects of the parametric and nonparametric spectral techniques can be combined if a parametric model is used to prewhiten the time series prior to the application of a smoothed periodogram analysis. In most prewhitening situations, one is limited to using the first-difference filter in which the current data value is subtracted from the next value multiplied by some weighting coefficient, $0 \leq \alpha \leq 1$. That is $y'(t) = y(t) - \alpha y(t + \Delta t)$. The weighting coefficient can be taken as equal to the correlation coefficient of the initial data series with a shift of one time step, Δt. The filter suppresses low frequencies and stresses high frequencies and has a transform

$$H(f) = [1 - \alpha\, e^{-i2\pi f \Delta t}]^2 = 1 - 2\alpha \cos(2\pi f \Delta t) + \alpha^2 \tag{5.6.100}$$

Prewhitening reduces leakage and increases the effectiveness of frequency averaging of the spectral estimate (reduces the random error). The reduced leakage gives rise to a greater dynamic range of the analysis and allows us to examine weak spectral components. Notice that, if $Y(f)$ is the Fourier transform of $y(t)$, then the Fourier transform of $y'(t)$ is

$$Y'(\omega) = \int_t y'(t)e^{-i\omega t}\,dt \approx \omega \cdot Y(\omega) \tag{5.6.101}$$

so that *first differencing* is like a linear high-pass filter with amplitude $|H(\omega)| = |\omega|$. This effect shows up quite well in the processing of satellite-tracked drifter data. Spectra of the drifter positions (longitude, $x(t)$; latitude, $y(t)$) as functions of time, t, are generally "red" whereas the spectra of the corresponding drifter velocities (zonal, $u = \Delta x/\Delta t$; meridional, $v = \Delta y/\Delta t$) are considerably "whiter" (Figure 5.6.28).

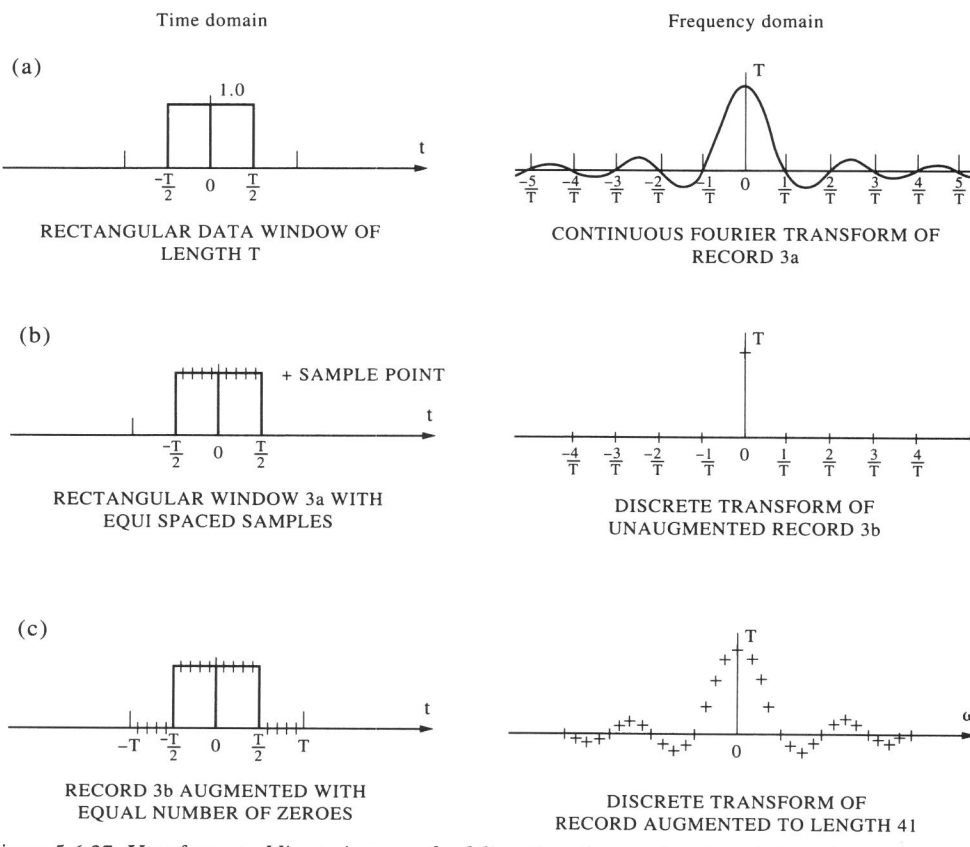

458 *Data Analysis Methods in Physical Oceanography*

Figure 5.6.27. Use of zero padding to improve the delineation of spectral peaks. (a) A continuous box-car window of length T and its continuous Fourier transform; (b) a discrete sample of (a) at equally spaced sampling intervals and its discrete Fourier transform; (c) same as (b) but with zero padding of 2T data points. (From Henry and Graefe, 1971.)

5.6.10 Spectral analysis of unevenly spaced time series

Most discrete oceanographic time-series data are recorded at equally spaced time increments. However, some situations arise where the recorded data are spaced unevenly in time or space. For example, positional data obtained from satellite-tracked drifters are sampled at irregular time intervals due to the eastward progression in the swaths of polar-orbiting satellites and to the advection of the drifters by surface currents. Repeated time-series oceanic transects are typically spaced at irregular intervals due to the vagaries of ship scheduling and weather. In addition, instrument failure and data drop-outs generally lead to "gappy", irregularly spaced time series.

As noted in Section 3.17, a common technique for dealing with irregularly sampled or gappy data is to interpolate data values to a regular grid. This works well as long as there are not too many gaps and the gaps are of short duration relative to the signals of interest. Long data gaps can lead to the creation of erroneous low-frequency oscillations in the data at periods comparable to the gap lengths. Only for the least-squares method for harmonic analysis described in Section 5.5 is unevenly sampled data perfectly acceptable. Vaníček (1971), Lomb (1976) and others have devised a

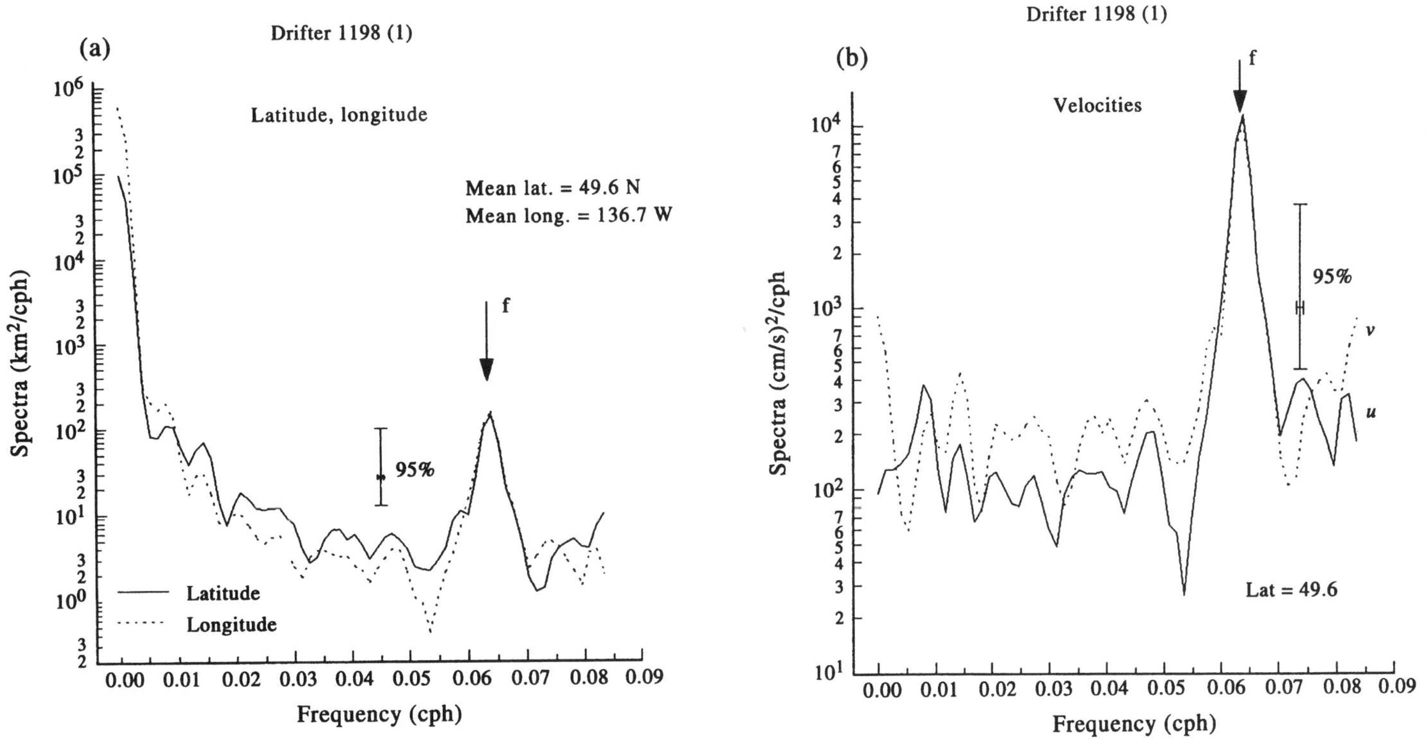

Figure 5.6.28. Effect of a first-difference (high-pass) filter on resulting spectra. (a) Spectra of longitude (Δx) and latitude (Δy) displacements of a satellite-tracked drifter launched in the northeast Pacific in September 1990 ($\Delta t = 3$ h; duration $T = 90$ days); (b) as with (a) but for the zonal ($u = \Delta x/\Delta t$) and meridional velocity ($v = \Delta y/\Delta t$). Mean position of the drifter was 49.6°N, 136.7°W. f denotes the mean inertial frequency; vertical line is the 95% confidence interval.

least-squares spectral analysis method for unevenly spaced time series. The Lomb method described by Press *et al.* (1992) evaluates data, and associated sines and cosines, at the times t_n that the data are measured. For the N data values $x(t_n) = x_n, i = 1, \ldots, N$, the Lomb-normalized periodogram is defined as

$$P(\omega) = \frac{1}{2\sigma^2} \left\{ \frac{\left[\sum_{n=1}^{N}(x_n - \bar{x})\cos[\omega(t_n - \tau)]\right]^2}{\sum_{n=1}^{N}\cos^2[\omega(t_n - \tau)]} + \frac{\left[\sum_{n=1}^{N}(x_n - \bar{x})\sin[\omega(t_n - \tau)]\right]^2}{\sum_{n=1}^{N}\sin^2[\omega(t_n - \tau)]} \right\}$$

(5.6.102)

where as usual

$$\bar{x} = \frac{1}{N}\sum_{n=1}^{N} x_n, \quad \sigma^2 = \frac{1}{N-1}\sum_{n=1}^{N}(x_n - \bar{x})^2 \quad (5.6.103)$$

are the mean and standard deviation of the time series, and the time offset, τ, is defined by

$$\tan(2\omega\tau) = \frac{\sum_{n=1}^{N}\sin(2\omega t_n)}{\sum_{n=1}^{N}\cos(2\omega t_n)} \quad (5.6.104)$$

The offset τ renders equation (5.6.102) identical to the equation we would derive if we attempted to estimate the harmonic content of a data set at frequency ω using the linear least-squares model

$$x(t) = A\cos\omega t + B\sin\omega t \quad (5.6.105)$$

In fact, Vaníček's founding paper on the technique refers to it as a least-squares spectral analysis method. The method, which gives superior results to FFT methods, weights the data on a per point basis rather than on a time-interval basis. By not using weights that span a constant time interval, the method reduces errors introduced by unevenly sampled data. For further details on the Lomb periodogram, including the introduction of significance testing of spectral peaks, the reader is referred to Press *et al.* (1992; pp. 569–577).

5.6.11 General spectral bandwidth and Q of the system

Once the power spectral density, $S(\omega)$, has been computed, the general spectral bandwidth BW may be determined from the three moments, m_k, of the spectra

$$m_k = \int_0^\infty \omega^k S(\omega)\, d\omega, \quad k = 0, 1, 2$$

$$\approx \sum_{i=0}^{N/2} \omega_i^k S(\omega_i) \Delta\omega$$

(5.6.106)

where $N/2$ is the number of spectral estimates and $\Delta\omega$ is the frequency resolution of the spectral estimates (cf. Masson, 1996). In particular

$$BW = (m_2 m_0/m_1^2 - 1)^{1/2} \qquad (5.6.107)$$

The bandwidth, $\Delta\omega_{BW}$, of a particular spectral peak within an oscillatory system can be used to estimate the dissipation of the system at the peak (resonant) frequency, ω_r. Specifically, the "Q" or *Quality* factor of the system measures the amount of energy stored in a linear oscillator compared to the amount of energy lost per cycle through frictional dissipation. The Q-factor characterizes the sharpness of the resonant frequency and is commonly used as a direct measure of tidal dissipation in the ocean. Suppose that the energy of a simple linear system passes through a maximum at resonance frequency and that the energy of the system falls to 50% of its maximum value at frequencies $\omega \approx \omega_r \pm \Delta\omega_{BW}/2$. The Q of the system is then given by

$$Q = \omega_r/\Delta\omega_{BW} \qquad (5.6.108)$$

For example, Wunsch (1972) finds $Q \approx 3.3$ for an apparent resonant period of 14.8 h for the North Atlantic Ocean while Garrett and Munk (1971) obtain an global-wide lower bound of 25 for normal modes near the semidiurnal frequency.

5.6.12 Summary of the standard spectral analysis approach

In summary, PSD estimates for time series $y(t)$ can be obtained as follows using the standard autocorrelation and periodogram approaches:

(1) Remove the mean and trend from the time series. If block averaging is to be used to improve the statistical reliability of the spectral estimates (i.e. to increase the number of degrees of freedom), divide the data series into M sequential blocks of N' data values each, where $N' = N/M$ (see Section 5.6.7).
(2) To partially reduce end effects (Gibbs' phenomenon) or to increase the series length to a power of two for FFT analysis, pad the data with $K \leq N$ zeros. Also pad the record with zeros if you wish to increase the frequency resolution or center spectral estimates in specific frequency bands. To further reduce end effects and side-lobe leakage, taper the time series using a Hanning (raised cosine) window, Kaiser–Bessel window, or other appropriate window (see Section 5.6.6).
(3) Compute the Fourier transforms, $Y(f_k), k = 0, 1, 2, ..., N - 1$, for the time series (for convenience, we have taken $K = 0$). For block-segmented data, calculate the Fourier transforms $Y_m(f_k)$ for each of the M blocks ($m = 1, ..., M$) where $k = 0, 1, ..., N' - 1$ and $N' < N$. To reduce the variance associated with the tapering in step 2, the transforms can be computed for overlapping segments.
(4) Re-scale the spectra to account for the loss of "energy" during application of the window. That is, adjust the scale factor of $Y(f_k)$ (or $Y_m(f_k)$ in the case of smaller block size partitioning) to account for the reduction in spectral energy due to the tapering in step 2. For the Hanning window, multiply the amplitudes of the Fourier transforms by $\sqrt{(8/3)}$. The rescaling factors for other windows are listed in the right-hand column of Table 5.6.4.

(5) Compute the raw power spectral density for the time series (or for each block) where for the two-sided spectral density estimates

$$S_{yy}(f_k) = \frac{1}{N\Delta t}[Y^*(f_k)Y(f_k)], \quad k = 0, 1, 2, ..., N-1$$

(no block averaging)

$$S_{yy}(f_k; m) = \frac{1}{N\Delta t}[Y_m^*(f_k)Y_m(f_k)], \quad k = 0, 1, 2, ..., N'-1 \qquad (5.6.109a)$$

(block averaging)

and for the one-sided spectral density estimates

$$G_{yy}(f_k) = \frac{2}{N\Delta t}[Y^*(f_k)Y(f_k)], \quad k = 0, 1, 2, ..., N/2$$

(no block averaging)

$$G_{yy}(f_k; m) = \frac{2}{N\Delta t}[Y_m^*(f_k)Y_m(f_k)], \quad k = 0, 1, 2, ..., N'/2 \qquad (5.6.109b)$$

(block averaging)

(6) In the case of the block-segmented data, average the raw spectral density estimates from the m blocks of data, frequency-band by frequency-band, to obtain the smoothed periodogram for $S_{yy}(f_k)$ or $G_{yy}(f_k)$. Remember, the trade-off for increased smoothing (more degrees of freedom) is a decrease in frequency resolution.

(7) Incorporate 80, 90, and/or 95% confidence limits in spectral plots to indicate the statistical reliability of spectral peaks. Most authors use the 95% confidence intervals.

We can illustrate some of the points in the above summary using the log–log spectra of sea-level oscillations (Figure 5.6.29) recorded over 14 days (20,160 min) in 1991 at Malokurilsk Bay on the west coast of Shikotan Island in the western Pacific. The main spectral peak is centered at a period of 18.6 min and corresponds to a wind-generated seiche amplitude of about 25 cm (Rabinovich and Levyant, 1992). All spectra have been obtained using segmented versions of the 14-day time series. Each time-series segment has been smoothed using a Kaiser–Bessel window with 50% overlap between segments and each segment has been treated as an independent time series. An FFT algorithm was used to calculate the spectrum for each segment. The smoothest spectrum (Figure 5.6.29a) is based on block averaged spectral estimates from roughly 157 overlapping segments (~20,160 min/128 min), the moderately smooth spectrum (Figure 5.6.29b) from the average of 39 overlapping segments, and the noisiest spectrum (Figure 5.6.29c) from the average of 10 overlapping segments. Taking into account the 50% overlap between segments and the fact that there are two degrees of freedom (DOF) per raw spectral estimate, there are 628 (= 157 × 4), 154, and 36 DOF for the three spectra, respectively. The smoothed spectrum in Figure 5.6.29(d) is derived using a slightly different approach. Although the segment lengths are the same as those in Figure 5.6.29(c) (i.e. 2048 min), the number of DOF is increased with increasing frequency, ω. In this sliding scale, the lowest frequency range uses 36 DOF (as with Figure 5.6.29c), the next frequency band averages together the spectra for

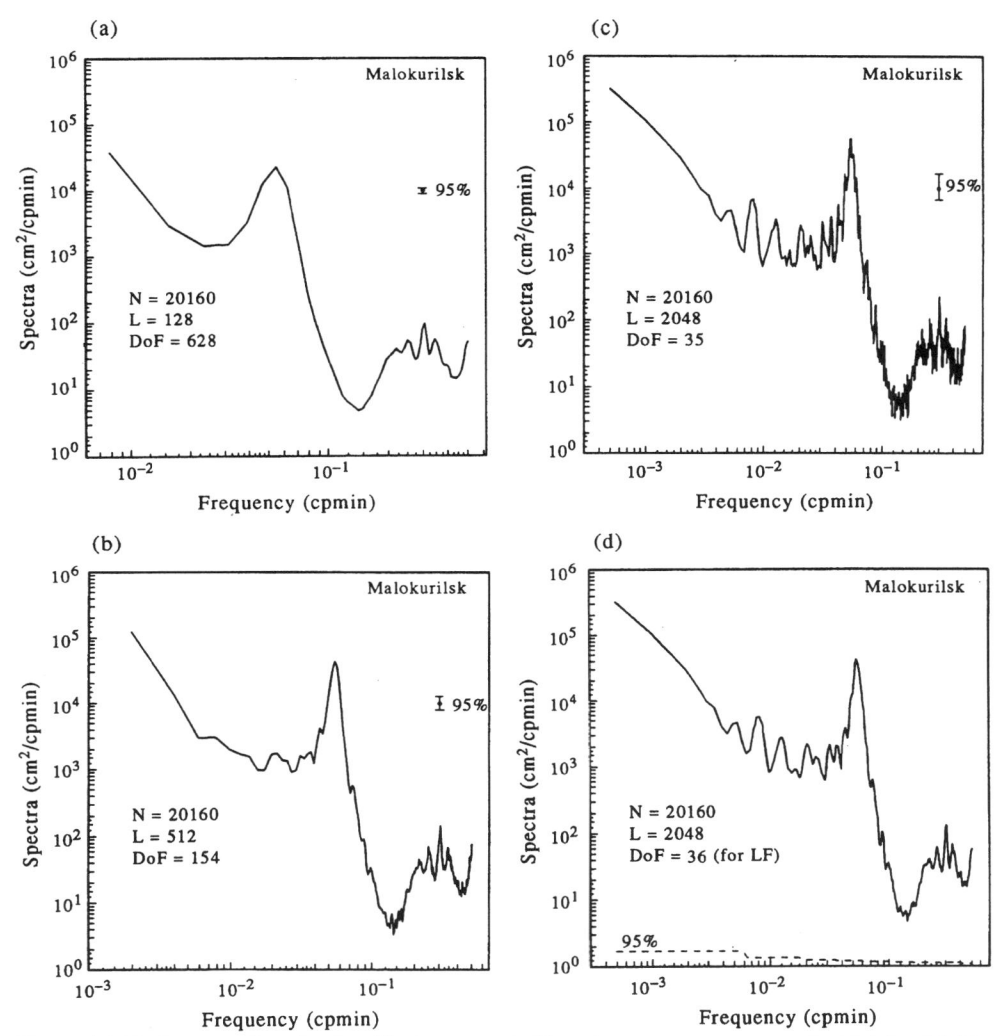

Figure 5.6.29. Spectra of sea-level oscillations recorded by a bottom-pressure gauge in Malokurilsk Bay on the west coast of Shikotan Island. Time-series length $T = N\Delta t$, where $N = 20,160$ and $\Delta t = 1$ min. Segment lengths are $T_s = M\Delta t$, $M \ll N$. Each time-series segment has been smoothed with a Kaiser–Bessel window with 50% overlap between segments. Block averaging has been used to smooth the spectral estimates. (a) Highly smoothed spectrum with $M = 128$ (2^7), DOF = 628; (b) moderately smoothed spectrum with $M = 512$ (2^9), DOF = 154; (c) weakly smoothed spectrum with $M = 2048$ (2^{11}), DOF = 36; (d) same as (c) except that DOF = 36 applies to the lowest frequency range only. For $f \geq 6 \times 10^{-2}$ cycles/min, the number of spectral estimates averaged together increases as 3×36, 5×36, and 7×36, for each of the next three frequency ranges. (Courtesy of A. Rabinovich.)

three adjacent frequencies to give 108 DOF, the next averages together the spectra for five adjacent frequencies to give 180 DOF, and so on.

As indicated by Figure 5.6.29, increasing the number of frequency bands averaged in each spectral estimate enhances the overall smoothness of the spectrum and improves the statistical reliability for specific spectral peaks. The number of DOFs increases and the confidence interval narrows. The penalty we pay for improved statistical confidence is reduced resolution of the spectral peaks. As in Figure 5.6.29(a), too much smoothing diminishes our ability to specify the frequency of spectral peaks and washes out peaks linked to some of the weaker seiches. Because each time-series segment is so short, we also lose definition at the low-frequency end of the spectrum. As indicated by Figure 5.6.29(c), too little smoothing leads to a noisy spectrum for which few spectral peaks are associated with any physical processes. The sliding DOF scale in Figure 5.6.29(d) is a useful compromise.

Covariance function: Since the covariance function, $C_{yy}(\tau)$, and the autospectrum are Fourier transform pairs, the above analysis can be used to obtain a smoothed or unsmoothed estimate of the covariance function. To do this, first calculate the Fourier transform $Y(f)$ of the time series, and determine the product $S_{yy}(f) = N^{-1}\Delta t[Y^*(f)Y(f)]$. Then take the inverse Fourier transform (IFT) of the autospectrum, $S_{yy}(f)$, to obtain the covariance function, $C_{yy}(\tau)$. If the spectrum is unsmoothed prior to the IFT (or IFFT if the FFT was used), we obtain the raw covariance function. If, on the other hand, the autospectrum is smoothed prior to the above integral using one of the spectral windows, such as the Hanning window, the covariance function also will be a smoothed function.

A word of caution: Although everyone agrees on the basic formulation for the discrete Fourier transform (DFT) and the inverse discrete Fourier transform (IDFT), there are several ways to normalize the relations using the number of records, N. In our definitions, (5.6.10) and (5.6.12), N appears in the denominator of the inverse discrete Fourier transform. Some authors normalize using $1/N$ in the DFT only while others insist on symmetry by using $1/\sqrt{N}$ in both DFT and its inverse. User alert: When using "canned" programs to obtain DFTs and IDFTs, ensure that you know how the transforms are defined and adapt your analysis to fit the appropriate processing routines.

5.7 SPECTRAL ANALYSIS (PARAMETRIC METHODS)

If the analytical model for a time series was known exactly, a sensible spectral estimation method would be to fit the model spectrum to the observed spectrum and determine any unknown parameters. In general, however, oceanic variability is too complex to admit simple analytical models and parametric spectral estimates over the full frequency range of the data series. In addition, the imposition of an overly simplified spectral model could seriously degrade any estimation. On the other hand, it is reasonable that relatively simple spectral models might adequately reflect the system dynamics over limited frequency bands. Under some very general conditions, any stationary series can be represented in closed form by a statistical model in which the corresponding spectrum is a rational function of frequency (i.e. a ratio of two polynomials in ω).

If the time series under investigation is long relative to the time scales of interest, and if the spectrum is not overly complicated and does not have too large a dynamic range, the simple smoothed periodogram technique will probably yield adequate results. At a minimum, it will identify the major features in the spectrum. For shorter time series or in studies of fine spectral structure, other techniques may be more applicable. One such spectral analysis technique was developed by Burg (1967, 1972) who showed that it was possible to obtain the power spectrum by requiring the spectral estimate to be the most random or have the maximum entropy of any power spectrum which is consistent with the measured data. This leads to a spectral estimate with a high frequency resolution since the method uses the available lags in the autocovariance function without modification and makes a nonzero estimate (prediction) of the autocorrelation function beyond those which are routinely calculated from the data. Because the spectral values are computed using a maximum entropy condition, the resulting spectral estimates are not accurate in terms of spectral amplitude.

The most popular of the "modern" parametric techniques is the *Autoregressive power spectral density* (AR PSD) model whose origins are in economic time series forecasting and statistical estimation. Autoregressive estimation was introduced to the earth sciences in the 1960s where it was originally applied to geophysical time-series data under the name *maximum entropy method* (MEM). The duality between AR and MEM estimation has been thoroughly explored by Ulrych and Bishop (1975). Autoregressive spectral estimation is attractive because it has superior frequency resolution compared to conventional FFT techniques. As an example of the frequency resolution capability, consider the 14-year time series of average monthly air temperature for New York city (Figure 5.7.1a). The unsmoothed periodogram and three smoothed periodograms reveal a broad spectral peak centered at a period of one year (Figure 5.7.1b). This compares to the much sharper annual peak obtained via AR estimation (Figure 5.7.1c). The results reveal another important difference between the two methods. With the nonparametric periodogram approach, we can determine confidence limits for the spectral peaks while for the parametric method the significance of the peaks is unknown. For example, the maximum entropy method is good for finding the location of spectral peaks but is not reliable for computing the correct spectral energy at those peaks. (The periodogram smoothing in Figure 5.7.1(b) was performed using a Parzen window with truncation values $N = 16, 32$, and 64; the weights for these windows are $w(n) = 1.0 - |2n/N|^2$, with $0 \leq |n| \leq \frac{1}{2}N$.)

In general, autoregressive and maximum entropy PSD estimation are not as widely used in oceanography as traditional spectral analysis methods. The former find their greatest application in analytical climate modeling and in wavenumber spectral estimation. Modern parametric techniques are good so long as the model is good. On the other hand, if the model is false, the resulting spectrum estimate can be highly misleading. It follows that if you have no reason for believing a specific model you are better served using a nonparametric model. For this reason, we limit our presentation to the essential elements of the two methods. The reader is directed to Marple (1987) for a thorough discussion of the topic, including an introduction to Fourier transform methods of spectral analysis.

5.7.1 Some basic concepts

Many deterministic and stochastic discrete-time series processes encountered in oceanography are closely approximated by a rational transfer model in which the input sequence $\{x_n\}$ and the output sequence $\{y_n\}$, which is meant to model the input data, are related by the linear difference relation

$$y_n = \sum_{k=0}^{q} b_k x_{n-k} - \sum_{m=1}^{p} a_m y_{n-m} \qquad (5.7.1)$$

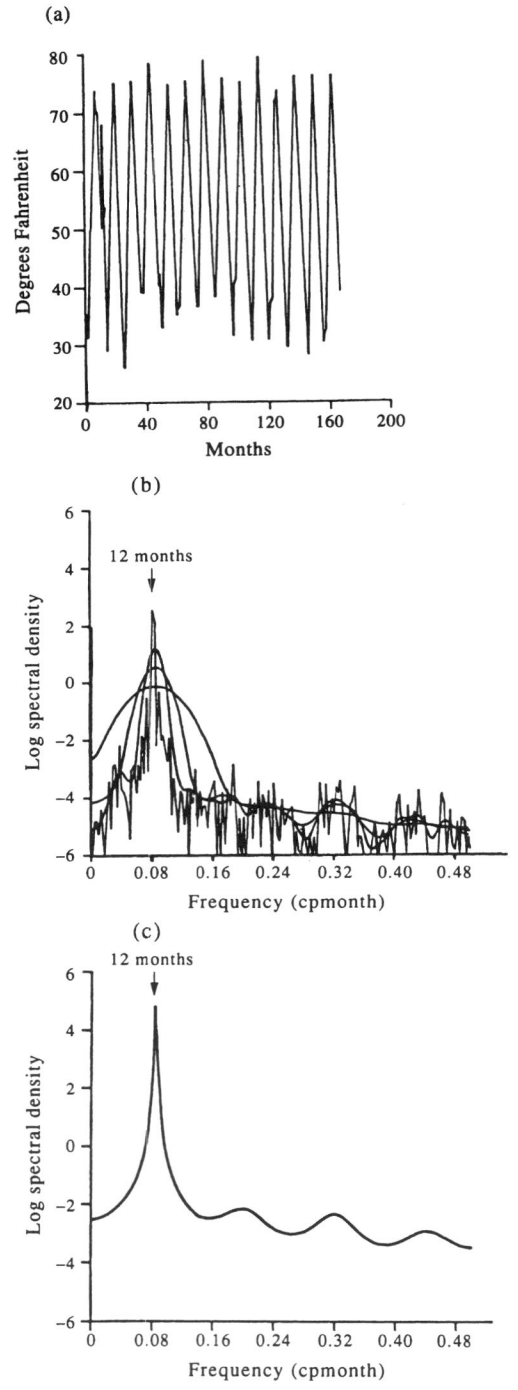

Figure 5.7.1. (a) Time series of monthly average air temperature for New York city (1946–1959); (b) the unsmoothed (raw) periodogram and three smoothed periodograms for Parzen windows with truncation lengths of 16, 32, and 64 months; and (c) an autoregressive spectral estimate of (a) showing the sharp peak at 12-months period. (From Pagano, 1978.)

Here, y_n is shorthand notation for $y(n\Delta t)$, also written as $y(n)$. In its most general form, the linear model (5.7.1) is termed an *autoregressive moving average* (ARMA) model. The power spectral density (PSD) of the ARMA output process is

$$P_{ARMA}(f) = \sigma^2 \Delta t [A(f)/B(f)]^2 \tag{5.7.2}$$

where σ^2 is the variance of the applied white-noise driving mechanism and $\sigma^2 \Delta t$ is the PSD of the noise for the Nyquist interval $-1/(2\Delta t) \leq f \leq 1/(2\Delta t)$. Here

$$A(f) = \alpha[\exp(i2\pi f \Delta t)], \quad B(f) = \beta[\exp(i2\pi f \Delta t)] \tag{5.7.3}$$

where the coefficients α, β are defined in terms of the z-transform, $X(z)$, of the variable $z = \exp(i2\pi f \Delta t)[= \exp(i2\pi n k/N)$ in discrete form] where $k, n = 0, 1, \ldots, N-1$

$$X(z) = \sum_{n=0}^{N-1} x_n z^{-n} \tag{5.7.4}$$

which maps a real-valued sequence into a complex plane. Note that equation (5.7.4) is defined through negative powers of z, the convention used in electrical engineering. Geophysicists expand in positive powers of z (z^{+n}) but define $z = \exp(-iz\pi f \Delta t)$ so the results are the same. The z-transform of the autoregressive branch is

$$\alpha(z) = \sum_{n} a_n z^{-n} \tag{5.7.5a}$$

while that of the moving average branch is

$$\beta(z) = \sum_{n} b_n z^{-n} \tag{5.7.5b}$$

Specification of the parameters $\{a_k\}$, termed the autoregressive coefficients, the parameters $\{b_k\}$, termed the moving-average coefficients, and the variance σ^2 is equivalent to specifying the spectrum of the process $\{y_n\}$. Without loss of generality, one can assume $a_o = 1$ and $b_o = 1$ since any gain of the system (5.7.1) can be incorporated into σ^2. If all the $\{a_k\}$ terms except $a_o = 1$ vanish then

$$y_n = \sum_{k=0}^{q} b_k x_{n-k} \tag{5.7.6}$$

and the process is simply a moving average of order q, and

$$P_{MA}(f) = \sigma^2 \Delta t |A(f)|^2 \tag{5.7.7}$$

This model is sometimes called an *all-zero model* since spectral peaks and valleys are formed through zeros of the function $A(f)$. If all the $\{b_k\}$ terms except $b_o = 1$ vanish then

$$y_n = \sum_{m=1}^{p} a_m x_{n-m} + \varepsilon_n \tag{5.7.8}$$

and the process is strictly an autoregressive model of order p. The process is called AR in the sense that the sequence x_n is a linear regression on itself with ε_n representing the error. With this model, the present value y_n is expressed as a weighted sum of past values plus a noise term. The PSD is

$$P_{AR}(f) = \frac{\sigma^2 \Delta t}{|B(f)|^2} \qquad (5.7.9)$$

In the engineering literature, this model is sometimes called an *all-pole model* since narrow spectral peaks can be sharply delineated through zeros in the denominator.

5.7.2 Autoregressive power spectral estimation

The discrete form of an autoregressive model $y(t)$ of order p is represented by the relationship

$$y(n) = a_1 y(n-1) + a_2 y(n-2) + \ldots + a_p y(n-p) + \varepsilon(n) \qquad (5.7.10)$$

where time $t = n\Delta t$, the a_k ($k = 1, \ldots, p$) are constant coefficients, and $\varepsilon(t)$ is a white-noise series (usually called the innovation of the AR process) with zero mean and variance σ^2. Another interpretation of the AR process links $y(t)$ with a value that is predicted from the previous $p - 1$ values of the process with a prediction error equal to $\varepsilon(t)$. Thus, the a_k ($k = 1, \ldots, p$) represent a p-point prediction filter. If $Y(z)$ is the z-transform of $y(n)$ then

$$Y(z) = \sum_{n=0}^{p} y(n) z^n \qquad (5.7.11)$$

and

$$Y(z) - Y(z)(a_1 z + a_2 z^2 + \ldots + a_p z^p) = D(z) \qquad (5.7.12)$$

so that

$$|Y(z)|^2 = \frac{|D(z)|^2}{|1 - a_1 z - a_2 z^2 \ldots - a_p z^p|^2} \qquad (5.7.13)$$

Substituting $z = \exp(-i2\pi f \Delta t)$ we obtain half of the true power spectrum. If the autoregression is a reasonable model for the data, then the autoregressive power spectral density estimate based on (5.7.9) is

$$P_{AR}(f) = \frac{\sigma^2 \Delta t}{\left|1 + \sum_{k=1}^{p} a_k \exp(-i2\pi f k \Delta t)\right|^2} \qquad (5.7.14)$$

To find the PSD we need only estimate three things: (1) the autoregressive parameters $\{a_1, a_2, \ldots, a_p\}$; (2) the variance, σ^2, of the white-noise process that is assumed to be driving the system; and (3) the order, p, of the process. The limitations of the AR model are the degrading effect of observational noise, spurious peaks, and some anomalous effects which occur when the data are dominated by sinusoidal components. Unlike conventional Fourier spectral estimates, the peak amplitudes in AR spectral estimates are not linearly proportional to the power when the input process consists of sinusoids in

Time-series Analysis Methods 469

noise. For high signal-to-noise ratios, the peak is proportional to the square of the power with the area under the peak proportional to power.

5.7.2.1 Autoregressive parameter estimation

Yule–Walker equations: If the autocorrelation function, $R_{yy}(k)$, is known exactly, we can find the $\{a_k\}$ by the Yule–Walker equations. This method relates the AR parameters to the known (or estimated) autocorrelation function of $y(n)$

$$R_{yy}(k) = \frac{1}{N}\sum_{n=1}^{N-k}[[x(n+k) - \bar{x}][x(n) - \bar{x}]; \quad \bar{x} = \frac{1}{N}\sum_{n=1}^{N}x(n) \tag{5.7.15}$$

There are other methods of estimating R_{yy} but this estimator has the attractive property that its mean-squared error is generally smaller than that of other estimators (Jenkins and Watts, 1968). Since it is generally assumed that the mean $\bar{x}$ has been removed from the data, the autocovariance and autocorrelation functions are identical. To get the AR parameters, one need only choose p equations from the Yule–Walker equations for $k > 0$, solve for $\{a_1, a_2, \ldots, a_p\}$, and then find σ^2 from (2.39) for $k = 0$. The matrix equation to find the a_is and σ^2 is

$$\begin{vmatrix} R_{yy}(0) & R_{yy}(-1) & \cdots & R_{yy}(-p) \\ R_{yy}(1) & R_{yy}(0) & \cdots & R_{yy}[-(p-1)] \\ \cdots & \cdots & \cdots & \cdots \\ \cdots & \cdots & \cdots & \cdots \\ R_{yy}(p) & R_{yy}(p-1) & \cdots & R_{yy}(0) \end{vmatrix} \begin{vmatrix} 1 \\ a_1 \\ \cdots \\ \cdots \\ a_p \end{vmatrix} = \begin{vmatrix} \sigma^2 \\ 0 \\ \cdots \\ \cdots \\ 0 \end{vmatrix} \tag{5.7.16}$$

Thus, to determine the AR parameters and the variance σ^2 one must solve (5.7.16) using the $p + 1$ autocorrelation lags, $R_{yy}(0), \ldots, R_{yy}(p)$, where $R_{yy}(-k) = R_{yy}^*(k)$.

Solutions to the Yule–Walker matrix equation can be found via the computationally efficient Levinson–Durbin algorithm which proceeds recursively to compute the parameter sets $\{a_{11}, \sigma_1^2\}, \{a_{21}, a_{22}, \sigma_2^2\}, \ldots, \{a_{p1}, a_{p2}, \ldots, a_{pp}, \sigma_p^2\}$. The final set at order p (the first subscript) is the desired solution. The algorithm requires p^2 operations as opposed to the $O(p^3)$ operations of Gaussian elimination. More specifically, the recursion algorithm gives

$$a_{11} = \frac{-R_{yy}(1)}{R_{yy}(0)} \tag{5.7.17a}$$

$$\sigma_1^2 = (1 - |a_{11}|^2)R_{yy}(0) \tag{5.7.17b}$$

with the recursion for $k = 2, 3, \ldots, p$ given by

$$a_{kk} = \frac{-1}{\sigma_1^2}\left[R_{yy}(k) + \sum_{j=1}^{k-1}a_{k-1,j}R_{yy}^{(k-j)}\right] \tag{5.7.18a}$$

$$a_{ki} = -a_{k-1,i} + a_{kk}(a_{k-1,k-i})^* \tag{5.7.18b}$$

$$\sigma_k^2 = (1 - |a_{kk}|^2)\sigma_{k-1}^2 \tag{5.7.18c}$$

Burg algorithm: Box and Jenkins (1970) point out that the Yule–Walker estimates of the AR coefficients are very sensitive to rounding errors, particularly when the AR process is close to becoming nonstationary. The assumption that $y(k) = 0$, for $|k| > p$ leads to a discontinuity in the autocorrelation function and a smearing of the estimated PSD. For this reason, the most popular method for determining the AR parameters (prediction error filter coefficients) is the Burg algorithm. This algorithm works directly on the data rather than on the autocorrelation function and is subject to the Levinson recursion (5.7.18b). As an illustration of the differences in the YW and the Burg estimates, the respective values of a_{11} for the series $y(t_k) = y(k)$ are

$$a_{11} = \frac{\sum_{k=2}^{p} y(k)y(k-1)}{\sum_{k=1}^{p} y(k)^2}, \text{ for the Yule-Walker estimate}$$

$$a_{11} = \frac{\sum_{k=2}^{p} y(k)y(k-1)}{\frac{1}{2}x_1^2 + \sum_{k=1}^{p} y(k)^2 + \frac{1}{2}x_p^2}, \text{ for the Burg estimate}$$

(5.7.19)

Detailed formulation of the Burg algorithm is provided by Kay and Marple (1981; p. 1392). Again, there are limitations to the Burg algorithm, including spectral line splitting and biases in the frequency estimate due to contamination by rounding errors. Spectral line splitting occurs when the spectral estimate exhibits two closely spaced peaks, falsely indicating a second sinusoid in the data.

Least squares estimators: Several least squares estimation procedures exist that operate directly on the data to yield improved AR parameter estimates and spectra than the Yule–Walker or Burg approaches. The two most common methods use forward linear prediction for the estimate, while a second employs a combination of forward and backward linear prediction. Ulrych and Bishop (1975) and Nuttall (1976) independently suggested this least squares procedure for forward and backward prediction in which the Levison recursion constraint imposed by Burg is removed. The least squares algorithm is almost as computationally efficient as the Burg algorithm requiring about 20 more computations. The improvement by the LS approach over the Burg algorithm is well worth the added computation time. Improvements include less bias in the frequency estimates, and absence of observed spectral line splitting for short sample sinusoidal data.

Barrodale and Erickson (1978) provide a FORTRAN program for an "optimal" least-squares solution to the linear prediction problem. The algorithm solves the underlying least-squares problem directly without forcing a Toeplitz structure on the model. Their algorithm can be used to determine the parameters of the AR model associated with the maximum entropy method and for estimating the order of the model to be used. As illustrated by the spectra in Figure 5.7.2, this approach leads to a more accurate frequency resolution for short sample harmonic processes. In this case, the test data were formed by summing 0.03 and 0.2 Hz sine waves generated in single precision and sampled 10 times per second. The reader is also referred to Kay and Marple (1981; p. 1393) for additional details.

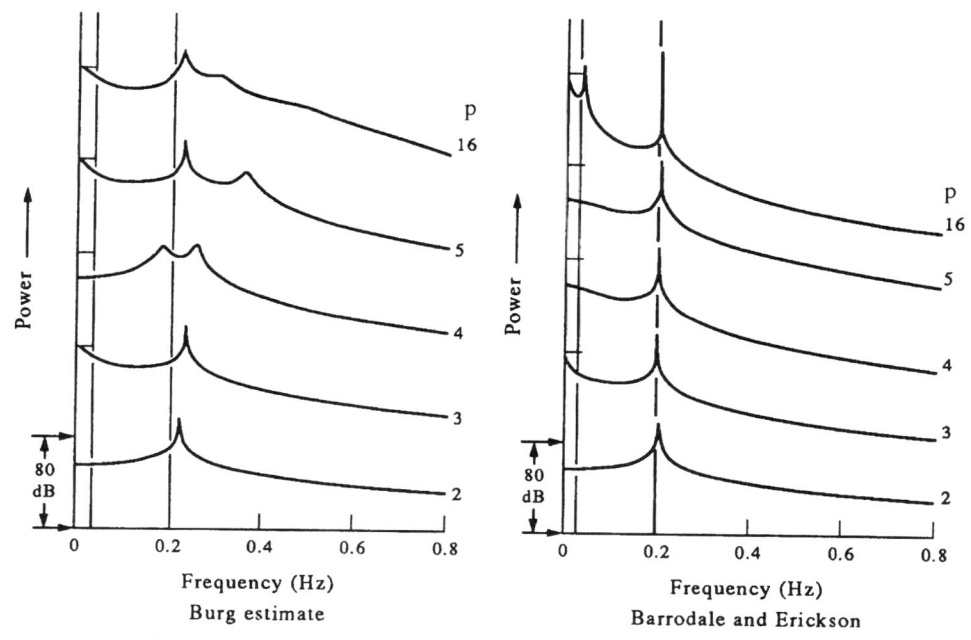

Figure 5.7.2. Maximum entropy method spectra obtained using (a) the Burg and (b) the Barrodale and Erickson algorithms. Signal consists of a combined 0.2 and 0.03 Hz (cps) sine wave. Spectra are plotted for increasing numbers of coefficients, p. (From Barrodale and Erickson, 1978.)

5.7.2.2 Order of the autoregressive process

The order p of the autoregressive filter is generally not known *a priori* and is acknowledged as one of the most difficult tasks in time series modeling by parametric methods. The choice is to postulate several model orders then compute some error criterion that indicates which model order to pick. Too low a guess for the model order results in a highly smoothed spectral estimate. Too high an order introduces spurious detail into the spectrum. One intuitive approach would be to construct AR models with increasing order until the computed prediction error power σ_k^2 reaches a minimum. Thus, if a process is actually an AR process of order p, then $a_{p+1,k} = a_{pk}$ for $k = 1, 2, \ldots, p$. The point at which a_{pk} does not change would appear to be a good indicator of the correct model order. Unfortunately, both the Yule–Walker equations and Burg algorithm involve prediction error powers

$$\sigma_k^2 = \sigma_{k-1}^2[1 - |a_{kk}|^2] \tag{5.7.20a}$$

that decrease monotonically with increasing order p, so that as long as $|a_{kk}|^2$ is nonzero (it must be ≤ 1) the prediction error power decreases. Thus, the prediction error power is not sufficient to indicate when to terminate the search. Alternative approaches (Kay and Marple, 1981) have been proposed by Akaike (termed the final prediction error, FPE, and the Akaike information criterion, AIC), and by Parzen (termed the criterion autoregressive transfer function). The Akaike information criterion determines the model order by minimizing an information theoretic function. If the process has Gaussian statistics, the AIC is

Summary of algorithms

Method	Model applied	Advantages	Disadvantages
Periodogram method using FFT or direct Fourier transform	Sum of harmonics (sines and cosines). No specific model needed.	1. Uses harmonic least squares fit to the data; 2. output $S(f)$ directly proportional to power; 3. most computationally efficient; 4. well-established methodology; 5. confidence intervals easily computed; 6. integral of $S(f)$ over frequency band Δf is equal to the variance of the signal in that band. 7. easily generalized to cross-spectra and rotary spectra analysis.	1. Frequency resolution $\Delta f \approx 1/T$ dependent only on record length, T; 2. poor performance for short data records; 3. side-lobe leakage distorts spectra if appropriate windowing not done; windowing reduces frequency resolution, Δf; 4. must average spectral estimates to improve statistical reliability.
Autoregressive, Yule–Walker algorithm.	Autoregressive (all-pole) process. Specific model.	1. Improved spectral resolution over Fourier transform methods; 2. sharp spectral peaks; 3. no side-lobe leakage problems; 4. minimum phase (stable) linear prediction filter guaranteed if biased lag estimates computed; 5. related to linear prediction analysis and adaptive filtering.	1. Model order, p, must be specified; 2. spectral line splitting occurs; 3. implied windowing distorts spectra; 4. confidence intervals not readily computed.
Autoregressive, Burg algorithm.	Autoregressive (all-pole) process. Specific model.	1. Improved resolution over Fourier transform methods. Uses a constrained recursive least squares approach 2. no side-lobe leakage problems; 3. high resolution for low noise signals; 4. good spectral fidelity for short data series; 5. no windowing implied; 6. Stable linear prediction filter guaranteed.	1. Model order, p, must be specified; 2. spectral line splitting can occur; 3. confidence intervals not readily computed.
Autoregressive, least-squares method.	Autoregressive (all-pole) process. Specific model.	1. Sharper spectra than for other AR methods 2. no side-lobes; 3. good spectral fidelity for short data series; 4. no windowing; 5. no line splitting; 6. uses exact recursive least squares solution with no constraint.	1. Model order must be specified; 2. stable linear prediction filter not guaranteed, though stable filter results in most cases.

$$\text{AIC}(p) = \ln(\sigma_p^2) + 2(p+1)/N \tag{5.720b}$$

where σ_p^2 is the prediction error power and N is the number of data samples. The second term represents the penalty for the use of extra autoregressive coefficients that do not result in a substantial reduction in the prediction error power. The order p is the one that minimizes the AIC.

5.7.2.3 Maximum entropy method (MEM)

The only constraint on the AR method is that the data yield the known autocorrelation function $R_{yy}(k)$ for the interval $0 < k < p$. The assumption that $y(k) = 0$, for $|k| > p$ leads to a discontinuity in the autocorrelation function and a smearing of the estimated power spectral density. The MEM was designed, independently of autoregressive estimation, to eliminate the distortion of the spectrum caused by the truncated $R_{yy}(k)$. By adding a second constraint to improve the spectral estimation, the method gets away from the problems with the Yule–Walker algorithm. In essence, the MEM is a way of extrapolating the known autocorrelation function to lags $k > p$, which are not known. In words, we assume that $\{R_{yy}(0), \ldots, R_{yy}(p)\}$ are known and find a logical way to extend to lags $\{R_{yy}(p+1), \ldots\}$. As it turns out, the power spectral estimate for the MEM approach is equivalent to the power spectral estimate for the AR process.

In general, there exist an infinite number of possible extrapolations. Burg (1968) argued that preferred extrapolation should do two things: (1) Yield the known R_{yy} for $0 \leq k \leq p$; and (2) generate an extrapolated R_{yy} for $k > p$ that causes the time series to have maximum entropy under the constraint (1). The time series that results is the most random one which adheres to the known R_{yy} for the first $p + 1$ lags. Alternatively, we can say that PSD is the one with whitest noise (flattest spectrum) of all possible spectra for which $\{R_{yy}(0), \ldots, R_{yy}(p)\}$ is known. The reason for choosing the maximum entropy criterion is that it imposes the fewest constraints on the unknown time series by maximizing its randomness thereby causing minimum bias and operator intervention. For a Gaussian process, the entropy per sample is proportional to

$$\int_{-1/2\Delta t}^{1/2\Delta t} \ln[P_y(f)] \, df \tag{5.7.21}$$

where $P_y(f)$ is the PSD of y_n. The spectrum is found by maximizing (5.7.21) subject to the constraint that the $p + 1$ known lags satisfy the Wiener–Khinchin relation

$$\int_{-1/2\Delta t}^{1/2\Delta t} P_y(f) e^{-i2\pi f n \Delta t} \, df = R_{yy}(n), \quad n = 0, 1, \ldots, p \tag{5.7.22}$$

The solution is found using the Lagrange multiplier technique (see Ulrych and Bishop, 1975) as

$$P_y(f) = \frac{\sigma_p^2 \Delta t}{\left|1 + \sum_{k=1}^{p} a_{pk} \exp(-i2\pi f k \Delta t)\right|^2} \tag{5.7.23}$$

where $\{a_{p1}, ..., a_{pp}\}$ and σ_p^2 are just the order-p predictor parameters and prediction error power, respectively. With knowledge of $\{R_{yy}(0), R_{yy}(1), ..., R_{yy}(p)\}$ the power spectral density (PSD) of the maximum entropy method (MEM) is equivalent to the PSD of the autoregressive method. That is, the MEM spectral analysis is equivalent to fitting an AR model to the random process. It is indeed interesting that the representation of a stochastic process by an AR model is that representation that exhibits maximum entropy. The duality of the AR model and MEM has enabled workers to apply the large body of literature on AR time-series analysis to overcome shortcomings of the MEM.

The estimation of the MEM spectral density requires a knowledge of the order of the AR process that we use to model the data. The importance of correctly estimating the order p is illustrated using the AR process $y_n \equiv y(t_n)$ at times $t_n = n\Delta t$

$$y_n = 0.75 y_{n-1} - 0.5 y_{n-2} + \varepsilon_n \tag{5.7.24}$$

with noise variance $\sigma_\varepsilon^2 = 1$ (Figure 5.7.3a). Here $E[y(t)\varepsilon(t)] = \sigma_\varepsilon^2$, but $E[y(t)\varepsilon'(t)] = 0$ for any other additive noise, ε'. As indicated by Figure 5.7.3(b), which compares the theoretical power of a specified second-order AR process with the power spectral density computed from a realization of this process using $p = 2$ and $p = 11$ (Ulrych and Bishop, 1975), the correct choice of p is vital in obtaining a meaningful estimate of the power spectrum of the process. The peak value and the width of the spectral line of the MEM power spectral density estimate also may have considerable variance in the MEM estimates.

Although the MEM has numerous advantages over traditional nonparametric spectral techniques, especially for short data series, the usefulness of the approach is diminished by the lack of a straightforward criterion for choosing the length (order) of the prediction model. Too short a length results in a highly smoothed spectrum obviating the resolution advantages of the MEM, whereas an excessive length introduces spurious detail into the spectrum.

Confidence intervals: A major shortcoming of MEM is the lack of a mathematically consistent variance estimator (confidence interval) for the spectral density. One approach is to approximate the confidence bounds in the same way that we compute the bounds in traditional multivariate spectral analysis (i.e. using a chi-square variable with ν degrees of freedom) under the assumption that the equivalent number of degrees of freedom is given by $\nu = N/p$, where N the number of data points in the time series and p is the order of the model (Privalsky and Jensen, 1993, 1994). The order p should be chosen on the basis of objective criteria such as Akaike's information criterion, Parzen's criterion and so on (see Lütkepohl, 1985).

5.7.2.4 An autoregressive model of global temperatures

One way to determine the effect of initial conditions and random noise on the global temperature predictions of computer-simulated general circulation models (GCMs) is to obtain a control realization, modify the initial conditions and noise, obtain a second realization and compare results. Since this could take several months of supercomputing time, a more practical approach is to employ a model of the global air temperature series, $T(t)$, derived by Jones (1988) (Figure 5.7.4). If we assume that the sensitivity of GCMs to changing conditions is similar to that of a stationary

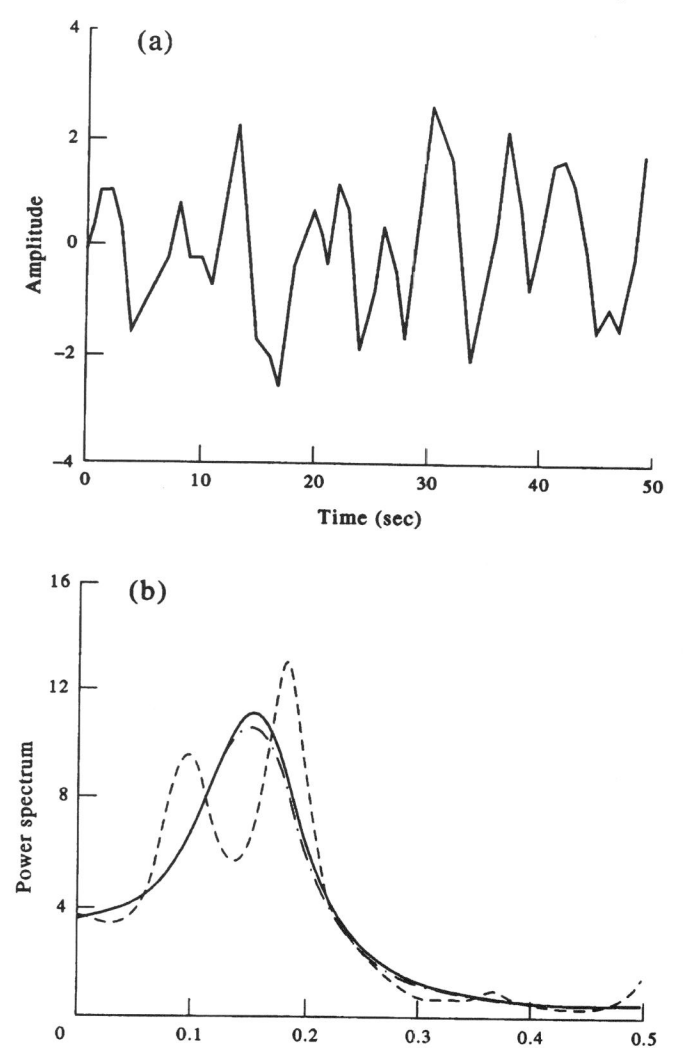

Figure 5.7.3. Maximum entropy spectra. (a) Time series for the second-order AR process $y_n = 0.75y_{n-1} - 0.5y_{n-2} + \varepsilon_n$ (5.7.24). (b) Spectral computation for the AR process. Solid line: the true power spectrum. Dot–dash line: maximum entropy method (MEM) estimate with 3-point ($p = 2$) prediction error filter. Dashed line: MEM estimate with 12-point ($p = 11$) error filter. (From Ulrych and Bishop, 1975.)

autoregressive model, then marked changes in the AR model that result from slight changes in the initial conditions or inherent noise are evidence that GCMs are too sensitive to these parameters to be reliable.

If $Z_n \equiv Z(t_n)$ represents the temperature deviation (departure from the long-term mean) at year t_n, then the maximum likelihood fourth order AR model for the temperature data in Figure 5.7.4 is

$$Z_n = 0.669Z_{n-1} - 0.095Z_{n-2} + 0.104Z_{n-3} + 0.247Z_{n-4} + \varepsilon_n \quad (5.7.25)$$

where $Z_n = T_n - \overline{T}$, and ε_n is an uncorrelated white-noise series with zero mean and variance equal to $0.0115°C^2$ (Tsonis, 1991; Gray and Woodward, 1992). In general, we can state that for any AR process, the initial values will have little effect on forecasts if the sample size is large relative to the order of the process. For this reason, AR processes are often known as "short memory" processes. In the above model, the correlation between $Z(t)$ and $Z(t + m\Delta t)$ is $0.9(0.96)^m$, for values of m greater than about five. For example, the correlation between $Z(t)$ and $Z(t + 30\Delta t)$ is 0.27, while that between $Z(t)$ and $Z(t + 50\Delta t)$ is 0.14. These correlations imply that, even if we started the model with the same initial values $Z_1, \ldots, Z_4$, different realizations of the model would typically have low cross-correlation after 30 years and possess very little similarity beyond 50 years (Figure 5.7.5a). The dissimilarity is associated with the stochastic nature of the noise $\varepsilon(t)$ which quickly decorrelates the present value of the model from its past values. The fact that the two series converge to a similar level near $t = 100$ years is not an indication that they are merging since extending these realizations causes them to depart from one another.

To show the importance of the noise, rather than the initial conditions, Gray and Woodward generated two samples with different starting values but with the same noise sequence. This was intended to mimic a specified set of random conditions driving the weather but having different starting values. As revealed by Figure 5.7.5(b), the realizations begin to merge by year 30, demonstrating their insensitivity to the initial conditions. A further point is that for stationary AR processes, the forecast function is only a function of the sample mean and the last four observations. Since the starting values are independent of the last four observations and small changes in the starting conditions have little effect on the sample mean for a long time series, the forecasts from such a model will be insensitive to changes. In closing their article, Gray and Woodward note that conventional autoregressive moving average (ARMA) modeling methodology indicates that the temperature time series should first be differentiated. Application of a variety of techniques suggests an order 10 (AR(10)) model as the "optimum" model for the differentiated data which gives rise to an AR(11) model for the original time series, not an AR(4) model used in the analysis.

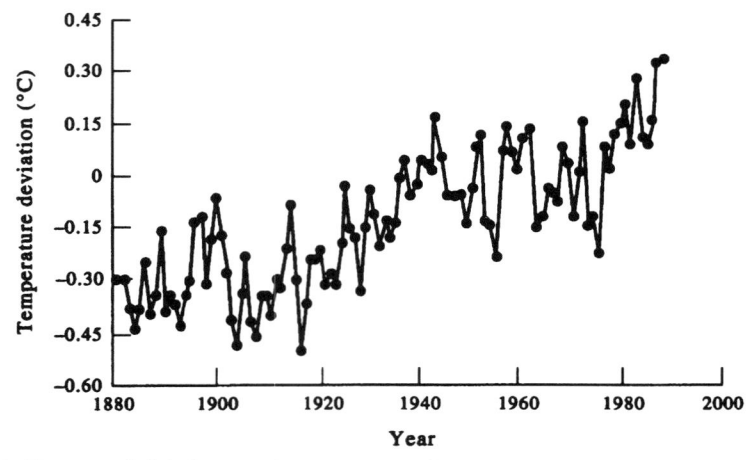

Figure 5.7.4. *The annual global mean air temperatures from 1881 to 1988 as deviations (°C) from the 1951–1970 average. (From Gray and Woodward, 1992.)*

Lastly, Tsonis (1992) replies that it is not appropriate to change the noise of the signal without also changing the initial conditions.

5.7.3 Maximum likelihood spectral estimation

As first demonstrated by Capon (1969), spectra can be defined using the maximum likelihood procedure. Instead of using a fixed window to operate on the autocorrelation function, the window shape is changed as a function of wavenumber or frequency. The window is designed to reject all frequency components in an optimal way, except for the one frequency component which is desired.

Rather than go through the details of defining the procedure for the maximum likelihood spectrum, we offer here comparisons between the traditional method (in this case, represented by a spectrum computed using a Bartlett window), a maximum likelihood spectrum, and a spectrum computing using the maximum entropy

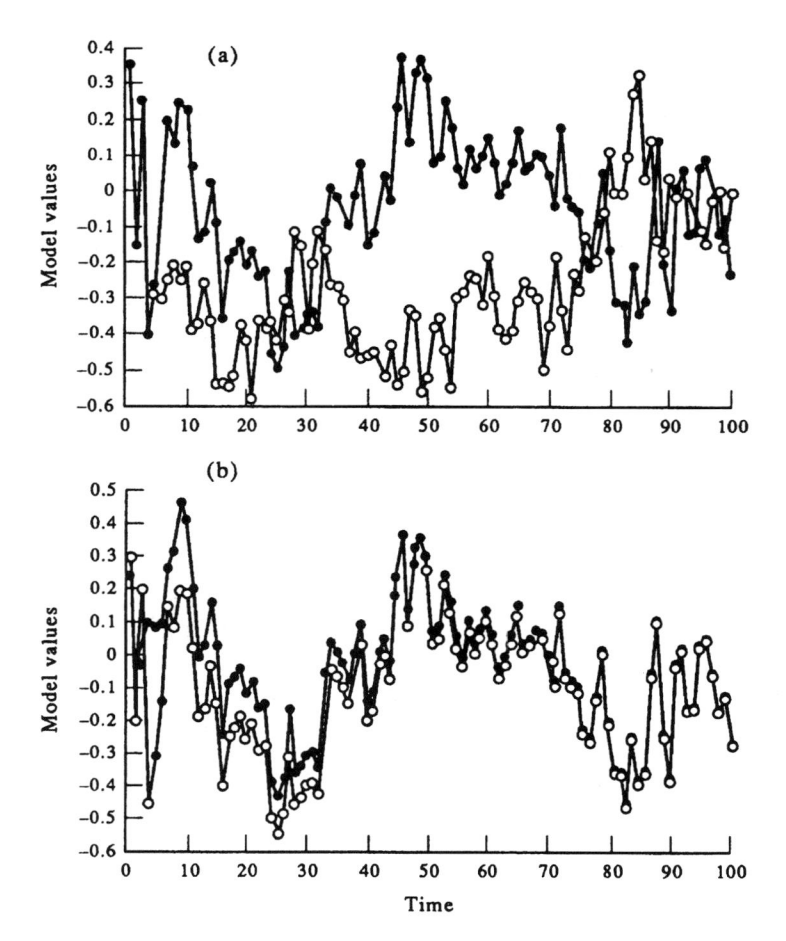

Figure 5.7.5. Two simulated realizations from the AR(4) model given by equation (5.7.25). (a) Same starting values but different and independently derived noise sequence; (b) different starting values but the same noise sequence. (From Gray and Woodward, 1992.)

procedure (Figure 5.7.6). As the figure illustrates, the maximum entropy spectrum has narrow peaks while both the Bartlett window and maximum likelihood method yield much broader spectral peaks. Note also that, except for the maximum spectral values, the maximum entropy spectrum significantly underestimates the spectral estimates for the 0.15 Hz signal and white noise. The maximum entropy spectrum also has small side-lobe energy that is dramatically less than the off-peak energy in either of these two spectra. The maximum likelihood spectral values are also systematically lower than those using the standard method with a Bartlett window. A similar comparison is shown in Figure 5.7.7, which first shows a time series of a 1 Hz (1 cps) sinusoid with 10% white noise added to it (Figure 5.7.7a). The power spectrum computed as the square of the Fourier coefficients is displayed in Figure 5.7.7(b). This can be compared with the narrow-peaked maximum entropy spectrum in Figure 5.7.7(c). The peaks are located at the same frequency representative of the 1 Hz, but the maximum entropy spectrum is extremely narrow while the Fourier power spectrum has a very wide peak. It is easy to see that the maximum entropy method seriously underestimates the spectral values at frequencies other than the main peak.

5.8 CROSS-SPECTRAL ANALYSIS

Estimation of autospectral density functions deals only with the frequency characteristics of a single scalar or vector time series, $x(t)$. Estimation of cross-spectral density functions performs a similar analysis but for two time series, $x_1(t)$ and $x_2(t)$, spanning concurrent times, $0 \leq t \leq T$. Although we often use time series from similar distri-

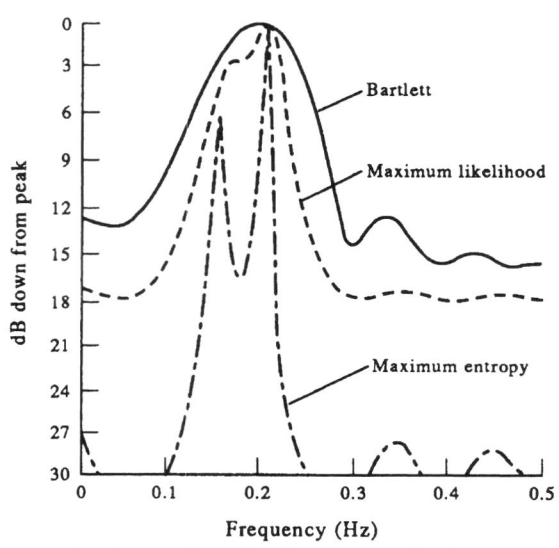

Figure 5.7.6. Power spectral estimates for a signal consisting of white noise plus two sine waves with frequencies 0.15 and 0.2 Hz (cps). Solid line: spectrum using the autocovariance method with a Bartlett smoothing window. Dashed line: Maximum likelihood spectral estimate. Dash–dot line: maximum entropy spectrum. (From Lacoss, 1971.)

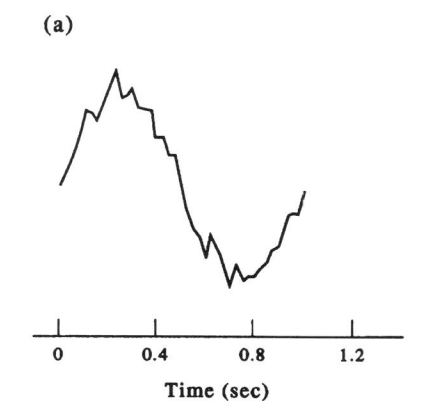

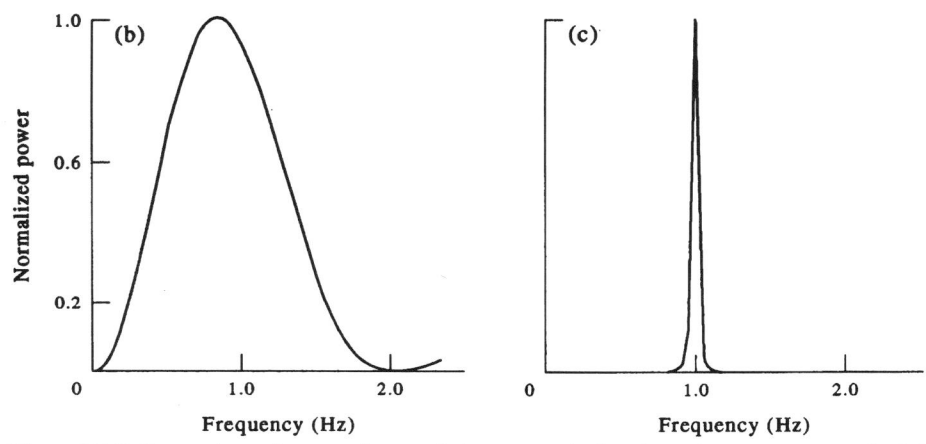

Figure 5.7.7. Comparison of spectra from periodogram method and maximum entropy method (MEM). (a) A sinusoid with 10% white noise and truncated with a 1 s window; (b) the power spectrum of (a) computed as the square of the modulus of the Fourier transform; (c) the MEM power spectrum of (a). Frequency in Hz (cps). (From Ulrych, 1972.)

butions, such as the velocity records from nearby moorings, cross-spectra may also be computed for two completely different quantities. In that sense, we *can* mix apples and oranges. For example, the cross-spectrum formed from the time-varying velocity fluctuations, $x_1(t) = u'(t)$, and the temperature fluctuations, $x_2(t) = T'(t)$, measured over the same time span at the same location gives an estimate of the local eddy heat flux, $q' = \rho C_p u' T'(t)$, as a function of frequency (ρ is the density and C_p the specific heat of seawater). Because autospectra involve terms like $x_1 x_1^*$, where the asterisk denotes complex conjugate, the spectra are real-valued and all phase information in the original signal is lost. Cross-spectra, on the other hand, involve terms like $x_1 x_2^*$ and are generally complex quantities whose real and imaginary parts take into account the correlated portions of both the amplitudes and relative phases of the two signals.

There are two ways to quantify the real and imaginary parts of cross-spectra. One approach is to write the cross-spectrum as the product of an amplitude function, called the *cross-amplitude spectrum*, and a phase function called the *phase spectrum*. The sample cross-amplitude spectrum gives the distribution of co-amplitudes with frequency while the sample phase spectrum indicates the angle (or time) by which one series leads or lags the other series as a function of frequency. Alternatively, the cross-spectrum can be decomposed into a *coincident spectral density function (or co-spectrum)*, which defines the degree of co-oscillation for those frequency constituents of the two time series that fluctuate in-phase, and a *quadrature spectral density function (or quad-spectrum)*, which defines the degree of co-oscillation for frequency constituents of the two series that co-oscillate but are out-of-phase by ±90°. Statistical confidence intervals can be provided for normalized versions of the cross-spectral estimates.

5.8.1 Cross-correlation functions

In Section 5.6.3.1, we showed that the autocovariance function, $C_{xx}(\tau)$, and the autospectrum, $S_{xx}(f)$, are Fourier transform pairs. Similarly, for separate time series $x_1(t)$ and $x_2(t)$, the cross-covariance function, $C_{x1x2}(\tau)$, and the cross-spectrum, $S_{x1x2}(f)$, are transform pairs. Thus, we can take the Fourier transform of the lagged cross-covariance function to obtain the cross-spectrum or we can take the inverse Fourier transform of the cross-spectrum to obtain the cross-covariance function. As a prelude to cross-spectral analysis, it is worth presenting a brief summary of cross-correlation functions commonly used in oceanography for scalar and vector time series. The cross-correlation functions tell us how closely two records are "related" in the time domain, whereas the cross-spectrum tells us how oscillations within specific frequency bands are related in the frequency domain.

Using the abbreviation $C_{12}(\tau)$ for $C_{x1x2}(\tau)$, the *cross-covariance function* is defined as

$$C_{12}(\tau) = \frac{1}{N-m} \sum_{N=0}^{N-m} x_1(n\Delta t) x_2(n\Delta t + \tau) \tag{5.8.1}$$

where $\tau = m\Delta t$ is the lag time for $m = 0, 1, \ldots, M$, $M \ll N$. Division of (5.8.1) by the product $C_{11}(0)C_{22}(0)$, corresponding to the autocovariance functions for each series at zero lag, gives the *cross-correlation coefficient function* for the data samples

$$\rho_{12}(\tau) = \frac{C_{12}(\tau)}{[C_{11}(0)C_{22}(0)]^{1/2}} \tag{5.8.2}$$

The time series $x_1(t)$ and $x_2(t)$ represent any two quantities we wish to compare. They also may represent quantities measured at different depths or locations for the same time period. For example, Kundu and Allen (1976) used the lagged covariance function

$$\rho(\mathbf{x}_1, \mathbf{x}_2, \tau) = \frac{\overline{v'(\mathbf{x}_1, t) v'(\mathbf{x}_2, t+\tau)}}{\left[\overline{(v'(\mathbf{x}_1, t))^2} \, \overline{(v'(\mathbf{x}_2, t))^2} \right]^{1/2}}$$

$$= \frac{\frac{1}{N-m} \sum_{n=1}^{N-m} v'(\mathbf{x}_1, n) v'(\mathbf{x}_2, n+m)}{\frac{1}{N} \left[\sum_{n=1}^{N} \left(v'(\mathbf{x}_1, n)\right)^2 \left(v'(\mathbf{x}_2, n)\right)^2 \right]^{1/2}}, \quad m = 0, 1, \ldots, M \ll N \tag{5.8.3}$$

to examine the correlation between the longshore (v) components of current for different coastal sites separated by a distance $d = |\mathbf{x}_1 - \mathbf{x}_2|$. Moreover, if $\tau_{\max}$ is the lag which gives the maximum correlation, then the speed of propagation, c, of the coherent signal in the direction $\mathbf{d} = \mathbf{x}_1 - \mathbf{x}_2$ is $c = |\mathbf{d}|/\tau_{max}$, the direction of propagation determined from the sign of τ_{max} (Figure 5.8.1). In Figure 5.8.1, the lagged correlations between time series of low-pass filtered longshore currents, $v(\mathbf{x}, t)$, at different sites along the continental shelf are used to examine the poleward propagation of low-frequency coastal-trapped waves. Results in the figure are based on currents at 60-m depth. Letters refer to pairs of stations used; e.g. C − P is the lag between the Carnation and Poinsettia stations.

A generalization of (5.8.3) is given by Kundu (1976). If $w = u + iv$ is the complex velocity, then the correlation between the rotating velocity vectors is given by the complex correlation coefficient

$$\rho(\mathbf{x}_1, \mathbf{x}_2, \tau) = \frac{\overline{w_1^*(t)w_2(t+\tau)}}{\left[\overline{w_1^*(t)w_1(t)}^{1/2}\overline{w_2^*(t)w_2(t)}^{1/2}\right]} \tag{5.8.4}$$

where subscripts denote locations 1 and 2, and the overbars denote the time or ensemble average. The correlation, ρ, which is independent of the choice of coordinate systems, is a complex quantity whose magnitude gives the overall measure of correlation and whose phase gives the average counterclockwise angle of the second vector with respect to the first.

5.8.2 Cross-covariance method

Following the Blackman–Tukey procedure for autospectral density estimation, the Fourier transform of the cross-covariance function, $C_{12}(\tau)$, can be used to find the cross-spectrum, $S_{12}(f)$. Although the cross-covariance method is straightforward to apply, the sample cross-covariance function, $C_{12}(\tau)$, suffers from the same disadvantage as the sample autocovariance function, $C_{11}(\tau)$, in that neighboring values tend to be highly correlated, thereby reducing the effective number of degrees of freedom. Moreover, the statistical significance falls off rapidly with increasing lag, τ, so that the number of lags, M, is much shorter than the record length ($M \ll N$).

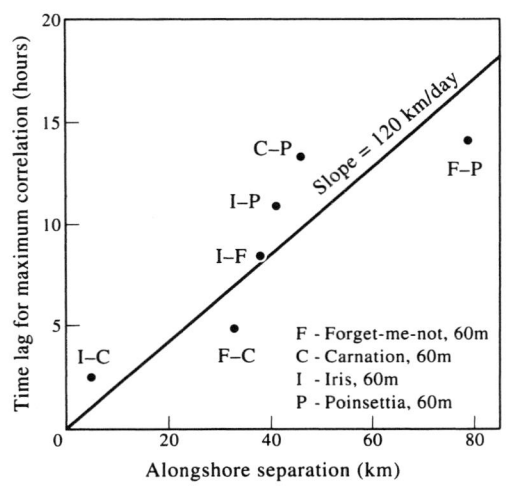

Figure 5.8.1. The lag time of maximum correlation of the longshore component of current at 60-m depth versus the distance of separation for the Oregon coast for 1973. Results indicate a mean northward signal propagation of 120 km/day. (From Kundu and Allen, 1976.)

Calculation of cross-spectra is best performed using the direct Fourier transform method. In fact, it is common practice these days to use the inverse Fourier transform of the cross-spectrum to get the cross-covariance function.

5.8.3 Fourier transform method

As with autospectral analysis, estimates of cross-spectral density functions are most commonly derived using Fourier transforms. The steps in calculating the cross-spectrum using standard Fourier transforms or FFTs are as follows (see also Bendat and Piersol, 1986):

(1) Ensure that the two time series $x_1(t)$ and $x_2(t)$ span the same period of time, t_n, where $n = 0, 1, ... , N - 1$, and $T = N\Delta t$ is the length of each record. Remove the means and trends from each of the two time series. If block averaging is to be used to improve the statistical reliability of the spectral estimates, divide the available data for each pair of time series into m sequential blocks of N' data values each, where $N' = N/m$
(2) To reduce side-lobe leakage, taper the time series $x_1(t)$ and $x_2(t)$ using a Hanning (raised-cosine) window, Kaiser–Bessel window, or other appropriate taper. Rescale the spectra calculated in step 4 to account for the loss of "energy" during application of the window (see Table 5.6.4).
(3) Compute the Fourier transforms, $X_1(f_k), X_2(f_k), k = 0, 1, 2, ... , N - 1$, for the two time series $x_1(t)$ and $x_2(t)$. For block-segmented data, calculate the Fourier transforms $X_{1m}(f_k)$ and $X_{2m}(f_k)$ for each of the m blocks, where $k = 0, 1, ... , N' - 1$. To reduce the variance associated with the tapering in step 2, the transforms can be computed for overlapping segments.
(4) Adjust the scale factor of $X_1(f_k)$ and $X_2(f_k)$ [or $X_{1m}(f_k), X_{2m}(f_k)$] for the reduction in spectral energy due to the tapering in step 2. For the Hanning window, multiply the amplitudes of the Fourier transforms by $\sqrt{(8/3)}$.
(5) Compute the raw cross-spectral power density estimates for each pair of time series (or each pair of blocks) where for the two-sided spectral density estimate

$$S_{12}(f_k) = \frac{1}{N\Delta t}[X_1^*(f_k)X_2(f_k)], \quad k = 0, 1, 2, ..., N - 1$$

(no block averaging)

$$S_{12}(f_k; m) = \frac{1}{N\Delta t}[X_{1m}^*(f_k)X_{2m}(f_k)], \quad k = 0, 1, 2, ... , N' - 1 \quad (5.8.5a)$$

(to be used for block averaging)

and for the one-sided spectral density estimates

$$G_{12}(f_k) = \frac{2}{N\Delta t}[X_1^*(f_k)X_2(f_k)], \quad k = 0, 1, 2, ..., N/2$$

(no block averaging)

$$G_{12}(f_k; m) = \frac{2}{N\Delta t}[X_{1m}^*(f_k)X_{2m}(f_k)], \quad k = 0, 1, 2, ..., N'/2 \quad (5.8.5b)$$

(for block averaging)

(6) In the case of the block-segmented data, average the raw cross-spectral density estimates from the m blocks of data to obtain the smoothed periodogram for $S_{12}(f_k)$, the two-sided cross-spectrum, or $G_{12}(f_k)$, the one-sided cross-spectrum.

Cross-covariance function: Since the cross-covariance function, $C_{12}(\tau)$ [$= R_{12}(\tau)$, the cross-correlation function, if the mean is removed from the record], and the cross-spectrum are Fourier transform pairs, equation (5.8.5) can be used to obtain a smoothed or unsmoothed estimate of the cross-covariance function. To do this, we first calculate the Fourier transforms $X_1(f)$ and $X_2(f)$ of the individual time series, and determine the product $S_{12}(f) = (N\Delta t)^{-1}[X_1^*(f)X_2(f)]$. We then take the inverse Fourier transform (IFT) of the cross-spectrum, $S_{12}(f)$, to obtain the cross-covariance function

$$C_{12}(\tau) = \int_{-\infty}^{\infty} S_{12}(f)e^{i2\pi f\tau}\,df \qquad (5.8.6)$$

If the spectrum is unsmoothed prior to the IFT (or IFFT if the number of spectral estimates is a power of 2), we obtain the raw cross-covariance function. If, on the other hand, the cross-spectrum is smoothed prior to (5.8.6) using one of the spectral windows, such as the Hanning window, the cross-covariance function also will be a smoothed function.

We can use the acoustic backscatter data in Table 5.1(a) to illustrate the direct and indirect methods for calculating the cross-covariance function. In Table 5.8.1, we present the normalized, unsmoothed cross-covariance function, $\rho_{12}(\tau) = C_{12}(\tau)/[C_{11}(0)C_{22}(0)]^{1/2}$, obtained directly from the definition (5.8.1). In this case, the lag τ is in 5-m depth increments. The indirect approach is based on the Fourier estimates presented in Table 5.8.2. Here, we first give the Fourier transforms, $X_1(f)$ and $X_2(f)$, of the two profile series as a function of wavenumber, f (Table 5.8.2a). We next calculate the cross-spectrum, $S_{12}(f) = (N\Delta t)^{-1}[X_1^*(f)X_2(f)]$, and then take the inverse transform of $S_{12}(f)$ to obtain the cross-covariance function $C_{12}(\tau)$ as a function of lag (Table 5.8.2b). No smoothing was applied to either data set, and the results obtained from the inverse Fourier transform method are identical to those listed in Table 5.8.1, within roundoff error. The advantage of the transform approach is that it is straightforward to derive a smoothed cross-covariance function by windowing the cross-spectral estimate prior to Fourier inversion.

5.8.4 Phase and cross-amplitude functions

Suppose that the constituents of the bivariate time series $\{x_1(t), x_2(t)\}$ have the same frequency, f_0, but different amplitudes (A_1, A_2) and different phases (ϕ_1, ϕ_2), respectively. In particular, let

$$x_k(t) = A_k \cos(2\pi f_0 t + \phi_k), \quad k = 1, 2 \qquad (5.8.7)$$

The Fourier transform of $x_k(t)$, over $-T/2 \leq t \leq T/2$ is

$$X_k(f) = \frac{A_k}{2}\left\{e^{i\phi_k}\frac{\{\sin[\pi(f-f_0)T]\}}{\pi(f-f_0)} + e^{-i\phi_k}\frac{\{\sin[\pi(f+f_0)T]\}}{\pi(f+f_0)}\right\}, \quad i = 1, 2 \qquad (5.8.8)$$

Hence, the sample cross-spectra of the two series is

484 *Data Analysis Methods in Physical Oceanography*

$$S_{12}(f) = \frac{1}{T}[X_1^*(f)X_2(f)] \qquad (5.8.9)$$

where X_1^* is the complex conjugate of X_1. From this expression, we obtain

$$S_{12}(f) = \frac{A_1 A_2}{4T} \left\{ e^{-i\phi_1} \frac{\sin[\pi(f-f_0)T]}{\pi(f-f_0)} + e^{i\phi_1} \frac{\sin[\pi(f+f_0)T]}{\pi(f+f_0)} \right\} \\ \times \left\{ e^{i\phi_2} \frac{\sin[\pi(f-f_0)T]}{\pi(f-f_0)} + e^{-i\phi_2} \frac{\sin[\pi(f+f_0)T]}{\pi(f+f_0)} \right\} \qquad (5.8.10)$$

where

Table 5.8.1. Unsmoothed, normalized cross-covariance function, $\rho_{12}(\tau)$ given by (5.8.2), as a function of lag τ in increments of 5 m for bin 1 of beams 1 and 2 of the acoustic backscatter spatial series (profiles) listed in Table 5.1(a)

Lag τ (m)	0	5	10	15	20	25	30	35
	0.96	0.94	0.85	0.71	0.57	0.48	0.40	0.31
Lag τ (m)	40	45	50	55	60	65	70	75
	0.23	0.14	0.02	−0.19	−0.24	−0.37	−0.46	−0.48

Table 5.8.2(a) Complex Fourier transforms of $X_1(f_k)$ and $X_2(f_k)$ for the profiles of acoustic backscatter listed in Table 5.1(a). For each wavenumber, f_k, the table lists the real part of the transform (top) followed by the imaginary part (bottom), where $X_j(f_k) = \text{Re}X_j(f_k) + i\text{Im}X_j(f_k)$, $j = 1, 2$. The vertical wavenumber $f_k = kf'$, $k = 0, 1, ..., 16$, where the fundamental vertical wavenumber, $f' = 1/155$ m $= 0.00645$ cpm (cycles per meter)

FFT $k = 0$	1	2	3	4	5	6	7
$X_1(f_k)$ 348.13	−289.32	71.17	15.52	55.16	97.59	−28.66	5.07
0.00	214.96	−16.35	−117.25	105.57	−16.98	−21.37	4.28
$X_2(f_k)$ 339.02	−226.53	119.54	55.84	−5.24	59.55	−36.39	4.22
0.00	227.88	38.22	−93.12	122.33	−24.13	−6.57	−19.09

$k = 8$	9	10	11	12	13	14	15	16
1.13	−6.16	41.11	24.03	−1.79	4.63	3.74	4.09	27.13
6.87	21.29	−2.96	−36.43	−4.60	1.08	3.54	18.45	0.00
11.90	5.68	23.89	13.85	3.96	7.37	11.27	2.34	27.79
−5.35	−4.63	−5.13	−18.72	−1.67	−4.93	−4.47	9.00	0.00

Table 5.8.2(b) The inverse fast Fourier transform (IFFT) of the cross-spectrum $S_{12}(f_k) = (N\Delta t)^{-1}[X_1^*(f_k)X_2(f_k)]$ using the values in Table 5.8.2(a). The values represent the raw (unnormalized) estimates of the cross-covariance function, $C_{12}(\tau)$, as a function of lag $\tau (0 \le \tau \le 16)$ in increments of 5 m for bin 1 of beams 1 and 2 of the acoustic backscatter spatial series (profiles) listed in Table 5.1(a)

	$\tau = 0$	1	2	3	4	5	6	7
$C_{12}(\tau)$	13483.7	12752.4	11151.5	9087.4	6992.3	5436.5	4589.9	3411.7

$\tau = 8$	9	10	11	12	13	14	15	16
2382.5	1393.8	160.6	−1103.6	−2096.0	−3103.5	−3610.5	−3623.8	−3222.1

$$S_{12}(f) \underset{T\to\infty}{\to} \frac{A_1 A_2}{4}\left[e^{-i(\phi_2-\phi_1)}\delta(f+f_0) + e^{i(\phi_2-\phi_1)}\delta(f-f_0)\right] \tag{5.8.11}$$

The phase difference, $(\phi_2 - \phi_1)$, in the above expressions determines the lead (or lag) of one cosine oscillation relative to the other for given frequency, f. The cross amplitude, $A_1 A_2$, gives the geometric mean amplitude of the co-oscillation for frequency f. From equation (5.8.2), the sample cross-spectrum is

$$S_{12}(f) = \frac{A_1(f)A_2(f)}{T}\left[e^{i[\phi_2(f)-\phi_1(f)]}\right] \tag{5.8.12}$$

or

$$S_{12}(f) = A_{12}(f)\left[e^{i\phi_{12}(f)}\right] \tag{5.8.13}$$

where the sample phase spectrum, $\phi_{12}(f) = \phi_2(f) - \phi_1(f)$, is an odd function of frequency, and the sample cross-amplitude spectrum, $A_{12}(f) = A_1(f)A_2(f)/T$, is a positive even function of f.

5.8.5 Coincident and quadrature spectra

An alternative description of this same information is to describe cross-spectra in terms of coincident (C) and quadrature (Q) spectra. In this case, we can write

$$S_{12}(f) = C_{12}(f) - iQ_{12}(f) \tag{5.8.14}$$

where

$$C_{12}(f) = A_{12}(f)\cos[\phi_{12}(f)]; \quad Q_{12}(f) = -A_{12}(f)\sin[\phi_{12}(f)] \tag{5.8.15}$$

and

$$A_{12}^2(f) = C_{12}^2(f) + Q_{12}^2(f); \quad \phi_{12}(f) = \tan^{-1}\left[\frac{-Q_{12}(f)}{C_{12}(f)}\right] \tag{5.8.16}$$

Here $C_{12}(f)$ is an even function of frequency and $Q_{12}(f)$ is an odd function. (The co-spectral density function $C_{12}(f)$ for frequency f is not to be confused with the covariance function $C_{12}(\tau)$ at time lag τ. Where confusion may arise, we use the cross-correlation $R_{12}(\tau)$ in place of $C_{12}(\tau)$.) If we consider the bivariate cosine example that we used in (5.8.7), we have

$$\begin{aligned}C_{12}(f) &= \frac{A_1 A_2}{4}\cos(\phi_2 - \phi_1)[\delta(f+f_0) + \delta(f-f_0)] \\ &= \left\{\frac{A_1\cos\phi_1 A_2\cos\phi_2}{4} + \frac{A_1\sin\phi_1 A_2\sin\phi_2}{4}\right\}[\delta(f+f_0) + \delta(f-f_0)]\end{aligned} \tag{5.8.17}$$

The sample co-spectrum, $C_{12}(f)$, measures the covariance between the two cosine components and the two sine components. That is, the contributions to the cross-spectrum from those components of the two time series that are "in phase" (phase differences of 0 or 180°). The sample quadrature spectrum, $Q_{12}(f)$, determines the contributions from those components of the time series that are coherent but "out of phase" (phase difference ±90°).

5.8.5.1 Relationship of co- and quad-spectra to cross-covariance

The inverse transform of the cross-spectrum gives the cross-covariance (cross-correlation)

$$R_{12}(\tau) = \int_{-\infty}^{\infty} [C_{12}(f) - iQ_{12}(f)] e^{i2\pi f \tau} \, df$$

$$= \int_{-\infty}^{\infty} C_{12}(f) \cos(2\pi f \tau) \, df + \int_{-\infty}^{\infty} Q_{12}(f) \sin(2\pi f \tau) \, df \qquad \varsigma(5.8.18)$$

Since $C_{12}(f)$ is an even function, $R_{12}(0) = \int_{-\infty}^{\infty} C_{12}(f) \, df$. If we define

$$C_{12}(f) = \int_{-T}^{T} R_{12}^{+}(\tau) \cos(2\pi f \tau) \, d\tau$$

$$Q_{12}(f) = \int_{-T}^{T} R_{12}^{-}(\tau) \sin(2\pi f \tau) \, d\tau \qquad (5.8.19)$$

then

$$\begin{aligned} R_{12}^{+}(\tau) &= \tfrac{1}{2}[R_{12}(\tau) + R_{12}(-\tau)] \text{ (the even part)} \\ R_{12}^{-}(\tau) &= \tfrac{1}{2}[R_{12}(\tau) - R_{12}(-\tau)] \text{ (the odd part)} \end{aligned} \qquad (5.8.20)$$

5.8.6 Coherence spectrum (coherency)

The *squared coherency, coherence-squared function,* or *coherence spectrum* between two time series $x_1(t)$ and $x_2(t)$ is defined for frequencies f_k, $k = 0, 1, \ldots, N-1$, as

$$\begin{aligned} \gamma_{12}^{2}(f_k) &= \frac{|G_{12}(f_k)|^2}{G_{11}(f_k) G_{22}(f_k)} \\ &= \frac{|S_{12}(f_k)|^2}{S_{11}(f_k) S_{22}(f_k)} \\ &= \frac{[C_{12}^{2}(f_k) + Q_{12}^{2}(f_k)]}{S_{11}(f_k) S_{22}(f_k)} \end{aligned} \qquad (5.8.21)$$

where $G_{11}(f_k)$ is the one-sided spectrum (confined to $f_k \geq 0$), $S_{11}(f_k) = \tfrac{1}{2} G_{11}(f_k)$ is the two-sided spectrum defined for all frequencies and $G_{12}(f_k)$ is the one-sided cross-spectrum. Here

$$0 \leq |\gamma_{12}^{2}(f_k)| \leq 1 \qquad (5.8.22)$$

and

$$\gamma_{12}(f_k) = |\gamma_{12}^{2}(f_k)|^{1/2} e^{-i\phi_{12}f_k} \qquad (5.8.23)$$

where $|\gamma_{12}^{2}(f_k)|^{1/2}$ is the modulus of the coherence function and $\phi_{12}(f_k)$ the phase lag between the two signals at frequency f_k (Figure 5.8.2). In the literature, both the squared coherency, γ_{12}^{2}, and its square root are termed "the coherence" so that there is often a confusion in meaning (Julian, 1975). To avoid any ambiguity, it is best to use squared-coherency when conducting coherence analyses once the sign of the coherence function is determined. This has the added advantage that squared coherency represents the fraction of the variance in x_1 ascribable to x_2 through a linear relationship between x_1 and x_2. Two signals of frequency f_k are considered highly coherent and in phase if $|\gamma_{12}^2(f_k)| \approx 1$ and $\phi_{12}(f_k) \approx 0$, respectively (Figure 5.8.2). The addition of random noise to the functions x_1 and x_2 of a linear system decreases the coherence-squared estimate and increases the noisiness of the phase associated with the system parameters. Estimation of $\gamma_{12}^2(f_k)$ is one of the most difficult problems in time-series analysis since it is so highly noise dependent. We also point out that phase estimates generally become unreliable where coherency amplitudes fall below the 90–95% confidence levels for a given frequency.

The real part of the coherence function, $\gamma_{12}(f_k)$, lies between -1 and $+1$ while the squared-coherency is between 0 and $+1$. If the noise spectrum, $S_{\varepsilon\varepsilon}(f_k)$, is equal to the output spectrum, then the coherence function is zero. This says that white noise is incoherent, as required. Also, when $S_{\varepsilon\varepsilon}(f_k) = 0$, we have $\gamma_{12}^2(f_k) = 1$; that is the coherence is perfect if there is no spectral noise in the input signal. It is important to note that, if no spectral smoothing is applied, we are assuming that we have no spectral noise. In this case, the coherency spectrum will be unity for all frequencies, which is clearly not physically realistic. Noise can be introduced to the system by

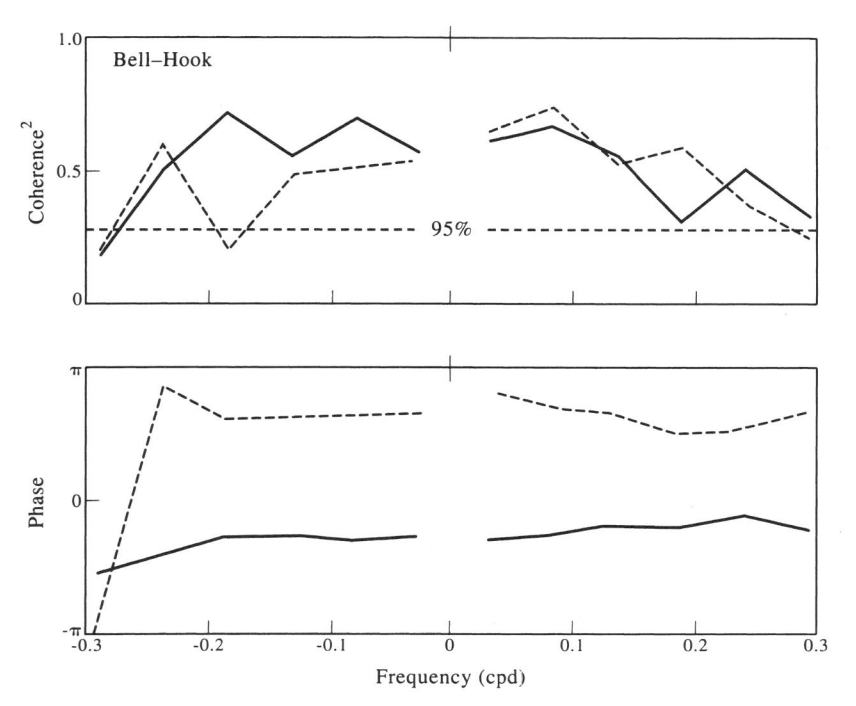

Figure 5.8.2. Coherence between current vector time series at sites Hook and Bell on the northeast coast of Australia (separation distance ≈ 300 km). (a) Coherence squared; (B) phase lag. Solid line: Inner rotary coherence (rotary current components rotating in the same sense). Dashed line: Outer rotary coherence (rotary current components rotating in the opposite sense). The increase in inner phase with frequency indicates equatorward phase propagation. Positive phase means that Hook leads Bell. (From Middleton and Cunningham, 1984.)

smoothing over adjacent frequencies. We also can overcome this problem by a prewhitening step that introduces some acceptable noise into the spectra.

5.8.6.1 Confidence levels

The final step in any coherence analysis is to specify the confidence limits for the coherence-square estimates. If $1 - \alpha$ is the $(1 - \alpha)100\%$ confidence interval we wish to specify for a particular coherence function, then, for all frequencies, the limiting value for the coherence-square (i.e. the level up to which coherence-square values can occur by chance) is given by

$$\gamma^2_{1-\alpha} = 1 - \alpha^{[1/(\text{EDOF}-1)]}$$
$$= 1 - \alpha^{[2/(\text{DOF}-2)]} \qquad (5.8.24)$$

where EDOF = DOF/2 (called the *equivalent* degrees of freedom) is the number of independent cross-spectral realizations in each frequency band (Thompson, 1979). The commonly used confidence intervals of 90, 95, and 99% correspond to $\alpha = 0.10$, 0.05, and 0.01, respectively. As an example, suppose that each of our coherence estimates is computed from an average over three adjacent cross-spectral Fourier components, then EDOF = 3 (DOF = 6). The 95% confidence level for the squared coherence would then be $\gamma^2_{95} = 1 - (0.05)^{0.5} = 0.78$. Alternatively, if the cross-spectrum and spectra were first smoothed using a Hamming window spanning the entire width of the data series, the equivalent degrees of freedom are EDOF = 2.5164 × 2 = 5.0328 (Table 5.6.4) and the 95% confidence interval $\gamma^2_{95} = 1 - (0.05)^{0.6595} = 0.86$. For EDOF = 2, $\gamma^2_{1-\alpha} = 1 - \alpha$ so that the confidence level is equal to itself.

A useful reference for coherence significance levels is Thompson (1979). In this paper, the author tests the reliability of significance levels $\gamma^2_{1-\alpha}$ estimated from (5.8.24) with the coherence-square values obtained through the summations

$$\gamma^2(f) = \frac{\left|\sum_{k=1}^{K} X_{1k}(f) X^*_{2k}(f)\right|^2}{\sum_{k=1}^{K} |X_{1k}(f)|^2 \sum_{k=1}^{K} |X_{2k}(f)|^2} \qquad (5.8.25)$$

In this expression, X_{1k} and X_{2k} are the Fourier transforms of the respective random time series $x_{1k}(t)$ and $x_{2k}(t)$ generated by a Monte Carlo approach, and the asterisk denotes the complex conjugate. The upper limit K corresponds to the value of EDOF in (5.8.24a). Because $\gamma^2(f)$ is generated using random data, it should reflect the level of squared coherency that can occur by chance. For each value of K, $\gamma^2(f)$ was calculated 1000 times and the resultant values sorted as 90th, 95th, and 99th percentiles. The operation was repeated 10 times and the means and standard deviations calculated. This amounts to a total of 20,000 Fourier transforms for each K (=EDOF). There is excellent agreement between the significance level derived from (5.8.24) and the coherence-square values for a white-noise Monte Carlo process (Table 5.8.3), lending considerable credibility to the use of (5.8.24) for computing coherence significance levels. The comparisons in Table 5.8.3 are limited to the 90 and 95% confidence intervals for $4 \leq K \leq 30$. Thompson (1979) includes the 99% interval and a wider range of K (EDOF) values.

Confidence intervals for coherence amplitudes, as well as for coherence phase, admittance, and other signal properties (see next section), can be derived using the data itself (Bendat and Piersol, 1986). Let $\hat{\varphi}$ be an estimator for φ, a continuous, stationary random process, and define the standard error or random error of sample values as

$$\text{random error} = \sigma[\hat{\varphi}] = (E[\hat{\varphi}^2] - E^2[\hat{\varphi}])^{1/2} \quad (5.8.26a)$$

and the root mean square (RMS) error as

$$\text{RMS error} = (E[(\hat{\varphi} - \varphi)^2])^{1/2} = (\sigma^2[\hat{\varphi}] + B^2[\hat{\varphi}])^{1/2} \quad (5.8.26b)$$

where B is the bias term $B[\hat{\varphi}] = E[\hat{\varphi}] - \varphi$ and $E[x]$ is the expected value of x. If we now divide each error term by the quantity φ being estimated, we obtain the normalized random error

$$\varepsilon_r = \frac{\sigma[\hat{\varphi}]}{\varphi} = \frac{(E[\hat{\varphi}^2] - E^2[\hat{\varphi}])^{1/2}}{\varphi} \quad (5.8.27a)$$

and the normalized RMS error

$$\varepsilon = \frac{(E[(\hat{\varphi} - \varphi)^2])^{1/2}}{\varphi} = \frac{(\sigma^2[\hat{\varphi}] + B^2[\hat{\varphi}])^{1/2}}{\varphi} \quad (5.8.27b)$$

where it is assumed that $\varphi \neq 0$. Provided ε_r is small, the relation

$$\hat{\varphi}^2 = \varphi^2(1 \pm \varepsilon_r) \quad (5.8.28)$$

yields

$$\hat{\varphi} = \varphi(1 \pm \varepsilon_r)^{1/2} \approx \varphi(1 \pm \varepsilon_r/2) \quad (5.8.29)$$

so that

$$\varepsilon_r[\hat{\varphi}^2] \approx 2\varepsilon_r[\hat{\varphi}] \quad (5.8.30)$$

Thus, for small ε_r the normalized error for squared estimates $\hat{\varphi}^2$ is roughly twice the normalized error for unsquared estimates.

Table 5.8.3 Monte Carlo estimates, $\gamma^2(f)$, of the significant coherence-squared and prediction of this value using (5.8.24) for intervals $\alpha = 0.05$ and 0.10 for EDOF = 4, 5, 6, 8, 10, 20, and 30. (After Thompson, 1979)

	EDOF = 4	EDOF = 5	EDOF = 6	EDOF = 8	EDOF = 10	EDOF = 20	EDOF = 30
$\alpha = 0.10$							
$\gamma^2(f)$	0.539	0.437	0.371	0.288	0.230	0.114	0.076
$\gamma^2_{0.90}$	0.536	0.438	0.369	0.280	0.226	0.114	0.076
$\alpha = 0.05$							
$\gamma^2(f)$	0.629	0.531	0.452	0.354	0.288	0.144	0.099
$\gamma^2_{0.95}$	0.632	0.527	0.451	0.348	0.283	0.146	0.098

490 *Data Analysis Methods in Physical Oceanography*

When the estimates $\hat{\varphi}$ have a small bias error, $B[\hat{\varphi}] \approx 0$, and a small normalized error, e.g. $\varepsilon \leq 0.2$, the probability density for the estimates can be approximated by a Gaussian distribution. The confidence intervals for the unknown true parameter φ based on a single estimate $\hat{\varphi}$ are then

$$\hat{\varphi}(1-\varepsilon) \leq \varphi \leq \hat{\varphi}(1+\varepsilon) \text{ with 68\% confidence} \quad (5.8.31a)$$

$$\hat{\varphi}(1-2\varepsilon) \leq \varphi \leq \hat{\varphi}(1+2\varepsilon) \text{ with 95\% confidence} \quad (5.8.31b)$$

$$\hat{\varphi}(1-3\varepsilon) \leq \varphi \leq \hat{\varphi}(1+3\varepsilon) \text{ with 99\% confidence} \quad (5.8.31c)$$

5.8.7 Frequency response of a linear system

We define the admittance (or transfer) function of a linear system as

$$H_{12}(f_k) = \frac{S_{12}(f_k)}{S_{11}(f_k)} = \frac{G_{12}(f_k)}{G_{11}(f_k)}, \quad f_k = k/T, \quad k = 1, ..., N$$
$$= |H_{12}(f_k)| e^{-i\phi_{12}(f_k)} \quad (5.8.32)$$

where $S_{11}(f_k)$ and $G_{11}(f_k)$ are, respectively, the two-sided and one-sided autospectrum estimates for the time series $x_1(t)$ selected here as the input time series. The gain (or admittance amplitude) function $H(f_k)$ behaves like a spectral regression coefficient at each frequency f_k. Using the definition $G_{12}(f_k) = C_{12}(f_k) - iQ_{12}(f_k)$, we obtain

$$|H_{12}(f_k)| = \frac{G_{12}(f_k)}{G_{11}(f_k)}$$
$$= \frac{|C_{12}^2(f_k) + Q_{12}^2(f_k)|^{1/2}}{G_{11}(f_k)} \quad (5.8.33)$$

and where $\phi_{12}(f_k) = \tan^{-1}[-Q_{12}(f_k)/C_{12}(f_k)]$ by (5.8.16). Figure 5.8.3 shows the complex admittance for the observed longshore component of oceanic wind velocity (time series 1) and the longshore component of wind velocity derived from pressure-derived geostrophic winds (time series 2). The geostrophic winds closely approximate the amplitude and phase of the actual winds up to a frequency of about 0.05 cph (period = 20 h) after which the two signals no longer resemble one another. It is also at this frequency that the coherence consistently begins to fall below the 90% confidence level.

5.8.7.1 Multi-input systems cross-spectral analysis

Many oceanographic time series are generated through the combined effects of several mutually coherent inputs. For example, low-frequency fluctuations in coastal sea level typically arise through the combined forcing of atmospheric pressure, along- and cross-shore wind stress, and surface buoyancy flux. Coherences between the forcing variables (e.g. pressure, longshore wind stress, and runoff) are generally quite high. Because of this, it would be physically incorrect to use ordinary cross-spectral analysis which simply examines the correlation functions, $\gamma_{y:x}^2$, between the output, $y(t)$, and each of the inputs, $x(t)$, individually without taking into account the mutual correlation among all the inputs. If this is not done, the sum of the individual correlation functions can exceed unity. Provided that long-term sea-level fluctuations

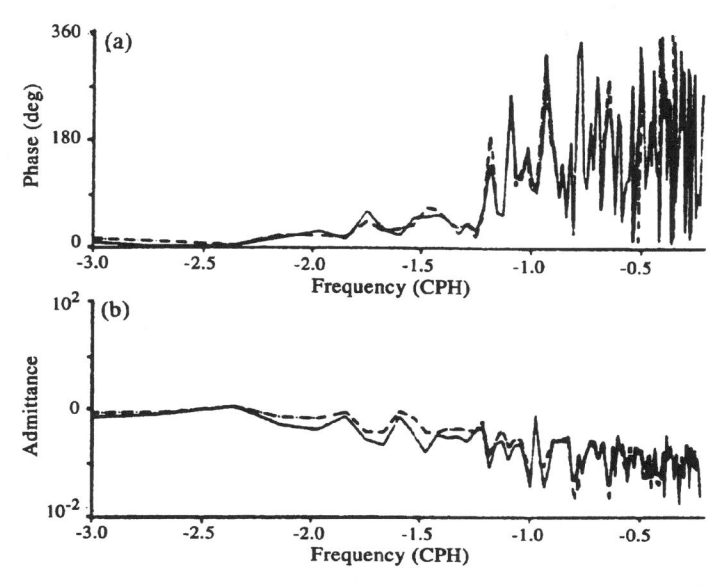

Figure 5.8.3. Complex admittance for observed (series 1) and calculated (series 2) longshore components of oceanic wind velocity. (a) Phase; (b) amplitude. Positive phase means that series 1 leads series 2. (From Thomson, 1983.)

(the output time series) are linearly related to the individual forcing functions (the input time series), we can use *multi-input systems cross-spectral analysis* to calculate the relative contribution each of the input terms makes to the output. The effective correlation function for the total system will then be less than unity, as required. This concept was pioneered in oceanography by Cartwright (1968), Groves and Hannan (1968), and Wunsch (1972). All three studies were concerned with sea-level variations.

The purpose of this section is to provide a brief overview of multiple systems analysis. For a thorough generalized presentation, the reader is directed to Bendat and Piersol (1986). Consider K constant-parameter linear systems associated with K stationary and ergodic input time series, $x_k(t)$, $k = 1, 2, \ldots, K$, a noise function, $\varepsilon(t)$, and a single output, $y(t)$, such that

$$y(t) = \sum_{k=1}^{K} y_k(t) + \varepsilon(t) \qquad (5.8.34)$$

where the $y_k(t)$ are the outputs generated by each of the measured inputs $x_k(t)$. We can only measure the accumulated response $y(t)$, not the individual responses, $y_k(t)$. In the present context, $y(t)$ represents the measured time series of coastal sea level, $x_k(t)$ the corresponding weather variables, and $\varepsilon(t)$ the deviations from the ideal response due to instrument noise, remotely generated subinertial waves, and other physical processes not correlated with the input functions. The Fourier transform of the output $y(t)$ is

$$Y(f) = \sum_{k=1}^{K} Y_k(f) + E(f)$$
$$= \sum_{k=1}^{K} H_k(f) X_k(f) + E(f) \qquad (5.8.35)$$

where

$$H_k(f) = \frac{Y_k(f)}{X_k(f)}, \quad k = 1, 2, \ldots, K \quad (5.8.36)$$

is the admittance (or transfer) function relating the kth input with the kth output at frequency f. The frequency-domain spectral variables $X_k(f)$ and $Y(f)$ can be computed from the measured time series $x_k(t)$ and $y(t)$. Using these variables, we can then determine the functions $H_k(f)$ and other properties of the system.

Multiplication of both sides of (5.8.35) by $X_j^*(f)$, the complex conjugate of $X_j(f)$, for any fixed $j = 1, 2, \ldots, K$, yields the power spectral relation

$$S_{jy}(f) = \sum_{k=1}^{K} H_k(f) S_{jk}(f) + S_{j\varepsilon}(f), \quad j = 1, 2, \ldots, K \quad (5.8.37)$$

in which

$$\begin{aligned}S_{jy}(f) &= \overline{X_j^*(f) Y(f)}, \quad j = 1, 2, \ldots, K \\ S_{jk}(f) &= \overline{X_j^*(f) X_k(f)}, \quad j, k = 1, 2, \ldots, K\end{aligned} \quad (5.8.38)$$

Here, the overbar denotes the average value, the $S_{jy}(f)$ are the cross-spectra between the K inputs and the single output, $S_{jk}(f)$ are the cross-spectra ($j \neq k$) and spectra ($j = k$) among the input variables, and $S_{j\varepsilon}(f)$ is the cross-spectrum between the input variables and the noise function. If the noise function $\varepsilon(t)$ is uncorrelated with each input x_k (as is normally assumed), the cross-spectral terms $S_{j\varepsilon}(f)$ will be zero and (5.8.37) becomes

$$S_{jy}(f) = \sum_{k=1}^{K} H_k(f) S_{jk}(f), \quad j = 1, 2, \ldots, K \quad (5.8.39)$$

This expression is a set of K equations in K unknowns—the $H_k(f)$ for $k = 1, 2, \ldots, K$—where all spectral terms can be computed from the measured records of $y(t)$ and $x_k(t)$. If the model is well defined, matrix techniques can be used to find the $H_k(f)$. Bendat and Piersol (1986) also define the problem in terms of the *multiple and partial coherence functions* for the system. The multiple coherence function is given by

$$\gamma_{y:x}^2 = \frac{S_{vv}(f)}{S_{yy}(f)} = 1 - \frac{S_{\varepsilon\varepsilon}(f)}{S_{yy}(f)} \quad (5.8.40)$$

where $S_{vv}(f)$ is the multiple coherent output spectrum, $S_{yy}(f)$ is the output spectrum and $S_{\varepsilon\varepsilon}(f)$ is the noise spectrum. As with any squared coherence function, $0 \leq |\gamma_{y:x}^2| \leq 1$. For any problem with multiple inputs, $\gamma_{y:x}^2$ takes the form of a matrix whose off-diagonal elements take into account the coherent interactions among the different input terms. Expressions (5.8.39) and (5.8.40) simplify even further if the inputs themselves are mutually uncorrelated. In that case

$$H_j(f) = \frac{S_{jy}(f)}{S_{jj}(f)}, \quad j = 1, 2, \ldots, K; \quad |H_j(f)|^2 S_{jj}(f) = \gamma_{jy}^2 S_{yy}(f) \quad (5.8.41)$$

Hence, the contribution of the input variable, $x_j(t)$, to the output variable, $y(t)$, occurs only through the transfer (admittance) function $H_j(f)$ of that particular input variable. No leakage of $x_j(t)$ takes place through any of the other transfer functions since $x_j(t)$ is uncorrelated with $x_k(t)$ for $k \neq j$.

In general, the output $y(t)$ is forced not only by the mutually coherent parts of the various inputs but also by the noncoherent portions of the inputs which go directly to the output through their own transfer functions without being affected by other transfer functions. This leads to the need for *partial coherence functions*. If part of one record causes part or all of a second record, then turning off the first record will eliminate the correlated parts from the second record and leave only that part of the second record that is not due to the first record. Because we do not want to incorporate the coherent portions of given forcing terms in the partial coherence functions, the partial coherences are found by first subtracting out the coherent parts of the various input signals. Bendat and Piersol (1986) state that, if any correlation between $x_1(t)$ and $x_2(t)$ is due to $x_1(t)$, then the optimum linear effects of $x_1(t)$ to $x_2(t)$ should be found. Denoting this mutual effect as $x_{2:1}(t)$, this should be subtracted from $x_2(t)$ to yield the conditioned (or residual) record, $x_{2:1}(t)$ representing that part of $x_2(t)$ not due to $x_1(t)$.

Multi-input systems cross-spectral analysis takes into account the fact that any input record $x_k(t)$ with nonzero correlations between other inputs will contribute to variations in the output $y(t)$ by passage through any of the K linear systems, $H_k(f)$. The conditioned portion of $x_k(t)$ will contribute directly to the output through its own response function only. The problem is to determine what percentage contribution each input function makes to the total variance of $y(t)$ for a specified frequency band. The simplest case is a two-input system consisting of inputs $x_1(t)$ and $x_2(t)$ for which

$$Y(f) = H_1(f)X_1(f) + H_2(f)X_2(f) + E(f) \tag{5.8.42}$$

and, provided $\gamma_{12}^2 \neq 0$

$$H_1(f) = \frac{S_{1y}(f)\left[1 - \dfrac{S_{12}(f)S_{2y}(f)}{S_{22}(f)S_{1y}(f)}\right]}{S_{11}(f)[1 - \gamma_{12}^2(f)]} \tag{5.8.43a}$$

$$H_2(f) = \frac{S_{2y}(f)\left[1 - \dfrac{S_{21}(f)S_{1y}(f)}{S_{11}(f)S_{2y}(f)}\right]}{S_{22}(f)[1 - \gamma_{12}^2(f)]} \tag{5.8.43b}$$

What is important to note here is the nonzero coupling between the different input variables when the cross-coherence, $\gamma_{12}^2(f)$, is nonzero. The product $H_1(f)S_{11}(f)$ in (5.8.43a) still represents the ordinary coherent spectrum between the input x_1 and the output y. However, when $|\gamma_{12}| \neq 0$, $x_1(t)$ influences $y(t)$ through the transfer function $H_2(f)$ as well as through its own transfer function $H_1(f)$. Similarly, $x_2(t)$ influences $y(t)$ through the transfer function $H_1(f)$ as well as through its transfer function $H_2(f)$ (5.8.43b). In general, the sum of $\gamma_{1y}^2(f)$ and $\gamma_{2y}^2(f)$ can be greater than unity when the outputs are correlated. The contributions from the conditioned records of $x_1(t)$ and $x_2(t)$ must also be taken into account when estimating the output response, $y(t)$. Once this is done, it becomes possible to construct reliable forecasting models for y.

Cartwright (1968) used the multiple input method to study tides and storm surges around east and north Britain. He expanded the tide height, ζ, at each of the ports studied as a Taylor series of the atmospheric pressure, p, about the port location ($x =$

$0, y = 0)$

$$\zeta(x,y,t) = p_{00}(t) + xp_{10}(t) + yp_{01}(t) + x^2p_{20}(t) + 2xyp_{11}(t) + y^2p_{02}(t) + \dots \quad (5.8.44)$$

in which the pressure gradient terms $(p_{10}, p_{01}) = (\partial p/\partial x, \partial p/\partial y)$ are proportional to the geostrophic wind stress, the second derivatives $(p_{20}, p_{02}) = (\partial^2 p/\partial^2 x, \partial^2 p/\partial y^2)$ are related to wind stress gradients, and so on. As indicated by Table 5.8.4, the variances in different frequency bands for the sea level at Aberdeen, Scotland are significantly reduced relative to the original values as the pressure, first derivatives, and second derivatives are successively included. Consequently, all of the mutually correlated weather variables are considered relevant to the predictability of sea level. In a more recent study, Sokolova *et al.* (1992) used the multiple spectral analysis technique to study sea-level oscillations measured from July to September 1986 at different locations around the perimeter of the Sea of Japan. According to their analysis for both the multiple and partial coherences, 46–77% of the sea-level variance was coherent with atmospheric pressure and 5–37% was coherent with the wind stress.

5.8.8 Rotary cross-spectral analysis

As outlined in Section 5.6.4, the decomposition of a complex horizontal velocity vector, $w(t) = u(t) + iv(t)$, into counter-rotating circularly polarized components can aid in the analysis and interpretation of oceanographic time series. (Here, u and v typically represent the eastward and northward components of the current or wind.) Many of the fundamentals of this approach can be found in Fofonoff (1969), Gonella (1972), Mooers (1973), Calman (1978), and Hayashi (1979). In rotary spectral analysis, the different frequency components of the vector $w(t)$ are represented in terms of clockwise and counterclockwise rotating vectors (Figure 5.6.12). The counterclockwise component is considered to be rotating with positive angular frequency ($\omega \geq 0$) and the clockwise component with negative angular frequency ($\omega \leq 0$). Depending on which of the two components has the largest magnitude, the vector rotates clockwise or counterclockwise with time, with the tip of the vector tracing out an ellipse. If, for a given frequency, both components are of equal magnitude, the ellipse flattens to a line and the motions are *rectilinear* (back and forth along a straight line). Two one-sided autospectra and two one-sided cross-spectra can be computed for the rotary components. Mooers (1973) formulated these as two two-sided rotary autospectra called, respectively, the *inner* and *outer rotary autospectra*, the terminology originating from the resemblance of the inner and outer rotary autocovariance functions derived from the autospectra to the inner (dot) and outer (cross) products in mathematics. (A note

Table 5.8.4 Residual variances (cm^2) for different frequency bands for Aberdeen, Scotland sea-level oscillations. The predictive model explains increasingly more of the variance as additional weather variables are incorporated in the analysis. (Modified after Cartwright, 1968)

	0–0.5 cpd	0.5–0.8 cpd	1.1–1.8 cpd	2.1–2.8 cpd
Variables included				
Original variance	181	16	9.6	4.1
p_{00}	88	13	9.1	3.9
p_{00}, p_{10}, p_{01}	49	9	7.1	3.6
$p_{00}, p_{10}, p_{01}, \dots p_{02}$	38	6	5.3	3.3

on terminology: Mooers (1973) uses A and C for counterclockwise (+) and clockwise components (−) while Gonella (1972) uses +/− subscripts for these components of the form u_+ and u_-. In this text, we use +/− superscripts where, for example, the amplitude of the two vector components is written as A^+ and A^-.)

To simplify the mathematics, we assume that u and v are continuous, stationary processes with zero means and Fourier integral representations. The velocity vector $w(t)$ can then be written in terms of its Fourier transform

$$w(t) = u(t) + iv(t) = \sum_p W_p e^{i\omega_p t}$$

$$= \sum_p \{[A_{1p} \cos(\omega_p t) + B_{1p} \sin(\omega_p t)] + i[A_{2p} \cos(\omega_p t) + B_{2p} \sin(\omega_p t)]\} \quad (5.8.45)$$

in which the Fourier transform component, W_p, is a complex quantity, the A and B are constants, and ω_p is the frequency of the pth Fourier component. As outlined in Section 5.6.4, each Fourier component of frequency $\omega = \omega_p$ can be expressed as a combination of two circularly polarized components having counterclockwise ($\omega \geq 0$) and clockwise ($\omega \leq 0$) rotation. Each of two components has its own amplitude and phase, and the tip of the vector formed by the combination of the two oppositely rotating components traces out an ellipse over a period, $T = 2\pi/\omega$. The semi-major axis of the ellipse has length $L_M = A^+(\omega) + A^-(\omega)$ and the semi-minor axis has length $L_m = |A^+(\omega) - A^-(\omega)|$. The angle, θ, of the major axis measured counterclockwise from the eastward direction gives the ellipse orientation.

If we specify $A_1(\omega)$ and $B_1(\omega)$ to be the amplitudes of the cosine and sine terms for the eastward (u) component in equation (5.8.45) and $A_2(\omega)$ and $B_2(\omega)$ to be the corresponding amplitudes for the northward (v) component, the amplitudes of the two counter-rotating vectors for a given frequency are

$$A^+(\omega) = \tfrac{1}{2}\left\{[B_2(\omega) + A_1(\omega)]^2 + [A_2(\omega) - B_1(\omega)]^2\right\}^{1/2} \quad (5.8.46a)$$

$$A^-(\omega) = \tfrac{1}{2}\left\{[B_2(\omega) - A_1(\omega)]^2 + [A_2(\omega) + B_1(\omega)]^2\right\}^{1/2} \quad (5.8.46b)$$

and their phases are

$$\tan(\theta^+) = [A_1(\omega) - B_1(\omega)]/[A_1(\omega) + B_2(\omega)] \quad (5.8.47a)$$

$$\tan(\theta^-) = [B_1(\omega) + A_2(\omega)]/[B_2(\omega) - A_1(\omega)] \quad (5.8.47b)$$

The eccentricity of the ellipse is

$$\varepsilon(\omega) = 2[A^+(\omega)A^-(\omega)]^{1/2}/[A^+(\omega) + A^-(\omega)] \quad (5.8.48)$$

where the ellipse traces out an area $\pi[(A^+)^2 - (A^-)^2]$ during one complete cycle of duration $2\pi/\omega$. The use of rotary components leads to two-sided spectra; i.e. defined for both negative and positive frequencies. If $S^+(\omega)$ and $S^-(\omega)$ are the rotary spectra for the two components, then $A^\pm(\omega) \propto [S^\pm(\omega)]^{1/2}$ can be used to determine the ellipse eccentricity. The sense of rotation of the vector about the ellipse is given by the rotary coefficient (see Section 5.6.4.2)

$$r(\omega) = [S^+(\omega) - S^-(\omega)]/[S^+(\omega) + S^-(\omega)] \quad (5.8.49)$$

where $-1 \leq r \leq 1$. Values for which $r > 0$ indicate counterclockwise rotation while values of $r < 0$ indicate clockwise rotation; $r = 0$ is rectilinear motion.

If u, v are orthogonal Cartesian components of the velocity vector, $w = (u, v)$, then the rotary spectra can be expressed as

$$\begin{aligned} S^+(\omega) &= [A^+(\omega)]^2, \quad \omega \geq 0 \\ &= \tfrac{1}{2}[S_{uu} + S_{vv} + 2Q_{uv}] \end{aligned} \quad (5.8.50a)$$

$$\begin{aligned} S^-(\omega) &= [A^-(\omega)]^2, \quad \omega \leq 0 \\ &= \tfrac{1}{2}[S_{uu} + S_{vv} - 2Q_{uv}] \end{aligned} \quad (5.8.50b)$$

where S_{uu} and S_{vv} are the autospectra for the u and v components, and Q_{uv} is the quadrature spectrum between the two components. The stability of the ellipse is given by

$$\begin{aligned} \mu(\omega) &= \frac{|\langle (A^-(\omega)A^+(\omega) \exp[i(\theta^+ - \theta^-)]\rangle|^2}{\langle (A^-)^2\rangle\langle (A^+)^2\rangle}, \quad \omega \geq 0 \\ &= \frac{|Y|}{[S^+(\omega)S^-(\omega)]^{1/2}} \end{aligned} \quad (5.8.51)$$

where

$$Y = \tfrac{1}{2}[S_{uu} - S_{vv} + i2S_{uv}] \quad (5.8.52)$$

and the ellipse has a mean orientation

$$\phi = \tfrac{1}{2}\tan^{-1}[2S_{uv}/(S_{uu} - S_{vv})] \quad (5.8.53)$$

where ϕ is measured counterclockwise from east (the function ϕ is not coordinate invariant). The brackets $\langle \cdot \rangle$ denote an ensemble average or a band average in frequency space. The ellipse stability, $\mu(\omega)$, resembles the magnitude of a correlation function and is a measure of the confidence one might place in the estimate of the ellipse orientation (Gonella, 1972).

5.8.8.1 Rotary analysis for a pair of time series

Having summarized the rotary vector analysis for a single location, we now want to consider the coherence and cross-spectral properties for two time series measured simultaneously at two spatial locations. The object of the rotary spectral analysis is to determine the "similarity" between the two time series in terms of their circularly polarized rotary components. For two vector time series, the inner and outer rotary cross-spectra can be computed. As the spectra are complex, they have both amplitude and phase. Hence, coherence and phase spectra can be computed, just as with the cross-spectra of two scalar time series. *Inner* functions describe co-rotating components and *outer* functions describe counter-rotating components. We could, of course, use standard Cartesian components for this task. Unfortunately, the Cartesian vectors and their derived relationships generally are dependent on the selected orientation of the coordinate system. The advantages of the rotary type of analysis are: (1) The coherence analysis is independent of the coordinate system (i.e. is coordinate invariant); and (2) the results encompass the coherence and phase of oppositely rotating, as well as like-rotating components, for motions that may be highly nonrectilinear. Because the counter-rotating components have circular symmetry, invariance under coordinate rotation follows for coherence.

We consider two vector time series defined by the relations

$$w_1(t) = (u_1, v_1); \quad w_2(t) = (u_2, v_2) \tag{5.8.54}$$

where, as before, $(u, v) = u + iv$ are complex quantities. If $W_1(\omega)$ and $W_2(\omega)$ are components of the Fourier transforms of these time series, then the transforms can be expressed in the form

$$W(\omega) = \begin{cases} A^+ \exp(-i\theta^+), & \omega \geq 0 \\ A^- \exp(-i\theta^-), & \omega \leq 0 \end{cases} \tag{5.8.55}$$

with the same definitions for amplitudes and phases as in the previous subsection. These expressions equate the negative frequency components from the Fourier transform with the clockwise rotary components and the positive frequency components from the transform with the counterclockwise components.

Inner-cross spectrum: The inner cross-spectrum, $S_{w_j w_k}(\omega)$, provides an estimate of the joint energy content of two time series for rotary components rotating in the same direction (e.g. the clockwise component of series 1 with the clockwise component of series 2; Figure 5.8.4). For all frequencies, $-\omega_N < \omega < \omega_N$

$$S_{w_j w_k}(\omega) = \langle W_j^*(\omega) W_k(\omega) \rangle, \quad j, k = 1, 2$$
$$= \begin{cases} A_j^+(\omega) A_k^+(\omega) \exp[-i(\theta_j^+ - \theta_k^+)], & \omega \geq 0 \\ A_j^-(\omega) A_k^-(\omega) \exp[i(\theta_j^- - \theta_k^-)], & \omega \leq 0 \end{cases} \tag{5.8.56}$$

where, as before, $\langle \cdot \rangle$ denotes an ensemble average or a band average in frequency space, and the asterisk denotes the complex conjugate. It follows that the inner-autospectrum for each time series is

$$S_{w_j w_j}(\omega) = \begin{cases} [A_j^+(\omega)]^2, & \omega \geq 0 \\ [A_j^-(\omega)]^2, & \omega \leq 0 \end{cases} \tag{5.8.57}$$

Thus, $S_{w_j w_j}(\omega)$ ($j = 1, 2$) is the power spectrum of the counterclockwise component of the series j for $\omega \geq 0$, and the power spectrum for the clockwise component for $\omega \leq 0$. The area under the curve of $S_{w_j w_k}(\omega)$ versus frequency equals the sum of the variance of the eastward (u) and northward (v) components. For $\omega \geq 0$, $S_{w_1 w_2}(\omega)$ is the cross-spectrum for the counterclockwise component of series 1 and 2, while for $\omega \leq 0$, $S_{w_1 w_2}(\omega)$ represents the cross-spectrum for the clockwise rotary component.

Inner-coherence squared: The two-sided inner-coherence squared, $\gamma_{12}^2(\omega)$, between the two time series at frequency ω is defined in the usual manner. Specifically, using the previous definitions for the rotary components, we find

$$\gamma_{12}^2(\omega) = \begin{cases} \{\langle A_1^+ A_2^+ \cos(\theta_1^+ - \theta_2^+)\rangle)^2 + \langle A_1^+ A_2^+ \sin(\theta_1^+ - \theta_2^+)\rangle)^2\}/\langle A_1^{+2}\rangle\langle A_2^{+2}\rangle, & \omega \geq 0 \\ \{\langle A_1^- A_2^- \cos(\theta_1^- - \theta_2^-)\rangle)^2 + \langle A_1^- A_2^- \sin(\theta_1^- - \theta_2^-)\rangle)^2\}/\langle A_1^{-2}\rangle\langle A_2^{-2}\rangle, & \omega \leq 0 \end{cases}$$
(5.8.58)

where $0 \leq |\gamma_{12}^2| \leq 1$. A coherence of near zero indicates a negligible relationship between the two like-rotating series while a coherence near unity indicates a high degree of variability between the series. The inner-phase lag, ϕ_{12}, between the two vectors is

$$\phi_{12}(\omega) = \tan^{-1}[-\text{Im}(S_{w_1 w_2})/\text{Re}(S_{w_1 w_2})]$$
(5.8.59)

or, in terms of the clockwise and counterclockwise components

$$\tan(\phi_{12}) = \begin{cases} \langle A_1^+ A_2^+ \sin(\theta_1^+ - \theta_2^+)\rangle/\langle A_1^+ A_2^+ \cos(\theta_1^+ - \theta_2^+)\rangle, & \omega \geq 0 \\ \langle -A_1^- A_2^- \sin(\theta_1^- - \theta_2^-)\rangle/\langle A_1^- A_2^- \cos(\theta_1^- - \theta_2^-)\rangle, & \omega \leq 0 \end{cases}$$
(5.8.60)

The phase, which is the same for both the inner cross-spectrum and the inner coherence, is a measure of the phase lead of the rotary component of time series 1 with respect to that of time series 2. Figure 5.8.4(a) shows the inner rotary coherence and phase for five years of monthly winter (November through February) wind data measured off Alaska at Middleton Island (59.4°N, 146.3°W) and Environmental Weather Buoy EB03 (56.0°N, 148.0°W). Co-rotating wind vectors were generally

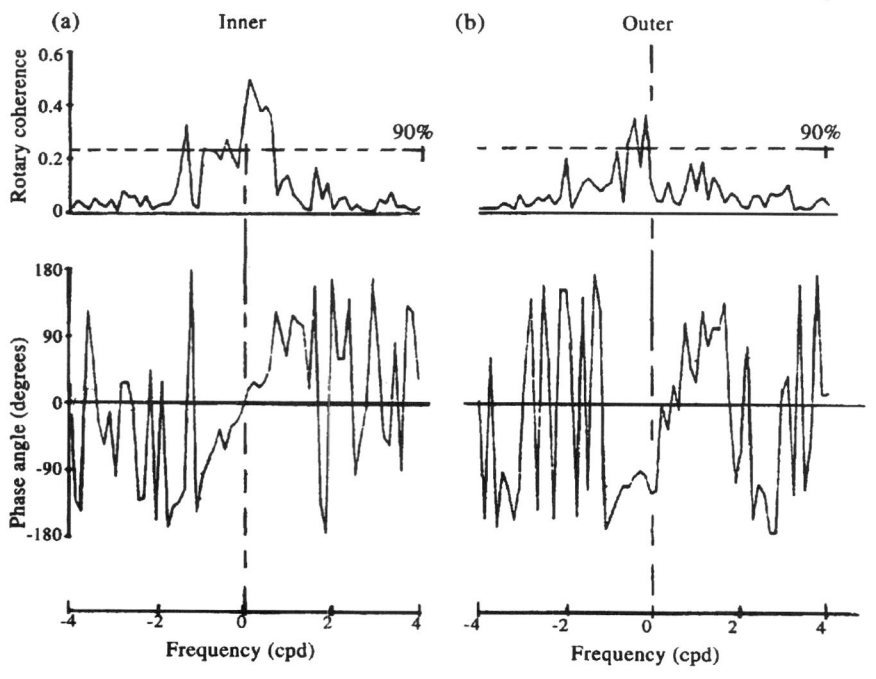

Figure 5.8.4. Rotary coherence and phase for five-year time series of monthly mean winter (November through February) wind velocity from two sites off Alaska. (a) Co-rotating (inner) coherence and phase with 90% confidence level; (b) counter-rotating (outer) coherence and phase. (From Livingstone and Royer, 1980.)

coherent above the 90% confidence level for frequencies $-1 < f < 1$ cpd, with greater coherence at positive frequencies (Livingstone and Royer, 1980). The inner phase was nearly a straight line in the frequency range $-1 < f < 0$ cpd, increasing by 120° over this range.

Outer-cross spectrum: The outer cross-spectrum, $Y_{w_j w_k}(\omega)$, provides an estimate of the joint energy content between rotary components rotating in opposite directions (e.g. between the clockwise component of time series 1 and the counterclockwise component of time series 2). For frequencies in the Nyquist frequency range, $-\omega_N < \omega < \omega_N$

$$Y_{wjwk}(\omega) = \langle W_j(-\omega) W_k(\omega) \rangle, \quad j, k = 1, 2$$
$$= \begin{cases} A_j^-(\omega) A_k^+(\omega) \exp[i(\theta_k^+ - \theta_j^-)], & \omega \geq 0 \\ A_j^+(\omega) A_k^-(\omega) \exp[i(\theta_j^+ - \theta_k^+)], & \omega \leq 0 \end{cases} \quad (5.8.61)$$

(Middleton, 1982). These relations resemble those for the inner-cross spectra but involve a combination of oppositely rotating vector amplitudes and phases. For the case of a single series, j, the outer rotary autospectrum is then

$$Y_{wjwj}(\omega) = A_j^-(\omega) A_j^+(\omega) \exp[i(\theta_j^+ - \theta_j^-)], \quad \omega \geq 0 \quad (5.8.62)$$

and is symmetric about $\omega = 0$, and so is defined for only $\omega \geq 0$. Hence, $Y_{w_j w_j}(\omega)$ is an even function of frequency; i.e. $Y_{w_j w_j}(\omega) = Y_{w_j w_j}(-\omega)$. As noted by Mooers, $Y_{w_j w_j}(\omega)$ is not a power spectrum in the ordinary physical sense because it is complex valued. Rather it is related to the spectrum of the uv-Reynolds stress.

Outer-coherence squared: After first performing the ensemble or band averages in the brackets $\langle \cdot \rangle$, the outer-rotary coherence squared between series j and k is expressed in terms of the Fourier coefficients as

$$\lambda_{jk}^2(\omega) = \begin{cases} \langle A_j^- A_k^+ \rangle^2 \left[\langle \cos(\theta_k^+ - \theta_j^-) \rangle^2 + \langle \sin(\theta_k^+ - \theta_j^-) \rangle^2 \right] / \langle A_k^{+2} \rangle \langle A_j^{-2} \rangle, & \omega \geq 0 \\ \langle A_j^+ A_k^- \rangle^2 \left[\langle \cos(\theta_j^+ - \theta_k^-) \rangle^2 + \langle \sin(\theta_j^+ - \theta_k^-) \rangle^2 \right] / \langle A_j^{+2} \rangle \langle A_k^{-2} \rangle, & \omega \leq 0 \end{cases}$$
(5.8.63)

The phase lag, $\psi_{jk}(\omega)$ between the two oppositely rotating components of the two time series is then the same for the coherence and the cross-spectrum and is given by

$$\tan(\psi_{12}) = \begin{cases} \langle A_j^- A_k^+ \sin(\theta_j^- - \theta_k^+) \rangle / \langle A_j^- A_k^+ \cos(\theta_j^- - \theta_k^+) \rangle, & \omega \geq 0 \\ \langle A_j^+ A_k^- \sin(\theta_k^- - \theta_j^+) \rangle / \langle A_j^+ A_k^- \cos(\theta_k^- - \theta_j^+) \rangle, & \omega \leq 0 \end{cases} \quad (5.8.64)$$

If the values of

$$A_j^- A_k^+ \quad \text{and} \quad A_j^+ A_k^-$$

change little over the averaging interval covered by the angular brackets, then

$$\psi_{jk}(\omega) = \begin{cases} \theta_j^- - \theta_k^+, & \omega \geq 0 \\ \theta_k^- - \theta_j^+, & \omega \leq 0 \end{cases} \quad (5.8.65)$$

Figure 5.8.4(b) shows the outer rotary coherence and phase for five-year records of winter winds off Alaska. Counter-rotating vectors were coherent at negative frequencies in the range $-1 < f < 0$ cpd and exhibited little coherence at positive frequencies. In this portion of the frequency band, the linear phase gradient was similar to that for the co-rotating vectors (Figure 5.8.4a).

Complex admittance function: If we think of the wind vector at location 1 as the source (or input) function and the current at location 2 as the response (or output) function, we can compute the complex inner admittance, Z_{12}, between two co-rotating vectors as

$$Z_{12}(\omega) = S_{w_1 w_2}(\omega)/S_{w_1 w_1}(\omega), \quad -\omega_N < \omega < \omega_N \quad (5.8.66)$$

The amplitude and phase of this function are

$$|Z_{12}(\omega)| = |S_{w_1 w_2}(\omega)|/S_{w_1 w_1}(\omega) \quad (5.8.67a)$$

$$\Phi_{12}(\omega) = \tan^{-1}\{\operatorname{Im}[S_{w_1 w_2}(\omega)]/\operatorname{Re}[S_{w_1 w_2}(\omega)]\} \quad (5.8.67b)$$

For frequency ω, the absolute value of $Z_{12}(\omega)$ determines the amplitude of the clockwise (counterclockwise) rotating response one can expect at location 2 to a given clockwise (counterclockwise) rotating input at location 1. The phase, $\Phi_{12}(\omega)$, determines the lag of the response vector to the input vector.

The corresponding expressions for the complex outer admittance, Z_{12}, between two opposite-rotating vectors are

$$Z_{12}(\omega) = Y_{w_1 w_2}(\omega)/S_{w_1 w_1}(\omega), \quad -\omega_N < \omega < \omega_N \quad (5.8.68)$$

with amplitude and phase

$$|Z_{12}(\omega)| = |Y_{w_1 w_2}(\omega)|/S_{w_1 w_1}(\omega) \quad (5.8.69a)$$

$$\Phi_{12}(\omega) = \tan^{-1}\{\operatorname{Im}[Y_{w_1 w_2}(\omega)]/\operatorname{Re}[Y_{w_1 w_2}(\omega)]\} \quad (5.8.69b)$$

For frequency ω, the absolute value of $Z_{12}(\omega)$ yields the amplitude of the clockwise (counterclockwise) rotating response one can expect at location 2 to a given counterclockwise (clockwise) rotating input at location 1. The phase, $\Phi_{12}(\omega)$, determines the lag of the response vector to the input vector.

5.9 WAVELET ANALYSIS

The terms "wavelet transform" and "wavelet analysis" are two recent additions to the lexicon of time-series analysis. First introduced in the 1980s for processing seismic data (cf. Goupillaud et al., 1984), the technique has begun to attract attention in meteorology and oceanography where it has been applied to time-series measurements of turbulence (Farge, 1992; Shen and Mei, 1993), surface gravity waves (Shen et al., 1994), low-level cold fronts (Gamage and Blumen, 1993), and equatorial Yanai waves (Meyers et al., 1993).

As frequently noted in the literature, Fourier analysis does a poor job of dealing with signals of the form $\phi(t) = A(\tau)\cos(\omega t)$, where the amplitude, A, varies on the slow time scale, τ. Wavelet analysis has a number of advantages over Fourier analysis that are particularly attractive. Unlike the Fourier transform, which generates record-averaged values of amplitude and phase for each frequency component or harmonic, ω, the wavelet transform yields a localized, "instantaneous" estimate for the amplitude and phase of each spectral component in the data set. This gives wavelet analysis an advantage in the analysis of nonstationary data series in which the amplitudes and phases of the harmonic constituents may be changing rapidly in time or space. Where a Fourier transform of the nonstationary time series would smear-out any detailed information on the changing processes, the wavelet analysis attempts to track the evolution of the signal characteristics through the data set. As with other transform techniques, problems can develop at the ends of the time series, and steps must be taken to mitigate these effects. Similar to other transform techniques involving finite length data, steps also must be taken to minimize the distortion of the transformed data caused by the nonperiodic behavior at the ends of the time series. Lastly, we note that increasing the temporal resolution, Δt, of the wavelet analysis decreases the frequency resolution, Δf, and vice versa, such that $\Delta t \Delta f < \frac{1}{4}\pi$, reminiscent of the Heisenberg uncertainty relation. The more accurately we want to resolve the frequency components of a time series, the less accurately we can resolve the changes in these frequency components with time.

5.9.1 The wavelet transform

Wavelet analysis involves the convolution of a real time-series, $x(t)$, with a set of functions $g_{a\tau}(t) = g(t:\tau, a)$ that are derived from a "mother wavelet" or analyzing wavelet, $g(t)$, which is generally complex. In particular

$$g_{a\tau}(t) = \frac{1}{\sqrt{a}} g[a^{-1}(t-\tau)] \qquad (5.9.1)$$

where τ (real) is the *translation* parameter corresponding to the central point of the wavelet in the time series and a (real and positive) is the *scale dilation* parameter corresponding to the width of the wavelet. For the Gaussian-shaped Morlet wavelet (Figure 5.9.1) described in detail later in this section, the dilation parameter can be related to a corresponding Fourier frequency (or wavenumber).

The continuous wavelet transform, $X(t)$, of the time series with respect to the analyzing wavelet, $g(t)$, is defined through the convolution integral

$$X_g[\tau, a] = \frac{1}{\sqrt{a}} \int_{-\infty}^{\infty} g^*[a^{-1}(t-\tau)] x(t) \, dt \qquad (5.9.2)$$

in which g^* denotes the complex conjugate of g and variables τ, a are allowed to vary continuously through the domain $(-\infty, \infty)$. Wavelet analysis provides a two-dimensional unraveling of a one-dimensional time series into position, τ, and amplitude scale, a, as new independent variables. The wavelet transformation (5.9.2) is a sort of mathematical microscope, with magnification $1/a$, position τ, and optics given by the choice of the specific wavelet, $g(t)$ (Shen et al., 1994). Whereas Fourier analysis provides an average amplitude over the entire time series, wavelet analysis yields a measure of the localized amplitudes a as the wavelet moves through the time series with increasing values of τ. Although wavelets have a definite scale, they typically do not bear any resemblance to the sines and cosines of Fourier modes. Nevertheless, a correspondence between wavelength and scale a can sometimes be achieved.

To qualify for mother wavelet status, the function $g(t)$ must satisfy several properties (Meyers et al., 1993):

(1) Its amplitude $|g(t)|$ must decay rapidly to zero in the limit $|t| \to \infty$. It is this feature that produces the localized aspect of wavelet analysis since the transformed values, $X_g[\tau, a]$ are generated only by the signal in the cone of influence about $t = \tau$. In most instances, the wavelet $g[(t - \tau)/a]$ is assumed to have an insignificant effect at some time $|t| = \tau_c$.
(2) $g(t)$ must have zero mean. Known as the *admissibility condition*, this ensures the invertability of the wavelet transform. The original signal can then be obtained from the wavelet coefficients through the inverse transform

$$x(t) = \frac{1}{C} \int_{-\infty}^{\infty} \int_{-\infty}^{\infty} \{X_g[\tau, a] a^{-2} g_{a\tau}\} \, d\tau \, da$$

where

$$C^{-1} = \int_{-\infty}^{\infty} (\omega^{-1} |G(\omega)|^2) \, d\omega \qquad (5.9.4)$$

in which $G(\omega)$ is the Fourier transform of $g(t)$. For $1/C$ to remain finite, $G(0) = 0$.
(3) Wavelets are often regular functions, such that $G(\omega < 0) = 0$. These are also called *progressive* wavelets. Elimination of negative frequencies means that wavelets need only be described in terms of positive frequencies.
(4) Higher-order moments (such as variance and skewness) should vanish allowing the investigation of higher-order variations in the data. This requirement can be relaxed, depending on the application.

One of most extensively used wavelets is the standard (admissible and progressive) Morlet wavelet

$$g(t) = e^{-t^2/2} e^{+ict} \qquad (5.9.5)$$

consisting of a plane wave of frequency $c = \omega$ (or wavenumber $c = k$ in the spatial domain) which is modulated by a Gaussian envelope of unit width. Another possible wavelet which is applicable to a signal with two frequencies c_1 and c_2 is

$$g(t) = e^{-t^2/2} e^{ic_1 t} e^{ic_2 t} \qquad (5.9.6)$$

while the wavelet

$$g(t) = e^{-t^2/2} e^{ict} e^{ikt^2/2} \qquad (5.9.7)$$

is applicable to short data segments with linearly increasing frequency ("chirps").

5.9.2 Wavelet algorithms
The choice of $g(t)$ is dictated by the analytical requirements. More specifically, the wavelet should have the same pattern or signal characteristic as the pattern being

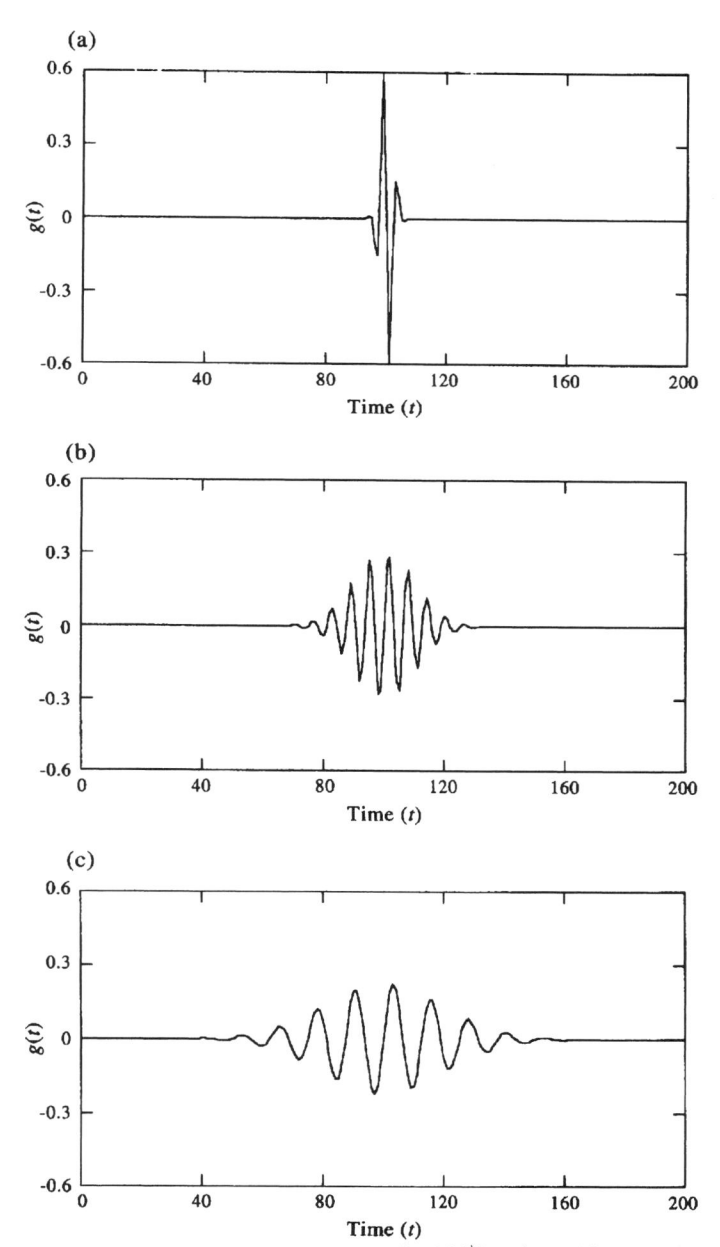

Figure 5.9.1. The Morlet wavelet, $g(t) = (1/\sqrt{a})e^{-[(t-\tau)/a]^2/2}\sin[c(t-\tau)/a]$, where t is time in arbitrary units ($t = t_n$; $n = 1, \ldots, 200$). The example is for $c = 10$ and time lag $\tau = 100$ so that the wavelet is seen midway through the time series. (a) $a = 2$; (b) $a = 10$; (c) $a = 20$.

sought in the time series. Large values of the transform $X_g(\tau, a)$ will then indicate where the time series $x(t)$ has the desired form. The simplest—and most time-consuming—method for obtaining the wavelet transform is to compute the transform at arbitrary points in parameter (τ, a) space using the discrete form of equation (5.9.2) for known values of $x(t)$ and $g(t)$. If one integrates from $0 < a \leq M$ and $0 < \tau \leq N$, the integration time goes as MN^2. An alternate method is to use the convolution theorem and then obtain the wavelet transform in spectral space

$$X_g[\tau, a] = \frac{1}{\sqrt{a}} \int_{-\infty}^{\infty} e^{i\tau\omega} G^*(a\omega) X(\omega) \, d\omega \qquad (5.9.8)$$

where $G(\omega)$ and $X(\omega)$ are the Fourier transforms of $g(t)$ and $x(t)$, respectively. Since FFT transforms can now be exploited, the analysis time drops to $MN\log_2 N$. To use this method, $G(\omega)$ should be known analytically and the data must be preprocessed to avoid errors from the FFT algorithms. For example, if $x(t)$ is aperiodic, the discrete form of (5.9.7) will generate an artificial periodicity in the wavelet transform that greatly distorts the results for the end regions. Methods have been devised to work around this problem. Aliasing and bias in FFT routines must also be taken into account.

Meyers et al. (1993) used the standard Morlet wavelet (5.9.5), for which $g(t) = e^{-t^2/2} e^{ict}$, to examine a signal that changes frequency halfway through the measurement. Here, we have followed tradition and used c for frequency ω. After considerable attempts (including use of raw data, cosine weighted data and other variations), the authors decided that the best approach was to taper or buffer the original time series with added data points that attenuate smoothly to zero past the ends of the time series. "The region of the transform corresponding to these points is then discarded after the transform. Without this buffering, a signal whose properties are different near its ends will result in a wavelet transform that has been forced to periodicity at all scales through a distortion (in some cases severe) of the end regions. The greater the aperiodicity of the signal, the greater the distortion."

For the Morlet wavelet, the dilation parameter a giving the maximum correlation between the wavelet and a plane Fourier component of frequency ω_o (i.e. a wave of the form $e^{i\omega_o t}$) is

$$a_o = \frac{[c + (2+c^2)^{1/2}]}{4\pi} T_o \qquad (5.9.10)$$

where $T_o = 2\pi/\omega_o$ is the Fourier period. (In wavenumber space, T_o is replaced by wavelength λ_o and ω_o by k_o.) We note that any linear superposition of periodic components results in separate local maxima. Consequently, the wavelet transform of any function $x(t) = \sum A_j e^{ik_j t}$ will have modulus maxima at $a_j = [c + (2+c^2)^{1/2}]/(2k_j)$.

5.9.3 Oceanographic examples

In this section, we will consider two oceanographic wavelet examples (surface gravity wave heights and zonal velocity from a satellite-tracked drifter) using the standard Morlet wavelet

$$g(t) \rightarrow g[(t-\tau)/a] = \frac{1}{\sqrt{a}} e^{-\frac{1}{2}[(t-\tau)/a]^2} \sin\left[c(t-\tau)/a\right] \qquad (5.9.11)$$

In this real expression, the Gaussian function determines the envelope of the wavelet while the sine function determines the wavelengths that will be preferentially weighted by the wavelet. The wavelet function progresses through the time series with increasing τ, its cone of influence centered at times $t = \tau$. As a increases, the width of the Gaussian spreads in time from its center value (Figure 5.9.1a–c). Increasing c increases the number of oscillations over the span of the function. The processing procedure is as follows: (1) read in the time series $x(n)$ ($n = 0, \ldots, N - 1$) to be analyzed, where $N = 2^m$ (m is an integer). To reduce ringing, extend each end of the time series by adding a trigonometric taper, tap $= 1 - \sin\phi$, where tap $= 1.0$ at the end values $x(0)$ and $x(N - 1)$. The total length of the buffered time series must remain a power of two; (2) remove the mean of the new record and then take the FFT of the time series to obtain $X(\omega)$; (3) take the Fourier transform of the wavelet $g(t)$ at given length scales, a, to obtain $G(a\omega)$; (4) calculate the integral (5.9.8) by convolving the product $G^*(a\omega)X(\omega)$ in Fourier space; (5) take the inverse FFT of the result to obtain $\sqrt{a}X_g[\tau, a]$ as a function of time dilation τ and amplitude, a.

In Figure 5.9.2(a) we have plotted a 300 s record of surface gravity wave heights measured off the west coast of Vancouver Island in the winter of 1993. Maximum wave amplitudes of around 3 m occurred mid-way through the time series. The Morlet wavelet transform of the record yields an estimate of the wave amplitude (Figure 5.9.2b) and phase (Figure 5.9.2c) as functions of the wave period (T) and time (t). Also plotted is the value of the wave period (T = scale a) at peak energy (Figure 5.9.2d). Comparison of Figures 5.9.2(b) and 5.9.2(d) reveals that the larger peaks near times of 75, 150, and 210 s all have about the same wavelet scale, a, corresponding to a peak wave period of around 8 s. Also, as one would expect, the 2π changes in phase between crests (Figure 5.9.2c) increases with increasing wave period (scale, a).

In our second example, we have applied a standard Morlet wavelet transform to a 90-day segment of 3-hourly sampled east–west (u) current velocity (Figure 5.9.3a) obtained from a satellite-tracked drifter launched in the northeast Pacific in August 1990 as part of the World Ocean Circulation Experiment (WOCE). The drifter was drogued at 15-m depth and its motion indicative of currents in the surface Ekman layer. The 90-day velocity record has been generated from positional data using a cubic spline interpolation algorithm. We focus our attention on the high-frequency end of the spectrum, $0 < a < 1.5$ days. As indicated by Figures 5.9.3(b) and (c), the first 30 days of the record, from Julian day (JD) 240 to 270, were dominated by weak semidiurnal tidal currents with periods of 0.5 days. Beginning on JD 270, strong wind-generated inertial motions with periods around 16 h ($f \approx 1.5$ cpd) dominated the spectrum. These energetic motions persisted through the record, except for a short hiatus near JD 295. A blow-up of the segment from JD 240 to 270 shows a rapid change in signal phase associated with the shift from semidiurnal tidal currents to near-inertial motions. The contribution from the beat frequency between the M_2 tidal signal and the inertial oscillations, $fM_2 = 0.0805 + 0.0621$ cph $= 0.1426$ cph can also be seen in the transformed data at period $T \approx 0.29$ days. Examination of the longer period motions ($2 < a < 30$ days) suggests the presence of a long-period modulation of the high-frequency motions associated with the near-inertial wave events.

5.9.4 The S-transformation

Wavelet transforms are not the only method for dealing with nonstationary oscillations with time-varying amplitudes and phases. The S-transformation (Stockwell et al., 1994) is an extension of the wavelet transform that has been used by Chu (1994) to

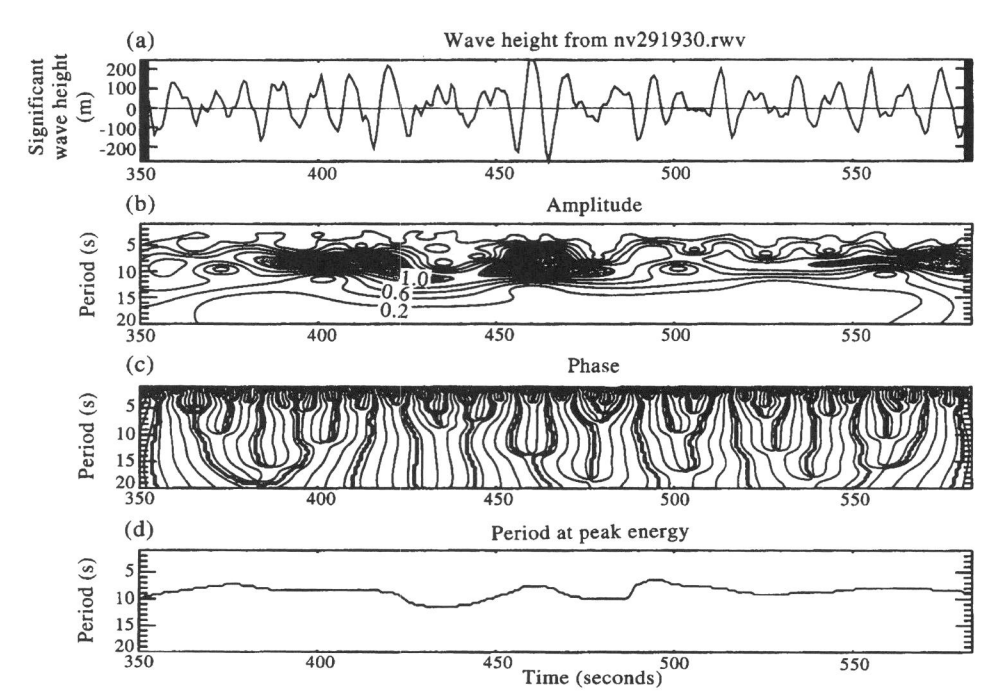

506   *Data Analysis Methods in Physical Oceanography*

Figure 5.9.2. Morlet wavelet transform of surface gravity waves measured from a waverider buoy moored off the west coast of Vancouver Island. (a) Original five-minute time series of significant wave height for the winter of 1993. (b) Wave amplitude (m) and (c) phase (deg.) as a functions of time; (d) the value of a (wave period) at peak wave amplitude. (Courtesy, D. Masson.)

examine the localized spectrum of sea level in the TOGA data sets. For this particular transform, the relationship between the S-transform, $S(\omega, \tau)$, and the data, $x(t)$, is given by

$$S(\omega, \tau) = \int_{-\infty}^{\infty} H(\omega + \alpha) e^{-(2\pi^2 \alpha^2/\omega^2)} e^{i2\pi\alpha\tau} d\alpha \qquad (5.9.12)$$

$$x(t) = \int_{-\infty}^{\infty} \int_{-\infty}^{\infty} S(\omega, \tau) e^{i2\pi\alpha\tau} \, d\omega \, d\tau \qquad (5.9.13)$$

where

$$H(\omega + \alpha) = \int_{-\infty}^{\infty} x(t) e^{-i2\pi(\omega+\alpha)\tau} \, dt \qquad (5.9.14a)$$

$$= \int_{-\infty}^{\infty} S(\omega + \alpha, \tau) \, d\tau \qquad (5.9.14b)$$

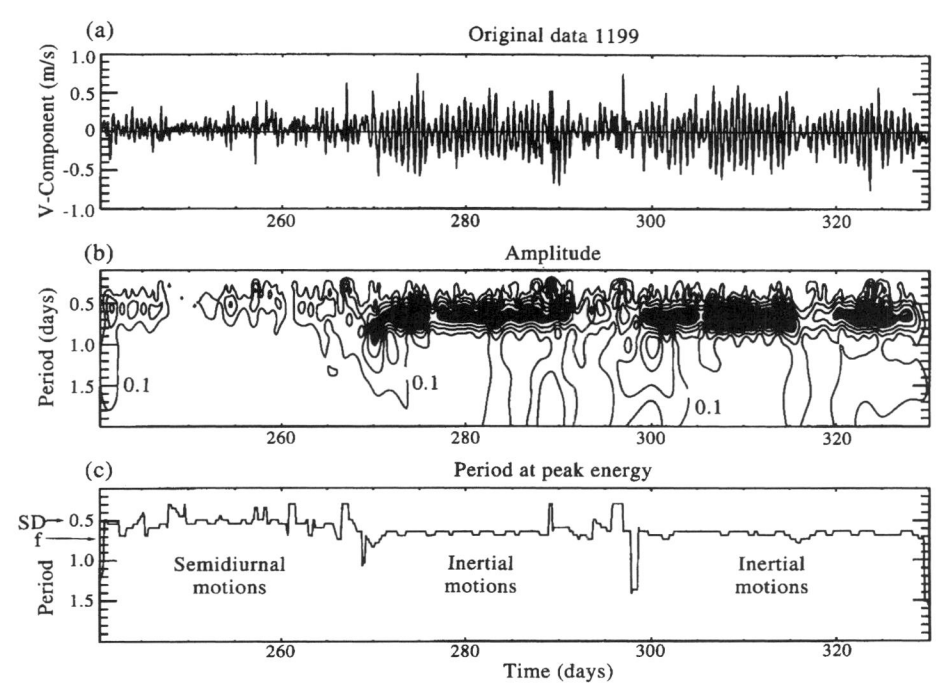

Figure 5.9.3. The Morlet wavelet transform of a 90-day record of the east–west velocity component from the trajectory of a satellite-tracked drifter in the northeast Pacific, September 1990. (a) Original 3-hourly time series; (b) amplitude (cm/s) versus time as a function of period, T, in the range $0 < T < 2.0$ days; (c) period (days) of the current oscillations at peak amplitude. (Courtesy, J. Eert.)

is the standard Fourier transform of the input time series data. As indicated by (5.9.14b), the Fourier transform is the time average of the S-transform, such that $|H(\omega)|^2$ provides a record-averaged value of the localized spectra $|S(\omega)|^2$ derived from the S-transform. Equation (5.9.13) can also be viewed as the decomposition of a time series $x(t)$ into sinusoidal oscillations which have time-varying amplitudes $S(\omega, \tau)$.

The discrete version of the S-transformation can be obtained as follows. As usual, let $x(t_n) = x(n\Delta t)$, $n = 0, 1, \ldots, N-1$ be a discrete time series of total duration $T = N\Delta t$. The discrete version of (5.9.12) is then

$$S(0, \tau_q) = \frac{1}{N}\sum_{m=0}^{n-1} x(m/T), \quad p = 0 \tag{5.9.15a}$$

$$S(\omega_p, \tau_q) = \sum_{m=0}^{N-1}\left\{H[(m+p)/T]e^{-(2\pi^2 m^2/p^2)}e^{i2\pi mq/N}\right\}, \quad p \neq 0 \tag{5.9.15b}$$

where $S(0, \tau_q)$ is the mean value for the time series, $\omega_p = p/N\Delta t$ is the discrete frequency of the signal, and $\tau_q = q\Delta t$ is the time lag. The discrete Fourier transform is given by

$$H(p/T) = \frac{1}{N}\sum_{k=0}^{N-1} x(k/T)e^{-i2\pi pk/N} \tag{5.9.16}$$

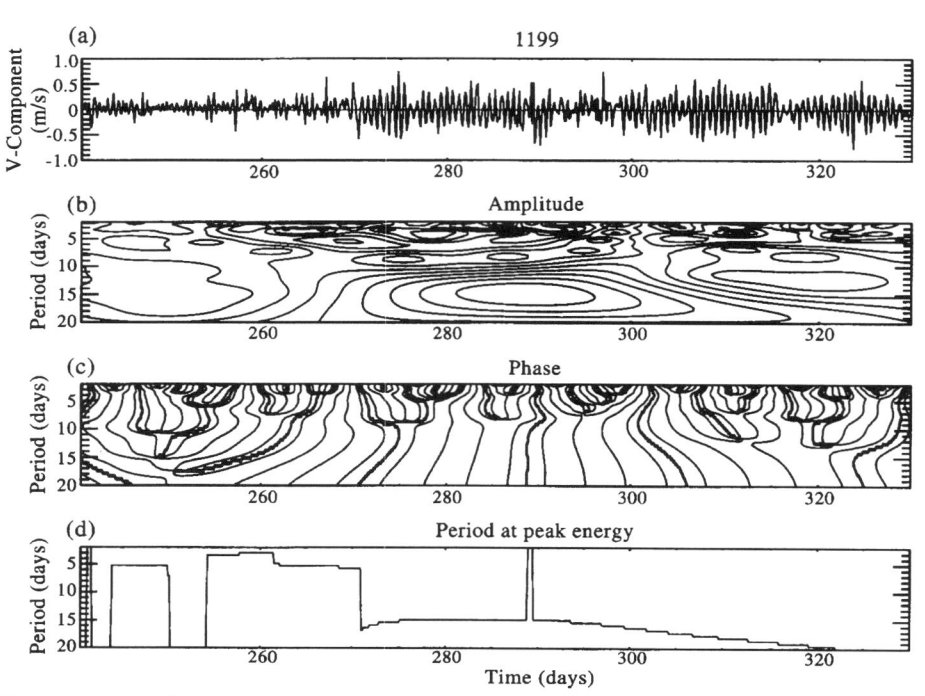

508 *Data Analysis Methods in Physical Oceanography*

Figure 5.9.4. As for Figure 5.9.3 but for a larger range of periods. (a) Original 3-hourly velocity time series; (b) amplitude (cm/s) and (c) phase (degrees) versus time as a function of period, T, in the range $2 < T \leq 20$ days; (d) period (days) of the current oscillations at peak amplitude.

The S-transform is a complex function of frequency ω_p and time τ_q, with amplitude and phase defined by

$$A(\omega_p, \tau_q) = |S(\omega_p, \tau_q)| \qquad (5.9.17a)$$

$$\Phi(\omega_p, \tau_q) = \tan^{-1}\{\text{Im}[S(\omega_p, \tau_q)]/\text{Re}[S(\omega_p, \tau_q)]\} \qquad (5.9.17b)$$

For a sinusoidal function of the form

$$X(\omega_p, \tau) = A(\omega_p, \tau) \cos[2\pi\omega_p \tau + \Phi(\omega_p, \tau)] \qquad (5.9.18)$$

the function X at frequency ω_p is called the "voice".

Chu (1994) applied the S-transform to the nondimensionalized sea-level records, $x(t)$, collected at Nauru (0°32'S, 166°54'W) in the western equatorial Pacific and La Libertad (2°12'S, 80°55'W) in the eastern equatorial Pacific. Here

$$x(t) = \frac{[\eta(t) - \bar{\eta}]}{\bar{\eta}} \qquad (5.9.19)$$

and $\bar{\eta}(t)$ represents the mean value of the sea level, $\eta(t)$. A Fourier spectral analysis of the time series revealed a strong annual sea-level oscillation in the western Pacific and a weak annual oscillation in the eastern Pacific. Both stations had strong quasi-biennial oscillations with periods of 24–30 months. The S-transformation was then used to examine the temporal variability in these components throughout the 16 and

18-year time series. For example, the voices for the annual oscillation ($\omega_{16} = 16/T$; $T = 192$ months) were similar at the two locations with higher amplitudes in the late 1970s than in the late 1980s (Figure 5.9.5). At La Libertad, the annual cycle became weak after 1979. The temporally varying quasi-biennial oscillations ($\omega_8 = 8/T$) were out-of-phase between the western and eastern Pacific (Figure 5.9.6).

5.9.5 The multiple filter technique

The multiple filter technique is a form of signal demodulation that uses a set of narrow-band digital filters (windows) to examine variations in the amplitude and phase of dispersive signals as functions of time, t, and frequency, ω (or f). Originally designed to resolve complex transient seismic signals composed of several dominant frequencies (Dziewonski et al., 1969), the technique has recently been modified for the analysis of clockwise and counterclockwise rotary velocity components (Thomson et al., 1997) and in investigations of tsunami wave dispersion (Gonzalez and Kulikov, 1993).

The multiple filter technique relies on a series of band-pass filters centered on a range of narrow frequency bands to calculate the instantaneous signal amplitude or phase. Dziewonski et al. (1969) filter in the frequency domain rather than the time domain, although the results are equivalent to within small processing errors. The filtering algorithm generates a matrix (grid) of amplitudes or phases with columns representing time and rows representing frequency (or period). The gridded values can then be contoured to give a three-dimensional plot of the demodulated signal amplitude (or phase) as a function of time and frequency. Gonzalez and Kulikov (1993) used the technique to examine the evolution of tsunami waves generated by an undersea earthquake in the Gulf of Alaska on 6 March 1987 (Figure 5.9.7). Sea-level heights measured by two bottom-pressure recorders deployed in the deep ocean to the south of Kodiak Island show that the tsunami waves were highly dispersive (low frequencies propagated faster than high frequencies) and that the arrival times of the waves closely followed the theoretical predictions for shallow-water wave motions. Peak spectral amplitudes were centered around a period of roughly 5 min and the signal duration was about 40 min.

5.9.5.1 Theoretical considerations

Since the technique is used to examine signal energy as a function of time and frequency, it is desirable that the filtering function has good resolution in the immediate vicinity of each center frequency and time value of the f–t diagram. The Gaussian function was chosen to meet these requirements since the frequency–time resolution is greater for this function than any other type of nonband-limited function. A system of Gaussian filters with constant relative response leads to a constant resolution on a $\log(\omega)$ scale. If $\omega_n = 2\pi f_n$ denotes the center frequency of the nth row, the Gaussian window function can be written

$$H_n(\omega) = \exp\left\{-\alpha[(\omega - \omega_n)/\omega_n]^2\right\} \qquad (5.9.20)$$

The Fourier transform of H_n, which bears a close resemblance to the Morlet wavelet (5.9.11), is

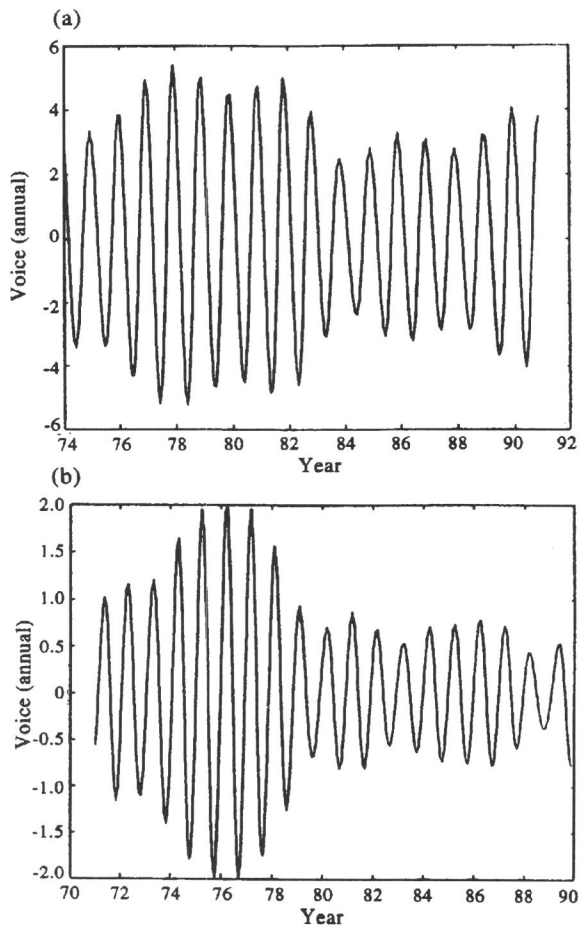

510 *Data Analysis Methods in Physical Oceanography*

Figure 5.9.5. The "voices" for the annual oscillation ($\omega_{16} = 16/T$; $T = 192$ months) for (a) Nauru; and (b) La Libertad. Higher amplitudes were recorded in the late 1970s than in the late 1980s. (Chu, 1994.)

$$h_n(t) = \frac{\sqrt{\pi}}{2\alpha} \omega_n \exp\left[-(\omega_n^2 t^2/4\alpha)\right] \cos(\omega_n t) \qquad (5.9.21)$$

The resolution is controlled by the parameter, α. The value of α that we choose depends on the dispersion characteristics in the original signal and, as the user of this method will soon discover, improved resolution in time means reduced resolution in frequency, and vice versa. We also need to truncate the filtering process. Dziewonski et al. (1969) used a filter cut-off where the filter amplitude was down 30 dB from the maximum.

If we let BAND be the relative bandwidth, then the respective lower and upper limits of the symmetrical filter, denoted $\omega_{L,n}$ and $\omega_{U,n}$, are

$$\omega_{L,n} = (1 - \text{BAND})\omega_n \qquad (5.9.22a)$$

$$\omega_{U,n} = (1 + \text{BAND})\omega_n \qquad (5.9.22b)$$

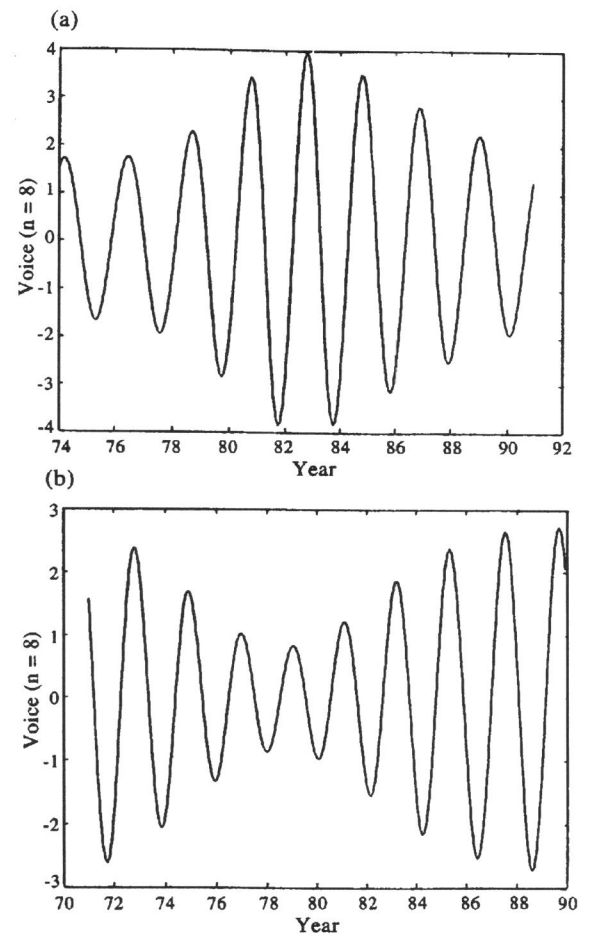

Figure 5.9.6. The "voices" for the quasi-biennial oscillations ($\omega_8 = 8/T$) for (a) Nauru; and (b) La Libertad. The oscillations were out-of-phase between the western and eastern Pacific. (Chu, 1994.)

The parameter α in (5.9.20) and (5.9.21) is expressed in terms of the bandwidth and the function β

$$\alpha = \beta/\text{BAND}^2 \tag{5.9.23}$$

where

$$\beta = \ln[H_n(\omega_n)/H_n(\omega_{L,n})] = \ln[H_n(\omega_n)/H_n(\omega_{U,n})] \tag{5.9.24}$$

describes the decay of the window function, $H_n(\omega)$. The window function then takes the form

$$H_n(\omega) = \begin{cases} 0 & \text{for } \omega(1 - \text{BAND})\omega_n \\ \exp\{-\alpha[(\omega - \omega_n)/\omega_n]^2\} & \text{for } (1 + \text{BAND})\omega_n \leq \omega \leq (1 + \text{BAND})\omega_n \\ 0 & \text{for } \omega > (1 + \text{BAND})\omega_n \end{cases}$$

$$\tag{5.9.25}$$

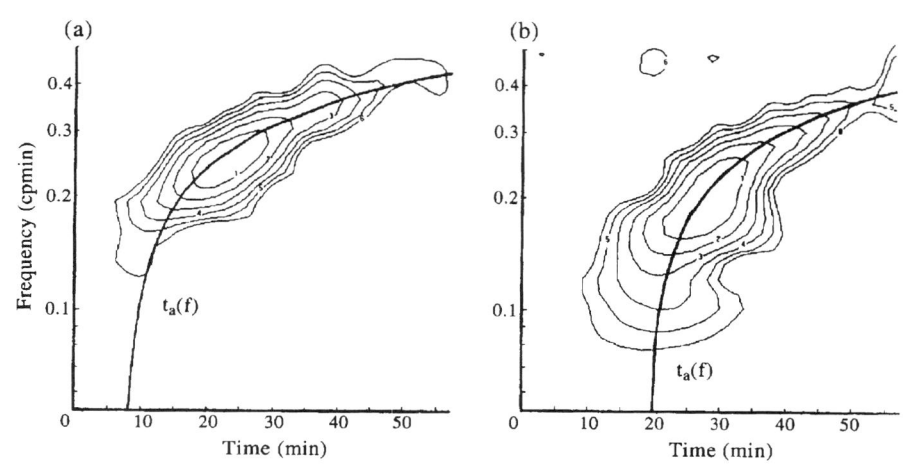

512   *Data Analysis Methods in Physical Oceanography*

Figure 5.9.7. Multiple filter technique applied to sea-level heights measured in 5 km of water near 53°N, 156°W in the Gulf of Alaska on 6 March 1988. Amplitude contours in the f–t diagram are normalized by the maximum value and drawn with a step of 1 dB. Solid curve denotes the theoretical arrival time for these highly dispersive waves. (From Gonzalez and Kulikov, 1993.)

In their analysis of seismic waves, Dziewonski et al. (1969) used BAND = 0.25, $\beta = 3.15$, and $\alpha = \beta/\text{BAND}^2 = 50.3$.

The *f–t* diagram for the Alaska tsunamis (Figure 5.9.7) was obtained by windowing in the frequency domain with the truncated Gaussian function (5.9.25). In the time domain, the traces represent the convolution of the original data series with the Gaussian weighting function. The authors first set $\alpha = 25$ and chose $\beta = 1$, so that BAND = 0.20. The choice of β in (5.9.24) is arbitrary and can be set to unity, whereupon the bandwidth is determined by the e^{-1} values of the Gaussian function. For $\alpha = 25$ but $\beta = 2$, we have BAND = 0.28, and so on.

The flow chart for the analysis (Figure 5.9.8) is as follows:

(1) Remove the mean and trend (linear or other obvious functional trend) from the digital time series, $y(t)$.
(2) Fourier transform the time series. If an FFT algorithm is to used for this purpose, augment the time series with zeros to the nearest power of 2.
(3) Evaluate the center frequencies, $\omega_n = \omega_{n-1}/\text{BAND}$, for the array of narrow-band filters. The filters have a constant relative bandwidth, BAND, with the total width of each filter occupying the same number of rows in the log (frequency) scale. As noted on numerous occasions in the text, it is the length of the time series and the sampling rate which determine the frequency of the Fourier components. Since it often is difficult to get the frequencies obtained from the Fourier analysis to line up exactly with the center frequencies of the filters, select those components of the Fourier analysis which are closest to each member of the array and use these as the center frequencies.
(4) Select equally spaced times (columns) for calculation of amplitude or phase, focusing mainly on the times following the arrival of the waves.
(5) Filter the wave spectrum (sine and cosine functions of the Fourier transform) in the frequency domain with the Gaussian filter $H_n(\omega)$. This filter is symmetric about the center frequencies, ω_n.

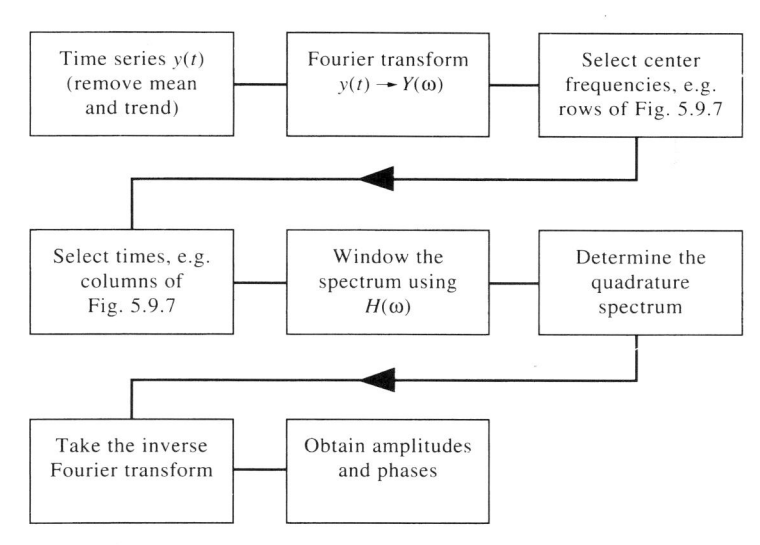

Figure 5.9.8. Flow chart for application of the multiple-filter technique. (Adapted from Dziewonski et al., 1969.)

(6) Take the inverse Fourier transform of the spectra using the same Fourier transform used in step 2. Since the inverse Fourier transform for the wave spectrum as windowed by the function $H_n(\omega)$ yields only the in-phase component of the filtered signal for each ω_n, knowledge of the quadrature spectrum is also required for evaluation of the instantaneous spectral amplitudes and phases. The quadrature spectrum is found from the in-phase spectrum using

$$Q_n(\omega) = H_n(\omega) e^{i\pi/2} \qquad (5.9.26)$$

The amplitude and phase of the signal for each center frequency for each time are derived from the inverse Fourier transforms of the spectra and quadrature spectra.

(7) Instantaneous spectral amplitudes and phases are computed for each time step. The procedure (5)–(7) is repeated for each center frequency.

The multiple filter technique can be used to examine rotary components of current velocity fields. In this case, the input is not a real variable, as it is for scalar time series, but a complex input, $w(t) = u(t) + iv(t)$. Figure 5.9.9 is obtained from the analysis of a 90-day time series of surface currents measured by a 15 m drogued satellite-tracked drifter launched off the Kuril Islands in the western North Pacific on 4 September 1993 (Thomson et al., 1997). The 3-hourly sampling interval used for this time series was made possible by the roughly eight position fixes per day by the satellite-tracking system. Plots show the variation in spectral amplitude of the clockwise and counterclockwise rotary velocity components as functions of time and frequency. For illustrative purposes, we have focused separately on the high and low frequency ends of the spectrum (periods shorter and longer than two days). Several interesting features quickly emerge from these *f–t* diagrams. For example, the motions are entirely dominated by the clockwise rotary component except within the narrow channel (Friza Strait) between the southern Kuril Islands where the motions become more rectilinear. The burst of clockwise rotary flow encountered by the drifter over the Kuril-Kamchatka Trench starting on day 28 was associated with wind-generated

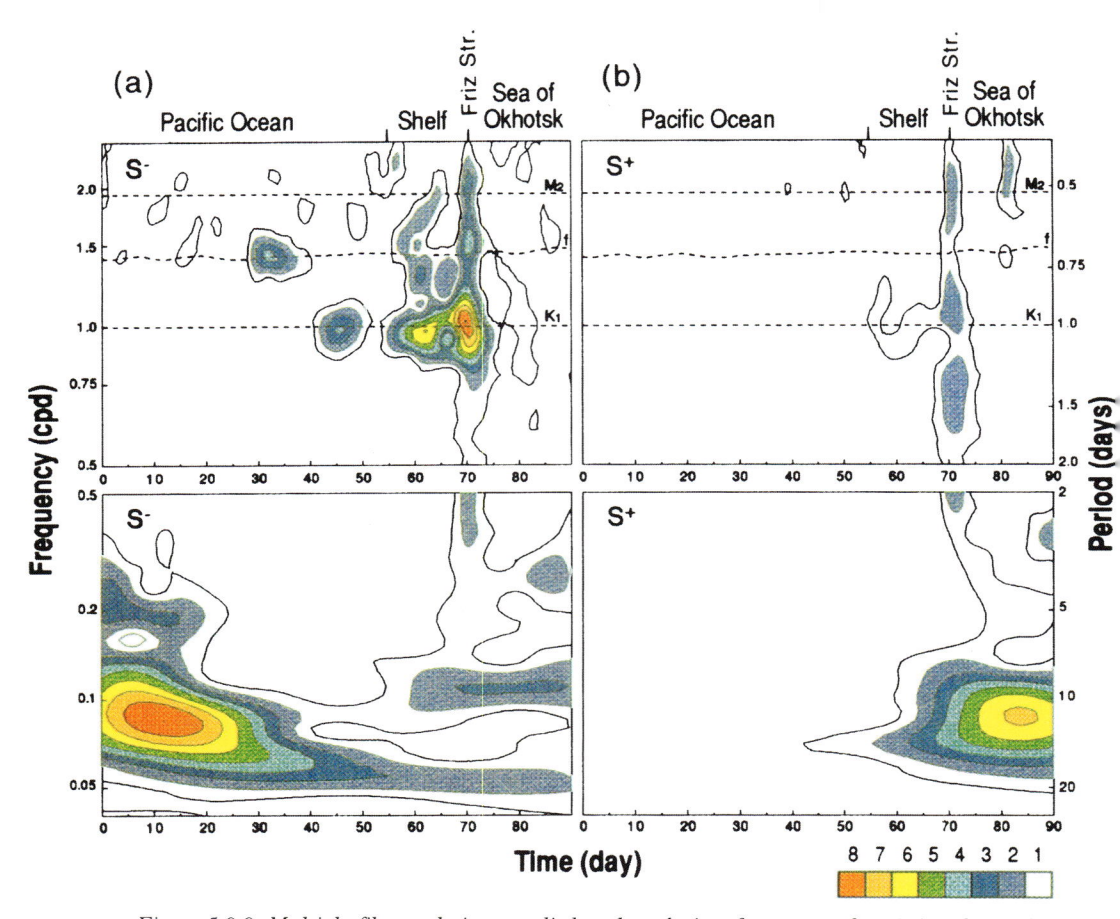

Figure 5.9.9. Multiple-filter technique applied to the velocity of a near-surface (15 m drogued) satellite tracked drifter launched off the Kuril Island in 1993. S^- denotes the spectral amplitude (cm/s) of the clockwise rotary component versus frequency (cpd) and time (day); S^+ denotes the spectral amplitude of the counterclockwise component. (From Thomson et al., 1997.)

inertial waves whereas the strong clockwise rotary diurnal currents first encountered on day 40 and then again on day 55 were associated with diurnal-period continental shelf waves propagating along the steep continental slope of the Kuril Islands.

5.10 DIGITAL FILTERS

5.10.1 Introduction

Digital filtering is often an important step in the processing of digital oceanographic data. Applications include smoothing and decimation of time series, removal of fluctuations in selected frequency bands, and the alteration of signal phase. The term "decimation" originally meant the removal of every tenth point but is now commonly used for values other than 10. Digital filtering facilitates data processing by preconditioning the frequency content of the record. For example, filters are commonly used in studies of inertial waves to isolate current variability centered near the local Coriolis frequency, to remove background sea-level fluctuations in investigations of tsunamis, and to eliminate tidal frequency fluctuations in studies of low-frequency current oscillations (Figure 5.10.1). The terms "detided" or "residual" time series are

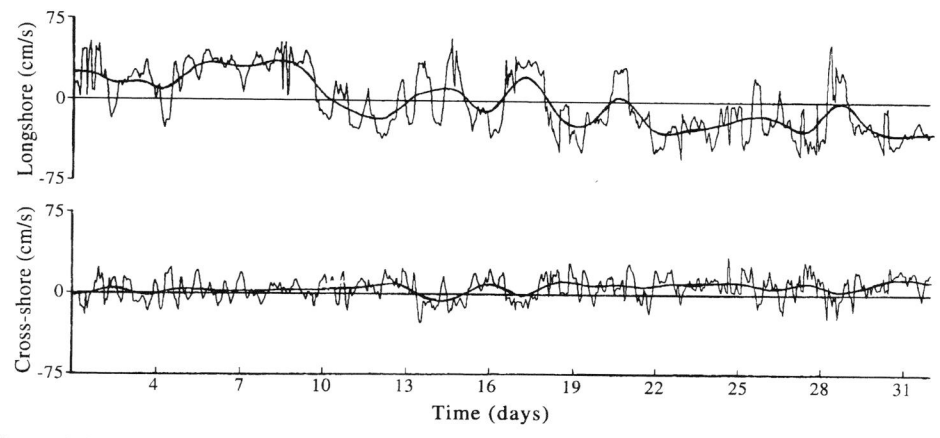

Figure 5.10.1. Time series of hourly longshore (top) and cross-shore (bottom) components of current velocity at 53-m depth on the continental shelf of northern Vancouver Island for March 1980. Thin line: original hourly data. Thick line: Hourly data filtered with a low-pass Godin tide-elimination filter, $A_{25}^2 A_{24}/(25^2 24)$. (From Huggett, Crawford, Thomson, and Woodward, 1987.)

used to describe time series that have been filtered to remove tidal components. Filters also provide algorithms for data interpolation, for integration and differentiation of recorded signals, and for linear prediction models.

There is no single type of digital filter for general oceanographic use. Selection of an appropriate filter depends on a variety of factors, including the frequency content of the data and the kind of analysis to be performed on the filtered record. Personal preference or familiarity with one type of filter also can be deciding factors. However, in certain instances, one type of filter may be superior to another for a specific task, and proper filter selection involves some forethought. Often the type of filter must be tailored to the job at hand. For example, some of the so-called "tide-elimination" or "tide-killer" filters once used extensively in oceanography are inadequate for time series with marked diurnal period variability (Walters and Heston, 1982; Thompson, 1983). These filters permit leakage of unwanted diurnal tidal energy into the nontidal (residual) frequency bands of the filtered record. Elimination of this problem is possible through proper filter selection.

This chapter begins with a brief outline of basic filtering concepts then proceeds to descriptions of some of the more useful digital filters presently used in oceanographic research. We use the term "filter" to cover any linear operation on the data. In *optimal estimation* applications, the term applies specifically to an optimal estimate of the last measurement point. *Smoothing* is reserved for estimates spanned by the observations. Much of the emphasis in this chapter is on the design of low-pass digital filters that remove high-frequency oscillations from a given oceanographic time series. These filters can be used to construct other types of filters. The running-mean filter, the cosine Lanczos-window (or Lanczos–cosine) filter, and the Butterworth filter are among those most commonly used in oceanography.

5.10.2 Basic concepts

From a practical standpoint, a good low-pass filter should have five essential qualities: (1) a sharp cut-off, so that unwanted high-frequency components are effectively removed; (2) a comparatively flat pass-band that leaves the low-frequency components unchanged; (3) a clean transient response, so that rapid changes in the signal do not result in spurious oscillations or "ringing" within the filtered record; (4) zero phase shift; and (5) acceptable computation time. As a rule, many of these desirable features are mutually exclusive and there are severe limitations to achieving the desired filter. We are invariably faced with a trade-off between the ability of the filter to produce the required results and the amount of filter-induced data loss we can afford to tolerate. For example, improved statistical reliability (increased degrees of freedom) for specified frequency bands decreases the frequency resolution of a filter while more sharply defined frequency cut-offs lead to greater ringing and associated data loss.

Suppose we have a time series consisting of the sequence

$$x(t_n) = x_n, \quad n = 0, 1, \ldots, N-1 \tag{5.10.1}$$

with observations at discrete times $t_n = t_o + n\Delta t$ in which t_o marks the start time of the record and Δt is the sampling increment. A digital filter is an algebraic process by which a sequential combination of the input $\{x_n\}$ is systematically converted into a sequential output $\{y_n\}$. In the case of linear filters, for which the output is linearly related to the input, the time domain transformation is accomplished through convolution (or "blending") of the input with the weighting function of the filter. Filters having the general form

$$y_n = \sum_{k=-M}^{M} h_k x_{n-k} + \sum_{j=-L}^{L} g_j y_{n-j}, \quad n = 0, 1, \ldots, N-1 \tag{5.10.2}$$

(in which M, L are integers and h_k, g_j are nonzero weighting functions) are classified as *recursive* filters since they generate the output by making use of a feed-back loop specified by the second summation term. Such filters "remember" the past in the sense that all past output values contribute to all future output values. Filters based on the input data only ($g_j = 0$), are classified as *nonrecursive* filters. Any filter for which $-M \leq k \leq M$ is said to be physically unrealizable (in the sense of any real-time output) because both past and future data are needed to calculate the output. Filters of this type have widespread application in the analysis of pre-recorded data for which all digital values are available beforehand. Filters for which $0 \leq k \leq M$ are said to be physically realizable or causal, and are used in real-time data acquisition and in forecasting procedures.

Impulse response: The output $\{y_n\}$ of a nonrecursive linear filter is obtained through the convolution

$$y_n = \sum_{k=-M}^{M} h_k x_{n-k} = \sum_{k=-M}^{M} h_{n-k} x_k, \quad n = 0, 1, \ldots, N-1 \tag{5.10.3}$$

where h_k are the time invariant weights and there are N data values $x_o, x_1, \ldots, x_{N-1}$. For a symmetric filter, the time domain convolution becomes

$$y_n = \sum_{k=0}^{M} h_k (x_{n-k} + x_{n+k}), \quad n = 0, 1, \ldots, N-1 \tag{5.10.4}$$

in which $h_k = h_{-k}$. The set of weights $\{h_k\}$ is known as the *impulse response* function and is the response of the filter to a spike-like impulse. To see this, we set $x_n = \delta_{0,n}$

where $\delta_{m,n}$ is the Kronecher delta function

$$\delta_{m,n} = 0, \quad m \neq n$$
$$= 1, \quad m = n \tag{5.10.5}$$

Equation (5.10.1) then becomes

$$y_n = \sum_{k=-M}^{M} h_k \delta_{0,n-k} = h_n \tag{5.10.6}$$

The summations in equations (5.10.3) and (5.10.4) are based on a total of $2M + 1$ specified weights with individual values of h_k labeled by subscripts $k = -M, -M + 1, \ldots, M$. To make practical sense, the number of weights is limited to $M \ll N/2$ where $N\Delta t$ is the record length. In reality, it is not possible to use equation (5.10.3) to calculate an output value y_n for each time t_n. Because the response function spans a finite time (equal to $2M\Delta t$), difficulties arise near the ends of the data record and we are forced to accept the fact that there are always fewer output data values than input values. There are three options: (1) We can make do with $2M$ fewer estimates of y_n (resulting from time losses of $M\Delta t$ at each end of the record); (2) we can create values of $x(t_n)$ for times outside the observed range $0 \leq t < (N-1)\Delta t$ of the time series; or (3) we can progressively decrease the filter length, M, in accordance with the number of remaining input values. In the first approach, x_n is defined for $n = 0, 1, \ldots, N-1$ whereas y_n is defined for the shortened range $n = M, M+1, \ldots, N-(M+1)$. In the second approach, the appendaged estimates of x_n should qualitatively resemble the data at either end of the record. For example, we could use the "mirror images" of the data reflected at the end points of the original time series. In the third approach, the values y_{M-1} and $y_{N-(M-1)}$ are based on $(M-1)$ weights, the values y_{M-2} and $y_{N-(M-2)}$ on $(M-2)$ weights, and so on.

Frequency response: The Fourier transform of $y(t_n)$ in (5.10.3) is

$$Y(\omega) = \sum_{n=-M}^{M} y_n e^{-i\omega_n \Delta t}$$
$$= \sum_{k=-M}^{M} h_k e^{-i\omega_k \Delta t} \sum_{n=-M}^{M} x_{n-k} e^{-i\omega_{(n-k)} \Delta t} \tag{5.10.7}$$
$$= H(\omega) X(\omega)$$

so that convolution in the time domain corresponds to multiplication in the frequency domain. The function

$$H(\omega) = \frac{Y(\omega)}{X(\omega)} = \sum_{k=-M}^{M} h_k e^{-i\omega k \Delta t}, \quad \omega \equiv \omega_n = 2\pi n/N\Delta t \tag{5.10.8}$$

$n = 0, \ldots, N/2$ is known as the *frequency response* (or *admittance function*; see Section 5.8.7) since it determines how a specific Fourier component $X(\omega)$ is modified as it is transformed from input to output. For the symmetric filter (5.10.4), the transfer function reduces to

$$H(\omega) = h_o + 2\sum_{k=1}^{M} h_k \cos(\omega k \Delta t) \tag{5.10.9}$$

Once $H(\omega)$ is specified, the weights h_k are found through the inverse Fourier transform

$$h_k = \sum_{n=-N/2}^{N/2} H(\omega) e^{i\omega_n k \Delta t} \tag{5.10.10}$$

In general, $H(\omega)$ is a complex function that can written in the form

$$H(\omega) = |H(\omega)| \, e^{i\phi(\omega)} \tag{5.10.11}$$

where the amplitude $|H(\omega)|$ is called the *gain* of the filter (a term originating with electrical circuitry) and $\phi(\omega)$ is the *phase lag* of the filter. The power $P(\omega)$ of the transfer function is given by

$$P(\omega) = H(\omega)H(-\omega) = H(\omega)H^*(\omega) = |H(\omega)|^2 \tag{5.10.12}$$

where, as usual, the asterisk denotes the complex conjugate.

5.10.3 Ideal filters

An ideal filter is one that has unity gain, $|H(\omega)| = 1$, at all frequencies within the specified *pass band(s)* and zero gain at frequencies within the *stop band(s)* (Figure 5.10.2). When processing oceanographic data, it is generally advantageous to have $\phi(\omega) = 0$ for all ω so that the filter produces no alteration in the phase of the frequency components. As we discuss in conjunction with recursive filters, zero phase shift can be guaranteed by first passing the input forward then backward (after inversion) through the same set of weights. In the case of nonrecursive filters, zero phase is accomplished using symmetric filters (i.e. those with no imaginary components).

Digital filters commonly used in processing oceanographic data can be classified under the general headings of *low-pass*, *high-pass*, or *band-pass* filters. Although impossible to achieve, we would like the amplitudes of our ideal filters to satisfy the following relations (see Figure 5.10.2)

$$\text{Low-pass:} \; |H(\omega)| = 1 \; \text{for} \; |\omega| \leq \omega_c \\ = 0 \; \text{for} \; \omega_c \leq \omega \tag{5.10.12a}$$

$$\text{High-pass:} \; |H(\omega)| = 0 \; \text{for} \; |\omega| \leq \omega_c \\ = 1 \; \text{for} \; \omega_c \leq \omega \tag{5.10.12b}$$

$$\text{Band-pass:} \; |H(\omega)| = 1 \; \text{for} \; \omega_{c1} \leq |\omega| \leq \omega_{c2} \\ = 0 \; \text{otherwise} \tag{5.10.12c}$$

The *cut-off frequency*, $\omega_c (= 2\pi f_c)$, marks the transition from the pass-band to the stop-band. For ideal filters, the transition is step-like while for practical filters, the transition has a finite width. In the latter case, ω_c is defined as the frequency at which the mean filter amplitude in the pass-band is decreased by a factor of $\sqrt{2}$ and should roughly coincide with spectral minima in the time series being analyzed; the power of the filter is down by a factor of 2 (-3 dB) at the cut-off frequency. As its name implies, a low-pass filter lets through (or is "transparent" to) low-frequency signals but strongly attenuates high-frequency signals (cf. Figures. 5.10.3a, b). High-pass filters let through the high-frequency components and strongly attenuate the low-frequency components (cf. Figures. 5.10.3a, c). Band-pass filters permit only frequencies in a limited range (or band) to pass unattenuated.

Low-pass filters are the most common filters used in oceanographic data analysis. It is through these filters that low-frequency, long-term variability of oceanographic signals is determined. The running-mean filter, which involves a moving average over an odd number of values, is the simplest form of low-pass filter. More complex filters with better frequency responses, such as the low-pass Kaiser–Bessel window used in Figure 5.10.3(b), also are commonly used. High-pass filtered data are readily obtained by subtracting the low-pass filtered data from the original record from which the low-pass data were derived. One does not need to create a separate high-pass filter. Similarly, band-pass filters can be formed by an appropriate combination of low-pass and high-pass filters. In the ocean, seawater acts as a form of natural low-pass filter, attenuating high-frequency wave or acoustic energy at a much more rapid rate than low-frequency energy. Acoustic waves of a few hertz (cycles per second) can propagate thousands of kilometers in the ocean whereas acoustic waves of hundreds of kilohertz are strongly attenuated over a few hundred meters.

High-pass filters are less frequently used than low-pass filters. Applications include the delineation of high-frequency, high-wavenumber fluctuations in the internal wave band (roughly $2f < \omega < N$, where N is the Brunt–Väisälä frequency) and the isolation of seiche or tsunami motions in closed or semi-enclosed basins. Band-pass filters are used to isolate variability in relatively narrow frequency ranges such as the near-inertial frequency band or, in North America, the electronic-induced 60-cycle noise in high-frequency oceanic data caused by AC power supplies.

The maximum range of frequencies that can be covered by a digital filter is determined at the high-frequency end by the Nyquist frequency, $\omega_N = \pi/\Delta t$ (radians/unit time), and at the low-frequency end by the fundamental frequency, $\omega_1 = 2\pi/T$, where $T = N\Delta t$ is the length of the record. The corresponding range in cycles/unit time are determined by $f_N = 1/(2\Delta t)$ and $f_1 = 1/T$. Provided that the cut-off frequencies are sufficiently far removed from the ends of the intervals, digital filters can be applied throughout the range, $\omega_1 < |\omega| < \omega_N$ ($f_1 < |f| < f_N$).

5.10.3.1 Bandwidth

The difference in frequency between the two ends of a pass-band defines an important property known as the *bandwidth* of the filter. To illustrate the relevance of this property, we consider an ideal band-pass filter with constant gain, linear phase, and cut-off frequencies ω_{c1}, ω_{c2} such that

$$H(\omega) = H_o \exp(-i\omega t_o), \quad \omega_{c1} \leq |\omega| < \omega_{c2}$$
$$= 0, \quad \text{otherwise} \tag{5.10.13}$$

From (5.10.12c), the impulse response is

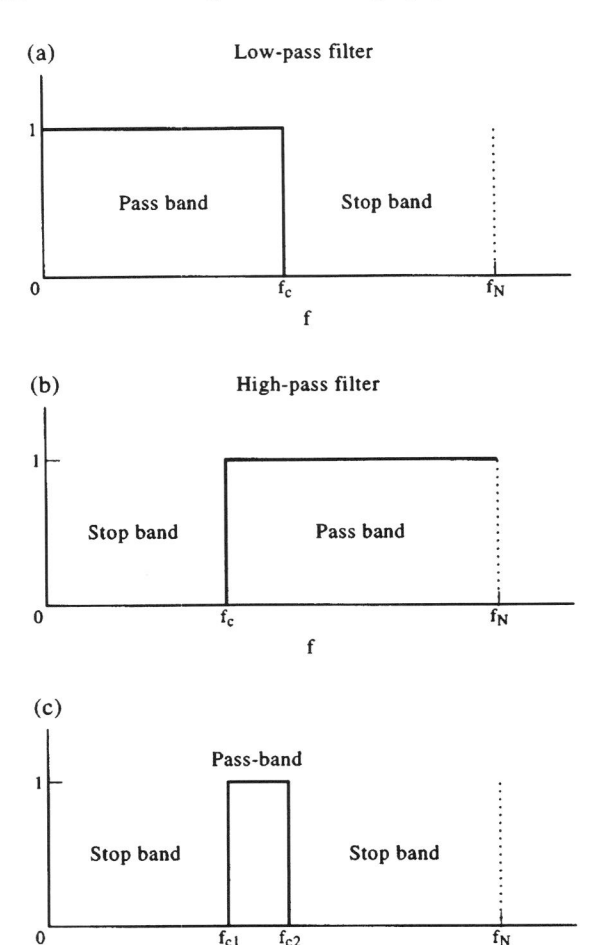

Figure 5.10.2. Frequency response functions, $|H(f)|$, for ideal filters. (a) Low pass; (b) high pass; and (c) band pass. The band-pass filter has been constructed from the combined low-pass and high-pass filters. f_N and f_c are the Nyquist and cut-off frequencies, respectively.

$$h_k = \frac{1}{2\pi} H_o \left(\int_{\omega_{c1}}^{\omega_{c2}} e^{-i\omega t_o} e^{i\omega k \Delta t} d\omega + \int_{\omega_{c1}}^{\omega_{c2}} e^{i\omega t_o} e^{-i\omega k \Delta t} d\omega \right) \quad (5.10.14)$$

$$= \frac{2H_o}{\pi} \Delta\omega \cos[\Omega(k\Delta t - t_o)] \frac{\sin[\Delta\omega(k\Delta t - t_o)]}{\Delta\omega(k\Delta t - t_o)}$$

in which $\Omega = \frac{1}{2}(\omega_{c1} + \omega_{c2})$ is the center frequency and $\Delta\omega = \omega_{c2} - \omega_{c1}$ is the bandwidth. For high or low-pass filters, the bandwidth is equal to the cut-off frequency.

Using the fact that $\sin p/p \to 1$ as $p \to 0$, we find that the peak amplitude response of the filter (5.10.14) is directly proportional to the bandwidth $\Delta\omega$ as $\Delta\omega(k\Delta t - t_o) \to 0$. Note also that a narrow-band filter (one for which $\Delta\omega \to 0$) will

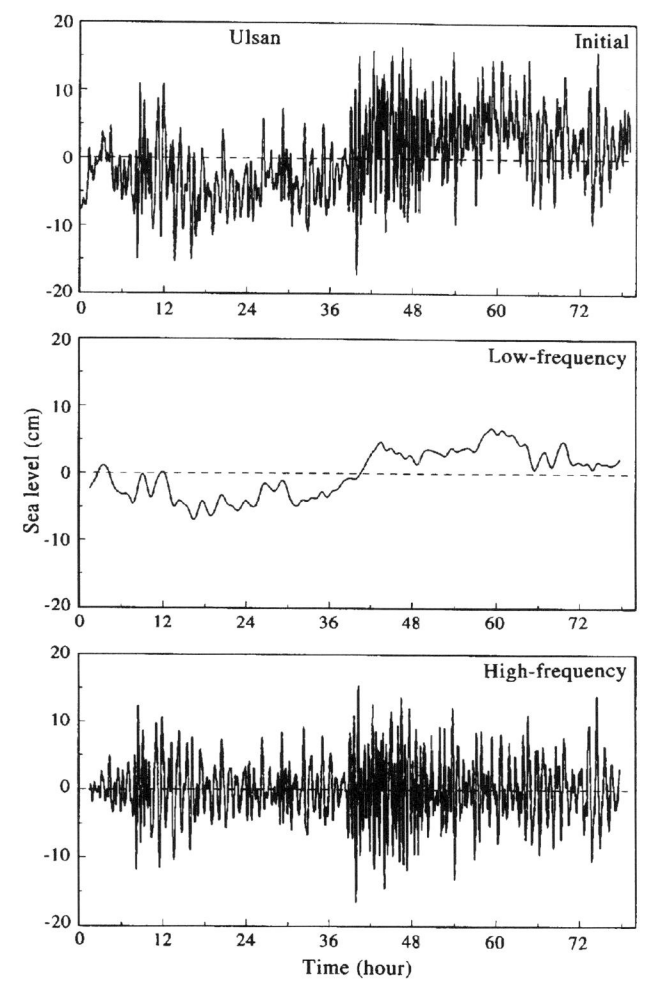

Figure 5.10.3. Filtering of a tide gauge record for Ulsan, Korea using low and high-pass Kaiser–Bessel filters (windows) with length T/27 = 3 h; T = 81 h is the record length and $\Delta t = 0.5$ min the sampling increment. (a) Original record; (b) low-pass filtered record; (c) high-pass filtered record. (Courtesy, A. Rabinovich.)

oscillate longer (i.e. persist to higher values of k) than a broad-band filter when subjected to a transient loading. Put another way, the persistence of the ringing that follows the application of the filter to a data set increases as the bandwidth decreases. From a practical point of view, this means that the ability of a filter to resolve sequential transient events is inversely proportional to the bandwidth. The narrower the bandwidth (i.e. the finer the resolution in frequency), the longer the time series needed to resolve individual events. For example, if we use a band-pass filter to isolate inertial frequency motions in the range 0.050–0.070 cph, the bandwidth $\Delta f = \Delta\omega/2\pi = 0.020$ cph and the filter could accurately resolve inertial events that occurred about $1/\Delta f = 50$ h apart. If we now reduce the bandwidth to 0.010 cph, the filter is only capable of resolving transient motions that occur more than 100 h apart.

522 Data Analysis Methods in Physical Oceanography

(The need to have long records to resolve closely spaced frequencies is exactly the problem we faced in Section 5.6.5.4 regarding the Rayleigh criterion for tidal analysis.)

Another way of stating the above relationship is that the uncertainty in frequency, Δf (or $\Delta \omega$), is inversely proportional to the length of time T over which the signal oscillates (i.e. $\Delta f \approx 1/T$) so that $T \Delta f \approx 1$ for a given filter. If we wish to use a filter with a very narrow bandwidth, we need to analyze long time-series records in which the signals of interest, such as the tides, have a high degree of persistence. In terms of observed data, the measured bandwidth of an oscillation in current speed, sea-level elevation, or other oceanic parameter is directly related to the persistence of the signal. For example, a wind-generated clockwise rotary inertial current having an observed bandwidth $\Delta f \approx 0.10$ cpd implies that the burst of inertial energy had a duration $T \approx 1/\Delta f = 10$ days.

5.10.3.2 Gibbs' phenomenon

In practice, step-like transfer functions such as described by equation (5.10.12) are not possible. Digital filters invariably possess finite-slope transition zones between the stop and pass-bands. To illustrate some of the fundamental impediments to creating ideal filters, consider the step-like transfer function

$$H(\omega) = 1 \qquad 0 < \omega \leq \omega_N$$
$$= 0 \qquad -\omega_N \leq \omega < 0 \tag{5.10.15}$$

(Figure 5.10.2a) where, for convenience, we specify a cut-off frequency $\omega_c = 0$. Assuming that $H(\omega)$ is repeated over multiples of the basic interval $(-\omega_N, \omega_N)$, the appropriate Fourier series expansion for equation (5.10.15) is given in the usual manner by

$$H(\omega) = \frac{1}{2}a_o + \sum_{n=1}^{\infty} [a_n \cos(\omega n \Delta t) + b_n \sin(\omega n \Delta t)] \tag{5.10.16}$$

with coefficients

$$a_n = \frac{1}{\omega_N} \int_{-\omega_N}^{\omega_N} H(\omega) \cos(\omega n \Delta t) \, d\omega \tag{5.10.17a}$$

$$b_n = \frac{1}{\omega_N} \int_{-\omega_N}^{\omega_N} H(\omega) \sin(\omega n \Delta t) \, d\omega \tag{5.10.17b}$$

The fact that $a_n = 1$, for all n, suggests reformulation of the problem in terms of the function

$$H_c(\omega) = H(\omega) - 1/2 \tag{5.10.18}$$

centered about $H(\omega) = 1/2$, the mean functional value at the discontinuity. Since $H(\omega)$ is then an odd function, cosine terms in (5.10.16) can be eliminated immediately. Moreover, H_c is symmetric about $\omega = \pm \frac{1}{2}\omega_N = \pm \pi/(2\Delta t)$ so that there are no even sine terms. For odd n, (5.10.17b) yields $b_n = 2/n\pi$ and (5.10.16) becomes

$$H(\omega) = \frac{1}{2} + \frac{2}{\pi}\left[\sin(\omega\Delta t) + \frac{\sin(3\omega\Delta t)}{3} + \frac{\sin(5\omega\Delta t)}{5} + \cdots\right] \quad (5.10.19)$$

which must be truncated after a finite number of terms.

Successive approximations to the series (5.10.19), and hence to the function (5.10.15), are not convergent near discontinuities such as that for the step-like transition region of the ideal high-pass filter shown in Figure 5.10.4. In this example, the filter amplitude $|H(\omega)|$ is zero for $\omega < \omega_c$ (the stop band) and unity for $\omega_c < \omega < \omega_N$ (the pass band). The succession of overshoot ripples, or ringing, is known as *Gibbs' phenomenon*. The ripple period, $T = p\pi t$ (p is an integer), is fixed but increasing the number of terms in the Fourier series for $H(\omega)$ decreases the distortion due to the overshoot effects. However, even in the limit of infinitely many terms, Gibbs' phenomenon persists as the amplitude of the first overshoot diminishes asymptotically to about 0.18 or about 9% of the pass-band amplitude. The first minimum decreases asymptotically to about 5% of the pass-band amplitude. In the limit of large $N \to \infty$, it can be shown (Godin, 1972; Hamming, 1977) that

$$H_\infty(0) \to \frac{1}{\pi}\int_0^\pi \sin u/u \, du \quad (5.10.20)$$

The values of $H_\infty(0)$ can be found in tables of the sine integral function. In the case of Figure 5.10.4, the value for the first maximum is 1.08949 ($= 1.0 + 0.08949$) while that for the first undershoot is 0.9514 ($= 1.0 - 0.04858$).

Gibbs' phenomenon has considerable importance in that it occurs whenever a function has a discontinuity. For example, suppose that we want to use equation (5.10.19) to remove spectral components near a cut-off frequency, ω_c. Unless the spectral components in the stop and pass-bands are well separated relative to the width of the transition zone, the finite ripples will cause leakage of unwanted energy into the filtered record. Noise from the stop-band will not be completely removed and certain frequencies in the pass-band will be distorted. A critical aspect of filter design is the attenuation of the overshoot ripples using smoothing or tapering functions (windows). As discussed in Section 5.6.6, windows are important in reducing side-lobe leakage in spectral estimates.

Further difficulties arise when we apply the weights $\{h_k\}$ of an ideal filter in the time domain. Consider the nonrecursive, low-pass filter (positive frequency only)

$$\begin{aligned} H(\omega) &= 1, \quad 0 \le \omega \le \omega_c \\ &= 0 \quad \text{otherwise} \end{aligned} \quad (5.10.21)$$

for which the impulse function is, for $k = -N, \ldots, N$

$$\begin{aligned} h(t_k) = h_k &= \frac{1}{\omega_N}\sum_{\omega=0}^{\omega_c} \cos(\omega k \Delta t)\Delta\omega \\ &= \frac{\sin(\omega_c k \Delta t)}{\omega_N k \Delta t} \\ &= \frac{f_c}{f_N}\frac{\sin(2\pi f_c k \Delta t)}{2\pi f_c k \Delta t} \end{aligned} \quad (5.10.22)$$

524 Data Analysis Methods in Physical Oceanography

in which $h_o = f_c/f_N$. The weights h_k attenuate slowly, as $1/k$, so that a large number of terms are needed if the filter response $H(\omega)$ is to be effectively carried over to the time domain. In addition to being computationally inefficient, filters constructed from a large number of weights lead to considerable loss of information at the ends of the data sequence. Practical considerations force us to truncate the set of weights thereby enhancing the overshoot problem associated with Gibbs' phenomenon in the frequency domain. Moreover, if we truncate the length of the data set (5.10.1), we are unable to accurately replicate (5.10.21) in the frequency domain. This leads to a finite slope between the stop and pass-bands of the filter.

The situation is similar for high-pass filters

$$H(\omega) = 0, \quad 0 \leq \omega \leq \omega_c$$
$$= 1, \quad \text{otherwise} \tag{5.10.23a}$$

In this case

$$h_k = \frac{1}{\omega_N} \sum_{\omega=\omega_c}^{\omega_N} \cos(\omega k \Delta t) \Delta \omega$$
$$= -\frac{f_c}{f_N} \frac{\sin(2\pi f_c k \Delta t)}{2\pi f_c k \Delta t}, \quad k = -N, ..., N \tag{5.10.23b}$$

where $h_o = 1 - f_c/f_o$. Notice that, except for the central term h_o, the weights h_k of the high-pass filter (5.10.23b) are equal to minus the weights h_k of the low-pass filter

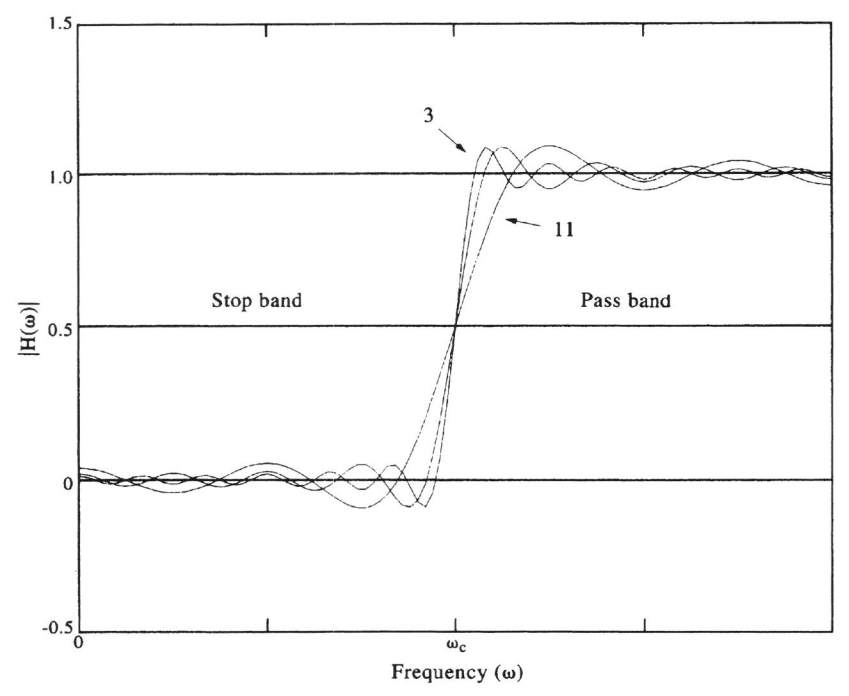

Figure 5.10.4. Gibbs' phenomenon (overshoot ripples) arising from successive approximations to the step-like function $|H(\omega)| = 1$, $\omega_c < \omega \leq \omega_N$, and zero otherwise. $\omega_c = 2\pi f_c$ is the cut-off frequency. Curves are derived from (5.10.19) using M = 3, 7, and 11 terms.

(5.10.22). The center value, h_o of the high-pass filter is found from h_o of the low pass filter by: $h_o(high\ pass) = 1 - h_o\ (low\ pass)$.

The difficulties that arise with Gibbs' phenomenon are somewhat alleviated by applying smoothing functions that attenuate the overshoot ripples. As usual, the price we pay for improved decay of the weighting terms is a broadening of the main lobe centered at the frequency being filtered. As we remarked earlier, the fact that the transition from the pass to the stop-band takes place over a finite range of frequencies necessitates a working definition for the cut-off frequency, ω_c. Here, ω_c is defined as the frequency at which the power $|H(\omega)|^2$ of the filter is attenuated by a factor of 2 (-3 db) from its mean pass-band value. (Power in dB = $20\ \log(A/A_o)$ where A_o is a reference level for the signal amplitude, A, having power proportional to A^2.) Alternatively, the cut-off frequency marks the frequency at which the amplitude $|H(\omega)|$ of the filter is reduced by a factor of $\sqrt{2}$ of the pass-band amplitude (amplitude in dB = $10\ \log(A/A_o)$).

5.10.3.3 Recoloring

The transfer function amplitude $|H(\omega)|$ defines the effectiveness of a particular filter in transmitting or blocking power within specific frequency bands. Since no filter is perfect, in the sense that its transfer function is exactly unity throughout the pass band(s) and zero in the stop band(s), it is often necessary to "re-color" (re-scale) the output $Y(\omega)$ so that the total variance in the pass-band spectral estimates equals the total variance of the input data for that frequency range. The need to re-color stems from practical considerations involving the choice of filter, cut-off frequency, and filter steepness through the transition band. For a pass-band of width $\Delta\omega$, multiplication of the filter output $|Y(\omega)|$ by a frequency-independent correction factor γ given by

$$\gamma(\Delta\omega) = \frac{\text{input variance within bandwidth}}{\text{output variance within bandwidth}}$$

ensures that the output power is adequately re-scaled.

We can illustrate the re-coloring process using the Hanning (von Hann) and Hamming windows. If $x(t)$ is any scalar time series of length N, and $y(t)$ is the filtered output of this series following application of one of these windows, then the Fourier transform of the output, $Y(f_k)$, for discrete frequencies $f_k = (k/T), k = 0, 1, \ldots, (N/2)$ is given by

$$Y(f_k) = 0.50X(f_k) - 0.25X(f_{k-1}) - 0.25X(f_{k+1}) \quad \text{(Hanning)} \quad (5.10.24a)$$

$$Y(f_k) = 0.54X(f_k) - 0.23X(f_{k-1}) - 0.23X(f_{k+1}) \quad \text{(Hamming)} \quad (5.10.24b)$$

where $X(f_k)$ is the Fourier transform of the original time series. The corresponding expected values for $|Y(f_k)|^2$ in (5.10.24a, b) are

$$E[|Y(f_k)|^2] = (0.50)^2 + (0.25)^2 + (0.25)^2 = 0.3750 \quad (5.10.24c)$$

$$E[|Y(f_k)|^2] = (0.54)^2 + (0.23)^2 + (0.23)^2 = 0.3974 \quad (5.10.24d)$$

so that the spectral density estimates $S(f_k) \approx |Y(f_k)|^2]$ for each frequency component of a time series smoothed by a Hanning window should be rescaled by the exact factor

$(0.375)^{-1} = 8/3$ to correct for the loss of power due to the filter. For the Hamming window, the factor is roughly $(0.397)^{-1} \approx 5/2$. Note that, according to equation (5.10.24), we can easily obtain spectral estimates $S(f_k)$ for each windowed time series by summing up the squared amplitudes $|X(f)|^2$] of three adjacent Fourier components of the original time series

$$S(f_k) = C_o|X(f_k)|^2 + C_{-1}|X(f_{k-1})|^2 + C_{+1}|X(f_{k+1})|^2 \qquad (5.10.25)$$

where $C_o = 0.50$ and $C_{-1} = C_{+1} = -0.25$ for the Hanning window and $C_o = 0.54$ and $C_{-1} = C_{+1} = -0.23$ for the Hamming window.

5.10.4 Design of oceanographic filters

The isolation of signal variability within specific frequency bands requires filters with well-defined frequency characteristics. The design of application-specific filters can proceed in two basic ways. The first approach is to assemble a combination of simple filters, such as moving averages of variable length, and from them construct a filter with the required characteristics. This is referred to as *cascading* since the output from the lead-off filter is used as input to the second filter, output from the second filter is used as input to the third, and so on. Filter cascading is used in the design of Godin's (1972) tide-elimination filters and the squared Butterworth filters described later in this section. The second approach is to specify the desired characteristics of the filter precisely and then use poles and zeros of mathematical functions to design a filter that meets these requirements as closely as possible. As an example, we might wish to eliminate the annual cycle from a long time-series of upper-ocean variability, such as sea surface temperature, so that weaker fluctuations are no longer overwhelmed by the dominant seasonal changes. The filter properties are then directly tailored to the processing requirements and to the data specific to the region of interest. (In this example, we could also use least-squares analysis to determine the annual cycle and then subtract this cycle from the original data.)

Regardless of which approach is taken, it is important that the impulse and frequency response functions of the filter have a number of fundamental properties: (1) The frequency response function should have reasonably sharp transitions between adjacent stop and pass bands, especially if the data do not have wide "spectral-gaps" between dominant frequencies within the two bands. At the same time, the transition should not be so steep as to introduce large side-lobe effects or cause the filter output to become unstable; (2) the transfer function should have nearly constant amplitude and zero phase (even symmetry) within the pass and stop bands so that corrections to amplitude and phase are easily applied. Linear phase change as a function of frequency is acceptable but requires corrective work at the end of the processing; and (3) the impulse response should have as short a span as possible to both minimize the number of points lost (or that are need to be appended at the ends of the data) and to reduce the amount of computation.

5.10.4.1 Frequency versus time domain filtering

In most instances, filters are designed to precondition the frequency content of the data prior to further analysis. This immediately suggests that the design of a filter begin with specification of the transfer function, $H(\omega)$. Once $H(\omega)$ has been determined there are two ways to proceed. The standard time-domain approach (e.g. Hamming, 1977) is to Fourier transform $H(\omega)$ to obtain the time-domain filter weights, h_k, which are then used in the convolution (5.10.3) to determine the output $\{y_n\}$. The output is subsequently Fourier transformed to determine $Y(\omega)$. The frequency domain approach (e.g. Walters and Heston, 1982; Middleton, 1983) makes use of the fact that $Y(\omega) = H(\omega)X(\omega)$, where $X(\omega)$ is the Fourier transform of the data $x(t)$. In this approach, the data are Fourier transformed to obtain $X(\omega_i)$, $i = 1, 2, ..., N/2$, where $X(\omega)$ consists of a set of $N/2$ frequency-dependent amplitudes and phases $[A(\omega_i), \phi(\omega_i)]$ at discrete frequencies. The filtered record is obtained by multiplying $X(\omega)$ by $H(\omega)$. The time-domain series $\{y_n\}$ can be derived from the inverse Fourier transform of $Y(\omega)$.

There are pros and cons for both approaches. The time-domain approach uses the actual recorded data and filtering consists of simple sums and products. Moreover, the filtered series $\{y_n\}$ can be immediately plotted against the original input $\{x_n\}$ to see directly the effectiveness of the filter. Discontinuities in the time series, which lead to transient filter ringing effects, can be dealt with on the spot. However, if the calulation of $Y(\omega)$ and its associated spectral estimate $|Y(\omega)|^2$ are the ultimate goals, the time-domain approach requires application of two Fourier transforms: First, we use $H(\omega)$ to define the filter weights $\{h_k\}$ and then transform $y_n \rightarrow Y(\omega)$ to obtain the Fourier components. This can lead to roundoff and computational errors.

In the frequency-domain analysis, only one Fourier transform, $x_n \rightarrow X(\omega)$, is required. On this basis, it seems preferable to use the Fourier transform method and just set to zero all those frequency components outside the range of interest. The filtered data $\{y_n\}$ are then found through an inverse transform of the modified Fourier components, $Y(\omega) = H(\omega)X(\omega)$. One obvious difficulty with this procedure is that the discrete frequencies of the Fourier estimates may not be properly positioned relative to the required cut-off frequency of the filter; that is, the cut-off frequency may fall mid-way between two discrete Fourier components. Walters and Heston (1982) also pointed out that the sharp cut-off associated with this process causes ringing through the entire data set (Figure 5.10.5). For this reason, the Fourier coefficients must be reduced gradually to zero over a range of frequencies. For example, Nowlin et al. (1986) used a trapezoidal-shaped band-pass filter to study inertial oscillations in data collected in Drake Passage. In this particular instance, "Fourier coefficients within 0.03 cpd of the local inertial frequency were retained undiminished, and this central portion was flanked by two tapered sections 0.06-cpd wide in which the coefficients were reduced linearly to zero." The smooth filter transition results in a substantial reduction in ringing in the filtered data but is certainly reminiscent of data tapering required in the time-domain analysis. A more detailed discussion of frequency domain filtering is presented in Section 5.10.9.

5.10.4.2 Filter cascades

In some instances, a desired filter $H(\omega)$ can be constructed from a series or *cascade* of basis filters $H_j(\omega)$ such that

$$H(\omega) = H_1(\omega) \times H_2(\omega) \times ... \times H_q(\omega) \quad (5.10.26)$$

where "×" denotes successive applications of individual transfer functions, beginning with H_1. That is, the data are first processed with $H_1(\omega)$ and the output from this filter passed through $H_2(\omega)$; the output from $H_2(\omega)$ is then passed through $H_3(\omega)$, and so on until the last filter, $H_q(\omega)$. The final output from $H_q(\omega)$ corresponds to the sought-after output from $H(\omega)$. Although the technique is straightforward and helps to minimize roundoff error, it has a number of major drawbacks, including the need for extended computations and the possibility of repeated ringing as one filter after another is applied in succession.

A high-pass filter $H_H(\omega)$ is obtained from its low-pass counterpart $H_L(\omega)$ by the relation $H_H(\omega) = 1 - H_L(\omega)$ where, in theory, the combined output from the two filters simply recreates the original data, since $H_L(\omega) + H_H(\omega) = 1$. This has advantages in situations where $H_L(\omega)$ is easily derived or is already available. In the time domain, the high-pass filtered record $\{y'_n\}$ is obtained by subtracting the output $\{y_n\}$ from the low-pass filter form the input time series $\{x_n\}$. Care is needed to ensure that the times of y_n and x_n are properly aligned so that $y'_n = x_n - y_n, n = M, M+1, \ldots, N - 2M$.

A band-pass filter can be constructed from an appropriate high and low-pass filter using the method illustrated in Figure 5.10.2(c). Here, the cut-off frequency of the low-pass filter becomes the high-frequency cut-off of the band-pass filter; similarly, the cut-off frequency of the high-pass filter becomes the low-frequency cut-off of the band-pass filter. The cascade then has the form $H_B(\omega) = H_L(\omega) \times H_H(\omega)$.

Because nonrecursive filters are symmetric ($H(\omega)$ is a real function), there is no shift in phase between the input and output signals. This feature of the filters, as well as their general mathematical simplicity, has contributed to their popularity in oceanography. Recursive filters, on the other hand, are typically nonsymmetric. This introduces a frequency-dependent phase shift between the input and output variables and adds to the complexity of these filters for oceanic applications. Despite these difficulties, recursive filters are useful additions to any processing repertoire. The good news is that we can remove phase shifts introduced through the "forward" application of the filter by reversing the process and passing the data "backward"

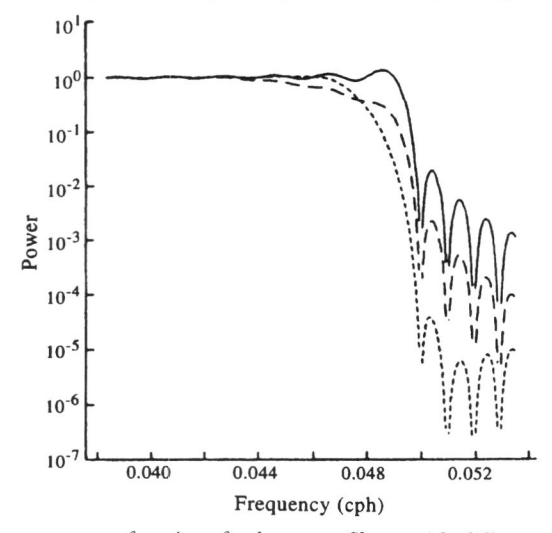

Figure 5.10.5. Frequency-response functions for low-pass filters with different transition bands. Solid line: A step-like transition band. Long-dashed line: A nine-point cosine-tapered transition band. Short-dashed line: A three-point optimally designed transition band. The cut-off associated with each filter causes ringing through the entire data set. (From Elgar, 1988.)

through the filter. In performing the latter step, we must be careful to invert the order of the record values between the forward and backward passes. Specifically, if the recursive filter introduces a phase shift $\phi(\omega)$ at frequency ω (or equivalently, a time shift $\phi/\omega = \phi/2\pi f$), it will introduce a compensating shift $-\phi(\omega)$ when passed in the reverse order through the filter. To show this sequence let $x_1, x_2, \ldots, x_n$ be the original data sequence used as input to a given filter with nonzero phase characteristics, and $y_1, y_2, \ldots, y_n$ the output from the filter (Figure 5.10.6). If we now invert the order of the output and pass the inverted signal through the filter again, we obtain a new output $z_1, z_2, \ldots, z_n$. The order of the z-output is then inverted to form $z_n, z_{n-1}, \ldots, z_1$, which returns us to the proper time sequence. For simplicity we can rewrite this later sequence as $y'_1, y'_2, \ldots, y'_n$. The act of applying the filter a second time cancels any phase change from the first pass through the filter. Note that this corresponds to squaring the transfer function so that the final transfer function for the recursive filter is $|H(\omega)|^2$.

As an example of a phase-dependent recursive filter, consider the high-pass *quasi-difference filter*

$$y(n\Delta t) = x(n\Delta t) - \alpha x[(n-1)\Delta t] \qquad (5.10.27\text{a})$$

where α is a parameter in the range $0 < \alpha \leq 1; \alpha = 1$ corresponds to the simple difference filter (Koopmans, 1974). The transfer function for this filter is

$$H(\omega) = 1 - \alpha e^{-i\omega \Delta t} \qquad (5.10.27\text{b})$$

and the phase function is

$$\phi(\omega) = \tan^{-1}[\alpha \sin(\omega \Delta t)/(1 - \alpha \cos(\omega \Delta t))] \qquad (5.10.27\text{c})$$

Reversing the order of the output from the first pass of the data through the filter and then running the time-inverted record through the filter again is tantamount to passing the data through a second filter $H(\omega)^*$. This introduces a phase change $-\phi(\omega)$ which cancels the phase change $\phi(\omega)$ from the first filter (Figure 5.10.7). The symmetric filter obtained from this cascade is then

$$\begin{aligned}|H(\omega)|^2 &= H(\omega) \times H(\omega)^* \\ &= (1 - \alpha e^{-i\omega \Delta t})(1 - \alpha e^{+i\omega \Delta t}) = [1 - 2\alpha \cos(\omega \Delta t) + \alpha^2]^{1/2}\end{aligned} \qquad (5.10.27\text{d})$$

5.10.5 Running-mean filters

The *running-mean* or *moving-average filter* is the simplest and one of the most commonly used low-pass filters in physical oceanography. In a typical application, the filter (which is simply a moving rectangular window) consists of an odd number of $2M + 1$ equal weights, h_k, $k = 0, \pm 1, \ldots, \pm M$, having constant values

$$h_k = \frac{1}{2M+1} \qquad (5.10.28\text{a})$$

where h_k resembles a uniform probability density function in which all occurrences are equally likely. The running-mean filter produces zero phase alteration since it is symmetric about $k = 0$, it satisfies the normalization requirement

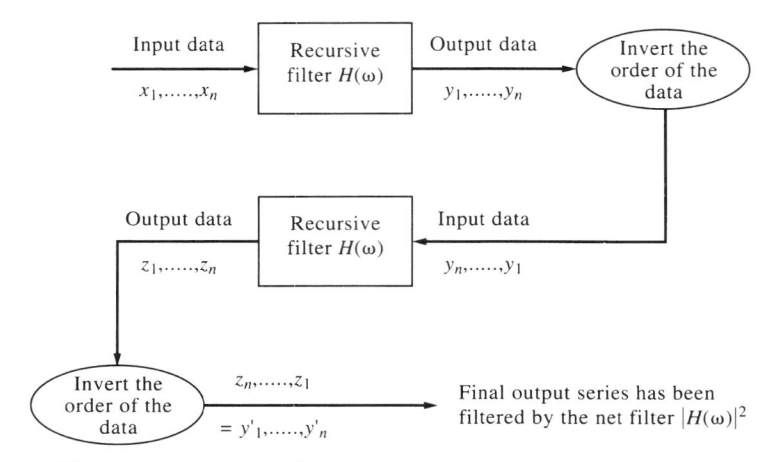

530 *Data Analysis Methods in Physical Oceanography*

Figure 5.10.6. The processing sequence for a nonsymmetric recursive filter $H(\omega)$ which removes phase changes $\phi(\omega)$ introduced to the data sequence x_i ($i = 1, \ldots, n$) by the filter. This cascade produces a symmetric squared-filter response $|H(\omega)|^2$.

$$\sum_{k=-M}^{M} h_k = 1 \qquad (5.10.28b)$$

and is straightforward to apply. To obtain the output sequence $\{y_m\}$ for input sequence $\{x_n\}$, the first $2M + 1$ values of x_n (namely $x_0, x_1, \ldots, x_{2M}$) are summed and then divided by $2M + 1$, yielding the first filtered value $y_M = y(2M\Delta t/2)$. The subscript M reminds us that the filtered value replaces the original data record x_M at the appropriate location in the time series. The next value, $y_M + 1$, is obtained by advancing the filter weights one time step Δt and repeating the process over the data sequence $x_1, x_2, \ldots, x_{2M+1}$ and so on up to $N - 2M$ output values. The $\{y_m\}$ consist of a "smoothed" data sequence with the degree of smoothing, and associated loss of information from the ends of the input, dependent on the number of filter weights. Mathematically

$$y_{M+i} = \frac{1}{2M+1} \sum_{j=0}^{2M} x_{i+j}, \quad i = 0, \ldots, N - 2M \qquad (5.10.29)$$

A high-pass running-mean filter can be generated by subtracting the output $\{y_m\}$ from the original data. The output $\{y'_m\}$ for the high-pass filter is

$$y'_m = x_m - y_m, \quad m = M, M+1, \ldots, N - 2M \qquad (5.10.30)$$

where we make certain we subtract data values for the correct times. This technique of obtaining a high-pass filtered record from a low-pass filtered record will also be applied to other types of filters.

The transfer function $H(\omega)$ for the running-mean filter is given by equation (5.10.8). Using equation (5.10.27) and the fact that $\Delta t = \pi/\omega_N$, we find that

$$H(\omega) = \frac{1}{2M+1} \left\{ \frac{1 + 2\sin\left[(\pi/2M)(\omega/\omega_N)\right]\cos\left[\pi/2(M+1)(\omega/\omega_N)\right]}{\sin(\pi/2\,\omega/\omega_N)} \right\} \qquad (5.10.31a)$$

$$= \frac{1}{2M+1} \frac{\sin\left[\pi/2(2M-1)(\omega/\omega_N)\right]}{\sin\left[(\pi/2)(\omega/\omega_N)\right]} \qquad (5.10.31b)$$

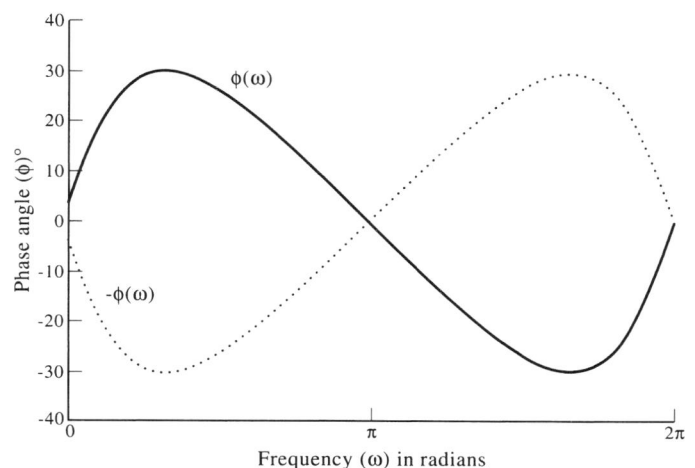

Figure 5.10.7. The phase change $\phi(\omega)$ for a quasi-difference filter (with $\alpha = 0.5$ as a function of frequency, ω).

where $H(\omega) \to 1$ as $\omega/\omega_N \to 0$. As M increases, the central lobe of the transfer function narrows (Figure 5.10.8) and the cut-off frequency (at which $|H(\omega)| = e^{-1}|H(0)|$) moves closer to zero frequency. The filter increasingly isolates the true mean of the signal. Unfortunately, the filter has considerable contamination in the stop-band due to the large, slowly attenuating side-lobes. Reduction of these side-lobe effects requires a long filter which means severe loss of data at either end of the time series. The running-mean filter should therefore only be used with long data sets ("long" compared with the length of the filter). Accurate filtering requires use of more sophisticated filters.

For the three-point weighted average, $h_k = 1/3$ and equation (5.10.31) yields

$$H(\omega; 3) = \frac{1}{3}[1 + 2\cos(\pi\omega/\omega_N)]$$
$$= \frac{1}{3} \frac{\sin[(3\pi/2)(\omega/\omega_N)]}{\sin[(\pi/2)(\omega/\omega_N)]} \quad (5.10.32)$$

while for five-point weighted average, $h_k = 1/5$ and

$$H(\omega; 5) = \frac{1}{5} \frac{\sin[(5\pi/2)(\omega/\omega_N)]}{\sin[(\pi/2)(\omega/\omega_N)]} \quad (5.10.33)$$

(Figure 5.10.8). Numerous examples of running-mean filters appear in the oceanographic literature. A common use of running-mean filters is to convert data sampled at times t to an integer multiple of this time increment for use in standard analysis packages. Data collected at intervals Δt of 5, 10, 15, 20, or 30 min are usually converted to hourly data for use in tidal harmonic programs, although the least-squares algorithms used in these programs also work with unequally spaced time-series data (e.g. Foreman, 1977, 1978). Running-mean filters also are commonly used to create weekly, monthly, or annual time series (Figure 5.10.9).

5.10.6 Godin-type filters

For the low-pass filtering of sub-hourly sampled tidal records prior to decimation to "standard" hourly values, Godin (1972) recommends the use of cascaded running-mean filters with response functions of the form

$$\frac{A_n^2 A_{n+1}}{n^2(n+1)}, \quad \frac{A_n A_{n+1}^2}{n(n+1)^2} \tag{5.10.34}$$

Here, A_n and A_{n+1} are the average values of n and $n+1$ consecutive data points, respectively. Each filter smoothes the data three times. In the first version in (5.10.34), the smoothing is performed twice using the n-point average and once using the $\{n+1\}$-point average. The alternative version uses the $\{n+1\}$-point average twice and the n-point average once. Following the filter operation, the smoothed records can then be sub-sampled at hourly intervals without concern for aliasing by higher-frequency components. For the second version in (5.10.34), the response function is

$$H(\omega) = \frac{1}{n^2(n+1)} \frac{\sin^2[(\pi/2)(n\omega/\omega_N)] \sin[(\pi/2)(n+1)\omega/\omega_N]}{\sin^3[(\pi/2)(\omega/\omega_N)]} \tag{5.10.35}$$

Godin filters $(A_{12}^2 A_{14})/(12^2 14)$ are used routinely to smooth oceanographic time series sampled at multiples of 5-min increments prior to their use in tidal analysis programs. On the other hand, 30-min data would first be smoothed using the filter $(A_2^2 A_3)/(2^2 3)$ (Figure 5.10.10) and then decimated to hourly data. Conversion of 30-min data from Aanderaa RCM4 current meters to hourly data requires such a three-stage running-average filter. The filter is needed to convert the instantaneous directions and average speeds from the current meter to quantities more closely resembling vector-averaged currents. Application of the moving low-pass filter (5.10.34) removes high-frequency components and helps avoid the aliasing errors that would occur if the raw data were simply decimated to hourly values without any form of prior smoothing. Simply picking out a value each hour is, of course, akin to not having recorded the higher frequency variability in the first place. Some care is required in that the smoothing process reduces the amplitude of various Fourier components outside the tidal band. As a result, amplitudes of Fourier components derived after application of the filter must be corrected (recolored) in inverse proportion to the amplitude of the filter at the particular frequency. Phases of the Fourier components are unaltered by this symmetric filter.

The formulation (5.10.34) also can be used to generate low-pass filters to remove diurnal, semidiurnal, and shorter period fluctuations from the hourly records. Although these filters have been criticized in recent years because of their slow transition through the high-frequency end of the "weather band" (periods longer than two days), they are easy to apply, have good response in the daily tidal band and consume relatively little data from the ends of the time series. The most commonly used version of the low-pass Godin filter is $(A_{24}^2 A_{25})/(24^2 25)$ in which the hourly data are smoothed twice using the 24-point (24-h) average and once using the 25-point average. The filter response is

$$\begin{aligned} H(\omega) &= \frac{1}{24^2 25} \sin^2[24(\pi/2)(\omega/\omega_N)] \frac{\sin[25(\pi/2)(\omega/\omega_N)]}{\sin[(\pi/2)(\omega/\omega_N)]} \\ &= \frac{1}{24^2 25} \sin^2(24\pi f \Delta t) \frac{\sin(25\pi f \Delta t)}{\sin^3(\pi f \Delta t)} \end{aligned} \tag{5.10.36}$$

where as before $\omega = 2\pi f$ (f is in cycles per hour), $\omega_N = \pi/\Delta t$ and $\Delta t = 1$ h. Note that a

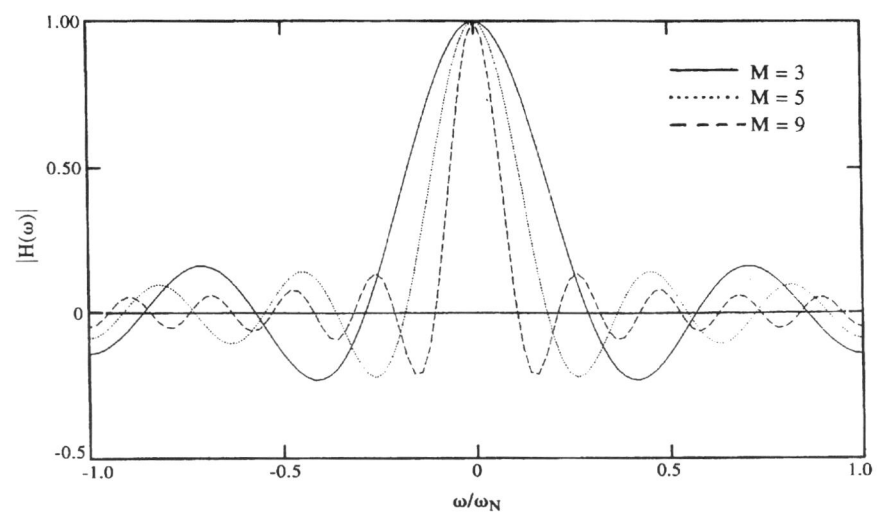

Figure 5.10.8. The frequency response functions, $|H(\omega)|$, for running-mean (weighted average) filters for $M = 3, 5, 9$. ω_N = Nyquist frequency.

total of 35 data points (i.e. 35 h) are lost from each end of the time series and that the filter has a half-amplitude point near 67 h (Figure 5.10.11). The weights of this symmetric 71-h-length filter are (Thompson, 1983)

$$h_k = \frac{1/2}{24^2 25}[1200 - (12-k)(13-k) - (12+k)(13+k)], \quad 0 \leq k \leq 11$$
$$= \frac{1/2}{24^2 25}(36-k)(37-k), \quad 12 \leq k \leq 35 \quad (5.10.37)$$

The Godin low-pass filter (5.10.37) effectively removes all daily tidal period energy except for slight leakage in the diurnal frequency band. More precisely, the filter eliminates variability due to the principal mixed diurnal constituent, K_1, for which the amplitude is down by 3.2×10^{-3}, and is only slightly less effective in removing variability due to the declinational diurnal constituent, O_1. The filter represents a marked improvement over the simple A_{24} and A_{25} running-mean filters and Doodson filter commonly used earlier for tidal analysis (cf. Groves, 1955). The principal failing of the Godin filter is its relatively slow transition between the pass and stop-bands which leads to significant attenuation of nontidal variability in the range of two to three days. This shortcoming of the filter has inspired a number of authors to investigate more efficient techniques for removing the high-frequency portion of oceanographic signals. The cosine–Lanczos filter, the transform filter, and the Butterworth filter are often preferred to the Godin filter, or earlier Doodson filter, because of their superior ability to remove tidal period variability from oceanic signals.

5.10.7 Lanczos-window cosine filters

As mentioned in Section 5.10.3.2, transfer functions for ideal (rectangular) filters are formulated in terms of truncated Fourier series. This leads to overshoot ripples (Gibbs' phenomenon) near the cut-off frequency with subsequent leakage of unwanted signal energy into the pass band. *Lanczos-window cosine filters* are reformulated rectangular filters which incorporate a multiplicative factor (the *Lanczos window*) to

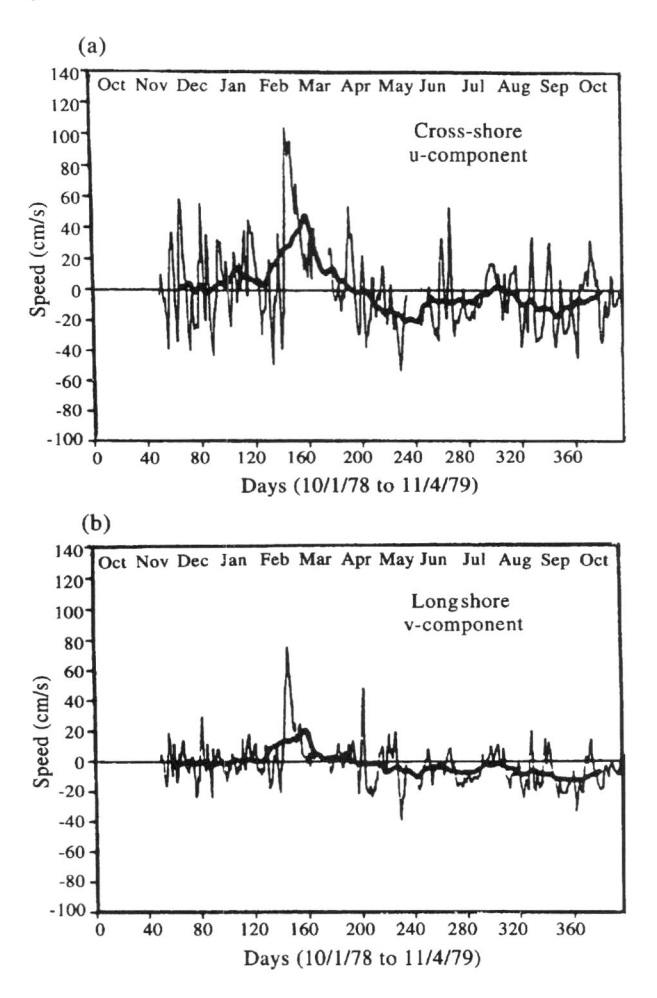

534 *Data Analysis Methods in Physical Oceanography*

Figure 5.10.9. Daily mean time series of cross-shelf (top) and longshelf (bottom) near-surface currents off Cape Romain in the South Atlantic Bight for the period 10 January 1979 to 11 April 1979. Thin line: Daily average data. Thick line: 30-day running-mean values. (From McClain et al., 1988.)

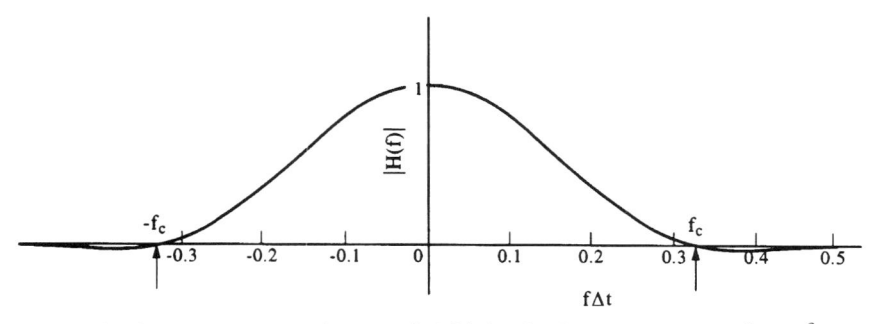

Figure 5.10.10. The frequency-response function, $|H(f)|$, for the Godin-type filter $A_2^2 A_3/(2^2 3)$ used to smooth 30-min data to hourly values. The horizontal axis has units $f \Delta t$, with $f_N \Delta t = 0.5$; f_c is the cut-off frequency. (From Godin, 1972.)

rectangular filters which incorporate a multiplicative factor (the *Lanczos window*) to ensure more rapid attenuation of the overshoot ripples. A variety of other windows can also be used. The terms *Lanczos–cosine filter* and *cosine–Lanczos filter* are commonly used names for a family of filters using windows to reduce the side-lobe ripples. Owing to their simplicity and favorable characteristics, these filters have gained considerable popularity among physical oceanographers over the years (Mooers and Smith, 1967; Bryden, 1979; Freeland et al., 1986).

5.10.7.1 Cosine filters

We start with an ideal, low-pass filter with transfer function

$$H(\omega) = 1 \quad 0 \leq |\omega| \leq \omega_c$$
$$= 0 \quad \text{elsewhere} \quad (5.10.38)$$

and assume that the function $H(\omega)$ is periodic over multiples of the Nyquist frequency domain $(-\omega_N, \omega_N)$. Written as Fourier series, the response function is

$$H(\omega) = \frac{a_0}{2} + \sum_{k=1}^{M} [a_k \cos(\omega k \Delta t) + b_k \sin(\omega k \Delta t)] \quad (5.10.39)$$

where we have truncated the series at $M \ll N$; as usual, N is the number of data points to be processed by the filter. To eliminate any frequency-dependent phase shift, we insist that $H(\omega) = H(-\omega)$, whereby $b_k = 0$. The resulting *cosine filter* has the transfer function

$$H(\omega) = h_o + \sum_{k=1}^{M} h_k \cos(\pi k \omega / \omega_N) \quad (5.10.40)$$

where coefficients $h_k \, (= \frac{1}{2} a_k)$ given by

$$h_k = \frac{1}{\omega_N} \int_0^{\omega_N} H(\omega) \cos(\pi k \omega / \omega_N) \, d\omega \quad (5.10.41)$$

with $k = 0, 1, \ldots, M$. The weighting terms h_k are those which determine the output series $\{y_n\}$ for given $\{x_n\}$. We assume that M is sufficiently large that $H(\omega)$ is close to unity in the pass-band and near zero in the stop-band.

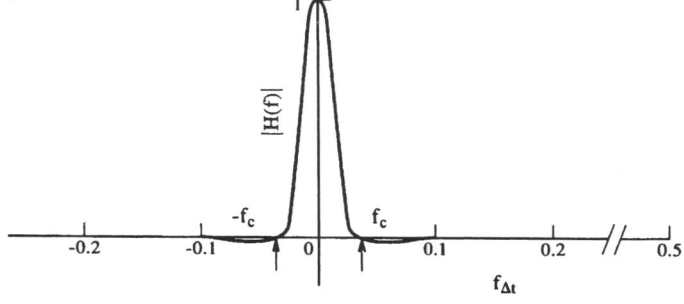

Figure 5.10.11. Same as Figure 5.10.10 but for the Godin-type low-pass filter $A_{25}^2 A_{24}/(25^2 24)$ used to eliminate tidal oscillations in hourly data. (From Godin, 1972.)

536 *Data Analysis Methods in Physical Oceanography*

For a low-pass cosine filter, $0 \leq |\omega| \leq \omega_c$ defines the bounds of the integral (5.10.41) and the weights are given by

$$h_k = \frac{\omega_c}{\omega_N} \frac{\sin(\pi k \omega_c/\omega_N)}{\pi k \omega_c/\omega_N}, \quad k = 0, \pm 1, ..., \pm M \qquad (5.10.42)$$

for which $h_o = \omega_c/\omega_N$. The corresponding weights for a high-pass filter, $|\omega| > \omega_c$, are

$$h_o = 1 - \omega_c/\omega_N, \quad k = 0 \qquad (5.10.43)$$

$$h_k = \frac{-\omega_c}{\omega_N} \frac{\sin(\pi k \omega_c/\omega_N)}{\pi k \omega_c/\omega_N}, \quad k = \pm 1, ..., \pm M \qquad (5.10.44)$$

That is, h_o (*high pass*) $= 1 - h_o$ (*low pass*) while for $k \neq 0$, the coefficients h_k are simply of opposite sign. The functions (5.10.42) and (5.10.44) are identical to those discussed in context of Gibbs' phenomenon. Thus, the cosine filter is a poor choice for accurately modifying the frequency content of a given record based on preselected stop and pass-bands. As an example of the response of this filter, Figure 5.10.12 presents the transfer function

$$H(\omega) = 0.4 + 2 \sum_{k=1}^{9} [\sin(0.4 k \pi)/k \pi] \cos(k\omega)$$

for a low-pass cosine filter with $\omega_c/\omega_N = 0.4$ and $M = 10$ terms. This filter response is compared to the ideal low-pass filter response and to the modified cosine filter using the Lanczos window (with sigma factors) discussed in the next section.

5.10.7.2 The Lanczos window

Lanczos (1956) showed that the unwanted side-lobe oscillations of the form $\sin(p)/p$ in equations (5.10.42) and (5.10.44) could be made to attenuate more rapidly through use of a smoothing function or window. The window consists of a set of weights that successively average the (constant period) side-lobe fluctuations over one cycle, with the averaging period determined by the last term kept or the first term ignored in the Fourier expansion (5.10.44). In essence, the window acts as a low-pass filter of the weights of the cosine filter. The Lanczos window is defined in terms of the so-called *sigma-factors* (cf. Hamming, 1977)

$$\sigma(M,k) = \frac{\sin(\pi k/M)}{\pi k/M} \qquad (5.10.45)$$

in which M is the number of distinct filter coefficients, h_k, $k = 1, ..., M$ and $\omega_M = (M-1)/M$ is the frequency of the last term kept in the Fourier expansion. Multiplication of the weights of the cosine filter by the sigma factors yields the desired weights of the Lanczos-window cosine filter. Thus, the weights of the low-pass cosine–Lanczos filter become, using $\sigma(M, 0) = 1$

$$h_o = \omega_c/\omega_N, \quad \text{for } k = 0 \qquad (5.10.46a)$$

$$h_k = (\omega_c/\omega_N) \frac{\sin(\pi k \omega_c/\omega_N)}{\pi k \omega_c/\omega_N} \sigma(M,k) \qquad (5.10.46b)$$

for $k = \pm 1, \ldots, \pm M$ and $M \ll N$. The corresponding weights for the high-pass Lanczos–cosine filter are

$$h_o = 1 - \omega_c/\omega_N, \qquad \text{for } k = 0 \qquad (5.10.47a)$$

$$h_k = -(\omega_c/\omega_N) \frac{\sin(\pi k \omega_c/\omega_N)}{\pi k \omega_c/\omega_N} \sigma(M,k) \qquad (5.10.47b)$$

The transfer function (5.10.39) for a low-pass cosine–Lanczos filter is then

$$H_L(\omega) = \frac{\omega_c}{\omega_N} \left[1 + 2 \sum_{k=1}^{M-1} \sigma(M,k) \frac{\sin(\pi k \omega_c/\omega_N)}{\pi k \omega_c/\omega_N} \cos(\pi k \omega/\omega_N) \right] \qquad (5.10.48)$$

while for the high-pass cosine–Lanczos filter

$$H_H(\omega) = 1 - H_L(\omega) \qquad (5.10.49)$$

Examination of the transfer functions in Figure 5.10.12 reveals that the side-lobe ripples are considerably reduced by the sigma factors of the Lanczos window. Again, the tradeoff is a broadened central lobe, so that, although there is much less contamination from frequencies within the stop-band, the transition of the filter amplitude at the pass-band is less steep than that for the cosine filter. The effect of this smoothing,

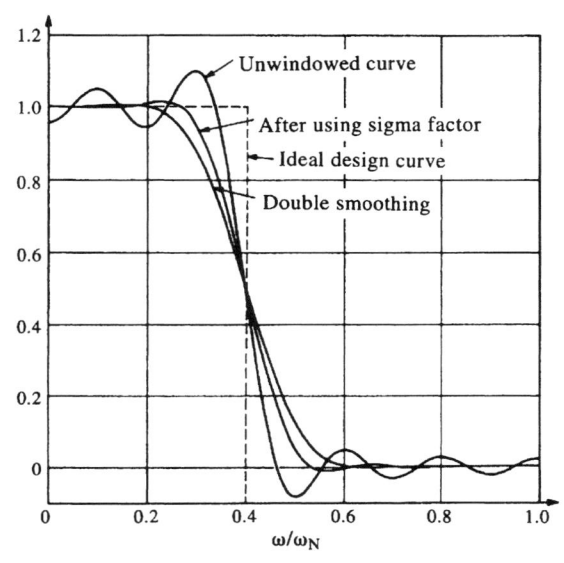

Figure 5.10.12. Approximations to the frequency response of an ideal low-pass filter (dashed line). Solid curves give: The frequency response for an unwindowed cosine filter, a Lanczos–cosine filter that uses sigma factors, and the response after double application of the Lanczos–cosine filter. Filters use $M = 10$ Fourier terms and $\omega_c = 0.4\omega_N$; $\omega_N =$ Nyquist frequency. Gibbs' effect is reduced by the sigma factors of the Lanczos window. (From Hamming, 1977.)

538 Data Analysis Methods in Physical Oceanography

which represents a long period modulation of the weighting terms h_k in (5.10.42), can be illustrated numerically by taking a record length $N = 25$ and calculating the filter response $H(\omega/\omega_N)$ with and without the sigma factors. This exercise is instructive in other ways in that it emphasizes the effect of truncation errors during the calculations and indicates what happens if ω_c/ω_N is too near to the ends of the principal interval $0 \leq \omega/\omega_N \leq 1$. Consider the case $\omega_c/\omega_N = 0.022$, $N = 25$, and filter truncation at the fourth decimal place. For a high-pass cosine-type filter with no Lanczos window (which we want to have zero amplitude near zero frequency), we find $H(0) = 0.0740$ whereas use of the sigma factors (Lanczos window) yields $H(0) = 0.4015$. With the cut-off frequency so close to the end of the frequency range, the sigma factors clearly degrade the usefulness of the filter. Increasing the record length to $N = 50$ for the same cut-off frequency improves matters considerably; in this case, $H(0) = 0.0527$ and $H(1) = 0.9997$ using the sigma factors.

5.10.7.3 Practical filter design

Design of a low or high-pass cosine–Lanczos filter begins with specification of: (1) The cut-off frequency; and (2) the number M of weighting terms required to achieve the desired roll-off between the stop and pass-bands. The cut-off frequency is then normalized by the Nyquist frequency, ω_N, obtained from the sampling interval Δt of the time series. As with other types of filters, it is advantageous to keep the normalized cut-off frequency away from the ends of the principal interval

$$0 \leq \omega/\omega_N \leq 1 \tag{5.10.50}$$

The weights h_k are then derived via (5.10.46) and (5.10.47).

Using (5.10.4) and (5.10.8), and assuming an input $\{x_n\}$, $n = 0, 1, \ldots, N-1$, the output for a low-pass cosine–Lanczos filter with $M + 1$ weights is

$$y_n = \frac{2\omega_c}{\omega_N}\left[x_n + \sum_{k=1}^{M} F(k)(x_{n-k} + x_{n+k})\right] \tag{5.10.51a}$$

in which

$$F(k) = \frac{\frac{1}{2}\sin(\pi k/M)}{\pi k/M} \frac{\sin(\pi k\omega_c/\omega_N)}{\pi k\omega_c/\omega_N} \tag{5.10.51b}$$

The output time series begins with $y_M = y(M\Delta t)$ corresponding to the first calculable value for the given filter length, M, and the assumption that the input data begin at $x_n = x_o$. That is

$$y_M = \frac{2\omega_c}{\omega_N}\bigg[x_M + \frac{1}{2}F(1)(x_{M-1} + x_{M+1}) \\ + \frac{1}{2}F(2)(x_{M-2} + x_{M+2}) + \ldots \\ \ldots + \frac{1}{2}F(M)(x_o + x_{2M})\bigg] \tag{5.10.52}$$

The chosen number of filter coefficients, M, is always a compromise between the desired roll-off of the filter at the cut-off frequency and the acceptable number of data points (= 2M) that are lost from the two ends of the record. The greater the number M, the sharper the filter cut-off and the greater the data loss. Repeated (q times) processing of a given record by the same filter generates an increasingly sharper cascade filter response $[H(\omega/\omega_q)]^q$ with an corresponding greater loss (qM) of data values from each end of the record. For a high-pass filter, M should be large enough that, in the time domain, the 2M weights for the corresponding low-pass filter span "many" periods of the higher frequency oscillations one is attempting to isolate using the filter.

The sum S of the weights h_k in (5.10.46) and (5.10.47)

$$S = \sum_{k=0}^{M} h_k$$

gives a qualitative measure of the filter performance. An ideal low-pass filter (i.e. one with no truncation or numerical roundoff effects) should give $S = 1$ while an ideal high-pass filter would have $S = 0$. Close proximity to these values indicates a numerically reliable filter.

5.10.7.4 The Hanning (von Hann) window

A variety of cosine-type filters are presented in the recent oceanographic literature under the general term of Lanczos–cosine or cosine–Lanczos filters. A popular formulation having widespread application is the five-day low-pass filter proposed by Mooers and Smith (1967) in a study of continental shelf waves off Oregon. In this study, a Hanning or raised cosine window defined by

$$\omega_k = \tfrac{1}{2}[1 + \cos(\pi k/M)], \quad |k| < M$$
$$= 0, \quad |k| > M \quad (5.10.54)$$

replaces the sigma factors in (5.10.47).

Let x_n, $n = 1, 2, \ldots, N$ denote an hourly digital time series and $2M + 1 = 120$ be the total number of weights spanning a period of 120 h (five days). The hourly output $\{y_n\}$ from the filter is then

$$y_n = \frac{1}{A}\left[x_n + \sum_{k=1}^{60} F(k)(x_{n-k} + x_{n+k})\right] \quad (5.10.55a)$$

where

$$F(k) = \frac{1}{2}[1 + \cos(\pi k/60)]\frac{\sin(p\pi k/12)}{p\pi k/12} \quad (5.10.55b)$$

and

$$A = 1 + 2\sum_{k=1}^{60} F(k) \quad (5.10.55c)$$

is the normalization factor. Once the number of filter weights k is specified (here, $k = 60$), the transfer function $H_L(\omega)$ is determined by the parameter p, the half-amplitude frequency of the filter in cycles per day (cpd). Specifically, we find

$$H_L(\omega) = \frac{1}{A}\left[1 + 2\sum_{k=1}^{60} F(k) \cos(\pi k\omega/\omega_N)\right] \quad (5.10.56)$$

in which F and A are given by (5.10.55b) and (5.10.55c).
Comparison of (5.10.55b) with (5.10.51b) shows that

$$p = 12(\omega_c/\omega_N) = 24 f_c \text{ (in cpd)} \quad (5.10.57)$$

where $f_c = \omega_c/2\pi$ is the cut-off frequency in cycles per hour (cph) and where we have used the Nyquist frequency $f_N = 0.5$ cph for the hourly sampled data. The arguments of the angles in (5.10.51) and (5.10.55) are, therefore, identical. Where the filters differ is in the use of the sigma factors. Whereas the oscillations of $[1 + \cos(\pi k/M)]$ are uniform with k, those of $\sin(\pi k/M)/(\pi k/M)$ decay with increasing k, similar to the way we have seen the term, $\sin(\pi \omega_c/\omega_N)/(\pi k\omega_c/\omega_N)$, decay in amplitude (e.g. Figure 5.6.1b). In this regard, the raised cosine window provides a more severe weighting of the truncated Fourier series than the sigma factors.

The value $p = 0.7$ cpd, corresponding to a cut-off period of 34.29 h, has been commonly used in the design of low-pass Lanczos–cosine filters (cf. Bryden, 1979). Although this produces an acceptable filter response for periods of two days and longer (where two days is generally the central period of the oceanic "spectral gap"), it has been shown to pass an unacceptable amount of high-frequency energy from the diurnal band, particularly from the O_1 and Q_1 tidal constituents (Walters and Heston, 1982). In an attempt to further reduce the leakage from the diurnal band, Mooers and Smith (1967) applied a separate filter to the low-pass filtered data from the $p = 0.7$ cpd filter or "Lancz7" filter (Thompson, 1983; Figure 5.10.13). Walters and Heston (1982) passed the data twice through the filter to produce the 10-day (Lancz7) filter. This results in a significantly improved filter amplitude throughout the diurnal band but also doubles the amount of data lost from the ends of the time series. Thompson (1983) suggested the use of a Lanczos–cosine filter with $p = 0.6$ cpd (the "Lancz6" filter) which equates to a cut-off period of 40 h. The Lancz6 filter essentially removes the leakage from the diurnal band but simultaneously shifts the low-pass portion of the filtered record to periods somewhat in excess of two days. The difference in the filters is quite subtle. For the Lanczos–cosine filter with $p = 0.7$ (Lancz7 filter), the first zero of the transfer function occurs at 15.4°/h (at 0.0428 cph), which is past the diurnal band (Figure 5.10.13); for the Lancz6 filter, the first zero is shifted to 14°/h (at 0.0389 cph) near the O_1 frequency of 13.9°/h.

5.10.8 Butterworth filters

The windowed cosine filters described in the previous section attempt to approximate an ideal rectangular transfer function using truncated Fourier cosine series. For nonrecursive filters, the output is a simple linear combination of the data and the role of the window is to attenuate the overshoot ripples created by truncation in the time domain (Gibbs' phenomenon). We now turn to a specific type of recursive filter for which the transfer function is created using a rational function in sines and cosines. Because this is a recursive filter, the output consists of both input data and past values of the output.

Let $w = w(\omega)$ be a monotonically increasing rational function of sines and cosines in the frequency, ω. The monotonic function

$$|H_L(\omega)|^2 = 1/[1 + (w/w_c)^{2q}] \qquad (5.10.58)$$

(Figure 5.10.14) generates a particularly useful approximation to the squared gain of an ideal low-pass recursive filter with frequency cut-off w_c. (Our filter design will eventually require $w(0) = 0$ so that the final version of $H_L(\omega)$ will closely resemble (5.10.58).)

Butterworth filters of the form (5.10.58) have a number of desirable features (Roberts and Roberts, 1978). Unlike the transfer function of a linear nonrecursive filter constructed from a truncated Fourier series, the transfer function of a Butterworth filter is monotonically flat within the pass and stop-bands, and has high tangency at both the origin ($\omega = 0$) and the Nyquist frequency, ω_N. The attenuation rate of $H_L(\omega)$ can be increased by increasing the *filter order*, q. However, too steep a transition from the stop-band to the pass-band can lead to ringing effects in the output due to Gibbs' phenomenon. Since it has a squared response, the Butterworth filter produces zero phase shift and its amplitude is attenuated by a factor of two at the cut-off frequency, for which $w/w_c = 1$ for all q. In contrast to nonrecursive filters, such as the Lanczos–cosine filter discussed in the previous section, there is no loss of output data from the ends of the record; N input values yield N output values. However, we do not expect to get something for nothing. The problem is that ringing distorts the data at the ends of the filtered output. As a consequence, we are forced to ignore output values near the ends of the filtered record, in analogy with the loss of data associated with nonrecursive filters. In effect, the loss is comparable to that from a nonrecursive filter of similar smoothing performance. A subjective decision is usually needed to determine where, at the two ends of the filtered record, the "bad" data end and the "good" data begin.

Butterworth filters fall into the category of physically realizeable recursive filters having the time-domain formulation (5.10.2) with $k = 0, \ldots, M$. They may also be classified as infinite impulse response filters since the effects of a single impulse input can be predicted to an arbitrary time into the future. To see why we expect $w(\omega)$ to be

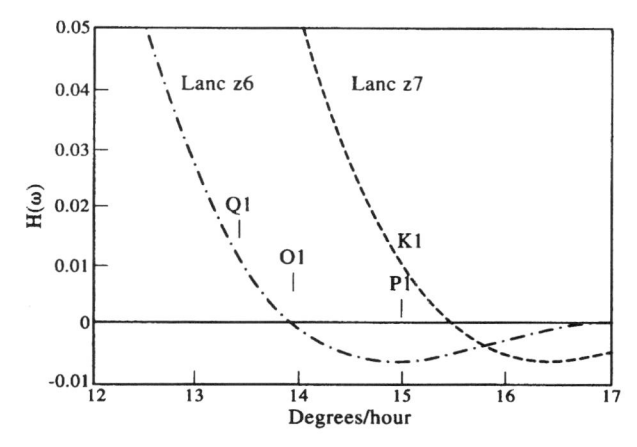

Figure 5.10.13. Expanded views of the filter responses for two tide-elimination filters for the diurnal frequency band. The Lancz6 and Lancz7 filters are low-pass Lanczos–cosine filters. $15°/h = 1.0$ cpd. (Modified from Thompson, 1983).

542   Data Analysis Methods in Physical Oceanography

a rational function in sines and cosines, we use (5.10.2) and the fact that $H(\omega)$ is the ratio of the output to the input. We can then write

$$H(\omega) = \frac{\text{output}}{\text{input}} = \frac{\sum_{k=0}^{M} h_k e^{-i\omega k \Delta t}}{1 - \sum_{j=1}^{L} g_j e^{-i\omega k \Delta t}} \qquad (5.10.59)$$

where the summations in the numerator and denominator involve polynomials in powers of $\exp(-i\omega k\Delta t)$ which can in turn be expressed through the variable w. The substitution $z = \exp(i\omega kt)$ leads to expression of the filter response $H(\omega)$ in terms of the z-transform and zeros of poles.

5.10.8.1 High-pass and band-pass filters

High-pass and band-pass Butterworth filters can be constructed from the low-pass filter (5.10.58). For example, to construct a high-pass filter with cut-off, w_c, we use the transformation $w/w_c \rightarrow -(w/w_c)^{-1}$ in (5.10.58). The square transfer function of the high-pass filter is then

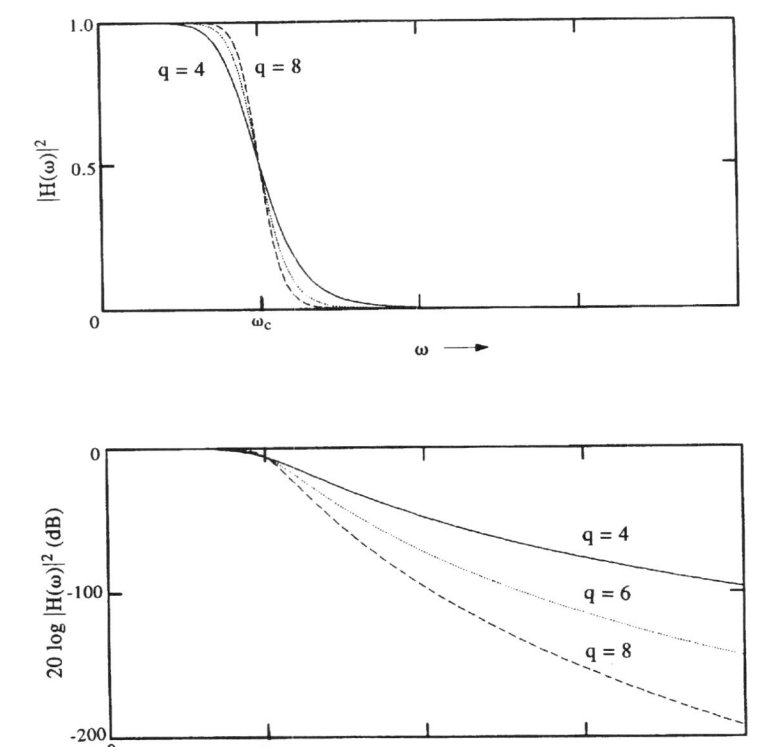

Figure 5.10.14. The frequency response functions $|H_L(\omega)|^2$ for an ideal squared, low-pass Butterworth filter for filter orders q = 4, 6, 8. Bottom panel gives response in decibels (dB). Power = 0.5 at the cut-off frequency, ω_c.

Time-series Analysis Methods 543

$$|H_H(w)|^2 = (w/w_c)^{2q}/[1 + (w/w_c)^{2q}] \qquad (5.10.60)$$

where, as required

$$|H_H(w)|^2 = 1 - |H_L(w)|^2 \qquad (5.10.61)$$

Band-pass Butterworth filters (and their counterparts, *stop-band* Butterworth filters) are constructed from a combination of low-pass and high-pass filters. For instance, the appropriate substitution in (5.10.58) for a band-pass filter is $w/w_c = w_*/w_c - (w_*/w_c)^{-1}$ which leads to the quadratic equation

$$(w_*/w_c)^2 - (w/w_c)(w_*/w_c) - 1 = 0 \qquad (5.10.62a)$$

with roots

$$w_{*1,2}/w_c = (w/w_c)/2 \pm [(w/w_c)^2/4 + 1]^{1/2} \qquad (5.10.62b)$$

Substitution of $w/w_c = \pm 1$ (the cut-off points of the low-pass filter) yields the normalized cut-off functions $w_{*1}/w_c = 0.618$ and $w_{*2}/w_c = 1.618$ of the band-pass filter based on the cut-off frequency $\pm w_c$ of the associated low-pass filter. The corresponding band-pass cut-off functions for the cut-off frequency $-w_c$ of the low-pass filter are $w_{*1}/w_c = -1.618$ and $w_{*2}/w_c = -0.618$. Specification of the low-pass cut-off determines w_{*1}/w_{*2} of the band-pass filter. The bandwidth $\Delta w/w_c = -(w_{*1} - w_{*2})/w_c = 1$ and the product $(w_{*1}/w_c)(w_{*2}/w_c) = 1$. Note that specification of w_{*1} and w_{*2} gives the associated function w_c of the low-pass filter

$$w_{*1}w_{*2} = w_c^2 \qquad (5.10.63)$$

5.10.8.2 Digital formulation

The transfer functions (5.10.58)–(5.10.61) involve the continuous variable w whose structure is determined by sines and cosines of the frequency, ω. To determine a form for $w(\omega)$ applicable to digital data, we seek a rational expression with constant coefficients a to d such that the component $\exp(i\omega\Delta t)$ in (5.10.59) takes the form

$$\exp(i\omega\Delta t) = \frac{aw + b}{cw + d} \qquad (5.10.64)$$

(Here, we have replaced $-i\omega\Delta t$ with $+i\omega\Delta t$ without loss of generality.) As discussed by Hamming (1977), the constants are obtained by requiring that $\omega = 0$ corresponds to $w = 0$ and that $\omega \to \pi/\Delta t$ corresponds to $w \to \pm\infty$. Constants b and d (one of which is arbitrary) are set equal to unity. The final "scale" of the transformation is determined by setting $(\omega/2\pi)\Delta t = 1/4$ for $w = 1$. This yields

$$\exp(i\omega\Delta t) = \frac{1 + iw}{1 - iw} \qquad (5.10.65)$$

or, equating real and imaginary parts

$$w = \frac{2}{\Delta t}[\tan(\tfrac{1}{2}\omega\Delta t)]$$
$$= \frac{2}{\Delta t}[\tan(\pi\omega/\omega_s)], \quad -\omega_N < \omega < \omega_N \tag{5.10.66}$$

where $\omega_s/(2\pi) = f_s$ is the sampling frequency ($f_s = 1/\Delta t$). We note that the derivation of (5.10.66) is equivalent to the conformal mapping

$$w = i\frac{2}{\Delta t}\frac{1-z}{1+z} \tag{5.10.67a}$$

where

$$z = e^{2\pi i f \Delta t} = e^{i\omega\Delta t} \tag{5.10.67b}$$

is the standard z-transform.

The transfer function of the (discrete) low-pass Butterworth filter is then (Rabiner and Gold, 1975)

$$|H_L(\omega)|^2 = \frac{1}{1 + [\tan(\pi\omega/\omega_s)/\tan(\pi\omega_c/\omega_s)]^{2q}} \tag{5.10.68a}$$

and that of the high-pass Butterworth filter

$$|H_H(\omega)|^2 = \frac{[\tan(\pi\omega/\omega_s)/\tan(\pi\omega_c/\omega_s)]^{2q}}{1 + [\tan(\pi\omega/\omega_s)/\tan(\pi\omega_c/\omega_s)]^{2q}} \tag{5.10.68b}$$

The sampling and cut-off frequencies in these expressions are given by $\omega_s = 2\pi/\Delta t$ and $\omega_c = 2\pi/T_c$ in which $T_c = 1/f_c$ is the period of the cyclic cut-off frequency f_c. Plots of (5.10.68a) for various cut-off frequencies and filter order q are presented in Figure 5.10.15.

Use of the bilinear z-transform, $i(1-z)/(1+z)$, in (5.10.67a) eliminates aliasing errors that arise when the standard z-transform is used to derive the transfer function; these errors being large if the digitizing interval is large. Mathematically, the bilinear z-transform maps the inside of the unit circle ($|z| < 1$, for stability) into the upper half plane. A thorough discussion of the derivation of pole and zeros of Butterworth filters is presented in Kanasewich (1975) and Rabiner and Gold (1975).

We note that the above relationships define the square of the response of the filter $H(\omega)$ formed by multiplying the transfer function by its complex conjugate, $H(\omega)^* = H(-\omega)$. (In this instance, $H(\omega)^*$ and $H(-\omega)$ are equivalent since $i = \sqrt{-1}$) always occurs in conjunction with ω. The product $H(\omega)H(-\omega)$ eliminates any frequency-dependent phase shift caused by the individual filters and produces a squared, and therefore sharper, frequency response than produced $H(\omega)$ alone. The sharpness of the filter (as determined by the parameter q) is limited by filter ringing and stability problems. When q becomes too large, the filter begins to act like a step and Gibbs' phenomenon rapidly ensues.

Equations (5.10.68a, b) are used to design the filter in the frequency domain. In the time domain, we first determine the filter coefficients h_k and g_j for the low-pass filter (5.10.2) and then manipulate the output from the transfer function $H(\omega)$ to generate the output $|H(\omega)|^2$. To obtain the output for a high-pass Butterworth filter, $|H_H(\omega)|^2$, the output from the corresponding low-pass filter, $|H_L(\omega)|^2$, is first obtained and the resulting data values subtracted from the original input values on a data point-by-data point basis.

5.10.8.3 Tangent versus sine filters

Equations (5.10.68a, b) define the transfer functions of *tangent* Butterworth low-pass filters. Corresponding transfer functions for *sine* Butterworth low-pass filters are given by

$$|H_L(\omega)|^2 = \frac{1}{1 + [\sin(\pi\omega/\omega_s)/\sin(\pi\omega_c/\omega_s)]^{2q}} \quad (5.10.69)$$

where we have simply replaced tan$x$ with sin$x$ in (5.10.68). Although this book deals only with the tangent version of the filter, there are situations where the sine-version may be preferable (Otnes and Enochson, 1972). The tangent filter has "superior" attenuation within the stop-band but at a cost of doubled algebraic computation (the sine version has only recursive terms while the tangent version has both recursive and nonrecursive terms).

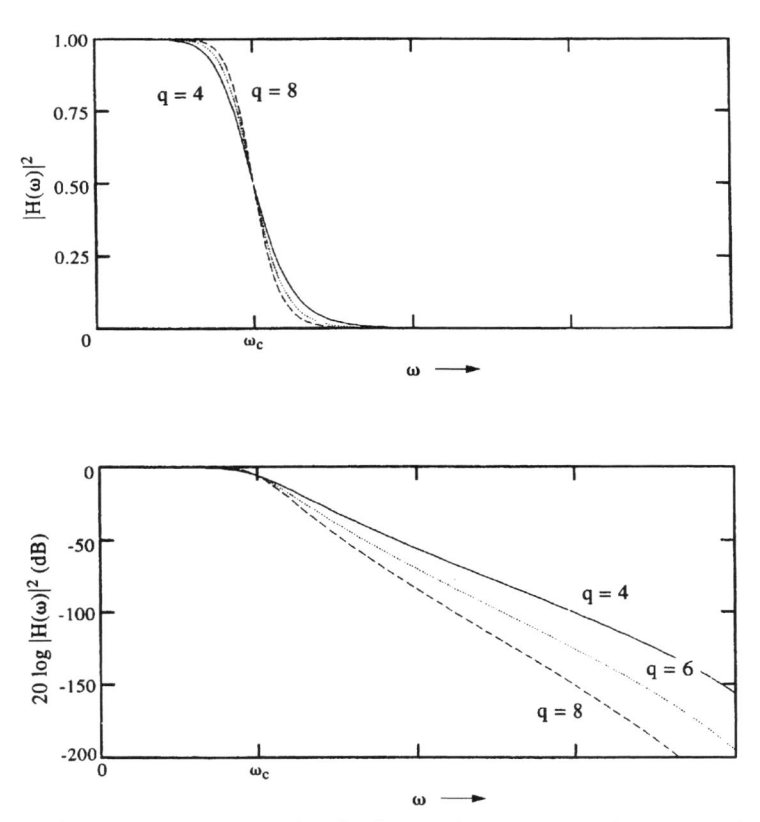

Figure 5.10.15. Same as Figure 5.10.14 but for discrete, low-pass squared Butterworth filters. (After Rabiner and Gold, 1975.)

5.10.8.4 Filter design

The design of Butterworth filters is discussed in Hamming (1977). Our approach is slightly different but uses the same general concepts. We begin by specifying the sampling frequency $\omega_s = 2\pi f_s = 2\pi/\Delta t$ based on the sampling interval Δt for which

$$0 < \omega/\omega_s < 0.5 \tag{5.10.70}$$

and where the upper limit denotes the normalized Nyquist frequency, ω_N/ω_s. We next specify the desired cut-off frequency ω_c at the half-power point of the filter. For best results, the normalized cut-off frequency of the filter, ω_c/ω_s, should be such that the transition band of the filter does not overlap to any significant degree with the ends of the sampling domain (5.10.70). Once the normalized cut-off frequency (or frequencies) is known, specification of the filter order q fully determines the characteristics of the filter response. Our experience suggests that the parameter q should be less than 10 and probably not larger than eight. Despite the use of double precision throughout the calculations, runoff errors and ringing effects can distort the filter response for large q and render the filter impractical.

There are two approaches for Butterworth filter design once the cut-off frequency is specified. The first is to specify q so that the attenuation levels in the pass and stop-bands are automatically determined. The second is to calculate q based on a required attenuation at a given frequency, taking advantage of the fact that we are working with strictly monotonic functions. Suppose we want an attenuation of $-D$ decibels at frequency ω_a in the stop-band of a low-pass filter having a cut-off frequency $\omega_c < \omega_a$. Using the definition for decibels and (5.10.48), we find that

$$\begin{aligned} q &= 0.5 \frac{\log(10^{D/10} - 1)}{\log(w_a/w_c)} \\ &\approx \frac{D/20}{\log(w_a/w_c)}, \quad \text{for } D > 10 \end{aligned} \tag{5.10.71}$$

where D is a positive number measuring the decrease in filter amplitude in decibels (dB) and w is defined by (5.10.66). The nearest integer value can then be taken for the filter order provided that the various parameters (w_a, D) have been correctly specified and q is less than 10. If the latter is not followed, the imposed constraints are too severe and new parameters need to be specified. The above calculations apply equally to specification of q based on the attenuation $-D$ at frequency $\omega_a < \omega_c$ in the stop-band of a high-pass filter, except that $\log(w_a/w_c)$ in (5.10.71) is replaced by $\log(w_c/w_a)$. Since $\log(x) = -\log(1/x)$, we can simply apply (5.10.71) to the high-pass filter, ignoring the minus sign in front of $\log(1/x)$.

5.10.8.5 Filter coefficients

Once the characteristics of a transfer response have been specified, we need to derive the filter coefficients to be applied to the data in the time domain. We assume that the transfer function $H_L(\omega; q)$ of the low-pass filter can be constructed as a product, or cascade, of second-order ($q = 2$) Butterworth filters $H_L(\omega; 2)$ and, if necessary, one first-order ($q = 1$) Butterworth filter $H_L(\omega; 1)$. For example, suppose we required a filter of order $q = 5$. The transfer function would then be constructed via the cascade

$$H_L(\omega; 5) = H_L(\omega; 1) \times H_{L,1}(\omega; 2) \times H_{L,2}(\omega; 2) \tag{5.10.72}$$

in which the two second-order filters, $H_{L,1}$ and $H_{L,2}$, have different algebraic structure. Use of the cascade technique allows for variable order in the computer code for Butterworth filter programs without the necessity of computing a separate transfer function $H_L(\omega; q)$ each time. This eliminates a considerable amount of algebra and reduces the roundoff error that would arise in the "brute-force calculation" of H_L for each order.

The second-order transfer functions for a specified filter order q are given by

$$H_L(\omega; 2) = \frac{[w_c^2(z^2 + 2z + 1)]}{a_k z^2 + 2z(w_c^2 - 1) + \{1 - 2w_c \sin[\pi(2k+1)/2q] + w_c^2\}} \tag{5.10.73a}$$

where w and z are defined by (5.10.66) and (5.10.67b)

$$a_k = 1 + 2w_c \sin[\pi(2k+1)/2q] + w_c^2 \tag{5.10.73b}$$

and k is an integer that takes on values in the range

$$0 \le k < 0.5(q-1) \tag{5.10.73c}$$

When q is an odd number, the first-order filter $H_L(\omega; 1)$ must also be used where

$$H_L(\omega; 1) = \left(\frac{w_c}{1+w_c}\right) \frac{z+1}{z - \left(\frac{1-w_c}{1+w_c}\right)} \tag{5.10.74}$$

Again, suppose that $q = 5$. The transfer function H_L is then composed of the lead filter $H_L(\omega; 1)$ given by (5.10.74) and two second-order filters, for which k takes the values $k = 0$ and 1 in (5.10.73). Note that we have strictly adhered to the inequality in (5.10.73c). The first second-order filter is obtained by setting $k = 0$ in (5.10.73); the second second-order filter is obtained by setting $k = 1$. For $q = 7$, a third second-order for $k = 2$ would be required, and so on.

The next step is to recognize that the first-order function (5.10.74) has the general form

$$H_L(\omega) = \frac{d_0 z + d_1}{z - e_1} \tag{5.10.75}$$

and that the second-order function (5.10.73a) has the general form

$$H_L(\omega) = \frac{c_0 z^2 + c_1 z + c_2}{z^2 - b_1 z - b_2} \tag{5.10.76}$$

where the sine terms in the coefficients of (5.10.73a) change with filter order q. The coefficients d, e in (5.10.74) are obtained by direct comparison with (5.10.73) while the coefficients b, c in (5.10.76) are obtained through comparison with (5.10.73a).

The recursive digital filters (5.10.2), whose time-domain algorithms have the transfer functions (5.10.75) and (5.10.76) are, respectively

$$y_n = d_0 x_n + d_1 x_{n-1} + e_1 y_{n-1} \tag{5.10.77}$$

and

$$y_n = c_0 x_n + c_1 x_{n-1} + c_2 x_{n-2} + b_1 y_{n-1} + b_2 y_{n-2} \qquad (5.10.78)$$

Direct comparison of (5.10.75) with (5.10.77) yields the time domain coefficients for the first-order filter; comparison of (5.10.76) with (5.10.78) yields the corresponding coefficients for the second-order filters for each value of k beginning with $k = 0$. In particular, we find, for the first-order filter

$$d_o = d_1 = \frac{w_c}{1 + w_c}; \quad e_1 = \frac{1 - w_c}{1 + w_c} \qquad (5.10.79)$$

and for the second-order filter

$$\begin{aligned} b_1 &= -2w_c/a_k; \quad b_2 = [a_k^{-2}(1 + w_c^2)]/a_k \\ &= c_o = w_c^2/a_k; \quad c_1 = 2c_o; \quad c_2 = c_o \end{aligned} \qquad (5.10.80)$$

where the coefficients in (5.10.80) change with the parameter k according to the number of second-order filters needed to create the filter of order q.

To apply the $q = 5$ filter, we process the input data x_n ($n = 0, 1, \ldots, N$) by the first-order filter (5.10.77). We then take the output from the first-order filter and process it by the first of the second-order filters (5.10.78) with $k = 0$. The resultant output is then processed by the next second-order filter (5.10.78) with $k = 1$. The sequence y'_n ($n = 0, 1, \ldots$) derived from the three filter applications is the low-pass output for the fifth order Butterworth filter $H_L(\omega; 5)$, as indicated by (5.10.72).

The task is only half complete since our ultimate goal is to remove any filter-induced phase shift by smoothing the data with the squared-response of the filter $|H_L|^2$, given by (5.10.50). The sequence we require is: $\{x_n\}$ yields $\{y'_n\}$ as the output from $H_L(\omega)$; $\{y'_n\}$ yields $\{y_n\}$ as the output from $|H_L(\omega)|^2$. To obtain the output $\{y_n\}$ for the square response of the filter, $|H_L(\omega)|^2$, we need to process the output $\{y'_n\}$, obtained from $H_L(\omega)$, with the filter $H_L(-\omega)$. There are three options: (1) We can separately design $H(-\omega)$, a relatively straightforward task involving some sign changes in (5.10.79) and (5.10.80); (2) we can invert the order of the calculations such that the output $\{y'_n\}$ from $H_L(\omega)$ is passed through the inverted version of this filter. That is, the data from $H_L(\omega)$ are first run through the second-order filter ($k = 1$ for $q = 5$), with the output from this filter passed through the second-order filter ($k = 0$) and finally through the first-order filter; or (3) we can simply invert the chronological order of the data $\{y'_n\}$ and pass the inverted sequence through the original filter $H_L(\omega)$. Since all the data are recorded beforehand, we recommend approach (3). The one caution is that the sequence of the final output must be inverted to regain the original chronological order of the data. In all cases, passing the inverted version of $\{y'_n\}$ through the filter cascade removes any phase shift associated with the first pass which produced $\{y'_n\}$ from $\{y_n\}$. A phase shift $\phi(\omega)$ caused by the first sequence of filters $H_1(\omega) \times H_2(\omega) \times \ldots$ is canceled by the phase shift $-\phi(\omega)$ caused by the second sequence of filters $H_L(-\omega)$.

Computer programs designed to carry out the Butterworth filter operations should assign the output $\{y'_n\}$ from each filter as the new input $\{x_n\}$ to the next filter in the cascade until the output corresponding to the filter $|H_L(\omega)|^2$ is achieved. The last set of output is then chronologically inverted and re-run through the same filter. Following the final set of calculations, the output sequence is inverted to ensure correct ordering in time.

Time-series Analysis Methods 549

To obtain the results for a *high-pass* Butterworth filter, one further operation is required. The final output $\{y_n\}$ ($n = 0, 1, \ldots$) from the low-pass filter is subtracted point-for-point from the original input, $\{x_n\}$ ($n = 0, 1, \ldots$) to create the high-pass filtered data $y_{*n} = x_n - y_n$. The procedure to obtain the low and high-pass Butterworth filters is illustrated schematically in Figure 5.10.16.

5.10.9 Frequency-domain (transform) filtering

The type of digital filtering discussed in the previous sections involves convolution of the time-series data with weighting functions called *impulse response functions* that eliminate selected ranges of frequencies from the data. In the case of Fourier transform filtering, the weights are defined in terms of a Fourier transform window or *frequency response function*, $H(\omega)$, and filtering involves: (1) taking the FFT of the original data set; (2) multiplying the FFT output by the appropriate form of $H(\omega)$ that lets through the frequencies of interest and blocks all the others; and (3) taking the inverse FFT of the result to get back a filtered data set in the time domain. These steps are shown schematically in Figure 5.10.17. As an example, $H(\omega)$ might be a low-pass filter designed to eliminate frequency components with periods $2\pi/\omega$ that are longer than 40 h. Alternatively, $H(\omega)$ could be a "notch" filter used to isolate oscillations centered near the local Coriolis frequency, or a two-notch filter designed to remove energy in the diurnal and semidiurnal tidal bands. Transform methods have been discussed from an oceanographic perspective by Walters and Heston (1982), Evans (1985), and Forbes (1988). As these papers indicate, the choice of an "appropriate" form for $H(\omega)$ is critical to the success of the method. Filtering in the frequency domain is attractive because of its simplicity compared to convolution in the time domain and because it is conceptually more in accord with our objective in filtering; namely, to remove specific periodicities in the data while retaining those of interest. Perhaps contrary to expectation, multiplication of the Fourier transform by a window is not always more computationally efficient than convolution of filter weights with the data (Evans, 1985).

We can outline use of the Fourier transform filtering as follows. Suppose we have a time series $x(t)$ with discrete values $x(n\Delta t) = x_n$, where n is an integer in the range $-N < n \leq N$. The Fourier transform of this time series is

$$X_k = \frac{1}{T} \sum_{n=-N+1}^{N} x_n \exp(-i\omega_k n \Delta t) \qquad (5.10.81)$$

where $T = 2N\Delta t$ is the record length and the Fourier frequencies are

$$\omega_k = 2\pi f_k = \frac{2\pi k}{T}, \quad -N < k \leq N. \qquad (5.10.82)$$

Let $w(r\Delta t) = w_r$, $-s \leq r \leq s$, represent a set of filter weights whose sum is unity to preserve the series mean and whose distribution is symmetric about $r = 0$ to preserve the phase information in the data. The number of weights, $S = 2s + 1$, is called the span of the filter. Since s points are lost from each end of the input data series, the filtered output series

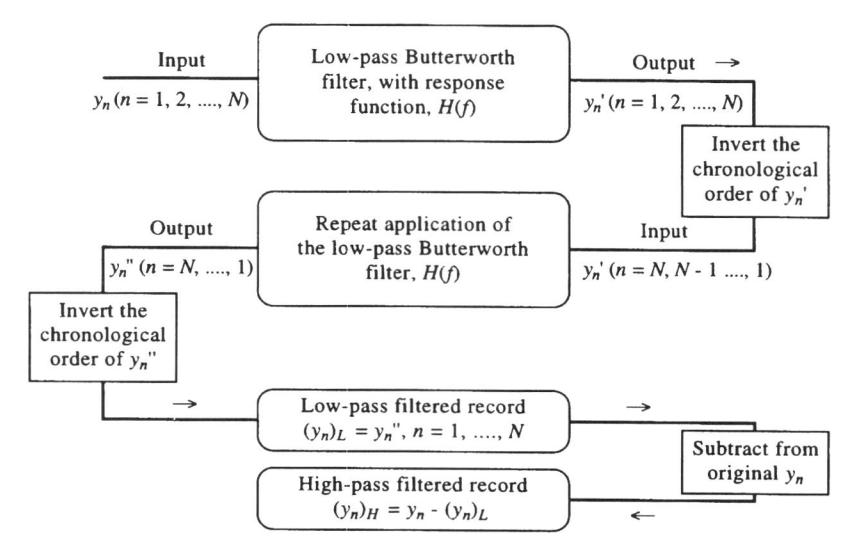

550 *Data Analysis Methods in Physical Oceanography*

Figure 5.10.16. The procedure for obtaining low and high-pass Butterworth filters.

$$y_n = \sum_{r=-s}^{s} w_r x_{n-r} = \sum_{r=-s}^{s} w_{n-r} x_r \qquad (5.10.83)$$

is shorter than the original series by 2s values. The effect of the convolution is to smear the signal $x(t)$ according to the weighting imposed by the impulse response function (IRF), $w(t)$. The frequency response function (FRF)

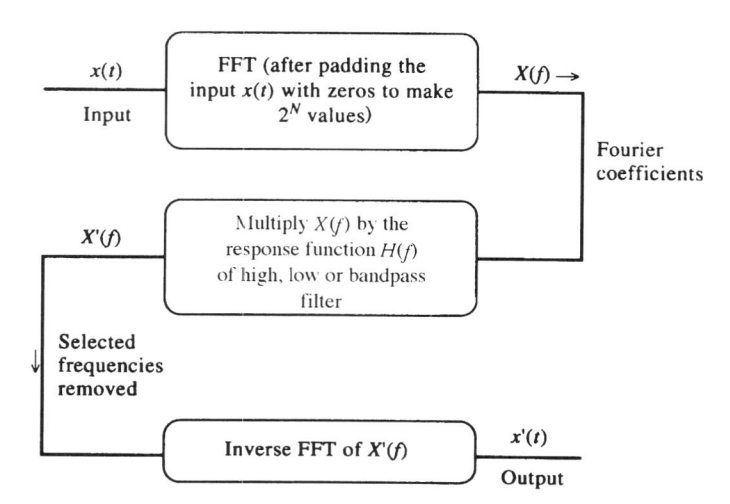

Figure 5.10.17. The procedure for obtaining discrete Fourier transform filters for application in the frequency domain.

$$H(\omega) = \sum_{r=-s}^{s} w_r \exp(-i\omega r \Delta t) = |H(\omega)|e^{-i\phi(\omega)} \quad (5.10.84)$$

gives the effect of the impulse response function on the transform of a sinusoid of unit amplitude and frequency ω ($= 2\pi f$). As stated earlier, the absolute value $|H(\omega)|$ is the *gain factor* of the system and the associated phase angle, $\phi(\omega)$, the *phase factor* of the system. If a linear system is subjected to a sinusoidal input with a frequency ω and produces a sinusoidal output at the same frequency, then $|H(\omega)|$ is the ratio of the output amplitude to the input amplitude and $\phi(\omega)$ is the phase shift between the output and input. The FRF is viewed as a window or transfer function that lets through some frequencies and stops others. Note that H is defined at all frequencies such that $-\pi/\Delta t < \omega \leq \pi/\Delta t$, and not just at the Fourier frequencies, ω_k.

The key to Fourier transform filtering is that, for a constant-parameter linear system, the Fourier transform of the filtered data, $Y(\omega)$, is related to the Fourier transform of the input data, $X(\omega)$, through the product

$$Y(\omega) = H(\omega)X(\omega) \quad (5.10.85)$$

In other words, convolution in the time domain, defined by (5.10.83), translates to multiplication in the frequency domain. The merits of a filter are judged by its FRF (frequency domain) and its IRF (time domain). We would like the magnitude of the FRF to be near unity in the frequency bands to be passed by the filter and near zero in the bands to be stopped; i.e. $|H(\omega)| \approx 1$ and 0, respectively. The transition band between the stop and pass-bands should be as narrow as possible since a broad transition band results in a filtered time series whose frequency content may be contaminated by unwanted frequencies. Similarly, the span of the IRF should be short so that the magnitude of weights decay to zero rapidly as r increases toward $\pm s$. If convolution is used, short filters are computationally more efficient and, moreover, result in less data loss. Unfortunately, the two criteria are at odds with one another. In general, the narrower the transition band in the frequency domain, the slower is the decay rate of the IRF in the time domain. Also, the steeper the maximum slope of the transition band, the larger are the side initial side-lobes of the IRF that arise from the well-known Gibbs' phenomenon. In the limit of a step function-type FRF, in which the transition zone has zero width, the resulting IRF decays very slowly and has large side-lobes (ringing). Thus, one must always compromise in specifying a FRF.

In all time-domain filtering (convolution), data are lost from each end of the original digital time series. For example, in the case of nonrecursive filters, in which the output is based on input time series alone, a known segment of the record of length $T/2$ is lost from either end of the time series (T is the filter length). The same applies to recursive filters in which the present output from the filter is based on the original data series as well as previous values of the output. Here, the difficulty is that the amount of data we must discard from either end is not well defined because of ringing effects associated with the convolution and abrupt data discontinuities at the ends of the record. Transform windowing typically results in exactly the same amount of data loss as the equivalent time-domain filter (Walters and Heston, 1982). The Fourier transform treats the data outside the record as if it were zero, so that the ringing at the ends is introduced by the abrupt changes in the series from nonzero to zero and to the circular convolution of the window's IRF with the data (see Section 5.10.10). Ringing

(Gibbs' phenomenon) occurs throughout the entire time series and becomes evident when the filtered FFT data are inverted to recover the desired filtered time-series data. The effects of Gibbs' phenomenon are mitigated by tapering the frequency-domain filter using a linear or cosine function.

According to Thompson (1983), careful construction of weighting functions in the time domain can more effectively remove tidal components than Fourier transform filtering. This is because tidal frequencies do not generally coincide with Fourier frequencies of the record length. Design of IRF weights to minimize the squared deviation from some specified norm (least squares filter design) offers more control over the FRF at particular nonFourier frequencies. On the other hand, broad-band signals are best served by the FRF approach. Evans (1985) suggests that the ratio of convolution cost to windowing cost is $E = S/[2 \log_2(N)]$, where S is the filter span. If $E > 1$, then windowing in the frequency domain is more efficient method. Forbes (1988) addressed the problem of removing tidal signals from the data while retaining the near-inertial signal and argues that Fourier transform filtering is effective provided that careful consideration is given to the filter bandwidth and the amount of tapering of the sides of the filter. Note that, in trying to remove strong tidal signals from a data series, it is sometimes beneficial to first calculate the tidal constituents and then subtract the harmonically predicted tidal signal from the data prior to filtering. This is time consuming and not an advantage if the filter is properly designed.

Figure 5.10.18(a) shows the energy-preserving power spectrum for a mid-depth current meter record from a Cape Howe mooring site (37°35'S, 150°25'E) off the coast of New South Wales. To remove the strong tidal motions from this record, Forbes first used an untapered discrete Fourier transform (DFT) with 12 and 17 adjacent Fourier coefficients set to zero in the diurnal and semidiurnal bands, respectively (Figure 5.10.18b). The greatest improvement in the Fourier transform filtering came from setting only three Fourier terms to zero but tapering the filter with a nine-point cosine taper in the frequency domain at the diurnal and semidiurnal frequencies (Figure 5.10.18c). Thus, tapering the time series, not widening the filter by using more zero frequencies, is a better way to improve filter characteristics. Perhaps, the most important conclusion from Forbes' work is that DFT filters are effective if the number of Fourier coefficients set to zero is sufficient to cover the unwanted frequency band and if the filter is cosine-tapered in the frequency domain to ensure a smooth transition to nonzero Fourier coefficients. In the nonintegral single-frequency case presented here (Forbes was looking at near-inertial motions) this amounted to a three-point filter with a nine-point cosine taper. The widths of the filter and taper must be determined for each application by a careful examination of the spectrum for leakage into adjacent frequencies, but once this is done, the technique is fast and simple to apply.

To summarize the use of Fourier transform filtering:

(1) Remove any linear trend (or nonlinear trend if it is well defined) from the data prior to filtering but do not be too concerned with cosine tapering the first and last 10% of the data. Fast Fourier-transform the data.
(2) Define the Fourier transform filter $H(\omega)$ for both positive and negative frequencies with the extreme frequencies given by $\pm 1/2\Delta t$.
(3) If the measured data are real, and the filtered output is to be real, the filter should obey $H(-\omega) = H(\omega)^*$, where the asterisk denotes complex conjugate. The easiest way to satisfy this condition is to pick $H(\omega)$ real and symmetric in frequency.

(4) If $H(\omega)$ has sharp vertical edges then the impulse response of the filter (the response arising from a short impulse as input) will have damped ringing at frequencies corresponding to these edges. If this occurs, pick a smoother $H(\omega)$. You can take the FFT inverse of $H(\omega)$ to see the impulse response of the filter. The more points used in the smoothing the more rapid the fall of the impulse response.

(5) Multiply the transformed data series $X(\omega)$ by $H(\omega)$ and invert the resultant data series, $Y(\omega)$, to obtain the filtered data in the time domain. To eliminate ringing effects, discard $T/2$ data points from either end of the filtered time series, where T is the span of the IRF for the transform filter.

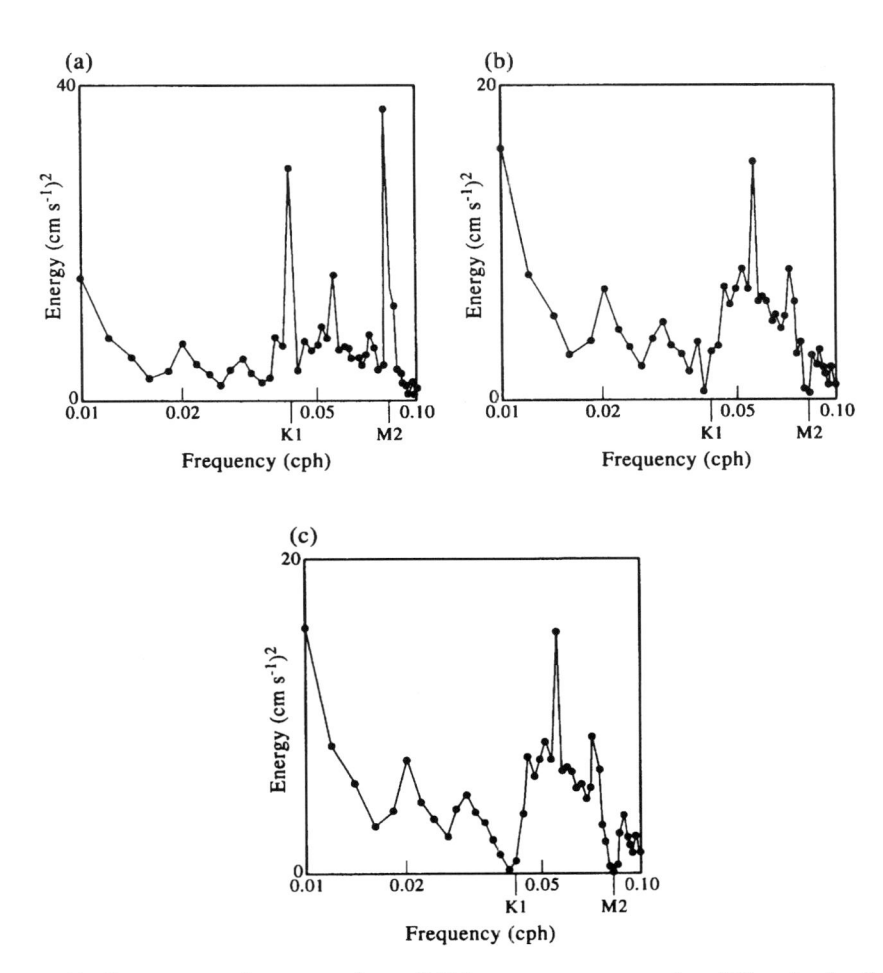

Figure 5.10.18. Energy-preserving spectra for a 4000-h current meter record at 720-m depth off Cape Howe, Australia. (a) Raw hourly data; (b) after applying a discrete Fourier transform (DFT) filter with 12 and 17 adjacent Fourier coefficients set to zero in the diurnal and semidiurnal bands (no tapering); (c) after applying a DFT filter with three Fourier coefficients set to zero and nine Fourier coefficients cosine-tapered on each side of the zero coefficients. (From Forbes, 1988.)

5.10.10 Truncation effects

For all digital filters, a percentage of the end values from the filtered record must be omitted prior to further analysis. This loss of information from the ends of the output is linked to ringing effects associated with discontinuities at the ends of the input and to the nonexistence of integrable data prior to the start of the record. The ringing decays toward the interior of the data sequence after the end effects have been smoothed by a sufficient number of filter integrations (Figure 5.10.19). In the case of the squared Butterworth filter, both ends of the data are affected twice since the data are passed forward and backward through the filter. One approach is to assume that 10% of output data at each end of the filter output is contaminated and remove these points from the final output. However, each case is different and data elimination should be based on a trial and error approach using visual inspection to estimate the extent of the data removal. Padding the ends of the input with zeros appears to serve no useful purpose. In some cases the ringing effect can be substantially reduced by using the zero cross-over points (for input centered about the mean record value) as the first record of the input.

5.11 FRACTALS

The term "fractal" was coined by Mandelbrot (1967) to describe the bumpiness of geometrical curves and surfaces. Regardless of how closely we examine a fractal object it fails to become smooth and its degree of fluctuation remains unchanged. Fractal objects are uneven at all scales and possess no characteristic length scales. Fractals are ubiquitous features whose presence has been reported in a wide variety of fluid dynamical settings including the mixing of turbulent flows (Sreenivasan *et al.*, 1989), the trajectories of oceanic drifters (Osborne *et al.*, 1989; Sanderson *et al.*, 1990) and the paths of atmospheric cyclones (Fraedrich *et al.*, 1990). More everyday examples involve the fractal dimensionality of coastlines, the shapes of clouds, and the forms of lightning strikes. The fractal curve in Figure 5.11.1(a), called a *Koch curve*, resembles a coastline or the outline of a snowflake that would be mapped at ever-increasing spatial resolution. In this case, one begins with an equilateral triangle of side-length L and then successively attaches smaller and smaller equilateral triangles of size $L/3$, $L/3^2$, and so on to the middle of every straight-line segment. After N iterations, the perimeter consists of N segments of length r, where $r = L/3^N$ and

$$N = \alpha(L/r)^D \quad (5.11.1)$$

where $\alpha = 3$ and $D = \log 4/\log 3 \approx 1.262$ is called the fractal dimension. This dimension lies between $D = 1$ for a true one-dimensional curve and $D = 2$ for a true surface area. Figure 5.11.1(b) is an example of an area fractal called the *Sierpinski gasket* which finds use in studies of sediment porosity. Again, one begins with a triangle of side L but then cuts out successively smaller triangles of lengths $L/2$, $L/2^2$, and so on. After N iterations, the "pore" space between the sides of the triangles consists only of triangles of size $r = L/2^N$. The number of such triangles is given by equation (5.11.1) but with $\alpha = 1$ and $D = \log 3/\log 2 \approx 1.585$.

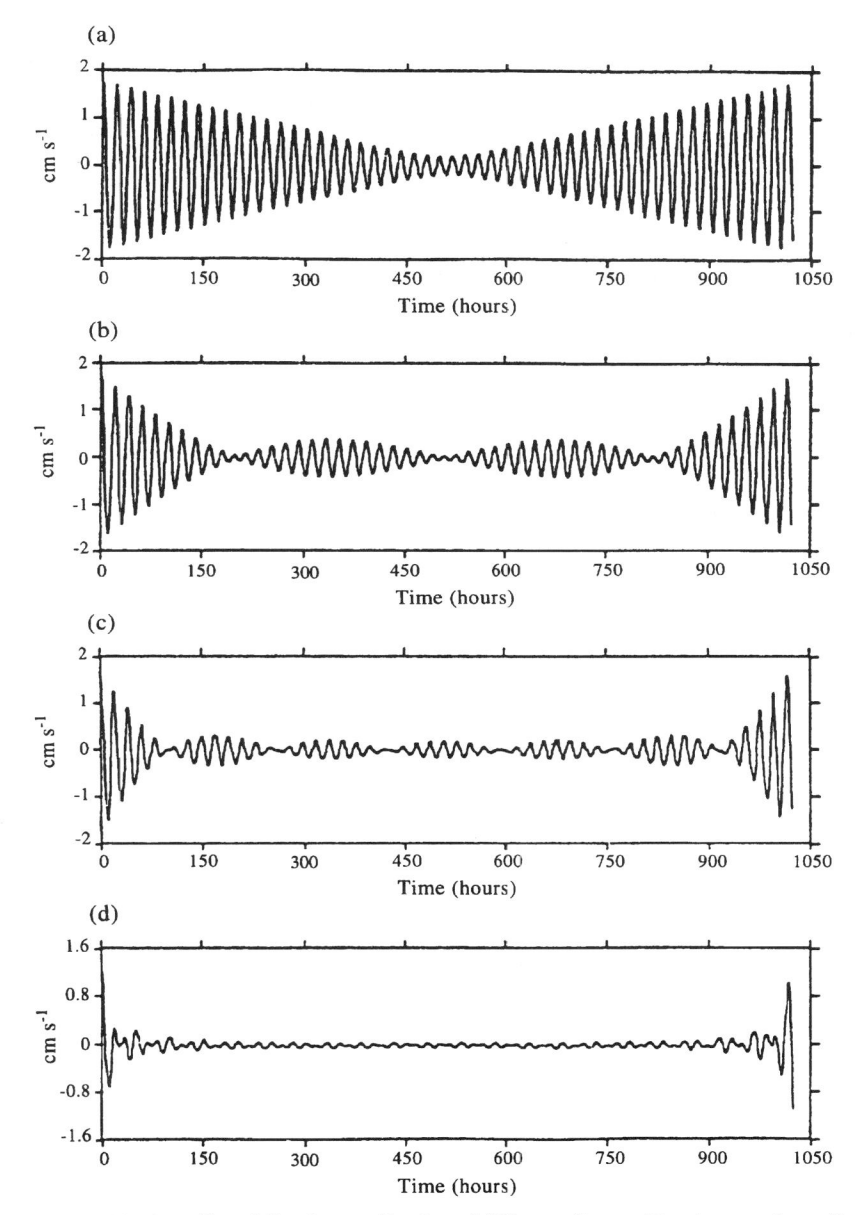

Figure 5.10.19. Ringing effects following application of different discrete Fourier transform filters to an artificial time series with frequency f = 0.05 cph and then inverting the transform. (a) Single Fourier coefficient at 0.05 cph set to zero; (b) three Fourier coefficients set to zero; (c) five Fourier coefficients set to zero; (d) 21 coefficients set to zero. (From Forbes, 1988.)

556   *Data Analysis Methods in Physical Oceanography*

The study of fractal geometry is related to the problem of predictability and propagation of order in nonequilibrium, frictionally dependent dynamical systems, such as turbulent flow in real fluids. In fluid systems, predictability is related to the rate at which initially close fluid particles diverge and the sensitivity of this divergence to initial conditions. Since low predictability implies a highly irregular dynamical system with sensitive dependence on initial conditions, the dispersion of tagged fluid parcels is related to the ultimate skill that can be achieved by deterministic numerical prediction models.

The fractal (or Hausdorff) dimension, D, provides a measure of the roughness of a geometrical object. For example, drifter trajectories confined to a horizontal plane can have a fractal dimension somewhere between that of a topological curve ($D = 1$) and that of random Brownian motion ($D = 2$). The case $D = 1$ is for a smooth differentiable curve whose length remains constant regardless of how the measurements are made. For fractal curves ($D > 1$), the length of the curve increases without bound for decreasing segment length. In the absence of a stationary mean flow, the track of a fluid parcel undergoing Brownian (random-walk) motion will eventually occupy the

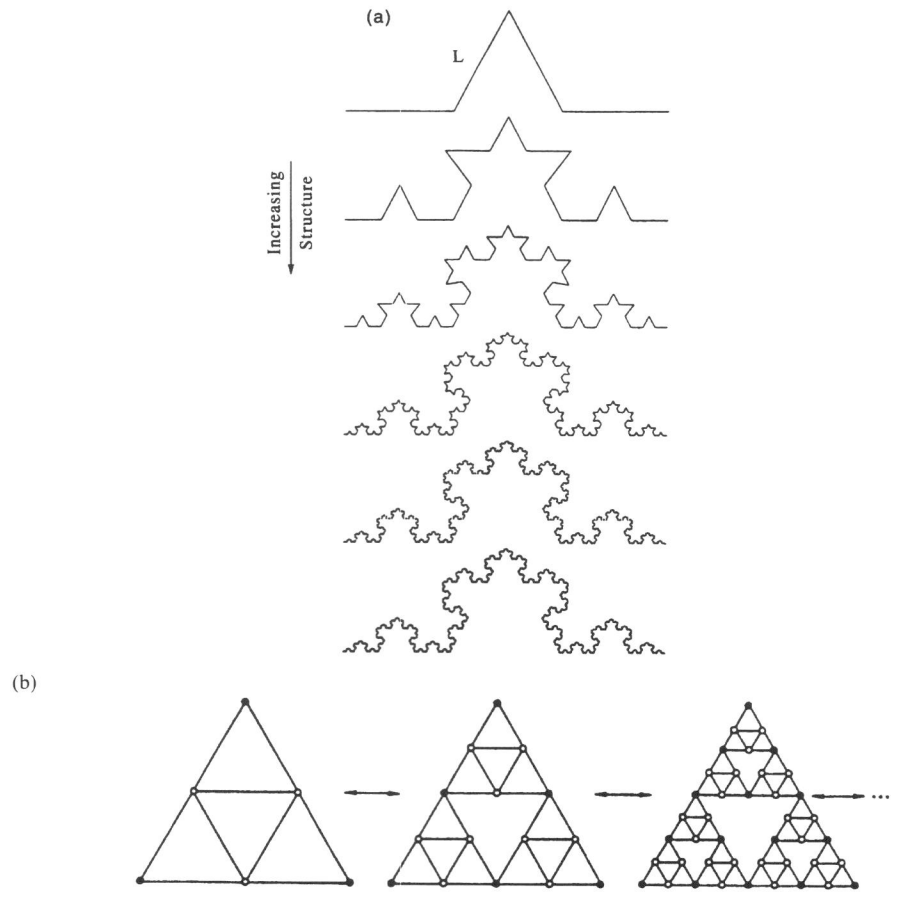

Figure 5.11.1. Examples of common fractals. (a) Generation of the Koch curve fractal by successive attachment of equilateral triangles; $D = 1.262$; (b) generation of the Sierpinski gasket fractal by successive removal of smaller triangles; $D = 1.585$.

entire horizontal plane available to it, whereas a parcel displaying fractal Brownian motion will not. The case $D < 2$ implies that the motion has inherent "memory" in the sense that a given incremental displacement in the fluid path is not independent of all previous displacements. In terms of dynamical systems, this means that there are a finite number of variables required to explain the dynamics of the fluid motions.

Osborne et al. (1989) examined the scaling properties of drifter trajectories for the upper ocean using year-long tracks of three satellite-tracked drifters deployed within the Kuroshio Extension region in 1977. Based on results from four fundamentally different fractal analysis methods, the Lagrangian trajectories were found to exhibit fractal behavior with dimension $D = 1.27 \pm 0.11$ over spatial scales of 20–150 km and temporal scales of 1.5 days to one week. These scales are thought to be representative of two-dimensional geophysical fluid dynamical turbulence within the inertial subrange—the eddy cascade region of self-similar turbulence which separates short-period current motions (daily tidal oscillations and inertial currents) from long-period oscillations such as Rossby waves and mean flows. Sanderson et al. (1990) have reported fractal dimensions at scales of 0.1–4 km for clusters of drifters deployed in Lake Erie, the Atlantic Equatorial Undercurrent, and in coastal waters off the south shore of Long Island. In a related study, the degree of chaotic behavior and predictability of the atmosphere has been studied using tropical and mid-latitude maritime cyclone tracks (Fraedrich and Leslie, 1989; Fraedrich et al., 1990). Results suggest that the atmosphere has an e-folding error growth rate of about 24 h and an ultimate predictability of eight to 14 days.

In this section, we provide several methods for determining the fractal characteristics of oceanic variability using particle track motions.

5.11.1 The scaling exponent method

Consider a particle track sampled at times (t) along the path $\mathbf{x}(t) = (x(t), y(t))$ in longitude–latitude (x–y) coordinates. Displacements along each of the two orthogonal horizontal axes are assumed to be independent self-affine (self-scaling) scalar functions. The scaling exponent H (which may be different for the two axes) is positive, less than or equal to unity and related to the fractal dimension of the function by $D = \min[1/H, 2]$. Brownian motions have scaling exponent $H = 1/2$ ($D = 2$) while monofractal scalar displacements exhibit fractional Brownian motions with $H > 1/2$ ($D < 2$). If the scalar series are sampled at equal time intervals, the exponents H_x, H_y are given by the *structure functions*

$$\overline{[x(t+\alpha\Delta t) - x(t)]^2} = \overline{[\Delta x(\alpha\Delta t)]^2}$$
$$= \alpha^{2H_x}\overline{[\Delta x(\Delta t)]^2} \qquad (5.11.2a)$$
$$= \alpha^{2H_x}\overline{[x(t+\Delta t) - x(t)]^2}$$

$$\overline{[y(t+\alpha\Delta t) - y(t)]^2} = \overline{[\Delta y(\alpha\Delta t)]^2}$$
$$= \alpha^{2H_y}\overline{[\Delta y(\Delta t)]^2} \qquad (5.11.2b)$$
$$= \alpha^{2H_y}\overline{[y(t+\Delta t) - y(t)]^2}$$

where overbars denote averages over time and the α are assigned integer values. The scaling exponents also can be found using the absolute value of the above functions (Osborne et al., 1989)

$$\overline{|y(t+\alpha\Delta t)-y(t)|} = \alpha^{2H_y}\overline{|y(t+\Delta t)-y(t)|} \qquad (5.11.2c)$$

$$\overline{|x(t+\alpha\Delta t)-x(t)|} = \alpha^{2H_x}\overline{|x(t+\Delta t)-x(t)|} \qquad (5.11.2d)$$

Figure 5.11.2 provides examples of the scaling exponents, H_y, derived from (5.11.2b) using one-year time series of 6-hourly meridional displacements of 120-m-drogued satellite-tracked drifters launched in the northeast Pacific in 1987. Part (a) of the figure is the log of the structure function

$$\overline{\{[y(t+\alpha\Delta t)-y(t)]^2\}}^{1/2}$$

versus $\log(\alpha)$. The slopes of these curves, H_y, are presented in part (b). Figure 5.11.3 is the same as Figure 5.11.2 except that it uses artificial drifter tracks generated from a Brownian motion (random-walk) algorithm. For the real drifter data, all four tracks had a constant fractal dimension $D_y = 1/H_y \approx 1.18 \pm 0.07$ over time scales of about 0.5–10 days. At longer time scales, motions were strongly affected by mesoscale eddies (cf. Thomson et al., 1990) and fractal analysis is no longer valid. For the pseudo-drifters, $D_y \approx 2$, which is what we would expect for a random-walk regime in which the drifters can occupy the entire two-dimensional space available to them.

Although confined to monofractal functions, the scaling dimension approach is attractive because it is computationally fast and defined in terms of simple scaling properties. The principal drawback is that irregularly sampled particle trajectories, such as those of satellite-tracked drifters, must be converted to equally spaced data using a spline or other interpolation scheme. For isotropic monofractal trajectories, a single fractal dimension is sufficient to define the overall scaling properties of the motions including scaling properties of the mean, variance, and higher moments. Anisotropy in the drifter motions may lead to significantly different values for the scaling exponents H_x, H_y, and associated fractal dimensions. Where these differences are small, fractal dimensions can be expressed through a mean scaling exponent, $\overline{H} = \frac{1}{2}(H_x + H_y)$.

5.11.2 The yardstick method

The fractal dimension of a drifter trajectory of length $L(\Delta)$ can be measured in the usual sense using a ruler (or *yardstick*) with variable length, Δ. As the length of the ruler is decreased and the yardstick estimation of the total length becomes more precise, the length of the trajectory will follow a power-law dependence

$$L(\Delta) \approx \Delta^{1-D_L}; \quad \lim \Delta \to 0 \qquad (5.11.3)$$

The divider dimension D_L, which closely approximates the fractal dimension D, is found from the slope of log-transformed $L(\Delta)$ for small length scales Δ (Figure 5.11.4). The case $D_L = 1$ is the topological dimension for a smooth differential curve. For fractal dimensions, $D > 1$ and the length of the curve increases without bound for decreasing segment length.

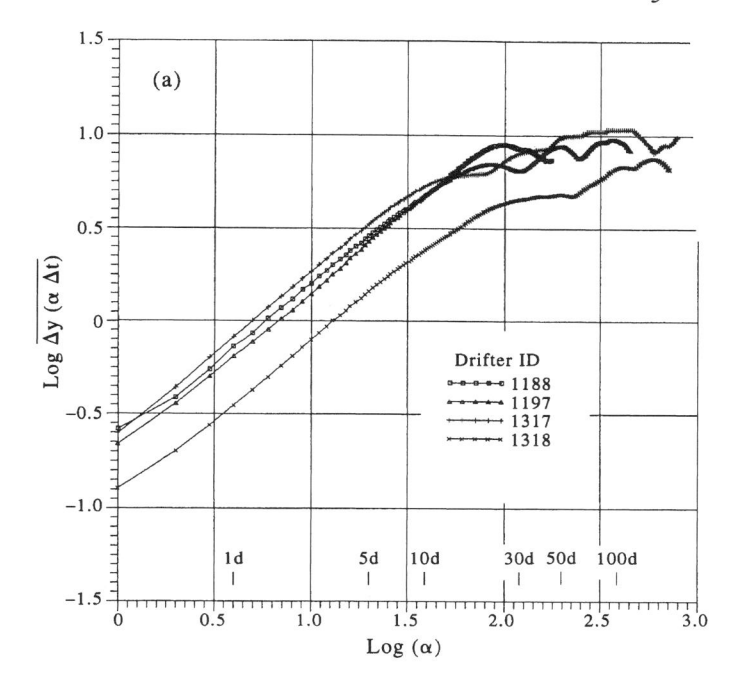

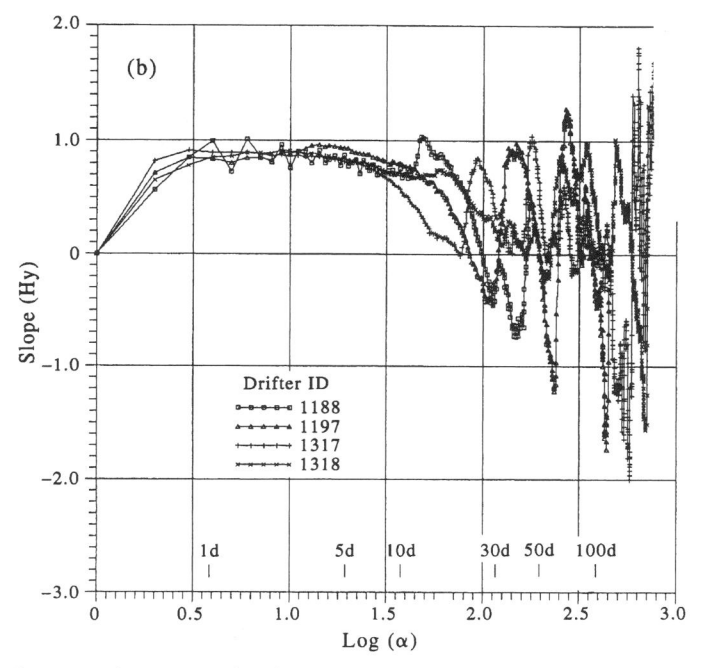

Figure 5.11.2. Structure functions and scaling exponents for trajectories of four 6-hourly sampled, 120-m-drogued satellite-tracked drifters launched in the northeast Pacific in 1987. (a) Absolute values of the structure functions versus the scaling factor, α, plotted on a log–log scale. (b) Slopes, H_y, of the curves in (a) versus scaling factor. Slopes were roughly equal and constant over time scales of one to 10 days.

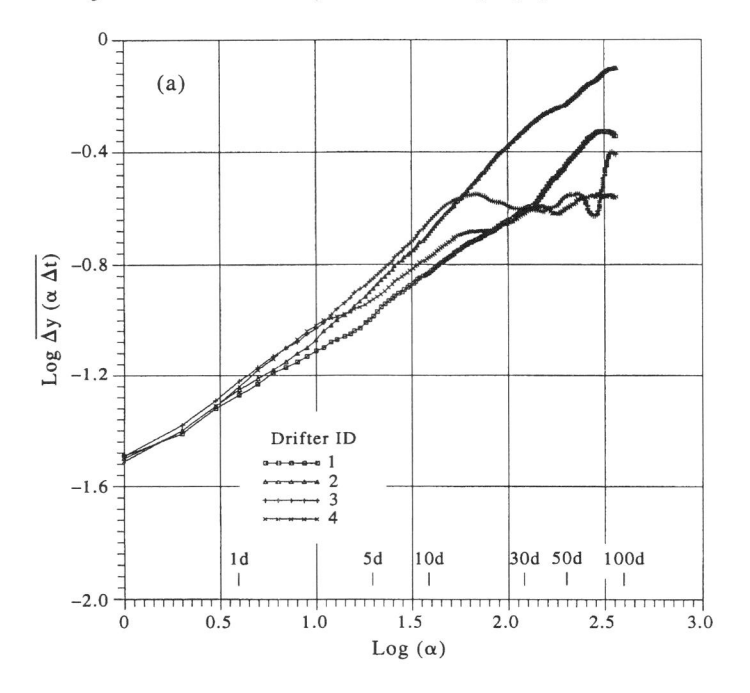

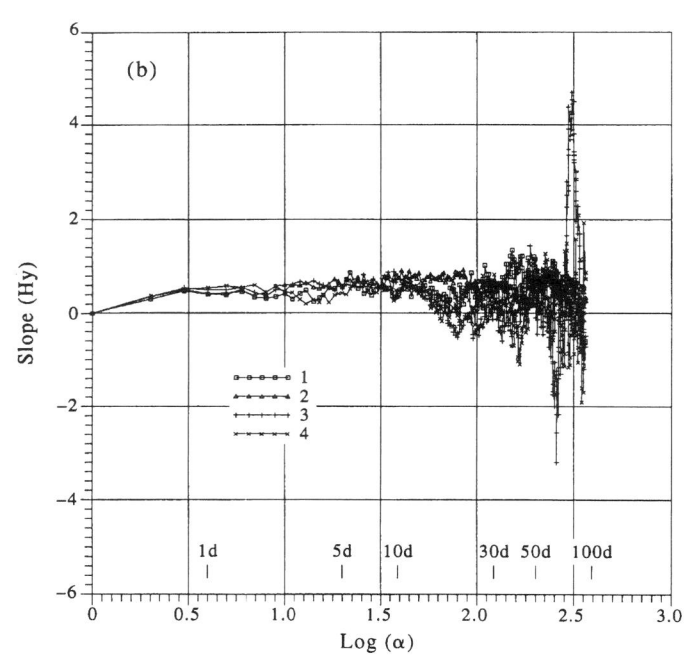

Figure 5.11.3. As in Figure 5.11.2 except for pseudo-drifter tracks generated using a random number generator. In this case, $H_y \approx 0.5$ and drifters perform a non-fractal random walk with dimension $D \approx 2$.

A problem with applying equation (5.11.3) to irregularly sampled drifter records is that the data are unequally sampled both in time and space. Although it makes sense to use a spline-interpolation scheme to generate scalar coordinate data with equally spaced time increments, it is less meaningful to generate coordinate series with equally sampled positional increments. The reason is simple enough: Time is single-valued whereas location is not. Drifters often loop back on themselves. If the data are not equally spaced, we cannot define a sequence of fixed-length yardsticks but must measure the curve $L(\Delta)$ as a function of the average yardstick length, Δ_{av}. This averaging is valid provided the errors introduced by the averaging process are no worse than those arising from other sources (cf. Osborne et al., 1989). Another problem with the yardstick method is that it is based on the slope of (5.11.3) for small spatial scales. The measurement of these scales is often difficult in practice due to limitations in the response and/or positioning of the drifters, cyclone, or other Lagrangian particle.

5.11.3 Box counting method

In this method, one counts the number $N_m(L)$ of boxes of length L in m-dimensional space that are needed to cover a "cloud" or set of points in the space. The Hausdorff–Besicovich dimension, D, of this set can be estimated by determining the number of cubes needed to cover the set in the limit as $L \to 0$. For a fractal curve, the number of boxes increases without bound as $L \to 0$. That is

$$N_m(L) \to L^{-D}, \quad L \to 0 \qquad (5.11.4)$$

If the original series is random, then $D = n$ for any dimension n (a random process embedded in an n-dimensional space always fills that space). If, however, the value of D becomes independent of n (i.e. reaches a saturation value, D_0, say), it means that the system represented by the time series has some structure and should possess an attractor whose Hausdorff–Besicovitch dimension is equal to D_0. Once saturation is reached, extra dimensions are not needed to explain the dynamics of the system.

As an example, if we were to measure the area of surfaces embedded in three-dimensional space, we would count the number $N_3(L)$ of cubic boxes of size L required to cover the surface. The area S is then of order

$$S \approx N_3(L)L^2 \qquad (5.11.5)$$

For a nonfractal surface, the area asymptotes to a constant value independent of L, which is the true area of the surface. In general

$$N_3(L) \approx L^{-D}, \quad S \approx L^{2-D} \qquad (5.11.6)$$

5.11.4 Correlation dimension

An important method for determining the self-similarity of monofractal curves has been proposed by Grassberger and Procaccia (1983). The technique also has found widespread use in studies of chaos and the dimensionality of strange attractors. Specifically, one determines the number of times that the computed distances d_{ij} between points in a time series $x(t_i)$ (or pair of time series $x_i(t)$ and $x_j(t)$) are less than a prescribed length scale, ε. That is, one finds what fraction of the total number of

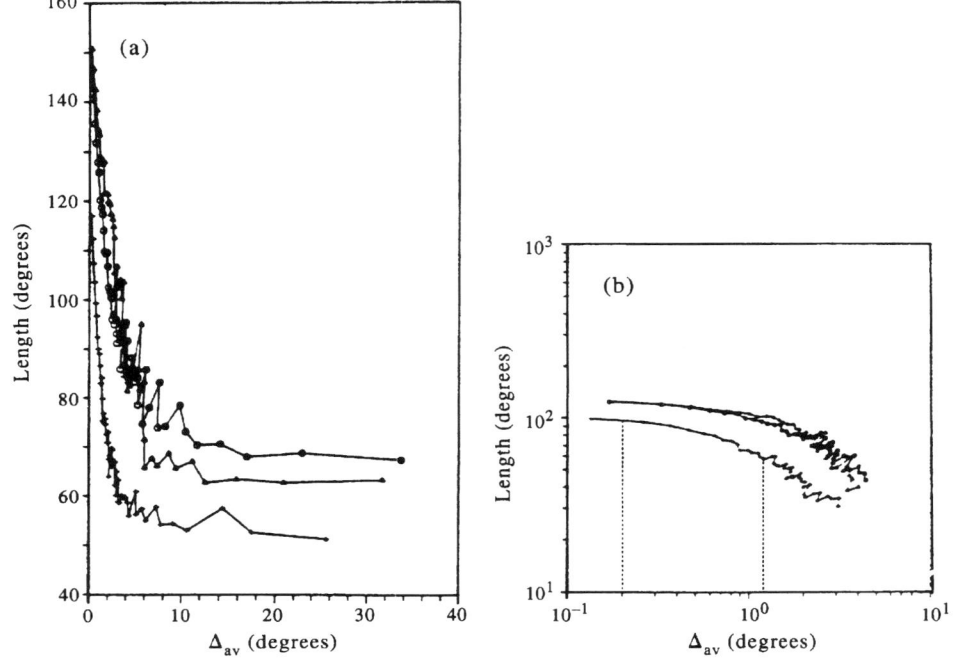

Figure 5.11.4. Yardstick length $L(\Delta)$ measured using a ruler with variable average yardstick length, Δ_{av} (in degrees of latitude), for three drifters launched in the Kuroshio Extension in 1977. (a) Linear coordinates; and (b) log–log coordinates. Note the divergence of the lengths for small Δ. (From Osborne et al., 1989.)

possible estimates of the distance $d_{ij} = |x(t_i) - x(t_j)|$ that are less than ε. For a single discrete vector time series, the Grassberger–Procaccia correlation function is defined as

$$C(\varepsilon) = \frac{1}{M(M-1)} \sum_{i,j}^{M} H\big[\varepsilon - |x(t_i) - x(t_j)|\big], \quad M \to \infty \qquad (5.11.7)$$

where $H(\varepsilon, r_{ij})$ is the Heavyside step function ($= 0$ for $\varepsilon < r$; $= 1$ for $\varepsilon > r$) and M is the number of points in the time series. In (5.11.7), the vertical bars denote the norm of the vector $d_{ij} = [(x(t_i) - x_j)^2 + (y(t_i) - y_j)^2]^{1/2}$. The fractal dimension for a self-affine curve is then obtained as the correlation dimension defined by

$$C(\varepsilon) \approx \varepsilon^{\nu}, \quad \varepsilon \to 0 \qquad (5.11.8)$$

The fractal dimension is obtained from the log-transformed version of this equation (Figure 5.11.5). According to Osborne et al. (1989), the correlation method gives the least uncertainty in the estimate of the fractal dimension whereas largest errors are associated with the exponent scaling method.

5.11.5 Dimensions of multifractal functions

The various techniques discussed above will (within statistical error) give the same fractal dimension provided that the series being investigated exhibits self-similar monofractal behavior. However, because the techniques rely on different assumptions and measure different scaling properties of the series, the calculated dimensions will be different if the series has a multifractal structure. Multifractal properties are related to multiplicative random processes and are associated with different scaling properties at different scales.

A form of box-counting can be used to study the multifractal properties of ocean drifters (Osborne et al., 1989). Given a fractal curve on a plane, the plane is covered with adjacent square boxes of size Δ and the probability, $p_i(\Delta)$, is computed that the ith box contains a piece of the fractal curve

$$p_i(\Delta) = \frac{n_i(\Delta)}{N} \tag{5.11.9}$$

where n_i is the number of data points falling in the ith box and N is the total number of points in the time series. For fractal curves for small Δ

$$\sum_i [p_i(\Delta)]^q \approx \Delta^{(q-1)D} \tag{5.11.10}$$

where the sum is extended over all nonempty boxes. The quantities $D = D_q$ are the generalized fractal dimensions. A fundamental difference between monofractals and multifractals is that for monofractals D_q is the same for all q while for multifractals the different generalized dimensions are not equal. In general, $D_q < D_{q'}$ for $q > q'$.

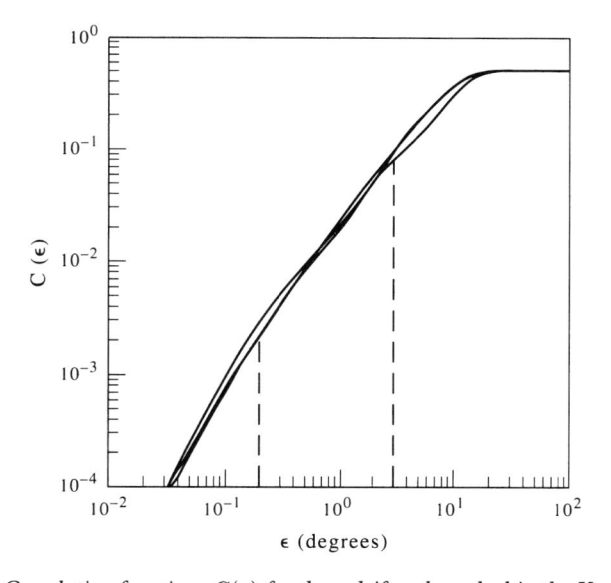

Figure 5.11.5. Correlation functions $C(\varepsilon)$ for three drifters launched in the Kuroshio Extension in 1977. The slope of the function in log–log coordinates is a measure of the correlation dimension of the signal. The two vertical lines indicate the approximate limits of the scaling range. (From Osborne et al., 1989.)

564 *Data Analysis Methods in Physical Oceanography*

5.11.6 Predictability

A box-counting method can be used to investigate the degree of chaotic behavior associated with the Lagrangian motions such as those of drifters and tropical cyclones. In this method, one counts the number $N_n(\Delta)$ of boxes of dimension Δ in n-dimensional space needed to cover a "cloud" or set of points in the space in the limit $\Delta \to 0$. In practice, the box-counting method is difficult to apply. Estimates of the predictability of drifter trajectories are more readily obtained using the correlation integral technique of Grassberger and Pocaccia (1983). In this case, the degree of predictability is found from the dimension of the attractor derived from an embedded phase space created from all possible pairs of "drifters". The phase space serves, in turn, as a substitute for the state space needed to study the dynamics of a system (Tsonis and Elsner, 1990).

The analysis takes the following steps: (1) we first consider a pair of independent tracks of length $m\Delta t$, where m is the embedding dimension and Δt the sampling increment. Specifically, consider the cyclone tracks for Australia for July 1982 and 1983 (Figure 5.11.6a) examined by Fraedrich *et al.* (1990). For convenience, the start times and positions of the tracks are reinitialized so that they begin at the same time and location. Fraedrich and Leslie (1989) found that the errors introduced by reinitializing are less than those from other sources; (2) we next examine the divergence of the paths by calculating the multiple track correlation function (or correlation integral) $C_m(\varepsilon)$ for the particular embedding dimension m and path separation scale, ε. To this end, we count the number tracks $N_m(\varepsilon)$ of length $m\Delta t$ for which the track length remains less than the great circle distance ε for all the segments in the track. For $m = 1$, each individual data point forms a unit-length segment of the drifter track. One then counts the number of times, $N_1(\varepsilon)$, that the distance between the drifter positions is less than ε for the $N = m$ possible drifter tracks. The distance between each drifter pair is considered; hence, for 10 drifters or cyclone tracks there would be $10 \times 10 = 100$ pairs. This process is repeated for all values of m to create a cloud of points in m-dimensional space which then approximate the dynamics of the system from which the observations $x(t)$ are drawn. The correlation integral is defined by

$$C_m(\varepsilon) = \frac{N_m(\varepsilon)}{[N_m - 1]^2} \qquad (5.11.11)$$

where $N_m(\varepsilon)$ is the number of pairs of trajectories of dimension m that remain less than a distance ε from one another. Note that the numerator in the above expression is a squared quantity since it is based on the number of drifter pairs; (3) we then plot $\log[C_m(\varepsilon)]$ versus $ln(\varepsilon)$ to find the slope D_2 of the curve

$$C_m(\varepsilon) \approx \varepsilon^{D_2}, \quad \varepsilon \to 0 \qquad (5.11.12)$$

The subscript "2" indicates that pairs of points are used to create the phase space.

If both original time series are random, then $D_2 = 2m$. A random process embedded in a $2m$-dimensional space always fills that space. On the other hand, if D_2 becomes independent of m at some saturation value, D_0, it means that the system represented by the time series has some structure (i.e. predictability) and should possess an attractor whose Hausdorff–Besicovitch dimension is equal to D_0 (Figure 5.11.6b). The need to calculate D_0 from the observations arises because we do not know the value of

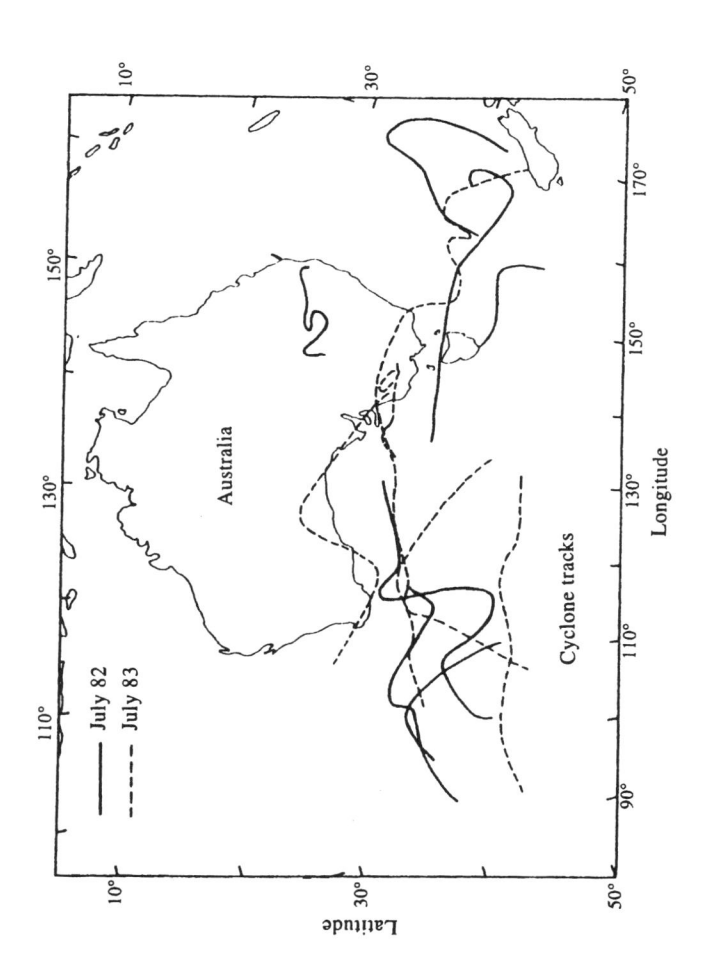

Figure 5.11.6. Use of fractals to study the predictability of cyclone tracks. (a) Cyclone tracks for Australia in July 1982, 1983. (From Fraedrich et al., 1990.)

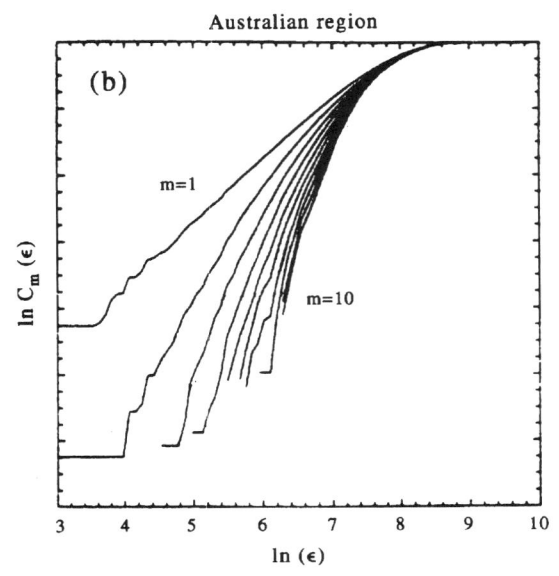

Figure 5.11.6. Use of fractals to study the predictability of cyclone tracks. (b) Correlation integral or cumulative distance distributions $C_m(\varepsilon)$ of pairs of independent cyclone trajectory pieces versus $\ln(\varepsilon)$. Each curve is for m times the data time step of 24 h (m = 1–10 from left to right in the figure). For increasing m, structure eventually becomes invariant at highest embedding dimensions, an indication that extra variables are not needed to account for the dynamics of the system. (From Fraedrich et al., 1990.)

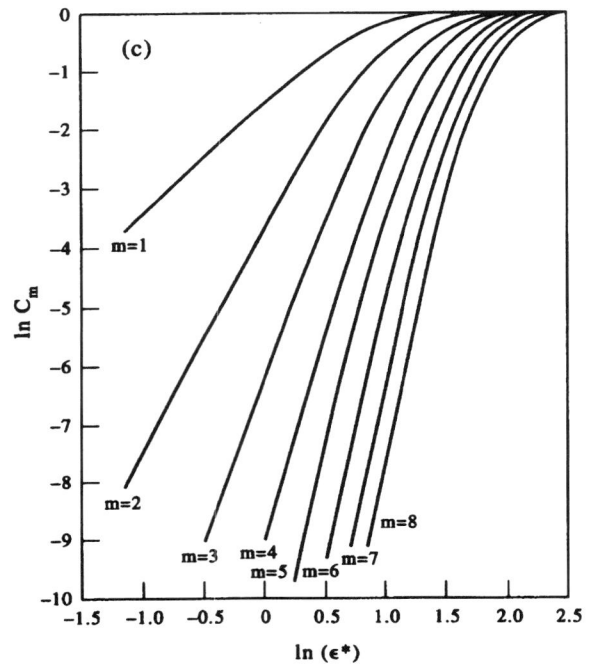

Figure 5.11.6. Use of fractals to study the predictability of cyclone tracks. (c) Same as (b) but for a random-walk pseudo-cyclone generated using a random-number generator. The slopes approach $D_2 \to 2m$ for decreasing distance threshold, ε. (From Fraedrich et al., 1990.)

m a priori. We, therefore, calculate D_0 for increasing m until we approach a structure that becomes invariant at higher embedding dimensions, an indication that extra variables are not needed to account for the dynamics of the system. The attractor can be a topological structure such as a point, limit cycle or torus, or a nontopological submanifold with fractal structure. For a random-walk regime, D_2 approaches $2m$ so that there is no corresponding limiting value, D_0.

The independent segments of the paired drifter trajectories of sufficiently long duration embed the attractor in a substitute phase space spanned by the time-lagged coordinates provided by the data. The correlation dimension D_2 measures the spatial correlation of the points that lie on the attractor. For a random time series there will be no such spatial correlation in any embedding dimension and thus no saturation will be observed in the exponent D_2. We note that the dimensionality of an attractor, whether fractal or nonfractal, indicates the minimum number of variables present in the evolution of the corresponding dynamical system. In other words, the attractor must be embedded in a state space of at least its dimension. Therefore, the determination of the Hausdorff dimension of an attractor sets a number of constraints that should be satisfied by any numerical or analytical model used to predict the evolution of the system. The main concern is that we do not extend the interpretation when going from a densely populated low-dimensional space to a sparsely occupied high-dimensional space. We cannot go beyond the critical embedding dimension above which the scaling region cannot be accurately determined (Essex *et al.*, 1987; Tsonis and Elsner, 1990).

Appendices

APPENDIX A: UNITS IN PHYSICAL OCEANOGRAPHY

Length	1 micrometer (μm; micron) = 10^{-3} millimeter (mm) = 10^{-6} meter (m) 1 centimeter (cm) = 10 mm = 10^{-2} m = 0.3937 inches (in) 1 meter (m) = 10^2 cm = 39.37 in = 3.281 feet (ft) = 1.094 yards (yd) 1 kilometer (km) = 10^3 m = 0.5396 nautical mile (naut mi) = 0.6214 statute mile (mi) 1 nautical mile = 1 minute latitude = 6080 ft = 1.152 mi = 1.8532 km 1° latitude = 111.19 km; 1° longitude = 111.19 cos|(latitude)|km At 45°; 1° longitude = 78.62 km 1 cable = 0.1 naut mi = 608 ft = 185.3 m 1 fathom (fm) = 6 ft = 1.8288 m 1 league = 3040 fathoms = 3 naut mi 1 inch (in) = 2.54 cm; 12 in = 1 ft; 36 in = 1 yard
Area	1 square kilometer (km^2) = 10^6 m^2 = 100 hectares (ha) = 0.386 mi^2 *1 ha = 2.471 acres (ac)* 1 square mile (mi^2) = 640 ac = 259 ha
Volume	1 cubic meter (m^3) = 264.2 US gallon (gal) = 220.0 imperial gal = 35.314 ft^3 1 litre (l) = 10^3 ml = 10^{-3} m^3 = 0.264 US gal = 0.220 imperial gal 1 barrel (oil; bbl) = 42 US gal = 0.159 m^3 = 158.987 litres (l) 1 US gal = 0.83 imperial gal
Time	*1 hour (h) = 3.6 $\times$ 10^3 seconds (s)* *1 solar day = 24 h = 8.64 $\times$ 10^4 s* *1 sidereal day = 23 h, 56 min, 4 s*
Mass	1 gram (g) = 0.03527 ounces (oz) = 0.03215 troy ounces 1 kilogram (kg) = 10^3 g = 2.205 lb; 1 pound (lb) = 0.4536 kg 1 metric ton (tonne) = 10^6 g = 2205 lb = 1.1025 ton 1 ton = 2000 lb
Pressure	*1 pascal (Pa) = 1 newton/m^2 (N/m^2) = 10^{-5} bar = 10^{-4} decibar (dbar)* *1 atmosphere (atm) = 1.01325 $\times$ 10^5 Pa* *1 bar = 0.98692 atm = 10^5 Pa = 1.02 kg/cm^2* *1 millibar (mb) = 10^{-3} bar = 10^{-1} kPa* *1 kPa = 10^3 Pa = 10^{-1} dbar = 10 millibar (mb)*
Stress	1 dyn/cm^2 = 10^{-1} N/m^2 1 kg/cm^2 = 0.96784 atm = 14.2233 lb/in^2
Speed	*1 knot (nautical mi/h; kn) = 0.5148 m/s = 51.48 cm/s = 44.48 km/day* *1 meter per second (m/s) = 2.24 statute mi/h = 1.943 knots = 86.4 km/day* *1 (statute) mi/h = 1.609 km/h = 0.868 knots*
Temperature	°F (Fahrenheit) = 9/5° $\times$ Celsius + 32 °C (Celsius) = (°F − 32) $\times$ 5/9 K (kelvin) = °C + 273.15 (0 K = absolute zero)
Dissolved O$_2$	*1 milliliter per liter (ml/l) = 1.43 mg/l* *1 ml/l $\approx$ 43.3 micromole/kg (μmol/kg) for S = 34.7 psu, T = 3.5°C, σ_t = 27.96*
Earth rotation	ω = 0.72921 $\times$ 10^{-4} rad s^{-1} = 0.04178 cycles/h (cph)

Appendix A: *Continued.*

Gravitational acceleration	$g = 9.81 \text{ m/s}^2 = 981 \text{ cm/s}^2 = 32.1722 \text{ ft/s}^2$
Force	1 newton (N) = 1 kg m/s^2 = 10^5 dynes (dyn) = 2.2 lb
Energy and power	1 (thermochemical) calorie (cal) = 4.184 joules (J) = 3.968×10^{-3} British thermal units (BTU; Btu; @ 60°F) 1 J = 1 newton meter (N m) = 10^7 ergs = 0.2390 cal = 2.78×10^{-7} kW h 1 watt (W) = 1 joule per second (J/s) = 1.341×10^{-3} horsepower (hp) 1 kilowatt hour (kW h) = 3.6×10^6 J = 3.41×10^3 Btu 1 hp = 7.457×10^2 W
Geopotential	1 dynamic centimeter (dyne cm) = 10^3 cm^2 s^{-2} ≈ 1.02 cm 1 joule/kg (J/kg) = 1 m^2 s^{-2} = 10 dyne cm Specific volume anomaly: $\delta = \alpha_{S,T,P} - \alpha_{35,0,P}$ where $\alpha = \rho^{-1}$ m^3/kg. Geopotential anomaly (or *dynamic height anomaly* in the older literature): $$\Delta\Phi \text{ (or } \Delta D\text{)} = \Phi_{P_2} - \Phi_{P_1} = -\int_{P_1}^{P_2} \delta \, dP \text{ joules/kg}$$ Potential energy: $$PE = g^{-1} \int_{P_1}^{P_2} \delta(P) P \, dP \text{ joules/m}^2$$
Transport	1 sverdrup (Sv) = 10^6 m^3 s^{-1}

APPENDIX B: GLOSSARY OF STATISTICAL TERMINOLOGY

Alternative hypothesis: Value of a parameter of a population other than the value hypothesized or believed to be true by the investigator.

Asymptotically normal distribution: A distribution of values which is not truly normal, but which approaches a normal distribution as the number of samples becomes very large.

Autocorrelation: In a time series, $x(t)$, with zero mean, the statistical relationship between values of a variable taken at certain times in the series and values of a variable taken at other times (a function of the time lag, τ, between the two series) written as

$$R_{xx}(\tau) = E[x(t)x(t+\tau)]$$

where E is the expected value. Autocovariance is similar except that the mean of the record is not subtracted prior to the analysis.

Biased estimator: An estimator $\hat{x}$ which results in the expected value $E[\hat{x}]$ of a sample having a systematic error with respect to the true expected value; i.e. $E[\hat{x}] \neq \mu_x$.

Bin interval: A specified arbitrary interval which partitions a quantity whose number of occurrences are being measured; used for constructing a histogram (frequency of occurrence distribution) of the data set.

Central limit theorem: States that the distribution of sample means taken from a large population approaches a normal (Gaussian) distribution.

Chi-squared distribution: The distribution function generated by the random variable

$$\chi_n^2 = X_1^2 + X_2^2 + X_3^2 + \ldots + X_n^2$$

where $X_1, X_2, \ldots, X_n$ are n independent random variables drawn from a normal population with zero mean. The chi-squared variable χ_n^2 has an expected value (mean) of n and a variance of $2n$.

Confidence interval: An interval which has a specified probability of containing a given parameter or characteristic.

Continuous random variable: A random variable which may be characterized as a continuous function.

Correlation (covariance): A quantitative measure of the interdependence or association between two variables.

Countable: Either finite or denumerable.

Cross-correlation: The correlation between corresponding members of two or more different series of the same duration: if $x(t) = (x_1, x_2, \ldots, x_n)$ and $y(t) = (y_1, y_2, \ldots, y_n)$ are two series, the cross-correlation is the correlation between $x(t)$ and $y(t)$, or between $x(t)$ and $y(t+t)$ for lag t such that

$$R_{xy}(\tau) = E[x(t)y(t+\tau)]$$

(see *Autocorrelation*).

Cumulative distribution function: The integral of the probability density function, $p(x)$, over some specified interval, (x_1, x_2). The integral or sum of $p(x)$ from $-\infty$ to x gives the cumulative total of all values whose value is less than or equal to x.

Degrees-of-freedom: The number of truly independent samples used to estimate a parameter; when each sample of a series of length N is independent of all other values, the degrees of freedom, $\nu = N - 1$.

Discrete random variable: A random variable which may be characterized as a discrete function.

Ensemble average: An average over several realizations of a random variable taken over times of equal duration.

Ergodic hypothesis: The hypothesis that replaces statistical replications (ensemble averages) with samples in space or time. Allows us to compute averages in space or time as ensemble averages.

Estimator bias: The difference between an estimate and the true value of the parameter being estimated.

Expected value: For a random variable x with a probability function $f(x)$, this is the integral from $-\infty$ to ∞ of $xf(x)$; also known as the expectation and written as

$$E(x) = \int_{-\infty}^{\infty} xf(x)\,dx = \mu$$

Gamma distribution: A normal distribution whose probability distribution function involves the gamma function as

$$f(x) = \frac{x^{\alpha-1} e^{-x/\beta}}{\beta^\alpha \Gamma(\alpha)}, \quad \alpha, \beta > 0;\ 0 \leq x \leq \infty$$
$$= 0 \quad \text{elsewhere}$$

where α and β are parameters of the distribution and $\Gamma(\alpha)$ is the gamma function

$$\Gamma(\alpha) = \int_0^\infty x^{\alpha-1}\,e^{-x}\,dx$$

Gauss–Markov theorem: An unbiased linear estimator of a parameter has the minimum variance, that is, the best estimator when it is determined by the least squares method.

Hypothesis testing: The branch of statistics which considers the problem of choosing between two actions on the basis of the observed value of a random variable whose distribution depends on a parameter, the value of which would indicate the correct action.

Independent random variables: The discrete random variables, $X_1, X_2, ..., X_n$ are independent if for arbitary values $x_1, x_2, ..., x_n$ of the variables the probability that $X_1 = x_1, X_2 = x_2, ...$ is equal to the product of the probabilities that $X_i = x_i$, for $i = 1, 2, ..., n$; in this case, the random variables are unrelated.

Inference: The act of passing from statistical sample data to generalizations (as of the value of true population parameters) with calculated degrees of certainty (confidence intervals at selected significance levels).

Joint probability density function: The distribution which gives the probability that $X_i = x_i$, for $i = 1, 2, ..., n$ for all values x_i of the random variable X_i.

Jointly sufficient statistics: Let $X_1, X_2, ..., X_n$ be a random sample from a probability density function $p(X; \theta)$, where θ is an unknown statistical parameter such as the mean or variance; the statistics $S_1, ..., S_r$ are defined to be jointly sufficient if and only if the conditional distribution of $X_1, X_2, ..., X_n$, given $S_1 = s_1, ..., S_r = s_r$ does not depend on θ.

Likelihood $L(x)$: The likelihood of occurrence of a sample of independent values of $x_1, x_2, ..., x_n$, with a probability function $f(x)$, is the product $f(x_1)f(x_2), ..., f(x_n)$.

Likelihood ratio: The probability of a random drawing of a specified sample from a population, assuming a given hypothesis about the parameter of the population, divided by the probability of a random drawing of the same sample assuming that the parameters of the population are such that this probability is maximized; i.e. $L(x_1)/L(x_2)$.

Least-squares method: A technique for fitting a line, polynomial or other curve to a given distribution of points which minimizes the sum of the squares of the deviations of the given points from the fitted curve.

Linear regression: The straight line running through the points of a scatter diagram about which the amount of scatter is a minimum in the least squares sense.

Maximum likelihood estimation: A method whereby the likelihood distribution is maximized to produce an estimate of the random variable.

Mean square error: A measure of the extent to which a collection of numbers, $x_1, x_2, ..., x_n$, is unequal and defined by the expression

$$\text{mean square error} = E[(\hat{x} - x)^2] = E[(\hat{x} - E[\hat{x}])^2] + E[(E[\hat{x}] - x)^2]$$

where $E[x]$ is the expected (mean) value and $\hat{x}$ is the estimator for x.

Method of moments: A procedure for estimating the parameters of a distribution.

Minimal sufficient statistic: A set of jointly sufficient statistics is defined to be minimal sufficient if and only if it is a function of every other set of sufficient statistics.

Moments: The nth moment of a distribution $f(x)$ about a point x_0 is the expected value $E[f^n(x - x_0)]$; the first moment is the mean of the distribution, while the variance is a function of the first and second moments (this definition only applies to moments about the mean and not the origin).

Moment generating function: Let x be a random variable with probability density function $p(x)$; the expected value of some function $f(rx)$, $E[f(rx)]$, is defined to be the moment generating function of f if the expected value exists for every value of r in some interval $-\eta < r < \eta; \eta > 0$

$$m(r) = E[f(rx)] = \int_{-\eta}^{\eta} f(rx)p(x)\,dx$$

Moment of a power spectrum, $S(f)$: Defined as $m_n = \int_0^\infty S(f)f^n\,df$ where f is the frequency. In the case of surface gravity waves, the significant wave height $H_s = 4m_0^{1/2}$ and the *mean zero crossing period* is $T = (m_0/m_2)^{1/2}$.

Multivariate analysis: The study of random variables which are multidimensional.

Normal (or Gaussian) probability distribution function: A normally distributed frequency distribution of a random variable x with a mean μ and standard deviation σ given by

$$p(x) = \frac{1}{\sigma\sqrt{2\pi}} \exp\left[\frac{(x - \mu)^2}{2\sigma^2}\right]$$

Null hypothesis: The hypothesis that there is no validity to the specific claim that two variations (treatments) of the same thing can be distinguished by a specific procedure.

Pivotal statistic: The statistic that allows one to compute a confidence interval for a specific estimate.
Population: Any finite or infinite collection of individuals or elements that can be specified or labeled.
Population distribution: The distribution that characterizes a population; may be displayed using a histogram or frequency of occurrence diagram.
Population mean and variance: The population mean, μ, is the arithmetic average of values x_i ($i = 1, ..., N$) obtained from all members in a population of size N based on the measurement of some quantity, x, associated with each member; population variance is the arithmetic average of the numbers $(x_i - \mu)^2$.
Population moment: The rth moment associated with a particular population.
Probability: The probability of an event is the ratio of the number of times it occurs to the (larger) number of trials that take place.
Probability density function (PDF): A real-valued function whose integral over any set gives the probability that a random variable has values in this set.
Random variable: A well-defined function that allows us to assign a real number to any outcome of an experiment. Specifically, the random outcome of a particular experiment, indexed by k, can be represented by a real number x_k called the random variable.
Relative frequency distribution: A frequency distribution in which the individual class frequencies are expressed as a fraction of the total frequency range.
Sample: A selection of values from a larger collection of values.
Sample mean and variance: The mean value $\bar{x}$ and variance s of a sample taken from a given data set, X. In general these differ from the true mean, μ, and variance, σ, of the population.
Sample moment: The moment of a sample taken from a given set of samples.
Significance level: The probability of a false rejection of the null hypothesis.
Standard error: A measure of the variability any statistical constant would be expected to show in taking repeated random samples of a given size from the same universe of observations.
Standard normal variable: A normal (Gaussian) distributed random variable with specified mean and standard deviation which has been transformed to a random variable with a mean of zero $(\bar{X} = 0)$ and a standard deviation of unity $(s = 1)$.
Stationarity: The property by which the statistics of a random variable do not change with time. For a stationary time series, quantities such as the mean and variance are nearly identical for different segments of the record.
Student's t-distribution: A probability distribution used to test the hypothesis that a random sample of n observations comes from a normal population with a given mean.
Sufficiency: Condition of an estimator that uses all the information about the population parameter contained in the sample observations.
Sufficient statistics: Let $X = (x_1, x_2, ..., x_n)$ be a random sample from the probability density function $p(x; \theta)$; a statistic $S = s(x_1, x_2, ..., x_n)$ is defined to be a sufficient statistic if and only if the conditional distribution of X given S, does not depend on θ for any statistic $R = r(x_1, x_2, ..., x_n)$.
Tschebysheff's theorem: Given a nonnegative random variable $f(x)$, and $k > 0$, the probability that $f(x) \geq k\sigma$ is less than or equal to the expected value of f divided by k^2.
Unbiased estimator: An estimate $\hat{\theta}$ for a parameter θ whose expected value is $E[\hat{\theta}] = \theta$.
Uniform probability density function: The distribution of a random variable in which each value has the same probability of occurrence.

APPENDIX C: MEANS, VARIANCES AND MOMENT-GENERATING FUNCTIONS FOR SOME COMMON CONTINUOUS VARIABLES

Distribution	Probability function	Mean	Variance	Moment-generating function
Uniform	$f(x) = \dfrac{1}{\theta_2 - \theta_1}; \; \theta_1 \leq x \leq \theta_2$	$\dfrac{\theta_2 + \theta_1}{2}$	$\dfrac{(\theta_2 - \theta_1)^2}{12}$	$\dfrac{e^{t\theta_2} - e^{t\theta_1}}{t(\theta_2 - \theta_1)}$
Normal	$f(x) = \dfrac{1}{\sigma\sqrt{2\pi}} \exp\left[-\left(\dfrac{1}{2\sigma^2}\right)(x-\mu)^2\right]$	μ	σ^2	$\exp\left(\mu t + \dfrac{t^2 \sigma^2}{2}\right)$
Gamma	$f(x) = \left[\dfrac{1}{\Gamma(\alpha)\beta^\alpha}\right] x^{\alpha-1} e^{-x/\beta}$	$\alpha\beta$	$\alpha\beta^2$	$(1 - \beta t)^{-\alpha}$
Chi-squared	$f(\chi^2) = \dfrac{(\chi^2)^{(\nu/2)-1} e^{-\chi^2/2}}{2^{\nu/2}\Gamma(\nu/2)}$	ν	2ν	$(1 - 2t)^{-\nu/2}$
Beta	$f(x) = \left[\dfrac{\Gamma(\alpha + \beta)}{\Gamma(\alpha)\Gamma(\beta)}\right] x^{\alpha-1}(1-x)^{\beta-1}$	$\dfrac{\alpha}{\alpha + \beta}$	$\dfrac{\alpha\beta}{(\alpha+\beta)^2(\alpha+\beta+1)}$	Does not exist in closed form

APPENDIX D: STATISTICAL TABLES

Table D.1. Cumulative normal distribution. The area or cumulative distribution, F(z), under the standardized normal distribution curve for $z \leq z_F$ such that the probability $P(z < z_F) = F(z)$. For example, $P(z < z_F = 1.21) = 0.8869$, and $P(z > z_F = 1.21) = 1 - 0.8869 = 0.1131$. [Adapted from Introductory Statistical Analysis by D. L. Harnett and J. L. Murphy, 1976, Addison-Wesley.]

$$F(z) = \int_{-\infty}^{z} \frac{1}{\sqrt{2\pi}} e^{-z^2/2} \, dz$$

z	0.00	0.01	0.02	0.03	0.04	0.05	0.06	0.07	0.08	0.09
0.0	0.5000	0.5040	0.5080	0.5120	0.5160	0.5199	0.5239	0.5279	0.5319	0.5359
0.1	0.5398	0.5438	0.5478	0.5517	0.5557	0.5596	0.5636	0.5675	0.5714	0.5753
0.2	0.5793	0.5832	0.5871	0.5910	0.5948	0.5987	0.6026	0.6064	0.6103	0.6141
0.3	0.6179	0.6217	0.6255	0.6293	0.6331	0.6368	0.6406	0.6443	0.6480	0.6517
0.4	0.6554	0.6591	0.6628	0.6664	0.6700	0.6736	0.6772	0.6808	0.6844	0.6879
0.5	0.6915	0.6950	0.6985	0.7019	0.7054	0.7088	0.7123	0.7157	0.7190	0.7224
0.6	0.7257	0.7291	0.7324	0.7357	0.7389	0.7422	0.7454	0.7486	0.7517	0.7549
0.7	0.7580	0.7611	0.7642	0.7673	0.7704	0.7734	0.7764	0.7794	0.7823	0.7852
0.8	0.7881	0.7910	0.7939	0.7967	0.7995	0.8023	0.8051	0.8078	0.8106	0.8133
0.9	0.8159	0.8186	0.8212	0.8238	0.8264	0.8289	0.8315	0.8340	0.8365	0.8389
1.0	0.8413	0.8438	0.8461	0.8485	0.8508	0.8531	0.8554	0.8577	0.8599	0.8621
1.1	0.8643	0.8665	0.8686	0.8708	0.8729	0.8749	0.8770	0.8790	0.8810	0.8830
1.2	0.8849	0.8869	0.8888	0.8907	0.8925	0.8944	0.8962	0.8980	0.8997	0.9015
1.3	0.9032	0.9049	0.9066	0.9082	0.9099	0.9115	0.9131	0.9147	0.9162	0.9177
1.4	0.9192	0.9207	0.9222	0.9236	0.9251	0.9265	0.9279	0.9292	0.9306	0.9319
1.5	0.9332	0.9345	0.9357	0.9370	0.9382	0.9394	0.9406	0.9418	0.9429	0.9441
1.6	0.9452	0.9463	0.9474	0.9484	0.9495	0.9505	0.9515	0.9525	0.9535	0.9545
1.7	0.9554	0.9564	0.9573	0.9582	0.9591	0.9599	0.9608	0.9616	0.9625	0.9633
1.8	0.9641	0.9649	0.9656	0.9664	0.9671	0.9678	0.9686	0.9693	0.9699	0.9706
1.9	0.9713	0.9719	0.9726	0.9732	0.9738	0.9744	0.9750	0.9756	0.9761	0.9767
2.0	0.9772	0.9778	0.9783	0.9788	0.9793	0.9798	0.9803	0.9808	0.9812	0.9817
2.1	0.9821	0.9826	0.9830	0.9834	0.9838	0.9842	0.9846	0.9850	0.9854	0.9857
2.2	0.9861	0.9864	0.9868	0.9871	0.9875	0.9878	0.9881	0.9884	0.9887	0.9890
2.3	0.9893	0.9896	0.9898	0.9901	0.9904	0.9906	0.9909	0.9911	0.9913	0.9916
2.4	0.9918	0.9920	0.9922	0.9925	0.9927	0.9929	0.9931	0.9932	0.9934	0.9936
2.5	0.9938	0.9940	0.9941	0.9943	0.9945	0.9946	0.9948	0.9949	0.9951	0.9952
2.6	0.9953	0.9955	0.9956	0.9957	0.9959	0.9960	0.9961	0.9962	0.9963	0.9964
2.7	0.9965	0.9966	0.9967	0.9968	0.9969	0.9970	0.9971	0.9972	0.9973	0.9974
2.8	0.9974	0.9975	0.9976	0.9977	0.9977	0.9978	0.9979	0.9979	0.9980	0.9981
2.9	0.9981	0.9982	0.9982	0.9983	0.9984	0.9984	0.9985	0.9985	0.9986	0.9986
3.0	0.9987	0.9987	0.9987	0.9988	0.9988	0.9989	0.9989	0.9989	0.9990	0.9990
3.1	0.9990	0.9991	0.9991	0.9991	0.9992	0.9992	0.9992	0.9992	0.9993	0.9993
3.2	0.9993	0.9993	0.9994	0.9994	0.9994	0.9994	0.9994	0.9995	0.9995	0.9995
3.3	0.9995	0.9995	0.9995	0.9996	0.9996	0.9996	0.9996	0.9996	0.9996	0.9997
3.4	0.9997	0.9997	0.9997	0.9997	0.9997	0.9997	0.9997	0.9997	0.9997	0.9998

Table D.2. Cumulative chi-square distribution. The area or cumulative distribution, $F(\chi^2)$ under the χ^2 distribution curve for different degrees of freedom, ν, such that the probability $P(\chi^2_\nu < \chi^2_{\nu;F}) = F(\chi^2)$. For example, for $\nu = 16$ the probability $P(\chi^2_{16} < \chi^2_{16;F} = 26.3) = F(26.3) = 0.950$. Consequently, $P(\chi^2_{16} > \chi^2_{16;F} = 26.3) = 1 - F(26.3) = 0.050$. [Adapted from Introductory Statistical Analysis by D. L. Harnett and J. L. Murphy, Addison-Wesley, 1976; abridged from Tables of percentage points of the incomplete beta function and of the chi-square distribution (C.M. Thompson, Biometrika, Vol. 32 (1941).]

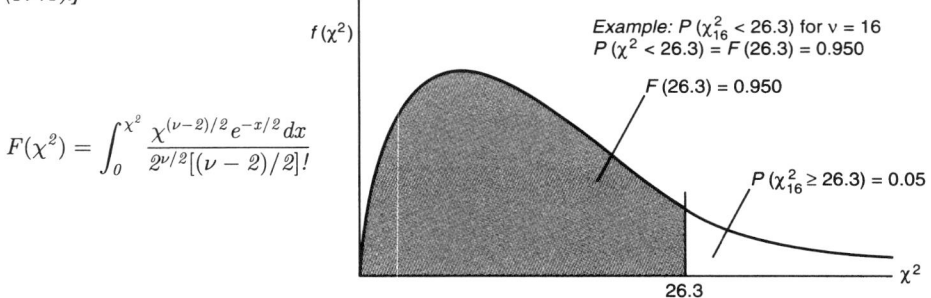

$$F(\chi^2) = \int_0^{\chi^2} \frac{\chi^{(\nu-2)/2} e^{-x/2} dx}{2^{\nu/2}[(\nu-2)/2]!}$$

Example: $P(\chi^2_{16} < 26.3)$ for $\nu = 16$
$P(\chi^2 < 26.3) = F(26.3) = 0.950$
$F(26.3) = 0.950$
$P(\chi^2_{16} \geq 26.3) = 0.05$

						F							
ν	0.005	0.010	0.025	0.050	0.100	0.250	0.500	0.750	0.900	0.950	0.975	0.990	0.995
1	0.0⁴393	0.0³157	0.0³982	0.0²393	0.0158	0.102	0.455	1.32	2.71	3.84	5.02	6.63	7.88
2	0.0100	0.0201	0.0506	0.103	0.211	0.575	1.39	2.77	4.61	5.99	7.38	9.21	10.6
3	0.0717	0.115	0.216	0.352	0.584	1.21	2.37	4.11	6.25	7.81	9.35	11.3	12.8
4	0.207	0.297	0.484	0.711	1.06	1.92	3.36	5.39	7.78	9.49	11.1	13.3	14.9
5	0.412	0.554	0.831	1.15	1.61	2.67	4.35	6.63	9.24	11.1	12.8	15.1	16.7
6	0.676	0.872	1.24	1.64	2.20	3.45	5.35	7.84	10.6	12.6	14.4	16.8	18.5
7	0.989	1.24	1.69	2.17	2.83	4.25	6.35	9.04	12.0	14.1	16.0	18.5	20.3
8	1.34	1.65	2.18	2.73	3.49	5.07	7.34	10.2	13.4	15.5	17.5	20.1	22.0
9	1.73	2.09	2.70	3.33	4.17	5.90	8.34	11.4	14.7	16.9	19.0	21.7	23.6
10	2.16	2.56	3.25	3.94	4.87	6.74	9.34	12.5	16.0	18.3	20.5	23.2	25.2
11	2.60	3.05	3.82	4.57	5.58	7.58	10.3	13.7	17.3	19.7	21.9	24.7	26.8
12	3.07	3.57	4.40	5.23	6.30	8.44	11.3	14.8	18.5	21.0	23.3	26.2	28.3
13	3.57	4.11	5.01	5.89	7.04	9.30	12.3	16.0	19.8	22.4	24.7	27.7	29.8
14	4.07	4.66	5.63	6.57	7.79	10.2	13.3	17.1	21.1	23.7	26.1	29.1	31.3
15	4.60	5.23	6.26	7.26	8.55	11.0	14.3	18.2	22.3	25.0	27.5	30.6	32.8
16	5.14	5.81	6.91	7.96	9.31	11.9	15.3	19.4	23.5	26.3	28.8	32.0	34.3
17	5.70	6.41	7.56	8.67	10.1	12.8	16.3	20.5	24.8	27.6	30.2	33.4	35.7
18	6.26	7.01	8.23	9.39	10.9	13.7	17.3	21.6	26.0	28.9	31.5	34.8	37.2
19	6.84	7.63	8.91	10.1	11.7	14.6	18.3	22.7	27.2	30.1	32.9	36.2	38.6
20	7.43	8.26	9.59	10.9	12.4	15.5	19.3	23.8	28.4	31.4	34.2	37.6	40.0
21	8.03	8.90	10.3	11.6	13.2	16.3	20.3	24.9	29.6	32.7	35.5	38.9	41.4
22	8.64	9.54	11.0	12.3	14.0	17.2	21.3	26.0	30.8	33.9	36.8	40.3	42.8
23	9.26	10.2	11.7	13.1	14.8	18.1	22.3	27.1	32.0	35.2	38.1	41.6	44.2
24	9.89	10.9	12.4	13.8	15.7	19.0	23.3	28.2	33.2	36.4	39.4	43.0	45.6
25	10.5	11.5	13.1	14.6	16.5	19.9	24.3	29.3	34.4	37.7	40.6	44.3	46.9
26	11.2	12.2	13.8	15.4	17.3	20.8	25.3	30.4	35.6	38.9	41.9	45.6	48.3
27	11.8	12.9	14.6	16.2	18.1	21.7	26.3	31.5	36.7	40.1	43.2	47.0	49.6
28	12.5	13.6	15.3	16.9	18.9	22.7	27.3	32.6	37.9	41.3	44.5	48.3	51.0
29	13.1	14.3	16.0	17.7	19.8	23.6	28.3	33.7	39.1	42.6	45.7	49.6	52.3
30	13.8	15.0	16.8	18.5	20.6	24.5	29.3	34.8	40.3	43.8	47.0	50.9	53.7

Appendices 579

Table D.3a. Cumulative t-distribution. The area or cumulative distribution, F(t), under the t-distribution curve for different degrees of freedom, ν, such that the probability $P(t_\nu < t_{\nu;F}) = F(t)$. For example, for $\nu = 9$ the probability $P(t_9 < t_{9;F} = 2.262) = F(2.262) = 0.975$ and $P(t_9 > t_{9;F} = 2.262) = 1 - F(2.262) = 0.025$, corresponding to the 95% confidence interval ($F = F_{0.025}$). Note that $F_{0.100}$, $F_{0.050}$, and $F_{0.005}$ correspond to the 80, 90, and 99% levels, respectively. [Adapted from Introductory Statistical Analysis by D.L. Harnett and J.L. Murphy, Addison-Wesley, 1976; abridged from the Statistical tables of R.A. Fisher and Frank Yates, Oliver & Boyd, Edinburgh and London, 1938.]

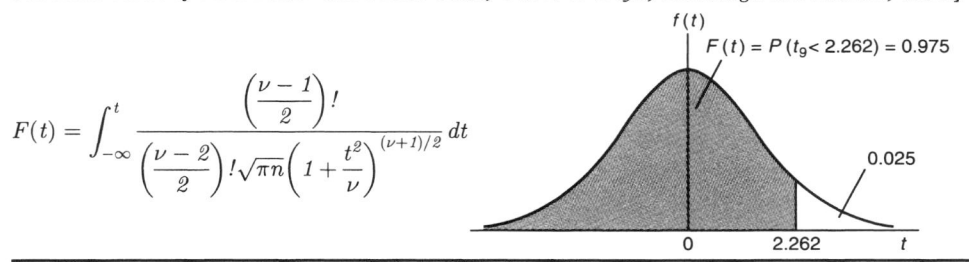

$$F(t) = \int_{-\infty}^{t} \frac{\left(\frac{\nu-1}{2}\right)!}{\left(\frac{\nu-2}{2}\right)!\sqrt{\pi n}\left(1+\frac{t^2}{\nu}\right)^{(\nu+1)/2}} dt$$

$F(t) = P(t_9 < 2.262) = 0.975$

ν	0.75	0.90	0.95	0.975	0.99	0.995	0.9995
1	1.000	3.078	6.314	12.706	31.821	63.657	636.615
2	0.816	1.886	2.920	4.303	6.965	9.925	31.598
3	0.765	1.638	2.353	3.182	4.541	5.841	12.941
4	0.741	1.533	2.132	2.776	3.747	4.604	8.610
5	0.727	1.476	2.015	2.571	3.365	4.032	6.859
6	0.718	1.440	1.943	2.447	3.143	3.707	5.959
7	0.711	1.415	1.895	2.365	2.998	3.499	5.405
8	0.706	1.397	1.860	2.306	2.896	3.355	5.041
9	0.703	1.383	1.833	2.262	2.821	3.250	4.781
10	0.700	1.372	1.812	2.228	2.764	3.169	4.587
11	0.697	1.363	1.796	2.201	2.718	3.106	4.437
12	0.695	1.356	1.782	2.179	2.681	3.055	4.318
13	0.694	1.350	1.771	2.160	2.650	3.012	4.221
14	0.692	1.345	1.761	2.145	2.624	2.977	4.140
15	0.691	1.341	1.753	2.131	2.602	2.947	4.073
16	0.690	1.337	1.746	2.120	2.583	2.921	4.015
17	0.689	1.333	1.740	2.110	2.567	2.898	3.965
18	0.688	1.330	1.734	2.101	2.552	2.878	3.922
19	0.688	1.328	1.729	2.093	2.539	2.861	3.883
20	0.687	1.325	1.725	2.086	2.528	2.845	3.850
21	0.686	1.323	1.721	2.080	2.518	2.831	3.819
22	0.686	1.321	1.717	2.074	2.508	2.819	3.792
23	0.685	1.319	1.714	2.069	2.500	2.807	3.767
24	0.685	1.318	1.711	2.064	2.492	2.797	3.745
25	0.684	1.316	1.708	2.060	2.485	2.787	3.725
26	0.684	1.315	1.706	2.056	2.479	2.779	3.707
27	0.684	1.314	1.703	2.052	2.473	2.771	3.690
28	0.683	1.313	1.701	2.048	2.467	2.763	3.674
29	0.683	1.311	1.699	2.045	2.462	2.756	3.659
30	0.683	1.310	1.697	2.042	2.457	2.750	3.646
40	0.681	1.303	1.684	2.021	2.423	2.704	3.551
60	0.679	1.296	1.671	2.000	2.390	2.660	3.460
120	0.677	1.289	1.658	1.980	2.358	2.617	3.373
∞	0.674	1.282	1.645	1.960	2.326	2.576	3.291

Table D.3b. Cumulative t-distribution (two-tailed tests). Similar to Table D.3a except that values give cumulative distribution, $F(t)$, under the t-distribution curve for different degrees of freedom, ν, regardless of sign, such that the probability $P(|t_\nu| > |t_{\nu,F}|) = F(t)$. For example, for $\nu = 9$ the probability $P(|t_9| > |t_{9,F}| = 2.262) = F(2.262) = 0.05$ and $P(|t_9| < |t_{9,F}| = 2.262) = 1 - F(2.262) = 0.95$, corresponding to the 95% confidence interval. Note that $F_{0.200}$, $F_{0.100}$, and $F_{0.010}$ correspond to the 80, 90, and 99% levels, respectively

ν	\multicolumn{10}{c}{F Probability of a larger value, sign ignored}
	0.500	0.400	0.200	0.100	0.050	0.025	0.010	0.005	0.001
1	1.000	1.376	3.078	6.314	12.706	25.452	63.657		
2	0.816	1.061	1.886	2.920	4.303	6.205	9.925	14.089	31.598
3	0.765	0.978	1.638	2.353	3.182	4.176	5.841	7.453	12.941
4	0.741	0.941	1.533	2.132	2.776	3.495	4.604	5.598	8.610
5	0.727	0.920	1.476	2.015	2.571	3.163	4.032	4.773	6.859
6	0.718	0.906	1.440	1.943	2.447	2.969	3.707	4.317	5.959
7	0.711	0.896	1.415	1.895	2.365	2.841	3.499	4.029	5.405
8	0.706	0.889	1.397	1.860	2.306	2.732	3.355	3.832	5.041
9	0.703	0.883	1.383	1.833	2.262	2.685	3.250	3.690	4.781
10	0.700	0.879	1.372	1.812	2.228	2.634	3.169	3.581	4.587
11	0.697	0.876	1.363	1.796	2.201	2.593	3.106	3.497	4.437
12	0.695	0.873	1.356	1.782	2.179	2.560	3.055	3.428	4.318
13	0.694	0.870	1.350	1.771	2.160	2.533	3.012	3.372	4.221
14	0.692	0.868	1.345	1.761	2.145	2.510	2.977	3.326	4.140
15	0.691	0.866	1.341	1.753	2.131	2.490	2.947	3.286	4.073
16	0.690	0.865	1.337	1.746	2.120	2.473	2.921	3.252	4.015
17	0.689	0.863	1.333	1.740	2.110	2.458	2.898	3.222	3.965
18	0.688	0.862	1.330	1.734	2.101	2.445	2.878	3.197	3.922
19	0.688	0.861	1.328	1.729	2.093	2.433	2.861	3.174	3.883
20	0.687	0.860	1.325	1.725	2.086	2.423	2.845	3.153	3.850
21	0.686	0.859	1.323	1.721	2.080	2.414	2.831	3.135	3.819
22	0.686	0.858	1.321	1.717	2.074	2.406	2.819	3.119	3.792
23	0.685	0.858	1.319	1.714	2.069	2.398	2.807	3.104	3.767
24	0.685	0.857	1.318	1.711	2.064	2.391	2.797	3.090	3.745
25	0.684	0.856	1.316	1.708	2.060	2.385	2.787	3.078	3.725
26	0.684	0.856	1.315	1.706	2.056	2.379	2.779	3.067	3.707
27	0.684	0.855	1.314	1.703	2.052	2.373	2.771	3.056	3.690
28	0.683	0.855	1.313	1.701	2.048	2.368	2.763	3.047	3.674
29	0.683	0.854	1.311	1.699	2.045	2.364	2.756	3.038	3.659
30	0.683	0.854	1.310	1.697	2.042	2.360	2.750	3.030	3.646
35	0.682	0.852	1.306	1.690	2.030	2.342	2.724	2.996	3.591
40	0.681	0.851	1.303	1.684	2.021	2.329	2.704	2.971	3.551
45	0.680	0.850	1.301	1.680	2.014	2.319	2.690	2.952	3.520
50	0.680	0.849	1.299	1.676	2.008	2.310	2.678	2.937	3.496
55	0.679	0.849	1.297	1.673	2.004	2.304	2.669	2.925	3.476
60	0.679	0.848	1.296	1.671	2.000	2.299	2.660	2.915	3.460
70	0.678	0.847	1.294	1.667	1.994	2.290	2.648	2.899	3.435
80	0.678	0.847	1.293	1.665	1.989	2.284	2.638	2.887	3.416
90	0.678	0.846	1.291	1.662	1.986	2.279	2.631	2.878	3.402
100	0.677	0.846	1.290	1.661	1.982	2.276	2.625	2.871	3.390
120	0.677	0.845	1.289	1.658	1.980	2.270	2.617	2.860	3.373
∞	0.6745	0.8416	1.2816	1.6448	1.9600	2.214	2.5758	2.8070	3.2905

Appendices 581

Table D.4. *Critical values of the F-distribution for* (a) $\alpha = 0.05$ *and* (b) 0.01. *The distributions represent the area exceeding the value of* $F_{0.05,\nu_1,\nu_2}$ *and* $F_{0.01,\nu_1,\nu_2}$ *as shown by the shaded area in the figure for different degrees of freedom,* ν. *For example, if* $\nu_1 = 15$ *and* $\nu_2 = 20$, *then the critical value for* $\alpha = 0.05$ *is 2.20.* [Adapted from Introductory Statistical Analysis by D. L. Harnett and J. L. Murphy, Addison-Wesley, 1976, abridged from tables of percentage points of the inverted beta (F) distribution by M. Merrington and C. M. Thompson Biometrika, Vol. 33 (1943).]

$P(F > 2.20) = 0.05,$
$P(F < 2.20) = 0.95.$

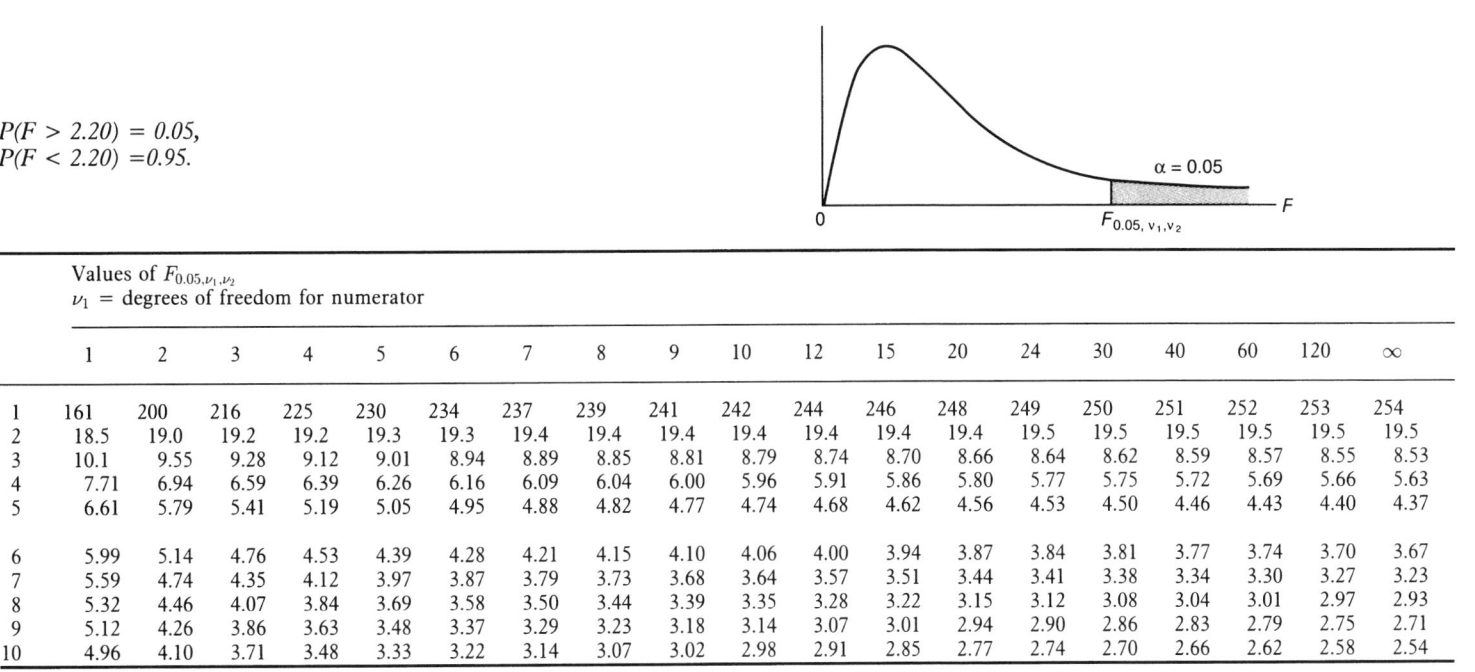

Values of $F_{0.05,\nu_1,\nu_2}$
ν_1 = degrees of freedom for numerator

ν_2	1	2	3	4	5	6	7	8	9	10	12	15	20	24	30	40	60	120	∞
1	161	200	216	225	230	234	237	239	241	242	244	246	248	249	250	251	252	253	254
2	18.5	19.0	19.2	19.2	19.3	19.3	19.4	19.4	19.4	19.4	19.4	19.4	19.4	19.5	19.5	19.5	19.5	19.5	19.5
3	10.1	9.55	9.28	9.12	9.01	8.94	8.89	8.85	8.81	8.79	8.74	8.70	8.66	8.64	8.62	8.59	8.57	8.55	8.53
4	7.71	6.94	6.59	6.39	6.26	6.16	6.09	6.04	6.00	5.96	5.91	5.86	5.80	5.77	5.75	5.72	5.69	5.66	5.63
5	6.61	5.79	5.41	5.19	5.05	4.95	4.88	4.82	4.77	4.74	4.68	4.62	4.56	4.53	4.50	4.46	4.43	4.40	4.37
6	5.99	5.14	4.76	4.53	4.39	4.28	4.21	4.15	4.10	4.06	4.00	3.94	3.87	3.84	3.81	3.77	3.74	3.70	3.67
7	5.59	4.74	4.35	4.12	3.97	3.87	3.79	3.73	3.68	3.64	3.57	3.51	3.44	3.41	3.38	3.34	3.30	3.27	3.23
8	5.32	4.46	4.07	3.84	3.69	3.58	3.50	3.44	3.39	3.35	3.28	3.22	3.15	3.12	3.08	3.04	3.01	2.97	2.93
9	5.12	4.26	3.86	3.63	3.48	3.37	3.29	3.23	3.18	3.14	3.07	3.01	2.94	2.90	2.86	2.83	2.79	2.75	2.71
10	4.96	4.10	3.71	3.48	3.33	3.22	3.14	3.07	3.02	2.98	2.91	2.85	2.77	2.74	2.70	2.66	2.62	2.58	2.54

ν_2 = degrees of freedom for denominator

Table D.4 (a) Continued

Values of $F_{0.05,\nu_1,\nu_2}$
ν_1 = Degrees of freedom for numerator

ν_2	1	2	3	4	5	6	7	8	9	10	12	15	20	24	30	40	60	120	∞
11	4.84	3.98	3.59	3.36	3.20	3.09	3.01	2.95	2.90	2.85	2.79	2.72	2.65	2.61	2.57	2.53	2.49	2.45	2.40
12	4.75	3.89	3.49	3.26	3.11	3.00	2.91	2.85	2.80	2.75	2.69	2.62	2.54	2.51	2.47	2.43	2.38	2.34	2.30
13	4.67	3.81	3.41	3.18	3.03	2.92	2.83	2.77	2.71	2.67	2.60	2.53	2.46	2.42	2.38	2.34	2.30	2.25	2.21
14	4.60	3.74	3.34	3.11	2.96	2.85	2.76	2.70	2.65	2.60	2.53	2.46	2.39	2.35	2.31	2.27	2.22	2.18	2.13
15	4.54	3.68	3.29	3.06	2.90	2.79	2.71	2.64	2.59	2.54	2.48	2.40	2.33	2.29	2.25	2.20	2.16	2.11	2.07
16	4.49	3.63	3.24	3.01	2.85	2.74	2.66	2.59	2.54	2.49	2.42	2.35	2.28	2.24	2.19	2.15	2.11	2.06	2.01
17	4.45	3.59	3.20	2.96	2.81	2.70	2.61	2.55	2.49	2.45	2.38	2.31	2.23	2.19	2.15	2.10	2.06	2.01	1.96
18	4.41	3.55	3.16	2.93	2.77	2.66	2.58	2.51	2.46	2.41	2.34	2.27	2.19	2.15	2.11	2.06	2.02	1.97	1.92
19	4.38	3.52	3.13	2.90	2.74	2.63	2.54	2.48	2.42	2.38	2.31	2.23	2.16	2.11	2.07	2.03	1.98	1.93	1.88
20	4.35	3.49	3.10	2.87	2.71	2.60	2.51	2.45	2.39	2.35	2.28	2.20	2.12	2.08	2.04	1.99	1.95	1.90	1.84
21	4.32	3.47	3.07	2.84	2.68	2.57	2.49	2.42	2.37	2.32	2.25	2.18	2.10	2.05	2.01	1.96	1.92	1.87	1.81
22	4.30	3.44	3.05	2.82	2.66	2.55	2.46	2.40	2.34	2.30	2.23	2.15	2.07	2.03	1.98	1.94	1.89	1.84	1.78
23	4.28	3.42	3.03	2.80	2.64	2.53	2.44	2.37	2.32	2.27	2.20	2.13	2.05	2.01	1.96	1.91	1.86	1.81	1.76
24	4.26	3.40	3.01	2.78	2.62	2.51	2.42	2.36	2.30	2.25	2.18	2.11	2.03	1.98	1.94	1.89	1.84	1.79	1.73
25	4.24	3.39	2.99	2.76	2.60	2.49	2.40	2.34	2.28	2.24	2.16	2.09	2.01	1.96	1.92	1.87	1.82	1.77	1.71
30	4.17	3.32	2.92	2.69	2.53	2.42	2.33	2.27	2.21	2.16	2.09	2.01	1.93	1.89	1.84	1.79	1.74	1.68	1.62
40	4.08	3.23	2.84	2.61	2.45	2.34	2.25	2.18	2.12	2.08	2.00	1.92	1.84	1.79	1.74	1.69	1.64	1.58	1.51
60	4.00	3.15	2.76	2.53	2.37	2.25	2.17	2.10	2.04	1.99	1.92	1.84	1.75	1.70	1.65	1.59	1.53	1.47	1.39
120	3.92	3.07	2.68	2.45	2.29	2.18	2.09	2.02	1.96	1.91	1.83	1.75	1.66	1.61	1.55	1.50	1.43	1.35	1.25
∞	3.84	3.00	2.60	2.37	2.21	2.10	2.01	1.94	1.88	1.83	1.75	1.67	1.57	1.52	1.46	1.39	1.32	1.22	1.00

ν_2 = Degrees of freedom for denominator

Table D.4 (b) Critical values of the F-distribution

$P(F > 3.09) = 0.01$,
$P(F < 3.09) = 0.99$.

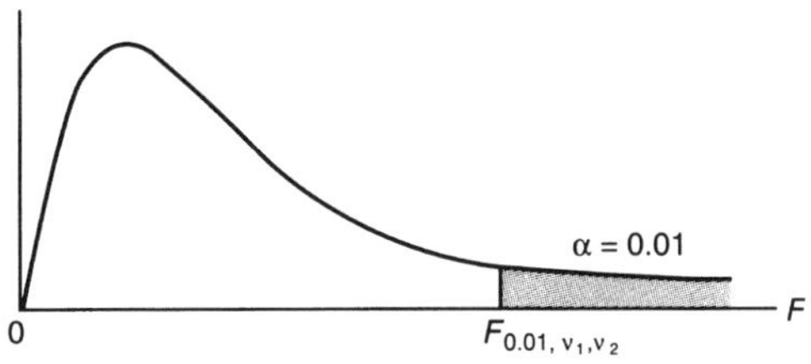

Values of $F_{0.01,\nu_1,\nu_2}$
ν_1 = Degrees of freedom for numerator

ν_2	1	2	3	4	5	6	7	8	9	10	12	15	20	24	30	40	60	120	∞
1	4052	5000	5403	5625	5764	5859	5928	5982	6023	6056	6106	6157	6209	6235	6261	6287	6313	6339	6366
2	98.5	99.0	99.2	99.2	99.3	99.3	99.4	99.4	99.4	99.4	99.4	99.4	99.4	99.5	99.5	99.5	99.5	99.5	99.5
3	34.1	30.8	29.5	28.7	28.2	27.9	27.7	27.5	27.3	27.2	27.1	26.9	26.7	26.6	26.5	26.4	26.3	26.2	26.1
4	21.2	18.0	16.7	16.0	15.5	15.2	15.0	14.8	14.7	14.5	14.4	14.2	14.0	13.9	13.8	13.7	13.7	13.6	13.5
5	16.3	13.3	12.1	11.4	11.0	10.7	10.5	10.3	10.2	10.1	9.89	9.72	9.55	9.47	9.38	9.29	9.20	9.11	9.02
6	13.7	10.9	9.78	9.15	8.75	8.47	8.26	8.10	7.98	7.87	7.72	7.56	7.40	7.31	7.23	7.14	7.06	6.97	6.88
7	12.2	9.55	8.45	7.85	7.46	7.19	6.99	6.84	6.72	6.62	6.47	6.31	6.16	6.07	5.99	5.91	5.82	5.74	5.65
8	11.3	8.65	7.59	7.01	6.63	6.37	6.18	6.03	5.91	5.81	5.67	5.52	5.36	5.28	5.20	5.12	5.03	4.95	4.86
9	10.6	8.02	6.99	6.42	6.06	5.80	5.61	5.47	5.35	5.26	5.11	4.96	4.81	4.73	4.65	4.57	4.48	4.40	4.31
10	10.0	7.56	6.55	5.99	5.64	5.39	5.20	5.06	4.94	4.85	4.71	4.56	4.41	4.33	4.25	4.17	4.08	4.00	3.91

ν_2 = Degrees of freedom for denominator

Table D.4 (b) Critical values of the F-distribution

Values of $F_{0.01, \nu_1, \nu_2}$
ν_1 = Degrees of freedom for numerator
ν_2 = Degrees of freedom for denominator

	1	2	3	4	5	6	7	8	9	10	12	15	20	24	30	40	60	120	∞
11	9.65	7.21	6.22	5.67	5.32	5.07	4.89	4.74	4.63	4.54	4.40	4.25	4.10	4.02	3.94	3.86	3.78	3.69	3.60
12	9.33	6.93	5.95	5.41	5.06	4.82	4.64	4.50	4.39	4.30	4.16	4.01	3.86	3.78	3.70	3.62	3.54	3.45	3.36
13	9.07	6.70	5.74	5.21	4.86	4.62	4.44	4.30	4.19	4.10	3.96	3.82	3.66	3.59	3.51	3.43	3.34	3.25	3.17
14	8.86	6.51	5.56	5.04	4.70	4.46	4.28	4.14	4.03	3.94	3.80	3.66	3.51	3.43	3.35	3.27	3.18	3.09	3.00
15	8.68	6.36	5.42	4.89	4.56	4.32	4.14	4.00	3.89	3.80	3.67	3.52	3.37	3.29	3.21	3.13	3.05	2.96	2.87
16	8.53	6.23	5.29	4.77	4.44	4.20	4.03	3.89	3.78	3.69	3.55	3.41	3.26	3.18	3.10	3.02	2.93	2.84	2.75
17	8.40	6.11	5.19	4.67	4.34	4.10	3.93	3.79	3.68	3.59	3.46	3.31	3.16	3.08	3.00	2.92	2.83	2.75	2.65
18	8.29	6.01	5.09	4.58	4.25	4.01	3.84	3.71	3.60	3.51	3.37	3.23	3.08	3.00	2.92	2.84	2.75	2.66	2.57
19	8.19	5.93	5.01	4.50	4.17	3.94	3.77	3.63	3.52	3.43	3.30	3.15	3.00	2.92	2.84	2.76	2.67	2.58	2.49
20	8.10	5.85	4.94	4.43	4.10	3.87	3.70	3.56	3.46	3.37	3.23	3.09	2.94	2.86	2.78	2.69	2.61	2.52	2.42
21	8.02	5.78	4.87	4.37	4.04	3.81	3.64	3.51	3.40	3.31	3.17	3.03	2.88	2.80	2.72	2.64	2.55	2.46	2.36
22	7.95	5.72	4.82	4.31	3.99	3.76	3.59	3.45	3.35	3.26	3.12	2.98	2.83	2.75	2.67	2.58	2.50	2.40	2.31
23	7.88	5.66	4.76	4.26	3.94	3.71	3.54	3.41	3.30	3.21	3.07	2.93	2.78	2.70	2.62	2.54	2.45	2.35	2.26
24	7.82	5.61	4.72	4.22	3.90	3.67	3.50	3.36	3.26	3.17	3.03	2.89	2.74	2.66	2.58	2.49	2.40	2.31	2.21
25	7.77	5.57	4.68	4.18	3.86	3.63	3.46	3.32	3.22	3.13	2.99	2.85	2.70	2.62	2.53	2.45	2.36	2.27	2.17
30	7.56	5.39	4.51	4.02	3.70	3.47	3.30	3.17	3.07	2.98	2.84	2.70	2.55	2.47	2.39	2.30	2.21	2.11	2.01
40	7.31	5.18	4.31	3.83	3.51	3.29	3.12	2.99	2.89	2.80	2.66	2.52	2.37	2.29	2.20	2.11	2.02	1.92	1.80
60	7.08	4.98	4.13	3.65	3.34	3.12	2.95	2.82	2.72	2.63	2.50	2.35	2.20	2.12	2.03	1.94	1.84	1.73	1.60
120	6.85	4.79	3.95	3.48	3.17	2.96	2.79	2.66	2.56	2.47	2.34	2.19	2.03	1.95	1.86	1.76	1.66	1.53	1.38
∞	6.63	4.61	3.78	3.32	3.02	2.80	2.64	2.51	2.41	2.32	2.18	2.04	1.88	1.79	1.70	1.59	1.47	1.32	1.00

APPENDIX E: CORRELATION COEFFICENTS AT THE 5% AND 1% LEVELS OF SIGNIFICANCE FOR VARIOUS DEGREES OF FREEDOM ν

Degrees of freedom	5%	1%	Degrees of freedom	5%	1%
1	0.997	1.000	24	0.388	0.496
2	0.950	0.990	25	0.381	0.487
3	0.878	0.959	26	0.374	0.478
4	0.811	0.917	27	0.367	0.470
5	0.754	0.874	28	0.361	0.463
6	0.707	0.834	29	0.355	0.456
7	0.666	0.798	30	0.349	0.449
8	0.632	0.765	35	0.325	0.418
9	0.602	0.735	40	0.304	0.393
10	0.576	0.708	45	0.288	0.372
11	0.553	0.684	50	0.273	0.354
12	0.532	0.661	60	0.250	0.325
13	0.514	0.641	70	0.232	0.302
14	0.497	0.623	80	0.217	0.283
15	0.482	0.606	90	0.205	0.267
16	0.468	0.590	100	0.195	0.254
17	0.456	0.576	125	0.174	0.228
18	0.444	0.561	150	0.159	0.208
19	0.433	0.549	200	0.138	0.181
20	0.423	0.537	300	0.113	0.148
21	0.413	0.526	400	0.098	0.128
22	0.404	0.515	500	0.088	0.115
23	0.396	0.505	1000	0.062	0.081

APPENDIX F: APPROXIMATIONS AND NONDIMENSIONAL NUMBERS IN PHYSICAL OCEANOGRAPHY

Beta parameter, β^*: A nondimensionalized form of β (the beta parameter) defined as the ratio of the horizontal gradient in relative vorticity, $\nabla_h \zeta$, to the horizontal gradient in planetary vorticity, $\nabla_h f$

$$\beta^* \equiv \frac{\beta L^2}{U}$$

Here, $\zeta = \partial v/\partial x - \partial u/\partial y$ is the relative vorticity (for velocity components u, v in the x, y directions, respectively), f is the local Coriolis parameter, U is a horizontal velocity scale, L is a horizontal length scale (*see also Rhines length*) and β is defined as

$$\beta \equiv \nabla_h f \approx 10^{-11} \text{ m}^{-1} \text{ s}^{-1}$$

In the *beta-plane approximation*, the curved surface of the earth is approximated by a flat plane tangent to the earth for which $f = f_o + \beta y$, where f_o is a reference value for f and y is the latitude. For this case, $\beta = df/dy$ is a constant.

Boussinesq approximation: Assumes that density changes in the fluid can be neglected except where density, $\rho = \rho_o + \rho'$, is multiplied by the acceleration of gravity, g. That is, the effects of ρ' can be neglected in terms of the form $\rho F = (\rho_o + \rho')F$ for any variable F except for those involving g (i.e. ρg). Here $\rho_o = \rho_o(z)$ is the mean density and $\rho'/\rho_o \approx 10^{-3}$. At large *Mach* numbers ($U/c > 1$) the compressibility of the fluid becomes important and large density changes can occur. Since the speed of sound in water, $c \approx 1500$ m/s, is almost always large compared to flow speeds, U, the approximation is good for normal oceanic conditions.

Brunt–Väisälä frequency, $N(z)$ (also *Väisälä* or *Buoyancy frequency*): The natural frequency of oscillation of a parcel of water displaced vertically (z-direction upward) from its level of equilibrium

$$N(z) = \sqrt{-\left(\frac{g}{\rho_o}\frac{d\rho}{dz} + \frac{g^2}{c^2}\right)} \cong \sqrt{-\frac{g}{\rho_o}\frac{d\rho}{dz}}$$

where $g = 9.81$ m s^{-1} is the acceleration due to gravity, c is the speed of sound, $d\rho/dz \leq 0$ is the vertical *in situ* density gradient, and ρ_o is a reference density. The compressibility term, $g^2/c^2 \approx 5 \times 10^{-5}$ s^{-1}, associated with adiabatic displacement of the fluid can generally be ignored in the upper few thousand meters of the ocean where $10^{-4} < N < 10^{-2}$ s^{-1} (periods of 17.5 h to 10.5 min). However, this is not the case in the deep ocean where $d\rho/dz$ is small and N can be of order 10^{-5} s^{-1}. Derivation of N using CTD data usually requires considerable low-pass filtering to eliminate erroneously high or negative values of the density gradient.

Burger number, B (also *Stratification parameter*, S): The squared ratio of the internal (Rossby) radius of deformation, r_i, to a long-wave scale, L (such as the wavelength of a Rossby or coastal-trapped wave)

$$B \equiv \frac{N^2 H^2}{f^2 L^2}$$

where N is a characteristic Brunt–Väisälä frequency for the water depth H. For continental shelf-slope regions influenced by coastal-trapped waves, motions are baroclinic when $B \gg 1$ and barotropic when $B \ll 1$. For internal waves over a sloping bottom tilted at an angle θ to the horizontal, $H = L \sin\theta$ and

$$B \equiv \frac{N^2 \sin^2\theta}{f^2}$$

Coriolis parameter (also Inertial frequency, Coriolis frequency, Planetary vorticity): The local vertical component of the earth's rate of rotation given by $f = 2\Omega \sin\theta$, where θ is the latitude and $\Omega = 0.72921 \times 10^{-4}$ s^{-1} is the angular rate of rotation based on a *sidereal day* of 23 h, 56 min and 4 s. A sidereal day is the time for the earth to complete one rotation relative to an absolute reference point in space. Because of the movement of the earth about the sun, a sidereal day differs slightly from the solar day of 24 h. Latitude is positive for the northern hemisphere and negative for the southern hemisphere. At 50°N, $f = 1.117 \times 10^{-4}$ s^{-1} = 0.0640 cph, corresponding to a period of 15.6 h.

Cox number, C_θ: A relative measure of high vertical wavenumber temperature structure (temperature "noisiness") defined as the ratio of the mean vertical gradient squared to the mean-square vertical gradient for temperature, $T(z)$

$$C_\theta \equiv \frac{\langle \mathrm{d}T/\mathrm{d}z \rangle^2}{\langle (\mathrm{d}T/\mathrm{d}z)^2 \rangle}$$

Ekman number, E: A nondimensional number giving the relative importance of frictional forces at a boundary to the Coriolis force

$$E \equiv \frac{\text{frictional force}}{\text{Coriolis force}} = \frac{\nu}{fD^2}$$

where ν is the turbulent eddy viscosity, f is the Coriolis parameter and D is the depth of the fluid. The characteristic thickness, δ, of the Ekman layer is given by

$$\delta = \sqrt{\frac{2\nu_v}{f}}$$

where ν_v is the vertical component of eddy viscosity. For ν_v of order 10^{-2} m^2 s^{-1}, $\delta \approx 20$ m at mid-latitudes ($f \approx 1 \times 10^{-4}$ s^{-1}).

Froude number, F_r: The square root of the ratio of the inertial force to the gravitational force for barotropic motions with a free surface

$$F_r \approx \left[\frac{\text{inertial force}}{\text{gravitational force}}\right]^{1/2} = \frac{U}{\sqrt{gH}}$$

where U is the flow velocity and $c = \sqrt{gH}$ is the phase speed of a surface wave in a fluid of depth H. The flow is *supercritical* if $F_r > 1$ and *subcritical* if $F_r < 1$. The Froude number is analogous to the *Mach Number* used for compressible flow.

Froude number (internal), F'_r: The square root of the ratio of the inertial force to the buoyancy force for baroclinic motions in a stratified fluid

$$F'_r \equiv \left[\frac{\text{inertial force}}{\text{buoyancy force}}\right]^{1/2} = \frac{U}{\sqrt{g'H_n}}$$

where $g' = g(\rho_2 - \rho_1)/\rho_2$ is the reduced gravity and $c'_n = \sqrt{g'H_n}$ is the phase speed of a mode n internal wave in a fluid with an effective depth H_n (see *Internal Rossby radius*). The flow is *supercritical* if $F'_r > 1$ and *subcritical* if $F'_r < 1$.

Geostrophic approximation: Assumes that the Rossby number is small ($Ro \ll 1$) so that horizontal motions are mainly a balance between the Coriolis force and the horizontal pressure gradient.

Hydrostatic approximation: Assumes that the vertical velocity, w, can be ignored in the vertical component of the momentum balance and that the vertical pressure gradient $\partial p/\partial z$ is proportional to the density, ρ

$$\frac{\partial p}{\partial z} = -g\rho$$

Integration from depth z to the ocean surface $z = \eta$ gives, for near-uniform density $\rho \approx \rho_o$

$$p = p_o + g\rho_o(\eta - z)$$

where p_o is the atmospheric pressure at the ocean surface. The approximation cannot be used to study high-frequency internal wave dynamics.

Inertial period, T_f: The period of oscillation for the Coriolis frequency, f (see *Coriolis frequency*)

$$T_f = \frac{2\pi}{|f|} = \frac{\pi}{\Omega|\sin\theta|}$$

$T_f \approx 15.62$ h for inertial motions at latitude $\theta = 50°$.

Intrinsic frequency: If ω_o is the frequency of a wave measured at a fixed point and $\mathbf{k}$ the wavenumber vector of the wave, the intrinsic frequency, ω, of the wave as seen by an observer in a coordinate system moving with the mean flow, $\mathbf{U}$, is given by

$$\omega = \omega_o - \mathbf{k}\cdot\mathbf{U}$$

Thus, the frequency of the wave measured at fixed point is Doppler shifted by the amount $+\mathbf{k}\cdot\mathbf{U}$ relative to the intrinsic frequency. For most oceanic motions, the Doppler shift measured at fixed point is within a few percent of the intrinsic frequency, ω.

Kolmogorov microscale, η: The length scale at which turbulent motions begin to be damped out by small-scale molecular viscosity, ν

$$\eta \equiv 2\pi\left(\frac{\nu^3}{\epsilon}\right)^{1/4}$$

in which $\epsilon = 2\nu\langle(\partial u_i/\partial x_j)^2\rangle$ is the mean rate of dissipation of turbulent kinetic energy (see *Ozmidov scale*). In the upper ocean, η is a few centimeters.

Mach number, M: The relative importance of fluid compressibility defined by the relation

$$M \equiv \left[\frac{\text{inertial force}}{\text{compressibility force}}\right]^{1/2} = \frac{U}{c}$$

where c is the speed of sound (≈ 1500 m/s in water) and U is the velocity of the fluid. Flows are subsonic if $M < 1$ and supersonic if $M > 1$. Compressibility effects can be ignored if $M < 0.3$.

Monin–Obukov length, L_M: The height above a heated boundary at which mechanical (shear) production of turbulent kinetic energy equals the buoyant (convective) destruction of turbulent kinetic energy

$$L_M \equiv \frac{\text{shear production}}{\text{buoyant destruction}} = \frac{u_*^3}{k\alpha g \overline{wT'}}$$

where u_*, k, and α are, respectively, the friction velocity, von Karman constant, and coefficient of thermal expansion, and $\overline{wT'}$ is the mean heat flux for vertical velocity fluctuations w and temperature fluctuations T'.

Ozmidov (buoyancy) scale, η_b (or L_R): The ratio of nonlinear to buoyancy scales in a turbulent fluid; the scale above which eddy-like motions are damped by stratification

$$\eta_b \equiv 2\pi \left(\frac{\epsilon}{N^3}\right)^{1/2}$$

Here, $\epsilon = 2\nu\langle(\partial u_i/\partial x_j)^2\rangle$ is the rate of dissipation of turbulent kinetic energy, and u_i is the ith component of velocity in the jth direction, x_j ($i, j = 1, 2, 3$ corresponding to the x, y, z directions, respectively). In the upper ocean, η_b can be up to a few meters.

Péclet Number, Pe: The diffusivity analog to the Reynolds number

$$Pe \equiv \frac{UL}{\kappa} = Pr \cdot Re$$

where κ is the diffusivity of heat or salt. In geophysical fluid dynamics, κ corresponds to the turbulent eddy diffusivity (see *Prandtl number*, Pr and *Reynolds number*, Re).

Prandtl number, Pr: The ratio of momentum to heat (or salt) diffusivity

$$Pr \equiv \frac{\text{momentum diffusivity}}{\text{heat diffusivity}} = \frac{\nu}{\kappa}$$

For typical values of molecular viscosity $\nu \approx 10^{-2}$ cm^2 s^{-1} and molecular heat diffusivity $\kappa \approx 10^{-3}$ cm^2 s^{-1}, $Pr \approx 10$. For salt, $\kappa \approx 10^{-5}$ cm^2 s^{-1} and $Pr \approx 1000$. A turbulent *Prandtl number* can be defined in terms of the turbulent eddy viscosity and turbulent diffusivities of heat and salt. The *Schmidt number* is similar to the Prandtl number with momentum diffusivity replaced by mass diffusivity.

Rayleigh number, Ra: The ratio of the destabilizing effect of the buoyancy force to the stabilizing effect of the viscous force

$$Ra \equiv \frac{g\alpha\Gamma d^4}{\kappa\nu}$$

where α is the coefficient of thermal expansion, $\Gamma = -d\langle T \rangle/dz$ is the vertical gradient of the background temperature $\langle T \rangle$ (the adiabatic temperature gradient, also known as the "lapse rate" by meteorologists), d is the depth of the layer, is the thermal diffusivity and κ is the kinematic viscosity. The "lapse rate" is the fastest rate at which the temperature can decrease with height without causing instability.

Reynolds number, Re: The ratio of the inertial (nonlinear) force to the viscous force

$$Re \equiv \frac{\text{inertial force}}{\text{viscous force}} = \frac{UL}{\nu}$$

where U is the flow velocity, L is a characteristic length scale, and ν is the kinetic viscosity ($\nu \approx 0.01$ cm^2 s^{-1} for molecular processes). Viscous effects become important at small Reynolds numbers, $Re \ll 1$. In geophysical fluid dynamics, ν corresponds to the turbulent eddy viscosity.

Rhines length, ℓ: The scale at which mesoscale eddies transform from individual features to Rossby wave packets (the scale at which the planetary β-effect becomes comparable to nonlinear effects). Rossby wave propagation causes anisotropic elongation of the eddies in the zonal (east–west) direction and the eddy size in the meridional (north–south, y) direction stops growing at the scale

$$\ell = \sqrt{\frac{u}{\beta}}$$

where u is the RMS velocity and β is the north–south gradient of the Coriolis parameter, $f = f_o + \beta y$ (see *Beta parameter*).

Richardson number, Ri: A measure of the dynamic stability of the water column. In a two-layer fluid with reduced gravity g', mean flow U and horizontal length scale, L, the *local Richardson number* is defined as

$$Ri \equiv \frac{g'L}{U^2}$$

while for a continuously stratified fluid with buoyancy frequency $N(z)$

$$Ri \equiv \frac{N^2 L^2}{U^2}$$

The above expressions also are known as the *bulk Richardson number* since they define the overall stability characteristics of the water column. In both cases, $Ri \propto 1/Fr'^2$, where Fr' is the internal Froude number. For $Ri > 0$, the stratification is stable; for $Ri = 0$ it is neutral and for $Ri < 0$ it is unstable. The *gradient Richardson number*

$$Ri \equiv \frac{N^2}{(dU/dz)^2}$$

is a measure of the localized stability of the water column in which the stabilizing effect of the density gradient, or buoyancy N, competes with the destabilizing effect of turbulent mixing due to the vertical shear, dU/dz. Shear instability typically can be expected for $Ri \leq 1$ (the often used $Ri \leq 1/4$ criterion is a necessary, but not sufficient condition for instability). The *flux Richardson number*, which is the ratio of the rate of increase in fluid potential energy due to entrainment (buoyant destruction of turbulent kinetic energy) to the rate of production of turbuent energy associated with the velocity shear, may be defined as

$$Rf \equiv \frac{-g\alpha \overline{wT'}}{-\overline{uw}(dU/dz)} \approx \frac{\nu_v N^2}{\epsilon}$$

where $g\alpha\overline{wT'}$ is the production of turbulent kinetic energy by the vertical heat flux $\overline{wT}$, $-\overline{uw}(dU/dz)$ is the production of turbuent kinetic energy by the Reynolds stress $\overline{uw}$ working against the mean shear dU/dz, ν_v is the vertical diffusion coefficient, N is the buoyancy frequency and ϵ is turbulent energy production.

Rigid-lid approximation: For surface displacement $\eta(t)$, the rigid-lid approximation requires that the vertical velocity $w = \partial\eta/\partial t + \mathbf{u}\cdot\nabla\eta = 0$ at the surface ($z = 0$) and that vertical baroclinic motions within the fluid greatly exceed those at the surface. One implication of the rigid-lid approximation is that the external Rossby radius, r_o, becomes infinite; hence, a measure of the validity of the approximation is, that for motions of length scale L, $L/r_o \ll 1$. The rigid-lid approximation allows surface pressure in the ocean to vary spatially but eliminates surface gravity waves. If one could put a rigid cover on top of the ocean, the upward pressure beneath the cover would vary in space but gravity waves would be eliminated. Application of the rigid lid approximation removes barotropic Kelvin waves from the coastal trapped wave problem and simplifies calculation of baroclinic modes.

Rossby number, Ro: The ratio of nonlinear to Coriolis forces, and the ratio of the relative vorticity to the planetary vorticity, defined by

$$Ro \equiv \frac{\text{nonlinear accelerations}}{\text{Coriolis force}} = \frac{U^2/L}{fU} = \frac{U}{fL}$$

For common oceanic scales $U \approx 10$ cm/s, $L \approx 100$ km and $f \approx 10^{-4}$ s^{-1}, we find $Ro \approx 0.01$ so that nonlinear terms are of second order in the equations of motion.

Rossby radius of deformation (external; barotropic), r_o: The natural e-folding scale for barotropic currents in the sea defined as

$$r_o \equiv \frac{\sqrt{gH}}{f} = \frac{c}{f}$$

where $c = \sqrt{gH}$ is the propagation speed of long gravity waves (e.g. the tide) in water of depth H. For a mid-latitude ocean of depth 1000 m, $r_o \approx 1000$ km.

Rossby radius of deformation (internal; baroclinic), r_i: The natural e-folding scale for baroclinic motions which, for a continuously stratified ocean, is normally written as

$$r_i \equiv \frac{NH}{f}$$

where H is the local water depth and N a representative value for the local buoyancy frequency. We may also define the baroclinic Rossby radius as

$$\pi r_i \equiv \frac{\sqrt{gH_n}}{f} = \frac{c_n}{f}$$

where H_n is the "equivalent depth"

$$H_n = H^2 N^2 / g n^2 \pi^2$$

and

$$c_n = NH/n\pi, \quad n = 1, 2, \ldots$$

are the horizontal phase speeds of the different vertical wave modes. For first mode ($n = 1$) wave propagation in a mid-latitude region of depth $H \approx 1000$ m, buoyancy frequency $N \approx 3 \times 10^{-3}$ s^{-1}, and Coriolis frequency $f \approx 10^{-4}$ s^{-1}, we find $c_1 \approx 1.0$ m/s and $r_i \approx 60$ km. For a two-layer fluid with upper and lower layer densities and thicknesses ρ_1, H_1 and ρ_2, H_2 we have

$$r_i \equiv f^{-1} \left[\frac{g(\rho_2 - \rho_1)}{\rho_2} \cdot \frac{H_1 H_2}{H_1 + H_2} \right]^{1/2} = f^{-1} \left[g' \frac{H_1 H_2}{H_1 + H_2} \right]^{1/2}$$

Strouhal number, S: The ratio of the boundary-imposed frequency of fluid oscillation, n, to "natural" frequency of oscillation, U/D, based on the flow velocity U and length scale D of an obstacle

$$S \equiv \frac{uD}{U}$$

In the case of an obstacle in a steady flow, n is the vortex shedding frequency of the leeward flow.

Thorpe scale, L_T: An objective measure of the vertical overturning scale in a turbulent stratified fluid. First proposed by Thorpe (1977) to describe overturning structures within turbulent mixing events in a Scottish loch, the scale is obtained by rearranging an observed density profile, which may contain inversions, into a profile in which density increases monotonically with depth. Heat and mass are conserved during the rearrangement process. Consider an observed profile of n density values, ρ_n, sampled at depths z_n. If a given sample with density ρ_n must be moved to a depth z_m in generating the stable profile, then the Thorpe displacement for the sample is $z_m - z_n$. In general, a unique displacement will result from each density sample and n Thorpe displacements will be generated from the original profile. The Thorpe scale, L_T, is the root-mean-square of these displacements (Dillon, 1982; Libe Washburn, personal communication). Typical values are of the order of 1 m.

APPENDIX G: CONVOLUTION

Convolution and Fourier transforms

Consider the time-dependent functions $g(t)$ and $h(t)$ and their respective frequency-dependent Fourier transforms $G(f)$ and $H(f)$. The convolution of the two original functions (written $g * h$) is defined as

$$g * h \equiv \int_{-\infty}^{\infty} g(t) h(t - \tau) dt \qquad (G1)$$

where $g * h$ is a function of the time lag, τ, and $g * h = h * g$. There is a one-to-one relationship between the function $g * h$ and the product of the Fourier transforms of the two functions such that

$$g * h \leftrightarrow G(f) \cdot H(f) \qquad (G2)$$

Known as the convolution theorem, (G2) states that the Fourier transform of the convolution is the product of the Fourier transforms of the individual functions. In other words, convolution in one domain equates to multiplication in the other domain. We further note that the correlation of g and h [corr(g, h); see Section 5.3] is written as

$$\text{corr}(g, h) \equiv \int_{-\infty}^{\infty} g(t + \tau) h(t) dt \qquad (G3)$$

which is also a function of the lag τ. As with convolution, we can form the transform pair

$$\text{corr}(g, h) \leftrightarrow G(f) \cdot H(f)^* \qquad (G4)$$

called the correlation theorem, where $H(f)^*$ is the complex conjugate of $H(f)$ and $H(f)^* = H(-f)$ since we are restricting discussion to the usual case in which g and h are real functions. As this relationship indictates, multiplying the Fourier transform of one function by the complex conjugate of the Fourier transform of the other function yields the Fourier transform of their correlation. The correlation of a function with itself is called its autocorrelation (Section 5.3).

Convolution of discrete data

The analysis of geophysical data commonly involves the convolution of specially designed "data windows" (convolution functions or filters) with time series records in order to smooth the spectral estimates obtained from these data and to improve the statistical reliability of spectral peaks. Good filters are those that minimize unwanted spectral leakage associated with the filter's side lobes in the frequency domain. Consider a filter $h(t_k)$ applied to a discrete data series $g(t_j)$, where the t_j and t_k ($j, k = 0, ...$) are discrete times in the data series. The filter will have non-zero values over a short segment of the data to which it is being applied and will be zero elsewhere, yielding a single value for the central time of the filter for that specific piece of the data. The filter $h(t_k)$ typically has a central peak and falls off to zero on either side of the maximum.

The convolution theorem can be extended to discrete time series as follows. Assume that the time series, $g(t_j)$, has duration N and is completely determined by the N values $g(t_0), \ldots, g(tN - 1)$. The convolution of this function with the window, $h(t_k)$, is a member of the discrete Fourier transform pair

$$\sum_{k=-N/2+1}^{N/2} g(t_{j-k})h(t_k) \leftrightarrow G_n H_n \quad (G5)$$

where G_n ($n = 0, \ldots, N - 1$) is the discrete Fourier transform of the time series $g(t_j)$ ($j = 0, \ldots, N - 1$), and H_n ($n = 0, \ldots, N - 1$) is the discrete Fourier transform of the function $h(t_k)$, ($k = 0, \ldots, N - 1$). The values of $h(t_k)$ typically span a small fraction of the full data range $k = -N/2 + 1, \ldots, N/2$.

In Figure G1, the original time series, $g(t_j)$ (we have chosen normalized monthly values of the Southern Oscillation Index, SOI) is shown at the top and the convolution function, $h(t_k)$, used to filter the time series is presented in the middle panel. Here, we have used a simple five-year long Hamming window [see equation (5.6.78)]. The window (filter) is symmetrical, uses 61 monthly weights (with non-zero first and last weights), and begins with the first month of the time series.

The bottom panel in Figure G1 shows the convolution of $h(t_k)$ with $g(t_j)$. As the filtered result clearly demonstrates, $h(t_k)$ acts as a smoothing function that flattens out the "bumpiness" of $g(t_j)$, reducing sharp year-to-year changes in the normalized SOI. This smoothing depends on the duration and the shape of the window, $h(t_k)$. A more sharply peaked $h(t_k)$ would produce less time series smoothing, leaving more of the large year-to-year variability. The function, $h(t_k)$, has exactly the same purpose as a moving average, except that the weights of the filter (the filter coefficients) are specially designed to reduce side lobe spectral leakage problems. For the moving average, all weights are of equal value. The convolved data (bottom panel of Figure G1) consists of variations longer than five years. Note the extended period of El Niño events (negative SIO) in the 1980s and 1990s.

Convolution as truncation of an infinite time series

An observed time series, $x(t)$, can be considered a subset of an unlimited duration time series $g(t)$, obtained by convolving $g(t)$ with a rectangular window $h(t)$ of the form

$$h(t) = \begin{cases} 1 & 0 \leq t \leq T \\ 0 & \text{otherwise} \end{cases} \quad (G6)$$

$$x(t) = h(t)g(t) \quad (G7)$$

As illustrated in Figure G2, the series $x(t)$ can be defined as

$$x(t) = h(t)g(t) \quad (G7)$$

It follows that the Fourier transform of $x(t)$ is the convolution of the Fourier transforms of $h(t)$ and $g(t)$, namely

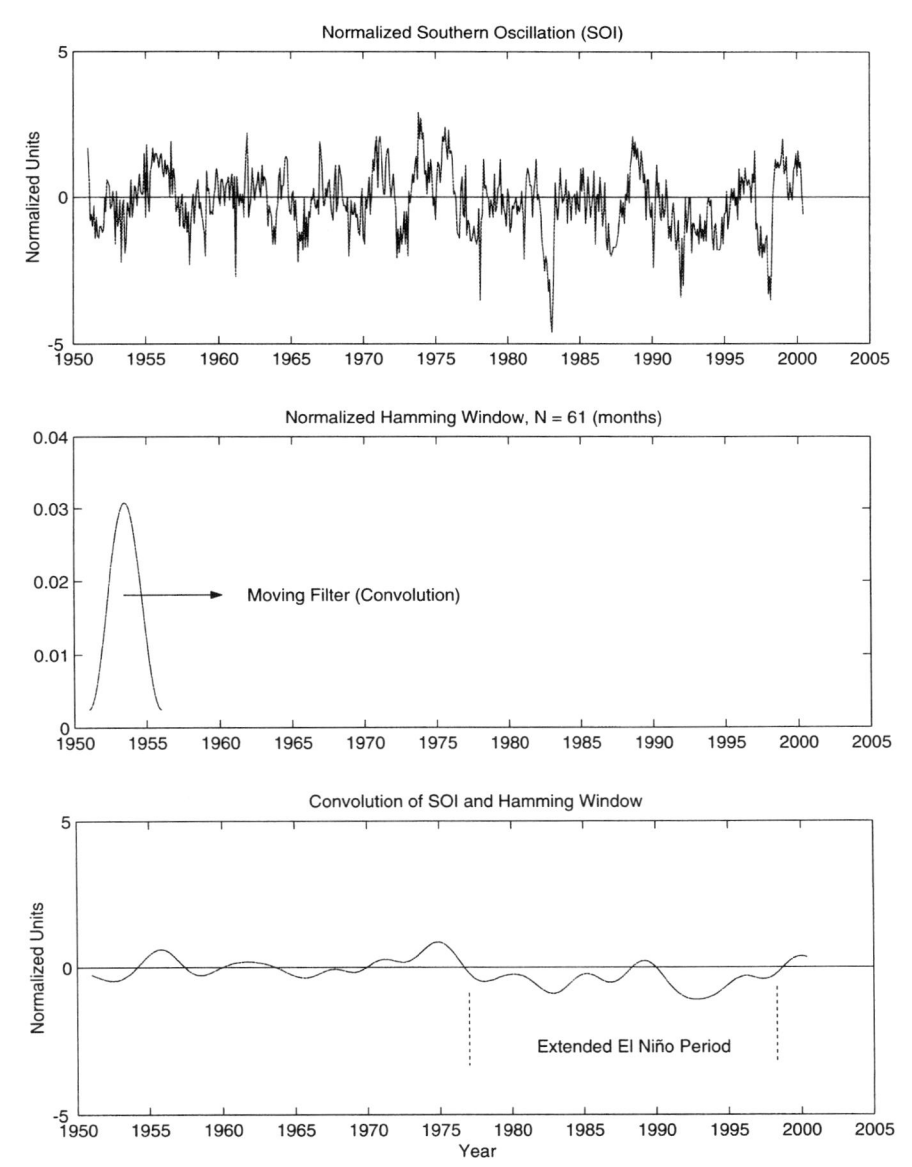

Figure G1. Convolution of the monthly time series of normalized Southern Oscillation Index (see U.S. government website ftp.necp.noaa.gov/pub/cpc/wd52dg/data/indicies/...) using a 61-month Hamming window (filter). Negative (positive) values of the index are associated with El Niño (La Niña) events. The convolution emphasizes the low-frequency variability of the El Niño–La Niña phenomenon in the equatorial Pacific.

$$X(f) = \int_{-\infty}^{\infty} H(\zeta)G(f-\zeta)\mathrm{d}\zeta \tag{G8}$$

In this case, multiplication in the time equals convolution in the frequency domain,

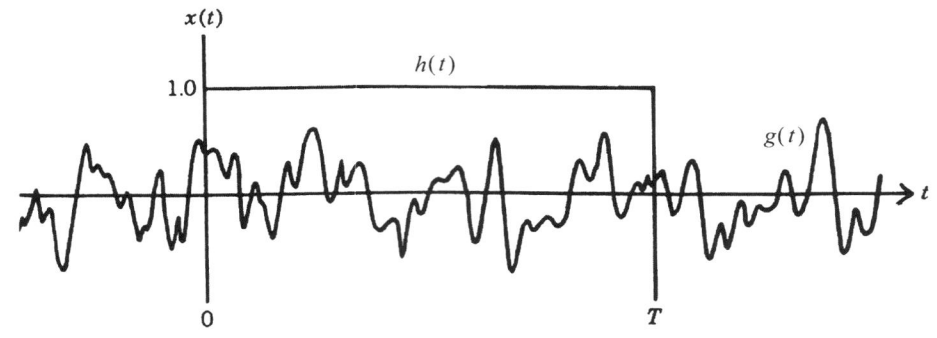

596 *Data Analysis Methods in Physical Oceanography*

Figure G2. Sampling a time series segment of duration T. The measurement is analogous to application of a rectangular window, h(t), of amplitude 1.0 and duration T to an extensive time series g(t).

whereas in the previous case we examined convolution in the time domain (as with a running average) and multiplication in the frequency domain. These concepts are essential for the application of data windows in both the time and frequency domains.

Deconvolution

Deconvolution is the process of reversing (undoing) the smoothing that took place during application of the "data window", either in the time or frequency domains. It is assumed in this case that the response function is known and the process of deconvolution requires only a reverse of the process described above. Thus, the equation for deconvolution follows from that for convolution presented in equation (G1).

References

Allen, S.E. and R.E. Thomson 1993: Bottom-trapped subinertial motions over midocean ridges in a stratified rotating fluid. *J. Phys. Oceanogr.*, 23, 566–581.

Anderson, D.L. 1993: ^{3}He from the mantle; primordial signal or cosmic dust? *Science*, 261, 170–176.

Arnault, S., Y. Menard, and J. Merle 1990: Observing the Tropical Atlantic Ocean in 1986–87 from altimetry. *J. Geophys. Res.*, 95, 17,921–17,946.

Baker, D.J., Jr 1981: Ocean instruments and experiment design. In *Evolution of Physical Oceanography* (eds B.A. Warren and C. Wunsch). MIT Press, Cambridge, Mass., pp. 396–433.

Baker, E.T. and J.E. Lupton 1990: Changes in submarine hydrothermal ^{3}He/heat ratios as an indicator of magmatic/tectonic activity. *Nature*, 346, 556–558.

Baker, E.T. and G.J. Massoth 1986: Hydrothermal plume measurements: A regional perspective. *Science*, 234, 980–982.

Baker, E.T. and G.J. Massoth 1987: Characteristics of hydrothermal plumes from two vent fields on the Juan de Fuca Ridge, northeast Pacific Ocean. *Earth Planet. Sci. Lett.*, 85, 59–73.

Baker, E.T., J.W. Lavelle, R.A. Feely, G.J Massoth, S.L. Walker, and J.E. Lupton 1989: Episodic venting of hydrothermal fluids from the Juan de Fuca Ridge. *J. Geophys. Res.*, 94, 9237–9250.

Baker, E.T. and J.W. Lavelle 1984: The effect of particle size on the light attenuation coefficient of natural suspensions. *J. Geophys. Res.*, 89, 8197–8203.

Baker, E.T., C.R. German, and H. Elderfield 1995: Hydrothermal plumes over spreading-center axes: Global distributions and geological inferences. In *Seafloor Hydrothermal Systems: Physical, Chemical, Biological and Geological Interactions* (eds S.E. Humphris, R.A. Zierenberg, L.S. Mullineaux, and R.E. Thomson). AGU, Geophysical Monograph 91, pp. 47–71.

Bakun, A. 1973: Coastal upwelling indices, west coast of North America, 1946–71. NOAA Technical Report NMFS SSRF-671. US Department of Commerce.

Bard, E., M. Arnold, H.G. Östlund, P. Maurice, P. Monfray, and J.-C. Duplessy 1988: Penetration of bomb radiocarbon in the tropical Indican Ocean measured by means of accelerator mass spectrometry. *Earth & Planet. Sci. Lett.*, 87, 379–389.

Barkley, R.A. 1968: *Oceanographic Atlas of Pacific Ocean*. University of Hawaii Press, Honolulu.

Barnett, T.P. 1983: Recent changes in sea level and their possible causes. *Climate Change*, 5, 15-38.

Barnett, T.P., W.C. Patzert, S.C. Webb, and B.R. Brown 1979: Climatological usefulness of satellite determined sea-surface temperatures in the tropical Pacific. *Bull. Am. Met. Soc.*, 60, 197–205.

Barrodale, I. and R.E. Erickson 1978: Algorithms for least-squares linear prediction and maximum entropy spectral analysis. MS report, DM-142-IR, University of Victoria.

Bartz, R., J. Zaneveld, and H. Pak. 1978: A transmissometer for profiling and moored observations in water. *Soc. Photo-Opt. Instrum. Engrs J.*, 160(V), 102–108.

Batchelor, G.K. 1967: *An Introduction to Fluid Dynamics*. Cambridge University Press, Cambridge.

Bates, J.J and W.L. Smith 1985: Sea surface temperatures from geostationary satellites. *J. Geophys. Res.*, 90, 11,609–11,618.

Bates, J.J. and H.F. Diaz 1991: Evaluation of multichannel sea surface temperature product quality for climate monitoring. *J. Geophys. Res.*, 96, 20,613–20,622.

Beardsley, R.C. 1987: A comparison of the vector-averaging current meter and new Edgerton, Germeshausen, and Gier, Inc., vector-measuring current meter on a surface mooring in Coastal Ocean Dynamics Experiment 1. *J. Geophys. Res.*, 92, 1845–1859.

Beardsley, R.C. and W.C. Boicourt 1981: On estuarine and continental-shelf circulation in the Middle Atlantic Bight. In *Evolution of Physical Oceanography* (eds B.A. Warren and C. Wunsch). MIT Press, Cambridge, Mass., pp. 198–233.

Beardsley, R.C., W.C. Boicourt, L.C. Huff, J.R. McCullough, and J. Scott 1981: CMICE: a near-surface current meter intercomparison experiment. *Deep-Sea Res.*, 28A, 1577–1603.

Bendat, J.S., and A.G. Piersol 1986: *Random Data: Analysis and Measurement Procedures*. John Wiley, New York.

Bennett, A.F. 1976: Poleward heat fluxes in southern hemisphere oceans. *J. Phys. Oceanogr.*, 4, 785–798.

Bennett, A.F. 1992: *Inverse Methods in Physical Oceanography*. Cambridge University Press, Cambridge.

Bernstein, R.L. 1982: Sea surface temperature estimation using the NOAA-6 satellite advanced very high resolution radiometer. *J. Geophys. Res.*, 87, 9455–9466.

Bernstein, R.L. and D.B. Chelton 1985: Large-scale sea surface temperature variability from satellite and shipboard measurements. *J. Geophys. Res.*, 90, 11,619–11,630.

Berteaux, H.O. 1990: *Program SSMOOR: Users' Instructions*. Cable Dynamics and Mooring Systems (CDMS), Woods Hole, Mass.

Blackman, R.B. and J.W. Tukey 1958: *The Measurement of Power Spectra*. Dover, New York.

Bograd, S.J., A.B. Rabinovich, R.E. Thomson, and A.J. Eert 1999: On sampling strategies and interpolation schemes for satellite-tracked drifters. *J. Atmos. amd Oceanic Technol.*, 16, 893–904.

Boicourt, W.C. 1982: The recent history of ocean current meter measurement. *Proceedings of IEEE Second Workshop Conference on Current Measurement*, pp. 9–13.

Born, G.H, M.A. Richards, and G.W. Rosborough 1982: An empirical determination of the effects of sea-state bias on the SEASAT altimeter. *J. Geophys. Res.*, 87, 3221–3226.

Born, G.H., J.L. Mitchell, and G.A. Heyler 1987: GEOSAT-ERM mission design. *J. Astron. Sci.*, 35, 119–134.

Bowditch, N. 1977: American practical navigator. Defense Mapping Agency Hydrographic Center, DMA No. NVPUB9V1.

Box, G.E.P. and G.M. Jenkins 1970: *Time Series Analysis: Forecasting and Control*. Holden-Day, San Francisco, Calif.

Bretherton, F.P., R.E. Davis, and C.B. Fandry 1976: A technique for objective analysis and design of oceanographic experiments applied to MODE-73. *Deep-Sea Res.*, 23, 559–582.

Bretherton, F.P and J.C. McWilliams 1980: Estimations from irregular arrays. *Rev. Geophys. Space Phys.*, 18, 789-812.

Brink, K.H. and D.C. Chapman 1987: Programs for computing properties of coastal-trapped waves and wind-driven motions over the continental shelf and slope, 2nd edn. Woods Hole Oceanographic Institution, Technical Report No. WHOI-87-24.

Briscoe, M.G. 1975: Internal waves in the ocean. *Rev. Geophys. Space Res.*, 13, 591–598.

Broecker, W.S. and T.-H. Peng 1982: Tracers in the sea. *Lamont–Dohery Geological Observatory*. Eldigio Press, New York.

Broecker, W.S., T. H. Peng, G. Östlund, and M. Stuiver 1985: The distribution of bomb radiocarbon in the ocean. *J. Geophys. Res.*, 90, 6953–6970.

Broecker, W.S., T. H. Peng, and G. Östlund 1986: The distribution of bomb Tritium in the ocean. *J. Geophys. Res.*, 91, 14,331–14,344.

Broecker, W.S., A. Virgilio, and T-H. Peng 1991: Radiocarbon age of waters in the deep Atlantic revisited. *Geophys. Res. Lett.*, 18, 1–3.

Brooks, C.F. 1926: Observing water-surface temperatures at sea. *Mon. Wea. Rev.*, *54*, 241–254.
Brower, R.L., G.S. Gohrband, W.G. Pichel, T.L. Signore, and C.C. Walton 1976: Satellite derived sea-surface temperatures from NOAA spacecraft. NOAA Technical Memo, NESS No. 79, Washington, DC.
Brown, N.L. 1974: A precision CTD microprofiler. In *Ocean 74 Record, 1974 IEEE Conference on Engineering in the Ocean Environment*. IEEE Publication 74 CHO873-0 OEC, Institute of Electrical and Electronics Engineers, New York, Vol. 2, 270–278.
Brown, R.A. 1983: On a satellite scatterometer as an anemometer. *J. Geophys. Res.*, *88*, 1663–1673.
Brown, N.L. and G.K. Morrison 1978: WHOI/Brown conductivity, temperature and depth profiler. WHOI Report No. 78-23.
Brown, O.B., J.W. Brown, and R.H. Evans 1985: Calibration of advanced very high resolution radiometer infrared observations. *J. Geophys. Res.*, *90*, 11,667–11,678.
Bruland, K.W., J.R. Donat, and D.A. Hutchins 1991: Interactive influences of bioactive trace metals on biological production in oceanic waters. *Limnol. Oceanogr.*, *36*, 1555–1577.
Bryden, H. 1979: Poleward heat flux and conversion of available potential energy in Drake Passage. *J. Mar. Res.*, *37*, 1–22.
Bullister, J.L. 1989: Chlorofluorocarbons as time-dependent tracers in the ocean. *Oceanography*, *2*(2), 12–17.
Bullister, J.L. and R.F. Weiss 1988: Determination of CCl_3F and CCl_2F_2 in seawater and air. *Deep-Sea Res.*, *35*, 839–853.
Burd, B.J. and R.E. Thomson 1993: Flow volume calculations based on three-dimensional current and net orientation data. *Deep-Sea Res.*, *40*, 1141–1153.
Burd, B.J. and R.E. Thomson 1994: Hydrothermal venting at Endeavour Ridge: Effect on zooplankton biomass throughout the water column. *Deep-Sea Res.*, *41*, 1407–1423.
Burg, J.P. 1967: Maximum entropy spectral analysis, paper presented at the 37th Annual International Meeting, Soc. of Explor. Geophys., Oklahoma City, Okla., Oct. 31, 1967.
Burg, J.P. 1968: A new analysis technique for time series data, In NATO Advanced Study Institute on Signal Processing Emphasis on Underwater Acoustics, Enschede, The Netherlands, Aug. 12–23.
Burg, J.P. 1972: The relationship between maximum entropy spectra and maximum likelihood spectra. *Geophysics*, *37*, 375–376.
Busalacchi, A.J. and J.J. O'Brien 1981: Interannual variability of the equatorial Pacific in the 1960's. *J. Geophys. Res.*, *86*, 10,901–10,907.
Cahill, M.L., J.H. Middleton, and B.R. Stanton 1991: Coastal-trapped waves on the west coast of the South Island, New Zealand. *J. Phys. Oceanogr.*, *21*, 541–557.
Calder, M. 1975: Calibration of echo sounders for offshore sounding using temperature and depth. *International Hydrographic Review*, *LII*(2), 13–17.
Calman, J. 1978: On the interpretation of ocean current spectra. *J. Phys. Oceanogr.* *8*, 627–652.
Cannon, G.A. and R.E. Thomson 1996: Characteristics of 4-day oscillations trapped by the Juan de Fuca Ridge. *Geophys. Res. Lett.*, *23*, 1613–1616.
Capon, J. 1969: High resolution frequency-wavenumber spectral analysis. *Proc. IEEE*, *57*, 1408–1418.
Cartwright, D.E. 1968: A unified analysis of tides and surges round north and east Britain. *Phil. Trans. Roy. Soc. London A263*, 1–55.
Cartwright, D.E. and A.C. Edden 1973: Corrected tables of tidal harmonics. *Geophys. J. R. Astron. Soc.*, *33*, 253–264.
Chapman, D.C. 1983: On the influence of stratification and continental shelf and slope topography on the dispersion of subinertial coastally trapped waves. *J. Phys. Oceanogr.*, *13*, 1641–1652.
Charnock, H., J.E. Lovelock, P.S. Liss, and M. Whitfield 1988: Tracers in the Ocean. Proceedings of a Royal Society Meeting Held on 21 and 22 May, 1987. The Royal Society, London.

Chave, A.D. and D.S. Luther 1990: Low-frequency, motionally induced electromagnetic fields in the ocean. *J. Geophys. Res., 95*, 7185–7200.

Chave, A.D., D.S. Luther, and J.H. Filloux 1992: The barotropic electromagnetic and pressure experiment. 1. Barotropic current response to atmospheric forcing. *J. Geophys. Res., 97*, 9565–9593.

Chelton, D.B. 1982: Statistical reliability and the seasonal cycle: comments on "Bottom pressure measurements across the Antarctic Circumpolar Current and their relation to the wind". *Deep-Sea Res., 29*, 1381–1388.

Chelton, D.B. 1983: Effects of sampling errors in statistical estimation. *Deep-Sea Res., 30*, 1083–1101.

Chelton, D.B. 1988: WOCE/NASA Altimeter algorithm workshop. US WOCE Technical Report, No. 2.

Chelton, D.B. and R.E. Davis 1982: Monthly mean sea level variability along the west coast of North America. *J. Phys. Oceanogr., 6*, 757–784.

Chelton, D.B., P.A. Bernal and J.A. McGowan 1982: Large-scale interannual physical and biological interaction in the California Current. *J. Mar. Res., 40*, 1095–1125.

Cheney, R.E., J.G. Marsh and B.D. Becley 1983: Global mesoscale variability from collinear tracks of SEASAT altimeter data. *J. Geophys. Res. 88*, 4343–4354.

Chisholm, S.W. and F.M.M. Morel (eds) 1991: What controls phytoplankton production in nutrient-rich areas of the open sea? *Limnol. Oceanogr., 36*, 1507–1970.

Chiswell, S. 1992: Inverted echo sounders at the WOCE deep-water station. WOCE Notes 4(4) October 1992, 1–6. US Office, Department of Oceanography. Texas A&M University, College Station.

Chiswell, S.M., M. Wimbush and R. Luckas 1988: Comparison of dynamic height measurements from an inverted echo sounder and an island tidge gauge in the central Pacific. *J. Geophys. Res., 93*, 2277–2283.

Christensen, E.J., B.J Haines, S.J. Keihm, C.S. Morris, R.A. Norman, G.H. Purcell, B.G. Williams, B.D. Wilson, G.H. Born, M.E. Parke, S.K. Gill, C.K. Shum, B.D. Tapley, R. Kolenkiewicz, and R.S. Nerem 1994: Calibration of TOPEX/POSEIDON at platform Harvest. *J. Geophys. Res., 99*, 24,465–24,487.

Chu, P.C. 1994: Localized TOGA sea level spectra obtained from the S-transformation. *TOGA Notes, 17*, 5–8.

Church, J.A., H.J. Freeland, and R.L Smith 1986: Coastal-trapped waves on the East Australian continental shelf. Part I: Propagation of modes. *J. Phys. Oceanogr., 16*, 1929–1943.

Clayson, C.A., W.J. Emery and R. Savage 1993: Wind speed comparisons between GEOSAT, SSM/I, buoy and ECMWF wind speeds. *J. Geophys. Res. 19*, 558-563.

Collins, C.A., L.F. Giovando, and K.A. Abbott-Smith 1975: Comparison of Canadian and Japanese merchant ship observations of sea surface temperature in the vicinity of the present Ocean Station P, 1927–53. *J. Fish. Res. Bd. Canada, 32*, 253–258.

Cooley, J.W. and J.W. Tukey 1965: An algorithm for machine calculation of complex Fourier series. *Math. Comput. 19*, 297–301.

Cox, R.A. 1963: The salinity problem. *Progress in Oceanography, 1*, 243–261.

Cox, R.A., F. Culkin and J.P. Riley 1967: The electrical conductivity/chlorinity relationship in natural sea water. *Deep-Sea Res., 14*, 203-220.

Craig, H. 1994: Retention of Helium in subducted interplanetary dust particles. *Science, 265*, 1892–1893.

Crawford, A.B. 1969: Sea surface temperatures, some instruments, methods and comparison. Tech. Note No. 103, WMO, pp. 117–129.

Crawford, W.R. and R.E. Thomson 1984: Diurnal-period continental shelf waves along Vancouver Island: A comparison of observations with theoretical models. *J. Phys. Oceanogr., 14*, 1629–1646.

Cresswell, G.R. 1976: A drifting buoy tracked by satellite in the Tasman Sea. *Aust. J. Mar. Freshwat. Res., 27*, 251–262.

Dahlen, J.M. and N.K. Chhabra 1983: Slippage errors and dynamic response of four drogued buoys measured at sea. Report P-1729, C.C. Draper Laboratory. Presented at Marine Technology Society/NOAA Data Buoy Center, 1983 Symposium on Buoy Technology, New Orleans, April 1983.

Danielson, G.C. and C. Lanczos 1942: Some improvements in practical Fourier analysis and their application to X-ray scattering from liquids. *J. Franklin Inst.*, 233, 365–380 and 435–452.

Dantzler, H.L. 1974: Dynamic salinity calibrations of continuous salinity/temperature/depth data. *Deep-Sea Res.*, 21, 675–682.

Datawell bv 1992: Directional waverider. Datawell bv, Haarlem, The Netherlands.

Davis, R.E. 1976: Predictability of sea surface temperature and sea level pressure anomalies over the North Pacific Ocean. *J. Phys. Oceanogr.*, 6, 249–266.

Davis, R.E. 1977: Techniques for statistical analysis and prediction of geophysical fluid systems. *Geophys. Astrophys. Fluid Dyn.*, 8, 245–277.

Davis, R.E. 1978: Predictability of sea level pressure anomalies over the North Pacific Ocean. *J. Phys. Oceanogr.*, 8, 233–246.

Davis, R. and L. Regier 1977: Methods for estimating directional wave spectra from multi-element arrays. *J. Mar. Res.*, 35, 453–477.

Davis, R.E., D.C. Webb, L.A. Regier, and J. Dufour 1992: The autonomous Lagrangian circulation explorer (ALACE). *J. Atmos. Oceanic Technol.*, 9, 264–285.

Defant, A. 1936: Schichtung und Zirkulation des Atlantischen Ozeans. Die Troposphare. In *Wissenschaftliche Ergebnisse der Deutschen Altantishcen Expedition auf dem Forschungs- und Vermessungsschiff "Meteor" 1925-1927.*

Defant, A. 1937: Stratification and circulation of the Atlantic Ocean. The troposphere, scientific results of the German Atlantic expedition of the Research Vessel, *Meteor*, 1925–27, English translation (ed. W.J. Emery). Amerind, New Delhi, 1981.

Defant, A. 1961: *Physical Oceanography, VI*. Pergamon Press, New York.

Denbo, D.W. and J.S. Allen 1984: Rotary empirical orthogonal function analysis of currents near the Oregon Coast. *J. Phys. Oceanogr.* 14, 35–46.

Denbo, D.W. and J.S. Allen 1986: Reply to "Comments on: Rotary empirical orthogonal function analysis of currents near the Oregon Coast". *J. Phys. Oceanogr.* 16, 793–794.

Denman, K.L. and H.J. Freeland 1985: Correlation scales, objective mapping and a statistical test of geostrophy over the continental shelf. *J. Mar. Res.* 43, 517–539.

Diaconis, P. and B. Efron 1983: Computer-intensive methods in statistics. *Sci. American*, 248, 116–130.

Dillon, T.D. 1982: Vertical overturns: A comparison of Thorpe and Ozmidov length scales. *J. Geophys. Res.*, 87, 9601–9613.

Dodimead, A.J., F. Favorite, and T. Hirano 1963: Salmon of the North Pacific. II: Review of oceanography of the subarctic Pacific region. *Int. North Pacific Fish. Commun. Bull.*, 13, 195.

Dodson, S.I. 1990: Predicting diel vertical migration of zooplankton. *Limnol. Oceanogr.* 35, 1195–1200.

Doodson, A.T. and H.D. Warburg 1941: *Admiralty Manual of Tides*. Hydrographic Department, Admiralty, London.

Druffel, E.R.M. 1989: Decade time scale variability of ventilation in the North Atlantic: High-resolution measurements of bomb radiocarbon in banded corals. *J. Geophys. Res.*, 94, 3271–3285.

Druffel, E.R.M. and P.M. Williams 1991: Radiocarbon in seawater and organisms from the Pacific coast of Baja California. *Radiocarbon*, 33, 291–296.

Dziewonski, A., S. Bloch, and M. Landisman 1969: A technique for the analysis of transient seismic signals. *Bull. Seismological Soc. Am.*, 59, 427–444.

Edmond, J.M., C.M. Measures, R.E. McDuff, L.H. Chan, R. Collier, B. Grant, L.I. Gordon, and J.B. Corliss 1979: Ridge crest hydrothermal activity and the balance of the major and minor elements in the ocean: The Galapagos data. *Earth Planet. Sci. Lett.*, 46, 1–18.

Efron, B. and G. Gong 1983: A leisurely look at the bootstrap, the jackknife, and cross-validation. *Am. Statistician*, *37*, 36–48.

Egbert, G.D., A.F. Bennett, and M.G.G. Foreman 1994: TOPEX/POSEIDON tides estimated using a global inverse model. *J. Geophys. Res.*, *99*, 24,821–24,852.

Ekman, V.W. 1905: On the influence of the earth's rotation on ocean currents. *Arkiv für mathematik astronomi, cch fysik*. 2.

Ekman, V.W. 1932: On an improved type of current meter. *J. Cons. Int. Explor. Mer.*, *7*, 3–10.

Elgar, S. 1988: Comment on "Fourier transform filtering: A cautionary note" by A.M. Forbes. *J. Geophys. Res.*, *93*, 15,755–15,756.

Elsner, J.B. and A.A. Tsonis 1991: Comparison of observed Northern Hemisphere surface air temperature records. *Geophys. Res. Lett.*, *18*, 1229–1232.

Emery, W.J. 1983: On the geographical variability of the upper level mean and eddy fields in the North Atlantic and North Pacific. *J. Phys. Oceanogr.*, *12*, 269–291.

Emery, W.J. and J. Meincke 1986: Global water masses: summary and review. *Oceanoglogica Acta*, *9*, 383–391.

Emery, W.J., T.C. Royer, and R.W. Reynolds 1985: The anomalous tracks of North Pacific drifting buoys 1981–83. *Deep-Sea Res.*, *32*, 315–347.

Emery, W.J., A.C. Thomas, M.J. Collins, W.R. Crawford, and D.L. Mackas 1986: An objective procedure to compute surface advective velocities from sequential infrared satellite images. *J. Geophys. Res.*, *91*, 12,865–12,879.

Emery, W.J., G.H. Born, D.G. Baldwin, and C.L. Norris 1989a: Satellite derived water vapor corrections for GEOSAT altimetry. *J. Geophys. Res.*, *95*, 2953–2964.

Emery, W.J., J. Brown, and V.P. Novak 1989b: AVHRR image navigation; summary and review. *Photog. Eng. and Rem. Sens.*, *8*, 1175–1183.

Emery, W.J.. C.W. Fowler, and C.A. Clayson 1992: Satellite image derived Gulf Stream currents. *J. Oceanic and Atm. Sci. Tech.*, *9*, 285–304.

Emery, W.J., Y. Yu, G. Wick, P. Schluessel, and R.W. Reynolds 1994: Improving satellite infrared sea surface temperature estimates by including independent water vapor observations. *J. Geophys. Res.*, *99*, 5219–5236.

Emery, W.J., C.W. Fowler, and J. Maslanik 1994: Arctic sea ice concentrations from special sensor microwave imager and advanced very high resolution radiometer satellite data. *J. Geophys. Res.*, *99*, 18,329–18,342.

Enfield, D.B. and L. Cid 1990: Statistical analysis of El Niño/Southern oscillation over the last 500 years. *TOGA Notes*, *1*, 1–4.

Essex, C., T. Lookman, and M.A.H. Nerenberg 1987: The climate attractor over short timescales. *Nature*, *326*, 64–66.

Evans, J.C. 1985: Selection of a numerical filtering method: Convolution or transform windowing? *J. Geophys. Res. 90*, 4991–4994.

Farge, M. 1992: Wavelet transforms and their applications to turbulence. *Ann. Rev. Fluid Mech.*, *24*, 395–457.

Farmer, D. and J.D. Smith 1980: Nonlinear internal waves in a fjord. In *Fjord Oceanography* (eds H.J. Freeland, D.M. Farmer, and C.D. Levings). Plenum, New York, pp. 465–493.

Farmer, D.M., S.F. Clifford, and J.A. Verral 1987: Scintillation structure of a turbulent tidal flow. *J. Geophys. Res.*, *92*, 5369–5382.

Favorite, F., A.J. Dodimead, and K. Nasu 1976: Oceanography of the subarctic Pacific region. *Int. North Pac. Fish. Commission Bull.*, *33*, 187.

Feely, R.A., J.F. Gendron, E.T. Baker, and G.T. Lebon 1994: Hydrothermal plumes along the East Pacific Rise, 8°40′ to 11°50′N: Particle distribution and composition. *Earth. Planet. Sci. Lett.*, *128*, 19–36.

Fine, R.A. 1985: Direct evidence using tritium data for throughflow from the Pacific into the Indian Ocean. *Nature*, *315*, 478–480.

Flagg, C.N. and S.L. Smith 1989: On the use of the acoustic Doppler current profiler to measure zooplankton abundance. *Deep-Sea Res.*, *36*, 455–479.

Flierl, G. and A.R. Robinson 1977: XBT measurements of thermal gradient in the MODE eddy. *J. Phys. Oceanogr.*, 7, 300–302.

Fofonoff, N.P. 1960: Transport computations for the North Pacific Ocean 1955–1958. Fish. Res. Board Canada, Manuscript Report Oceanogr. and Limnol., No. 77-80.

Fofonoff, N.P. 1969: Spectral characteristics of internal waves in the ocean. Frederick C. Fuglister Sixtieth Anniversary volume. *Deep-Sea Res.*, 16(suppl.), 58–71.

Fofonoff, N.P. and C. Froese 1960: Programs for Oceanographic computations and data processing on the electronic digital computer ALWAC III-E. M-1 miscellaneous programs. Fish. Res. Board Canada, Manuscript Report Oceanogr. and Limnol., No. 72.

Fofonoff, N.P. and S. Tabata 1966: Variability of oceanographic conditions between Ocean Station P and Swiftsure Bank off the Pacific coast of Canada. *J. Fish. Res. Bd. Canada*, 23, 825–868.

Fofonoff, N.P., S.P. Hayes, and R.C. Millard 1974: WHOI/Brown CTD microprofiler: methods of calibration and data handling. WHOI-74-89.

Forbes, A.M.G 1988: Fourier transform filtering: a cautionary note. *J. Geophys. Res.*, 93, 6958–6962.

Forch, C., M. Knudsen and S.P.L. Sorensen 1902: Berichte ueber die Konstantenbestimungen zur Aufstellunt der hydrographischen Tabellen. Kgl. Danske Vedenskab. Selskab Skrifter, 7 Taekke, Naaturvidensk, og Mex.

Foreman, M.G.G. 1977: Manual for tidal height analysis and prediction. Pacific Mar. Sci. Rep. 77-10. Institute of Ocean Sciences, Sidney, BC.

Foreman, M.G.G. 1978: Manual for tidal currents analysis and prediction. Pacific Mar. Sci. Rep. 78-6. Institute of Ocean Sciences, Sidney, BC.

Fraedrich, K., and L.M. Leslie 1989: Estimates of cyclone track predictability. I: tropical cyclones in the Australian region. *Q. J. R. Meteorol. Soc.*, 115, 79–92.

Fraedrich, K., R. Grotjanh, and L.M. Leslie 1990: Estimates of cyclone track predictability. II: Fractal analysis of mid-latitude cyclones. *Q. J. R. Meteorol. Soc.*, 116, 317–335.

Freeland, H., J.A. Church, R.L. Smith, and F.M. Boland 1985: Current meter data from the Australian Coastal Experiment: a data report. CSIRO Marine Laboratories, Report 169, Hobart, Australia.

Freeland, H.J. and W. J. Gould 1976: Objective analysis of meso-scale ocean circulation features. *Deep-Sea Res.*, 23, 915–923.

Freeland, H.J., F.M. Boland, J.A. Church, A.J. Clarke, A.M.G. Forbes, A., Huyer, R.L. Smith, R.O.R.Y. Thompson, and N.J. White 1986: The Australian coastal experiment: A search for coastal-trapped waves. *J. Phys. Oceanogr.*, 16, 1230–1249.

Freeland, H.J., P.B. Rhines, and T. Rossby 1975: Statistical observations of the trajectories of neutrally buoyant floats in the North Atlantic. *J. Marine Res.*, 33, 383–404.

Friehe, C.A. and S.E. Pazan 1978: Performance of an air-sea interaction buoy. *J. Appl. Meteor.*, 17, 1488–1497.

Fuglister, F.C. 1960: Atlantic Ocean Atlas of temperature and salinity profiles and data from the International Geophysical Year of 1957–1958. Woods Hole Oceanographic Institution Atlas Series 1.

Gamage, N. and W. Blumen 1993: Comparative anlaysis of low-level cold fronts: Wavelet, Fourier, and empirical orthogonal function decompositions. *Month. Weather Rev.*, 121, 2867–2878.

Gammon, R.H., J. Cline and D. Wisegarver 1982: Chlorofluoromethanes in the northeast Pacific Ocean: Measured vertical distributions and application as transient tracers of upper ocean mixing. *J. Geophys. Res.*, 87, 9441–9454.

Gandin, L.S. 1965: *Objective Analysis of Meteorological Fields.* Israel Program for Scientific Translations, Jerusalem.

Gargett, A.E. 1994: Observing turbulence with a modified acoustic Doppler current profiler. *J. Atmospheric Oceanic Technol.*, 11, 1592–1610.

Gargett, A.E., G. Östlund, and C.S. Wong 1986: Tritium time series from Ocean Station P. *J. Phys. Oceanogr., 16*, 1720–1726.

Garrett, C. and W.H. Munk 1979: Internal waves in the ocean. *Ann. Rev. Fluid Mech. 11*, 339–369.

Garrett, C. and B. Petrie 1981: Dynamical aspects of the flow through the Strait of Belle Isle. *J. Phys. Oceanogr., 11*, 376–393.

Garrett, C.J.R. and W.H. Munk 1971: The age of the tide and the "Q" of the oceans. *Deep-Sea Res., 18*, 493–503.

Gendron, J.F., J.P. Cowen, R.A. Feely, and E.T. Baker 1993: Age estimate for the 1987 megaplume on the southern Juan de Fuca Ridge using excess radon and manganese partitioning. *Deep-Sea Res., 40*, 1559–1567.

Georgi, D.T., J.P. Dean, and J.A. Chase 1980: Temperature calibration of expendable bathythermographs. *Ocean Engng, 7*, 491–499.

Godfrey, J.S. and T.J. Golding 1981: The Sverdrup relation in the Indian Ocean and the effect of Pacific–Indian Ocean throughflow on the Indian Ocean circulation and on the East Australia Current. *J. Phys. Oceanogr., 11*, 771–779.

Godin, G. 1972: *The Analysis of Tides*. University of Toronto Press.

Goldstein, R.M., T.P. Barnett, and H.A. Zebker 1989: Remote sensing of ocean currents. *Science, 246*, 1282–1285.

Gonella, J. 1972: A rotary component method for analyzing meteorological and oceanographic vector time series. *Deep-Sea Res., 19*, 833–846.

Gonzalez, F.I. and Ye. A. Kulikov 1993: Tsunami dispersion observed in the deep ocean. In *Tsunamis in the World*, Kluwer, pp. 7–16.

Gooberlet, M.A., C.T. Swift and J.C. Wilkerson 1990: Ocean surface wind speed measurements of the special sensor microwave imager (SSM/I). *IEEE Trans. Geosci. Rem. Sens., 28*, 823–828.

Gordon, A.L. 1986: Interocean exchange of thermocline water. *J. Geophys. Res., 91*, 5037–5046.

Gordon, R.L. 1996: *Acoustic Doppler Current Profilers. Principles of Operation: A Practical Primer*, 2nd edn. RD Instruments, San Diego, Calif.

Gould, W.J. 1973: Effects of non-linearities of current meter compasses. *Deep-Sea Res., 20*, 423–427.

Gould, W.J. and E. Sambuco 1975: The effect of mooring type on meaured values of ocean currents. *Deep-Sea Res., 22*, 55–62.

Goupillaud, P., A. Grossmann, and J. Morlet 1984: Cycle-octave and related transforms in seismic signal analysis. *Geoexploration, 23*, 85–105.

Grassberger, P. and I. Procaccia 1983: Measuring the strangeness of strange attractors. *Physica 9D*, 189–208.

Grasshoff, K. 1983: *Methods of Seawater Analysis*. Verlag Chemie, Weinheim.

Grassl, H. 1976: The dependence of the measured cool skin of the ocean on wind stress and total heat flux. *Boundary Layer Meteorol., 10*, 465–474.

Gray, H.L. and W.A. Woodward 1992: Autoregressive models not sensitive to initial conditions. *EOS Trans., AGU, 73*(25), 267–268.

Green, A. 1984: Bulk dynamics of the expendable bathythermograph (XBT). *Deep-Sea Res., 31*, 415–426.

Groves, G.W. 1955: Numerical filters for discrimination against tidal periodicities. *Trans. Am. Geophys. Union, 36*, 1073–1084.

Groves, G.W. and E.J. Hannan 1968: Time series regression of sea level on weather. *Rev. Geophys., 6*, 129–174.

Gruza, G.V., E.Ya. Ran'kova, and E.V. Rocheva 1988: Analysis of global data variations in surface air temperature during instrument observation period. *Meteor. Gridr.*, 16–24.

Guymer, T.H., J.A. Businger, W.L. Jones, and R.H. Stewart 1981: Anomalous wind estimates from SEASAT scatterometer. *Nature, 294*, 735–737.

Halpern, D. 1978: Mooring motion influences on current measurements. In Proceedings of a Workshop Conference on Current Measurement, Technical Report DEL-SG-3-78, College of Marine Studies of Deleware, Newark.

Halpern, D., R.A. Weller, M.G. Briscoe, R.E. Davis, and J.R. McCullough 1981: Intercomparison tests of moored current measurements in the upper ocean. *J. Geophys. Res.*, 86, 419–428.

Halpern, D., W. Knauss, O. Brown, and F. Wentz 1993: An atlas of monthly mean distributions of SSMI surface wind speed, ARGOS buoy drift, AVHRR/2 sea surface temperature, and ECMWF surface wind components during 1991. JPL Publications 93-10, Jet Propulsion Laboratory, Pasadena.

Hamming, R.W. 1977: *Digital Filters*. Prentice-Hall, Englewood Cliffs, N.J.

Hamon, B.V. 1955: A temperature–salinity–depth recorder. Conseil Permanent International pour le Exploration de la Mer. *Journal du Conseil*, 21, 22–73.

Hamon, B.V. and N.L. Brown 1958: A temperature–chlorinity–depth recorder for use at sea. *J. Scientific Instru.*, 35, 452–458.

Hanawa, K. and H. Yoritaka 1987: Detection of systematic errors in XBT data and their correction. *J. Oceanogr. Soc. Japan*, 32, 68–76.

Hanawa, K., and Y. Yoshikawa 1991: Re-examination of depth error in XBT data. *J. Atmos. Oceanic Technol.*, 8, 422–429.

Hanawa, K. and T. Yasuda 1992: New detection method for XBT depth error and relationship between the depth error and coefficients in the depth-time equation. *J. Oceanogr.*, 48, 221–230.

Hansen, J. and S. Lebedeff 1987: Global trends of measured surface air temperature. *J. Geophys. Res.* 92, 13,345–13,372.

Harada, K. and S. Tsunogai 1986: Ra-226 in the Japan Sea and the residence time of the Japan Sea water. *Earth Planet. Sci. Lett.*, 77, 236–244.

Harnett, D.L. and J.L. Murphy 1975: *Intoductory Statistical Analysis*. Addison-Wesley, Reading Mass.

Harris, F.J. 1978: On the use of windows for harmonic analysis with the discrete Fourier transform. *Proc. IEEE*, 66, 51–83.

Haxby, W.F. 1985: Gravity field of the world's oceans, chart scale 1:51,400,000. Office of Naval Research, Washington, D.C.

Hayashi, Y. 1979: Space–time spectral analysis of rotary vector series. *J. Atmos. Sci.*, 36, 757–766.

Hayne, G.S. and D.W. Hancock 1982: Sea state-related altitude errors in the Seasat altimeter. *J. Geophys. Res.*, 87, 3227–3231.

Heinmiller, R.H. 1968: Acoustic release systems. WHOI Technical Report 68-48, Woods Hole, Mass.

Heinmiller, R.H., C.C. Ebbesmeyer, B.A. Taft, D.B. Olson, and G.P. Nitkin 1983: Systematic errors in expendable bathythermograph (XBT) profiles. *Deep-Sea Res.*, 30, 1185–1197.

Helland-Hansen, B. 1918: Nogen hydrografiske metoder. *Forh. Skand. Naturforskeres*, 16, Kristiania.

Hellerman, S. and M. Rosenstein 1983: Normal monthly wind stress over the World Ocean with error estimates. *J. Phys. Oceanogr.* 13, 1093–1104.

Hendry, R. 1993: Canadian Technical Report of Hydrography and Ocean Sciences, Bedford Institute of Oceanography CTD Trials. BIO, Dartmouth, Nova Scotia, B2Y 4A2.

Henry, R.F. and P.W.U. Graefe 1971: Zero padding as a means of improving definition of computed spectra. Manuscript Series Report Series No. 20, Canadian Department of Energy, Mines and Resources, Ottawa.

Hichman, M.L. 1978: *Measurement of Dissolved Oxygen*. John Wiley, New York.

Hill, M.N. (ed.) 1962: *The Sea, Volume 1, Physical Oceanography*. Interscience, New York.

Hilland, J.E., D.B. Chelton, and E.G. Njoku 1985: Production of global sea surface temperature fields for the Jet Propulsion Laboratory workshop. *J. Geophys. Res.*, 90, 11,642–11,650.

Hiller, W. and R.H. Käse 1983: Objective analysis of hydrographic data sets from mesoscale surveys. *Ber. Inst. Meereskd. Univ. Kiel*, 116.

Hiyagon, H. 1994: Retention of solar helium and neon in IDPs in deep sea sediment. *Science*, 263, 1257–1259.

Hogg, N. 1977: Topographic waves along 70°W on the continental rise. *J. Mar. Res.*, 39, 627–649.

Holl, M.M. and B.R. Mendenhall 1972: Fields by information blending, sea-level pressure version, Technical Note 72-2, Fleet Numerical Weather Central, Monterey, Calif.
Hollinger, J. 1989: DMSP Special Sensor Microwave/Imager calibration/validation. Final report, Vol. 1, Navy Research Laboratory, Washington, DC.
Horel, J.D. 1984: Complex principal component analysis: Theory and examples. *J. Climate Appl. Meteorol. 23*, 1660–1673.
Horne, E.P.W. and J.M. Toole 1980: Sensor response mismatches and lag correction techniques for temperature-salinity profilers. *J. Phys. Oceanogr. 10*, 1122–1130.
Hsieh, W.W. 1986: Comments on: "Rotary empirical orthogonal function analysis of currents near the Oregon Coast". *J. Phys. Oceanogr. 16*, 791–792.
Huang, N.E., C.D. Leitao, and CG. Parra 1978: Large-scale Gulf Stream frontal study using Geos 3 radar altimeter data. *J. Geophys. Res., 83*, 4673–4682.
Huggett, W.S., W.R. Crawford, R.E. Thomson and M.V. Woodward 1987: Data record of current observations Volume XIX. Coastal Ocean Dynamics Experiment (CODE), Institute of Ocean Sciences, Part 1.
Iler, R.K. 1979: *The Chemistry of Silica*. John Wiley, New York.
Jackson, D.D. 1972: Interpretation of inaccurate, insufficient and inconsistent data. *Geophys. J. R. Astron. Soc., 28*, 97–109.
Jacobsen, A.W. 1948: An instrument for recording continuously the salinity, temperature and depth of sea water. *Trans. Am. Meteorol. Soc.*, 1057–1070.
James, R.W. and P.T. Fox 1972:. Comparative sea-surface temperature measurements. Report No. 5, Report on Marine Science Affairs, WMO, Geneva, pp. 117–129.
Jamous, D., L. Mémery, C. Andrié, P. Jean-Baptiste, and L. Merlivat 1992: The distribution of helium 3 in the deep western and southern Indian Ocean. *J. Geophys. Res., 97*, 2243–2250.
Jenkins, F.A. and H.E. White 1957: *Fundamentals of Optics*. McGraw-Hill, New York.
Jenkins, G.M. and D.G. Watts 1968: *Spectral Analysis and its Applications*. Holden-Day, San Francisco, Calif.
Jenkins, W.J. 1988. The use of anthropogenic tritium and helium-3 to study subtropical gyre ventilation and circulation. *Phil. Trans. R. Soc. London A325*, 43–61.
Jenkins, W.J., J.M. Edmond and J. B Corliss 1978: Excess ^{3}He and ^{4}He in Galapagos submarine hydrothermal waters. *Nature, 272*, 156–158.
Jerlov, N.G. 1976: *Marine Optics*. Elsevier, New York.
Jones, P.D. 1988: Hemispheric surface air temperature variations: Recent trends and an update to 1987. *J. Climate, 1*, 654.
Jones, P.D., T.M.L. Wigley, and P.B. Wright 1986: Global temperature variations between 1861 and 1984. *Nature, 322*, 430–434.
Julian, P.R. 1975: Comments on the determination of significance levels of the coherence statistic. *J. Atmos. Sci., 32*, 836–837.
Kadko, D., N.D. Rosenburg, J.E. Lupton, R. Collier, and M. Lilley, 1990: Chemical reaction rates and entrainment within the Endeavour Ridge hydrothermal plume. *Earth Planet. Sci. Lett., 99*, 315–335.
Kanasewich, E.R 1975: *Time Series Analysis in Geophysics*. University of Alberta Press, Edmonton.
Kautsky, H. 1939: Quenching of luminescence by oxygen. *Trans. Faraday Soc., 35*, 216–219.
Kay, S. M., and S.L. Marple, Jr 1981: Spectrum analysis—a modern perspective. *Proc. IEEE, 69*, 1380–1417.
Kelly, K.A. 1988: Comment on "Empirical orthogonal function analysis of advanced very high resolution radiometer surface temperature patterns in Santa Barbara Channel" by G.S.E. Lagerloef and R.L. Berstein. *J. Geophys. Res., 93*, 15,753–15,754.
Kelly, K.A. and P.T. Strub 1992: Comparison of velocity estimates from advanced very high resolution radiometer. *J. Geophys. Res., 97*, 9653–9668.
Keyte, F.K. 1965: On the formulae for correcting reversing thermometers. *Deep-Sea Res., 12*, 163–172.

Kipphut, G.W. 1990: Glacial meltwater input to the Alaska Coastal Current: evidence from oxygen isotope measurements. *J. Geophys. Res.* 95, 5177–5181.

Kirwan, A.D. Jr and M.S. Chang 1976: On the micropolar Ekman problem. *Int. J. Engng Sci.*, 14, 685–692.

Kirwan, A.D., Jr, G. McNally, M.-S. Chang, and R. Molinari 1975: The effect of wind and surface currents on drifters. *J. Phys. Oceanogr.*, 5, 361–368.

Kirwan, A.D., Jr, G.J. McNally, E. Reyna and W.J. Merrell, Jr 1978: The near-surface circulation of the eastern North Pacific. *J. Phys. Oceanogr.*, 8, 937–945.

Kirwan, A.D., Jr, G. McNally, S. Pazan, and R. Wert 1979: Analysis of surface current response to wind. *J. Phys. Oceanogr.*, 9, 401–412.

Knudsen, M. (ed.) 1901: *Hydrographical Tables*. GEC, Copenhagen.

Konyaev, K.V. 1990: *Spectral Analysis of Physical Oceanographic Data*. National Science Foundation, Washington, D.C.

Koopmans, L.H. 1974: *The Spectral Analysis of Time Series*. Academic Press, New York.

Krauss, W. 1993: Ekman drift in homogenous water. *J. Geophys. Res.*, 98, 20,187–20,209.

Krauss, W. and R.H. Käse 1984: Mean circulation and eddy kinetic energy in the eastern North Atlantic. *J. Geophys. Res.*, 84, 3407–3415.

Kremling, K. 1972: Comparison of specific gravity in natural sea-water from hydrographical tables and measurements by a new density instrument. *Deep-Sea Res.*, 19, 377–383.

Kuhn, H., D. Quadfasel, F. Schott, and W. Zenk 1980: On simultaneous measurements with rotor, wing and acoustic current meters, moored in shallow water. *Deutsche Hydrographische Zeitschrift*, 33, 1–18.

Kulikov, E.A. and F.I. Gonzalez 1995: *On Reconstruction of the Initial Tsunami Signal from Distant Bottom Pressure Records*. Doklady Akademii Nauk.

Kundu, P.K. 1976: An analysis of inertial oscillations observed near the Oregon coast. *J. Phys. Oceanogr.* 6, 879–893.

Kundu, P.K. 1990: *Fluid Mechanics*. Academic Press, San Diego, Calif.

Kundu, P.K. and J.S. Allen 1976: Some three-dimensional characteristics of low-frequency current fluctuations near the Oregon coast. *J. Phys. Oceanogr.*, 6, 181–199.

Kundu, P.K. J.S. Allen, and R.L. Smith 1975: Modal decomposition of the velocity field near the Oregon coast. *J. Phys. Oceanogr.*, 5, 683–704.

Labrecque, A.M., R.E. Thomson, M.W. Stacey, and J.R. Buckley 1994: Residual currents in Juan de Fuca Strait. *Atmosphere-Ocean*, 32, 375–394.

Lacoss, R.T. 1971: Data adaptive spectral analysis methods. *Geophys.*, 36, 661–675.

LaFond, E.C. 1951: Processing oceanographic data, HO Publication No. 614, US Hydrographic Office.

Lanczos, C. 1956: *Applied Analysis*. Prentice-Hall, Englewood Cliffs, NJ. (Reprinted in 1988, Dover, New York.)

Langdon, C. 1984: Dissolved oxygen monitoring system using a pulsed electrode: Design, performance and evaluation. *Deep-Sea Res.*, 31, 1357–1367.

Laxon, S. and D. McAdoo 1994: Arctic Ocean gravity field derived from ERS-1 satellite altimetry. *Science*, 265, 621–625.

LeBlond, P.H. and L.A. Mysak 1979: *Waves in the Ocean*. Elsevier, Amsterdam.

Lemon, D.D. and D.M. Farmer 1990: Experience with a multi-depth scintillation flowmeter in the Fraser Estuary. Proceedings IEEE Fourth Working Conference on Current Measurement, Clinton, MD, April 3–5, 1990, pp. 290–298.

Lemon, D.D., R.E. Thomson, J.R. Delaney, D.M. Farmer, F. Rowe, and R.A.J. Chave 1996: Acoustic scintillation velocity measurement of a buoyant hydrothermal plume. Preliminary report, ASL Environmental Science.

Levitus, S. 1982: Climatological atlas of the world ocean. NOAA Professional Paper 13, US Department of Commerce, Rockville, Md.

Lewis, E.L. and R.G. Perkin 1981: The practical salinity scale 1978: conversion of existing data. *Deep-Sea Res.*, 28A, 307–328.

Lewis. E.L. 1980: The practical salinity scale 1978 and its antecedents. *J. Oceanic Engineering*, 5, 3–8.
Lilley, M.D., R.A. Feely, and J.H. Trefry 1995: Chemical and biochemical transformations in hydrothermal plumes. In *Seafloor Hydrothermal Systems: Physical, Chemical, Biological and Geological Interactions* (eds S.E. Humphris, R.A. Zierenberg, L.S. Mullineaux, and R.E. Thomson). AGU, Geophysical Monograph 91, 369–391.
Lipps, F.B. and R.S. Hemler 1992: On the downward transfer of tritium to the ocean by a cloud model. *J. Geophys. Res.*, 97, 12,889–12,900.
Livingstone, D. and T.C. Royer 1980: Eddy propagation determined from rotary spectra. *Deep-Sea Res.*, 27A, 883–835.
Llewellyn-Jones, D.T., P.J. Minett, R.W. Saunders, and A.M. Zavody 1984: Satellite multi-channel infrared measurements of sea surface temperature of the northeast Atlantic Ocean using AVHRR/2. *Q. J. R. Meteorol. Soc.*, 110, 613–631.
Lomb, N.R. 1976: Least-squares frequency-analysis of unequally spaced data. *Astrophys. Space Sci.*, 39, 447–462.
Lorenz, E. 1956: Empirical orthogonal functions and statistical weather prediction. Scientific Report No. 1, Air Force Cambridge Research Center, Air Research and Development Command. Cambridge Mass.
Lueck, R.G. 1990: Thermal inertia of conductivity cells: Theory. *J. Atmos. Oceanic Technol.*, 7, 741–755.
Lueck, R.G., O. Hertzman, and T.R. Osborn 1977: The spectral response of thermistors. *Deep-Sea Res.*, 24, 951–970.
Lukas, R. 1994: HOT results show interannual variability of Pacific Deep and Bottom waters. *WOCE Notes*, 6(2), 4.
Lupton, J.E. and H. Craig 1981: A major helium-3 source at 15°S on the East Pacific Rise. *Science*, 214, 13–18.
Lupton, J.E., J.R. Delaney, H.P. Johnson and M.K. Tivey 1985: Entrainment and vertical transport of deep-ocean water by buoyant hydrothermal plumes. *Nature*, 316, 621–623.
Lupton, J.E., E.T. Baker and G.J. Massoth 1989: Variable ^{3}He/heat ratios in submarine hydrothermal systems: evidence from two plumes over the Juan de Fuca Ridge. *Nature*, 337, 161–164.
Lupton, J.E., E.T. Baker, M.J. Mottl, F.J. Sansone, C.G. Wheat, J.A. Resing, G.J. Massoth, C.I. Measures, and R.A. Feely 1993: Chemical and physical diversity of hydrothermal plumes along the East Pacific Rise, 8°45′N to 11°50′N. *Geophys. Res. Lett.*, 20, 2913–2916.
Lütkepohl, H. 1985: Comparison criteria for estimating the order of a vector autoregressive process. *J. Time Series Anal.*, 6, 35–52.
Lynn, R.J. and J.L. Reid 1968: Characteristics and circulation of deep and abyssal waters. *Deep-Sea Res.*, 15, 577–598.
Lynn, R.J. and J. Svejkovsky 1984: Remotely sensed sea surface temperature variability off California during a "Santa Ana" clearing. *J. Geophys. Res.*, 89, 8151–8162.
Macdonald, R.W., F.A. McLaughlin, and C.S. Wong. 1986: The storage of reactive silicate samples by freezing. *Limnol. Oceanogr.*, 31, 1139–1142.
Mackas, D.L., K.L. Denman, and A.F. Bennett 1987: Least squares multiple tracer analysis of water mass composition. *J. Geophys. Res.*, 92, 2907–2918.
Mackenzie, K.V. 1981: Nine term equation for sound speed in the oceans. *J. Acoust. Soc. Am.*, 70, 807–812.
Macklin, S.A., P.J. Stabeno and J.D. Schumacher 1993: A comparison of gradient and observed over-the-water winds along a mountainous coast. *J. Geophys. Res.*, 98, 16,555–16,569.
MacPhee, S.B. 1976: Acoustics and echo sounding instrumentation. Canadian Hydrographic Service Technical Report 76-1.
Mandelbrot, B.B. 1967: How long is the coast of Britain? Statistical self-similarity and fractional dimension. *Science*, 155, 636–638.

Mantyla, A.W. and J.L. Reid 1983: Abyssal characteristics of the World Ocean waters. *Deep-Sea Res., 30*(8A), 805–833.

Marks, K.M., D.C. McAdoo, and W.H.F. Smith. 1993:. Mapping the southeast Indian Ridge with Geosat. *EOS, 74*(8), 81, 86.

Marple, Jr, S.L. 1987: *Digital Spectral Analysis*. Prentice-Hall, Englewood Cliffs, N.J.

Marsden, R.F. 1987: A comparison between geostrophic and directly measured surface winds over the northeast Pacific Ocean. *Atmosphere-Ocean, 25*, 387–401.

Martin, J.H. and G.A. Knauer 1973: The elemental composition of plankton. *Geochem. Cosmochim. Acta, 37*, 1639–1653.

Martin, M., L.D. Taley, and R.A. de Szoeke 1987: Physical, chemical and CTD data from Marathon Expedition R/V *Thomas Washington* 261, 4 May–4 June 1984. Oregon State University, College of Oceanography, Data Report 131, Ref. 87-15, May 1987.

Masson, D. 1996: A case study of wave–current interaction in a strong tidal current. *J. Phys. Oceanogr., 26*, 359–372.

Maul, G. and N.J. Bravo 1983: Fitting of satellite and in-situ ocean surface temperatures: results for Polymode during the winter of 1977–1978. *J. Geophys. Res., 88*, 9605–9616.

McClain, C.R., J.A. Yoder, L.P. Atkinson, J.O. Blanton, T.N. Lee, J.J.Singer, and F. Müller-Karger 1988: Variability of surface pigment concentration in the South Atlantic Bight. *J. Geophys. Res., 93*, 10,675–10,697.

McClain, E.P. 1981: Multiple atmosphere-window techniques for satellite sea surface temperatures. In *Oceanography from Space* (ed.) J.F.R. Gower. Plenum, New York, pp. 73–85.

McClain, E.P., W.G. Pichel, C.C. Walton, Z. Ahmad, and J. Sutton 1983: Multi-channel improvements to satellite-derived global sea surface temperatures. *Adv. Space Res., 2*, 43–47.

McDougall, R.J. 1985a: Double-diffusive interleaving/Part 1: Linear stability analysis. *J. Phys. Oceanogr., 15*, 1532–1541.

McDougall, R.J. 1985b: Double-diffusive interleaving/Part 2: Finite amplitude, steady state interleaving. *J. Phys. Oceanogr., 15*, 1542–1556.

McDuff, R.E. 1988: Effects of vent fluid properties on the hydrography of hydrothermal plumes. *EOS Trans., AGU, 69*, 1497.

McDuff, R.E. 1995: Physical dynamics of deep-sea hydrothermal plumes. In *Seafloor Hydrothermal Systems: Physical, Chemical, Biological, and Geological Interactions* (eds S.E. Humphris, R.A. Zierenberg, L.S. Mullineaux, and R.E. Thomson). American Geophysical Union, Geophysical Monograph 91, Washington, D.C., pp. 357–368.

McManus, J., R.W. Collier, C.-T.A. Chen, and J. Dymond 1992: Physical properties of Crater Lake, Oregon: A method for the determination of conductivity- and temperature-dependent expression of salinity. *Limnol. Oceanogr., 37*, 41–53.

McNally, G.J. 1981: Satellite-tracked drift buoy observations of the near-surface flow in the eastern mid-latitude North Pacific. *J. Geophys. Res., 86*, 8022–8030.

McNally, G.J. and W.B. White 1985: Wind driven flow in the mixed layer observed by drifting buoys during autumn-winter in midlatitude North Pacific. *J. Phys. Oceanogr., 15*, 684–694.

McNally, G.J., W.C. Patzert, A.D. Kirwan, Jr, and A.C. Vastano. 1983: The near-surface circulation of the North Pacific using satellite tracked drifting buoys. *J. Geophys. Res., 88*, 7507–7518.

McWilliams, J.C. 1976: Maps from the Mid-Ocean Dynamics Experiment: Part I. Geostrophic streamfunction. *J. Phys. Oceanogr., 6*, 810–827.

Meisel, D.D. 1978: Fourier transforms of data sampled at unequaled observational intervals. *Astron. J., 83*, 538–545.

Meisel, D.D. 1979: Fourier transforms of data sampled in unequally spaced segments. *Astron. J., 84*, 116–126.

Memery, L. and C. Wunsch 1990: Constraining the North Atlantic circulation with tritium data. *J. Geophys. Res., 95*, 5239–5256.

Mero, T.M. 1982: Performance results for the EG7G vector-measuring current meter (VMCM), In Proceedings of IEEE Second Working Conference on Current Measurement (eds M. Dursi and W. Woodward). Institute of Electrical and Electronics Engineers, New York, pp. 159–164.

Merrifield, M.A. and R.T. Guza 1990: Detecting propagating signals with complex empirical orthogonal functions: A cautionary note. *J. Phys. Oceanogr.*, 20, 1628–1633.

Meyers, S.D., B.G. Kelly, and J.J. O'Brien 1993: An introduction to wavelet analysis in oceanography and meteorology: With application to the dispersion of Yanai waves. *Monthly Weather Rev.*, 121, 2858–2866.

Middleton, J.H. 1982: Outer rotary cross spectra, coherences and phases. *Deep-Sea Res.*, 29(10A), 1267–1269.

Middleton, J.H. 1983: Low-frequency trapped waves on a wide, reef-fringed continental shelf. *J. Phys. Oceanogr.*, 13, 1371–1382.

Middleton, J.H. and A. Cunningham 1984: Wind-forced continental shelf waves from geographical origin. *Cont. Shelf Res.*, 3, 215–232.

Miller, L. and R. Cheney 1990: Large-scale meridional transport in the tropical Pacific Ocean during the 1986–1987 El Niño from Geosat. *J. Geophys. Res.*, 95, 17,905–17,919.

Miyake, Y. and K. Saruhashi 1967: A study on the dissolved oxygen in the ocean. In *Geochemistry Study of the Ocean and the Atmosphere, Yasuo Miyake Seventieth Anniversary*, Geochemical Laboratory, Meteorological Research Institute, Tokyo, 1978, pp. 91–98.

Miyakoda, K. and A. Rosati 1982: The variation of sea surface temperature in 1976 and 1977, I: The data analysis. *J. Geophys. Res.*, 87, 5667–5680.

Montgomery, R.B. 1938: Circulation in the upper layer of the southern North Atlantic deduced with the aid of isentropic analysis. *Papers Phys. Oceanogr. Meteorol.*, 6(2), 55.

Montgomery, R.B. 1958: Water characteristics of Atlantic Ocean and of World Ocean. *Deep-Sea Res.*, 5, 134–148.

Mooers, C.N.K. 1973: A technique for the cross spectrum analysis of pairs of complex-valued time series, with emphasis on properties of polarized components and rotational invariants. *Deep-Sea Res.*, 20, 1129–1141.

Mooers, C.N.K. and R.L. Smith 1967: Continental shelf waves off Oregon. *J. Geophys. Res.*, 73, 549–557.

Muench, R.D. and J.D. Schumacher 1979: Some observations of physical oceanographic conditions on the northeast Gulf of Alaska continental shelf. NOAA Technical Memorandum ERL PMEL-17, Seattle, Wash.

Munk, W. H. and D.E. Cartwright 1966: Tidal spectroscopy and prediction. *Phil. Trans. R. Soc. London A259*, 533–581.

Munk, W. H. and C. Hasselman 1964: Upper resolution of tides. In *Studies in Oceanography*. Tokyo Geophysical Institute, University of Tokyo, pp. 339–344.

Murty, T.S. 1984: Storm surges: meteorological ocean tides. *Can. Bull. Fish. Aquatic Sci.*, 212, 897.

Nemac, A.F.L. and R.O. Brinkhurst 1988: Using the bootstrap to assess statistical significance in the cluster analysis of species abundance data. *Can. J. Fish. Aquat. Sci.*, 45, 965–970.

Niiler, P.P., R.E. Davis, and H.J. White 1987: Water-following characteristics of a mixed layer drifter. *Deep-Sea Res.*, 34, 1867–1882

Ninnis, R.N., W.J. Emery, and M.J. Collins 1986: Automated extraction of sea ice motion from AVHRR imagery. *J. Geophys. Res.*, 91, 10,725–10,734.

Nowlin, W.D., J.S. Bottero, and R.D. Pillsbury 1986: Observations of internal and near-inertial oscillations at Drake Passage. *J. Phys. Oceanogr.*, 16, 87–108.

Nuttall, A.H. 1976: Spectral analysis of a univariate process with bad data points, via maximum entropy, and linear predictive techniques. Naval Underwater systems Center, Technical Document 5419, New London, CT.

Nuttall, A.H. and G.C. Carter 1980: A generalized framework for power spectral estimation. *IEEE Trans. Acoustic Speech, Signal Process*, ASSP-28, 334–335.

Olbers, D.J., P. Müller, and J. Willebrand 1976. Inverse technique analysis of large data sets. *Phys. Earth Planet. Interiors*, 12, 248–252.

Olbers, D.J., M. Wenzel, and J. Willebrand 1985: The inference of North Atlantic circulation patterns from climatological hydrographic data. *Rev. Geophys.*, *23*, 313–356.

Oldenburg, D.W. 1984: An introduction to linear inverse theory. *IEEE Geosci. Rem. Sens.*, *GW-22*, 665–674.

Osborne, A.R., A.D. Kirwan, A. Provenzale, and L. Bergamasco 1989: Fractal drifter trajectories in the Kuroshio Extension. *Tellus*, *41A*, 416–435.

Östlund, H.G. and C.H. Rooth 1990: The North Atlantic tritium and radiocarbon transients 1972–1983. *J. Geophys. Res.*, *95*, 20,147–20,165.

Otnes, R.K. and L. Enochson 1972: *Digital Time Series Analysis*. John Wiley, New York.

Pagano, M. 1978: Some recent advances in autoregressive processes. In *Directions in Time Series* (eds D.R. and G.C. Tiao). Institute for Mechanical Statistics.

Paros, J.M. 1976: Digital pressure transducers. *Measurements and Data*, *10*(2), 74–79.

Parsons, T.R., Y. Maita, and C.M. Lalli 1984: *A Manual of Chemical and Biological Methods for Seawater Analysis*. Pergamon Press, Oxford.

Paulson, C.A. and J.J. Simpson 1981: The temperature difference across the cool skin of the ocean. *J. Geophys. Res.*, *86*, 11,044–11,054.

Pearson, C.A., J.D. Schumacher, and R.D. Muench 1981: Effects of wave-induced mooring noise on tidal and low-frequency current observations. *Deep-Sea Res.*, *28A*, 1223–1229.

Peltier, W.R. 1990: Glacial isostatic adjustment and relative sea-level change. In *Sea-level Change*. National Academy Press, Washington, D.C., pp. 73–87.

Peng, T.-H., W.S. Broecker, G.G. Mathieu, and Y.-H. Li 1979: Radon invasion rates in the Atlantic and Pacific Oceans as determined during the Geosecs Program. *J. Geophys. Res.*, *84*, 2471–2486.

Peterson, J.I., R.V. Fitzgerald, and D.K. Buckhold 1984: Fiber-optic probe for in vivo measurement of oxygen partial pressure. *Anal. Chem.*, *56*, 62–67.

Pettigrew, N.R. and J.D. Irish 1983: An evaluation of a bottom mounted Doppler acoustic profiling current meter. In Proceedings Oceans '83, September 1983.

Pettigrew, N.R., R.C. Beardsley, and J.D. Irish 1986: Field evaluations of a bottom mounted acoustic Doppler current profiler and conventional current meter moorings. Proceedings of the IEEE Third Working Conference on Current Measurement, January 22–24, 1986, Airlie, Virginia, pp. 153–162.

Phillips, O.M., D. Gu, and M. Donelan 1993: Expected structure of extreme waves in a Gaussian sea. Part I: Theory and SWADE buoy measurements. *J. Phys. Oceanogr.*, *23*, 992–1000.

Pickard, G.L. and W.J. Emery 1992: *Descriptive Physical Oceanography: An Introduction*, 5th edn. Pergamon Press, New York.

Pierson, W.J. 1981: The variability of winds over the ocean. In *Space-borne Synthetic Aperature Radar for Oceanography*, Johns Hopkins Oceanographic Studies, Vol. 7 (eds R. Beal, P.S. DeLeonibus, and I. Katz), Johns Hopkins University Press, Baltimore, Md.

Pillsbury, R.D., J.S. Bottero, R.E. Still, and W.E. Gilbert 1974: A compilation of observations from moored current meters. Vols VI and VII. Refs 74-2 and 74-7, School of Oceanography, Oregon State University, Corvallis, Oregon.

Piola, A.R. and A.L. Gordon 1984: Pacific and Indian Ocean upper-layer salinity budget. *J. Phys. Oceanogr.*, *14*, 747–753.

Poulain, P.-M. and P.P. Niiler 1989: Statistical analysis of the surface circulation in the California current system using satellite-tracked drifters. *J. Phys. Oceanogr.*, *19*, 1588–1603.

Pratt, J.H. 1859: see Vogt and Jung, 1991.

Pratt, J.H. 1871: see Vogt and Jung, 1991.

Preisendorfer, R.W. 1988: *Principal Component Analysis in Meteorology and Oceanography*. Developments in Atmospheric Science, *17*. Elsevier, Amsterdam.

Press, W.H., S.A. Teukolsky, W.T. Vetterling and B.P. Flannery 1992: *Numerical Recipes*, 2nd edn. Cambridge University Press, Cambridge.

Priestley, M.B. 1981: *Spectral Analysis and Time Series*. Academic Press, London.

Privalsky, V.E. and D.T. Jensen 1993: Time series analysis package. Utah Climate Center, Logan.

Privalsky, V.E. and D.T. Jensen 1994: Assessment of the influence of ENSO on annual global air temperatures. *Dyn. Atmosphere Oceans*, 22, 161–178.

Quadfasel, D. and F. Schott 1979: Comparison of different methods of current measurements. *Dt. hydrogr. Z.*, 32, 27–38.

Quay, P.D., M. Stuiver, and W.S. Broecker 1983: Upwelling rates for the equatorial Pacific Ocean derived from the bomb ^{14}C distribution. *J. Marine Res.*, 41, 769–792.

Quinn, W.H., V.T. Neal, and S. Antunez de Mayolo 1987: El Niño occurrences over the past four and a half centuries. *J. Geophys. Res.* 92, 14,449–14,461.

Rabiner, L. and B. Gold 1975: *Theory and Application of Digital Signal Processing*. Prentice-Hall, Englewood Cliffs, N.J.

Rabinovich, A.B. and A.S. Levyant 1992: Influence of seiche oscillations on the formation of the long-wave spectrum near the coast of the southern Kuriles. *Oceanology*, 32, 17–23.

Rao, P.K., W.L. Smith and R. Koffler 1972: Global sea surface temperature distribution determined from an environmental satellite. *Mon. Wea. Rev.*, 100, 10–14.

RD Instruments 1989: (see also Gordon R.L.) Acoustic Doppler current profilers. *Principles of Operation: A Practical Primer*. RD Instruments, San Diego, CA.

Redfield, A.C. 1958: The biological control of chemical factors in the environment. *Am. Scientist*, 46, 205–221.

Redfield, A.C., B.H. Ketchum, and F.A. Richards 1963: The influence of organisms on the composition of sea-water. In *The Sea*, Vol. 2 (ed. M.N. Hill). Interscience, New York, pp. 26–77.

Reid, J.L. 1965: Intermediate waters of the Pacific Ocean. *Johns Hopkins Oceanographic Studies*, No. 2.

Reid, J.L. 1982: Evidence of an effect of heat flux from the East Pacific Rise upon the characteristics of the mid-depth waters. *Geophys. Res. Lett.*, 9, 381–384.

Reid, J.L. and R.J. Lynn 1971: On the influence of the Norwegian–Greenland and Weddell Seas upon the bottom waters of the Indian and Pacific Oceans. *Deep-Sea Res.*, 18, 1063–1088.

Reid, J.L. and A.W. Mantyla 1978: On the mid-depth circulation of the North Pacific Ocean. *J. Phys. Oceanogr.*, 8, 946–951.

Reynolds, R.W 1982: A monthly averaged climatology of sea surface temperature. NOAA Technical Report NWS-31, Nat. Oceanic Atmos. Admin., Silver Springs, Md.

Reynolds, R.W. 1983: A comparison of sea surface temperature climatologies. *J. Clim. App. Meteorol.*, 22, 447–459.

Richardon, P.L. 1993: A census of eddies observed in North Atlantic SOFAR float data. *Prog. Oceanogr.*, 31, 1–50.

Richardson, W.S., P.B. Stimson, and C.H. Wilkins 1963: Current measurements from moored buoys. *Deep-Sea Res.*, 10, 369–388.

Richardson, P.L., J.F. Price, W.B. Owens, and W.J. Schmitz, Jr. 1981: North Atlantic subtropical gyre: SOFAR floats tracked by moored listening stations. *Science*, 213, 435–437.

Riser, S.C. 1982: The quasi-Lagrangian nature of SOFAR floats. *Deep-Sea Res.*, 29, 1587–1602.

Roache, P.J. 1972: *Computational Fluid Dynamics*. Hermosa, Albuquerque.

Roberts, J. and T.D. Roberts 1978: Use of the Butterworth low-pass filter for oceanographic data. *J. Geophys. Res.*, 83, 5510–5514.

Robinson, I.S. 1985: *Satellite Oceanography*. Ellis Horwood, Chichester.

Roemmich, D. and B. Cornuelle 1987: Digitization and calibration of the expendable bathythermograph. *Deep-Sea Res.*, 34, 299–307.

Roll, H.U. 1951: Wassertemperaturemessungen an Deck und in Maschinenraum. *Ann. Meteor.*, 4, 439–443.

Rørbæk, K. 1994: Comparison of Aanderaa Instruments DCM 12 Doppler current meter with RD Instruments Broadband direct reading 600 khz ADCP. Danish Hydraulic Institute, Copenhagen.

Rosenberg, N.D., J.E. Lupton, D. Kadko, R. Collier, M.D. Lilley, and H. Pak 1988: Estimation of heat and chemcial fluxes from a seafloor hydrothermal vent field using radon measurements. *Nature, 334,* 604-607.

Rossby, H.T. 1969: On monitoring depth variations of the main thermocline acoustically. *J. Geophys. Res., 74,* 5542-5546.

Rossby, H.T. and D. Webb 1970: Observing abyssal motions by tracking Swallow floats in the SOFAR channel. *Deep-Sea Res., 17,* 359-365.

Rossby, H.T., D. Dorson, and J. Fontaine 1986: The RAFOS systems. *J. Atmos. Oceanic Tech., 3,* 672-679.

Royer, T.C. 1981: Baroclinic transport in the Gulf of Alaska. Part II. A fresh water driven coastal current. *J. Mar. Res., 39,* 251-266.

Rual, P. 1991: XBT depth correction. Addendum to the Summary Report of the Ad Hoc Meeting of the IGOSS Task Team on Quality Control for Automated Systems, Marion, Mass., USA, June 1991, IOC/INF-888 Add, pp. 131-144.

Sanderson, B.G., A. Okubo, and A. Goulding 1990: The fractal dimension of relative Lagrangian motion. *Tellus, 42A,* 550-556.

Sandford, T.B. 1971: Motionally induced electric and magnetic fields in the sea. *J. Geophys. Res., 76,* 3476-3492.

Sarmiento, J.L., H.W. Feely, W.S. Moore, A.E. Bainbridge, and W.S. Broecker 1976: The relationship between vertical eddy diffusion and buoyancy gradient in the deep sea. *Earth Planet. Sci. Lett., 32,* 357-370.

Sarmiento, J.L., J.R. Toggweiller, and R. Najjar 1988: Ocean carbon cycle dynamics and atmospheric pCO_2. *Phil. Trans. R. Soc. A325,* 3-21.

Saunders, P.M. 1976: Near-surface current measurements. *Deep-Sea Res., 23,* 249-258.

Saunders, P.M. 1980: Overspeeding of a Savonious rotor. *Deep-Sea Res., 27A,* 755-759.

Saur, T. 1963: A study of the quality of sea water temperatures reported in logs of ships' weather observations. *J. Appl. Meteorol., 2,* 417-425.

Sayles, M.A., K. Aagaard, and L.K. Coachman 1979: *Oceanographic Atlas of the Bering Sea Basin.* University of Washington Press, Seattle, Wash.

Scarborough, J.B. 1966: *Numerical Mathematical Analysis.* Johns Hopkins Press.

Scarlet, R.I. 1975: A data processing method for salinity, temperature, depth profiles. *Deep-Sea Res., 27,* 509-515.

Schlosser. P., G. Bönisch, M. Rhein, and R. Bayer 1991: Reduction of deepwater formation in the Greenland Sea during the 1980s: Evidence from tracer data. *Science, 251,* 1054-1056.

Schluessel, P., H.Y. Shin, W.J. Emery and H. Grassl 1987: Comparison of satellite derived seas surface temperature with in situ skin measurements. *J. Geophys. Res., 92,* 2859-2874.

Schott, F. 1986: Medium-range vertical acoustic Doppler current profiling from submerged buoys. *Deep-Sea Res., 33,* 1279-1292.

Schott, F. and K.D. Leaman 1991: Observations with moored acoustic Doppler current profilers in the convection regime in the Golfe du Lion. *J. Phys. Oceanogr., 21,* 558-574.

Schumacher, J.D. and R.K. Reed 1986: On the Alaska Coastal Current in the western Gulf of Alaska. *J. Geophys. Res., 91,* 9655-9661.

Schuster, A. 1898: On the investigation of hidden periodicities with application to a supposed 26 day period of meteorological phenomena. *Terrestrial Magnetism, 3,* 13-41.

Seaver, G.A. and S. Kuleshov 1982: Experimental and analytical error of the expendable bathythermograph. *J. Phys. Oceanogr., 12,* 592-600.

Shen, Z. and L. Mei 1993: Equilibrium spectra of water waves forced by intermittent wind turbulence. *J. Phys. Oceanogr., 23,* 505-531.

Shen, Z., W. Wang, and L. Mei 1994: Finestructure of wind waves analyzed with wavelet transform. *J. Phys. Oceanogr., 24,* 1085-1094.

Shum, C.K., R.A. Werner, D.T. Sandwell, B.H. Zhang, R.S. Nerem, and B.D. Tapley 1990: Variations of global mesoscale eddy energy observed from Geosat. *J. Geophys. Res., 95,* 17,865-17,876.

Siemens, C.W. 1876: On determining the depth of the sea without the use of a sounding line. *Phil. Trans. R. Soc. London, 166*, 671–692.

Smith, W.L., P.K. Rao, R. Koffler and W.R. Curtis 1970: The determination of sea-surface temperature from satellite high resolution infrared window radiation mesurements. *Mon. Wea. Rev., 98*, 604–611.

Snedecor, G.W. and W.G. Cochran 1967: *Statistical Methods*. Iowa State University Press, Ames, Iowa.

Snodgrass, F.E. 1968: Deep sea instrument capsule. *Science, 162*, 78–87.

Sokolova, S.E., A.B. Rabinovich, and K.S. Chu 1992: On the atmosphere-induced sea level variations along the western coast of the Sea of Japan. *La Mer, 30*, 191–212.

Spencer, R.W., H.M. Hood, and R.E. Hood 1989: Precipitation retrieval over land and ocean with the SSM/I: identification and characteristics of the scattering signal. *J. Atmos. Oceanic Tech. 6*, 254–273.

Spencer, R.W., B.B. Hinton, and W.S. Olson 1989: Nimbus-7 37 GHz radiances correlated with radar rain rates over the Gulf of Mexico. *J. Climate Appl. Meteor., 22*, 2095–2099.

Spiess, F. 1928: *The Meteor Expedition—Research and Experiences During the German Atlantic Expedition 1925–27*. Amerind, New Delhi, 1985.

Sprent, P., and G.A. Dolby 1980: The geometric mean functional relationship. *Biometrics, 36*, 547–550.

Sreenivasan, K.R., R. Ramshankar, and C. Meneveau 1989: Mixing, entrainment and fractal dimensions of surfaces in turbulent fluids. *Proc. R. Soc. London A421*, 79–109.

Stacey, M.W., S. Pond and P.H. LeBlond 1988: An objective analysis of the low-frequency currents in the Strait of Georgia. *Atmosphere-Ocean, 26*, 1–15.

Stegen, G.R., D.P. Delisi and R.C. Von Collins 1975: A portable, digital recording, expendable bathythermograph (XBT) system. *Deep-Sea Res., 22*, 447–453.

Steinhart, J.C. and S.R. Hart 1968: Calibration curves for thermistors. *Deep-Sea Res., 15*, 497–503.

Stockwell, R.G., L. Mansinha, and R.P. Lowe 1994: Localization of the complex spectrum: The S transformation. *AGU Trans. 55*.

Stommel, H. and F. Schott 1977: The beta spiral and the determination of the absolute velocity field from hydrographic station data. *Deep-Sea Res., 24*, 325–329.

Strickland, J.D.H. and T.R. Parsons 1968: *A Practical Handbook of Seawater Analysis*. Bull. Fish. Res. Board Canada.

Strickland, J.D.H. and T.R. Parsons 1972: *A Practical Handbook of Seawater Analysis*, 2nd edn. Bull. Fish. Res. Board Canada.

Strong, A.E. and J.A. Pritchard 1980: Regular monthly mean temperatures of the Earth's oceans from satellites. *Bull. Am. Meteorol. Soc., 61*, 553–559.

Stuiver, M.P, P.D. Quay, and N.D. Östlund 1982: Abyssal water carbon-14 distribution and the age of the world oceans. *Science, 219*, 849–851.

Sturges, W. 1983: On interpolating gappy records for time-series analysis. *J. Geophys. Res., 88*, 9736–9740.

Sverdrup, H.U. 1947: Wind driven currents in a baroclinic ocean with applications to the equatorial currents in the eastern Pacific. *Proc. Natl. Acad. Sci., USA, 33*, 318–336.

Sverdrup, H.V., M.W. Johnson, and R.H. Fleming 1942: *The Oceans: Their Physics, Chemistry, and General Biology*. Prentice-Hall, Engelwood Cliffs, N.J.

Swallow, J.C. 1955: A neutrally-buoyant float for measuring deep current. *Deep-Sea Res., 3*, 74–81.

Sybrandy, A.L. and P.P. Niiler 1990: The WOCE/TOGA SVP Lagrangian Drifter Construction Manual. Scripps Institution of Oceanography, University of California, San Diego, SIO Reference 90-248.

Tabata, S. 1978a: On the accuracy of sea-surface temperatures and salinities observed in the northeast Pacific. *Atmosphere-Ocean, 16*, 237–247.

Tabata, S. 1978b: Comparison of observations of sea-surface temperatures at Ocean Station "P" and N.O.A.A. buoy stations and those made by merchant ships travelling in their vicinities, in the northeast Pacific Ocean. *J. Appl. Meteorol.*, 17, 374–385.

Tabata, S. and J.A. Stickland 1972: Summary of oceanographic records obtained from moored instruments in the Strait of Georgia, 1969–70. Current velocity and seawater temperature from Station H-06: Pacific Marine Science Report 72-7. Environment Canada.

Talley, L.D. and T.M. Joyce 1992: Double silica maximum in the North Pacific. *J. Geophys. Res.*, 97, 5465–5480.

Talley, L.D., M. Martin, and P. Salameth 1988: TransPacific section in the subpolar gyre (TPS47): Physical, chemical, and CTD data, R/V *Thomas Thompson* TT190, 4 August 1985–7 September 1985. SIO Ref. 88-9, Scripps Institute of Oceanography, La Jolla, Calif.

Talley, L.D., T.M. Joyce, and R.A. deSzoeke 1991: Trans-Pacific sections at 47°N and 152°W: Distribution of properties. *Deep-Sea Res.*, 38, 563–582.

Tapley, B.D., G.H. Born, and M.E. Park 1982: The Seastat altimeter data and its accuracy assessment. *J. Geophys. Res.*, 87, 3179–3188.

Tauber, G.M. 1969: The comparative measurements of sea surface temperature in the USSR, Technical Note 103, Sea Surface Temperature, WMO, pp. 141–151.

Tchernia, P. 1980: *Descriptive Regional Oceanography. Pergamon Marine Series*, Vol. 3. Pergamon Press, Oxford.

Thompson, R. 1971: Spectral estimation from irregularly spaced data. *IEEE Trans. Geosci. Electron.*, GE-9, 107–119.

Thompson, R.O.R.Y. 1979: Coherence significance levels. *J. Atmos. Sci.*, 36, 2020–2021.

Thompson, R.O.R.Y. 1983: Low-pass filters to suppress inertial and tidal frequencies. *J. Phys. Oceanogr.*, 13, 1077–1083.

Thompson, T.W., D.E. Weissman, and F.I. Gonzalez 1983: L-band radar backscatter dependence upon surface wind stress: A summary of new Seasat-I and aircraft observations. *J. Geophys. Res.*, 88, 1727–1735.

Thomson, R.E. 1977: Currents in Johnstone Strait, British Columbia: supplementary data on the Vancouver Island side. *J. Fish. Res. Bd. Canada*, 34, 697–703.

Thomson, R.E. 1981: *Oceanography of the British Columbia Coast*. Can. Special Pub. Fish. Aquat. Sci. 56, Ottawa.

Thomson, R.E. 1983: A comparison between computed and measured oceanic winds near the British Columbia coast. *J. Geophys. Res.*, 88, 2675–2683.

Thomson, R.E. and W.S. Huggett 1980: M2 baroclinic tides in Johnstone Strait, British Columbia. *J. Phys. Oceanogr.*, 10, 1509–1539.

Thomson, R.E., S. Tabata, and D. Ramsden 1985: Comparison of sea level variability on the Caribbean and the Pacific coasts of the Panama Canal. In *Time Series of Ocean Measurements*, Vol. 2, Unesco, IOC Technical Series 30, pp. 33–37.

Thomson, R.E., T.A. Curran, M.C. Hamilton, and R. McFarlane 1988: Time series measurements from a moored fluorescence-based dissolved oxygen sensor. *J. Atmos. Oceanic Technol.*, 5, 614–624.

Thomson, R.E, P.H. LeBlond, and W.J. Emery 1990: Analysis of deep-drogued satellite-tracked drifter measurements in the northeast Pacific. *Atmosphere-Ocean*, 28, 409–443.

Thomson, R.E., R.L. Gordon, and A.G. Dolling 1991: An intense acoustic back-scattering layer at the top of a mid-ocean ridge hydrothermal plume. *J. Geophys. Res.*, 96, 4839–4844.

Thomson, R.E., B.J. Burd, A.G. Dolling, R.L. Gordon, and G.S. Jamieson 1992: The deep scattering layer associated with the Endeavour Ridge hydrothermal plume. *Deep-Sea Res.* 39, 55–73.

Thomson, R.E., J.R. Delaney, R.E. McDuff, D.R. Janecky, and J.S. McLain, 1992: Physical characteristics of the Endeavour Ridge hydrothermal plume during July 1988. *Earth Planetary Science Lett.*, 111, 141–154.

Thomson, R.E., P.H. LeBlond, and A.B. Rabinovich 1997: Oceanic odyssey of a satellite-tracked drifter: North Pacific variability delineated by a single drifter trajectory, *J. Oceanogr.*, 53, 81–87.

Thomson, R.E. and H.J. Freeland 1999: Lagrangian meassurement of mid-depth currents in the eastern tropical Pacific. *Geophys. Res. Lett.*, 26, 3125–3128.

Thorpe, S.A. 1977: Turbulence and mixing in a Scottish loch. *Phil. Trans. R. Soc. London A286*, 125–181.

Tichelaar, B.W. and L.J. Ruff 1989: How good are our best models? Jackknifing, bootstrapping, and earthquake depth. *EOS*, 70(20), 593–605.

Toggweiler, J.R. and S. Trumbore 1985: Bomb-test ^{90}Sr in Pacific and Indian Ocean surface water as recorded by banded corals. *Earth Planet. Sci. Lett.*, 74, 306–314.

Tokamamkian, R., P.T. Strub, and J. McClean-Padman 1990: Evaluation of the maximum cross-correlation method of estimating sea surface velocities from sequential satellite images. *J. Atmos. Oceanic Technol.*, 7, 852–865.

Topham, D.R. and R.G. Perkins 1988: CTD sensor characteristics and their matching for salinity calculations. *IEEE J. Oceanic Engineering*, 13, 107–117.

Trenberth, K.E., and J.G. Olson 1988: ECMWF global analysis 1979–1986: Circulation statistics and data evaluation. NCAR Technical Report Note NCAR/TN-300+STR, National Center for Atmospheric Research.

Trumbore, S.E., S.S. Jacobs, and W.M. Smethie, Jr 1991: Chlorofluorocarbon evidence for rapid ventilation of the Ross Sea. *Deep-Sea Res.*, 38, 845–870.

Trump, W. 1983: Effect of ship's roll on the quality of precision CTD data. *Deep-Sea Res.*, 30(11A), 1173–1183.

Tsonis, A.A. 1991: Sensitivity of the global climate system to initial conditions. *EOS Trans. AGU*, 72, 313–328.

Tsonis, A.A. 1992: Autoregressive models not sensitive to initial conditions. Reply. *EOS Trans. AGU*, 25, 268.

Tsonis, A.A. and J.B. Elsner 1990: Comments on "Dimension analysis of climatic data". *J. Climate*, 3, 1502–1505.

Tushingham, A.M. and W.R. Peltier 1991: ICE-3-G: A new global model of late Pleistocene deglaciation based upon geophysical predictions of post glacial relative sea level change. *J. Geophys. Res.*, 96, 4497–4523.

Ulrych, T.J. 1972: Maximum entropy spectrum of truncated sinusoids. *J. Geophys. Res.*, 77, 1396–1400.

Ulrych, T.J. and T.N. Bishop 1975: Maximum entropy spectral analysis and autoregressive decomposition. *Reviews of Geophysics and Space Physics*, 13, 183–200.

UNESCO 1966: *International Oceanographic Tables*. Unesco, Place de Fontenoy, Paris. National UNESCO Office of Oceanography, Institute of Oceanography, Wormley.

Urick, R.J. 1967: *Principles of Underwater Sound*. McGraw-Hill, New York.

Vachon, W.A. 1973: Scale model testing of drogues for free drifting buoys. Technical Report, The Charles Stark Draper Laboratory, Inc., Cambridge, Mass.

Van Scoy, K.A., R.A. Fine and H.G. Östlund. 1991: Two decades of missing tritium into the North Pacific Ocean. *Deep-Sea Res.*, 38, S191–S219.

van Leer, J., W. Düing, R. Erath, E. Kennelly and A. Speidel 1974: The cyclesonde: an unattended vertical profiler for scalar and vector quantities in the upper ocean. *Deep-Sea Res.*, 21, 385–400.

Vaníček, P. 1971: Further development and properties of the spectral analysis by least-squares. *Astrophys. Space Sci.*, 12, 10–73.

Vazquez, J., V. Zlotnicki and L.-L. Fu 1990: Sea level variability in the Gulf Stream beween Cape Hateras and 50°N, a GEOSAT study. *J. Geophys. Res.*, 95, 17,957–17,964.

Vogt, P.R. and W.-Y. Jung 1991: Satellite radar altimetry aids seafloor mapping. *EOS*, 72(43), 465, 468–469.

von Arx, W.S. 1950: An electromagnetic method for measuring the velocities of ocean currents from a ship under way. *Pap. Phys. Oceanogr. and Meteor.*, 11, 1–61.

Walden, H. 1966: Zur messung der Wassertemperatur auf Handelsschiffen. *Dtsch Hydro. Z.*, 19, 21–28.

Waliser, D.E. and C. Gautier 1993: Comparison of buoy and SSM/I-derived wind speeds in the tropical Pacific. *TOGA Notes, 12*, 1–7.

Walker, E.R. and K.D. Chapman 1973: Salinity–conductivity formulae compared. Pacific Marine Science Report 73-5, Institute of Ocean Sciences, Sidney, British Columbia, Canada.

Wallace, D.W.R. and J.R.N. Lazier 1988: Anthropogenic chlorofluoromethanes in newly formed Labrador Sea water. *Nature, 332*, 61–63.

Wallace, J.M. 1972: Empirical orthogonal representation of time series in the frequency domain. Part II: Application to the study of tropical wave disturbances. *J. Appl. Meteor. 11*, 893–900.

Wallace, J.M. and R.E. Dickinson 1972: Empirical orthogonal representation of time series in the frequency domain. Part I: Theoretical considerations. *J. Appl. Meteor., 11*, 887–892.

Walters, R.A. and C. Heston 1982: Removing tidal-period variations from time-series data using low-pass digital filters. *J. Phys. Oceanogr., 12*, 112–115.

Wang, D.-P. and C.N.K. Mooers 1977: Long coast trapped waves off the west coast of the United States, summer 1973. *J. Phys. Oceanogr., 7*, 856–864.

Warren, B.A. 1970: General circulation of the South Pacific. In *Scientific Exploration of the South Pacific*, (ed.) W.S. Wooster. National Academy of Sciences, Washington, D.C., pp. 33–49.

Warren, B.A. 1983: Why is no deep water formed in the North Pacific? *J. Mar. Res., 41*, 327–347.

Watanabe, Y.W., S. Watanabe, and S. Tsunogai 1991: Tritium in the Japan Sea and the renewal time of Japan Sea deep water. *Mar. Chem., 34*, 97–108.

Watson, A.J. and M.I. Liddicoat 1985: Recent history of atmospheric trace gas concentrations deduced from measurements in the deep sea: Application to sulpher hexafluoride and carbon tetrachloride. *Atmos. Environ., 19*, 1477–1484.

Watson, A.J. and J.R. Ledwell 1988: Purposefully released tracers. *Phil. Trans. R. Soc. London A325*, 189–200.

Watts, D.R. and H.T. Rossby 1977: Measuring dynamic heights with inverted echo sounders: Results from MODE. *J. Phys. Oceanogr., 7*, 345–358.

Weare, B.C., and J.S. Nasstrom 1982: Examples of extended empirical orthogonal function analyses. *Mon. Weath. Rev., 110*, 481–485.

Wearn, R.B. and D.J. Baker, Jr 1980: Bottom pressure measurements across the Antarctic Circumpolar Current and their relation to the wind. *Deep-Sea Res., 27A*, 875–888.

Weinreb, M.P., G. Hamilton, S. Brown, and R.J. Koczor 1990: Nonlinearity corrections in calibration of Advanced Very High Resolution Radiometer infrared channels. *J. Geophys. Res., 95*, 7381–7388.

Weiss, W. and W. Roether 1980: The rates of tritium input to the world oceans. *Earth Planet Sci. Lett. 49*, 453–446.

Weller, R.A, and R.E. Davis 1980: A vector measuring current meter. *Deep-Sea Res., 27*, 565–582.

Wenner, F., E.H. Smith, and F.M. Soule 1930: Apparatus for the determination aboard ship of the salinity of sea water by the electrial conductivity method. Bureau of Standards Journal of Research, 5, pp. 711–732.

Wentz, F.J., LA. Mattox, and W. Peteherych 1986: New algorithms for microwave measurements of ocean winds: Applications to SEASAT and the special sensor microwave imager. *J. Geophys. Res., 91*, 2289–2307.

Wilkin, J.L. 1987: A computer program for calculating frequencies and modal structures of free coastal-trapped waves. Woods Hole Oceanographic Institution, Technical Report WHOI-87-53.

Willebrand, J.W., P. Müller, and D.J. Olbers 1977: Inverse analysis of the trimoored internal wave experiment (IWEX). *Berichte aus dem Institut für Meereskunde, 20a,b*.

Wilson, W. D. 1960: Speed of sound in sea water as a function of temperature, pressure and salinity. *J. Acoust. Soc. Am., 32*, 641–644.

Wimbush, M. 1977: An inexpensive sea-floor precision pressure recorder. *Deep-Sea Res., 24*, 493–497.

Wimbush, M., S.M. Chiswell, R. Lukas, and K.A. Donohue 1990: Inverted echo sounder measurement of dynamic height through an ENSO cycle in the Central Equatorial Pacific. *IEEE J. Oceanic Engng, 15*, 380–383.

Witter, D.L. and D.B. Chelton 1988: Temporal variability of sea-state bias in SEASAT altimeter height measurements. In Proceedings of the WOCE/NASA Altimeter Algorithm Workshop, Oregon 1987, (ed. D.B. Chelton). US WOCE Technical Report No. 2, WOCE (World Ocean Circulation Experiment) Implementation Plan, Vol 1. Detailed requirements. 1988. WOCE International Planning Office, Wormley.

Witter, D. and D.B. Chelton 1991: A Geosat altimeter wind speed algorithm and a method for altimeter wind speed algorithm development. *J. Geophys. Res., 96*, 8853–8860.

WOCE Science Steering Committe (SSC) 1991: SSC discusses WOCE priorities in Pacific, Indian and Atlantic Oceans. *WOCE Notes, 3*(3), 1, 4–5.

WOCE, 1988: World Ocean Circulation Implementation Plan, Vols 1 and 2, WOCE International Planning Office, Wormley.

Woods, J.D. 1985: The world ocean circulation experiment. *Nature, 314*, 501–511.

Woodward, M.J. and W.R. Crawford 1992: Loran-C drifters for coastal ocean measurements. *Sea Technology*, August, 24–27.

Woodward, M.J., W.S. Huggett, and R.E. Thomson 1990: Near-surface moored current meter intercomparisons. Can. Tech. Report Hydrogr. and Ocean Sci., No. 125, Dept. Fish. Oceans.

Woodworth, P.L. 1991: The permanent service for mean sea level and the global sea level observing system. *J. Coastal Res., 7*, 699–710.

Wooster, W.S., A.J. Lee, and G. Dietrich 1969: Redefinition of salinity. *Deep-Sea Res., 16*, 321–322.

Worcester, P.F., B.D. Cornuelle, and R.C. Spindel 1991: A review of ocean acoustic tomography: 1987–1990. *Reviews Geophys., Supp., 29*, 557–570.

Worthington, L.V. 1976: On the North Atlantic circulation. In *The Johns Hopkins Oceanographic Studies*, Johns Hopkins University Press, Baltimore, MD.

Worthington, L.V. 1981: The water masses of the world ocean: Some results of a fine-scale census. In *Evolution of Physical Oceanography*, Chapter 2 (eds B.A. Warren and C. Wunsch). MIT Press, Cambridge, Mass., pp. 42–69.

Wu, Q.X. 1991: Tracking evolving sea surface temperature features. In Proceedings of 6th New Zealand Image Processing Workshop, DSIR Physical Sciences, Lower Hutt, New Zealand.

Wu, Q.X. 1993: Computing velocity fields from sequential satellite images. In *Satellite Remote Sensing of the Oceanic Environment* (eds I.S.F. Jones, Y. Sugimore, and R.W. Stewart). Seibutsu Kensyusha.

Wunsch, C. 1972: Bermuda sea level in relation to tides, weather, and baroclinic fluctuations. *Rev. Geophys. Space Phys., 10*, 1–49.

Wunsch, C. 1977: Determining the general circulation of the oceans: a preliminary discussion. *Science, 196*, 871–875.

Wunsch, C. 1978: The North Atlantic general circulation west of 50°W determined by inverse methods. *Rev. Geophys. Space Phys., 16*, 583–620.

Wunsch, C. 1988: Transient tracers as a problem in control theory. *J. Geophys. Res., 93*, 8099–8110.

Wüst, G. 1935: Die Sratosphäre. Wissenshaftliche Ergebinesse der Deutschen Atlantischen Expedition Meteor 1925–27.

Wüst, G. 1957: Strömgeschwindigkeiten und Strommengen in den Tiefen des Atlantischen Ozeans. *Wissenschaftliche Ergebnisse der Deutschen Atlantischen Expedition* Meteor *1925–1927, 6*, 261–420.

Wyrtki, K. 1961: The oxygen minimum in relation to ocean circulation. *Deep-Sea Res., 9*, 11–23.

Wyrtki, K. 1971: *Oceanographic Atlas of the International Indian Ocean Expedition*. NSF, Washington, D.C.

Wyrtki, K. and G. Meyers 1975a, b: The trade wind field over the Pacific Ocean. Part I, the mean field and the mean annual variation. Hawaii Institute of Geophysics Report, HIG-75-1, University of Hawaii, Honolulu.

Zenk, W., D. Halpern, and R. Käse 1980: Influence of mooring configuration and surface waves upon deep-sea near-surface current measurements. *Deep-Sea Res.*, 27, 217–224.

Index

Aanderaa current meters
  data 161
  Doppler (DCM) 83
  RCM4 70–7, 133, 267–9
    calibration 77
    data processing 73
    problems 74–7
  RCM5 267–9
  RCM7 72, 76
absolute currents 361–6
absorption coefficient 51
accidental errors 264–6
accuracy 2–8
  levels of 275
ACM see acoustic current meters
acoustics
  backscatter 92–4
  current meters (ACM) 72, 78–9, 94–5, 98
  depth sounders 48–54
  Doppler current meters (ADCM) 83–92, 377–8, 379
  Doppler current profiler (ADCP) 65, 83–92
    backscatter 92–4
    limiting factors 89–90
    waves 67
  pingers 52, 87, 99
    schematic 53
  releases 99–102
  scattering 52
  scintillation 97
  sound speed 49, 50
  surface buoys, acoustically-tracked 102
  travel time 66–7
ADCM see acoustic Doppler current meter
ADCP see acoustic Doppler current profiler
ADEOS satellite 125
Advanced Very High Resolution Radiometer (AVHRR) 27–30
AGC see Automatic Gain Control
airborne surveys 170–1

ALACE see Autonomous Lagrangian Circulation Explorer
Alaska 71
  Coastal Current 143
  Gulf of 143, 154, 512
  Panhandle 143
Aleutian Low 339
algorithms 470–1
  AR parameters 471
  summary of spectral 472
  wavelet analysis 504–5
aliasing
  autospectrum 438
  time-series data 434
all-pole model 468
all-zero model 468
alternate hypothesis 249
altimeter bias
  errors 252–3
  estimates, confidence intervals 225–6
altimetric satellites 54–5
  GEOS-3 55, 62
  GEOSAT 55, 62–3
  SEASAT 54, 55, 62
ambient temperature 32
Amphitrite Point
  Vancouver Island 387–8, 395–7
    sea surface temperature (SST) data 419
amplitudes 333, 344, 347–8, 383–4, 416
analog-to-digital (AD) converters 1
analysis of variance (ANOVA) 254–7
anemometers 120
annihilator space 367–8
anomaly fields 339–40
ANOVA see analysis of variance
Antarctica
  Ross Sea 154
  Weddell Endbery Basin 130, 137
anthropogenic tritium 143

AOU *see* apparent oxygen utilization
apparent oxygen utilization (AOU) 129
approximations and nondimensional numbers 586–92
AR *see* autoregressive
AR PSD *see* autoregressive power spectral density model
Arctic Ocean 55
  Canada 403–4
  rotary current spectra 430
area, units of 570
ARGOS
  satellite 67
  tracking buoys 116, 117
ARMA *see* autoregressive moving average model
artificial skill 259
Atlantic Ocean 32, 128, 129, 130, 144, 149, 151
  basin 162
  equatorial 64
  Georges Bank 98
  Hatteras Plain, IWEX study 367
  north 36, 90, 107, 151, 153, 154, 157
    CTD stations 314
      temperature data analysis 364–6
      objective mapping example 314–19
  south
    heat transport 266
    near-surface currents 534
atmospheric pressure 120
attenuation, light 138–42
Australia
  current meter record 553, 554
  cyclone tracks 565–6
autocorrelation 244
  functions 375–7, 378, 411
  method 406
    spectral analysis 417–19
autocovariance 265, 375–7, 378
Automatic Gain Control (AGC) 93
Autonomous Lagrangian Circulation Explorer (ALACE) 117
autonomous underwater vehicles (AUVs) 371
autoregressive models
  global temperature 476–8
  moving average (ARMA) 467
  power spectral density (AR PSD) 465–7
    estimation 468–78
      AR parameters 469–71
autoregressive process 473, 474
  order 471–3
autotrophic zone 135–6
AUV *see* autonomous underwater vehicle

averaging, block 305–9
AVHRR *see* Advanced Very High Resolution Radiometer

backscatter, acoustic 92–4
band averaging *see* frequency band averaging
band-pass filters 511, 520–1, 530, 544–5
bandwidth (BW)
  ideal filters 522–3
  spectral 461
  windows 443
Bangladesh 55
baroclinic dynamic height 63
baroclinic modes 346, 347, 348, 349, 350
barotropic modes 346, 347, 348, 350
Bartlett windows *see* triangular windows
Basic Sealogger CTD, schematic 19
bathymetry 54
  satellite 54–5
bathythermograph *see* mechanical bathythermograph
Beaufort Scale 120
Beaufort Sea, satellite-tracked drifters 403–4
Bedford Institute of Oceanography 36, 98
bell distribution 196
Bering Sea 129, 183
Bernstein correction method 27
best linear unbiased estimators 276
beta-spiral method 361–4
bias
  estimation 214, 225–6
  satellite altimeter measurements 252–3
BIGDRV2, statistical computer package 352
BIGLOAD2, statistical computer package 351
bin interval 312
binomial coefficients 199
binomial probability function 206
bins
  bootstrapping 296
  histogram 269–70
biological oxygen demand (BOD) 129
biological studies 83, 92–4
bit-resolution, digital instruments 3
Black Sea 144
Blackman–Tukey method 417–18, 451
block averaging techniques 305–9, 451–3
BOD *see* biological oxygen demand
bomb-generated radiocarbon 149, 151
bootstrap method 294, 295–301
  biological application 301
  El Niño recurrence rates 296–8
  histograms 298–300
bottle cast sampling 7

bottom topography 172
bottom tracking 89
bottom-ballasted window blind drogue 110
boundary layer, oceanic 120
Bourdon tube sensor 43
Boussinesq approximation 352
box counting method, fractals 563–4, 567
box-car windows 406–7, 433–4, 443–8, 450
Brest, tide gauge data 57
British Columbia
  Cape St James 123
  coastal currents 71
  salmon 246
  Tofino 389–90, 395, 399–402
  Victoria, windowed spectra 450
  windowed spectra 450
Broadband ADCP 83, 86, 90
Brunt–Väisälä frequency 345, 352
BTCSW, statistical computer package 351
bubbler gauge 60–1
bucket method, sea surface temperature 24
bulk-averaging techniques 305–9
buoyancy frequency 345–6, 351
buoys, directional waverider 67–8
Burg algorithm 470–1, 472
burst sampling 7
  ECM 81–2
  VACM 73–4
Butterworth filters 543–51
  band-pass 544–5
  coefficients 548–51
  design 547–8
  digital formulation 545–7
  high-pass 544–5
  tangent vs sine 547

calibration 1–2, 160–1
  ACM 79
  CTD/STD 23–4
  RCM4 73
  thermometer 11
capacitance staff, waves 67
Cape Hatteras 63
Cape St James, British Columbia Coast 123
carbon-14: 149–53
Cartesian component rotary current spectra 426–7
cascade, filters 528, 529–32
Cascadia Basin 137
CCW *see* counterclockwise
central limit theorem 211–14
CFCs *see* chlorofluorocarbons
Challenger expedition 1873–6: 9

Chandler effect 56–7
Chebyshev
  polynomials 353
  theorem 224
chemical pumps 128
chemical sniffers *see* chemical pumps
chemical tracers 127–45
chi-square probability distribution 209–10, 441, 454, 578
  limits 219
  spectral estimators 424–5
chi-square test 222
chi-square variable 221
chlorinity 33, 34
chlorofluorocarbons (CFCs) 153–5
circularly polarized motions, Fourier analysis 384
Clark cell 131
class intervals 221
climatology, sea surface temperature (SST) 25
clockwise (CW), one-sided autospectra 432
Coastal Ocean Doppler Radar (CODAR) 96
coastal subsidence 56
Coastal Upwelling Experiment (CUE-II) 342
coastal-trapped waves (CTWs) 350–6
CODAR *see* Coastal Ocean Doppler Radar
coherence functions 493–4
coherence spectrum 486–90
  confidence levels 488–90
coherency *see* coherence spectrum
coincident and quadra spectra 485–6
color
  graphical representation 189–90
  vertical sections 172
combinations, probability 198
compasses
  calibration, RCM4 77
  Eulerian current meters 69
complex admittance function 500–1
complex demodulation 402–4
complex empirical orthogonal functions 342–3
computer graphics 187–91
computerized methods 294–304
conductivity–temperature–depth profiler (CTD) 3, 17–18, 35–9, 130, 169–70
  comparison with XBT 15–16, 254–6
  data processing 271
  depth error 43–8
  dynamic response 18–22
  EGGeneral Oceanics MK3C CTD 18, 22
  schematic 20
  Guildline 8737 CTD 22

conductivity–temperature–depth profiler (CTD) (*Continued*)
   IES calibration 64–7
   Neil–Brown CTD 23
   objective mapping example 314–19
   pressure measurement 43
   sensor response 22–3
   temperature calibration 23–4
confidence coefficient 216
confidence intervals 216–24
   altimeter bias estimates 225–6
   correlation 253
   MEM 475–6
   population variance 219–20
   spectra 454–5, 456
confidence levels, coherence spectrum 488–90
conic projections 177
conservative tracers 127
consistency, estimation 214
continuous process, definition 373
continuous sampling 6–7
contours, vertical sections 172
conventional tracers 127, 128–38
Copenhagen Water 34, 40
core-layer method 176
correlation
   cause and effect 246–9
   confidence intervals 253
   functions 374–80
      linear regression comparison 380
   random errors 244–5
   regression relationship 243–9
   significance levels 253
correlation coefficient (r) 243, 253, 585
cosine filters 537–8
counterclockwise (CCW), one-sided autospectra 432
covariance 243, 290–4
   multivariate distributions 293–4
covariance functions 203, 464
   analytical 377–9
   autocovariance 375–7, 378
   observed 379
   stochastic time series 375
covariance matrix 293–4, 310–11, 312–13, 318, 321–2, 328–32, 338, 369–70
   lagged 320
Crater Lake, Oregon 41
CROSS, statistical computer package 352
cross-channel wavenumber 345
cross-correlation functions 376, 480–2

cross-covariance
   function 376
   method 482
cross-shore direction 328
cross-shore velocity components 425–6
cross-spectral analysis 405, 480–501
   cross-correlation functions 480–2
   multi-input systems 491–5
   phase and cross-amplitude functions 484–5
   rotary 495–501
cross-validation 295
crossovers, depth recorder 52–3
CTD *see* conductivity–temperature–depth profiler
CTW *see* coastal-trapped waves
CTWEIG, statistical computer package 356
cumulative probability function 198, 200–1, 207, 208
current meters 68–99
   Aanderaa
      data 161
      RCM4 70–7, 133
   acoustic 72, 78–9, 94–5, 98
   acoustic Doppler 83–94
   comparisons 94–5
   compasses 69
   data
      daily velocities 227–8
      error sources 267–8
      hourly data 269
      time-series presentation 267
   Ekman 70
   electromagnetic (ECM) 72, 79–82, 95–6
   failure 199
   Geodyne 850 70
   Kaijo Denki 70
   Marsh–McBirney 512 ECM 79
   measurements 311
   Nerpic CMDR 70
   Niel–Brown ACM 79
   nonmechanical 78–82
   paddle wheel rotor 76
   RCM7 72, 76
   rotor-type 70–8
   SimTronix UCM 40 79
   vector averaging (VACM) 73–7, 94–5, 98
   vector measuring (VMCM) 77, 94–5, 97
   vertical normal mode analysis 347–50
currents
   absolute 361–6
   daily averages 250–2
   Eulerian 68–102
   geostrophic 114

Lagrangian 102–25
  measurement 68–99
  spectral analysis 425–32
  speed sensors 69
  velocity, wavelet analysis 507–8
curves, fractals 556, 557–8
CW *see* clockwise
Cyclesonde 72
cyclone tracks, Australia 565–6
cylindrical projection 177

Danish Hydraulic Institute 83
data
  interpolation 277–90
  presentation 159–91
  processing 159–91
    drifters 107–9, 118–19
    RCM4 73
  quality 193
  reduction, manual methods 193
Datawell, The Netherlands 67
datums, GPS 90–2
DCM *see* Aanderaa Doppler current meter
Defense Meteorological Satellite Program (DMSP) 123
degree volume, thermometer 10
degrees of freedom (DOF) 8, 424, 452
  correlation coefficients 585
  effective 257–64
  equivalent 454, 488–9
  goodness-of-fit test 221
demodulation
  complex 402–4
  harmonic analysis 392
density 31–3
deposition, sediment 56
depth 42–55
  difference 47
  error 43–8, 54
  geometric 42
  indicators 48–54
design
  Butterworth filters 547–8
  Lanczos-window cosine filters 540–1
  oceanographic filters 527–32
despiking techniques 264–77
  detection of errors 266–70
deterministic processes, spectra 409–13
DEVLSF, statistical computer software 330
DFT *see* discrete Fourier transform
digital data 8
digital filters *see* filters
digital formulation, Butterworth filters 545–7

digital image processing 191
digital thermometers 30–1
digital-signal-processors (DSPs) 1
Dirac delta function 377, 433
directional waverider buoys 67–8
Dirichlet kernel 444, 446
discrete Fourier transform (DFT) 443, 444, 449, 464, 554–5
discrete process, definition 373
discrete series, spectra 413
discrete-level profiles 3
dissolved oxygen 129–30
  measurement 130–3
  sensors
    flourescence-based 131–3
    polarographic 131
  units 570
DLSVRR, statistical computer software 332, 333
DMSP *see* Defense Meteorological Satellite Program
DO2KEF, statistical computer package 346
DOF *see* degrees of freedom
Dolph–Chebyshev windows 442
dominant wavelengths 7
Doppler frequency shift 85, 87, 106, 125
double precision 275
Drake Passage, Southern Ocean 374
drifters 103–7
  bottles 103–4
  cards 103–4
  data processing 107–9, 118–19
  external forces 109–15
  nonsatellite tracked 115–16
  pop-up 117–19
  quasi-Lagrangian 104–7
  radar-tracked 115
  remotely tracked 103
  satellite-tracked 104–7
    complex demodulation 403–4
    data 305–6
    objective mapping example 314–19
    Pacific Ocean 507, 558–60, 562, 563
  sub-surface 104, 116–19
  surface 104–7
drogues 109–10
  configurations 104
  loss 107, 115
  parachute 110
  Tristar 110
duty cycles 7
  satellite transmissions 287–8, 290
dynamic height 66–7

earth rotation, units 570
East Pacific Rise 144
echo sounders, inverted (IES) 63–7
echo sounding 48–54
ECM *see* electromagnetic current meter
ECMWF *see* European Centre for Medium-range Weather Forecasting
editing
  errors 160
  techniques 264–77
  two-step 271
EDOF *see* equivalent degrees of freedom
effective degrees of freedom 257–64
efficiency, estimation 214
eigenvalues 322–4, 325, 333
eigenvectors *see* empirical orthogonal functions (EOFs)
Ekman
  current meter 70, 111–13, 114
  slab model 113
El Niño 55, 63
  recurrence rates, bootstrap method 296–8
  Southern Oscillation (ENSO) 64
electric field, horizontal (HEF) 95–6
electrical conductivity, salinity 34–9
electrodes 79–80
electrokinetograph, geomagnetic 96
electromagnetic current meters (ECM) 72, 79–82, 95–6
  manufacturer's specifications 80–1
electromagnetic (EM) bias correction 63
electromagnetic sensors, Eulerian current speed sensors 69
electromagnetic velocity profiler (EMVP) 96
EM *see* electromagnetic
empirical orthogonal function (EOF) analysis
  interpretation 336–40
  methods 319–36
  variations 340–3
EMVP *see* electromagnetic velocity profiler
ENBW *see* equivalent noise bandwidth
Endeavour Ridge, Pacific Ocean 155, 431
energy
  EOF analysis 323
  units of 571
energy conservation theorem, Parseval 409, 414
energy spectral density (ESD) 404, 409, 414, 420
engine cooling water 217
EOF *see* empirical orthogonal function analysis

EOLE satellite 106
equal-area projection 179
equally spaced data *see* evenly spaced data
equivalent degrees of freedom 454, 488–9
equivalent depth 345, 346
equivalent noise bandwidth (ENBW) 443
ergodic records 203
ergodic theorem 374
ERM *see* exact repeat mission
erosion 56
errors
  calculation 273
  editing 160, 271
  EOF analysis 324, 336–40
  handling 264–304
  hypothesis testing 249
  identification and removal 266–72
  inverse methods 359–61
  mean square 258
  of measurement 264
  objective mapping procedure 312, 313
  propagation 273–4
  types 160, 264–6
ERS-1 satellite 55, 125
ESD *see* energy spectral density
estimate, standard error of 238
estimation 214–16
  linear 233–43
  methods 227–33
  trends 261–4
estimators 214
  best linear unbiased 276
  least squares 470–1
  maximum likelihood correlation 245–6
  unbiased 203–4, 214–15
Euler, Leonhard 68
Eulerian currents
  measurement 68–102
  mechanical speed sensors 69
  platinum resistors 69
  propellors 69
euphotic zone 135–6
European Centre for Medium-range Weather Forecasting (ECMWF) 125
eustatic changes 56
evenly spaced data 277, 278
exact repeat mission (ERM) 62
examples
  AR model of global temperature 476–8
  bootstrapping 295, 296–8
  confidence intervals 217–20, 225–6
  correlation and regression 246
  covariance 291–3

Index

daily-average current velocities 227–8
empirical orthogonal functions (EOFs) 334–6
error identification 266–7
Fourier analysis computational procedure 387–8
fractals 565
goodness-of-fit test 222–3
harmonic tidal analysis 399–402
hypothesis testing 250–2
interpolating gappy records 285–90
IWEX internal wave problem 366–70
jackknife method 303–4
least squares method 395–7
linear regression 236, 237
matrix regression 240–2
maximum likelihood 230–2
monthly mean sea-level 286–7
normal modes of semidiurnal frequency 347–50
objective mapping 314–19
probability density function (PDF) 207–11
satellite-tracked positional data 287–90
temperature differences 210
treatment effects 255
vibrating drum 320
wavelet analysis 505–8
expected frequency 221
expected values 201–4
expendable bathythermograph (XBT) 14–17, 63
  comparison with CTD 15–16
  data averaging 306–7
  depth error 43–8
expendable current profiler (XCP) 96
extrapolation 279

F-distribution 254
  critical values 581–4
factor analysis *see* empirical orthogonal function (EOF) analysis
fall-rate equation 46, 47
  coefficients 48
false color 190
Faraday, Michael 79, 95
Faraday's Law 80
fast Fourier transform (FFT) 390–2, 406, 420, 421
FFT *see* fast Fourier transform
fidelity, spectra 455
filters 516–57
  band-pass 511, 520–1, 530, 544–5
  Butterworth 542, 543–51

cascades 528, 529–32
design 527–32
Fourier transform 519, 529, 550, 551–2, 554–5
frequency response 519, 520, 528, 533, 534, 542, 551
frequency-domain 526, 529, 551–7
Godin-type 535–6
high-pass 520–2, 526, 530, 531, 544–5, 550
ideal 519–27
impulse response 518, 528, 551
Lanczos-window cosine 536–43
low-pass 520, 522, 525–6, 532–8, 540, 543
multiple filter technique 511–16
running-mean 532–5
selection 516
time domain 526, 529
truncation effects 557
finite dimensional inverse theory 359
finite record length, effect on spectral estimates 433–4
first differencing 458–9
fish finders 48–54
Fleet Numerical Ocean Centre, U.S. (FNOC) 121
float-gauges 58–61
fluorescence-based dissolved oxygen sensor 131–3
FNOC *see* Fleet Numerical Ocean Centre
folding frequency *see* Nyquist frequency
force, units 571
Formazin Turbidity Units (FTUs) 142
FORTRAN, statistical computer packages 352
Fourier analysis 380–92
  limitations 417
  trigonometric functions 324, 343
Fourier line spectra 410
Fourier series 382
Fourier transforms 409, 413, 433, 458, 480, 482–4, 501
  discrete (DFT) 443, 444, 449, 464, 554–5
  filters 519, 529, 550, 551–2, 554–5
  inverse discrete (IDFT) 464
  Singleton 406
fractals 557–68
  box counting method 563–4
  correlation dimension 564
  curves 556, 557–8
  multifractal functions 564–7
  predictability 567–8
  scaling exponent method 561–2
  yardstick method 562–3
France, altimetric satellites 63

Franklin, Benjamin 24
Fredholm equation 358–9, 366
free-fall velocity 43–8
freons see chlorofluorocarbons (CFCs)
frequency
  band averaging 451
  components
    amplitude 393
    phase lag 393
  domain 371–2
    EOF analysis 342
    smoothing 441, 450–3
  expected 221
  least squares analysis 398–9
  observed 221
  resolution 5, 439–41
  response, linear system 490–5
  response function (FRF) 552–4
  specific, Fourier analysis 388–90
frequency-domain filters 551–7
frequency-of-occurrence diagram 187
FRF see frequency response function
FTU see Formazin Turbidity Unit
fundamental frequency 5

Galapagos Rift 144
gamma density function, PDF 208–9
gappy records 277, 278
  interpolation 285–90
Garrett–Munk spectrum 189
gas exchange 153
Gauss–Markov
  smoothing 309–19, 366
  theorem 258, 276
Gaussian distribution 196
GCMs see general circulation models
GCP see ground-control-points
general circulation models (GCMs) 476
General Oceanics MK3C CTD 18, 22
  schematic 20
Geochemical Ocean Sections Study
  (GEOSECS: 1972–4) 146, 149, 151
geodetic datum 60
Geodyne 850 current meter 70
geoid 62
geomagnetic electrokinetograph 96
geometric depth 42
geometric mean functional regression
  (GMFR) 248–9
geopotential, units 571
GEOS-3 altimetric satellite 55
  data 62
GEOSAT altimetric satellite 55, 62–3, 123

GEOSECS see Geochemical Ocean Sections Study
Geostationary Orbiting Earth Satellite
  (GOES) 27, 28
geostrophic currents 114
geostrophic wind 120, 121
Gibbs' phenomenon, ideal filters 523–6, 543, 553, 554
glaciers 56
glacio-isostatic rebound 56
global positioning systems (GPS) 60, 89, 90–2, 116
global temperature, AR model 476–8
global tide gauge network 55–6
global warming 57
glossary 572–5
GMFR see geometric mean functional regression
GMT see Greenwich Mean Time
Godin-type filters 535–6
GOES see Geostationary Orbiting Earth Satellite
goodness-of-fit test 220–4
  calculation table 223
GPS see global positioning systems
graphical presentation, recent developments 187–91
gravitational acceleration
  oceanic 54
  units 571
Great Barrier Reef 183
Greenland 57
  Deep Sea Water 154
Greenwich Mean Time (GMT) 61
grid-value estimates 312
ground-control-points (GCP) 190
Guildline 8737 CTD 22
Guildline salinometers
  8410 Portable 34
  Autosal 8400A 37
Gulf of Alaska 143, 154, 512
Gulf Stream 24, 29, 63, 64, 119

Hamming windows 442, 447, 448
Hanning windows 442, 446–8, 541–3
harmonic analysis 392–404
Hatteras Plain, North Atlantic, IWEX study 367
Hausdorff–Besicovitch dimension 563, 568
HEF see horizontal electric field
Helium isotope 143–5
Hermitian matrix 340

high resolution infrared sounder (HIRS) 28–30
high-pass filters 520–2, 526, 530, 531, 544–5, 550
Hilbert transforms 343
HIRS *see* high resolution infrared sounder
histograms 187
  bins 194
  bootstrap method 298–300
  error detection 269–70
holey-sock drogues 110, 112
horizontal electric field (HEF) 95–6
horizontal maps 162, 172–6
humic acids 140
hydrocasts, error detection 268, 271–2
hydrographic sampling 7
hydrostatic pressure 31, 42–3
hydrothermal heat flux 157
hydrothermal plumes 164, 185
hydrothermal venting 32, 140, 143–5, 155
hypothesis testing 249–57
  errors 249
  probability of outcome 250

ice, melting 56, 57
ice-point, thermometer calibration 11
ICES *see* International Council for the Exploration of the Sea
ideal filters 519–27
  bandwidth 522–3
  Gibbs' phenomenon 523–6, 543, 553, 554
  recoloring 527
IDFT *see* inverse discrete Fourier transform
IES *see* inverted echo sounder
impulse response function (IRF) 552–4
index corrections, mercury thermometer 11
index of precision 272, 273
Indian Ocean 129, 137, 144, 147, 151, 158
Indonesian Archipelago 147, 158
injection temperatures 25
inner-coherence squared 499
inner-cross spectrum 498–9
instrument noise 168
integral time scale 263, 265, 379–80
Inter-Ocean
  acoustic release 99
  S4 ECM current meter 79
internal deformation radius, coastal trapped waves 350
International Council for the Exploration of the Sea (ICES) 33, 34
International Ice Patrol 34
International Indian Ocean Expedition 129

International Math and Science Library (IMSL) 330
interpolation 161–2, 277–90
  methods 279–85
inverse discrete Fourier transform (IDFT) 464
inverse methods 356–70
inverted echo sounder (IES) 63–7
IRF *see* impulse response function
iron compounds 135
irregularly sampled data 7–8
isentropic analysis 176
isopycnal surfaces 163–4
isotopes
  helium 143–5
  oxygen 142–3
isotropy 307
ITOS 1 satellite 26
IWEX internal wave problem 366–70

jackknife method 294, 301–4
  resampling 302
Japan, Kuroshio Extension 339
Japan Sea 149
Jet Propulsion Laboratory (JPL) 28
Johnstone Strait 76
Juan de Fuca Ridge
  Pacific Ocean 41, 137, 144, 145, 156
  EOF example 334–6

K-class intervals 221, 222, 223
Kaijo Denki current meter 70
Kaiser–Bessel windows 448–50, 451
Kelvin wave mode 350
kernel functions 357–9, 361, 367, 368, 444, 446
knot 70
Knudsen's equation 33, 34
Koch curve 556, 557–8
Korea, tide gauge record 521
Kriging 309
Kronecker delta 321
Kuril Islands, surface currents 515
Kuroshio Extension, Japan 339
kurtosis 196

La Niña 55
Labrador Sea 129, 154
lagged covariance matrix 320
Lagrange, Joseph L. 102
Lagrange's method, interpolation 280–1, 282
Lagrangian current measurements 102–25
Lagrangian data, complex demodulation 403–4
Lambert conformal projection 177

Lanczos window 538–40
  cosine filters 536–43
    design 540–1
large errors 266
  current meter data 267
  detection 269–70
Laser induced detection and ranging (LIDAR) 54
least squares
  linear fit 241
  maximum likelihood 242–3
  polynomial curve fitting 242
  quadratic expression 160–1
least squares method 234–7
  estimators 470–1, 472
  frequencies 398–9
  harmonic analysis 393
  internal wave example 368
  specific frequencies 388–90
  tidal analysis 397–402
least squares optimal interpolation 309–19, 313
length, units 570
level of significance 216
LIDAR see Laser induced detection and ranging
light attenuation and scattering 138–42
light-emitting diodes (LED) 140
  blue light 133
likelihood
  definition 227
  maximum 230–3
limits
  chi-square probability distribution 219
  normal probability distribution 216
line spectrum, Fourier 385
linear interpolation 279
linear regression 233–43
  correlation analysis comparison 380
linear system, frequency response 490–5
linearly polarized motions, Fourier analysis 384
liquid expansion sensors 9–12
liquid-in-metal thermometers 12–14
lobes, windows 442
Lomb-normalized periodogram 460–1
Long Lines survey (1983-5) 146, 149
Long Range Prediction Group 338
longshore direction 328
longshore velocity components 425–6
LORAN satellite navigation 115

low-pass filters 520, 522, 525–6, 532–8, 540, 543, 550
  qualities 517
lunar cycles 399
lunar day 389

McClain equation 27
manual data reduction 193
map projections 177
Marsh–McBirney 512 ECM current meter 79
mass, units 570
matrix regression, example 240–2
maximum cross-correlation (MCC) method 97, 119
maximum entropy method (MEM) 465, 473–6
  confidence intervals 475–6
  spectral analysis 471, 474, 475
maximum likelihood 230–3
  correlation estimator 245–6
  least squares 242–3
  spectral estimation 478–9
MBT see mechanical bathythermograph
MCC see maximum cross-correlation
mean 194, 215, 576
  see also population mean; sample mean
  spectral analysis 407–8
  stochastic time series 374–5
mean calibration speed 49
mean currents, satellite-tracked positional data 288
mean differences, global air temperatures 299–300
mean sea-levels 56, 57
mean square error (MSE) 258, 416
mean squares
  between (MSB) 256–7
  within (MSW) 256–7
mean tide height 57
mechanical bathythermograph (MBT) 12–14
  data averaging 306–7
  dynamic response 21
  pressure measurement 43
mechanical speed sensors, Eulerian currents 69
median 196, 215
Mediterranean Sea 154
melting ice 56, 57
MEM see maximum entropy method
Mercator projections 177
merchant ship data, gappy records 278
mercury thermometers 9–12, 23
meridional extent 306
Meteor Expedition 128, 162

meteorological convention 165
methodology
  see also data; errors; spectral analysis; time-series analysis
  editing 264–77
  estimation 227–33
  interpolation 279–85
  nonparametric statistics 294–304
  parametric 464–79
  statistics 193–264
Metonic cycle 56, 57
MGFs see moment generating functions
microwave scatterometer 123, 125
mid-ocean ridges 155, 185
Mie scattering theory 142
minimum variance unbiased estimation (MVUE) 227, 228–9, 230
Mitchell, J. 62
modal analysis 345
mode 196
modified Lambert conformal projection 177
moment generating functions (MGFs) 204–6, 576
moments 201–6
  method of 229
Monte Carlo method 489
Montreal Accord (1988) 149
moon 55
mooring 97–9
Morlet wavelet 503, 504, 505–8
MSB see mean squares between
MSE see mean square error
MSW see mean squares within
multi-input cross-spectral analysis 491–5
multifractal functions 564–7
multiple coherence functions 493
multiple filter technique, wavelet analysis 511–16
multiple linear regression 234
multivariate distributions 293–4
multivariate regression 239–40
MVUE see minimum variance unbiased estimation

National Marine Fisheries Service (NMFS) 121
NATO see North Atlantic Treaty Organisation
NATRE see North Atlantic Tracer Release Experiment
natural cubic spline 283–4, 285
NAVSTAR satellite navigation 115
Neil-Brown CTD 23
nephelometers 139, 142

Nerpic CMDR current meter 70
neutral regression 247, 248
New York, monthly average air temperature 466
Niel Brown ACM current meter 79
NIMBUS 6 satellite 106
nitrate 133–6
nitrite 133–6
NMC see U.S. National Meteorological Center
NMFS see National Marine Fisheries Service
NOAA polar-orbiting satellites 106–7
noise, instruments 168
nonsatellite, tracked drifters 115–16
nonconservative tracers 127
nondimensional numbers and approximations 586–92
nonmechanical current meters 78–82
nonmechanical speed sensors, Eulerian currents 69
nonparametric spectral methods 405
nonparametric statistical methods 294–304
nonpositive definite covariance functions 310
nonrandom variables 233
nonrecursive filters 517, 530
nonuniform data coverage 306
normal density function, PDF 208
normal distribution 196–7, 209
  central limit theorem 212
  cumulative 577
  limits 216
normal modes see vertical normal modes
NORPAX experiment 105, 151
North American Datum
  1927 (NAD-27) 90
  1983 (NAD-83) 90–1
North Atlantic Tracer Release Experiment (NATRE) 157
North Atlantic Treaty Organization (NATO) 70
nuation 56
null hypothesis 249, 252–3, 254
  rejection regions 251
nutrients 133–6
Nyquist frequency 382, 414, 416, 435–7
  sampling 437–9
Nyquist wavenumber estimates 7

objective analysis see objective mapping procedure
objective mapping procedure 309–19
oblique Mercator projection 177, 180
observed frequency 221
Ocean Station P 25, 126, 147

ocean ventilation 153
oceanic boundary layer 120
oceanic gravity 54
oceanic variability, temporal 182
oceanographic convention 165
one-sided autospectra 432
opportunity platforms 277, 278
optimal smoothing 309–19
Oregon, Crater Lake 41
orthogonality condition 321, 424
outcome, hypothesis testing 250
outer-coherence squared 500
outer-cross spectrum 499–500
outliers 270
oxygen 133–6, 176
  dissolved 129–33
  minimum layer 136
oxygen isotopes 142–3

Pacific Ocean 63, 97, 121, 129, 135, 147, 151
  coast 153
  current velocity 507–8
  East Pacific Rise 144
  El Niño/La Niña 55
  Endeavour Ridge 155, 431
  EOF analysis examples 334–40
  equatorial 64, 120, 125, 136, 137
    drifter trajectories 107
  Juan de Fuca Ridge 41, 137, 144, 145, 156
  north 25, 90, 110, 129, 133, 146, 149, 158
    satellite-tracked drifter data 263–4, 265
    sea surface temperature (SST) spectra 419
  northeast 137, 154
  Palmyra Lagoon 65
  sea-level oscillation 509–11
  south 54
  subarctic 136
  subpolar 149
paddle wheel rotor, current meter 76
Palmyra Lagoon, Pacific 65
Panama Canal, sea level heights 408
parachute drogues 110
parametric methods 294
  spectral analysis 405, 464–79
Parosscientific sensors 61
Parseval's energy conservation theorem 384–5, 409, 414
Parzen windows 445
PDF see probability density function
perfluorodeclin (PFD) 157–8
performance index (PI) 330, 334
  windows 443

periodogram 465–6, 472
  spectral analysis method 414, 419–22
    limitations 406
Permanent Service for Mean Sea-level (PSMSL) 57
permutations, probability 198–9
PFD see perfluorodeclin
phase and cross-amplitude functions 484–5
phosphate 133–6
photic zone 135
phytoplankton 133–6
PI see Performance Index
pingers see acoustic pingers
pivotal quantity 217
PL see propagation loss
planktonic organisms 83, 92–4
platform transmit terminal (PTT) 104
platforms of opportunity 277, 278
platinum resistance thermocouples (PRT) 29
platinum resistance thermometers 14–17, 18
platinum resistors, Eulerian current 69
Plessey Model 9040 (STD) 18, 35
plots, vertical section 163
pneumatic gauge 60–1
point estimator 251
polar stereographic projection 177
polaragraphic dissolved oxygen sensors 131
Pole Tide 56–7
polynomial curve fitting, least squares 242
polynomial interpolation 279–81, 285
pop-up float 117
population
  distributions 213
  mean 194, 202, 216–18
  standard deviation 196, 217–20
  variance 196, 204
potential temperature 31–3
power, units of 571
power spectra, discrete signals 415
power spectral density (PSD) 404, 410, 417, 418, 419–21, 467–8
  log plot 423
  periodic data 422
practical salinity scale (PSS 78) 33, 39–41
precipitation measurement 125–7
precision 2–8
pressure 42–55
  atmospheric 120
  hydrostatic 31, 42–3
  units of 570
pressure gauges 58–61
pressure measurement 43
prewhitening, spectra 455–9

principal axes, single vector time series 325–8
principal component analysis (PCA) see empirical orthogonal function (EOF) analysis
probability 197–201
  combinations 198
  concept 194
  functions 576
  permutations 198–9
probability density function (PDF) 198, 200, 202, 206, 374
  see also cumulative probability function
  error detection 270
  examples 207–11
  gamma 208–9
  normal (Gaussian) 208
  uniform 207
probability mass function 197–8, 199
probability of outcome, hypothesis testing 250
profiler support cable 18
progressive vector diagram (PVD) 165, 165–7, 186
propagation of errors formula 273–4
propagation loss (PL) 51
propellors, Eulerian current 69
protected thermometers 12
PRT see platinum resistance thermocouples
PSD see power spectral density
PSMSL see Permanent Service for Mean Sea-level
PSS 78 see practical salinity scale
PTT see platform transmit terminal
PVD see progressive vector diagram
Pytheus of Marseilles 55

Q-factor, spectra 461
quality, data 193
quasi-difference filter 531–2
quasi-Lagrangian current drifters 104–7

radar, synthetic aperture (SAR) 54
radar altimeters 54–5, 62–3
radar tracked drifters 115
radiation temperature, satellite-sensed 26–30
radio transmission 25
radioactive tracers 127, 145–53
radiocarbon 149–53
  bomb-generated 149, 151
radiometers 26, 27–8, 63
Radon 155–7
RAFOS float 116–17
  schematic 118
raised cosine windows see Hanning windows

random errors 264
  correlation 244–5
  detection 272
  theory 272
random telegraph signal 378
random variables 202–3, 204, 206, 228, 234
randomness concept 194
range 196
Rayleigh
  coefficient 400
  criterion 5, 397, 439, 441
  reference constituent 398
RCM4 see Aanderaa RCM4 current meter
RCM see rotor-type current meter
RD Instruments acoustic Doppler current profiler (ADCP) 83, 85–6
real-time data 185
reception time 49
recoloring, ideal filters 527
recording current Meter (RCM) 70
rectangular projection 179
rectangular windows see box-car windows
recursive filters 517, 530–1
refractometers, salinity 42
regression
  cause and effect 246–9
  correlation relationship 243–9
  estimation limitations 260
  linear 233–43
  multivariate 239–40
  neutral 247, 248
  straight line 247
regularly sampled data 7–8
rejection regions, null hypothesis 251
relative sea-level (RSL) measurements 56
remotely operated vehicles (ROVs) 371
remotely-tracked drifters 103
resampling methods 294, 302
residual component, time series 392
resistance thermometers 14
resolution 2–8, 3
  frequency 439–41
resonance 351
reversing thermometers 9–12, 18
  dynamic response 21
Revised Local Reference (RLR) data set, sea-levels 57
Reynold's climatology 25
rhodamine dye 157
ridges, mid-ocean 155, 185
ripple effcts 433
RLR see Revised Local Reference
roll-related error 22–3

Index 633

Rosette bottle sampler 23
Ross Sea, Antarctic Ocean 154
Rossby waves 55, 118
rotary component spectra 427–32
rotary cross-spectral analysis 495–501
   autospectra 495
   time-series pair 497–500
rotary current spectra
   Arctic Ocean 430
   cartesian components 426–7
rotary empirical orthogonal function analysis 342
rotor-type current meters (RCMs) 70–8
roundoff 274–6
ROV see remotely operated vehicle
RSL see relative sea-level
running-mean filters 532–5

S-transformation, wavelet analysis 508–11
Saanich Inlet 133
SAIL see Shipboard ASCII Interrogation Loop
salinity 33–42, 143, 176
   definition 33
   electrical conductivity 34–9
   nonconductive measurement methods 42
   refractometers 42
   titration 33
   tracers 128–9
salinity–temperature–depth (STD) profiler 17–18, 35
   dynamic response 18–22
   Plessey Model 9040 STD 18, 35
   sensor response 22–3
   temperature calibration 23–4
salinometers 34, 35
   Guildline 8410 Portable 34
   Guildline Autosal 8400A 37
   Plessey Model 9040 18, 35
sample distributions 194–7
sample mean 194–5, 215
sample median 215
sample size selection 224–5
sample standard deviation 196, 218
sample variance 195–6
sampling
   accuracy 6
   burst 6–7
   continuous 6–7
   distribution of populations 213
   duration 4–5
   effect on spectral estimates 432–41
   frequency see Nyquist frequency

   hydrographic 7
   increment 3–4
   irregular 7–8
   regular 7–8
   time 171
SAR see synthetic aperture radar
satellite, position fixing 106
satellite altimetry 54–5, 62–3
   bias 225–6, 252–3
   data corrections 63
satellite bathymetry 54–5
satellite sensing systems, gappy records 278
satellite-sensed sea surface temperature 26–30
satellite-tracked drifters 7, 104–7
   complex demodulation 403–4
   data 305–6
   data processing 107–9
   objective mapping example 314–19
   Pacific Ocean 507, 558–60, 562, 563
satellite-tracked positional data 195, 265
   example 287–90
   jumps 289
satellite-tracked surface buoys 102–3
Savonious rotors 70–4
   Eulerian current speed sensors 69
   problems 74–7, 94
SBE see Sea Bird Electronics
scaling exponent method, fractals 561–2
SCANNER see submersible chemical analyzer
scanning radiometer 26
   multichannel microwave (SMMR) 28–9
scatter diagram, error detection 266–7, 268
scatter matrix method, empirical orthogonal function (EOF) analysis 328–32
scatter plots 425–8
scattering
   acoustic 52
   light 138–42
scene animation 191
Scripps Institution of Oceanography 99
Sea Bird Electronics (SBE)
   9 CTD 36–9
   25 CTD 18, 22
   schematic 20
sea surface temperature (SST) 24–30, 97
   bucket method 24
   climatology 25
   EOF analysis example 336–40
   satellite-sensed 26–30
   spectra, northeast Pacific 419
   Vancouver Island 387–8
sea-level, Revised Local Reference (RLR) data set 57

sea-level heights 65
  mean 56, 57
  measurement 55–68
  Panama Canal 408
  recording 61
  relative 56
  steric 56
  variability 56–8
sea-level pressure (SLP), EOF analysis
    example 336–40
SEASAT
  altimetric satellite 54, 55, 62
  mission 123
  scatterometer data 125, 126
seawater standard 34, 40
Secchi disk 139–40
secular changes 56
sediment, deposition 56
selective availability 92
Sensoren-Instrumente-System (SIS) RTM
    4002 30–1, 37
Service ARGOS 107, 185
Shikotan Island, sea-level oscillations 463
Shipboard ASCII Interrogation Loop (SAIL)
    171
shooting method 347
significance
  correlation 253
  statistical 2–8
silicate 133, 137–8
SimTronix UCM 40 current meter 79
sine-squared windows see Hanning
    windows
single precision 275
Singleton Fourier transform 406
singular value decomposition (SVD)
    332–4
SIS see Sensoren-Instrumente-System
skewness 196
skill 258–9
  artificial 259
slant range 99
SMMR see scanning multichannel
    microwave radiometer
smoothing 271
  aliased frequencies 434
  frequency domain 450–3
  optimal 309–19
  periodograms 465–6
  spectral 441–50
SMOW see Standard Mean Ocean Water
SOFAR see Sound Fixing and Ranging floats
sonar scintillation 97

sound
  speed of 49–50
  speed in water 93
Sound Fixing and Ranging floats (SOFAR)
    102, 116, 118, 157
sound waves 50
sounders see acoustic depth sounders
sounding 49
Southern Ocean 129, 135, 136, 144
  Drake Passage 374
spacing of data 277–8
spatial resolution 2–8
Special Sensor Microwave Imager (SSMI)
    123–5, 126–7
spectra
  coherence 486–90
  coincident and quadra 485–6
spectral analysis 404–64
  density function 427
  estimation, comparison of methods 478–9
  parametric methods 464–79
  periodogram method 406, 414, 419–22
spectral energy 383
spectral gap 7
spectral power 433, 442
spectrum, definition 405–6
speed, units of 570
speed of sound 49–50
  water 93
spikes 160, 266
  see also errors
spline function 282
spline interpolation 281–5, 288
squares
  mean 256
  sum of 235–7, 255–7
SSB see sum of squares between
SSE see sum of squared errors
SSMI see Special Sensor Microwave Imager
SSMOOR, Cable Dynamics and Mooring
    Systems 97
SSR see sum of squares regression
SST see sea surface temperature; sum of
    squares total
SSW see sum of squares within
stability, spectra 455
standard deviation 196, 270
standard error, estimate 238
Standard Mean Ocean Water (SMOW) 143
standard seawater 34, 40
Standard Seawater Service 34
statistical methods 193–264
  importance 193–4

statistical significance 2–8
statistical tables 577–84
STD *see* salinity–temperature–depth profiler
steric sea-level 56
stick-plot 185
stilling well gauge *see* float stilling well gauge
stochastic processes 373–4, 409–13
Stokes drift 111–12
straight line regression 234, 247
stratification parameter, coastal trapped waves 350
stratosphere 163
stream function 311
stress, units 570
strontium-90 158
structure functions 291
Student's *t*-distribution 211
Sturm–Liouville equation 345–7
SUAVE *see* submersible chemical analyzer
sub-surface drifters 104, 116–19
submersible chemical analyzer (SCANNER/ SUAVE) 128
sufficiency 227
sulphur hexafluoride 157–8
sum of squares
   between (SSB) 255–7
   errors (SSE) 235, 238, 242
   regression (SSR) 235–7, 244
   total (SST) 235–7, 244
   within (SSW) 255–7
super-resolution, tidal frequency 441–2
surface buoys
   acoustically tracked 102
   satellite-tracked 102–3
surface drifters 104–7
surface gravity wave heights, wavelet analysis 506
surface gravity wave measurement 67–8
surface mooring 97–9
surveys, airborne 170–1
SVD *see* singular value decomposition
Sverdrup interior vorticity balance 362
Swallow floats *see* Sound Fixing and Ranging floats
synthetic aperture radar (SAR) 54, 97, 123

t-distribution, cumulative 579–80
Tasmania 61
techniques, editing and despiking 264–77
tectonic processes 56
temperature 176
   CTD calibration 23–4
   Fourier analysis example 387–8

global warming 57
injection 25
measurement techniques 8–33
potential 31–3
tracers 128–9
units 570
temperature profile, global distribution 308
temporal resolution 2–8
TGBM *see* tide gauge bench mark
thermal conductivity 50
thermal wind relation 362, 363
thermistors 14–17, 18
   dynamic response 21
thermocouples, platinum resistance (PRT) 29
thermometers 9–17
   digital 30–1
   ice-point calibration 11
   mercury 9–12, 23
   platinum resistance 14–17, 18
   protected 12
   resistance 14
thermometric depth 43
thermosteric anomaly, TS diagrams 182
three-box model, Japan Sea 149
three-dimensional representation 189–91
tidal periods 389
tide gauge bench mark (TGBM) 59
tide gauges 58–61
   global network 55–6
   station 59
tides 55–68
   harmonic analysis 372, 397–402
time
   sampling, effect on spectral estimates 433–4
   units 570
time domain
   EOF analysis 321–5, 342, 343
   weighting 441
time lagged correlation 265
time scale, integral 263, 265
time variabilty, TS diagrams 183
time-series analysis
   concepts 371–3
   correlation functions 374–80
   digital filters 516–57
   Fourier analysis 380–92
   fractals 557–68
   harmonic analysis 392–404
   spectral analysis 404–64
      parametric methods 464–79
   stochastic processes 373–4
   wavelet analysis 501–16

time-series data
  aliasing 434
  collection 371
  irregularly spaced 460–1
  unevenly spaced 460–1
time-series pair, rotary analysis 497–500
time-series presentation 185–6
  current meter data 267
timing errors 159–60
titration, salinity 33
TMR see TOPEX Microwave Radiometer
Tofino, British Columbia 389–90, 395, 399–402
TOGA see Tropical Ocean Global Atmosphere
TOPEX Microwave Radiometer (TMR) 63
TOPEX/POSEIDON altimetric satellite 55, 63
topography, bottom 172
tracers
  chemical 127–45
    conservative 127
    coventional 127, 128–38
    nonconservative 127
    salinity 128–9
    temperature 128–9
  inverse theory applications 361–4
  transient chemical 127, 145–58
    radiocarbon 127, 145–53
    tritium 146–9
transducers 52
transient chemical tracers 127, 145–58
transmission time 49
transmissometers 140–2
transport, units 571
transverse Mercator 177
treatment effects 254–5
trench waves see coastal-trapped waves (CTWs)
trends
  estimates 261–4
  spectral analysis 407–8
triangular windows 443–8
Tristar drogue 110
tritium, anthropogenic 143
tritium tracers 146–9
tritium units (TU) 146
Tropical Ocean Global Atmosphere (TOGA) research program 25
troposphere 163
truncation 274–6
  filtering 557
TS curves 271–2
TS diagram 165–6, 168, 181–5

TS relationship 267–8
Tshebysheff's theorem 215
tsunami warning system 55
TTO experiment 151
TU see tritium units
Tucker wave-recorder system 67
Tukey windows 446
turbidity 142
  meters 139
two-way travel time 48

unbiased estimators 203–4, 214–15
  best linear 276
uncorrelated time variability 321
unevenly spaced data 277–8
uniform density function, PDF 207
Unimak Pass 143
United Control Corporation 61
Universal Temps Coordinée (UTC) 61, 125
universal transverse Mercator (UTM) 179
unprotected thermometers 12
U.S. National Meteorological Center (NMC) 338
U.S. Navy 62
UTC see Universal Temps Coordinée

VACM see vector-averaging current meter
Van de Casteele test 60
Vancouver Island
  Amphitrite Point 387–8, 395–7, 419
  current velocity 516
  surface gravity wave heights 506
variability 195
  estimate 270
  uncorrelated time 321
variance 195–6, 576
  analysis of (ANOVA) 254–7
  F-distribution 254
  stochastic time series 374–5
variance-preserving spectra 422–3
vector averaging current meters (VACM) 73–7, 94–5, 98
  RCM7 72, 76
vector measuring current meters (VMCM) 77, 94–5, 97
vector series, spectra 425
velocity components, cross-shore 425–6
ventilation, ocean 153
venting, hydrothermal 32, 140, 143–5, 155
vertical diffusion 135, 157
vertical normal modes 344–50
vertical profiles 163, 164, 167–70
vertical sections 163, 170–2

vertical time-series plot 186
Vibraton 61
viscosity 50
Visible Infrared Spin Scan Radiometer (VISSR) 27–8
VISSR *see* Visible Infrared Spin Scan Radiometer
VMCM *see* vector measuring current meter
volume, units 570

wastewater plumes, buoyant 83
water, rivers 211–12
water column mixing 155
water level gauges *see* pressure gauges
waterfall plot 169
wave heights, K-class intervals 222–3
wavelengths, dominant 7
wavelet analysis 501–16
  admissibility condition 504
  advantages 501–2
  algorithms 504–5
  examples 505–8
  multiple filter technique 511–16
  parameters 502
  progressive wavelets 504
  S-transformation 508–11
  wavelet transform 502–4
wavenumber 7
  domain 371–2
waverider buoys, directional 67–8
waves
  capacitance staff 67
  direction 67–8
  height 67–8
  measurement, surface gravity 67–8
  pumping 74
  Rossby 55, 118
  sound 50
Weddell Endbery Basin, Antarctica 130, 137
Weddell Sea 151
Weibull distribution, El Niño recurrence rates 297–8
weighting coefficients 312
Wiener–Khinchin relation 410

Wilkin model 352–3, 356
windowing 441–50
windows
  box-car 406–7, 433–4, 443–8, 450
  desired qualities 442–3
  Dolph–Chebyshev 442
  filtering 538–40, 541–3
  Hamming 442, 447, 448
  Hanning 442, 446–8, 541–3
  Kaiser–Bessel 448–50, 451
  Lanczos 538–40
  Parzen 445
  triangular 443–8
  Tukey 446
winds 119–25
  fields 121
  geostrophic 120, 121
  measurement 120–5
  spectral analysis 425–32
Winkler method 131
WOCE *see* World Ocean Circulation Experiment
woce/toga near-surface velocity drifter 104–5
Woods Hole Oceanographic Institution 34
World Geodetic Survey
  1972 (WGS-72) 92
  1984 (WGS-84) 90
World Ocean Circulation Experiment (WOCE) 128, 129, 157, 506
  Implementation Plan 128
  Surface Velocity Program (WOCE-SVP) 110
wrap-around points 52–3
Wunsch's method 364–6

XBT *see* expendable bathythermograph
XCP *see* expendable current profiler

yardstick method, fractals 562–3
Yule–Walker equations 469–70, 472

zero-padding, spectra 455–9
zonal extent 306
zooplankton 83, 92–4, 140

DATE DUE